Evaporation

Benchmark Papers in Hydrology, 2

Evaporation

Selection, Introduction and Commentary by

John H. C. Gash
Centre for Ecology and Hydrology, Wallingford, UK

W. James Shuttleworth
SAHRA, University of Arizona, USA

Series sponsored by SAHRA, the Center for Sustainability of semi-Arid Hydrology and Riparian Areas, University of Arizona

International Association of Hydrological Sciences / Association Internationale des Sciences Hydrologiques

Published by the International Association of Hydrological Sciences 2007
IAHS Press, Centre for Ecology and Hydrology, Wallingford, Oxfordshire OX10 8BB, UK

IAHS Benchmark Papers in Hydrology Series, no. 2
ISBN 978-1-901502-98-5

British Library Cataloguing-in-Publication Data.
A catalogue record for this book is available from the British Library.

Production of this book was in part supported by the NSF Center for Sustainability of semi-Arid Hydrology and Riparian Areas (SAHRA) under the STC Program of the National Science Foundation, Agreement no. EAR-9876800 and NSF award DEB–0415977.

IAHS Publications are available from:
IAHS Press, Centre for Ecology and Hydrology, Wallingford, Oxfordshire OX10 8BB, UK
tel: +44 1491 692442; fax: +44 1491 692448; email: jilly@iahs.demon.co.uk

Printed in the UK by 4Word, Bristol.

Foreword to the Benchmark Papers in Hydrology Series

Hydrology is a young science. Following the seminal work by Henry Darcy in the mid-19th Century and Robert Horton's influential studies in the first half of the 20th Century, knowledge generation in the hydrological sciences exploded through the second half of the 20th Century with new discoveries in each of the components of the hydrological cycle. Despite scientific and technological advances, the field of hydrology has been highly fragmented across the engineering–science interface, between basic and applied studies, mathematical and descriptive work, field and laboratory investigations. Hydrological science is now taught in most of the world's major universities, but the teachers and students may be found just as commonly in civil engineering, as in physical geography, forestry, environmental engineering, geology, soil science, or agricultural engineering. Our key journals, textbooks, conference proceedings and monographs are likewise dispersed and scattered across myriad outlets. While Internet searching and .pdf versions of more recent published material make "keeping up" easier, there is a body of important early work that is often ignored by young researchers. As a result, we stand at the beginning of the 21st Century in danger of memory loss of our young science.

The objective of the *Benchmark Papers in Hydrology* series is to define and create a world repository of the benchmark papers in hydrology for education, research and practice. In so doing, it is hoped that this series will counter memory loss in our young and highly interdisciplinary science by chronicling the development of ideas leading to current theory in hydrology. Indeed, there are many examples where experiments, ideas, etc. are being "re-invented" because the key papers are not known to new generations of students, scientists and practitioners. This hydrological foundation is what the *Benchmark Papers in Hydrology* series aims to define, publish and distil: the key papers that contributed to the current thinking of the sub-fields of the hydrological sciences. As such, the benchmark papers are hydrological exemplars. They are the best of the contributions that our field has to offer.

The *Benchmark Papers in Hydrology* series interprets the hydrological sciences in the broadest sense. Each volume tackles a specific topic and is edited by a recognized scientific authority. The volume editor makes the initial selection of papers and prepares the Introduction and Commentary to the papers that are finally included in that volume. The series Editorial Board approves the published selection of papers. Articles are reproduced in whole (but sometimes in part) from the key papers of (mostly) the 20th Century. Those associated with the *Benchmark Papers in Hydrology* appreciate the International Association of Hydrological Science's lead role in publishing the series and their support of this major effort to disseminate the most important ideas in hydrology to libraries, institutes and individuals worldwide.

Jeffrey J. McDonnell
Series Editor

Acknowledgements

The various publishers who granted permission to reproduce copyright material in this volume are thanked. Cate Gardner at IAHS Press, and Marloes Bakker at Oregon State University, are thanked for their huge contributions to this volume. Sarah Anne Graham is thanked for the original cover design. The cover photograph was taken by Alessandro Carioca de Araújo at 07:05 on 1 June 2006 from the K34 flux observation tower in the Cuieiras Biological Reserve of the Instituto Nacional de Pesquisas da Amazônia, 80 km to the northwest of Manaus, Brazil. Production of this book was in part supported by the NSF Center for Sustainability of semi-Arid Hydrology and Riparian Areas (SAHRA) under the STC Program of the National Science Foundation, Agreement no. EAR-9876800 and NSF award DEB–0415977.

SAHRA
www.sahra.arizona.edu

SAHRA promotes sustainable management of water resources not only through research, but by improving dissemination of important science results and products, and by encouraging dialogue about important water issues, research needs and findings among scientists and water managers, policy makers, academics, students, and the general public. The sponsorship and distribution of the IAHS Benchmark Papers in Hydrology furthers the SAHRA mission of transferring knowledge to promote scientifically informed choices by present and future generations of policy makers, water managers, and the general public.

SAHRA is headquartered at the Department of Hydrology and Water Resources at the University of Arizona, USA, but draws on a broad range of academic expertise, disciplines, and institutions to address water science and management questions that are critical in semi-arid and arid regions. Partner institutions include universities in the southwest USA and northern Mexico, US federal agencies and national laboratories.

Contents

Introduction

On average, about 60% of the precipitation falling on the Earth's land surfaces evaporates. This fraction changes greatly from around 85% for the African continent where arid and semi-arid regions dominate, to around 15% for the Antarctic continent which is permanently covered with snow and ice. Isotope studies suggest that for extensive continents with moist, warm climates, perhaps half of the water falling as precipitation is water that already fell upwind and re-evaporated. Moreover, as a global average, around two-thirds of the energy that enters and drives the atmosphere is transferred as latent heat stored in water vapour by the evaporation process. Thus, evaporation is not only a major component of the land surface water balance, but also the primary means through which land surfaces return water and energy to the atmosphere.

Evaporation is an invisible process and difficult to measure. Perhaps this is why during much of the history of hydrology, evaporation was a poorly quantified term in the land surface water balance calculations, and estimated only as the residual between precipitation and runoff. Over the last 30–40 years there has been a huge growth in our capability to measure, understand theoretically and model evaporation. Hydrologists' capability to document and describe evaporation is now comparable with that for the other terms in the water balance. These days, evaporation can be measured in a variety of ways, including some that can separate the transpiration and rainfall interception components, and with automatic instrument systems that can log evaporation flux routinely, regularly and frequently, over long periods. Present-day models of evaporation now combine our understanding of the physics involved in the surface energy and mass exchange with our knowledge of the plant physiological controls on transpiration. By combining models and data, global models now even dare to predict evaporation for all the Earth's land surface simultaneously. The objective of this book is to chart the enormous progress in evaporation research by selecting and reproducing some of the most influential papers and by describing how different facets of the subject have evolved.

A purpose-driven interdisciplinary science

The field of evaporation is an excellent example of a branch of natural science in which technical development and basic understanding have grown in direct response to the need for such development and understanding, and where progress has occurred only as the result of the collaborative participation by many different scientific disciplines. It is important, in the context of a book which selects benchmark papers in this field, to recognize and reflect on the history that created this purpose-driven interdisciplinary science.

Evaporation involves a change of state from liquid to vapour, which is fundamentally a physical phenomenon. However, in the natural world, evaporation from the land surface to the atmosphere generally occurs from a complex interface that comprises plants, soil, and air, and within this interface evaporation can arise both from within and from the surfaces of the plants and soil, depending on the ambient meteorological conditions. Consequently, evaporation science is inherently interdisciplinary; it involves the participation not just of hydrologists, but also of scientists who understand the behaviour of plants and soil, and of the atmosphere which permeates and overlies the evaporating land surface.

It is revealing to look through the titles, journals and authors of the papers in this book. The earliest papers are written by those who, though they might have considered themselves as hydrologists, were mainly trained as physicists, meteorologists and geographers. Later, in the mid-20th Century, the motivation for evaporation research changed and this is reflected in papers that describe evaporation measurements and theory developed by authors from agricultural institutes and departments concerned with improving crop production. Hence, papers are included which describe the development of measurement techniques and models of plant behaviour written not just by physicists and meteorologists, but by ecologists and plant physiologists.

More recently, climate scientists have also contributed very substantially to the growth in understanding, and this has brought awareness that to understand global warming requires learning how to couple water vapour and carbon dioxide fluxes, which requires understanding of the process of photosynthesis and a contribution from plant biochemists. The interdisciplinary approach has been critical to the successful development of this important facet of hydrology, and it is only because of it that progress in measuring and understanding the process of evaporation has been so rapid and substantial.

Evolution of the field of evaporation research

Let us first consider the issues that have motivated the development of evaporation science. The oldest paper we reproduce, by Dalton, was probably motivated by simple scientific curiosity and the desire to understand the natural world. However, by the early part of the 20th Century, evaporation research had more practical motivations. Land-use change in general, and the effects of the change from forest to agriculture in particular, were issues, especially in the rapidly developing United States of America, and understanding the effects of planting forest continued to motivate much of evaporation research in the United Kingdom until around 1980. However, in the second half of the last century, new issues emerged and gained priority in addition to understanding the effects of land-use change. The need to feed a rapidly growing world population generated a massive growth in irrigated agriculture and during the 20th Century global water use increased 10-fold, with the rate per capita doubling, and about 70% of water used by agriculture. Efficient irrigation requires accurate estimates of crop evaporation and this further built the demand for better understanding of the evaporation process. In the past two decades, growing recognition of, and concern for, global warming and the consequent requirement to predict how the climate will change in response to increased levels of greenhouse gases further increased the need for better models of evaporation. Estimating the impacts of climate change with global climate models requires that they model the evaporation accurately, not just with the Earth's present vegetation cover, but also when vegetation cover itself changes in response to changing climate.

Some of the papers selected and presented in this volume demonstrate how the options available for evaporation measurement have developed, from monitoring the change in the mass balance of whole catchments or carefully selected, representative volumes of soil and vegetation, to measurements of the water flow through sample plants or the flux of water vapour through the atmosphere's turbulent boundary layer immediately above the evaporating land surface. As a result of these developments, there is now an assortment of methods available, each one having its own strengths and most appropriate application. Water balance sampling methods are most suitable for measuring total evaporation from plants and soil in plot-scale research applications, throughfall sampling and plant physiological methods when seeking separate information on the interception and transpiration processes, while micrometeorological techniques are preferred for patch and landscape-scale measurements.

However, area-average evaporation is not yet widely measured in a routine way. Despite the need for skilled operators with time to quality-control and process the large amounts of data the instruments produce, and despite the very real possibility that it may underestimate evaporation unless used with care, the eddy correlation method has probably come closest to providing the opportunity to make area-average measurements. Measurement systems are now at the point where they can be run continuously with a minimal power requirement, and many hundreds of eddy correlation systems are currently being used around the world. Although the primary motivation for nearly all these measurements has so far been to quantify the carbon budget, hydrologists are also now starting to use the technique for routine measurements of evaporation, and this trend looks set to continue. The scintillometer technique is also an exciting development in this context, with microwave scintillometer devices now being developed which promise to provide direct measurements of evaporation from large areas. As yet, no concerted effort has been made to try to combine present-day understanding and models with direct measurements of evaporation and associated variables that provide indirect (perhaps remotely sensed) information on evaporation, such as surface temperature, using modern data assimilation techniques to estimate area-average evaporation.

Some of the papers selected and presented in this volume also demonstrate how the theory of evaporation has developed, with key points being:

- imposition of the law of conservation of energy, a constraint which is not always available to hydrologists unless actively recognized and sought for;
- improved recognition and quantification of the meteorological controls on evaporation, initially through the concept of potential evaporation;
- the idea of representing a plant canopy at a single level, the "big leaf" model, which is the key to estimating the actual evaporation arising as transpiration;
- recognition that, in order to create a general theory that also applies to tall vegetation, methods based on simply modifying potential evaporation are not adequate, and wet canopy and dry canopy evaporation must be represented separately; and finally,
- the introduction of theoretical methods and models to describe sparse or still growing plant canopies.

This growing understanding, which came about by a combination of measurements and theory, explained why different rates of evaporation can be expected from different crops, especially from tall and short vegetation, and why this difference can alter with rainfall regime. It also resulted in models of evaporation that are a more applicable and accurate representation of actual evaporation in both hydrological and atmospheric models, including models of global circulation used for climate change prediction.

As described earlier, the fundamental driver of past research and technical development in evaporation science was population growth and growth in the expectations of the growing population, expressed in the form of understanding the consequences of land-use and climate change, and increasing irrigation efficiency. These remain the motivation for future progress in this field, albeit that the outstanding challenges have now become more complex and subtle.

There are many areas of Earth system science which require more research to further development of theoretical understanding and improved modelling of the evaporation process, but here we identify just three areas where we think progress is due.

First, the simultaneous measurement of water and carbon fluxes at many sites using the eddy correlation technique has provided a rich new source of multi-site data for many land covers, which has not yet been sufficiently exploited as a combined data set. New, integrated models of carbon–water fluxes should appear, with applications extending to modelling the production of biomass and biofuels, and perhaps incorporating the effects of nutrients and ambient carbon dioxide concentration.

Secondly, the effects of spatial variability resulting from convective rainfall or from topography, are poorly accounted for in models of evaporation at present. The influence of topography on soil moisture, and thus evaporation, is not represented either at the landscape scale or at the scale of a few tens of centimetres to hundreds of metres. Both these types of spatial variability are particularly important in semi-arid environments.

Finally, there is a great need for new measurements and models of the evaporation under freezing conditions. These conditions are complex and dynamic, and our ability to represent them is poor. But much of the world's surface water supply is at some time stored in the form of ice and snow, and the evaporative loss from this store is still inadequately measured and modelled. A paper on snow and ice is notably absent from this volume: a benchmark paper in this area is needed.

What is very clear is that the driving force of growth in population and human expectations and their impact through changing landscapes and climate, and an increasing demand for food and pure drinking water, will continue into the indefinite future. Improved understanding of water-related processes and management of water resources is usually at the heart of efforts to maintain ecosystem services, improve the efficiency of agriculture, and predict the changing climate. Better husbandry of water is the key, and knowledge of evaporation is an essential, perhaps the most essential, component.

About the book

This book is structured in five sections. Within each section the benchmark papers are mainly presented in chronological order, the purpose being to demonstrate how the topic has developed and how in many cases the ideas of one paper have led to the next. Hydrology is fundamentally an observational science, and we reflect this fact by first reproducing papers which describe most of the main methods available for measuring evaporation in the field. These are followed by a set of papers concerned with evaporation as controlled by the atmosphere: potential evaporation. The term potential evaporation has many subtly and confusingly different meanings and, in our view, it is usually best avoided. Nevertheless, it is an historically important concept, and one which was critical in the intellectual development of understanding evaporation as a physical process. Papers on actual evaporation follow. The evaporation of intercepted rainfall and transpiration are treated separately in this volume, recognizing their very different character and controls. Finally, there are three papers on evaporation and climate, to recognize that evaporation is not just a loss from the surface water balance, but also an extremely important input to the atmosphere and control on its general circulation.

In making this selection of papers, we have avoided simply generating a list of the 30 most cited papers on evaporation although, because many of the papers are seminal works, several are indeed highly cited. We considered that selecting papers purely on the basis of their citation would result in a soulless league table based in part on transient popularity rather than lasting influence. We have also sought to avoid rewriting history. Although many of the papers we include were the first to initiate progress in the topic they address, we have not included some papers which could, and perhaps should, have been influential. It is unfair, but some pioneering papers were overlooked and perhaps will always remain overlooked, while a later paper became well recognized and had a great influence on the scientific community. The selection is certainly biased towards papers written in English, although we do include one first written in French. But arguably this is acceptable since English has become the *lingua franca* of science.

While several of the included papers would doubtless appear on any sample list for this field, different editors would likely have chosen different papers. We apologise to the many authors, alive or dead, who could justifiably consider that one of their own papers should have been included in this book. For every paper we have included, there are five or six we would like to have included but could not; our excuse is the constraint of space and the need to create a balance across the subject. But "beauty is in the eyes of the beholder", so the saying goes, and what we have done is to choose those papers that have added most to our own understanding of the process of natural evaporation, and have steered our own work in exploring its measurement and modelling. In this respect, the choice is necessarily subjective, and the preferred list a personal one. Our sincere hope is that doing this will give the reader a glimpse of our own love and enthusiasm for this subject.

We hope you enjoy reading these papers!

Acknowledgements

We are grateful to the Series Editor, Jeff McDonnell, for the invitation to edit this collection of papers and for his advice and guidance. We would also like to thank Marloes Bakker, Holly Barnard, Cate Gardner and all the staff of IAHS Press involved in the editing, production and sale of the book, and the NSF Center for Sustainability of semi-Arid Hydrology and Riparian Areas (SAHRA) for providing partial financial support for its production. Although responsibility for the choice of papers is ours, we are also very grateful for the many suggestions, corrections and criticisms we have received from a large number of colleagues in the course of preparing this volume.

EVAPORATION MEASUREMENT

Natural evaporation can be measured either as the loss of liquid water from the surface, or as the flux of water vapour into the atmosphere. Methods based on measuring the mass budget of liquid water exploit the law of conservation of mass by creating a constrained system where evaporation can be deduced from a water balance equation. Each component in this system is either measured or assumed to be unchanging. Vapour transfer methods may be unconstrained or, since they are also measuring the flux of latent heat, can exploit the conservation of energy to also create a closed system. In the following sections we identify benchmark papers in the development of both these approaches.

A Mass Budget Methods – Commentary

In this section the first paper describes a pioneering study into the comparative evaporation from different land covers by measuring the total area-average evaporation for a pair of catchments. We then introduce two papers which describe methods that measure total evaporation, combining both transpiration and interception, by monitoring change in soil moisture content, firstly using a well-defined soil monolith contained in a lysimeter, and secondly by measuring the soil moisture depletion in natural soil. These are followed by a discussion of the sampling methodology necessary to measure evaporation of the rainfall intercepted by forest canopies as the difference between gross and net rainfall.

A1 Catchment water balance

Bates, C. G. (1921) First results in the streamflow experiment, Wagon Wheel Gap, Colorado.

The motivation for single catchment studies is normally to define the magnitude and timing of the relationship between runoff and precipitation for the catchment under study. Paired-catchment studies, on the other hand, are concerned with changes in the influence of vegetation on the catchment water balance, including vegetation-related changes in evaporation. There have been many paired catchment studies and, although the results have not been straightforward to interpret (see Section D), we have selected the first paper describing the Wagon Wheel Gap experiment, published in 1921 by C. G. Bates. In this paper he describes Wagon Wheel Gap as "*the first serious effort to obtain, under experimental conditions, a quantitative expression of forest influences on snow-melting, streamflow (and thus by implication, evaporation) and erosion*".

In the Wagon Wheel Gap study, runoff from two similar forested catchments was first compared for several years, the forest in one catchment was then removed, and the resulting observed difference in measured runoff investigated. Summarizing the results, McCulloch & Robinson (1993) state that, "*... cutting increased the annual streamflow (relative to the unchanged 'control' catchment) by 30 mm, mostly because of an increase in the spring snowmelt. However, there was also a small increase in summer low flows ...... attributed not only to lower summer transpiration (or evaporation) losses from short vegetation as compared with forest, but mainly to increased groundwater recharge in winter as a result of reduced winter evaporation (Bates & Henry, 1928).*" This experiment led to a whole series of subsequent studies based on essentially the same philosophy (see the review by McCulloch & Robinson, 1993, and other papers in the special edition edited by O'Loughlin & Dunin, 1993).

There are substantial practical problems associated with deriving evaporation from a catchment water balance, because the evaporation loss is deduced as the (often small) residual between incoming precipitation, outgoing runoff, and change in water stored within the catchment. Although river flow integrates runoff from the whole catchment and is generally accurately measured, the measurements of precipitation and storage over large areas are difficult because both are prone to sampling errors, which may be substantial. In addition the catchment must be assumed watertight. Nevertheless, paired-catchment studies have made a huge contribution to the growth of hydrological understanding. By itself, each paired-catchment study provided definitive results relevant to local climate and land cover. When taken together, they set a challenge, because the results were inconclusive about the overall effect of forests on water resources. The challenge was met by process studies being incorporated into the design of paired-catchment experiments. These studies led to the general understanding of forest evaporation based on the importance of the relative contributions of transpiration and evaporation of the intercepted precipitation (see Section D).

A2 Lysimetric measurements

Pruitt, W. O. & Angus, D. E. (1960) Large weighing lysimeter for measuring evapotranspiration.

The simplest form of lysimeter is the evaporation pan, which was probably first used in the 17th Century (see Brutsaert, 1982). The simplicity of pans is attractive and a large global network of evaporation pans, albeit of many different types, was developed in the 1950s. Gangopadhyaya *et al.* (1966) list 27 different designs. However, a recognition that evaporation pans were likely to over-estimate natural evaporation from real surfaces led to a need for measurements of actual evaporation from vegetation. With an increasing area of the world being used for irrigated agriculture, there was a particular need for research into the water use of irrigated crops. Simple lysimeters had been used in research before, for example by Penman (see C3) and Pasquill (1950) (see B3), but we consider that the University of California's Davis lysimeters constructed in 1958 provided a benchmark. We have therefore selected the paper by W. O Pruitt and D. E. Angus published in 1960, which describes the construction of these two instruments. When they were created, the Davis lysimeters were the largest in the world and indeed perhaps still are. They provided accurate, continuous measurements of evaporation from representative samples of agricultural vegetation. With soil inside the lysimeter being maintained at the same moisture content as in the field outside, they are widely regarded as having set a standard which others had then to meet. The instruments, which are still operational today, are 6.1 m in diameter, weigh up to 45 tonnes, and are associated with what amounts to an underground laboratory. They have been used in hundreds of studies in agricultural meteorology and hydrology, most notably the research in the 1960s and 1970s which established the descriptions of evaporation subsequently used operationally in irrigation management.

Although weighing lysimeters have a small spatial sample and are time consuming and expensive to construct, they are conceptually simple, accurate, have a fast time response and allow the study of how evaporation varies with soil moisture. They continue to play an important part in the study of evaporation. As well as agricultural crops, they have been used to measure evaporation from mountain areas (e.g. Wright, 1990), and from forest trees (e.g. Fritschen *et al.*, 1977; Dunin *et al.*, 1985). Other lysimetric approaches have also been used, including so-called natural lysimeters in which a plot with an impervious substrate is isolated, and measurements of drainage and soil moisture made to determine water loss (e.g. Law (see D1); Calder, 1976).

A3 Soil moisture depletion measurements

Geisel, W., Lorsch, S., Renger, M. & Strebel, O. (1970) Water flow calculations by means of gamma absorption and tensiometer field measurements in the unsaturated soil profile.

Measurements of the change in soil water content can be made using a variety of non-destructive methods based on the change in absorption of radioactivity or the electrical properties of the soil (see e.g. Marshall *et al.*, 1996; Hillel, 1998; Gardner *et al.*, 2001; IAEA, 2002). However, soil drying may be caused by water moving either upwards, as evaporation and transpiration, or downwards, as drainage. Consequently, measurements of soil water content alone cannot be used to derive evaporation – changes associated with evaporation cannot always be distinguished from those due to drainage. However, when the net precipitation input to the soil is less than the evaporation and transpiration from the top of the soil profile, soil tension measurements will reveal the soil water potential decreasing towards the surface. The turning point in the vertical profile of soil water potential separates the upper layer, where the water is moving upwards, from the depth below where it is moving downwards. This phenomenon was first observed by Richards (1954) who termed it the "static plane". The static plane is now better known as the zero-flux plane. When the change in soil water content above the zero-flux plane is also measured, the two measurements can be combined to give the evaporation. This extra step is usually attributed to Geisel *et al.* and, for this reason we select their 1970 paper as the benchmark in this field.

The zero-flux plane method has been discussed by Cooper *et al.* (1990), who point out that there is an implicit, although probably reasonable, assumption that root extraction below the zero-flux plane is negligible. Khalil *et al.* (2003) give an interesting and comprehensive review of the method and its applications, which include studies of evaporation from a crop of tea (Cooper, 1979), a chalk grassland (Wellings & Bell, 1980) and an agricultural crop (Arya *et al.*, 1975). Use of a defined zero-flux plane is not only an elegant way of measuring evaporation and transpiration using soil moisture depletion, it also allows insight into the process of soil water extraction: for example, when Shimada *et al.* (1999) found two zero-flux planes in the soil water potential profile under a forest, they were able to use the method to separate evaporation from the forest floor from transpiration by the deep rooted trees.

As well as being a technique to derive evaporation, the zero-flux plane method can also give the downward flux of water and thus be used to derive solute fluxes to groundwater (e.g. Wellings & Bell, 1980). Khalil *et al.* (2003) see this as having several potential applications in studies of pollution and its prevention.

A4 Throughfall measurements

Lloyd, C. R. & Marques, A. O. (1988) Spatial variability of throughfall and stemflow measurements in Amazonian rain forest.

Measurements of rainfall interception by forests have been made for more than a century (see also D3) as the difference between gross rainfall, and throughfall plus stemflow measured on the forest floor. Because this measurement is a relatively small difference between two large values, it is important that both are measured adequately. However, the spatial variability of throughfall presents a particular sampling problem which has been tackled in a variety of ways. Calder & Rosier (1976) avoided the problem by using a plastic sheet to give a 100% sample – a good strategy in a dense, young plantation forest; others (e.g. most recently Cuartas *et al.*, 2007) have used troughs. However, the most commonly used technique is a simple array of randomly positioned raingauges. Although this is a sound, simple methodology, in practice it has not always given useful results as experimentalists, seeking to reduce the costs of the equipment and the time spent in measurements, have collected inadequate samples.

Wilm (1943) examined the statistics of throughfall sampling and pointed out the need for random placement of gauges. Several others, notably Czarnowski & Olszewski (1970) and Kimmins (1973), analysed how the sampling error reduced with the number of gauges deployed and how the error in the accumulated total slowly reduces if throughfall gauges are periodically randomly relocated. Ford & Deans (1978) analysed the probability distribution of throughfall as a percentage of gross rainfall. Nevertheless, we select the paper by Colin Lloyd and Ari Marques as the benchmark in this field because, in addition to confirming the merit of randomly moving gauges as an effective alternative to using very many gauges, it also categorically demonstrated the need to include throughfall samples that are greater than the gross rainfall and thus apparently anomalous.

Lloyd & Marques also clearly showed the need to select a sampling strategy which is consistent with the scale of canopy variation. In their case, working in tropical rain forest with a heterogeneous canopy, they used a transect approach in an attempt to sample a number of dominant trees. The paper has been widely cited and was influential in the design of subsequent interception studies, and has led to further studies of the spatial variability of throughfall in tropical forest. Notable among these is the work by Loescher *et al.* (2002), which suggested that to account for the variation of the canopy associated with dominant trees and tree falls, the spatial separation of gauges should be more than 45 m, and the comprehensive measurements of Manfroi *et al.* (2006), who sampled subplots within a 4 ha plot. Studies of interception still occasionally attempt to measure throughfall with static gauges or troughs; but unless their representativeness is established by comparison with randomly relocated gauges (e.g. Cuartas *et al.*, 2007), there is the risk that the sample may be systematically biased by outliers and not close to the true mean throughfall. Such bias can lead to confusing results, as demonstrated by Holwerda *et al.* (2006).

References

Arya, L. M., Farrell, D. A. & Blake, G. R. (1975) A field study of soil water depletion patterns in presence of growing soybean roots. I. Determination of hydraulic properties of the soil. *Soil Sci. Soc. Am. Proc.* **39**, 424–430.

Bates, C. G. & Henry, A. J. (1928) Forest and streamflow experiment at Wagon Wheel Gap, Colorado. *Mon. Weather Rev. Suppl.* **30**, 1–79.

Brutsaert, W. (1982) *Evaporation into the Atmosphere*. Kluwer, Dordrecht, The Netherlands.

Calder, I. R. (1976) The measurement of water losses from a forested area using a "natural" lysimeter. *J. Hydrol.* **30**, 311–325.

Calder, I. R. & Rosier, P. T. W. (1976) The design of large plastic sheet net rainfall gauges. *J. Hydrol.* **30**, 403–405.

Cooper, J. D. (1979) Water use of a tea estate from soil moisture measurements. *East African Agricultural and Forestry J.* **43**, 102–121.

Cooper, J. D., Gardner, C. M. K. & MacKenzie, N. (1990) Soil controls on recharge to aquifers. *J. Soil Sci.* **41**, 613–630.

Cuartas, L. A., Tomasella, J., Nobre, A. D., Hodnett, M. G., Waterloo, M. J. & Múnerab, J. C. (2007) Interception water-partitioning dynamics for a pristine rainforest in Central Amazonia: marked differences between normal and dry years. *J. Hydrol.* (in press).

Czarnowski, M. S. & Olszewski, J. L. (1970) Number and spacing of rainfall gauges in a deciduous forest stand. *Oikos* **21**, 48–51.

Dunin, F. X., McIlroy, I. C. & O'Loughlin, E. M. (1985) A lysimeter characterization of evaporation by eucalypt forest and its representativeness for the local environment. In: *The Forest–Atmosphere Interaction* (ed. by B. A. Hutchison & B. B. Hicks), 271–291. Reidel, Dordrecht, The Netherlands.

Ford, E. D. & Deans, J. D. (1978) The effects of canopy structure on stemflow, throughfall and interception loss in a young Sitka Spruce stand. *J. Appl. Ecology* **15**, 905–917.

Fritschen, L. J., Hsia, J. & Dorwaiswamy, P. (1977) Evapotranspiration of a Douglas Fir determined with a weighing lysimeter. *Water Resour. Res.* **13**, 145–148.

Gangopadhyaya, M., Uryvaev, V. A., Omar, M. H., Nordenson, T. J. & Harbeck, G. E. (1966) Measurement and estimation of evaporation and evapotranspiration. *WMO Tech. Note 83*, Geneva, Switzerland.

Gardner, C. M. K., Robinson, D., Cooper, J. D. & Blyth, K. (2001) Soil water content. In: *Soil and Environmental Analysis: Physical Methods*, second edn (ed. by K. A. Smith & C. E. Mullins), 1–64. Marcel Dekker, New York, USA.

Hillel, D. (1998) *Environmental Soil Physics*. Academic Press, San Diego, USA.

Holwerda, F., Scatena, F. N. & Bruijnzeel, L. A. (2006) Throughfall in a Puerto Rican lower montane rain forest: a comparison of sampling strategies. *J. Hydrol.* **327**, 592–602.

IAEA (2002) *Neutron and Gamma Probes: Their Use in Agronomy*. IAEA-TCS-16, IAEA, Vienna, Austria.

Khalil, M., Sakai, M., Mizoguchi, M. & Miyazaki, T. (2003) Current and prospective applications of zero flux plane (ZFP) method. *J. Japan. Soc. Soil Phys.* **95**, 75–90.

Kimmins, J. P. (1973) Some statistical aspects of sampling throughfall precipitation in nutrient cycling studies in British Columbian coastal forests. *Ecology* **54**, 1008–1009.

Loescher, H. W., Powers, J. S. & Oberbauer, S. F. (2002) Spatial variation of throughfall volume in an old-growth tropical wet forest, Costa Rica. *J. Tropical Ecology* **18**, 297–407.

Manfroi, O. J., Kuraji, K., Suzuki, M., Tanaka, N., Kume, T., Nakagawa, M., Kumagai, T. & Nakashizuka, T. (2006) Comparison of conventionally observed interception evaporation in a 100-m^2 subplot with that estimated in a 4-ha area of the same Bornean lowland tropical forest. *J. Hydrol.* **329**, 329–349.

Marshall, T. J., Holmes, J. W. & Rose, C. W. (1996) *Soil Physics*, third edn. Cambridge University Press, Cambridge, UK.

McCulloch, J. S. G. & Robinson, M. (1993) History of forest hydrology. *J. Hydrol.* **150**, 189–216.

O'Loughlin, E. M. & Dunin, F. X. (eds) (1993) *Water Issues in Forests Today* (special issue) *J. Hydrol.* **150**(2-4), 189–788.

Pasquill, F. (1950) Some further considerations of the measurement and indirect evaluation of natural evaporation. *Quart. J. Royal Met. Soc.* **76**, 287–301.

Richards, L. A. (1954) Multiple tensiometer for determining vertical component of the hydraulic gradient in soil. *Soil Sci. Soc. Am. Proc.* **18**, 7–10.

Shimada, J., Yoneska, T. & Zhou, Q. Z. (1999) Quantitative evaluation of forest floor evaporation and transpiration revealed by soil water content observation in humid temperate pine forests. In: *Integrated Methods in Catchment Hydrology – Tracer, Remote Sensing and New Hydrometric Techniques* (ed. by C. Leibundgut, J. McDonnell & G. Schultz), 29–36. IAHS Publ. 258. IAHS Press, Wallingford, UK.

Wellings, S. R. & Bell, J. P. (1980) Movement of water and nitrate in the unsaturated zone of upper chalk near Winchester, Hants, England. *J. Hydrol.* **48**, 119–136.

Wilm, H. G. (1943) Determining net rainfall under a conifer forest. *J. Agric. Res.* **67**, 501–513.

Wright, I. R. (1990) A lysimeter for the measurement of evaporation from high altitude grass. In: *Hydrology in Mountainous Regions, I: Hydrological Measurements* (ed. by H. Lang & A. Musy), 79–88. IAHS Publ. 193. IAHS Press, Wallingford, UK.

FIRST RESULTS IN THE STREAMFLOW EXPERIMENT, WAGON WHEEL GAP, COLORADO

By Carlos G. Bates

Director Freemont Experiment Station

It is nearly ten years since the first outline of the Wagon Wheel Gap streamflow experiment, started jointly by the Weather Bureau and Forest Service in 1910, was given to the Society. These ten years have been years of suspense not only for me, but, I am sure, for many other members of the profession, who have awaited the results of the first serious effort to obtain, under experimental conditions, a quantitative expression of forest influences on snow-melting, streamflow, and erosion.

The period up to July 1, 1919, was devoted entirely to the refinement of the technique, and the establishment of relations existing between the two streams while *both* were influenced by forest cover. For these relations we have eight years' reliable records. Denudation of one area (Watershed B) was begun in July, 1919, and completed by the burning of most of the brush and débris in September, 1920. As the "streamflow year" begins October 1 and ends September 30, we now have a record for one whole year, in which the forest has been almost completely destroyed, but the absorbing capacity of the forest floor has been little changed. While the results in this year may not be as significant as those to follow, and the average results for several years must be taken for the basis of any broad calculations, it is felt that these first results will be extremely interesting, and therefore, with the consent of the Weather Bureau, the following unbiased statement is given.

It should be remarked that each of the following calculations is made according to formulæ (expressed by diagrams) derived from the eight years' record prior to denudation, and agreed upon by the two Bureaus as showing the most probable relation of the two streams under forest influences, for various climatic conditions. These formulæ are not perfect, and improvement in them may be made as time goes on. As at present accepted, they will shortly be published in the Monthly Weather Review.

With these formulæ, we calculate the "most probable" discharge of stream B for any period in the year, and for conditions existing with forest cover, and compare this quantity directly with the actual discharge after denudation.

For the general conditions of the experiment, the reader is referred to my article in the Proceedings of the Society of American Foresters, Vol. VI, No. 1, 1911. Some discussion of this project has also been given by Zon,[1] in comparing it with the Swiss Experiment, whose results are recently available.

CALCULATIONS BY PERIODS [2]

1. Relations for the whole year, October 1, 1919, to September 30, 1920:

The precipitation and run-off of both streams is rather larger than in the average year, especially in the spring flood period.

Under the conditions existing, the most probable discharge of stream B, with forest cover, would have been 7.608 inches over watershed.[3]

[3] Corresponding, say, to 22 inches of precipitation over the watershed. This expression permits us to disregard the area of the watershed. Hereafter the abbreviation "ins. O. W." will be used.

The actual discharge was 8.566 inches.

Gain from removal of forest .958 inches or 12.6 per cent.

2. Relations in the spring flood, 1920:

(*a*) *First day of flood on A:*

The streams rose steadily from their winter rate without any marked cessation of snow-melting.

Stream A first passed into the flood stage on April 8.

The most probable discharge of B, on this date, for the conditions existing would be .0114 ins. O. W.

The actual discharge was .0165 ins. O. W.

---

[1] Zon, R. The Effects of Forests Upon Streamflow. Review in JOURNAL OF FORESTRY, XVIII, 6, October, 1920, p. 625.

[2] The Weather Bureau concurs in the statement of the preliminary calculations of streamflow.

It wishes to state, however, that in a number of respects the winter snowfall cover, its melting, and the general flood conditions for 1919-20, place this year in a class by itself, to some extent, as giving conditions more or less unlike corresponding conditions prevailing in the first stage of the experiment and before Watershed B was denuded.

The rate for this day was increased 44.6 per cent by early snow melting induced by denudation.

(*b*) *End of flood on A (July 13):*

Under the conditions of subsidence of the flood the most probable discharge of stream B for the end of the flood (3 days, including one before and one after the last technnical day) would be .0421 ins. O. W.

The actual discharge was .0443 ins. O. W.

The stream was still 5.4 per cent higher than it probably would have been under the influence of forest.

(*c*) *Crest day on stream A:*

Under the conditions leading up to the crest of the flood, the discharge of stream B, on the highest day for A, should have been .2265 ins. O. W.

The actual discharge was .3105 ins. O. W.

At this point the rate of discharge was 36.9 per cent higher than under forest conditions.

(*d*) *Time elapsing between crests:*

Under forest conditions the crest of B was always somewhat later than that on stream A.

Under the conditions in 1920 it should have been, and *was,* one day later than the crest on A.

It would seem, then, that while earlier melting on watershed B had produced a relatively large early flow, the time of maximum melting was not appreciably affected.

(*e*) *Relative height of B crest:*

Considering the height of stream B on the crest day for A, it is found that on its own crest day (the day following) stream B is depressed about 5 per cent. This is the first evidence of any reduction in the supply to stream B resulting from early discharge.

(*f*) *Total volume in flood:*

The most probable total volume to be discharged by stream B, considering the character of the flood, was 5.195 ins. O. W.

The actual volume discharged was 5.946 ins. O. W., or a gain of 14.5 per cent. It is thus evident that greater flow in the early stages of the flood was not compensated by decreased flow in the later stages. In fact, we have already shown that at the end of the flood stream B was still 5 per cent high.

(*g*) *Volume to crest of flood on A:*

In this first period (46 days out of a total of 97 for the whole flood) the most probable discharge of B was 2.115 ins. O. W. and its actual discharge 2.727 ins. O. W., or an excess of 29 per cent.

Rough calculations indicate that on some days in the early stages of the flood stream B may have discharged 100 per cent more than would have been expected under forest conditions.

3. Relations in the summer rainy period:

(*a*) *From end of flood to July 31:*

In this 18-day period the flow of stream B was 0.242 ins. O. W., or only 0.8 per cent above that to be expected.

(*b*) *Month of August:*

The discharge of stream B was .341 ins. O. W., or 2.4 per cent in excess of the expected.

(*c*) *Month of September:*

The discharge of stream B was .331 ins. O. W., or 4.4 per cent in excess of that to be expected, all things considered.

(*d*) A check on the three preceding calculations, by one calculation for the whole period, shows no appreciable correction to be applied to the months. This shows an excess in the discharge of B of 2.8 per cent for the whole period.

(*e*) *Last five days of September:*

This period is almost always rainless and permits us to examine the relation of the streams when not influenced by any current precipitation. In 1920 all of the 5 days were without rain.

Calculations for 1920 indicate stream B was running 2 per cent above the usual, at this time.

4. Relations for the period October 1, 1919, to beginning of flood in 1920, the freezing period:

Admittedly, in the effort to cover this long period by one calculation, only a rough approximation to the most probable flow of stream B is possible. Unless a number of years' results give very consistent effects ascribable to denudation, it will be necessary to devise formulæ for different parts of the winter period. The whole, however, is to be thought of as one for the storing-up of water for the next growing season.

By the tentative formulæ, then, the most probable flow of stream B for the period was 1.754 ins. O. W.

The actual discharge was 1.707 ins., or a deficit of 2.7 per cent.

By deducting from the most probable for the whole year, the calculated amounts for the flood and summer periods, the result is entirely different. The most probable discharge of B is then 1.523 ins. O. W. and the actual of 1.706 indicates an excess of 12 per cent.

It might be remarked in defense of the latter figure that calculations by another set of formulæ, showed the flow of B from August, 1919, to January, 1920, to be steadily in excess of the most probable flow under forest conditions.

However, for the present, we are unable to make any positive statement as to the effect of denudation in this least important period of streamflow.

5. Silt deposits in basins as a measure of erosion.

The provisions for measuring the detritus carried by the streams are entirely adequate so far as mineral soil is concerned, but in flood a very appreciable quantity of organic silt may escape.

Individual silt measurements in the period of comparison of the two streams under forest, showed wide fluctuations, so that the calculated ratio of B silt to A silt for any single period has only approximate value.

However, the results in this first year are so striking as to leave little doubt. It should be remembered that they have been secured *without* destruction of the humus by fire, and represent, in all probability, results of disturbance in logging the watershed.

RECAPITULATION

The calculations of streamflow for one year after denuding the watershed known as "B," but before destroying the ground cover by

Period	Amount for A	Most probable for B	Actual for B	Excess
	Lbs.	*Lbs.*	*Lbs.*	*Per cent*
July 15 to October 15, 1919.......	75.6	130.	58.6	55
October 15, 1919, to April 15, 1920..	117.9	92.	131.3	43
April 15 to July 15, 1920.........	362.5	400.	1211.0	203
Whole year..................	556.0	622.	1400.9	125

SUMMARY

Datum.	Most probable discharge under forest protection	Actual discharge after denudation	Gain in flow from denudation	
	Inches	*Inches*	*Inches*	*Per cent*
1. Total for the year, 10/1/19 to 9/30/20	7.608	8.566	.958	12.6
2. Total for spring flood, 4/8 to 7/13/20	5,195	5.946	.751	14.5
3. To crest of flood, 4/8-5/23.....	2.115	2.727	.612	29.0
4. After crest of flood (by differences, 2-3)..............	3.080	3.219	.139	4.5
5. July, after end of flood.......	.240	.242	.002	0.8
6. August	.333	.341	.008	2.4
7. September	.317	.331	.014	4.4
8. Whole summer, by sum 5, 6, 7.	.890	.914	.024	2.7
9. Whole summer, single calculation	.888	.914	.026	2.9
10. Fall and winter freezing period, 10/1/19 to 4/7/20......	1.754	1.707	–.047	–2.7
11. Total for year, lines 2, 5, 6, 7, 10.......................	7.839	8.567	.728	9.3
12. Fall and winter, by differences (line 1 minus 2, 5, 6, and 7).....................	1.523	1.706	.183	12.0
13. Whole summer, June 1-September 30..................	2.282	2.362	.080	3.5
Critical rates:				
14. First day of flood, 4/8/20.....	.01140	.01650	.0051	44.6[a]
15. Day of crest on A, 5/23/20....	.2265	.3105	.084	36.9[a]
16. Last day of flood, 7/13/20....	.0421	.0443	.0022	5.4[a]
17. Last 5 days September, rainless	.0554	.0568	.0014	2.0[a]

[a] These percentages are obtained from the original calculations in terms other than ins. O. W., the transposition of figures slightly altering the results.

fire or removing a narrow strip of timber on either side of the stream-channel, when compared with the most probable flow of this same stream, for conditions of the year as evidenced in climatic records and the flow of a second stream, show the following facts which may or may not be duplicated in succeeding years.

1. Removal of the forest cover had the effect of causing earlier melting of the snow and a higher rate of flood discharge during the entire flood period from April 8 to July 13, 1920. The excess of flow was most marked in the early stage of the flood and least important at the end. It amounted to 14.5 per cent for the entire flood period.

2. During the latter half of July the stream from the deforested area was nearly normal. In August and September, however, its flow increased slightly, probably both on account of larger direct contributions from rains, and because the loss of a storage water through the transpiration from leaves was minimized. At the end of September the stream was flowing 2 per cent above its expected rate.

3. Whether the stream flowed more or less than it would have done under the influence of forest, during the fall and winter period, is problematical.

4. There was an excess of 125 per cent in the amount of measurable detritus carried by the stream after the forest removal. Most of this was secured during the flood quarter.

5. Expressing the late flood and summer conditions as one quantity, it may be said that the removal of the forest made available .08 inches over the watershed, or .0067 acre feet per acre of watershed, which would not otherwise have been available during the period June 1 to September 30. This is to be balanced against more than .05 acre feet which came down as flood excess too early to be of value in irrigation, and against the 125 per cent increase in detritus.

Large Weighing Lysimeter for Measuring Evapotranspiration

W. O. Pruitt and D. E. Angus
MEMBER ASAE

THE meteorological approach to evapotranspiration has been hampered in the past by lack of adequate independent means of measuring the flux of water vapor away from the surface. Such measurements are required over short-time intervals, so that variations in the meteorological parameters are sufficiently small for mean values to be used. Under most daytime conditions this interval should probably not be longer than about one hour. It is possible to use weighing lysimeters to obtain the flux of water, but care is required lest certain sources of error to which they are subject make any data obtained with them of doubtful value.

A large weighing lysimeter designed for short-period measurements of evapotranspiration was installed in 1958-59 by the department of irrigation of the University of California, Davis, and has been in operation since June of 1959. In this installation considerable care was taken to minimize the known source of error in use of lysimeters for measuring evapotranspiration. The following discussion indicates what steps were taken.

Advected Energy

An energy balance can be set up for an extensive homogenous area, in which the net radiant energy available at the surface is partitioned between sensible heat flow into the soil, sensible heat flow into the air, and latent heat flow into the air (i.e., evaporation). If air moving in from the surroundings is hotter and drier than the air over such a surface, there will be additional latent heat loss near the upwind edge, which will continue into the crop for some distance.

It is therefore obvious that, if the evapotranspiration from a lysimeter is to be compared with that computed from an energy-balance approach, or an aerodynamic approach, the lysimeter must be located a sufficient distance in from the upwind edge of the area for gradients and heat fluxes to be representative of that area.

The Davis lysimeter is set in a field 1165 ft long and 480 ft wide, in such a position that in the directions from which about 85 percent of the winds come, the air has moved over the field for a distance of between 550 and 650 ft before reaching it. The whole field is planted to the same crop as that in the lysimeter. Although the effect of advection at different distances in from the upwind edge of a field is not known quantitatively, empirical data from other parts of the world suggest that profiles over the lysimeter up to a height of about 4 m should be representative of the field.

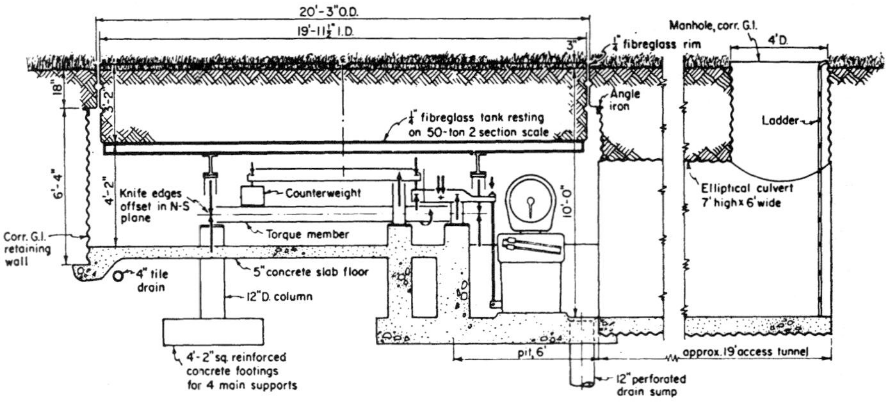

FIG. 1 Diagram of the 20-ft diameter weighing lysimeter installed by the irrigation department of the University of California at Davis, in 1958-59.

Lysimeter Wall Effects

In addition to the large-scale advection of heat on the field, there is small-scale advection from the area of surface occupied by the lysimeter walls. This effect can be reduced by making the ratio of the perimeter to the area of the lysimeter small, and for this reason a circular design was chosen. It can also be reduced by making the area of the annulus formed by the containing and retaining walls, plus the gap separating them, small compared to the enclosed area. A circular design is again an advantage since the walls of both the lysimeter and the retaining walls can be kept at a minimum thickness for a given strength of wall material. In certain types of construction this is not possible. For example, enclosing an undisturbed block of soil is very desirable where duplication of natural conditions of runoff, infiltration and percolation is important. To accomplish this, however, a relatively large surface area of retaining and containing walls is usually required which is undesirable if the lysimeters are to be used in evapotranspiration studies. Consequently the Davis lysimeter is a fill-type unit. The possible influence of soil disturbance is considered later.

The walls (and bottom) of the tank, and the top 20 in. of the outer retaining wall, are made of ¼-in. thick polyester resin reinforced with fiberglas. With a clearance between the two walls of only 1¼ in., the total wall plus gap area is less than 3 per cent of the enclosed soil area in the 20-ft-diameter tank.

Another wall effect which can be very important involves growth and transpiration differences of plants immediately adjacent to the inner and outer walls. Root restriction, soil fertility differences and unrepresentative heat flow in the walls are only some of the factors which might alter evapotranspiration in a given annular space around both inner and outer walls. For a given type of construction perhaps the most practical way this possible error can be minimized is to go to large sizes so that the affected area becomes a relatively small portion of the total surface area. This was another reason for choosing the rather large diameter of 20 ft.

Sensitivity

Since the weight of water evaporated over short periods is only a minute fraction of the total weight of the lysimeter, it is necessary for maximum sensitivity, to have a relatively large ratio of surface area to volume. Thus the lysimeter needs to be kept as shallow as possible, without causing unrepresentative root development. In the Davis lysimeter, once the area was decided on, the maxi-

Paper prepared expressly for publication by the American Society of Agricultural Engineers.
The authors—W. O. PRUITT and D. E. ANGUS—are, respectively, assistant research irrigation engineer, University of California, Davis, and physicist, division of meteorological physics, CSIRO, Melbourne, Australia.
* Numbers in parentheses refer to appended references.

mum depth was predetermined by the sensitivities of scales of various capacities. The chosen depth of 3 ft required a scales capacity of 100,000 lb, and the sensitivity obtainable (over the range from PWP to saturation in the soil mass) was ± 2 lb. This is equivalent to about ± 0.0012 in. of evaporation. Pressure fluctuations due to medium to strong winds cause an uncertainty in any given reading greater than this value. For this reason the automatic weight-printing mechanism is designed to record the dial reading every 4 min. The loss of weight over an hour's interval then is obtained with a fairly good degree of precision from the line of best fit through the individual values. Percentagewise, of course, the accuracy is best during periods of high evapotranspiration losses.

The 50-ton scale has sufficient sensitivity so that on a calm night the dial indicator responds to the application of water from a sprinkler jet, indicating an increase in weight of from ¼ to ½ lb during each sweep of the rotating sprinkler.

The dial is graduated in 2-lb intervals and the automatic printer records the dial reading to the nearest 2 lb. A counterweight in the scale system takes care of the dead load up to approximately 80,000 lb of soil and water (a condition somewhat below the wilting point). Above this load the dial with a 2,000-lb range plus a set of nine drop weights provide for a 20,000-lb total range of weight change.

At field capacity the dial indicates approximately 16,000 lb. Readings above 20,000 lb on up to saturated soil conditions can easily be obtained by adding some extra weights to the counterweight.

Crop

Because profile measurements (wind, temperature and moisture) are more easily taken over a very low-growing crop, it was decided that for the first phase of the project grass would be used. Desirable characteristics were: a grass that was turflike and not clumpy, that would stay green all winter, that has a rooting depth not much greater than 3 ft but not so shallow-rooted that it needs very frequent irrigation in summer, and that does not grow so rapidly that it requires too frequent mowing. The best compromise appeared to be perennial ryegrass.

Any discontinuity between the growth of the crop in the lysimeter and in the surrounding field will, by changing the turbulence of the air moving over it, cause unrepresentative evapotranspiration from the lysimeter. This error decreases as the lysimeter area increases, and is one more reason for going to a large unit. Also the whole field should be irrigated or cut in one operation in as short a period as possible. With four sprinkler laterals at 60-ft spacing it takes two days to apply a 2 to 2½-in. depth of water over the entire field (including the lysimeter). Irrigations are started at night with a lateral on both sides of the lysimeter to eliminate loss of daytime data. The grass is cut with a 5-gang reel mower in about 6 hr. An area of approximately 100-ft radius in and around the lysimeter is cut to the same height (approximately 4 in.) by a 22-in. rotary mower.

Disturbance of Soil

The most serious effect of using disturbed soil will probably be the altering of plant growth. This will not necessarily be a more serious problem with the Davis installation than where a monolithic construction is used. In the latter case a large area of soil around the lysimeter must be disturbed in order to lower the container down over the undisturbed soil block and make room for installation of retaining walls, scales, etc. To limit the effects of any differences in plant growth the grass is cut at 7 to 10-day intervals. By doing this the height of grass is kept between 4 and 8 in.

Another problem of using disturbed soil may be the altering of its moisture-retention characteristics, leading to differences in soil moisture tension and the soil heat flux component of the surface energy balance.

However, the field in which the lysimeter is located is a Yolo loam which exhibits virtually no change in structure within the top 3 ft, and is characterized by the absence of any clearly defined horizons. Because of this and the filling techniques used it is believed that alterations of the moisture-tension and thermal properties of this soil have not been appreciable. The effect on evapotranspiration should be negligible, especially if low moisture-tension values are maintained.

In excavating for the lysimeter, the soil from each one-foot layer was kept separate and was loaded back in the correct order. Approximately 16 in. of loose soil was replaced at a time. It was not mechanically packed but by saturating with water and draining several times, each 16-in. layer of loose soil settled approximately 4 in. The drainage system which greatly facilitated the filling and settling operations is described later.

Filling of the lysimeter was started early in March, 1959. By May 8 the soil was at a condition very close to the original field bulk density and sod from another part of the field was layed in place within the lysimeter.

Soil Heat Flux

Although the thermal properties of the soil in the lysimeter and in the field might be similar after settling the soil, any differences in the vertical temperature gradients will change the soil heat flux. (For this reason a metal container, which readily conducts some heat to the deeper soil layers, was not used.) Because the annual temperature wave in the soil will be cut off beneath the lysimeter by the underground room, the temperature of this room, and hence of the base of the lysimeter, need to be artificially maintained equal to that at the 3-ft depth in the field. Both heating and refrigerating equipment will be used to achieve this temperature control if necessary. With the soil temperature at the 3-ft depth maintained the same, and similar crop growth at the surface, it is unlikely that the surface soil temperature will be different in the lysimeter from that in the field. Hence the soil heat flux should be identical. The gap at the ground surface between the containing and retaining walls is sealed with a flexible strip of polyvinyl film to prevent warm air rising out of the underground room and influencing the evaporation or condensation processes near the rim. It also serves to keep out rain, dust, insects, and plant materials from the underground room.

Soil Moisture

As previously mentioned some alteration of moisture characteristics of the disturbed soil might be expected, but any direct effect on evapotranspiration can be reduced by maintaining low moisture-tension values.

During the summer, the sprinkler system is used to add only enough water to bring the soil in the lysimeter up to field capacity. There is then no trouble with excess water at the bottom of the lysimeter.

In winter, however, periods of heavy rain may cause the soil in the bottom of the lysimeter to become saturated. Although an open-drainage system would lower the water table to the level of the drain, the soil would remain well above field capacity while outside the lysimeter the soil would continue to drain. Special attention has therefore been given to this problem. The solution involved a drainage system consisting of three parts. Free water is drained through perforated plastic tubing placed in a 1-in. layer of sand on the bottom of the lysimeter. Water help up to about 0.05 bars of soil moisture tension is drained through a grid of about 150 ceramic tubes also placed in the sand, each 12 in. long, connected to a vacuum system. Finally, water held up to about 0.2 bars is

FIG. 2 (Left) Aerial view in November 1958 of the lysimeter installation. The fiberglas-plastic tank and upper retaining wall still remain to be assembled in place over the channel deck of the scales. FIG. 3 (Right) View of the lysimeter and some of the associated micrometeorological instrumentation. Wheel marks are from a mower which had just been used to cut the grass.

drained by vacuum through 250 additional ceramic tubes placed in the soil immediately above the sand. The system in the soil is never operated below 0.05 bars to prevent movement of the finer particles in the soil and possible clogging of the tubes. The whole system is used to drain the lysimeter at a rate as close as possible to the natural drainage rate in the field. So that the water drained will not affect the dial readings, it is collected in large vacuum tanks suspended from the scales platform. The tanks are emptied during a 10 to 15 min period at night. Thus there is no appreciable interruption of evapotranspiration records and the total amount drained is easily determined from dial records.

There are several recognized limitations in the use of this lysimeter. One of these concerns the type of crop that can be grown on it. If a crop with a rooting depth appreciably greater than 3 ft is used, more water will be withdrawn from the soil in the lysimeter than from the same depth of soil in the field, especially as the soil moisture tension near the surface increases. Although there is a considerable difference of opinion between workers in this field as to the effect of soil moisture tension on the rate of evapotranspiration, it seems preferable at this stage to ensure that as far as possible the moisture status is kept similar inside and outside the lysimeter.

During most of the early phases of this project, it is desired to maintain soil moisture conditions both within and outside of the lysimeter such that the losses will be near maximum for any given climatic condition. Irrigations are applied frequently enough to keep the soil moisture tension at a 1-ft depth below 0.4 to 0.5 bars. This required a frequency irrigation of from every 7 to 10 days during midsummer of 1959.

Inverted tensionmeters at five soil depths and three locations and at two depths at two other locations provide the measurement of the soil moisture tension within the lysimeter. All of the manometers for these units, as well as for twelve others located near the lysimeter, are located in the underground room. Electrical resistance type units indicate tension values above 0.8 of a bar for periods when the effect of soil moisture tensions on evapotranspiration is studied.

General Construction

General construction details are shown in the sketch of the installation in Fig. 1. Not shown are two drainage-collection tanks of 14-in. diameter by 10 ft long, suspended from the 5-in. channel beams which support the fiberglas-plastic tank. An automatic printer located on the right side of the dial is also missing in the diagram. All of the parts of the 50-ton scale are not shown but forces involved are indicated by arrows. The 19-ft long access tunnel leads out east of the lysimeter towards a road. The discontinuity in grass cover caused by the 4-ft diameter manhole should have no effect on evapotranspiration from the lysimeter, an air movement from the east is a rare occurrence. The tunnel has benches attached to either side for supporting various potentiometers and recorders, for the temperature and wind-profile registering systems located near the lysimeter.

From a depth of about 8 to 15 ft below the surface there is a natural stratum of sand in which the scales foundations are located. A 5-ft deep sump with a 2-ft section of perforated well casing serves as a drain for water from the drainage collection tanks and from the clay tile system underlying the floor of the underground room. It will also serve as a drainage well from which water can be pumped if necessary should a secondary water table develop under the installation.

Fig. 2 is an aerial picture taken in November, 1958, just before the five pieces of the tank and the four pieces of the upper retaining wall were assembled in place. In the center of the picture the channel-beam decking resting on the two main I beams is clearly visible. A plank rests across the 4-ft diameter manhole leading to the access tunnel.

Fig. 3 shows a view taken in November 1959 just after grass in the lysimeter was cut. One of the authors, D. E. Angus, is standing just beyond the south edge of the tank. Behind him but barely visible are anemometers on a 2-meter mast and a net exchange radiometer at a 2-me height. To the left the radiation shields of a temperature and humidity profile system are evident along with a 4-me wind and temperature mast. A number of soil temperature and soil heat flux plates are located in the general vicinity of the 4-me mast.

Most of this instrumentation is duplicated in an area just out of view north of the lysimeter. As indicated previously all of the recorders for all instruments are located underground in the access tunnel.

Five other lysimeters located within the same field of ryegrass supplement the data obtained from this installation.

The California State Department of Water Resources in 1958 installed three 2-ft diameter, 27-in. deep lysimeters. These units are weighed at 5 to 10-day intervals to the nearest pound using a portable tripod and beam balance.

Two floating-type lysimeters, similar to the one described by King, Tanner and Suomi (1)*, were installed in July

(Continued on page 18)

WEIGHING LYSIMETER

(Continued from page 15)

1959. They are 6 ft in diameter and 6 ft deep. Water stage recorders were altered to record on a 10 to 1 ratio the water level in a stilling well connected to the outer tank. The sensitivity of these lysimeters is better than ± 0.002 in. providing excellent opportunity for comparison, even on an hourly basis, with data from the larger unit. These two lysimeters which were installed primarily for studies of interception losses have been described previously by McMillan and Burgy (2).

References

1 King, K. M., Tanner, C. B., and Suomi, V. E. A floating lysimeter and its evaporation recorder. Trans. AGU 37 (6): 738-42. Dec. 1956.

2 McMillan, W. D. and Burgy, R. H. Interception loss from grass. J. Geophys. Research, (In press) 1960.

WATER-FLOW CALCULATIONS BY MEANS OF GAMMA-ABSORPTION AND TENSIOMETER FIELD MEASUREMENTS IN THE UNSATURATED SOIL PROFILE

W. GIESEL, S. LORCH, M. RENGER
Niedersächsisches Landesamt für Bodenforschung
and
O. STREBEL
Bundesanstalt für Bodenforschung,
Hannover-Buchholz,
Federal Republic of Germany

Abstract

WATER-FLOW CALCULATIONS BY MEANS OF GAMMA-ABSORPTION AND TENSIOMETER FIELD MEASUREMENTS IN THE UNSATURATED SOIL PROFILE.

By means of a double-probe densitometer, variations in the soil water content, and by means of tensiometers, soil moisture suctions, are determined as a function of time in different horizons of a soil profile. From these data are computed the water flow, in particular the evapotranspiration and the exchange of moisture between the soil and groundwater zones. By including the precipitation, the essential components of the water conservation equation are covered. In addition, it is possible to obtain from the field measurements the relations between soil-water content, soil-water suction and hydraulic conductivity for the individual horizons of a profile.

INTRODUCTION

The understanding of water movement in natural soil profiles under vegetation cover is of great importance for soil problems (water supply of plants, drainage, irrigation) and for problems in hydrology (groundwater recharge) and agrometeorology (evapotranspiration). However, in many cases, such water flow investigations are only partly carried out in the field because of difficulties in measuring and analysing. For instance, soil water contents at various depths are determined in the field, and the other data required for water movement calculations are derived from the relations between soil water content and soil water suction and between soil water content and hydraulic conductivity, which are determined on core samples in the laboratory. This procedure may cause several errors, so for a critical examination, experiments with simultaneous field measurements of soil water content and soil water suction as a function of time and depth are needed [1]. An analysis of these field data by computer gives evidence of the water flow, and also of the above-mentioned basic relations in the soil horizons or layers of a natural profile under non-equilibrium conditions.

The procedure and some preliminary results of such simultaneous soil water content and soil water suction measurements in a löss soil profile under arable cultivation are described in the present paper.

MEASURING METHODS

Determination of variations in soil water content

Since a high degree of accuracy in the moisture determination and good depth resolution were required, the double-probe densitometer [2, 3] was used for long-term measurements. This instrument operates on the principle of density determination by means of gamma-absorption measurement; density variations are attributed to variations in soil water content. Two access tubes were installed in the soil to a depth of up to 380 cm, about 50 cm apart and as parallel to each other as possible. In one of the tubes a gamma-source (^{137}Cs, 50 mCi) and in the other tube a detector (scintillometer) were moved up and down stepwise and synchronously by a special device. The step interval was 10 cm. For each measurement the source-to-detector arrangement was kept in each of the 35 measuring layers for 5 min. The output of the detector (the counted gamma-quantum) was punched on tape and the data were then analysed by computer. The drift of the instrument was almost completely eliminated by regular control measurements in some distinct layers below groundwater level.

An automatically operating gamma double-probe of this kind has proved its value in uninterrupted field measurements for more than three years. It allows exact determinations of soil water content variations in layers 5 - 10 mm thick. The measuring accuracy is 10-fold better than that of previous, conventional instruments. The total error in determinations of water content variations in long-term field measurements is approximately $\pm$ 0.0015 g/cm^3. The absolute water content is obtained from a previous calibration of double-probe density values with soil water content; for that purpose the water content is determined on core samples of the corresponding soil layers taken simultaneously with densitometer measurements. This calibration procedure leads to a result with an error in the absolute water content of $\sim$0.01 cm^3 water/cm^3 soil which is constant over the whole measuring period. However, the water content variation in time, which is the essential factor in this paper, is not influenced by this error.

A method to differentiate between variations in water content and variations in dry bulk density by means of two gamma-sources with different energies [4] is being tested.

Determination of soil water suction

The measurements were carried out by tensiometers with a Hg manometer. Two tensiometers were installed in eleven different measuring depths at about 200 cm apart to get an appropriate average value. The readings were taken daily. The use of conventional tensiometers was only provisional; later they will be replaced by a recording type of pressure transducer-tensiometer, which has a higher resolution and a lower thermal sensitivity [5].

Precipitation and groundwater level at the investigation site were measured with a rain-gauge and an automatic water-level recorder, respectively. In the laboratory, the pore size distribution of the various soil horizons and layers was determined on core samples according to Ref. [6].

FIELD SITE AND SOIL PROFILE

The investigated site near Hannover was on a grey-brown podzolic soil profile commonly found in large areas of the Federal Republic of Germany. The parent material is the silty loam of a löss layer 160 cm thick and underlain by gravelly sands of lower terrace sediments.

The pore size distributions in this two-layer profile (Fig. 1) show important differences between the upper and lower layers. These differences have a strong effect on the water movement (see Fig. 2). During the measuring period the groundwater level fluctuated between 255 and 285 cm below the soil surface (mean value 270 cm). Therefore, a significant upward flow from the groundwater in the gravelly sands into the upper löss layer is not to be expected.

The vegetation cover was a dense stand of sugar beets.

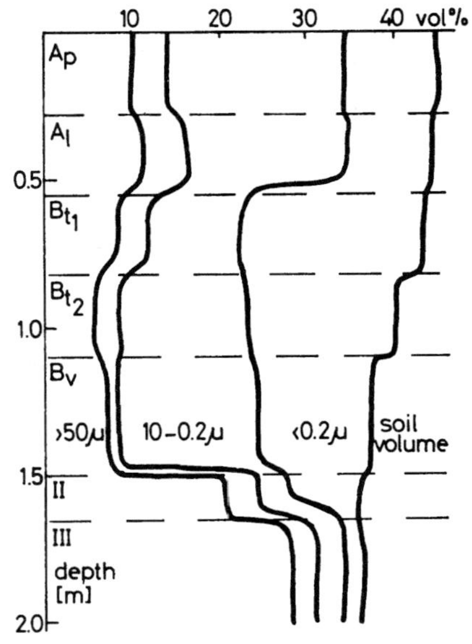

FIG. 1. Pore size distribution of a grey-brown podzolic soil profile.

THEORY

The equation of continuity

$$\text{div w} = -\frac{\partial \theta}{\partial t} \quad (1)$$

and Darcy's law can be applied to the water movement in the unsaturated soil zone between groundwater level and soil surface [7-11]. The flux w [cm^3 water/(cm^2 area × day)] is the sum of the water transport through pores v and the other flux components, which are expressed by r (vapour

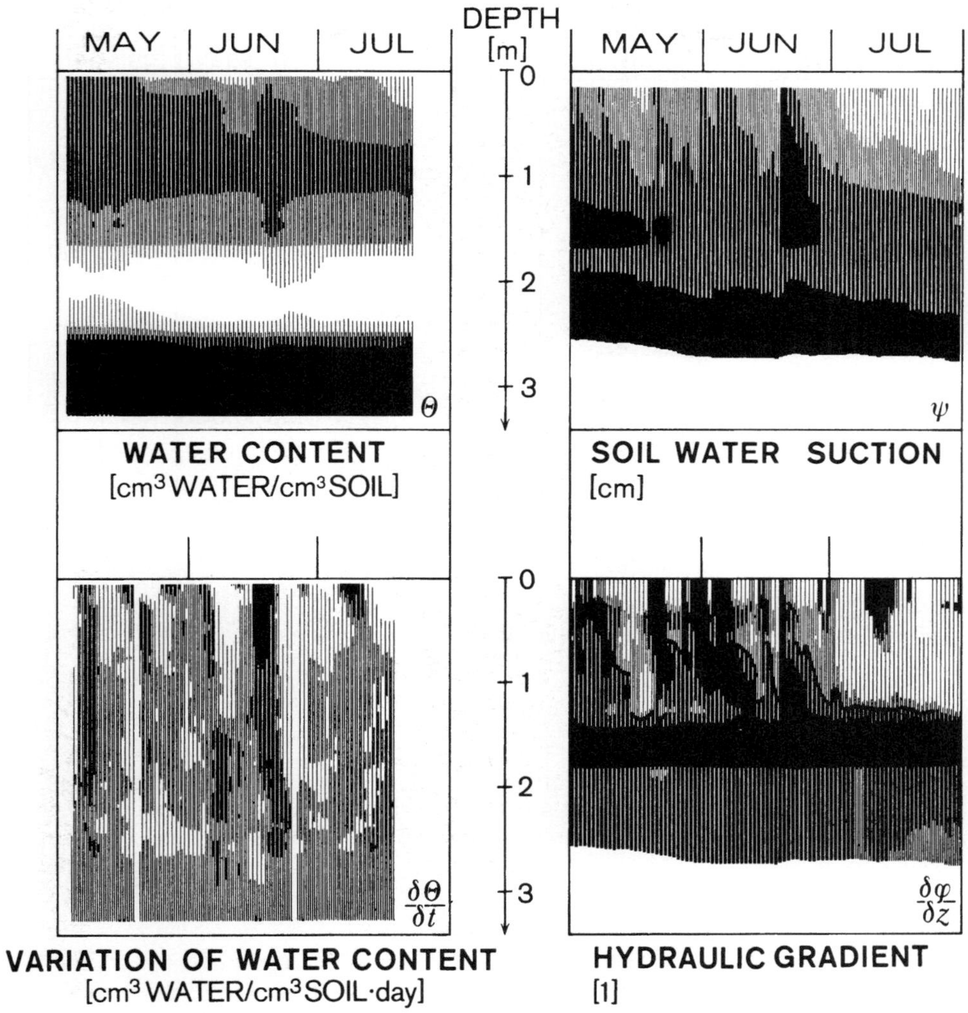
MAY
JUN
JUL
DEPTH [m]
0
1
2
3
θ
WATER CONTENT [cm3 WATER/cm3 SOIL]
ψ
SOIL WATER SUCTION [cm]
δθ/δt
VARIATION OF WATER CONTENT [cm3 WATER/cm3 SOIL·day]
δφ/δz
HYDRAULIC GRADIENT [1]

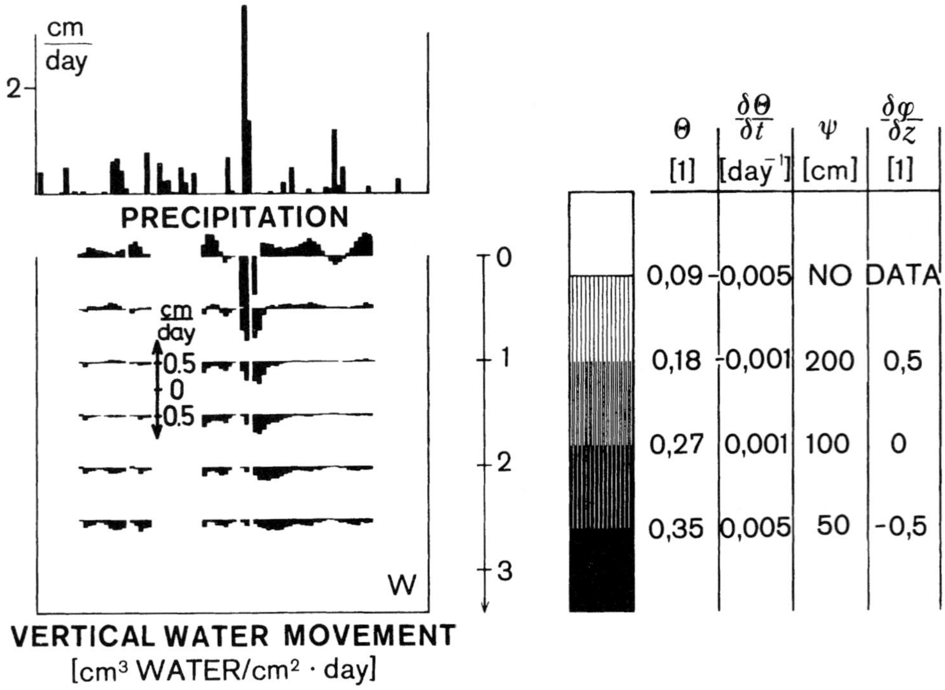

FIG. 2. Soil water content, soil water suction and vertical water movement as a function of depth and time.

transport, flux through roots and plants). The water content of the soil is designated as θ (cm^3 water/cm^3 soil). At flat sites the water movement in homogeneously rooted soils is practically vertical. So Eq. (1) can be reduced to

$$\frac{\partial w}{\partial z} = - \frac{\partial \theta}{\partial t} \tag{2}$$

and

$$\overline{w}_2 - \overline{w}_1 = - \int_{z_1}^{z_2} \frac{\partial \theta}{\partial t} \, dz \tag{2a}$$

respectively. The term on the right in Eq. (2a) can be calculated from the water content data, θ (z, t), which are determined with the gamma probe densitometer.

Figure 2 shows the water content data and the function:

$$\frac{\partial \theta}{\partial t} (z_1 t)$$

If the flux w_2 at any depth z_2 of the profile is known, then the flux w_1 at any other depth z_1 can be calculated according to Eq. (2a). The flux w_2 at distinct depths can be derived from tensiometer measurements.

The tensiometer readings give the capillary potential

$$\psi\ (z, t) \quad [\text{cm water}]$$

from which the gradient of the hydraulic potential

$$\frac{\partial\varphi}{\partial z} = \frac{\partial}{\partial t}(\psi + z)$$

can be calculated. The flow in the liquid phase follows from Darcy's law

$$v = -\,k\,(\theta)\,\frac{\partial\varphi}{\partial z} \quad [\text{cm/day}] \tag{3}$$

According to Eq. (3)

$$v = 0$$

if

$$\frac{\partial\varphi}{\partial z} = 0$$

The measuring data for

$$\psi\ (z, t)$$

and the resulting

$$\frac{\partial\varphi}{\partial z}(z, t)$$

are plotted in Fig. 2. The lower edges of these plottings signify the groundwater level. The gradient of the hydraulic potential is zero along the drawn line.
This depth

$$\zeta\,(t)$$

is used as a limit of integration when evaluating the water movement according to (2a)

$$w = \int_{z}^{\zeta} \frac{\partial\theta}{\partial t}\,dz \tag{2b}$$

since

$$w_2 = w_\zeta = 0; \quad w_1 = w$$

The formula can be applied only provided that no water (including vapour and flux through roots) penetrates the depth ζ. Therefore only ζ-values of sufficient depth can be used. The flux w calculated in several depths is shown in the lower part of Fig. 2. Upward and downward directed columns signify the amount and direction of water movement.

DATA PROCESSING

The daily tensiometer readings were punched on cards in the Institute, whereas the counting rates from the gamma probe-densitometer measurements were automatically punched on tape in the field. An IBM 1620 computer was fed with these data. From water content measurements, mean values were developed for each day. For computing the daily water content changes $\partial\theta/\partial t$, a quadratic approximation of daily mean θ-values of seven successive days was used at any given time. In horizons near the soil surface this procedure causes some errors in $\partial\theta/\partial t$ because of a moderate suppression of sudden water content changes which may happen there and which influence the computed data of daily water movement w in Fig. 2. But the flux w, averaged for a longer interval, say seven days, can be considered correct. It was plotted automatically as a column diagram in Fig. 2 after computation.

The relation between soil water content and water suction $\psi = f(\theta)$ is the soil moisture characteristic. It is shown in Fig. 3 for the A_1-horizon within a depth of 30-55 cm; only field data were used. The difference in water content at the same water suction, which amounts to ~ 0.08 cm^3 water cm^3 soil in the suction range of 100-200 cm, can be attributed to the effect of hysteresis.

Besides the water content, the gamma double probe also gives the air content of the soil layers which is an important value for aeration problems of the root zone. These air contents are the differences between the total pore volume and the water content. They are also plotted automatically as shown in Fig. 4.

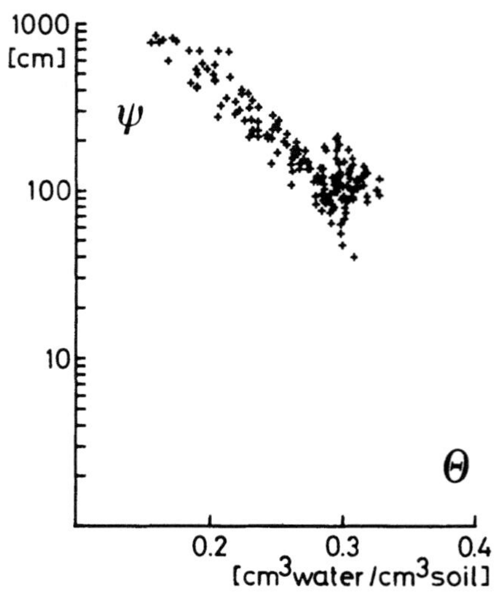

FIG. 3. Relation between soil water content and soil water suction of the A_1-horizon.

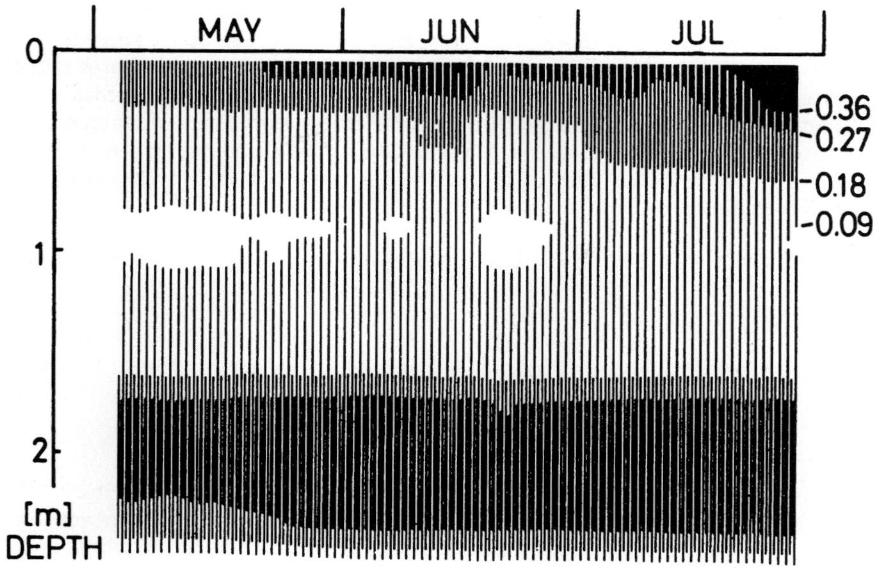

FIG. 4. Air content of the soil as a function of depth and time.

CONCLUSIONS

The results described here are from the authors' initial attempt to calculate water movement in soils without the use of laboratory data.

The calculated water fluxes show fairly good agreement with those which can be derived from some field experiments with tritiated water. For example, in a loamy sand soil under grass cover, Zimmermann et al. [12] found a downward displacement of the tritium peak of ~ 60 cm during 90 days in winter at a soil water content of about 0.3 cm^3 water/ cm^3 soil. These data yield an average daily downward movement of the bulk water of ~ 0.2 $cm^3/(cm^2$ day).

Some difficulties arose, especially with the numerical differentiation of the measuring data ($\partial\varphi/\partial z$, $\partial\theta/\partial t$). In future, a higher degree of accuracy of the basic values of soil water content, θ, and of soil water suction, ψ, should be achieved, as well as a better resolution in time, t, and measuring depth, z, of the tensiometer measurements. This resolution should correspond to that of the gamma double probe; this will be possible with the recording type of pressure transducer-tensiometer (see Section on Measuring Methods).

As soon as it is possible to measure the soil water content, $\theta = \theta(z, t)$, and the soil water suction, $\psi = \psi(z, t)$, more exactly and with better resolution in time, the water flow calculations will be more accurate. This, in turn, means that the hydraulic conductivities for each soil horizon can be determined according to Eq. (3) and can be given as a function of the known soil water content θ:

$$k = k(\theta, z)$$

However, this is true only for deeper soil layers for which water flux as vapour transport or flux through roots can be neglected. By using the hydraulic conductivities calculated in this way it is possible to determine the water flux also for time intervals without zero hydraulic potential gradients at any depth of the soil profile. Furthermore, quantitative data about the vapour transport and the water flux through roots, r, in the upper soil layers can be calculated in the following way. From Eq. (2b) the flux, at a distinct depth z, is

$$w = v + r = \int_z^{\zeta} \frac{\partial \theta}{\partial t}\, dz \qquad (2c)$$

and including Eq. (3)

$$k \frac{\partial \varphi}{\partial z} - r = - \int_z^{\zeta} \frac{\partial \theta}{\partial t}\, dz \qquad (2d)$$

For a flux r = 0, the term on the right divided by the gradient of the hydraulic potential gives the hydraulic conductivity $k = k(\theta)$. Systematic deviations from this previously known quotient must then be attributed to a flux $r \neq 0$.

With measurements that are improved and completed in the above manner it should be possible to reach a thorough understanding of the water movement and water balance in our investigation site. Moreover, the measurements should provide important indications of the possibilities and limitations of less intricate and more routine investigations on soil water problems. For instance, by comparing field and laboratory data of moisture characteristics and of the relation between the water content and hydraulic conductivity, the influence of hysteresis, non-equilibrium conditions and inhomogeneities in natural soil profiles can be evaluated.

ACKNOWLEDGEMENTS

The authors are greatly indebted to the "Deutsche Forschungsgemeinschaft" for their financial support of the investigations.

REFERENCES

[1] GARDNER, W.R., Present knowledge of the inter-relationships between soil moisture, irrigation, drainage, and water-use efficiency, Soil-Moisture and Irrigation Studies (Proc. Panel Vienna, 1966), IAEA, Vienna (1967) 77.

[2] LORCH, S., Messung zeitlicher Wassergehaltsänderungen im Boden, Z. Geophysik 33 (1967) 403.

[3] LORCH, S., "Bestimmung der Feuchtdichte und der Wassergehaltsänderung eines Bodens durch Messung der Absorption von γ-Strahlen", Section 4.4.3.3 in: SCHNEIDER, H.: Die Wassererschliessung, Hank-Druck (1970).

[4] SOANE, B.D., Double energy gamma transmission for moisture and density measurement in soil tillage studies, Int. Soil Water Symp, Prague, Discussions (1967) 197.

[5] STREBEL, O., GIESEL, W., RENGER, M., LORCH, S., Automatische Registrierung der Bodenwasserspannung im Gelände mit dem Druckaufnehmer-Tensiometer, Z. Pfl-Ernähr. Düng. Bodenk. (in press).

[6] RICHARDS, L.A., FIREMAN, M., Porous plate apparatus, Soil Sci. 56 (1943) 395.

[7] GARDNER, W.R., Dynamic aspects of water availability to plants, Soil Sci. 89 (1960) 63.

[8] GARDNER, W.R., EHLIG, C.F., Some observations on the movement of water to plant roots, Agron.J. 54 (1962) 453.
[9] ROSE, C.W., STERN, W.R., Determination of withdrawal of water from soil by crop roots as a function of depth and time, Aust.J. Soil Res. 5 (1967) 11.
[10] VAN BAVEL, C.H.M. et al., Hydraulic properties of a clay loam soil and the field measurement of uptake by roots, Soil Sci. Soc. Am. Proc. 32 (1968) 310.
[11] WESSELING, J., Principles of the unsaturated flow and their application to the penetration of moisture into the soil, Techn. Bull. No. 23, Wageningen (1961).
[12] ZIMMERMANN, U., EHHALT, D., MÜNNICH, K.O., "Soil-water movement and evapotranspiration: changes in the isotopic composition of the water", Isotopes in Hydrology (Proc. Symp. Vienna, 1966), IAEA, Vienna (1967) 567.

DISCUSSION

U. ZIMMERMANN: Are you able to determine lateral water movement with your method? Knowledge of this factor can be very important in studying phenomena such as evapotranspiration.

W. GIESEL: No, we are concerned at present only with flat, homogeneously rooted soil sites, where lateral water movement can be neglected.

U. ZIMMERMANN: Have you carried out tritium tracer experiments with a view to comparing the results with those obtained by gamma absorption and tensiometer measurements?

W. GIESEL: No, we have not. But it would be interesting to make such a comparison.

G. VACHAUD: It is necessary to have very good thermal insulation of the photomultiplier in order to avoid serious thermal drift. How is the photomultiplier insulated?

W. GIESEL: Perhaps Dr. Lorch, who designed the equipment, would care to answer this question.

S. LORCH: For thermal insulation we use relatively large masses of lead and Styropor (expanded polystyrene) around the photomultiplier and the crystals. The long-term drift is eliminated by regular control measurements below the groundwater level.

G. VACHAUD: What is the counting time and count-rate?

S. LORCH: The counting time is four minutes and the count-rate 10^5-10^6 counts/min.

E. SOMER: You mention in the paper that you have obtained a depth resolution of 5-10 cm. It is difficult to understand how you have been able to obtain such a good resolution. Could you give details concerning the size of the scintillation crystal, the collimators used and the discrimination level?

S. LORCH: The diameter of the scintillation crystal is 1 in. and the height 2 in. The crystal is surrounded by lead about 1 in. thick. The discrimination level is just below the ^{137}Cs peak.

Agricultural and Forest Meteorology, 42 (1988) 63–73
Elsevier Science Publishers B.V., Amsterdam — Printed in The Netherlands

SPATIAL VARIABILITY OF THROUGHFALL AND STEMFLOW MEASUREMENTS IN AMAZONIAN RAINFOREST

C.R. LLOYD

Institute of Hydrology, Wallingford, Oxon. (Gt. Britain)

A. DE O. MARQUES F°

Instituto Nacional de Pesquisas da Amazonia, Manaus, Amazonas (Brasil)

(Received February 10, 1987; revision accepted May 28, 1987)

ABSTRACT

Lloyd, C.R. and de O. Marques F°, A., 1988. Spatial variability of throughfall and stemflow measurements in Amazonian rainforest. Agric. For. Meteorol., 42: 63–73.

The spatial variability of throughfall and stemflow in Amazonian tropical rainforest is analysed. Statistical tests are used to assess the minimum sampling size and number of gauges required. Random relocation of throughfall gauges is shown to be better than fixed position gauges and random relocation of gauges on a transect line is better than over an area. During the study period, throughfall and stemflow, calculated as percentages of gross rainfall, were 91(±2) and 1.8(±1)%, respectively. No attempt was made in this study to investigate storm size dependency of these figures.

INTRODUCTION

Although there are still extensive areas of tropical rainforest, these are rapidly diminishing. Their influence on local and global climate is at present poorly understood, but an important aspect is likely to be the water balance of the forest canopy during and after rain. Interception loss in extensive rainforest not only affects streamflow and recharge, but through repeated rainfall–evaporation cycles, may significantly alter the rainfall climate (Lettau et al., 1979). Measurements of interception loss are, therefore, required and these are conventionally made as the difference between rainfall and throughfall plus stemflow. Equally important is a knowledge of the errors involved in such measurements.

Recent reviews of tropical rainforest interception results, (Clarke, 1987; Raich, 1983), have shown that estimates of interception loss have ranged from < 3% in Nigeria to > 63% in a Thailand teak forest. While some of this variation may reflect different climate, storm pattern and forest structure, it is possible that many of the measurements of throughfall and stemflow did not take full account of the high spatial variability present in a heterogeneous

0168-1923/88/$03.50

tropical rain forest. Jackson (1971) showed that, as a result of this spatial variability, errors in such data could be very high and the need for large samples in such studies was emphasised.

Although much research experience exists in the measurement and analysis of throughfall and stemflow in temperate forests, such forests are often plantations of one species, or natural forests in which one species dominates. This experience is not directly applicable to the varied canopy found in tropical rainforest and this has repercussions in defining the best measurement techniques and analysis procedures.

This paper presents an analysis of tropical rainforest throughfall and stemflow measurements with the objective of demonstrating the need for adequate sampling, and the importance of quantifying the measurement errors.

EXPERIMENTAL SITE

The experimental site is located close to the micrometeorological site described by Shuttleworth et al. (1984) at 02°57′S, 59°57′W in the Reserva Florestal Ducke, 25 km from Manaus, Amazonas, Brazil. It was selected as being representative of the natural vegetation and regional topography. The forest canopy around the experimental site is of varied species (see Moore and Fisch, 1986) and typical of undisturbed natural Amazonian terra firma rainforest. It extends with no obvious sub-storeys to a height of ~ 35 m, with occasional emergent trees reaching 40 m. The plant density is high, up to 3000 stems ha^{-1}, but $< 10\%$ have girths of $\geqslant 0.2$ m. The larger trees provide the bulk of the vegetation at the top of the canopy. The middle and lower canopy is made up of an increasing number of individually smaller contributions. Palms, some reaching 4–5 m, provide the lowest storey. Ground-level vegetation is sparse. A detailed analysis of Amazonian terra firma rainforest structure close to this experimental site can be found in Takeuchi (1961).

METHOD

Gross rainfall

Gross rainfall was measured by three 203 mm diameter tipping bucket raingauges (Rauchfuss Instruments Division, Australia). One of these gauges had its 203-mm funnel mounted at the top of a 45-m tower and the rainfall catch piped to its associated 0.25-mm tipping bucket gauge (i.e., with a sensitivity equivalent to 0.25 mm per tip), situated on the ground. This arrangement minimised the possibility of spurious counts during high winds or periods of activity on the tower. Tests indicated that the combined wetting error of the funnel and pipe was equivalent to ~ 0.02 mm per storm event. The raingauge had two independent outputs which were separately recorded on two cassette tape loggers (Microdata, U.K.).

Two similar tipping bucket gauges (with 0.5- and 0.25-mm buckets) were situated 100 m from the tower on a small platform erected at the top of a tree, the crown of which was missing. These gauges were sited at a height of 25 m and, although adequately exposed for rainfall measurement, were sheltered from the wind by neighbouring trees. The two independent outputs from each gauge were again separately recorded on two more cassette tape loggers. The double logging of six independent outputs from three separate raingauges provided continuity in the gross rainfall measurement.

Measurements of rainfall from elevated gauges are known to be prone to error due to the modification of airflow around the gauge. Errors are largest when windspeeds are high and raindrop size is small (Robinson and Rodda, 1969; Mueller and Kidder, 1972). Typically, storms recorded in this study were intense and accompanied by light wind; aerodynamic errors are, therefore, likely to be small. Nevertheless any such error should be apparent from comparison of the measurements of the raingauge exposed at the top of the tower and the two raingauges sheltered in the canopy. Analysis showed no significant difference between the three gauges: individual storm totals agreed within the resolution of the gauges, and the cumulative rainfall measurements agreed to within 0.5%. For each hour, the value of gross rainfall was taken as the arithmetic mean of all gauges reliably operational at the time. Figure 1 shows the weekly gross rainfall, used in this analysis, for the period 5 July 1984–21 August 1985.

Throughfall

Throughfall was measured in a 100 × 4-m plot which was situated close to and between gross rainfall gauges. This shape was designed to allow the inclusion of several dominant trees as well as to provide easy access with

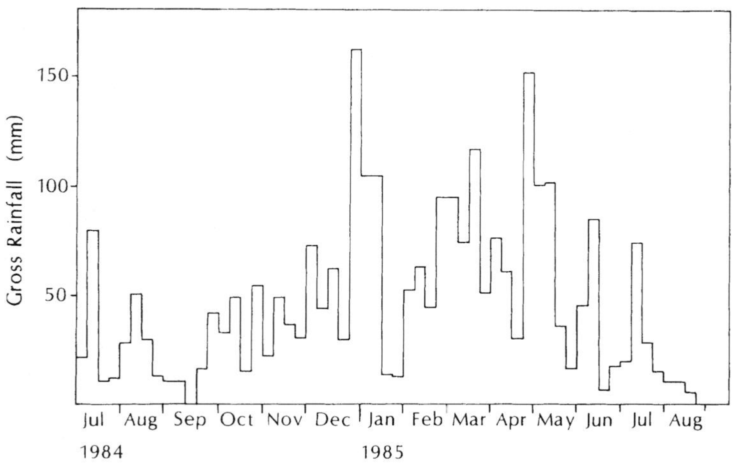

Fig. 1. Gross rainfall at the Ducke forest site for the period 5 July 1984–18 September 1985. Rainfall is grouped into weekly and 2-weekly totals.

minimal disturbance of the lower level vegetation. The plot area was marked out at 1-m intervals to give five lines and 505 sampling positions.

Thirty-six bottle gauges, each with a 127-mm diameter copper funnel, sampled throughfall in the area, each gauge being randomly relocated within the plot every week (except for four 2-week periods) from 5 July 1984 to 21 August 1985. For the period 14 March 1985–13 May 1985, the gauges were relocated after every storm. This intensive storm analysis was primarily used for estimating the canopy capacity, but has also been included in this analysis. Of the possible 505 sample positions, 494 were randomly chosen in 71 separate relocations during the period 5 July 1984–21 August 1985. The total throughfall collected in each bottle was measured with 0.5– and 1-l capacity measuring cylinders, the 1-l cylinder having 10-ml graduations. The estimation of any one of the 36 measurements is assumed to be uniformly distributed over a depth equal to one graduation. The variance of the sum of the 36 individual measurements is then such that the standard error, resulting from the resolution, in the sample mean of the 36 throughfall gauge amounts was < 0.02 mm. Steep angle funnels were used to minimise any "splash out" from the gauges during rain storms and, with the collecting area of each gauge 0.5 m above the ground, there was little opportunity for water to "splash in" after hitting the forest floor. Observation of the height to which soil particles had been deposited on the sides of the bottles showed a maximum splash height of ~ 0.25 m.

To estimate the evaporation loss from the bottle gauges, a test gauge was left for a month with a rain shield above it. The loss was equivalent to an error of < 0.06 mm per 2-week period.

Stemflow

Stemflow was measured on 19 adjacent trees within the area of the random grid. The trees ranged in girth from 0.12 to 1.64 m, and in height from ~ 5 m to canopy top; one tree had a buttressed lower trunk, two were tree palms and three trees had a liana which crossed from tree to tree at a height of 4 m. The bark texture ranged from very smooth to fibrous, often with a lichen or some other symbiotic plant cover.

The stemflow was collected in collars made from 0.2-m wide "flashing" (Evode, U.K.), a flexible, adhesive bitumen–lead building material. Six of the stemflows were measured using 0.5-mm tipping bucket raingauges similar to those used for gross rainfall. The outputs from these were accumulated on solid-state loggers (Computing Techniques, U.K.) with an integration time of 30 min. The remaining stemflows were collected in 20-l containers. The volumes were recorded weekly at the same time as the random grid throughfall measurements.

Data presented here are for the period 20 August 1984 – 21 August 1985. Logger faults and leaking or blocked stemflow gauges meant that data were not continuous, but measurements were available for > 17 gauges during 19 weekly periods and only 4 weeks had data available from less than nine gauges. In total, there were 47 consecutive weekly periods when stemflow and rainfall were simultaneously available.

RESULTS

Throughfall variability

The analysis of throughfall was divided into two sections: (1) an analysis of all the data to yield the percentage throughfall and to define the frequency distributions and required transformations necessary to allow the valid use of statistical tests; (2) to test whether sub-grid areas and transect lines were significantly different.

Total throughfall and its frequency distribution

The point throughfall catch expressed as a percentage fraction of the gross rainfall is shown in Fig. 2(a). Of the 494 unique sample points in Fig. 2(a), 29% recorded throughfall catches which were greater than gross rainfall measured during the same period. These accounted for 46% of the total throughfall catch. Again, while only 1.6% of the throughfall catches were greater than twice the gross rainfall, these accounted for 5% of the total throughfall catch. This compares with 3% of throughfall catches exceeding gross rainfall found by re-analysis of the data of Gash and Stewart (1977). These data, shown in Fig. 2(b), were taken in a plantation of Scots Pine (*Pinus sylvestris* L.) at Thetford Forest in Norfolk, U.K. Ford and Deans (1978) also indicate a similar pattern in a Sitka spruce (*Picea sitchensis* (Bong.) Carr.) plantation in Scotland. They found that throughfall catch ranged from 9 to 97% of incident rainfall. An analysis of separate storms indicated that the double peak, visible in Fig. 2(a), may have been related to storm size, but that the overall range of throughfall catch (0–410%) is present in both large and small storms.

The distribution shown in Fig. 2(a) is skew positive and, therefore, any attempt to assign standard errors, assuming a normal frequency distribution of individual period catches, will lead to erroneously large estimates of error. The logarithm and square-root transforms are commonly used for normalising skew-positive data and both were applied to the experimental distribution. A Chi-squared test showed the square-root transformation yielded a frequency distribution which was most nearly of Gaussian form, and it is with this transform that subsequent statistical tests were carried out.

The total throughfall was 2480 mm. This value is the arithmetic mean of all throughfall measurements over the observation period, as this is the best measure of the areal throughfall reaching the soil surface. This value is 91% of the total gross rainfall of 2721 mm.

Sub-grid samples

In many previous investigations, throughfall has been measured over a random grid on an area which is square. This method is most appropriate in

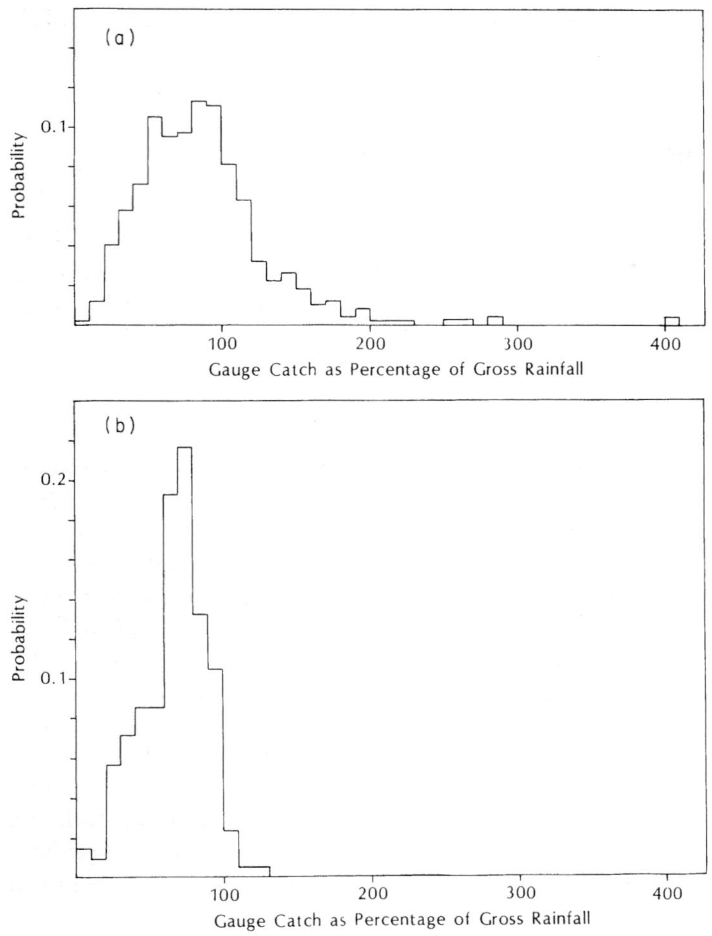

Fig. 2. (a) Amazon rainforest: probability distribution of throughfall gauge catch in the random grid expressed as a percentage of coincident gross rainfall. (b) Thetford forest: probability distribution of throughfall gauge catch expressed as a percentage of coincident gross rainfall.

homogeneous forests and where the trees are small compared to the size of the grid. On the other hand, phytogeographers have used a transect line as the reference to analyse forest structure.

To investigate the minimum "area" required to give a good estimation of throughfall consistent with the forest cover variability, the transformed random grid measurements were analysed by dividing the total area into five separate areas and into five line transects. A one-way analysis of variance was carried out on both of these different classes. The results of this analysis are shown in Table I. The F-test on the mean squares showed that mean throughfall on line transects were not significantly different at the 10% level. However, area mean throughfalls were significantly different, even at the 1% level. The

TABLE I

One-way analysis of variance for transect lines and sub-grid areas

Source of variation	Degrees of freedom	Sum of squares	Mean square	F
Between lines	4	6114.2	1528.6	0.729[a]
Within lines	489	1025454	2097.0	
Total	493	1031568		
Between areas	4	35915.0	8978.8	4.410[b]
Within areas	489	995653	2036.1	
Total	493	1031568		

[a] Not significant at the 10% level.
[b] Significant at the 1% level.

data were re-analysed by dividing the area into two halves and a simple difference of means test performed. Mean throughfalls of the two halves were significantly different at the 5% level. The ramifications of this result are discussed later.

Stemflow variability

Stemflow from each tree was summed over the 19 measurement periods for which data on 18 of the trees were available to give a stemflow volume per mm of gross rainfall (SVR) for each tree. Fifteen of the trees contributed 14% of the stemflow, while the three remaining trees contributed 7, 23 and 56%. The variation in the stemflow measurement for each tree had little discernible pattern from period to period, although there was a trend towards lower variability in trees with higher stemflow. Overall this temporal variability, expressed as a standard error of the mean stemflow, was 18% with a range of 9–26%. The spatial variability from tree to tree was calculated from the variance of the individual SVR for each tree about the mean SVR. This variance expressed as a standard error of the mean was 50%.

The percentage of gross rainfall which reached the soil surface as stemflow was calculated from the 19 periods analysed above as 3.3%. However, one tree, a palm which contributed 56% of the stemflow, although not rare, had a lower species density in the surrounding forest than in the sample area and this biased the stemflow sample. It was, therefore, necessary to adjust the contribution of this distinctive tree to reflect its relative species density in the surrounding forest. The overall stemflow percentage was consequently reduced to 1.8% of gross rainfall. Combining the estimates of spatial and temporal variability gives a standard error of $\pm$ 1% of gross rainfall on this value.

DISCUSSION

A measurement of throughfall at any point beneath the canopy is an independent random variable which depends on gauging errors, gross rainfall and spatial variability related to canopy structure. Gauging errors are assumed random, while gross rainfall variability is common to a set of measurements. Increasing the number of gauges will reduce both of these contributions to the error in the mean measurement, following conventional error propagation formulae. The effect of canopy structure is, to a first approximation, a unique function of position and increasing the number of gauges increases the definition of the canopy structure surface. The optimum number of gauges is that which adequately describes the canopy structure.

The measurements of throughfall obtained at each position in a random grid can be viewed as a coarse definition of the canopy structure. In this experiment, the canopy structure function is defined, within experimental error, at 494 individual points in a 100 × 4 m area. The mean value of the canopy structure function can only be defined within the variance of the 494 points. A value for this variance can be obtained from the inverse function of the variance of the transformed data. The result, expressed as a best estimate of the standard error of the mean throughfall, is 2.2% of gross rainfall. It is possible to use the data collected in this experiment to estimate the possible errors in other experimental programmes which involve different numbers of gauges, which may or may not be randomly relocated. The estimation, which is only valid for similar canopy type and gauge size, is illustrated in Fig. 3.

For the purposes of this exercise, the canopy structure function is assumed to be perfectly defined at the 494 possible sample points used in this experiment, and is assumed constant with time. Experimental programmes involving no relocation of (say) n gauges between the discrete sample points represent, in effect, only one n-point sample of the 494-point grid. The consequent error in the mean throughfall measurement in such an experiment is the sum of the variability in the canopy structure function together with an error term due to taking a small sample estimate of the canopy structure function variability. The total error is estimated as a percentage of gross rainfall by:

$$\sigma_\tau = 2.2\,[1 + (494/n)^{1/2}] \qquad (1)$$

Random relocation of these n gauges m times allows additional sampling of the canopy structure function and the error estimate is reduced according to the expression:

$$\sigma_\tau = 2.2\,[1 + (494/n \cdot m)^{1/2}] \qquad (2)$$

Ultimately, all the possible points in the grid are sampled and the error asymptotes towards the 2.2% error inherent in using a fixed point grid to sample a continuously varying function.

Although eqns. 1 and 2 are appropriate for continuously varying functions, it could be argued that throughfall is not continuously varying in space.

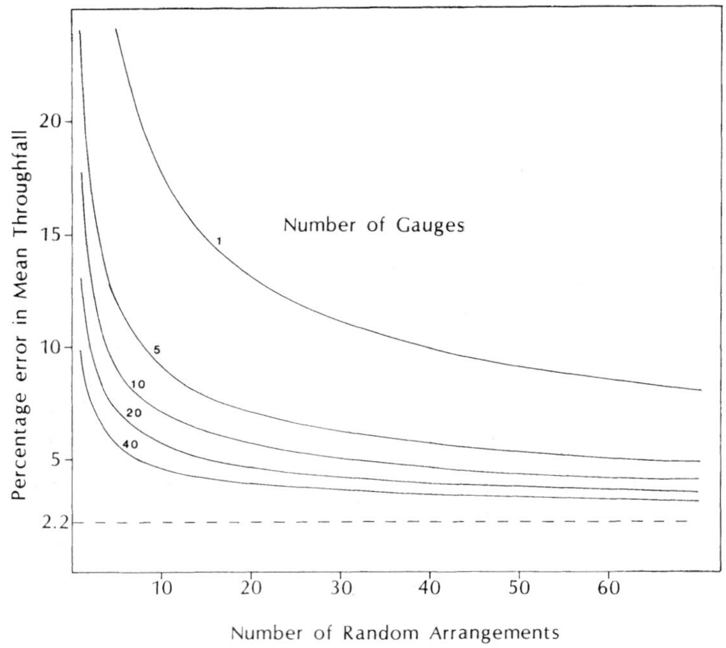

Fig. 3. Probable standard errors in measuring throughfall for increasing numbers of relocations and sample size.

However, raingauges are spatial integrators and they transform the possibly discontinuous distribution of throughfall into a continuous distribution, thus allowing the use of eqs. 1 and 2.

A previous experiment in the Ducke Forest by Franken et al. (1982), working nearby in the same forest, measured a mean throughfall value of 80% using 20 randomly located, but non-moving raingauges. From Fig. 3, it is apparent that such an experimental arrangement is prone to error in the order of 12% of gross rainfall due to variations in canopy structure. Their result, viz. throughfall is 80 ± 12% of gross rainfall, is not significantly different (at the 95% level) from the 91 ± 2% given in this study. In much the same way, a previous phase of the present experiment (not reported here) which used 16 gauges randomly relocated every 2 weeks over the period October 1983–March 1984 gave the throughfall as 87 ± 4% of gross rainfall. This result is also not significantly different (at the 95% level) to the present result.

The analysis of stemflow indicates that while certain trees are capable of very high stemflows, the overall contribution of stemflow to soil water is very small. Nevertheless, in terms of the interception loss, the stemflow is still an important term and for valid estimates of the stemflow contribution, stemflow samples must be large enough to take proper account of these high stemflow trees. It is also important, in the absence of well defined canopy areas for individual trees, to sample every tree within an area. The observations of

Jordan (1968, 1978), that trees with crowns in or below the main canopy often had greater stemflow than larger diameter emergent trees, is consistent with observations in this experiment.

CONCLUSIONS

The results presented here indicate that tropical rainforest is arguably the most difficult forest type in which to make measurements of interception loss, particularly when this loss is a small fraction of gross rainfall. The primary results of this study can be summarised as follows.

(i) The frequency distribution of the throughfall fraction for tropical forests is very much broader than for temperate plantation forests and this generates significantly greater "errors" in the measurements of interception fraction. The distribution is positively skewed and is optimally transformed to a normal distribution using the square root of the throughfall fraction.

(ii) Studies of sub-samples of the 100 × 4-m random grid used in the present experiment suggest that random relocations of gauges along a 100-m transect line with 1-m sample positions provide better estimates of mean throughfall than an area grid of 20 × 4 m with 1-m sample positions.

(iii) An investigation of the measurement errors generated by the variable canopy structure in tropical forest type indicates that much previous work may be in significant error, particularly if a fixed gauge pattern was used. Any predisposition on the part of observers, e.g., Manokaran (1979), to reduce or reject local throughfall measurements greater than gross rainfall (i.e., drip points) on the grounds that they were unrealistic, would generate a large and systematic overestimation of the interception loss. Although it is obvious from eq. 2 and Fig. 3 that an error reduction is attained by a random relocation of a small number of gauges, the same could be achieved using a very large number of fixed gauges, but in practice the number of fixed gauges required precludes this as a realistic alternative strategy.

(iv) The values of throughfall and stemflow, as a percentage of gross rainfall obtained in the experimental area, were 91(±2)% for throughfall and 1.8(±1)% for stemflow. These values are consistent with previous measurements in this type and region of Amazonian rainforest, if the size of the errors is taken into account.

ACKNOWLEDGEMENTS

The results presented here were obtained as part of an extensive programme of research into Amazonian micrometeorology supported by the Natural Environment Research Council, U.K., the Conselho Nacional de Desenvolvimento Cientifico et Tecnologico (CNPq), Brazil and the British Council. The authors are pleased to acknowledge this support.

We would also like to thank J.H.C. Gash and W.J. Shuttleworth for their constructive advice, and our colleagues in the Ciencas do Ambiente section of the Instituto National de Pesquisas da Amazonia for their help in data collection and equipment maintenance.

REFERENCES

Clarke, R.T., 1987. The interception process in tropical rain forests: A literature review and critique, Acta Amazonica, in press.

Ford, E.D. and Deans, J.D. 1978. The effects of canopy structure on stemflow, throughfall and interception loss in a young Sitka spruce plantation. J. Appl. Ecol., 15: 905–917.

Franken, W., Leopoldo, P.R., Matsui, E. and Ribeiro, M. de N.G., 1982. Estudo da interceptação da água de chuva em cobertura florestal amazônica do tipo terra firme. Acta Amazonica 12(2): 327–331.

Gash, J.H.C. and Stewart, J.B., 1977. The evaporation from Thetford forest during 1975. J. Hydrol., 35: 385–396.

Jackson, I.J., 1971. Problems of throughfall and interception assessment under tropical forest. J. Hydrol., 12: 234–254.

Jordan, C.F., 1968. Stem flow in the tropical rain forest. In: Puerto Rico Nuclear Center Rain Forest Project Annual Report, PRNC-119, Rio Piedras, Puerto Rico, pp. 35–40.

Jordan, C.F., 1978. Stem flow and nutrient transfer in a tropical rain forest. Oikos, 31: 257–263.

Lettau, H., Lettau, K. and Molion, L.C.B., 1979. Amazonia's hydrologic cycle and the role of atmospheric recycling in assessing deforestation effects. Mon. Weather Rev., 107: 227–238.

Manokaran, N., 1979. Stemflow, throughfall and rainfall interception in a lowland tropical rain forest in peninsular Malaysia. Malaysian Forester, 42: 174–201.

Moore, C.J. and Fisch, G., 1986. Estimating heat storage in Amazonian Tropical Forest. Agric. For. Meteorol., 38: 147–169.

Mueller, C.C. and Kidder, E.H., 1972. Rain gage catch variation due to airflow disturbances around a standard rain gage. Water Resour. Res., 8: 1077–1082.

Raich, J.W., 1983. Throughfall and stemflow in mature and year-old wet tropical forest. Trop. Ecol., 24: 234–243.

Robinson, A.C. and Rodda, J.C., 1969. Rain, wind and the aerodynamic characteristics of rain gauges. Meteorol. Mag., 98: 113–120.

Shuttleworth, W.J., Gash, J.H.C., Lloyd, C.R., Moore, C.J., Roberts, J., Marques, A. de O., Fisch, G., Silva, V. de P., Ribeiro, M.N.G., Molion, L.C.B., de Sa, L.D.A., Nobre, J.C., Cabral, O.M.R., Patel, S.R. and de Moraes, J.C., 1984. Eddy correlation measurements of energy partition for Amazonian forest. Q. J. R. Meteorol. Soc., 110: 1143–1162.

Takeuchi, M., 1961. The structure of the Amazonian vegetation. II. Tropical rain forest. J. Fac. Sci. Univ. Tokyo, III, 8(1): 1–26.

B Water Transfer Measurements – Commentary

This section introduces and comments on the measurement of evaporation and transpiration in the form of water moving through plants into the atmosphere and away from the surface. The first paper introduces the direct measurement of transpiration as water vapour from individual leaves and shoots using porometers, while the second paper pioneers the routine measurement of sap flow in stems and trunks using heat as a tracer in the thermal dissipation method. The third paper introduces the approach of estimating surface evaporation assuming the process of turbulent diffusion is similar for all atmospheric entities moving away from the ground, while the fourth describes the first attempt to measure turbulent fluxes directly and the fifth, an important step towards automating this latter approach as the now widely used eddy correlation technique. Most micrometeorological measurements sample the exchange from a source area upwind, and the sixth paper in this group introduced a simple way of defining this source area. The final paper describes an approach for estimating area-average atmospheric exchanges indirectly from fluctuations in the refractive index of the air that is now proven to be applicable even over heterogeneous surfaces.

B1 Porometry

van Bavel, C. H. M., Nakayama, F. S. & Ehrler, W. L. (1965) Measuring transpiration resistance of leaves.

The concept of stomatal resistance as a quantified measure of the restriction to water vapour flow through the stomata of leaves is attributed to Raschke (1956), but we have selected the 1965 paper by C. H. M. van Bavel, F. S. Nakayama and W. L. Ehrler as the benchmark in this section because it gives the first description of an instrument to measure the stomatal resistance of individual leaves in the field. The device, a diffusion porometer, temporarily encloses a leaf (or leaves) and, after the sample chamber has been flushed with dry air, water vapour from inside the leaf is allowed to diffuse into the chamber. The rate of increase in humidity within the chamber then allows the stomatal resistance to be calculated.

Porometers soon attained widespread use in physiological studies of plant water use, being further developed (e.g. Stiles, 1970) and becoming commercially available in the 1970s. Modern, steady-state porometers maintain the leaf chamber at ambient humidity by adjusting the input rate of dry air.

Although the initial motivation for developing this device was largely to synthesize whole plant transpiration by combining sample measurements of the transpiration of individual leaves and shoots, perhaps the most useful results of applying it were obtained in the form of better understanding of plant and ecosystem behaviour. The data that provided the understanding behind stomatal and surface resistance models (such as those described in D9 and D10) were collected with diffusion porometers. The devices were also used to allocate evaporation between the different components of ecosystems, for example between bare soil and a sparse crop (Boegh *et al.*, 1999), or the variation in transpiration with depth in forest canopies (e.g. Roberts *et al.*, 1990). In this way, data collected with porometers underpin many of the evaporation models which are in routine use today.

B2 Sap flow measurements

Granier, A. (1985) Une nouvelle méthode pour la mesure du flux de sèvre brute dans le tronc des arbres (A new method of sap flow measurement in tree trunks).

According to the comprehensive review by Swanson (1994), measuring the water use of plants by monitoring the movement of a tracer in their sap flow was first attempted in the 1930s. Stable isotopes such as tritium (e.g. Waring & Roberts, 1979) and deuterium (e.g. Calder *et al.*, 1992) have been used, but only rarely in comparison with the widespread use of heat as a tracer. Sap flow measurements based on some arrangement of electrical heaters embedded in plant stems, branches, or roots are now commonplace and the technique is now used as a routine component of numerous hydrological and ecological process studies. Sap flow monitoring has replaced scaling up leaf-level measurements of stomatal conductance as the preferred plant physiological method for deriving the transpiration. This is because it gives the integrated transpiration for a whole stem or plant and allows long-term, automatic measurements. It is also suitable for use in situations where micrometeorological methods cannot be applied, such as on isolated trees (e.g. David *et al.*, 2004) or in manipulative experiments (e.g. the droughting experiment described by Fisher *et al.*, 2006), for separating the components of transpiration from mixed vegetation (e.g. Tuzet *et al.*, 1997), and for locating the level from which transpired water is drawn (e.g. Hultine *et al.*, 2004). A further advantage is that there is no requirement for canopy access.

There are three main methods of using heat as a tracer: the heat-pulse velocity, the stem heat balance, and the thermal dissipation method. The theory and practice of these different techniques have been reviewed by Swanson (1994), and Smith & Allen (1996), with useful brief introductions to the subject also being given by Roberts (2005), and Verhoef & Campbell (2005). We have selected André Granier's 1985 paper describing the thermal dissipation method as the benchmark in this area, because it is our perception that it was only after its publication that sap flow monitoring was transformed from being a narrow specialization into being a routine measurement technique for studying evaporation from vegetation. According to Smith & Allen (1996) "*The major advantages of the thermal dissipation method appear to be easy installation, simple requirements for recording sensor outputs, simple sap flow calculations and lower costs…*".

Granier *et al.* (1990) suggested that the calibration parameters are not dependent on wood properties or tree species and that the technique may thus be used with standard calibrations. However, if a known portion of the sensor is not in contact with conducting wood, Clearwater *et al.* (1999) point out that a correction should be applied. If the cross-sectional distribution of the sapwood is unclear, or if a strong radial variation in the flux rate across the sapwood cannot be excluded, Herbst *et al.* (2007) advise the use of an independent calibration. Roberts (2005) shows a calibration rig in which heat dissipation can be measured in a cut section of tree trunk while water is passed through at different rates.

Note: to facilitate use of this paper, an English translation is provided.

B3 Atmospheric profile measurements

Thornthwaite, C. W. & Holzman, B. (1939) The determination of evaporation from land and water surfaces.

The 1939 paper by C. W. Thornthwaite and B. Holzman was the first to use a micrometeorological technique to give a continuous hourly-resolution measurement of evaporation. The required atmospheric measurements of humidity and wind speed were made at two levels over grassland. This approach assumes turbulent diffusion of water vapour and momentum are well described by a diffusion equation. At the time of the experiment, the diffusion coefficients were only known for near-neutral atmospheric conditions, but by making the measurements during three winter months, the problem associated with the method in unstable conditions was largely avoided.

Communication between scientists was likely to be neither very fast nor efficient during the 1940s, and we are unsure to what extent Thornthwaite and Holzman's paper, which was published in the USA, influenced the work of Pasquill (1949, 1950) in Britain, who adopted essentially the same method but also validated the results against measurements made with miniature lysimeters. The latter papers may also have been influenced by the development of the Monin-Obukov similarity theory in the Soviet Union. Nevertheless, together this combination of observational and theoretical development led to the parameterization of empirical flux–profile relationships becoming an international research priority in the 1950s and 1960s. Much of the work on this latter topic was carried out in Australia: see Dyer & Hicks (1970), and Webb (1970) for examples of functions describing unstable and stable atmospheric conditions, respectively. An overview is given by Panofsky & Dutton (1984). Although the aerodynamic method, which was pioneered in this benchmark paper, has now been replaced by other measurement techniques, the need for the use of the flux-profile methodology to which it led still persists in many of the models used to estimate evaporation (see Section D).

Subsequently, when robust net radiometers became commercially available, it became more popular to deduce evaporation from profile measurements of water vapour and (potential) temperature. If the turbulent diffusivities for sensible and latent heat are the same, the ratio of these two fluxes, the Bowen Ratio (Bowen, 1926) is proportional to the (measurable) gradients of potential temperature and water vapour content (or vapour pressure). The available radiant energy can then be partitioned into evaporation and sensible heat flux. The accuracy of this, the so-called Bowen Ratio–Energy Budget approach, is mainly dependent on how well the gradient of atmospheric humidity is measured. This accuracy can be improved (an essential procedure when working over forests where gradients are small) either by regularly interchanging the humidity sensor between two levels in the profile (e.g. McNeil & Shuttleworth, 1975), or by ducting air alternately to a single common sensor (http://www.campbellsci.com/bowen-ratio).

Verhoef & Campbell (2005) give a useful, brief introduction to profile measurement techniques.

B4 & B5 Eddy correlation

Swinbank, W. C. (1951) The measurement of vertical transfer of heat and water vapor by eddies in the lower atmosphere.

Dyer, A. J. (1961) Measurements of evaporation and heat transfer in the lower atmosphere by an automatic eddy correlation technique.

We have selected two papers describing the introduction of the eddy correlation (or eddy covariance) technique. Perhaps this is self-indulgence because it reflects our personal research interests, but given the current widespread use of the technique, we think the selection is justified. These two papers reflect the fact that in the 1950s and 1960s, CSIRO scientists in Australia led the world in micro-meteorology, and they are also fascinating papers to read.

Although the origin of the eddy correlation method can be traced back to Scrase (1930), we select the 1951 paper by W. C. Swinbank because it was the first to create "an apparatus" to successfully measure and record turbulent fluctuations of wind speed, temperature and humidity in the field, and to analyse these to derive estimates of the turbulent fluxes of sensible heat and water vapour. Measurements of temperature and humidity were made with a resistance thermometer and a fine wire, thermocouple wet bulb, respectively, but vertical wind speed was measured with a hot wire anemometer. To measure the vertical wind speed, the anemometer was mounted on a wind vane, so that the hot wire always pointed into the mean wind. This in itself was an original idea in that it took a sensitive wind tunnel system into a field situation. Analogue signals were recorded synchronously on a photographic paper chart using "mirror galvanometers". The five-minute runs were later analysed manually at one-second intervals, a laborious procedure but sufficient to prove the method.

The next significant step forward in the development of this technique occurred when A. J. Dyer published his 1961 paper, which we also select as a benchmark in this field. In this study, the basic sensors were similar to those used by Swinbank, but the need to record and later analyse data was removed by using an analogue multiplier circuit to derive the cross-products between the vertical wind speed, and the temperature and humidity of the air. Integration was achieved using circuitry which, interestingly, was based on a converted domestic Watt-hour meter. Surprisingly, when the performance of the instrument was evaluated by comparing the outgoing eddy fluxes with the incoming available energy, the instrument achieved 99% energy closure – a result which few modern systems can match!

This automation of the eddy correlation analysis pioneered by Dyer paved the way for the development of practical devices that could be used as research tools to investigate scientific problems: first the "Evapotron" (Dyer & Maher, 1965), and later the "Fluxatron" (Dyer *et al.*, 1967; Hicks, 1969), in which the vertical wind speed was measured by a more robust, and less power-hungry, propeller anemometer. The Fluxatron measured sensible heat only, with evaporation then being estimated as the residual in the surface energy balance – possible given the parallel development of field net radiometers. Useful results based on relatively long runs of data were obtained with Fluxatrons (e.g. Hicks *et al.*, 1975; Moore, 1976; Milne, 1979), although some problems were reported when measuring with this device, even over forests (McNeil & Shuttleworth, 1975).

Important subsequent developments in this field involved the development of simple one-dimensional sonic anemometry (Shuttleworth *et al.*, 1982) to measure vertical wind speed, together with digital electronics to compute the cross-products (Lloyd *et al.*, 1984), and the development of retrospective correction procedures to allow for some of the shortcomings in sensor performance (Moore, 1986). These technical developments also made the system require less power, which facilitated its routine unsupervised deployment in remote locations (e.g. paper E2). During the 1990s, as inexpensive three-dimensional sonic anemometers, laptop computers, and closed path, then later open path, infra-red gas analyzers became available, systems such as that described by Moncrieff *et al.* (1997) could be assembled from commercially available components. Several hundred such systems are in use today, but the origin of this work can be traced back to the pioneering work of Swinbank and Dyer. Bruce Hicks's (1988) tribute to A. J. Dyer gives a fascinating insight into the technical and human history of the development of micrometeorology in CSIRO.

B6 Micrometeorological source area

Gash, J. H. C. (1986) A note on estimating the effect of a limited fetch on micrometeorological evaporation measurements.

Although this paper by John Gash on estimating flux footprints does not describe a measurement technique or instrument, we include it as a benchmark paper in this section because the theory gave micrometeorologists new insight into what their instruments were actually measuring. Before this paper was published, assessing the fetch of micrometeorological sites was treated as a change in roughness problem, addressing the question: "How far from the nearest change in vegetation must a measurement be made?" This paper turned the question around and asked: "to what degree does the vegetation at varying distances upwind of the instrument contribute to the measurement?"

A whole literature subsequently developed, expanding this idea to include: a realistic wind profile, cross-wind diffusion, the effects of atmospheric stability and more realistic diffusion models (see the review of flux footprint modelling by Schmid, 2002). In his review Schmid states that "*Gash (1986) adopted Pasquill's (1972) idea to view an area of heterogeneous surface fluxes as an array of elemental sources in a two-dimensional flow approach. He estimated the effect of a limited fetch for evaporation measurements by considering an upwind array of spanwise elemental line sources, each occupying an infinitesimal strip of width δx. Gash (1986) is the first study that refers to a form of flux footprint in the way the term is used today in micro-meteorology*".

Gash's paper was only published after repeat submissions to two different journals. But we believe one of the reasons for its subsequent success is that, despite the simplifying assumptions made, the analytical approach adopted gave new insight into the process. Publication of the paper was also extremely timely. Two years after its publication the first FIFE experiment (Sellers *et al.*, 1988) took place. This experiment was designed to develop and test methods of estimating surface fluxes from remote sensing, thus creating the need to compare remote sensing estimates of fluxes with surface-based micrometeorological measurements. As a consequence, there was a requirement to know the micrometeorological flux footprint. Unfortunately, as pointed out by Schuepp *et al.* (1990), Equation 11 in Gash's paper is incorrect, because the integration should be between the measurement height, z_m, and z_0, not between z_m and zero. The effect is negligible when $z_m >> z_0$ but can be noticed for forest cover where z_0 is large. Schuepp *et al.* (1990) provide the corrected form of this and subsequent equations.

B7 Scintillometry

Lagouarde, J. P., McAneney, K. J. & Green, A. E. (1996) Scintillometer measurements of sensible heat flux over heterogeneous surfaces.

In the last decade much progress has been made in a turbulence theory-based approach to deducing area-average surface energy exchanges indirectly from long path-averaged fluctuations in temperature and humidity in the near-surface atmospheric boundary layer. The technique involves transmitting a beam of electro-magnetic radiation and viewing or measuring the intensity variations of the received signal due to air density differences altering the refractive index along the path between transmitter and receiver (Hill, 1992). The turbulent structure of these density changes are mainly due to temperature, but fluctuations in water vapour concentration also contribute, with the relative contribution of each depending on the wavelength of the radiation. In the visible and near-infrared waveband, the signal is much more sensitive to heat movement, while in the microwave to millimetre-wave band it is more sensitive to moisture fluctuations (Hill, 1997). Large and Extra-Large Aperture Scintillometers (see McAneney *et al.*, 1995), which can be deployed to estimate line-average sensible heat flux over distances of up to 10 km, have become popular as they are both robust and relatively inexpensive; in contrast millimetre-wave scintillometers are more expensive to build and currently less extensively used, but experiments have shown that they can make reasonable direct measurements of evaporation (Meijninger *et al.*, 2006).

Because it is based on the Monin-Obukov theory, in principle the scintillometry method is strictly only reliable when applied near (but not too near) a rough surface in the well-developed boundary layers over uniform vegetation. We consider that the breakthrough that made this branch of evaporation measurement unique and important is not its application as an alternative to other micrometeorological techniques over uniform cover, rather the demonstration that in practice the method provides a reasonable, worthwhile estimate of the line-average surface fluxes over extensive landscapes comprising heterogeneous vegetation cover. We therefore select the paper by Lagouarde *et al.* as the benchmark paper for this topic because it was the first to demonstrate this potential, although other papers by Chehbouni *et al.* (1999), and Mejininger *et al.* (2002a,b) also show similar results.

References

Boegh, E., Soegaard, H., Friborg, T. & Levy, P. E. (1999) Models of CO_2 and water vapour fluxes from a sparse millet crop in the Sahel. *Agric. Forest Met.* **93**, 7–26.

Bowen, I. S. (1926) The ratio of heat losses by conduction and by evaporation from any water surface. *Physical Rev.* **27**, 779–787.

Calder, I. R., Narayanswamy, M. N., Srinivasalu, N. V., Darling, W. G. & Lardner, A. J. (1986) Investigation into the use of deuterium as a tracer for measuring transpiration from eucalypts. *J. Hydrol.* **84**, 345–351.

Chehbouni, A., Watts, C., Lagouarde, J.-P., Kerr, Y. H., Rodriguez, J.-C., Bonnefond, J.-M., Santiago, F., Dedieu, G., Goodrich, D. C. & Unkrich, C. (2000) Estimation of heat and momentum fluxes over complex terrain using a large aperture scintillometer. *Agric. Forest Met.* **105**, 215–226.

Clearwater, M. J., Meinzer, F. C., Andrade, J. L., Goldstein, G. & Holbrook, N. M. (1999) Potential errors in measurement of nonuniform sapflow using heat dissipation probes. *Tree Physiology* **19**, 681–687.

David, T. S., Ferreira, M. I., Cohen, S, Pereira, J. S. & David J. S. (2004) Constraints on transpiration from an evergreen oak tree in southern Portugal. *Agric. Forest Met.* **122**, 193–205.

Dyer, A. J. & Hicks, B. B. (1970) Flux-gradient relationships in the constant flux layer. *Quart. J. Royal Met. Soc.* **96**, 715–721.

Dyer, A. J., Hicks, B. B. & King, K. M. (1967) The Fluxatron – a revised approach to the measurement of eddy fluxes in the lower atmosphere. *J. Appl. Met.* **6**, 408–413.

Dyer, A. J. & Maher, F. J. (1965) Automatic eddy-flux measurement with the Evapotron. *J. Appl. Met.* **4**, 622–625.

Fisher, R. A., Williams, M., do Vale, R. L., da Costa, A. L. & Meir, P. (2006) Evidence from Amazonian forests is consistent with isohydric control of leaf water potential. *Plant Cell and Environment* **29**, 151–165.

Granier, A., Bobay, V., Gash, J. H. C., Gelpe, J., Saugier, B. & Shuttleworth, W. J. (1990) Vapour flux density and transpiration rate comparisons in a stand of maritime pine (*Pinus pinaster* Ait.) in Les Landes forest. *Agric. Forest Met.* **51**, 309–319.

Herbst, M., Roberts, J. M., Rosier, P. T. W. & Gowing, D. J. (2007) Seasonal and interannual variability of canopy transpiration of a hedgerow in southern England. *Tree Physiology* **27**, 321–333.

Hicks, B. B. (1988) Some introductory notes to an issue of *Boundary-Layer Meteorology* dedicated to Arthur James Dyer. *Boundary-Layer Met.* **42**, 1–8.

Hicks, B. B. (1969) A simple instrument for the measurement of Reynolds stress by eddy correlation. *J. Appl. Met.* **8**, 825–827.

Hicks, B. B., Hyson, P. & Moore, C. J. (1975) A study of eddy fluxes over a forest. *J. Appl. Met.* **14**, 58–66.

Hill, R. J. (1992) Review of optical scintillation methods of measuring the refractive-index spectrum, inner scale and surface fluxes. *Waves in Random Media* **2**, 179–201.

Hill, R. J. (1997) Algorithms for obtaining atmospheric surface-layer fluxes from scintillation measurements. *J. Atmos. Oceanic Technol.* **14**, 456–467.

Hultine, K., Scott, R. L., Cable, W. L. & Williams, D. (2004) Hydraulic redistribution by a dominant, warm desert phreatophyte: seasonal patterns and response to precipitation pulses. *Functional Ecology* **18**, 530–538.

Lloyd, C. R., Shuttleworth, W. J., Gash, J. H. C. & Turner, M. (1984) A microprocessor system for eddy correlation. *Agric. Forest Met.* **33**, 67–80.

McAneney, K. J., Green, A. E. & Astill, M. S. (1995) Large-aperture scintillometry: the homogeneous case. *Agric. Forest Met.* **76**, 149–162.

McNeil, D. D. & Shuttleworth, W. J. (1975) Comparative measurements of the energy fluxes over a pine forest. *Boundary-Layer Met.* **9**, 297–313.

Meijninger, W. M. L., Hartogensis, O. K., Koshiek, W., Hoedjes, J. C. B., Zuurbier, R. M. & de Bruin, H. A. R. (2002a) Determination of area-averaged sensible heat fluxes with a large aperture scintillometer over a heterogeneous surface – Flevoland field experiment. *Boundary-Layer Met.* **105**, 37–62.

Meijninger, W. M. L., Green, A. E., Hartogensis, O. K., Kohsiek, W., Hoedjes, C. B., Zuurbier, R. M. & de Bruin, H. A. R. (2002b) Determination of area-averaged water vapour fluxes with large aperture and radio wave scintillometers over a heterogeneous surface – Flevoland field experiment. *Boundary-Layer Met.* **105**, 63–83.

Meijninger, W. M. L, Beyrich, F., Lüdi, A., Kohsiek, W. & de Bruin, H. A. R. (2006) Scintillometer-based turbulent fluxes of sensible and latent heat over a heterogeneous land surface – a contribution to LITFASS-2003. *Boundary-Layer Met.* **121**, 89–110.

Milne, R. (1979) Water loss and canopy resistance of a young Sitka spruce plantation. *Boundary-Layer Met.* **16**, 67-81.

Moncrieff, J. B., Massheder, J. M., de Bruin, H. A. R., Elbers, J., Friborg, T., Heusinkveld, B. G., Kabat, P., Scott, S., Sogaard, H. & Verhoef, A. (1997) A system to measure fluxes of momentum, sensible heat, water vapour and carbon dioxide. *J. Hydrol.* (HAPEX-Sahel Special Issue) **188–189**, 589–611.

Moore, C. J. (1976) Eddy flux measurements above a pine forest. *Quart. J. Royal Met. Soc.* **102**, 913–918.

Moore, C. J. (1986) Frequency response corrections for eddy correlation systems. *Boundary-Layer Met.* **37**, 17–45.

Panofsky, H. A. & Dutton, J. A. (1984) *Atmospheric Turbulence. Models and Methods for Engineering Applications*. Wiley, New York, USA.

Pasquill, F. (1949) Eddy diffusion of water vapour and heat near the ground. *Proc. Royal Soc. London, Series A, Mathematical and Physical Sciences* **198**, 116–140.

Pasquill, F. (1950) Some further considerations of the measurement and indirect evaluation of natural evaporation. *Quart. J. Royal Met. Soc.* **76**, 287–301.

Pasquill, F. (1972) Some aspects of boundary layer description. *Quart. J. Royal Met. Soc.* **98**, 469–494.

Raschke, K. (1956) Über die physikalischen Beziehungen zwischen Wärmeübergangszahl, Strahlungsaustausch. Temperature und Transpiration eines Blattes. *Planta* **48**, 200–238.

Roberts, J. (2005) Transpiration. In: *Encyclopedia of Hydrological Sciences, Part 4. Hydrometeorology* (ed. by M. J. Anderson), Chapter 42, 615–625. DOI: 10.1002/0470848944.hsa045. Wiley, Chichester, UK.

Roberts, J., Cabral, O. M. R. & de Aguiar, L. F. (1990) Stomatal and boundary-layer conductances measured in a terra firme rain forest, Manaus, Amazonas, Brazil. *J. Appl. Ecol.* **27**, 336–353.

Schmid, H. P. (2002) Footprint modeling for vegetation atmosphere exchange studies: a review and perspective. *Agric. Forest Met.* **113**, 159–183.

Schuepp, P. H., Leclerc, M. Y., MacPherson, J. I. & Desjardins, R. L. (1990) Footprint prediction of scalar fluxes from analytical solutions of the diffusion equation. *Boundary-Layer Met.* **50**, 355–373.

Scrase, F.J. (1930) *Some Characteristics of Eddy Motion in the Atmosphere.* UK Meteorological Office Geophysical Memoirs, no. 52.

Sellers, P. J., Hall, F. G., Asrar, G., Strebel, D. E. & Murphy, R. E. (1988) The First ISLSCP Field Experiment (FIFE). *Bull. Am. Met. Soc.* **69**, 22–27.

Shuttleworth, W. J., McNeil, D. D. & Moore, C. J. (1982) A switched continuous-wave sonic anemometer for measuring surface heat fluxes. *Boundary-Layer Met.* **23**, 425–448.

Smith, D. M. & Allen, S. J. (1996) Measurement of sap flow in plant stems. *J. Exp. Botany* **47**, 1833–1844.

Stiles, W. (1970) A diffusive resistance porometer for field use. *J. Appl. Ecol.* **7**, 617–622.

Swanson, R. H. (1994) Significant historical developments in thermal methods for measuring sap flow in trees. *Agric. Forest Met.* **72**, 113–132.

Tuzet, A., Castell, J.-F., Perrier, A. & Zurfluh, O. (1997) Flux heterogeneity and evapotranspiration partitioning in a sparse canopy: the fallow savanna. *J. Hydrol.* **188–189**, 482–493.

Verhoef, A. & Campbell, C. L. (2005) Evaporation measurement. In: *Encyclopedia of Hydrological Sciences* (ed. by M. J. Anderson), Chapter 40, 589–601. DOI: 10.1002/0470848944.hsa043. Wiley, Chichester, UK.

Waring, R. H. & Roberts, J. M. (1979) Estimating water flux through stems of Scots pine with tritiated water and phosphorus-32. *J. Exp. Botany* **116**, 459–471.

Webb, E. K. (1970) Profile relationships: the log-linear range, and the extension to strong stability. *Quart. J. Royal Met. Soc.* **96**, 67–90.

Measuring Transpiration Resistance of Leaves [1]

C. H. M. van Bavel, F. S. Nakayama, and W. L. Ehrler
United States Water Conservation Laboratory, Tempe, Arizona
Soil and Water Conservation Research Division, ARS, United States Department of Agriculture

Introduction

The transpiration process can be described physically, following Raschke (10), in terms of a resistance to diffusive and turbulent vapor flux in the external air, a similar diffusive resistance which results from the internal leaf geometry, inclusive of the stomata, and parallel to the latter, a resistance to vapor diffusion through the cuticle. In contrast, the last 2 resistances do not exist in the evaporation from an open water surface or a moist blotter paper. The resistance in these cases can be described using only external parameters.

In applying the resistance concept, the internal vapor concentration is generally computed from the leaf temperature, assuming that the state of the leaf water is such that its relative vapor pressure is substantially equal to one. This view is open to some objections; see, for example, Bange (2) and Heath (5). Nevertheless, it is applied by most plant physiologists as it will be in this paper.

The leaf resistance must be known when computing leaf or canopy evaporation from measured environmental parameters, although it may not always be of significant magnitude compared to the external vapor transport resistance. Thus, Bange (2) developed a procedure for calculating the resistance from measurements of stomatal dimensions and numbers, and substomatal cavity dimensions. This procedure does not allow for the contribution of cuticular transpiration to the total water loss, which still must be measured directly. Using an experimental correction for cuticular loss, Bange's work gave close agreement between measured and computed transpiration from leaf disks of *Zebrina pendula*, a hypostomatous plant.

Earlier, Penman and Schofield (9) suggested a method similar to that of Bange for estimating the stomatal resistance, the values to be applied in calculating canopy evaporation. For the latter, they proposed a combination method devised by Penman [see Pelton and Tanner (8) for a critical discussion] in which the energy balance and aerodynamics of the vegetative surface are simultaneously taken into account. Penman and Schofield did not give such detailed consideration to substomatal cavity dimensions and other geometrical details as did Bange, nor did they verify their results by direct experimentation.

Raschke (10) gave a method for computing the leaf resistance indirectly from simultaneous measurements of leaf transpiration, leaf temperature, leaf radiation balance, and ambient temperature and vapor pressure. Originally, he expressed the result by means of a dimensionless factor α, the wetness factor, being the ratio between the actual transpiration rate and the calculated evaporation rate of a hypothetical free water surface having the same dimensions, temperature, and exposure as the leaf. Raschke showed in a later publication (11) that α was identical with the ratio of the combined turbulent and laminar vapor flow resistance in the air outside the leaf (the external resistance, R_A) to the sum of R_A and the leaf resistance (R_L):

$$\alpha = R_A/(R_A + R_L). \qquad \text{I}$$

R_A can be approximated from observing the evaporation from properly exposed wet filter paper of leaf dimensions. R_A then equals the vapor concentration gradient from paper to bulk air divided by the evaporation rate per unit area.

Raschke's method is sound, but obviously not simple, nor always practical. On the other hand, the approaches taken by Bange, and by Penman and Schofield, do not account for cuticular water loss, and require a precise knowledge of stomatal aperture and internal leaf geometry. Whereas Bange showed agreement between theory and experiment, Pelton and Tanner (8) report unsatisfactory results from tests of the Penman-Schofield procedure. These tests were not stringent, and the disagreement may have originated in Penman and Schofield's proposal to use a generalized value for the stomatal resistance and apply it to a canopy as one would to a single leaf.

The stomatal aperture itself can be measured by various methods, such as visual observation of leaves or their replicas, measurements with pressure porometers, and infiltration techniques. Recent reviews of such techniques are found in Heath (5) and Eckhardt (3). Regardless of technical feasibility, none permits an exact calculation of the leaf resistance factor mentioned above.

Until now, the only available field method for measuring transpiration resistance directly was the classical cobalt chloride method, described in many textbooks. It is slow and subjective, and also likely to influence both stomatal aperture and leaf tempera-

[1] Received November 2, 1964.

ture. Classified by Heath (5) as a transpiration porometer, it has been recently discussed in some detail by Milthorpe (6). A refinement of the desiccant technique, suitable for laboratory studies, was proposed by Archer and La Mer (1) for measuring evaporation resistance caused by monomolecular films on open water surfaces. Their calibration and interpretation methods are quite similar to the ones we will describe in the following section.

Recently, Wallihan (13), although concerned with stomatal aperture rather than transpiration resistance, suggested a technique that can be the basis for a direct method to measure leaf resistance, one that is simple and rapid, and entails a minimum of disturbance to the plant.

In this article we describe improvements in the equipment and methods given by Wallihan. Further, we report on the calibration of the apparatus and the interpretation of the data. Finally, we show that the technique is physically sound and can lead to quantitative prediction of the transpiration rate of whole plants.

Materials and Methods

The method evolved from Wallihan's report

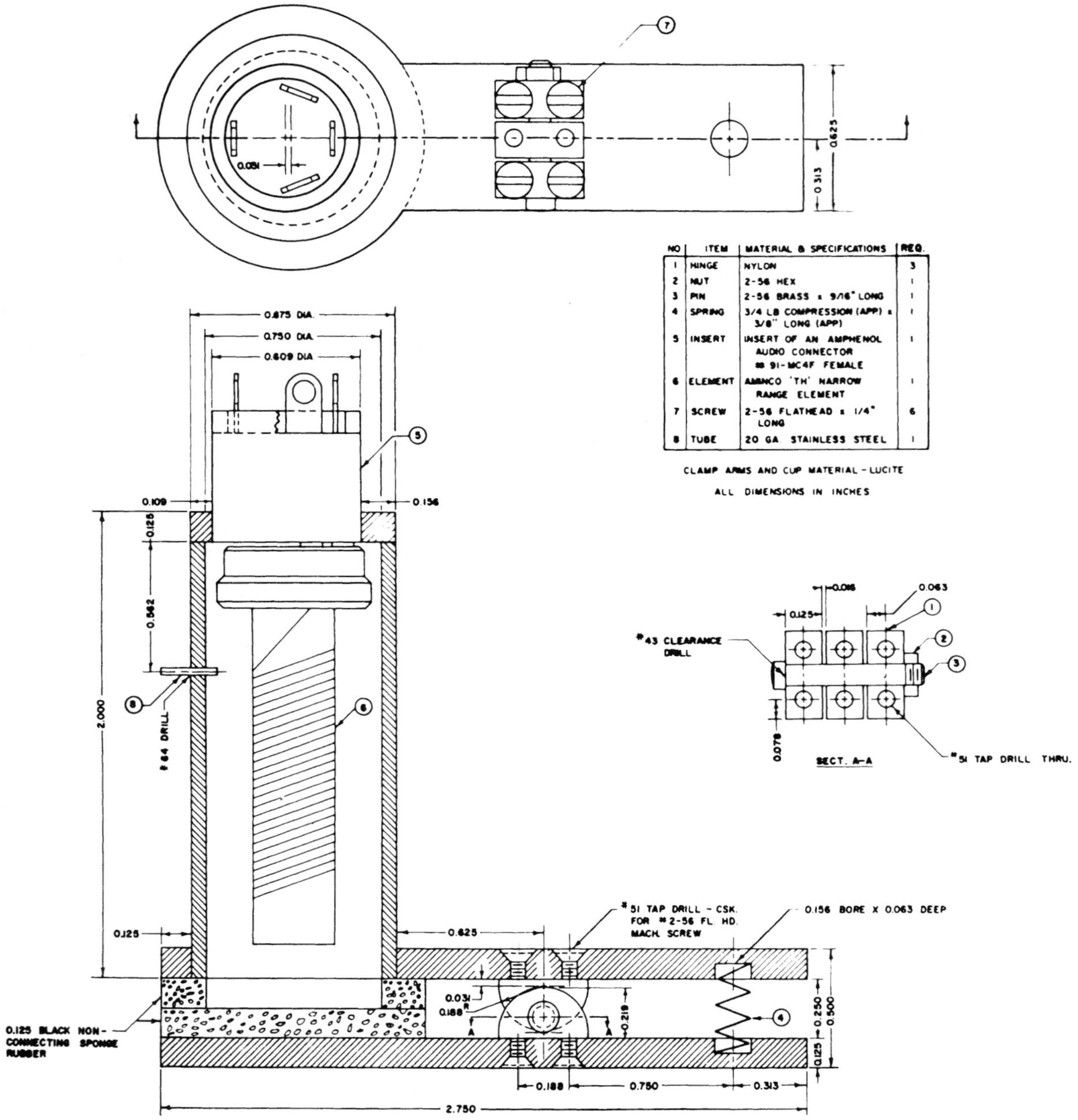

FIG. 1. Construction drawing of resistance meter leaf cup. Connecting leads are omitted.

consists of exposing a small portion of the upper or lower leaf surface to a hygroscopic surface, always in exactly the same spatial relationship and for only a short period, from 0.1 to 1.0 minute. The hygroscopic surface is a lithium chloride-impregnated resistor and, by calibration, the time rate of change in its electrical resistance can be interpreted in terms of the diffusion resistance, identified above as R_L and a known constant, typical of the device.

In effect, a portion of the leaf is allowed to transpire under known and prescribed conditions. The rate of transpiration when compared with the rate of evaporation from a free water surface of identical area under similar conditions, gives the value of R_L.

A similar approach has been followed using leaf chambers or plant enclosures. However, by using a small leaf cup as proposed here, the plant and its environment are not altered by the measurement, and the particular leaf spot is affected only very briefly. Also, the vapor transport in the cup appears to take place in a reproducible manner. Thus, the many objections to an enclosure technique do not apply. Further, as Wallihan already demonstrated, the equipment can be quite simple, inexpensive, and portable.

A construction drawing of the leaf resistance cup is given in figure 1. The sampling area is 2.84 cm^2 and the cup volume is 14.4 cm^3, excluding the measuring element. The latter is connected to an AC resistance meter with 30-gauge, stranded vinyl-insulated copper wire. An effective seal over veins and other surface elements of a leaf is obtained with low clamping pressure using sponge rubber gaskets. The entire assembly weighs 30 g, and it can be attached without further support to medium-size leaves, such as cotton, citrus, corn, or canna. If a leaf lacks adequate strength the cup must be held by hand.

The sensor is an Aminco-Dunmore[2] element (#4-4817) with an effective range from 14 to 27 % relative humidity. When used with an appropriate AC resistance meter (see Appendix), the relation between meter current and relative humidity is nearly linear from 0.10 to 0.80 full-scale. Before the cup is clamped onto the leaf, the air in the chamber is dried out with a stream of dry air to a reading of about 0.10 full-scale. After clamping the cup on, a stop watch is started when the reading attains a value of 0.20 full-scale and stopped when the value is 0.60 full-scale, thus giving a transit time Δt.

An appropriate analysis of the water vapor diffusion in the cup shows that the following relation should apply:

$$R_L = (S\Delta t - L_o)/D \qquad \text{II}$$

where R_L is the resistance of the leaf in min cm^{-1}, S is an instrument constant in cm min^{-1} which can

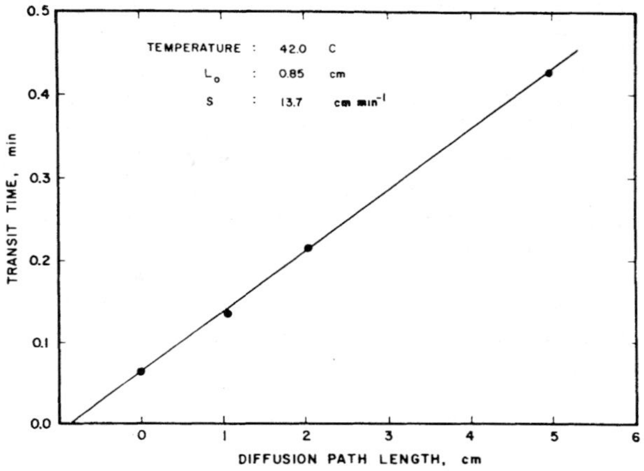

FIG. 2. Experimentally found relation between diffusion path length (L) and transit time (Δt) at 42.0°.

be designated as the sensitivity, L_o is a diffusion length or shape factor, characteristic for the cup, in cm, and D is the molecular diffusivity of water vapor in air in cm^2 min^{-1}.

Relation (II) is obtained by assuming that the temperature throughout the system is constant, that steady-state evaporation occurs, and that the relative humidity sensed by the element does not change appreciably with respect to unity. Experiments on the time rate of change of the element resistance, after exposure of the cup to a wet blotter, demonstrated the adequacy of the second assumption. With respect to the last assumption, a change in meter reading from 0.2 to 0.6 full-scale corresponds to a change in relative humidity of about 0.04, nearly independently of temperature. In fact, therefore, the condition assumed to be constant changes by 4 parts in 80, a change that may be ignored.

The effective validity of equation II was tested by measuring Δt at various temperatures and at varying values of a dummy R_L, obtained by interposing cylinders of the same internal diameter as the cup, but varying lengths L, between the cup and a wet blotter. The temperature of the blotter was measured with a fine thermocouple and was never more than 1° below that of the environment. Considering, in this case, that the equivalent R_L equals L/D, equation II can be written as

$$L = S\Delta t - L_o. \qquad \text{III}$$

By plotting L vs Δt, a straight line should be obtained, as exemplified in figure 2. Thus, not only is the validity of (II) established but, by measuring S and L_o at different temperatures, the absolute calibration of the instrument is obtained. This is necessary, since in equation II both S and D are temperature dependent, although L_o is not.

With our instrumentation, the average value for L_o was 0.72 cm. The dependence of S upon temperature is shown in figure 3. Values of D at 1000 mb pressure, taken from standard tables, are also plotted. Thus, from a measurement of Δt, and using the appropriate S and D from figure 3, R_L,

[2] Trade names are included for the benefit of the reader and do not infer any endorsement or preferential treatment of the product listed by the United States Department of Agriculture.

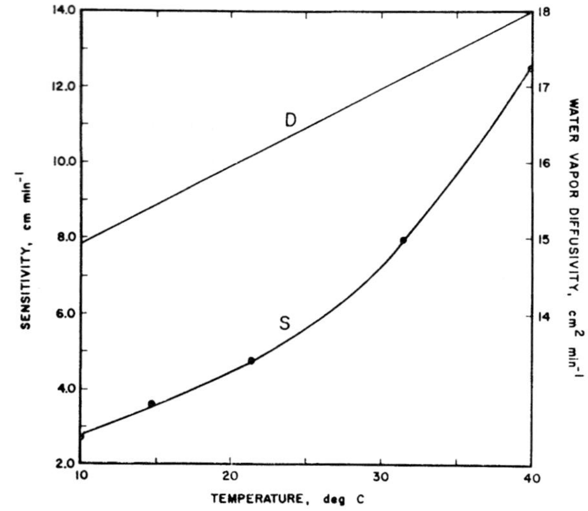

FIG. 3. Temperature dependence of resistance meter sensitivity (S) and of the water vapor diffusivity (D), the former having been experimentally determined, the latter from handbook values.

can be calculated from equation II. At pressures significantly different from 1000 mb, the value of D is to be multiplied by 1000/p, p being the ambient pressure in mb.

Results and Discussion

Experience with the leaf resistance meter showed that it could be used conveniently on various types of plants. The diurnal opening and closing of stomates, and differences between upper and lower epidermis, could be established reproducibly. Readings were substantially identical between different positions on any 1 leaf and between different leaves on 1 plant, provided exposure was the same. Read-

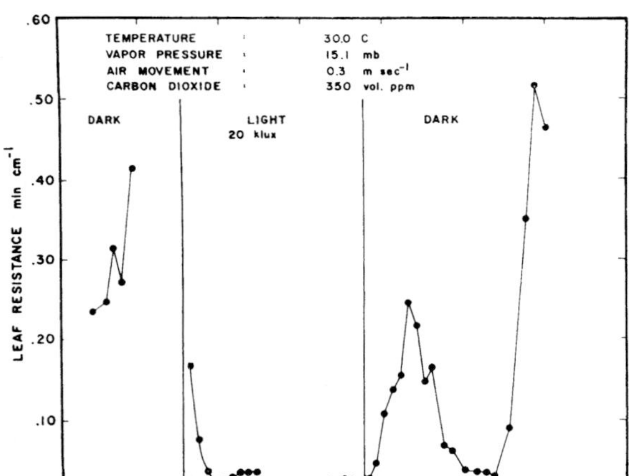

FIG. 4. Transpiration resistance of lower epidermis of a cotton leaf, as influenced by sudden illumination and darkening. Time scale reference is arbitrary.

ings could be taken up to 1 per minute by a single operator.

Figure 4 gives an example of data obtained in a controlled environment on a cotton plant (*Gossypium barbadense L.*, var. Pima S-2) grown in nutrient solution. Stomatal opening was induced by turning on fluorescent lights in an otherwise constant environment. Assuming that stomates were fully closed in the dark, cuticular resistance is approximately 20 times that of the combined (parallel) stomatal and cuticular resistance in the light. The time rate of change of leaf resistance can be documented quite precisely. Stomatal movement, as implied by the data, appears quite rapid in this case. Figure 4 also demonstrates that turning off the lights gave a reverse response of a rapidly increasing stomatal resistance, followed by an ephemeral decrease in the dark. The latter behavior has been observed repeatedly in our work.

A more rigorous test of the adequacy of the resistance measurement lies in its ability to predict absolute values of transpiration. Again, measurements were made on a cotton plant grown in nutrient solution and in a constant environment. Transpiration, leaf temperature, and resistances of upper and lower leaf epidermis were measured simultaneously on the same plant. The values, as shown in figure 5, were approximately constant with time, except the resistance of the upper epidermis, which fluctuated. Measurements of evaporation rate and temperature of wet blotter papers (5 × 10 cm rectangles), made in the same environment, gave a value of $R_A = 0.023$ min cm⁻¹. Transpiration was computed from

$$E = \frac{7.15 \times 10^{-7} \times \Delta e}{R_A + R_L} \text{ g cm}^{-2} \text{ min}^{-1} \qquad \text{IV}$$

where E is the water loss per unit leaf surface area of either upper or lower epidermis, Δe is the difference in vapor pressure between the air and a saturated atmosphere at leaf temperature in mb, and R_A and R_L are the air and leaf resistances, in min cm⁻¹. The conversion factor 7.15×10^{-7} g cm⁻³ mb⁻¹, taken at 30°, is required because of the use here of vapor pressure values rather than vapor concentrations.

From the average R_L value for the lower epidermis of 0.020 min cm⁻¹, the transpiration was calculated as 3.74×10^{-4} g cm⁻² min⁻¹. For the upper epidermis, with an average R_L of 0.148 min cm⁻¹, the transpiration was 0.94×10^{-4} g cm⁻² min⁻¹. For the leaf as a whole (2 sides) the transpiration rate would then be obtained by addition as 4.68×10^{-4} g cm⁻² min⁻¹, whereas the average measured value from water-loss determination of the entire plant was 3.75×10^{-4} g cm⁻² min⁻¹. This agreement is considered adequate considering the experimental errors involved in measuring R_L, as well as leaf temperature, short-time transpiration and, perhaps in particular, R_A.

Accepting the measured values for R_L as realistic, one computes the following values for Raschke's α:

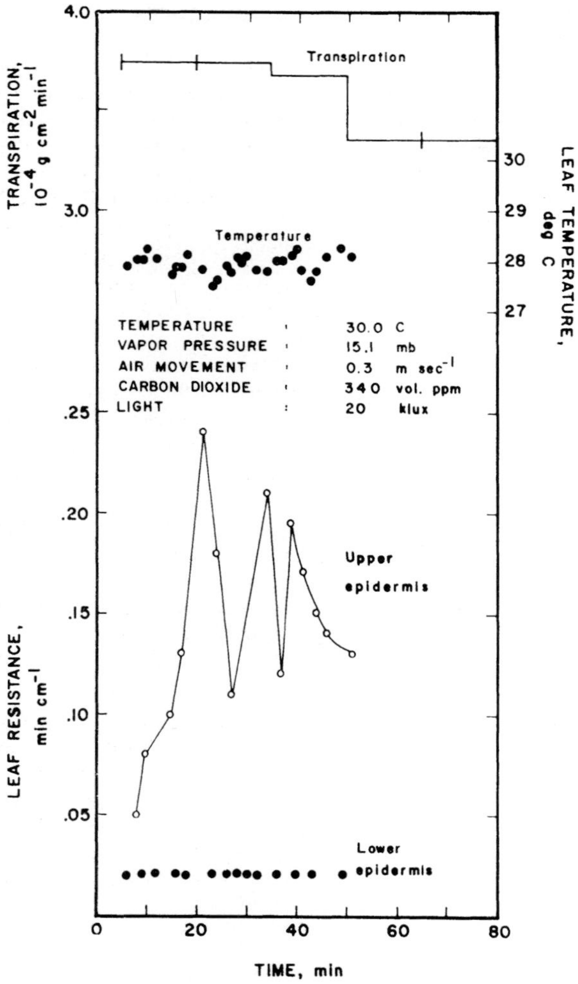

FIG. 5. Transpiration, leaf temperature, and leaf resistance of upper and lower epidermis of a cotton plant in a controlled, steady environment.

for the upper epidermis $\alpha = 0.13$, and for the lower epidermis $\alpha = 0.53$. Thus, the leaf as a whole evaporated 0.33 that of a hypothetical wet blotter at the same temperature and exposure to air movement. However, in reality, a similarly exposed blotter would have a lower temperature and the ratio would be greater than 0.33. Therefore, the disparity between leaf transpiration and blotter evaporation is less than suggested by the value of α. Raschke (10) cites values for α that are generally lower than 0.53, though he states that its value may lie between 0.004 and 0.6, the latter value applying to wide open stomates. Direct observation showed that the cotton stomates had a width of about 5 μ in our test, the length being about 15 μ. A report on simultaneous variation of R_L and transpiration in cotton is given by Ehrler, et al. (4).

From the stomatal dimensions and numbers on our cotton leaves, following Penman and Schofield's procedure, a value of R_L for the lower epidermis was found as 0.0041 min cm^{-1}. The generalized value of 1.56×10^3 cm^{-1} that Penman and Schofield have proposed for single stoma resistance would result in $R_L = 0.0062$ min cm^{-1} for the lower epidermis (15,000 stomates/cm^2). On the other hand, Bange's calculations for fully open stomates of *Zebrina* gave a value for R_L of 0.018 min cm^{-1}, again for the lower epidermis. This and our corresponding figure for R_L of 0.020 min cm^{-1}, suggests that the Penman-Schofield procedure considerably underestimates stomatal resistance, conceivably because the resistance of the substomatal cavities is not accounted for.

The physical significance of the measured R_L values depends upon their magnitude relative to the associated R_A values as is suggested by equation IV. One should bear in mind, however, that as R_L increases Δe will also increase and vice versa. A recent example of the application of equation IV is found in Slatyer and Bierhuizen (12), where additional discussion may be found. In this reference values for R_L for cotton were obtained in a leaf chamber. At 20 kilolux a value was found of 0.022 min cm^{-1} for cotton, virtually identical to our measurements.

In considering the evaporation from a vegetative canopy, as a problem in micrometeorology, an equation similar to equation IV can be written. Then, R_A or its analogue is determined by windspeed, roughness, and air stability, and R_L must refer to the canopy as a whole. As of now, it is not clear how R_L for a leaf is related to R_L for a canopy, though obviously the latter value must be smaller. For a detailed discussion reference is made to Monteith (7).

Summary

A method, suitable for field use, is given whereby the resistance offered by the leaf anatomy to diffusive loss of water can be measured directly, rapidly, and with simple, portable equipment.

A small cup, containing a sensitive humidity sensor, is clamped onto the leaf and the time rate of change of the sensor indication is measured over a short period, often less than 1 minute. Appropriate calibration permits the conversion of the time rate of change to the diffusion resistance, inclusive of a temperature correction. Construction details of suitable, portable equipment are supplied.

The diffusion resistance, thus measured, can be used to compute the absolute transpiration rate of leaves and plants, and to characterize the movements and responses of stomates with direct reference to the transpiration process.

The leaf resistance, as measured directly, can also be used in combination with a knowledge of atmospheric transport processes to predict evaporation from plant canopies.

Acknowledgment

The authors express thanks to their colleague, Mr. Ian McIlroy, for a critical reading of the manuscript.

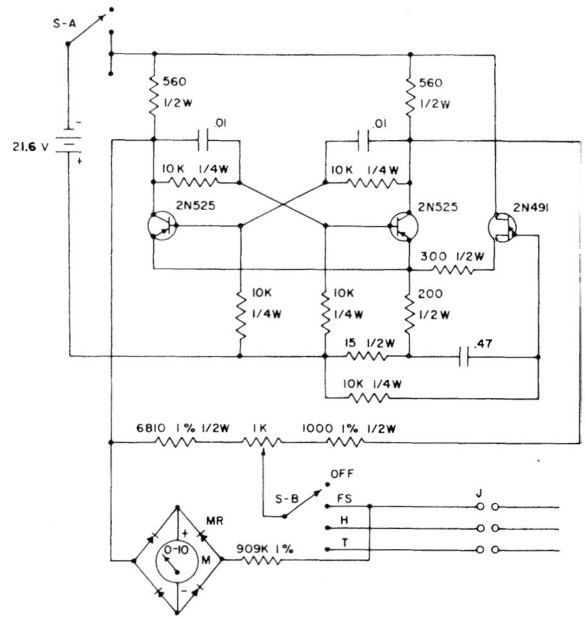

FIG. 6. Wiring diagram of portable AC resistance meter to be used with resistance meter leaf cup shown in figure 1.

Appendix

The changing resistance of the humidity sensor can be measured using commercially available equipment; however, the weight and bulk of this equipment is excessive for field work.

Figure 6 provides a wiring diagram of a lightweight (800 g), compact (8 × 10 × 16 cm) meter, that permits use of factory-supplied calibrations for the sensors. The meter consists of a unijunction trigger circuit and a flip-flop circuit, providing a pure and stable square wave voltage of amplitude 22 v and 250 cps frequency. Current consumption is 30 milliamps and maximum current through the sensor is 10 microamps.

Literature Cited

1. ARCHER, R. J. AND V. K. LAMER. 1954. The effect of monolayers on the rate of evaporation of water. Ann. New York Acad. Sci. 58: 807–29.
2. BANGE, G. G. J. 1956. On the quantitative explanation of stomatal transpiration. Acta Botan. Neerl. 2: 255–97.
3. ECKHARDT, F. E. 1960. Eco-physiological measuring techniques applied to research on water relations of plants in arid and semi-arid regions. In: Plant-water relationships in arid and semi-arid conditions. UNESCO Arid Zone Res. Ser. 15: 139–71.
4. EHRLER, W. L., F. S. NAKAYAMA, AND C. H. M. VAN BAVEL. 1965. Cyclic changes in water balance and transpiration of cotton leaves. In press.
5. HEATH, O. V. S. 1959. The water relations of stomatal cells and the mechanisms of stomatal movement. In: Plant Physiol. F. C. Steward, ed. Academic Press, New York. 2: 193–250.
6. MILTHORPE, F. L. 1955. The significance of the measurement made by the cobalt chloride paper method. J. Exptl. Botany 6: 17–19.
7. MONTEITH, J. L. 1963. Gas exchange in plant communities. In: Environmental Control of Plant Growth. Academic Press, New York. 95–111.
8. PELTON, W. L. AND C. B. TANNER. 1960. Potential evapotranspiration estimates by the approximate energy balance method of Penman. J. Geophys. Res. 65: 3391–3413.
9. PENMAN, H. L. AND R. K. SCHOFIELD. 1951. Some physical aspects of assimilation and transpiration. Symp. Soc. Exptl. Biology 5: 115–29.
10. RASCHKE, K. 1956. Über die physikalischen Beziehungen zwischen Wärmeübergangszahl, Strahlungsaustausch, Temperatur und Transpiration eines Blattes. Planta 48: 200–38.
11. RASCHKE, K. 1958. Über den Einflusz der Diffusionswiderstände auf die Transpiration und die Temperatur eines Blattes. Flora 146: 546–78.
12. SLATYER, R. O. AND J. F. BIERHUIZEN. 1964. Transpiration from cotton leaves under a range of environmental conditions in relation to internal and external diffusive resistances. Australian J. Biol. Sci. 17: 115–30.
13. WALLIHAN, E. F. 1964. Modification and use of an electric hygrometer for estimating relative stomatal apertures. Plant Physiol. 39: 86–90.

Ann. Sci. for., 1985, **42** (2), 193-200

Une nouvelle méthode pour la mesure du flux de sève brute dans le tronc des arbres

A. GRANIER

I.N.R.A.-C.R.F., Station de Sylviculture et de Production
Champenoux, F 54280 Seichamps

Résumé

La méthode décrite dans cet article utilise un capteur thermique composé de deux sondes insérées radialement dans le bois d'aubier du tronc. Une de ces sondes est chauffée à puissance constante, l'autre sert de référence de température. Une équation simple permet de calculer le flux de sève en fonction de la différence de température entre les deux éléments. Un étalonnage a été effectué sur des échantillons de tronc de différentes essences.

Le faible coût et la sensibilité de ce système doivent permettre d'aborder la mesure quantitative de la transpiration et de sa variabilité en forêt.

La mesure de la transpiration constitue un élément essentiel de la compréhension de la physiologie de l'arbre et de la dynamique des transferts d'eau dans les peuplements forestiers. Dans le cadre des recherches menées à la Station de Sylviculture et de Production du C.R.F., nous nous sommes intéressés depuis plusieurs années à la mesure du flux de sève brute dans le tronc des arbres. Il existe à ce jour de nombreuses méthodes de mesure du flux transpiratoire ; la méthode des impulsions de chaleur a eu, depuis HÜBER & SCHMIDT (1937) un développement très significatif, notamment grâce à SWANSON (1962, 1974). Nous avons largement utilisé cette technique, en la rendant automatique, et en modifiant la structure du capteur (GRANIER & GROSS, en cours de publication).

Toutefois, aucune de ces méthodes ne présente les spécifications requises pour permettre, à faible coût, de procéder à un échantillonnage de la transpiration des arbres en peuplements forestiers. La méthode des impulsions de chaleur a en particulier deux inconvénients : son caractère ponctuel dans le tronc et son imprécision en condition de transpiration faible.

Nous proposons ici une nouvelle méthode et un capteur permettant de mesurer le flux de sève suivant un axe radial dans le bois d'aubier du tronc.

Paper B2, is reprinted from *Annales des Sciences forestières* 42, 193–200, by permission of A. Granier.

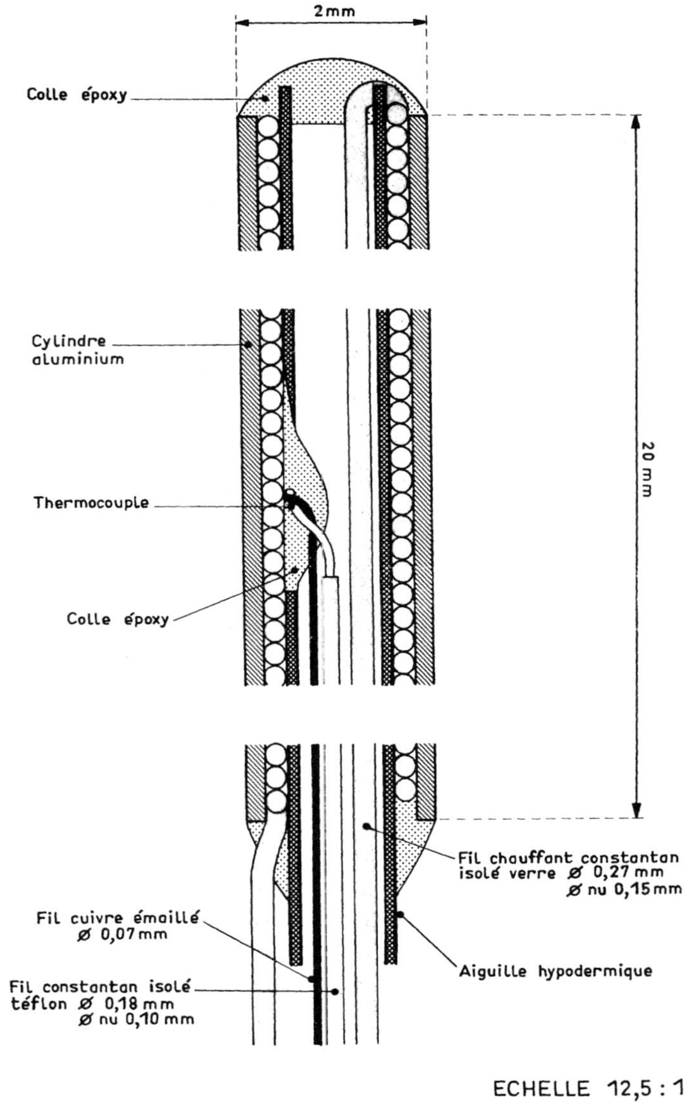

FIG. 1

Coupe longitudinale de l'élément chauffant.
Longitudinal sectional view of the heating probe.

1. Matériel et méthodes

1.1. *Matériel*

Le capteur (fig. 1) se compose de deux sondes cylindriques de 2 mm de diamètre et de 20 mm de longueur, insérées radialement dans le bois d'aubier du tronc, séparées d'une distance de 5 cm environ. Une de ces sondes est chauffée à puissance constante au moyen d'un enroulement de fil de constantan autour d'une aiguille hypodermique en acier inoxydable. La résistance chauffante est calibrée à 10 Ω et la tension d'alimentation est délivrée par une alimentation stabilisée réglable. Un thermocouple (cuivre - constantan) est disposé à mi-longueur de la résistance chauffante et une gaine en aluminium recouvre l'ensemble pour en uniformiser la température. La deuxième sonde, située dans le tronc au-dessous de la précédente, renferme un thermocouple identique et monté en opposition avec celui de l'élément chauffant. Le système permet ainsi de mesurer l'écart de température entre les deux sondes.

Un amplificateur opérationnel de type OP 27 PMI permet d'augmenter le signal des thermocouples (avec un gain de 1 000).

1.2. *Théorie*

En condition de régime thermique établi entre l'élément chauffant et le milieu (bois + sève), et pour un flux de sève constant, nous supposerons que l'apport de chaleur par effet Joule est égal à la quantité de chaleur dissipée au niveau de la paroi du capteur. Nous posons donc :

$$h\,S\,(T - T_\infty) = R\,i^2 \qquad (1)$$

avec :

- h = coefficient de transfert de la chaleur ($W \cdot m^{-2} \cdot {}^\circ C^{-1}$),
- S = aire de la surface d'échange (m^2),
- T = température du cylindre (°C),
- T_∞ = température du matériau bois en l'absence de chauffage (°C),
- R = résistance électrique (Ω),
- i = intensité du courant électrique (A).

Le coefficient h est supposé dépendre du débit de la sève u ($m \cdot s^{-1}$) sous la forme :

$$h = h_o\,(1 + \alpha \cdot u) \qquad (2)$$

où h_o est le coefficient d'échange lorsque $u = 0$ (transpiration nulle), qu'il est possible de calculer d'après (1) :

$$h_o = \frac{R\,i^2}{S\,(T_M - T_\infty)} \qquad (3)$$

T_M désigne la température à flux de sève nul ($u = 0$)

Lorsque u est constant et non nul, il vient :

$$u = \frac{1}{\alpha} \frac{T_M - T}{T - T_\infty} \qquad (4)$$

Le rapport $\frac{T_M - T}{T - T_\infty}$ est un nombre sans dimension, que nous appellerons index de flux K, proportionnel à u.

2. Résultats

2.1. *Etalonnage*

Le capteur a été testé et étalonné sur des fragments de troncs de diamètre compris entre 40 et 50 mm. Nous avons fait circuler de l'eau sous pression dans ces échantillons, et mesuré simultanément le débit de l'eau par pesée de l'exsudat et le signal

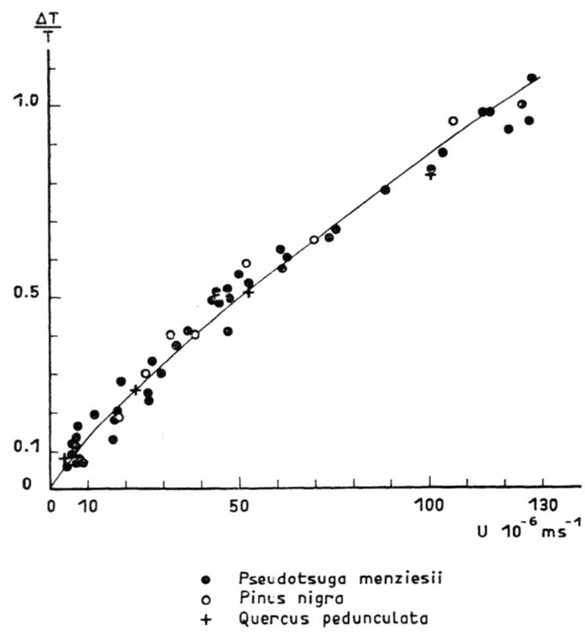

FIG. 2

Relation entre le rapport K (voir texte § 1.2) et le flux d'eau par unité de section de bois d'aubier pour différentes essences.

Relationship between K (see text § 1.2) and the water flux per unit of sapwood area, for different species.

délivré par le capteur thermique. Le débit pouvait être modifié en ajustant la pression de l'eau. Après les mesures, chaque échantillon était sectionné pour mesurer la section du bois d'aubier au niveau de l'élément chauffant. L'étalonnage a porté sur trois essences différentes : douglas, pin noir, et chêne pédonculé.

La figure 2 représente les résultats obtenus, avec en abscisse le flux d'eau par unité de surface u (en $m \cdot s^{-1}$) et en ordonnée le rapport K. Le calcul de K s'effectue d'après la relation (4) en connaissant pour chaque échantillon la température T_M atteinte lorsque le flux d'eau est nul. Il est intéressant de constater que la relation entre K et u est la même pour les trois espèces ; le coefficient α de l'équation (4) semble donc, dans les conditions expérimentales, indépendant de l'essence. Un ajustement non linéaire a conduit à la relation expérimentale :

$$K = 0.0206\ u^{0.8124} \qquad (5)$$

$r^2 = 0.96$
n = 53 points

où u est exprimé en $10^{-6}\ m \cdot s^{-1}$.

Signalons enfin que nous avons fixé l'intensité du courant dans la résistance chauffante à une valeur de 0,141 A, ce qui est un compromis entre la sensibilité du capteur (qui augmente avec l'intensité appliquée) et le risque d'un échauffement de la sonde de référence de la température (1).

2.2. *Mesures sur des arbres*

Durant l'été et l'automne 1983, nous avons effectué des enregistrements journaliers de flux de transpiration sur des douglas de la forêt domaniale d'Amance (à 15 km à l'est de Nancy). Nous avons choisi des arbres placés dans des situations différentes : des arbres de plein découvert, d'une hauteur de 5 m, et des arbres d'une plantation régulière âgée de 20 ans, pour une hauteur moyenne de 15 m. Quatre arbres (deux dans chacune des situations) ont été équipés chacun d'un capteur thermique inséré radialement dans leur bois d'aubier.

Le coefficient K repose sur l'évaluation de T_M (voir équation (4) au § 1.2). Nous avons supposé que cette température d'équilibre à flux de sève nul pouvait être mesurée lors de séquences nocturnes où l'humidité de l'air est voisine de la saturation.

La figure 3 reporte pour une journée (le 16-8-1983) la variation du coefficient K, observée pour un arbre de chaque traitement, ainsi que l'évolution de l'évapotranspiration potentielle (E.T.P.) calculée par la formule de Penman. Nous avons choisi une journée présentant des passages nuageux qui permettent de mettre en évidence la réponse des capteurs. Les deux arbres suivent une évolution similaire, en particulier au niveau des maximas et des minimas. Le retard important de K par rapport à l'E.T.P. le matin correspond à une phase d'évaporation de la rosée sur les aiguilles. Après cette phase, le facteur K des deux arbres suit la variation de l'E.T.P. Il est à noter que la chute brutale de l'E.T.P. intervenant en milieu de journée (passages nuageux) se manifeste de façon plus accentuée pour l'arbre situé en plein découvert. L'arbre de peuplement semble ainsi mieux tamponner les variations de l'E.T.P. En fin de journée, si les deux arbres voient diminuer K en même temps que l'E.T.P., un retard se manifeste pour le douglas du peuplement.

---

(1) D'après nos mesures, le coefficient α des relations (2) et (4) est dépendant de la quantité de chaleur apportée.

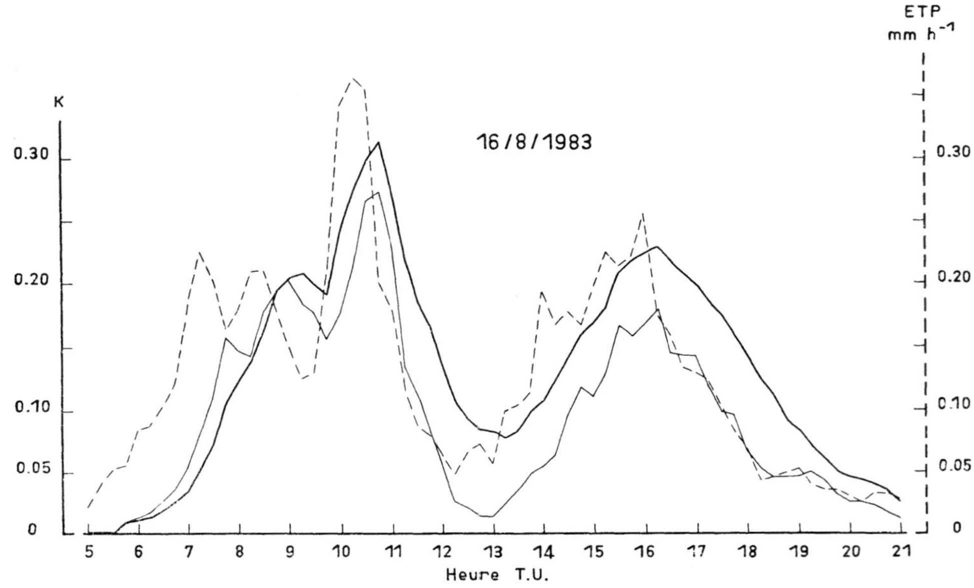

FIG. 3

Evolution du rapport K pour deux Douglas et de l'E.T.P. Penman au cours d'une journée.
Diurnal evolution of K for two Douglas fir trees and of Penman P.E.T.

——— Arbre en peuplement. *Tree in a closed stand.*

——— Arbre en découvert. *Open grown tree.*

– – – E.T.P. Penman. *Penman P.E.T.*

Un nombre important de journées a ainsi été étudié : nous avons représenté sur la figure 4, pour l'ensemble des journées, la valeur du coefficient K_{24} de l'arbre en peuplement, en fonction de l'E.T.P. journalière exprimée en mm. Le coefficient K_{24} est égal à la moyenne journalière de K défini d'après l'équation (4).

Pendant la durée de ces mesures, l'arbre étudié a été en permanence bien alimenté en eau ; la transpiration et donc le coefficient K_{24} sont donc sensiblement proportionnels à l'E.T.P. Nous avons aussi fait figurer les journées présentant des épisodes pluvieux diurnes, quelles que soient leur durée et leur intensité. Certaines de ces journées s'écartent du nuage de points : l'interception de l'eau par le feuillage provoque un blocage de la transpiration ; l'énergie incidente est alors utilisée pour évaporer l'eau interceptée.

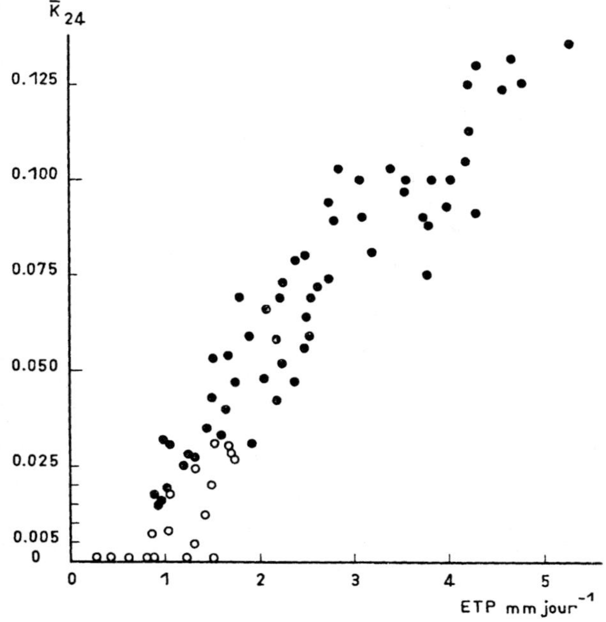

FIG. 4

Relation entre le coefficient K_{24} (voir texte § 2.2) et l'E.T.P. Penman pour un Douglas en peuplement fermé.

Relationship between K_{24} (see text § 2.2) and Penman P.E.T. for one Douglas-fir in a closed stand.

• Journées sans pluie.
Days without rain.

o Journées avec pluie.
Rainy days.

3. Conclusion

La méthode que nous avons décrite présente un certain nombre d'avantages, qui devraient permettre d'obtenir des résultats fiables, à deux niveaux :

— au niveau du fonctionnement hydrique de l'arbre, par l'étude fine des variations journalières du flux de sève, particulièrement en période de stress hydrique ;

— au niveau du fonctionnement hydrique des peuplements forestiers, sachant que la simplicité de mise en œuvre et le faible coût [(2)] de cette technique permettent d'aborder la mesure quantitative de la transpiration et de sa variabilité spatiale en forêt.

---

(2) Un capteur (deux sondes) revient à environ 30 FF, le système d'amplification à 80 FF par capteur.

L'utilisation de cette méthode suppose, lorsqu'on désire calculer le flux total, de connaître la section du bois d'aubier au niveau du point de mesure. Selon le type d'essence et la précision souhaitée, on pourra estimer la section du bois d'aubier grâce à un ou plusieurs sondages à la tarière de Pressler ou la mesurer directement après abattage de l'arbre.

Summary

A new method of sap flow measurement in tree stems

The method described in this paper is based on a thermal sensor composed of two probes radially inserted in the sapwood of the trunk. One of those probes is heated at a constant energy and the other considered as a temperature reference. A simple equation enables us to calculate the sapflow as a function of the difference of the temperature between the two elements. A calibration has been made on pieces of trunk of different species.

Owing to its sensitivity and its low cost, this system may fit for the quantitative measurement of forests transpiration.

Références bibliographiques

GRANIER A., GROSS P., 1984. Mesure du flux de sève brute chez le Douglas (*Pseudotsuga menziesii* Mirb. Franco) par la méthode thermo-électrique automatisée (à paraître).

HÜBER B.,SCHMIDT E., 1937. Eine Kompensationsmethode zur thermoelecktrischen Messung langsamer Stafströme. *Ber. dtsch. Bot. Ges.*, **55**, 514-529.

SWANSON R.H., 1962. An instrument for detecting sap movement in woody plants. Sta. Pap. Rocky Mt. For. Range Exp. Sta. n° 68.

SWANSON R.H., 1974. Velocitiy distribution patterns in ascending xylem sap during transpiration. In : Flow its Measurement and control in Science and Industry. Et. Rodger and Dowdell. Instrument Society of America. Vol. 1, 1425-30.

Translation of Paper B2 (by J. H. C. Gash and A. Granier)

A. Granier (1985) A new method of sap flow measurement in tree trunks.

Summary

This paper describes a thermal method of measuring sap flow in trees based on an instrument composed of two probes radially inserted in the sapwood of the trunk. One of those probes is heated at a constant energy and the other used as a temperature reference. A simple equation enables us to calculate the sap flow as a function of the difference in temperature between the two elements. Calibrations have been performed on samples of trunk from different species of tree.

Owing to its sensitivity and its low cost, this system is suitable for quantitative measurements of forest transpiration.

Introduction

Understanding the physiology of trees and the dynamics of water movement in forest stands requires measurements of transpiration. One method of measuring transpiration is by monitoring the rate of sap flow through tree trunks; application of this method has formed part of the research programme of the CRF Station de Sylviculture et de Production for many years. There are numerous methods of measuring transpiration flux; but thanks to the work of Swanson (1962, 1974), the heat pulse method of Hüber & Schmidt (1937) is the most important. Previously, we have mainly used this technique, but improving it by making it automatic and modifying the structure of the probe (Granier & Gross, 1984).

However, none of these existing methods are sufficiently inexpensive to allow the installation of a network of gauges as would be needed to measure transpiration from a forest stand. The heat pulse method, in particular, has two additional disadvantages: firstly it is a point measurement, and secondly, it is inaccurate when the transpiration is low.

Here we propose a new method with an instrument which can measure the sap flowing across the radial axis of trunk sapwood.

1 Material and Methods

1.1 Material

The instrument (see Fig. 1) comprises two cylindrical probes 2 mm in diameter and 20 mm in length, inserted radially in the trunk sapwood, separated by about 5 cm. One of the probes is heated at constant power by a coil of constantan wire wound round a stainless steel hypodermic needle. The resistance of the heater is set at 10 Ω, with the voltage being delivered by a stabilised power supply. A thermocouple (copper-constantan) is situated halfway along the heater. All this is then encased in an aluminium sheath, to create a probe with a uniform temperature. The second probe is placed in the trunk below the one described above, and contains an identical thermocouple acting as the cold junction. The system thus measures the difference in temperature between the two probes.

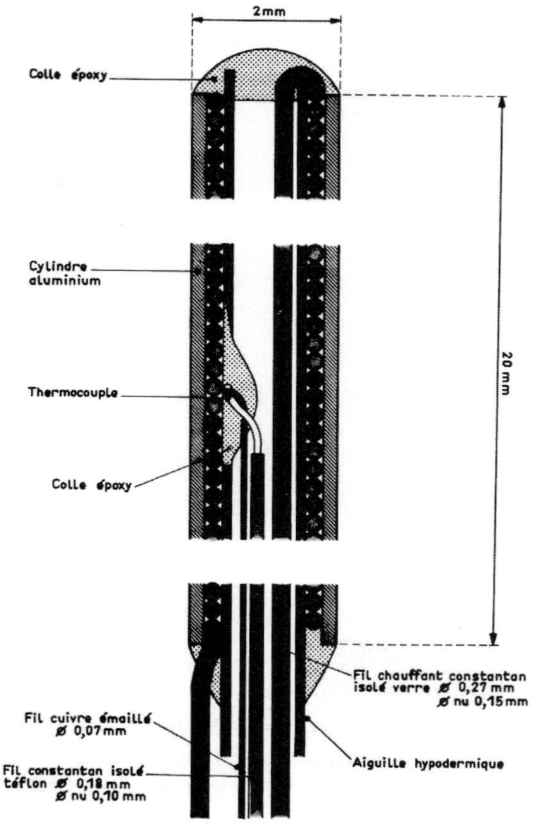

Fig. 1 Longitudinal sectional view of the heating probe.

An operational amplifier, type OP 27 PMI, increases the thermocouple signal (with a gain of 1000).

1.2 Theory

When thermal equilibrium has been established between the heated probe and its surroundings, and with a constant rate of sap flow, we can assume that the supply of energy by Joule heating equals the amount of heat dissipated at the level of the instrument. This can be expressed as:

$$hS(T - T_\infty) = Ri^2 \qquad (1)$$

where:

h = heat transfer coefficient (W m^{-2} °C^{-1})
S = heat exchange area (m^2)
T = temperature of the cylinder (°C)
T_∞ = temperature of the wood without heating (°C)
R = electrical resistance (Ω)
i = electrical current (A)

Assuming that the thermal conductivity, h, depends on the rate of sap flow, u (m s^{-1}), as:

$$h = h_0(1 + \alpha u) \qquad (2)$$

where h_0 is the heat transfer coefficient when $u = 0$ (zero transpiration), h_0 can be calculated from equation (1) as:

$$h_0 = \frac{Ri^2}{S(T_M - T_\infty)} \qquad (3)$$

where T_M represents the temperature for zero sap flow ($u = 0$).

When u is constant and non-zero, this becomes:

$$u = \frac{1}{\alpha}\frac{T_M - T}{T - T_\infty} \qquad (4)$$

The ratio $\frac{T_M - T}{T - T_\infty}$ is then a non-dimensional number proportional to u, that we will call the flux index, K.

2 Results

2.1 Calibration

The instrument has been tested and calibrated on sections of tree trunks of between 40 and 50 mm diameter. While water was forced through these samples under pressure, the flow rate was measured simultaneously by weighing the exudate and by using the thermal sensors. The rate of flow was varied by adjusting the water pressure. After measurement, a cross-section of each sample was taken at the position of the heating element to determine the area of sapwood. The calibration was completed on three different species of tree: Douglas Fir, Austrian Pine and Penduncular Oak.

The results are shown in Figure 2, where the flow rate, u (in m s^{-1}) is plotted against the ratio, K. K has been calculated for each sample using Equation 4 and T_M, the measurement of the temperature at zero flow rate. Interestingly it can be seen that the relationship between u and K is the same for each of the three species; the coefficient α in Equation 4 thus appears to be independent of species. A non-linear fit to the 53 data points gives the relationship:

$$K = 0.0206u^{0.8124}$$

where $r^2 = 0.96$ and u is expressed in 10^{-6} m s^{-1}.

Finally, we should state that the current in the heating element has been fixed at 0.141 A. This value was arrived at as a compromise between the sensitivity of the instrument (which increases with the current) and the risk of heating the probe containing the reference junction. Note that the coefficient α in Equations 2 and 4 is dependent on the amount of heating applied.

2.2 Measurements on trees

During the autumn of 1983, we carried out daily measurements of transpiration flux on Douglas Fir in the forest of

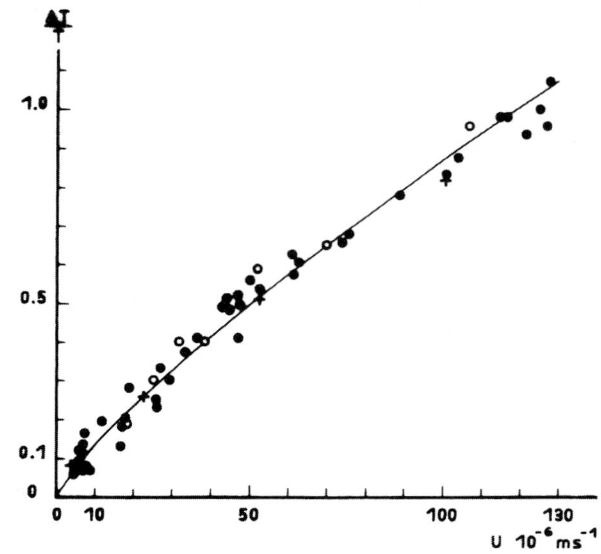

● *Pseudotsuga menziesii* ○ *Pinus nigra* + *Quercus pedunculata*

Fig. 2 Relationship between K and the water flux per unit of sapwood area for different species.

Amance (a state forest 15 km east of Nancy). The trees were chosen to be in two different situations: 5-m high, isolated trees, and in a 20 year old plantation with a mean height of 15 m. Four trees (two from each group) were each fitted with the thermal probes inserted radially into their sapwood.

The coefficient K relies on the evaluation of T_M (see Equation 4 in Section 1.2). We have assumed that this equilibrium temperature for zero sap flow could be measured at night when the air humidity is close to saturation.

Figure 3 shows the observed variation in K over one day (16 August 1983) for one tree in each group; also shown is the evolution of potential evaporation (ETP) calculated by the Penman formula. We have chosen a day characterised by broken cloud, as this demonstrates the time response of the instrument. The two trees show a similar pattern of variation, particularly their maxima and

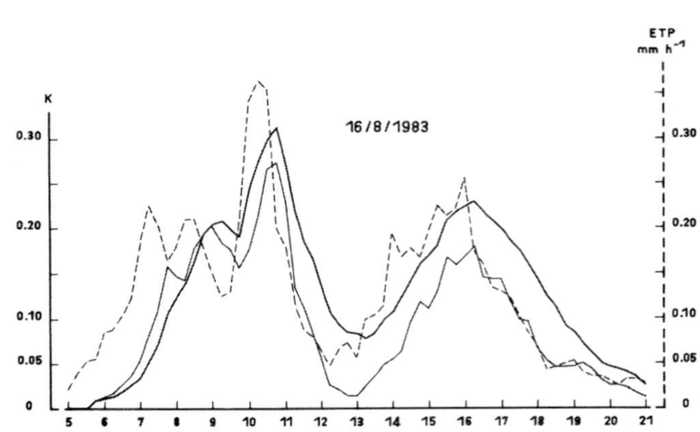

— Tree in a closed stand — Open grown tree - - Penman ETP

Fig. 3 Evolution of K for two Douglas fir trees and of Penman ETP.

minima. In the morning it can be seen that there is an important delay of K with respect to the ETP as dew evaporates from the needles. After this phase K follows the variation in ETP. Note that the rapid fall in ETP in the middle of the day (due to clouds) is followed more strongly by the isolated tree. The tree in the closed stand appears to be better able to absorb this variation in ETP. At the end of the day, although the two trees show a similar rate of decline in K as the ETP, the value of K in the Douglas Fir in the closed stand shows a delay.

A large number of other days have also been studied: these are shown in Figure 4, where the value of the daily coefficient, K_{24}, for the Douglas Fir stand is plotted against the daily total ETP in mm. K_{24} is the mean daily value of K as defined by Equation 4.

During the measurement period, the trees were permanently well supplied with water; the transpiration and thus the coefficient K were thus clearly proportional to the ETP. We have also plotted days with rainfall, without regard for the rainfall duration and intensity. Some of these days form a group of points where the interception of the rainfall by the canopy has created a barrier to transpiration; the incident energy is then used for evaporating intercepted water.

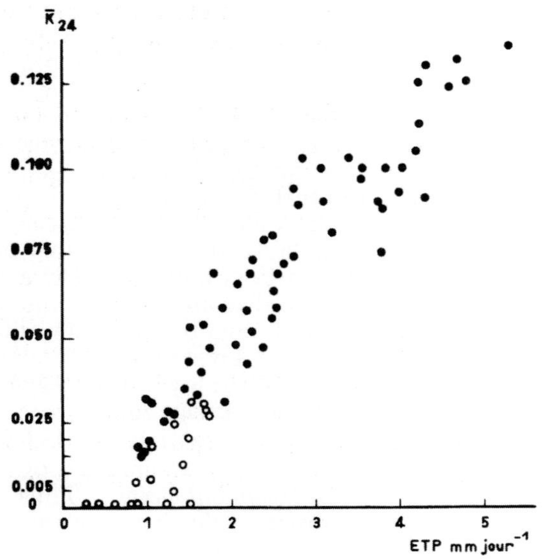

Fig. 4 Relationship between K_{24} and Penman ETP for one Douglas fir in a closed stand.

3 Conclusion

The method that we have described here has a number of advantages which should lead to reliable results in two scientific areas:

- research into tree water relations, through the detailed study of the daily variation in sap flow, particularly in periods of water stress;
- research into water use of forest stands, through the ability to install a network of low cost[1], simple instruments to measure transpiration and its spatial variability.

Using this method requires a knowledge of the sectional area of sapwood before it is possible to calculate the total transpiration. According to the species of tree and the desired precision, one can estimate the sectional area of sapwood using one or more Pressler augers or by measuring it directly after felling the tree.

[1]An instrument (two probes) costs about 30 French francs, the amplification system is about 80 French francs per instrument.

References

Granier, A. & Gross, P. (1984) Mesure du flux de sève brute chez le Douglas (*Pseudotsuga menziesii* Mirb. Franco) par la méthode thermo-électrique automatisée. INRA-Nancy, internal report.

Hüber, B. & Schmidt, E. (1937) Eine Kompensationmethode zur thermoelecktrischen Messung langsamer Stafströme. *Ber. dtsch. Bot. Ges.* **55**, 514–529.

Swanson, R.H. (1962) An instrument for detecting sap movement in woody plants. Sta. Pap. Rocky Mt. For. Range Exp. Sta. no. 68.

Swanson, R. H. (1974) Velocity distribution patterns in ascending xylem sap during transpiration. In: *Flow, Its Measurement and Control in Science and Industry* (Eds Rodger & Dowdell), vol. 1, 1425–1430. Instrument Society of America.

THE DETERMINATION OF EVAPORATION FROM LAND AND WATER SURFACES

By C. W. THORNTHWAITE and BENJAMIN HOLZMAN

[Soil Conservation Service, Washington, D. C., February 1939]

The problem of determining rate and amount of evaporation from land and water surfaces, or transpiration from vegetation, is rendered complex by the fact that the water becomes an invisible gas which rapidly mixes with the other gases of the atmosphere and is transported large distances and to great heights. A method for the accurate determination of water losses from lakes and reservoirs is of vital importance, but it is equally important to determine water losses from watersheds and from agricultural lands.

It is not possible to determine actual transpiration or evaporation from either free water surfaces or land surfaces by simply measuring the rate of loss of water from an exposed pan or atmometer. The amount of transpiration for any plant type is partly a function of leaf area, which varies seasonally with the growth cycle of the plant. Transpiration also varies from one plant species to another, depending on water requirements, the osmotic value in the leaves, and the number, nature, and size of the stomata. None of these factors is reflected by pan or atmometer measurements of evaporation.

Neither do these measurements indicate the evaporation from the soil. When the surface soil is moist the evaporation exceeds pan measurements because the soil with its minute irregularities presents a greater evaporating surface and because surface soil temperatures during the part of the day when most of the evaporation occurs are higher than water temperatures. When, however, the surface soil has become dry or partially dry, less evaporation occurs from the soil than from a pan. Even though the subsoil is moist, capillary action cannot supply the surface with water at a rate at all comparable to the evaporation from the surface of a body of water. Hence, water molecules can escape to the outer air only by a very slow diffusion process which takes place from the lower soil levels through the soil air, and results in a deceleration in the rate of moisture loss.

Since the rate of emission of water vapor from a land, water, or plant surface cannot be measured directly as can the rate of water accretion to the surface and since the measured rate from a pan or atmometer cannot give this rate of emission, the evaporation problem must be approached in a different manner.

Observations of evaporation from lakes, reservoirs, and pans have been used in the development of many empirical formulae in which evaporation is expressed as a function of various meteorological data such as temperature, relative humidity, barometric pressure, wind velocity, and solar radiation. Dalton (**5**) was the first to point out that evaporation is proportional to the difference between vapor pressure of the air at the water surface and that of the overlying air, although apparently he never expressed this relationship in mathematical terms.

In Rohwer's "Evaporation from Free Water Surfaces," published in 1931, (**20**) a number of evaporation formulae are presented and discussed. Most of these formulae contain the expression (e_s-e_d), e_s being the vapor pressure at the surface of the liquid and e_d the vapor pressure of the air, and a factor W expressing the influence of wind velocity. Rohwer's empirical formula is:

$$E=(0.44+0.118W)\ (e_s-e_d)$$

which, he says, is of the same form as Dalton's original formula:

$$E=C(e_s-e_d).$$

A different approach to the problem of determining evaporation from lakes and other bodies of water has been suggested by Ångström (**2**). He indicated that there must be a balance of energy between the insolation absorbed by a body of water and the energy exchanges due to radiation, convection, conduction, and the latent heat of evaporation, and that evaporation might be determined by measuring the necessary components of this equilibrium. Richardson (**18, 19**) and Cummings (**4**) have used this principle in determining evaporation losses but difficulties in obtaining various essential measurements, such as temperature gradients within the body of water and heat exchanges due to convection, prevent the determination of accurate short time measures of evaporation and prohibit the determination of moisture losses from land areas. Yamamoto (**39**) has made similar use of the conservation of energy principle in his studies of condensation of dew.

In recent years advances in aviation have stimulated the study of the structure of the lower atmosphere from which has come much essential information on turbulence and the intensity of turbulent mixing. In the free air there is a zone above the earth's surface in which the frictional effects of the ground are felt. This is the zone of turbulent mixing, or the turbulent layer. In this layer the mixing process is very important and depends on the shearing stresses associated with the roughness of the ground and with the wind structure, the thickness of the layer increasing as the value of these factors increases. The mixing process also depends upon the stability or density structure of the air, the mixing diminishing as stability increases. Beneath the turbulent layer there is a shallow layer usually measured in millimeters, called the laminar layer, in which air flow is laminar, where mixing proceeds by diffusion only. The vertical motion of the air in the turbulent layer tends to establish an adiabatic distribution of properties of the air, and thus to eliminate differences in moisture concentration. If moisture is neither added to nor withdrawn from the turbulent layer it quickly becomes uniform in moisture content throughout.

On the other hand, if water vapor is emitted from an evaporating surface it will be transported upward and scattered throughout the turbulent layer. Thus, as long as a stream of water vapor is flowing upward into the turbulent layer the moisture concentration will be highest at the base and will diminish upward, and a moisture gradient directed upwards will be established. Such a gradient could be maintained only as long as moisture continues to be added from below. When evaporation ceases the moisture will soon be distributed uniformly throughout the layer and the gradient will be destroyed. Similarly, if water vapor is abstracted from the base of the layer by condensation (dew or frost formation) the moisture concentration there will be reduced and as long as removal of moisture continues, a moisture gradient directed downwards will be maintained. The greater the intensity of turbulent mixing the greater will be the tendency toward the establishment of uniformity of moisture concentration, and the greater the evaporation or condensation required to maintain a constant gradient upward or downward. Similarly, the greater the moisture gradient which persists in an air column with a given rate of turbulent mixing, the greater will be the evaporation required to produce it.

From these facts it is evident that evaporation from any surface may be determined by taking into consideration the vertical distribution of moisture in the air and the intensity of turbulent mixing. The general form of an equation for evaporation based on these facts is $E=-A\,(dq/dh)$, in which A is the coefficient of turbulent mixing, or the Austausch coefficient of Willhelm Schmidt (**24**), and dq/dh is the rate of change of moisture concentration with respect to height above the evaporating surface. Observations required in order to measure evaporation are the moisture concentration at two levels within the turbulent layer, the heights of the two observations, so that moisture gradients may be obtained, and measurements of wind velocity at two or more levels to give the intensity of turbulent mixing. The moisture concentration in the air is measured as specific humidity. Within the height considered where air pressure is essentially constant the specific humidity is proportional to the vapor pressure.

Jeffreys (**9**) and Giblett (**7**) were among the earliest investigators who studied the problem of evaporation in terms of atmospheric turbulence. However, because of various nonrigorous assumptions regarding the nature of the Austausch coefficient, their results are limited and can be applied only to the evaporation from bodies of water.[1] Sutton (**26**), making use of Taylor's (**32**) researches, has extended Jeffrey's analysis by assuming that the Austausch coefficient varies with height. His theoretical work was found to be in very good agreement with experimental evaporation measurements for variously shaped areas. Schmidt (**25**) has made use of a formula devised by Ertel (**6**) for calculating the Austausch coefficient and claims to have measured the actual evaporation from a meadow.

During the last two decades great progress in the theoretical treatment of the mechanism of turbulent interchange has been made by various aerodynamicists, including Taylor (**30, 31, 32, 33, 34**), Prandtl (**16, 17**), von Kármán (**10, 11, 12, 13**), and others. In 1935 von Kármán (**12**) published an excellent summary of the problems associated with the various theories of turbulence. Rossby (**21, 22**) has made use of von Kármán's postulate of dynamic similarity and the concept of mixing length in his important researches on atmospheric turbulence. Sverdrup (**27**), working along lines explored by Rossby, has added much to our knowledge of the variation of the coefficient of eddy viscosity under changing conditions of thermal stability.

Although the theoretical treatment of problems of turbulence is still in the initial stage, sufficient has been accomplished to provide an understanding of the mechanism of evaporation and to permit the development of a formula by means of which evaporation from land and water surfaces alike may be determined.

A difficulty in the use of turbulent theory in the determination of evaporation is the fact that the laws of moisture transfer are different in the laminar and turbulent layers, being linear in the former and nonlinear in the latter. Leighly (**14**) avoided the turbulent layer entirely and confined his study to the laminar or boundary layer where the specific humidity gradient is linear and where the coefficient of diffusion can be used. Since the boundary layer is very thin if the air is in motion, Leighly's method would be useful only for determining evaporation from small areas (he thinks that the limit is a diameter not much greater than 4 feet). In fact, the making of moisture determinations at two levels in a layer as thin as the boundary layer would require a type of instrumentation not now available. With existing instruments it is quite impossible to recognize vapor pressure differences within a distance of a few millimeters. Thus, the use of Leighly's method must await the development of new instruments.

The problem of the difference in character of the moisture gradient and the process of moisture transfer in the boundary and turbulent layers confronted Sverdrup (**28, 29**) in his study of evaporation over the oceans. With the meager data at his disposal, consisting only of wind velocity and moisture concentration at one level in the turbulent layer, and the temperature and salinity of the ocean water, he was forced to consider the gradients in both the laminar and turbulent layers. Although this was accomplished with great ingenuity, the final computations can be considered only as rough indications of oceanic evaporation. Millar (**15**), in his study of evaporation from the Great Lakes, has approached the problem in much the same manner that Sverdrup did for the oceans, and his results are subject to the same limitations.

At this point it is possible to consider the physical limitations of the various empirical formulae based on Dalton's law, which have been presented in an earlier paragraph. These formulae are developed for determining evaporation from free water surfaces and make use of observations of the temperature of the water surface (from which the vapor pressure of the air in direct contact with the water is determined) and the vapor pressure and wind velocity at some level in the turbulent layer. Because of the previously mentioned difference in character of the moisture gradients and the processes of moisture transfer in the boundary and turbulent layers, it is impossible to derive simple empirical constants, to be used in connection with the difference between the vapor pressure of the water surface and the overlying air (e_w-e_d) and a wind factor, which will give more than a statistical approximation of the evaporation from free water surfaces. Using Sverdrup's method of integrating through the boundary layer, more satisfactory results, at least for large water bodies, could be expected. Under no circumstances could such observations be used for determining evaporation from land or transpiration from vegetation. Those formulae employing the so-called vapor pressure deficit (**1**), the difference between the saturation vapor pressure and the actual vapor pressure at some level in the air (E_d-e_d), are unrelated to the physical principles of evaporation, and cannot be expected to yield satisfactory results.

As an outgrowth of the researches on turbulence by von Kármán and Rossby and through discussion with them, a formula for determining evaporation from observations of moisture concentration and wind velocity in the turbulent layer has been devised and is here presented. The derivation of the formula follows directly from an expression for the Austausch coefficient obtained from the concepts of the mixing length and the shearing stress as developed by Prandtl and von Kármán. The derivation is straightforward and can readily be repeated by referring to the works of Rossby (**21, 22**) and Sverdrup (**27**). The formula itself is:

$$E=\frac{K_0^2\rho(q_1-q_2)u_2}{\log_e\frac{h_2}{h_1}\,\log_e\frac{h_2}{Z_0}}$$

in which K_0 = von Kármán's coefficient.
ρ = density of the air.
q_1 = moisture concentration at the lower level,
q_2 = moisture concentration at the upper level,
u_2 = wind velocity at the upper level,
h_2 = height of the upper instruments,
h_1 = height of the lower instruments,
and Z_0 = roughness coefficient.

[1] A brief critical discussion of the studies of Jeffreys and Giblett may be found in Brunt (3), pp. 264–267.

The roughness coefficient is determined from observations of wind velocity at two levels by means of the following formula:

$$\log_e Z_0 = \frac{u_2 \log_e h_1 - u_1 \log_e h_2}{u_2 - u_1}.$$

Since the intensity of turbulent mixing is dependent on the wind velocity and the roughness coefficient and since the latter can be determined from wind velocities at two levels, it is possible to simplify the evaporation formula to the following:

$$E = \frac{K^2{}_0 \rho (q_1 - q_2)\,(u_2 - u_1)}{\left(\log_e \frac{h_2}{h_1}\right)^2}.$$

In the above formulae all values are in *CGS* units. The formula giving evaporation in inches per hour for an installation where the upper observations are taken 28.6 feet and the lower ones 2 feet above the ground is:

$$E = \frac{0.0274\ P(q_1 - q_2)\,(u_2 - u_1)}{T + 459.4}$$

where P is pressure in inches *Hg*

q_1 and q_2 are in grams per kilogram
u is wind velocity in miles per hour
and T is temperature in °F.

Expressed in terms of vapor pressure the formula is simplified further, and becomes

$$E = \frac{17.1(e_1 - e_2)\,(u_2 - u_1)}{T + 459.4}$$

in which e_1 and e_2 are the vapor pressures in inches of *Hg* at the lower and upper levels, respectively.

The formulae as presented are rigorously correct for an adiabatic atmosphere but in cases of thermal stability a correction must be applied for a precise evaluation of evaporation. It is now known that when the temperature and wind structure of the air attain certain critical values (**23**) the effect of turbulent mixing can be completely suppressed and the moisture transfer must then proceed by diffusion alone. Under conditions of light winds and temperature inversions, therefore, the values for evaporation or condensation, as the case may be, determined by the use of the above formulae, are too large. This introduces no serious error in computations of evaporation for periods of a day or longer as the amounts of moisture evaporated under conditions of stability are negligible compared with those amounts evaporated during adiabatic conditions. For the determination of the amount of dew or frost deposition the adiabatic formula becomes inadequate when atmospheric stability is such as to bring about a marked suppression of turbulence. A correction for the stability factor devised along the lines suggested by Rossby (**21, 22**) and Sverdrup (**27**) is being tested at the present time. However, since the amounts of condensation are quite small, again no serious error is involved in the consideration of moisture losses from watersheds.

Two experimental evaporation stations have been established; one at the Muskingum Climatic Research Center in Ohio and the other on Arlington Farm in Virginia. At the Arlington installation the instruments are located in a hay field on the relatively flat top of a gentle ridge. The hay had been cut shortly before the installation was made, and being composed of perennial grasses, continued to grow and remained more or less green throughout the winter. In each case a tower has been erected in such a manner that an instrument shelter large enough to house a hygrothermograph may be raised to a height of 28 feet and lowered at will (fig. 1). Another shelter accommodates a second hygrothermograph 2 feet from the ground. Records from these hygrothermographs, together with records from a barograph, are the only requirements for determining the specific humidity at the two levels. Recording anemometers at these levels supply the other information required in the evaporation formula (fig. 2). At the present time three hygrothermographs and four anemometers, mounted at different elevations on a tower, are being operated at Arlington to supply information for testing the validity of an expression derived to correct the adiabatic formula for cases of atmospheric stability.

Despite the fact that the hair hygrometer as an instrument for measuring relative humidity has many defects, the chief of which are the uncertainty of registration of very high and very low values, the hygrothermograph is found to be the most satisfactory existing instrument for our purpose.[2] In order to be certain of the results, six instruments have been used on each installation. All six instruments were first carefully calibrated in a moisture chamber where, by means of saturated solutions of various salts,[3] constant humidities could be maintained at various levels, and a separate calibration chart was prepared for each instrument. They were then grouped in pairs and were used in succession on the tower. Before two instruments were used they were given a background run side by side in a Weather Bureau shelter. After use they were given another background run and were again calibrated in the moisture chamber. Three datum pens were added to each instrument to permit correction for shrinking and stretching of the chart paper. Recently nonhygroscopic metal charts have been used in place of the moisture-sensitive paper charts and the index pens are no longer needed. Safe-guarded in this manner the results are felt to be reasonably reliable.

The first usable records of moisture concentration were obtained on October 15, 1938. Continuous daily observations on the moisture gradients have been made since the latter part of November. From these records it has been possible to get some indication of the diurnal changes in moisture concentration and gradient for certain air mass types and of the magnitude of the moisture gradients. Records of wind velocity at the two levels have been available only since November 23, 1938, and not until December 10, 1938, were actual determinations of evaporation begun.

Figure 3 shows the march of moisture concentration at the two levels for two days in early autumn during which time a modified Polar Continental air mass remained over the area. The sky was clear and moderate temperatures prevailed. The air mass was relatively stagnant with light and variable winds. Significant radiational cooling occurred throughout the night and a moderate ground fog condition limiting visibilities to 1 to 2 miles was observed at the site of the experimental station.

During the first night the moisture concentration of the air dropped to a minimum of 5.4 and 5.8 grams in the lower and upper levels due to abstraction of moisture by condensation at the ground. As soon as the ground began to

---

[2] An automatic dewpoint recorder has been devised by the authors and a laboratory model has been built by Dr. L. R. Hafstad of the Carnegie Institution of Washington and J. C. Owen of our staff. If satisfactory field instruments can be made they should be more accurate than any type of instrument now available.

[3] The salts used and the relative humidities which they maintain at room temperatures are as follows: KNO_3, 95 percent; $NaCl$, 75 percent; $Mg(NO_3)_2.6H_2O$, 52 percent; $MgCl_2.6H_2O$, 32 percent; $LiCl$, 16 percent.

FIGURE 1.—Evaporation instrumental installation at Arlington Farm, Va.

FIGURE 2.—Recorder for wind velocity at two levels and wind direction at one level, designed especially for evaporation studies. A similar instrument giving a synchronized record of wind velocity at four levels and wind direction at one level has recently been installed.

TABLE 1.—*Data for the computation of evaporation at Arlington Farm, Va., for the 24-hour period Jan. 27, 1939*

	Hour ending—A. m.											
	1	2	3	4	5	6	7	8	9	10	11	Noon
Upper observations:												
Relative humidity	66.2	70.8	75.2	80.1	81.5	75.1	52.0	45.4	43.5	45.2	41.0	39.4
Temperature	20.8	20.3	20.1	18.9	18.7	19.3	22.8	23.3	24.0	25.1	26.5	28.5
Vapor pressure	.071	.074	.078	.079	.079	.075	.061	.055	.054	.059	.057	.061
Wind	5.0	3.8	4.0	1.8	1.5	3.5	9.0	5.8	13.5	12.8	19.2	12.5
Lower observations:												
Relative humidity	74.6	76.7	80.9	84.9	87.9	84.2	58.8	52.5	51.4	50.0	45.5	42.0
Temperature	19.4	18.5	17.5	16.5	16.0	16.5	20.5	21.0	22.4	24.9	26.9	29.4
Vapor pressure	.075	.074	.074	.074	.075	.073	.062	.057	.060	.065	.065	.067
Wind	3.2	2.0	1.8	1.0	.2	1.5	5.8	3.5	7.5	8.2	12.8	8.5
Difference:												
Vapor pressure	.004	.000	−.004	−.005	−.004	−.002	.001	.002	.006	.006	.008	.006
Wind	1.8	1.8	2.2	.8	1.3	2.0	3.2	2.3	6.0	4.6	6.4	4.0
Evaporation [1]	.0003	.0000	−.0003	−.0001	−.0002	−.0001	.0001	.0002	.0013	.0010	.0018	.0008
	Hour ending—P. m.											
	1	2	3	4	5	6	7	8	9	10	11	Midnight
Upper observations:												
Relative humidity	36.5	34.2	31.0	32.0	34.1	35.3	35.5	35.5	37.6	41.0	43.0	47.5
Temperature	30.5	32.5	34.0	34.0	34.0	33.0	32.3	31.7	31.0	30.0	28.3	26.5
Vapor pressure	.061	.063	.061	.062	.067	.066	.065	.063	.065	.068	.065	.067
Wind	13.5	15.2	14.8	12.2	14.5	11.2	8.0	7.2	5.0	3.5	2.8	4.2
Lower observations:												
Relative humidity	38.0	34.2	32.0	34.5	36.9	39.3	41.5	41.1	43.1	50.0	64.0	71.4
Temperature	31.9	33.9	34.9	33.6	32.9	31.4	30.4	29.4	29.4	26.9	22.9	21.4
Vapor pressure	.068	.067	.065	.066	.069	.069	.070	.066	.069	.071	.076	.079
Wind	9.2	10.5	9.8	9.8	9.0	8.0	5.0	4.2	2.8	1.8	.2	1.2
Difference:												
Vapor pressure	.007	.004	.004	.004	.002	.003	.005	.003	.004	.003	.011	.012
Wind	4.3	4.7	5.0	2.2	5.5	3.2	3.0	3.0	2.2	1.7	2.6	3.0
Evaporation [1]	.0011	.0007	.0007	.0003	.0004	.0003	.0005	.0003	.0003	.0002	.0010	.0013

[1] Negative evaporation is condensation. Totals for the period. Evaporation 0.0126 inches, Condensation 0.0007 inches.

warm up after sunrise and evaporation commenced, moisture concentration rose rapidly, beginning at the lower level and producing a reversal in the gradient. Shortly before noon thermal convection had developed to a point where the moisture was carried to the upper air faster than it was received from the ground and the concentration at both levels diminished about 0.8 grams. In late afternoon, with a diminution of convection, the evaporation again exceeded the rate of transport aloft and the moisture concentration rose to the high point of the day. Between 6 and 7 p. m. the cooling of the ground surface began, and soon the concentration at both levels was the same. With continued cooling condensation was initiated, and as moisture was abstracted from the lower level the gradient was reversed, and the concentration at both levels diminished throughout the night. In the morning there was a heavy dew deposit.

The same sequence of conditions was repeated the next day; first an upward shift of gradient followed by a rapid increase of moisture concentration as evaporation proceeded; then a diminution of concentration with a continued upward-directed gradient as thermal convection carried the moisture to the drier levels aloft; then as convection subsided a rapid increase in concentration; and finally, with the onset of ground radiation at the end of the day, condensation, a gradient directed downward, and a rapid diminution of moisture.

A few days later, with the passage of a well-marked front the transitional Polar Continental air mass was displaced by a fresh polar air mass. The moisture concentration dropped within the course of a few hours from about 10.5 to 4.0 grams. The moisture gradient pattern for the 2 days following is shown in figure 4. Except for scattered cirrus and a few scattered cumulus clouds during the afternoon the sky was clear throughout the 2-day period. For nearly the entire first day the wind direction was predominantly from the west and west-northwest with velocities ranging from 5 to 15 miles per hour. The moisture gradient, although small, was directed upward for approximately 28 hours after the invasion of the polar air mass. The moisture concentration remained almost stationary at about 4 grams until approximately 3 p. m., when a moderate amount of convection forced a small decline. It is of interest to note that scattered cumulus clouds were observed between 3 and 5 p. m., which coincided with the observed decline in moisture concentration. During the evening evaporation continued and the concentration rose more than a gram. Not until after 10 p. m. did the gradient turn downward and condensation commence.

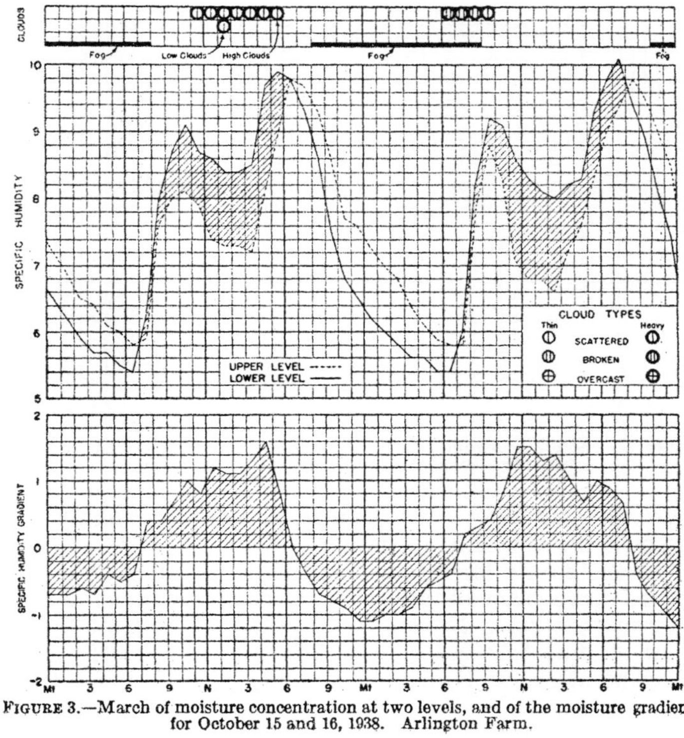

FIGURE 3.—March of moisture concentration at two levels, and of the moisture gradient, for October 15 and 16, 1938. Arlington Farm.

At this time the wind had subsided, becoming variable with velocities generally less than 2 to 3 miles per hour. Thereafter during the night, with moisture abstraction going on at the ground, the moisture concentration slowly declined. The gradient pattern of the second day resembled those of the 2 earlier days shown in figure 3,

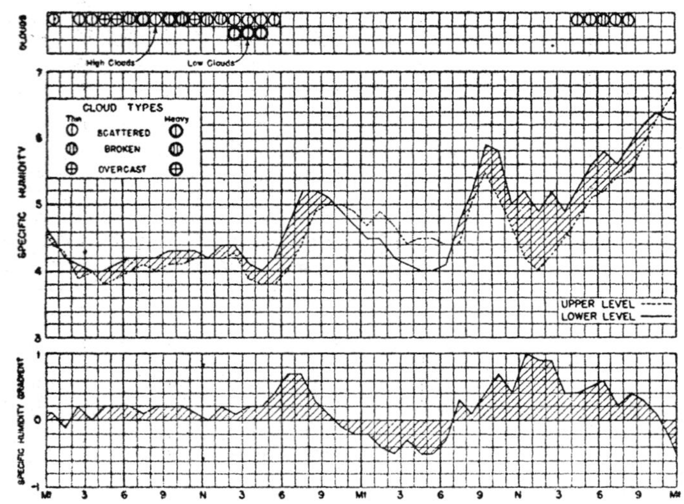

FIGURE 4.—March of moisture concentration at two levels, and of the moisture gradient, for October 21 and 22, 1938. Arlington Farm.

although the rate of evaporation was considerably reduced. The important difference in pattern is that the moisture gradient was directed upward until 11 p. m., with the moisture concentration continuing to rise until that time.

For the cases already discussed, wind velocity gradients

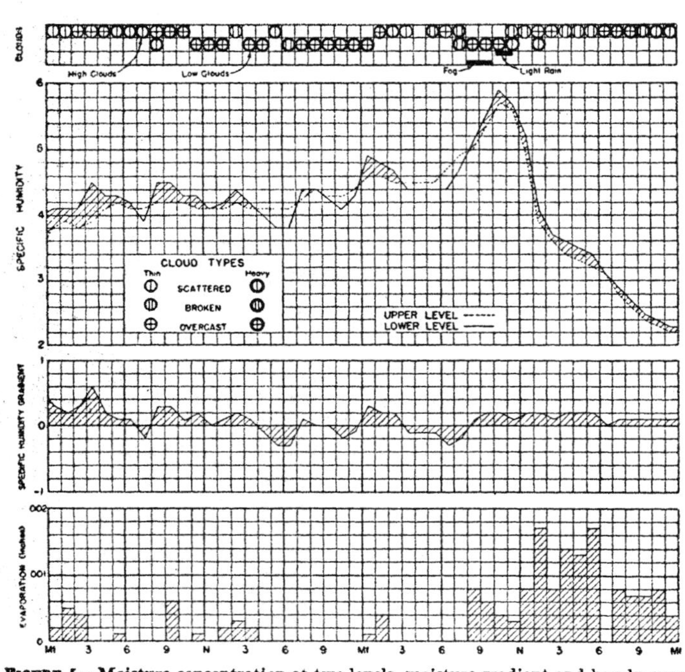

FIGURE 5.—Moisture concentration at two levels, moisture gradient and hourly evaporation for December 11 and 12, 1938. Arlington Farm.

were not available, consequently it was not possible to compute the actual evaporation.

Complete data for calculating the evaporation from the ground have been obtained with occasional unavoidable interruptions since November 23, 1938. Figures 5 and 6 show the moisture gradient pattern and the hourly rates of evaporation for December 11 to 14, during which time a Polar Continental air mass displaced a southerly current of well-modified polar air. On December 11, the synoptic weather chart indicated the presence of a well-modified returning flow of polar air over the eastern United States. The sky conditions as observed at the Washington airport adjoining the Arlington Farm, were quite variable from hour to hour. The thin, broken to overcast, altostratus clouds which predominated during the early morning and forenoon thickened to a nearly continuous overcast with a base at 6,000 to 7,000 feet during the afternoon and evening. During this time the specific humidity gradient was directed upward. Between 5 and 7 p. m. the clouds became scattered and broken and radiational cooling of the ground hastened the usual reversal of the specific humidity gradient. As the overcast thickened again, the gradient shifted upwards temporarily for an hour. Thereafter the gradient was directed downward except for a period between midnight and 3 a. m. of December 12. The upward gradient at this time was probably due to a

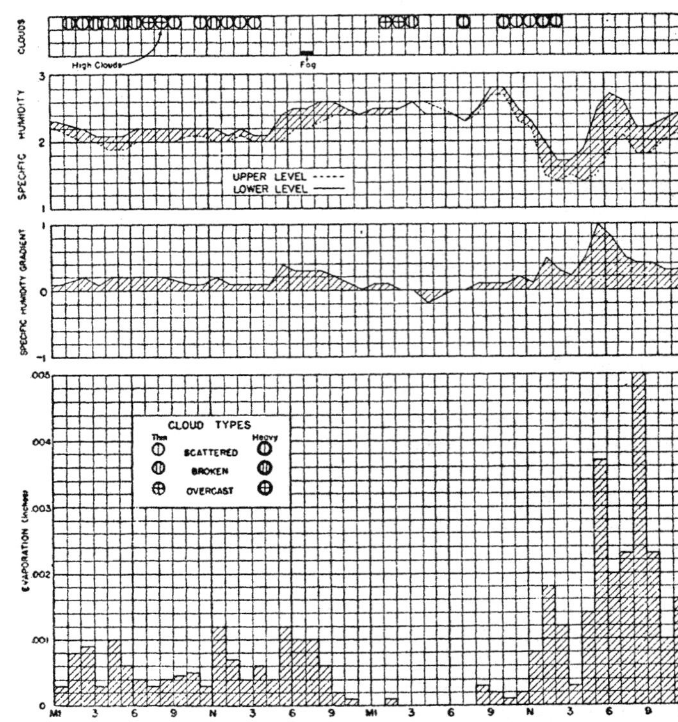

FIGURE 6.—Moisture concentration at two levels, moisture gradient and hourly evaporation for December 13 and 14, 1938. Arlington Farm.

slight increase in surface temperature caused by the downward transport of heat by turbulence induced by the increased wind velocities which prevailed during this period.

During the early morning of December 12, the clouds were, in general, high and scattered and calm or extremely light winds prevailed. The moisture gradient continued downward and a light dew formation resulted. At 9 a. m. the sky became overcast with upper altostratus and lower strato-cumulus decks heralding the approach of a cold front. Occasional sprinkles were noted during the hour ending at 9 a. m., and light rain was observed between 10:10 and 11:25 a. m. The amount of rainfall, however, was less than 0.01 inch. The front, which marked the boundary of Polar Continental air, passed the station at approximately 11 a. m. and was associated with a wind shift to the northwest and a dissipation of the strato-cumulus deck. During the ensuing 30 hours the wind remained in the northwest with velocities generally between 8 and 15 miles per hour. The specific humidity gradient was directed upward for 2 hours prior to the frontal passage and continued to be directed upward until the morning of the 14th. The gradient shifted downward at this time and a heavy frost ensued.

It is to be noted that on the 13th there was no drop in specific humidity resulting from large scale thermal or convective turbulence such as was illustrated in figure 3. This is explained by the fact that convective turbulence was limited to the thick layer of polar air, within which mixing by mechanical turbulence had already established complete uniformity of moisture concentration, thus precluding the possibility of replacement of the air at the level of the observations by drier air from aloft.

On December 14, however, the wind velocities in the polar air were light and variable and the thickness of the mechanically turbulent layer was much less than that on the 13th. From 11 a. m. to 3 p. m. the characteristic convective turbulence was manifested by a steady drop in the specific humidities in spite of the fact that evaporation was continuously adding moisture to the atmosphere. A secondary decrease in the specific humidities beginning at 8 p. m. is also due to mixing with drier air aloft. The wind velocities increased at this time, shifting from south at 3 miles per hour to west-northwest at 15 miles per hour, causing an increase in surface temperature from 38° to 42°.

The water loss from the land surface at Arlington Farm, combining evaporation from the soil, evaporation of dew and frost, and transpiration, for the 4 days from December 11 to December 14 was 0.0026 inches, 0.0137 inches, 0.0132 inches and 0.0239 inches respectively.

The station was operated without interruption from November 23, 1938 until February 7, 1939 and hourly evaporation has been determined for this entire period. Figure 7 shows the daily evaporation and condensation for the month of January. In table 1 the type of information obtained and hourly evaporation values for January 27, 1939 are presented merely for illustrative purposes.

During the first 4 days in January the ground surface was dry and the moisture was apparently removed chiefly by transpiration (**37**). Variable high scattered clouds and light variable winds prevailed during this period and appreciable amounts of dew condensed during the early morning hours. On January 5 a low stratus overcast prevailed throughout the greater portion of the day. Intermittent sprinkles and occasional light rain totaling 0.07 inches occurred during the 24-hour period. In spite of the cloudy conditions and the precipitation a measurable moisture gradient directed upward was detected and 0.0103 inches of evaporation occurred. The following day, January 6, the skies were relatively clear, moderate wind velocities were observed, and the evaporation was greatly increased, amounting to 0.0544 inches. On the 7th, 8th, and 9th variable high clouds to clear skies prevailed but the daily evaporation totals were, as was to be expected, less than that of the 6th when the ground surface was moistened by the previous day's rainfall. On the 8th the evaporation was slightly greater than on the 7th or 9th and may be explained by the moderate wind velocities, occasionally reaching 22 miles per hour. Light rain amounting to 0.07 inches fell between 7 a. m. and 9:30 a. m. on January 10. In spite of the moisture on the ground and a moderate south-southwest wind that varied between 8 and 12 miles per hour, the evaporation amounted to only 0.0276 inches. The suppression of evaporation on this day may be explained by the diminution of the moisture gradient due to the invasion of more moist air from the south, causing an increase in specific humidity from 3.5 grams to approximately 7 grams. On the following day, there was only slightly more evaporation. However, by the morning of the 12th fresh polar air had invaded the area and the specific humidity dropped to 2.2 grams. The wind blew prevailingly from the northwest with a velocity ranging between 5 and 15 miles per hour, and in spite of a high overcast the evaporation attained a value of 0.0535 inches. The increased evaporation into polar air masses, which has been found consistently in our observations, was anticipated on the basis of the cycle of air masses (**8, 35**).

The high alto-stratus overcast that was present throughout the 12th steadily lowered and thickened, heralding the approach of a warm front storm. On the 13th a mixed precipitation of freezing rain, sleet and snow fell throughout the greater part of the day. On the 16th a light snow fell between 6 a. m. and 11 a. m. On the 18th snow began falling shortly after midnight and continued throughout the day until 10 p. m. A light snow, associated with an upper cold front, occurred on the 20th between

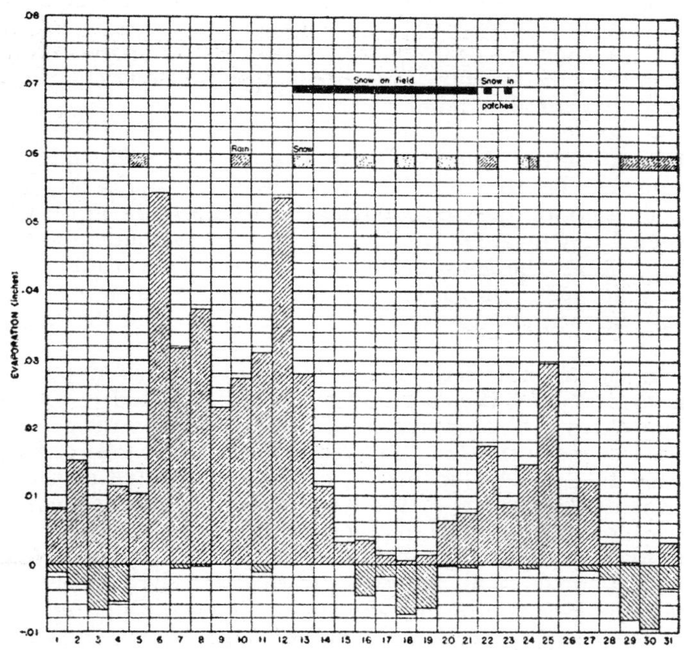

FIGURE 7.—Daily totals of evaporation and condensation for the month of January, 1939. Arlington Farm.

5 p. m. and 9 p. m. Thus from the 13th through the 20th the site of the evaporation station was covered with snow ranging in thickness from 1 to 6 inches.

During this period the blanket of snow was fundamentally responsible for the suppression of evaporation. A snow cover may suppress evaporation in a number of different ways, although it is by no means intended to imply that the moisture losses by evaporation from a snow field may not be large when suitable atmospheric conditions prevail. Because of the difference in the physical characteristics of a snow cover and a grass cover there must obviously be a difference in the roughness factor characterizing the type of surface and a consequent difference in the intensity of turbulent mixing.

Rossby and Montgomery (**22**, p. 10) using observations published by Shaw have determined the roughness factor for an open grassland. Sverdrup (**27**, p. 44) has made a similar determination for a snow field from observations taken in West Spitzbergen. Using these determinations of roughness it is found that under otherwise similar conditions the evaporation from the meadow would exceed by approximately 50 percent that from the snow field.

Similar calculations for Arlington Farm showed a reduction of the roughness factor during the period when the ground was snow covered that was in excellent agreement with this figure.

An important factor governing the rate of evaporation from the snow is the temperature of the snow surface. It is well known that a snow surface radiates energy as a black body (38) and on a clear calm night, excessive inversions of temperature may be established in the air next to the snow. During the daylight hours much of the incident radiation is reflected and the snow surface temperatures may remain well below the dewpoint temperatures of the air. In any event the temperature of the snow surface must remain at the freezing temperature until all of the snow is melted. Comparing a grass-covered field with a snow field, the former will also radiate energy like a black-body but during the daylight hours most of the incident energy will be absorbed and the surface temperatures may be greatly increased, resulting in increased transpiration of the plants. Transpiration may be regarded as the most effective mechanism whereby moisture is returned to the atmosphere from land areas (**37**). A snow cover obviously precludes the operation of the process of transpiration. Thus it will be evident that unless the air dew-point temperatures are less than the snow surface temperatures (vapor pressure gradient directed upwards) there can be no evaporation. If, as may frequently be the case, the air dew-point temperatures exceed that of the snow cover (vapor pressure gradient directed downwards) there must be a frost condensation occurring on the snow surface. This explains what took place between the 16th and 20th inclusive when an appreciable amount of frost was condensed upon the snow surface.

On the 22d and 23d only patches of snow remained on the field and by the morning of the 24th it had entirely disappeared. The increase of evaporation during the period between the 22d and 27th inclusive over that of the period 14th to 21st, inclusive, is to be attributed directly to the disappearance of the snow cover. However, even with the disappearance of the snow it is to be noticed that the average daily evaporation rates for the period from the 21st to the 27th, inclusive, were lower than those for the period prior to the first snowfall. This is accounted for by the greatly decreased temperature of the ground during the last half of the month.

Cold and dry Polar Continental air occupied the area on the 27th but was displaced by a southwesterly current of transitional tropical air on the 28th. The dew-point temperatures of the polar air averaged approximately 12° but increased to about 20° in the modified air. The ground temperatures were still quite low and with the invasion of the modified air with increased specific humidities, the moisture gradient was directed downward for a good portion of the day and appreciable condensation of frost occurred. The 29th was overcast and light rain began at 11 a. m. and continued through midnight. The cloud base which remained below 1,000 feet lowered to the ground in the evening, resulting in foggy conditions. On the 30th a low stratus overcast with ceilings varying between 200 and 600 feet and light rain or mist and fog persisted during the entire day. The rain ceased in the early morning of the 31st when a polar air mass again invaded eastern United States. Practically no evaporation occurred on the 29th or the 30th. On both of these days, however, significant amounts of condensation occurred. The small evaporation of the 31st was associated with the invasion of the polar air mass. The total amount of evaporation at the Arlington Farm for the month of January 1939 amounted to 0.4747 inches. The total amount of condensation as computed was 0.0632 inches. It is recognized that this latter figure may be somewhat exaggerated.

In general, most researches on the problem of evaporation have been directed toward the determination of moisture losses from bodies of water. This has led to a deficiency in basic and quantitative information regarding actual moisture losses from watersheds or land surfaces. Studies of the evaporation from free water surfaces alone have proved wholly inadequate from the standpoint of soil conservation. What is required is the actual moisture losses from all types of geographic surfaces, or in other words, a quantitative estimate of the evaporative phase of the hydrologic cycle. With this information the soil conservationist will then be able to interpret the interrelations between climatic forces and land utilization (**36**).

In an effort to supply this much-needed information a method for determinining the evaporation from either land or water areas has been presented in this preliminary report. The practicability of the technique has been completely demonstrated. It is hoped that with proper instrumental installation it will be possible to determine transpiration rates and moisture requirements of various field crops and forest trees, the effectiveness of various moisture conserving practices, and the relative importance of evaporation and transpiration in the hydrologic cycle. However, the question of satisfactory instrumentation has not been entirely solved and the validity of various theoretical assumptions has yet to be tested. These questions are expected to be discussed further in a detailed report on the problem of evaporation to be published in the near future.

LITERATURE CITED

(1) Anderson, Donald B.
1936. Relative humidity or vapor pressure deficit. Ecology 17: 277–282.

(2) Ångström, Anders.
1920. Applications of heat radiation measurements to the problems of the evaporation from lakes and the heat convection at their surfaces. Geogr. Ann. 2: 237–252.

(3) Brunt, David.
1934. Physical and dynamical meteorology. 411 pp. Cambridge, England.

(4) Cummings, N. W.
1929. Relation between evaporation and humidity as deduced quantitatively from rational equations based on thermodynamics and molecular theory. Natl. Research Council Bull. 68, pp. 47–56. (Amer. Geophys. Union Trans.)

(5) Dalton, John.
1802. Experimental essays on the constitution of mixed gases; on the force of steam of vapor from waters and other liquids in different temperatures, both in a torricellian vacuum and in air; on evaporation; and on the expansion of gases by heat. Mem. Manchester Lit. and Phil. Soc. 5: 535–602.

(6) Ertel, Hans.
1930. Eine methode zur berechung des austauschkoeffizienten aus den feinregistrierungen der turbulenten schwankungen. Gerlands Beitr. zur Geophysik 25: 279–289.

(7) Giblett, M. A.
1921. Some problems connected with evaporation from large expanses of water. Proc. Roy. Soc. (A) 99: 472–490.

(8) Holzman, Benjamin.
1937. Sources of moisture for precipitation in the United States. U. S. Dept. Agr. Tech. Bull. 589, 42 pp.

(9) Jeffreys, H.
1918. Some problems of evaporation. Phil. Mag. (6) 35: 270–280.

(10) Kármán, Th. von.
1930. Mechanische ähnlichkeit und turbulenz. Nachr.: Gesell. der Wiss. zur Göttingen (Math. Phys.) 1: 58–76.

(11) ———.
1934. Turbulence and skin friction. Jour. Aeronaut. Sciences 1: 1–20.

(12) ———.
1935. Some aspects of the turbulence problem. Fourth Internatl. Cong. Applied Mechanics Proc. 1934: 54–91.

(13) ———.
1937. Turbulence. Jour. Roy. Aeronaut. Soc. 41: 1109–1143.

(14) Leighly, John.
1937. A note on evaporation. Ecology 18: 180–198.

(15) Millar, F. Graham.
1937. Evaporation from free water surfaces. Canadian Met. Mem. 1 : 43–65.

(16) Prandtl, L.
1932. Meteorologische anwendung der strömungslehre. Beitr. zur physik der Freien Atmosphäre 19: 188–202.

(17) ———.
1935. The mechanics of viscous fluids. *In* Durand, W. F., ed., Aerodynamic Theory, v. 3, pp. 34–207. Berlin.

(18) Richardson, Burt.
1931. Evaporation as a function of insolation. American Soc. Civil Engineers Trans. 95: 996–1019.

(19) ——— and Montgomery, Carol.
1929. The measurement of insolation by means of a pan. Natl. Research Council Bull. 68, pp. 56–61. (Amer. Geophys. Union Trans.)

(20) Rohwer, Carl.
1931. Evaporation from free water surfaces. U. S. Dept. Agr. Tech. Bull. 271, 96 pp.

(21) Rossby, C. G.
1932. A generalization of the theory of the mixing length with applications to atmospheric and oceanic turbulence. Mass. Inst. Technol. Met. Papers 1 (4): 1–36.

(22) ——— and Montgomery, R. B.
1935. The layer of frictional influence in wind and ocean currents. Papers in Physical Oceanography and Meteorology, Mass. Inst. Technol. and Woods Hole Oceanographic Inst. 3 (3): 1–101.

(23) Schlichting, H.
1935. Turbulenz bei wärmeschichtung. Fourth Internatl. Cong. Applied Mechanics Proc. 1934: 245–246.

(24) Schmidt, Wilhelm.
1925. Der massenaustausch in freier luft und verwandte erscheinungen. Hamburg.

(25) ———.
1935. Turbulence near the ground. Jour. Roy. Aeronaut. Soc. 39: 355–376.

(26) Sutton, O. G.
1934. Wind structure and evaporation in a turbulent atmosphere. Proc. Roy. Soc. London (A) 146: 701–722.

(27) Sverdrup, H. U.
1936. The eddy conductivity of the air over a smooth snow field. Geofysiske Publikasjoner 11 (7): 1–69.

(28) ———.
1936. Das maritime verdunstungsproblem. Ann. der Hydrog. und Marit. Met. 64: 41–47.

(29) ———.
1937. On the evaporation from the oceans. Jour. of Marine Research 1: 3–14.

(30) Taylor, G. I.
1915. Eddy motion in the atmosphere. Roy. Soc. Phil. Trans. 215: 1–26.

(31) ———.
1918. Phenomena connected with turbulence in the lower atmosphere. Proc. Roy. Soc. London (A) 94: 137–155.

(32) ———.
1922. Diffusion by continuous movements. Proc. London Math. Soc. (2) 20: 196–212.

(33) ———.
1927. Turbulence. Quart. Jour. Roy. Met. Soc. 53: 201–211.

(34) ———.
1932. Transport of vorticity and heat through fluids in turbulent motion. Proc. Roy. Soc. (A) 135: 685–705.

(35) Thornthwaite, C. W.
1937. The hydrologic cycle re-examined. U. S. Dept. Agr. Soil Conserv. 3: [85]–91.

(36) ———, Holzman, Benjamin, and Blumenstock, David I.
1938. Climatic research in the soil conservation service. U. S. Monthly Weather Rev. 66: 351–368.

(37) Veihmeyer, F. J.
1938. Evaporation from soils and transpiration. Natl. Research Council, Amer. Geophys. Union Trans. 2: 612–619.

(38) Wexler, H.
1936. Cooling in the lower atmosphere and the structure of polar continental air. U. S. Monthly Weather Rev. 64: 122–136.

(39) Yamamoto, G.
1937. On the rate of condensation of dew. Geophys. Mag. 11: 91–96.

THE MEASUREMENT OF VERTICAL TRANSFER OF HEAT AND WATER VAPOR BY EDDIES IN THE LOWER ATMOSPHERE

By W. C. Swinbank

Commonwealth Scientific and Industrial Research Organization, Australia

(Manuscript received 7 October 1950)

ABSTRACT

Though the necessity of measuring the vertical flux of heat and water vapor, brought about by eddy movement in the lower atmosphere, has long been recognized, no direct method of determining these important quantities has previously appeared. The apparatus described here provides a continuous record, over a five-minute interval, of the detailed structure of temperature, vapor pressure, and total wind speed and its vertical component, of the air passing a fixed point. It is possible to derive the required fluxes from such records, and the method of analysis is discussed in some detail. Specimen records and measurements of the fluxes are presented.

1. Introduction

Though the following paragraphs may be modified to apply to the eddy transfer of other properties in the lower atmosphere, for definiteness the discussion will be restricted to heat flux. Suppose that the temperature of the air passing a fixed point is T, that its density is ρ, and that the upward vertical component of its velocity is w. Then, at that instant, heat is being transferred upwards at the point in question at a rate $c_p\rho wT$ per unit area, c_p being the specific heat of air at constant pressure.

If now we consider a period of time long enough to allow the passage of many eddies, some moving up, some down, the average rate at which heat has been transferred upwards at the point is $n^{-1}\Sigma c_p\rho wT$, where n is the number of observations. Alternatively, this may be written as $c_p\overline{\rho wT}$, where the bar denotes the mean value of the quantity beneath it over the time considered (c_p is constant).

The temperature of the air passing the point will fluctuate, and the fluctuations may be considered as taking place above some arbitrary level of temperature T_0. Thus, the instantaneous temperature T may be written as $T = T_0 + T'$. T' will be referred to as the temperature fluctuation.

The expression for the heat flux now becomes $c_p\overline{\rho w(T_0 + T')}$, which may be written

$$c_p(\overline{\rho w}T_0 + \overline{\rho wT'}).$$

The instrumental problem is to devise an apparatus to measure the quantities inside the parentheses.

Before going on to describe the apparatus that has been developed to achieve this, it would be well to consider what is meant when reference is made here to an "eddy." Attempts have been made to represent eddies, mathematically and mechanically, as if an eddy were a discrete sample of air. But the point-to-point variations of atmospheric properties are always more or less gradual, whether the property be temperature, water-vapor content, velocity, *etc.* The term "eddy" will be used as an economy of expression rather than an indication that a distinct mass of air is thereby envisaged. The method of measurement of flux of heat and water vapor to be described is independent of the nature of eddies, and it would, perhaps, be better to speak of the variations of the different properties as "fluctuations."

It seems likely that, ranging downwards from disturbances of synoptic size, there is no hiatus in the range of linear dimensions of eddies before the molecular scale is reached. Now, it is well known and easily demonstrated that molecular transfer is wholly inadequate to effect the observed changes of temperature and water-vapor structure that occur in the lower atmosphere. But at what point in the scale a limit can be set to the size of eddy that contributes materially to the flux of heat and water vapor (and other properties) cannot be decided *a priori*. This requires a detailed examination of eddy structure.

The recording of a fluctuating quantity always results in amplitude and phase distortion, more or less pronounced according to the period of the fluctuations and the characteristics of the recording apparatus. For the correct interpretation of such records, it is essential to know the degree of distortion introduced. It is also, of course, highly desirable to minimize this distortion, so that the range of frequencies of fluctuations that are adequately registered may be extended as far as possible.

Therefore, in the following sections, considerable attention is given to the amount of distortion introduced by the apparatus described, so that one may be fully aware of its limitations in recording the detailed atmospheric structure. In designing the apparatus, it has been assumed, with some justification from previous experience, that the fluctuations which are mainly responsible for the heat and water-vapor flux have a period of the order of a few seconds. Support for this assumption is given by an experiment described in an appendix to this paper.

2. The apparatus

Measurement of vertical-wind fluctuations.—The accurate measurement of a fluctuating vertical wind presents serious difficulties, particularly if mechanical methods are used. For example, from the behavior of a vane it is difficult to discriminate between horizontal and vertical fluctuations in wind velocity.

It seemed that the most satisfactory way of measuring wind fluctuations was by means of sensitive hot-wire anemometers, which do not suffer from this disability and, furthermore, have an important advantage over mechanical methods, as will be discussed later.

The techniques of hot-wire anemometry are well described by Ower (1949), and only a brief summary of the underlying principles is necessary here. King (1914), developing the theoretical treatment of Boussinesq (1901), established that the rate of heat loss H per unit length from a heated wire of diameter d, maintained at a temperature T above that of the fluid which flows normally across it with velocity V, could be represented by

$$H = KT + (2\pi KS\rho d)^{\frac{1}{2}} V^{\frac{1}{2}} T, \qquad (1)$$

where K is the thermal conductivity of the fluid, S its specific heat at constant volume and ρ its density. King showed that this relationship should hold down to a velocity of 0.25 ft/sec for a wire with diameter 0.001 in when T is 1000C, below which wind speed the free convection currents around the heated wire distort the air flow. For lower values of T, this effect is less important, and, for a value of T of about 500C, that commonly used in hot-wire anemometry, it is unimportant in the measurement of atmospheric turbulence.

Because of its thermal inertia, a hot wire is not able to follow fluctuations of velocity with exactitude, and this defect becomes more pronounced the shorter the period of the fluctuations. It is possible, using (1), to calculate the amplitude and phase distortion introduced by this lag into the recording of sinusoidal fluctuations of wind velocity of any period for any wind speed. Table 1 shows the results of such calculations for a platinum wire of diameter 0.001 in.[1]

It is seen that the distortion only becomes noticeable for fluctuations of less than one-second period, and then only in the lightest winds. It is to be noted that the distortions in the table refer only to the hot wire itself; the instrument recording the performance of the wire introduces further distortion.

[1] Similar computations were made by Dryden and Keuthe, N. A. C. A. Tech. Rep. 320, 1929.

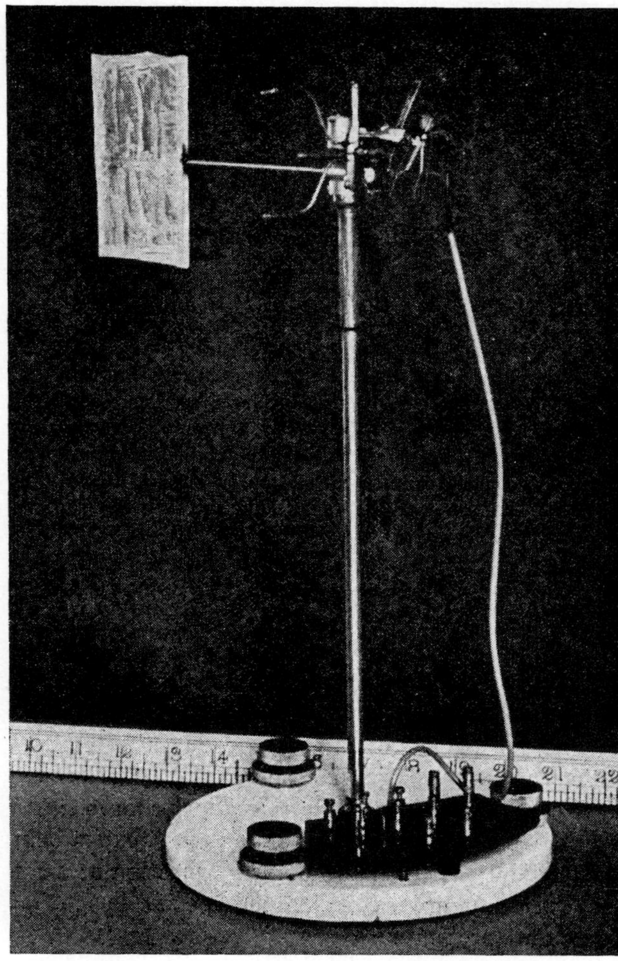

FIG. 1. General view of apparatus.

From (1), if the wire, of resistance R ohms per unit length, is heated by a current i, in the steady state

$$i^2R/J = T[K + (2\pi KSd)^{\frac{1}{2}}(V\rho)^{\frac{1}{2}}]. \qquad (2)$$

This expression shows that an electrically heated wire can be used to measure the speed of air flow in two ways: (a) by maintaining the wire at a constant temperature (and therefore constant resistance), in which case the current i provides a measure of the air flow, and (b) by keeping the current constant, in which case the temperature (and resistance) will vary with the air flow in accordance with (2).

The expression "air flow" is used deliberately here for, though the hot-wire instrument is customarily

TABLE 1. Amplitude and phase distortion caused by thermal inertia of platinum wire of diameter 0.001 in in recording sinusoidal wind fluctuations of various periods and speeds.

Period of fluctuations (sec)	Ratio of amplitude response to true amplitude, Wind speed (m/sec)			Phase lag (deg), Wind speed (m/sec)		
	0.5	1	4	0.5	1	4
0.5	0.955	0.97	0.99	17	15	9
1.0	0.99	0.995	0.998	9	7.5	4.5
2.0	0.998	1.0	1.0	4.5	4	2
4.0	1.0	1.0	1.0	2	2	1

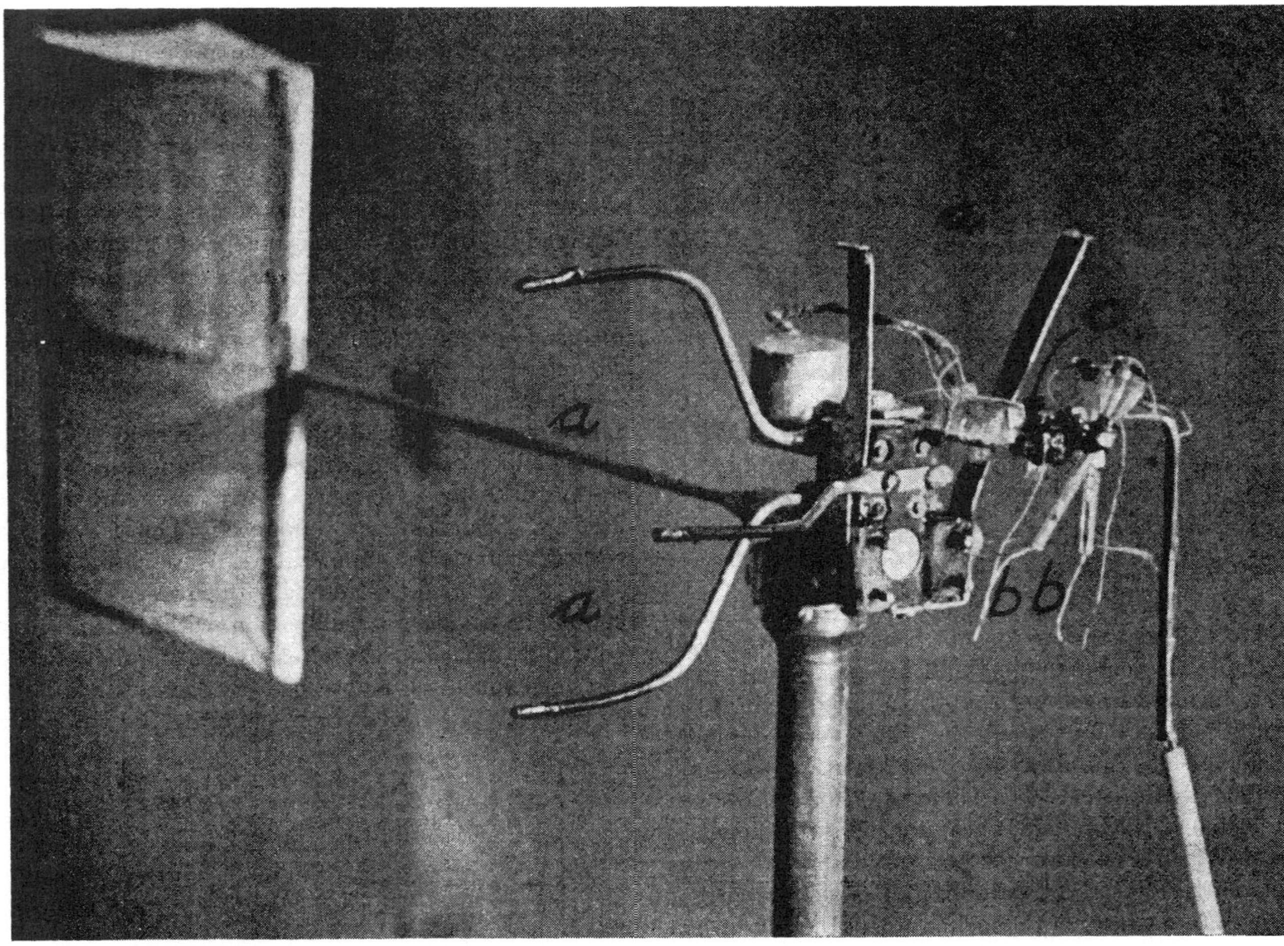

FIG. 2. Close-up view of head; a: 0.001 in platinum wire; b: fine thermocouple wet-bulb, fed from capillary tube; c: 50 SWG copper-constantan thermocouple.

used to measure air speed, (1) and (2) show that in reality it measures the product of speed and density; in the present investigation, this proves to be a particularly valuable property.

When the air flow is steady, the hot wire may be used, in either of the above ways, as an absolute instrument. But when the flow fluctuates, it is not practicable to keep either the current or the temperature constant, and the instrument must be calibrated against another anemometer.

The vertical component of the air flow cannot be measured with a single hot wire, for such a wire is not able to distinguish between variations in the air flow and of the angle at which the wind strikes the wire. To measure this component it is necessary to know (a) the total air flow and (b) its angle of inclination, and two hot-wire circuits are necessary for this.

The hot wires that are used are shown in figs. 1 and 2. The horizontal arm is supported on ball bearings on the vertical pillar, and carries a vertical light vane of wedge shape which swivels the head carrying the hot wires, so that the single, horizontal platinum wire of diameter 0.001 in is always held broadside to the wind. Therefore, this wire is always affected in the same way by a wind of given strength, no matter what its azimuth or inclination; thus, its behavior provides a measure of the total air flow. The wire forms one arm of a Wheatstone bridge net-work, and the galvanometer current provides a measure of the air flow when the wire has been calibrated. The circuit for this wire is shown in fig. 3.

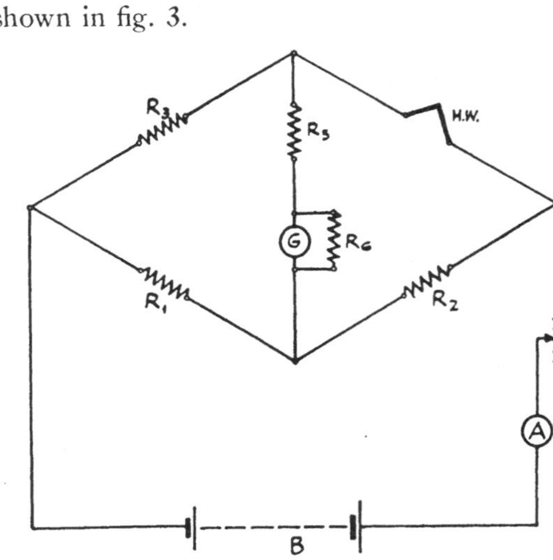

FIG. 3. Circuit diagram of total-airflow hot wire. H.W.: 0.001 in platinum wire, cold resistance 11 ohms; $R_1(=R_2)$: manganin wire, 100 ohms; R_3: 0.04 in platinum wire, 16 ohms; R_4: 200 ohms rheostat; R_5: 20,000 ohms; R_6: critical damping resistance of G mirror galvanometer; B: battery; A: ammeter.

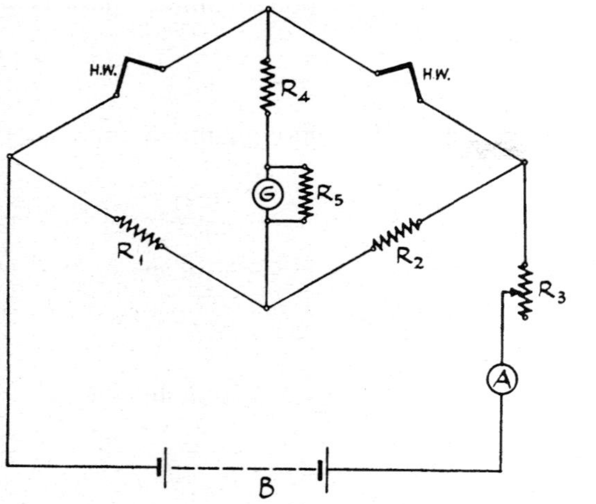

FIG. 4. Circuit diagram of inclination-of-airflow hot wires. H. W.: 0.001 in platinum wire, cold resistance 9 ohms; $R_1(=R_2)$: manganin wire, 100 ohms; R_3: 200 ohms rheostat; R_4: 20,000 ohms; R_5: critical damping resistance of G, mirror galvanometer; B: battery; A: ammeter.

The second hot-wire system consists of two equal platinum wires of diameter 0.001 in, equally and oppositely inclined to the horizontal, and mounted on the head so that they are in the same vertical plane, parallel to the vane, and hence are always in the same vertical plane as the wind. These wires form the opposite arms of the second Wheatstone bridge, shown in fig. 4. Winds inclined to the horizontal affect the wires differently, and the resultant galvanometer current provides a measure of the inclination of the wind when a calibration of the system has been made. This angle, combined with the total air flow provided by the horizontal wire, gives a measure of the vertical component of the air flow, *i.e.*, ρw in the expression given above for the heat flux.

The current in each circuit is adjusted by means of the rheostat until the current in each case, as indicated by the ammeter, is 0.2 amp in still air.

Calibration of the hot wires.—The two hot-wire systems are calibrated in a wind tunnel against a vane-type air meter as follows: the head carrying the hot wire is removed from its support and mounted in the wind tunnel on a horizontal rod fixed normal to the air flow, so that the horizontal wire is also normal to the flow. The inclined wires can be set at any inclination, in the line of the wind, as measured by a protractor fixed to the rod. Thus, different inclinations of the actual wind can be simulated. The bridge networks used for the calibration are exactly the same as those used for the operation of the hot wires, except for the substitution during calibration of resistances equal to the long leads to the instrument when used for recording actual winds.

A typical calibration curve for air flow is shown in fig. 5. The decrease of sensitivity of hot wires with increasing wind is well shown here, but it is possible, by modifying the resistances in the bridge arms, to reduce considerably this disability. Fig. 6 shows a calibration curve for determining the angle of inclination of the wind. (The ordinate here is the sine of the angle of inclination of the wind. This is used rather than the angle itself because, since ρw = air flow $\times$ sine of its inclination, the analysis is thereby simplified.) It will be seen that the determination of the angle of inclination requires that the total wind speed be known; it is hoped that, in the future development of the instrument, it will be possible to bring the curves of fig. 6 into practical coincidence, so that the output of the galvanometer will measure the angle of the wind independent of its speed.

The "ageing" effect of hot wires, referred to by King (1914), which causes a progressive change in the calibration of a hot wire, particularly when the wire is very fine, becomes less pronounced when the wire is worked at moderate temperatures, say 500C. This is about the temperature used in the present arrangement. The change of calibration with time is also reduced by annealing the wires, by heating them to red heat for some time before use. But even with these precautions, it is necessary to calibrate the wires fre-

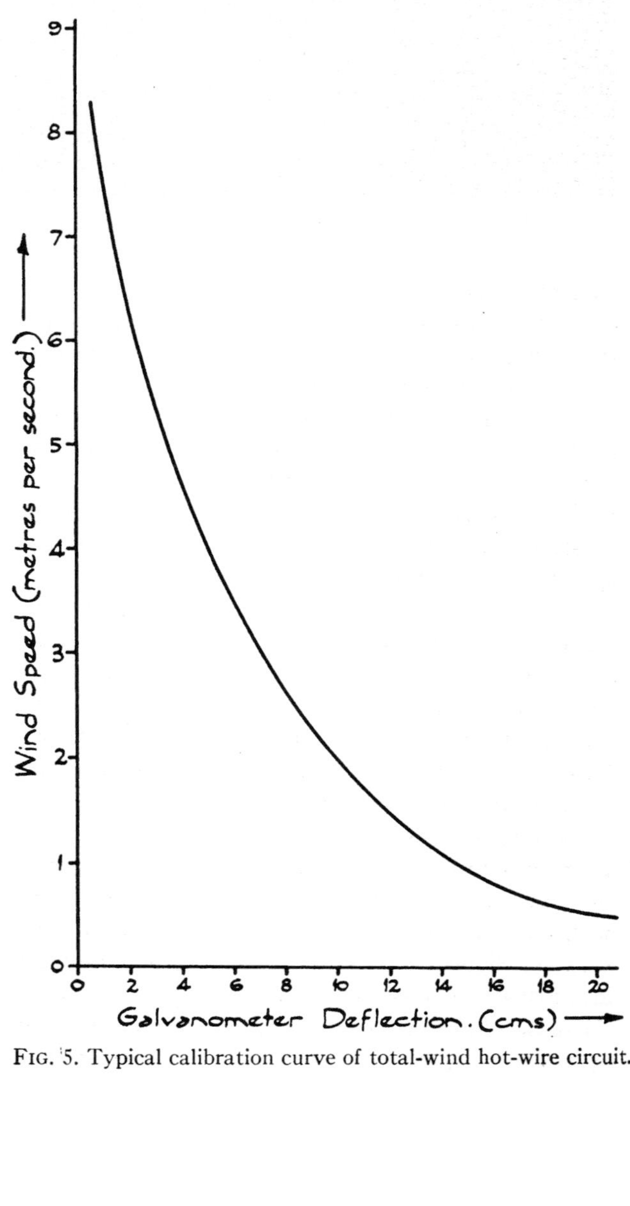

FIG. 5. Typical calibration curve of total-wind hot-wire circuit.

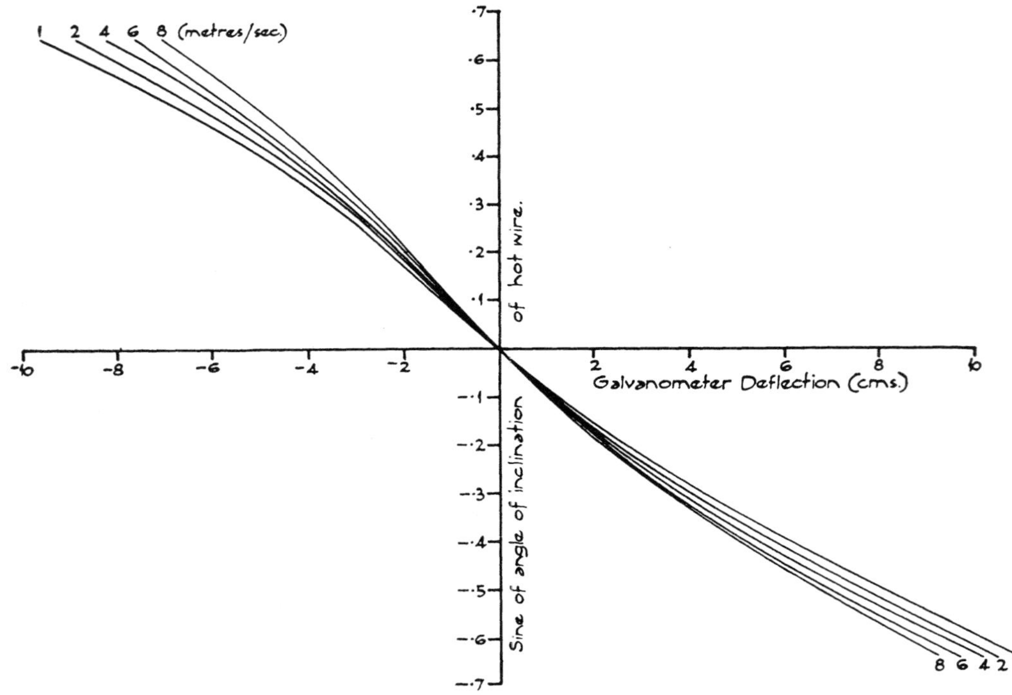

FIG. 6. Typical calibration curves of angle-of-inclination-of-wind hot-wire circuit.

quently. In the present investigation, measurements of the actual wind have not been made more than two days before or after a calibration of the hot wires.

Correction of zero drift caused by varying air temperatures.—Equation (1) relates the loss of heat from the wire to the excess of its temperature over that of the air. The short-period fluctuations of air temperature that accompany similar fluctuations in the wind are much too small to be of any consequence on a wire heated to some 500C, but the variations of temperature between one day and another, or between day and night, might easily exceed 20C. Therefore, it is necessary to provide compensation for the influence of such changes on the calibration of the hot wire. This is accomplished by making the arm of the Wheatstone bridge opposite to the hot wire of thick platinum wire which, having the same temperature coefficient of resistance as the hot wire itself, is equally affected by changes in air temperature. Such compensation is not necessary for the second hot-wire system, each wire in this being equally affected by temperature changes.

Measurement of temperature fluctuations.—Thermocouples of 50 SWG copper and constantan wire, soldered to form a junction whose effective diameter is less than 0.002 in, are used for the measurement of the fluctuations of air temperature. They are attached to the head carrying the hot wire, but sufficiently away and to windward from the hot wires to avoid thermal influence from them. The reference junctions are buried at about one-foot depth in the ground where, for the period during which recordings are made, the temperature is virtually constant. It is not necessary to know the temperature of the reference junction, as the description of the analysis of the records, to be given presently, will show. To increase the sensitivity, two pairs of junctions in series are used. The position of the exposed junctions on the head of the instrument is indicated in fig. 2.

The thermocouple junctions, though small, still have sufficient thermal capacity to prevent the faithful recording of short-period fluctuations of temperature, and will introduce an amplitude and phase distortion more or less pronounced according to the period of the fluctuations and the speed of the wind. Table 2 shows the results of calculations of the distortions introduced by such a junction (of assumed diameter 0.002 in) for various wind speeds and periods of sinusoidal temperature fluctuations.

The table shows that though, naturally, the distortion introduced by the thermojunctions is somewhat greater than in the case of the finer hot wire, it is still, for fluctuations of greater than one-second period and for winds of 1 m/sec or more, entirely allowable in the present investigation.

TABLE 2. Amplitude and phase distortion caused by thermal inertia of thermojunction (copper-constantan) of diameter 0.002 in in recording sinusoidal temperature fluctuations with varying periods and wind speeds.

Period of fluctuations (sec)	Ratio of amplitude response to true amplitude, Wind speed (m/sec)			Phase lag (deg), Wind speed (m/sec)		
	0.5	1	4	0.5	1	4
0.5	0.7	0.78	0.92	45	38	24
1.0	0.89	0.93	0.98	27	21	12
2.0	0.97	0.98	0.99	14	11	6
4.0	0.99	0.99	1.00	7	5	3

It was at first thought unnecessary to protect such small thermojunctions from the sun's radiation, but calculation shows that the effect may not be negligible. To calculate the order of magnitude of the influence of this radiation on the performance of the thermocouples, assume the intensity of solar radiation to be 1 cal cm^{-2} min^{-1} and the reflectivity of the junction to be 50 per cent. Then, in a wind of $\frac{1}{2}$ m/sec, the sun's radiation will be sufficient to maintain the junction at a temperature about 0.2C above the mean air temperature, while in a wind of 2 m/sec the temperature excess will be about 0.13C. Thus, since the fluctuations of air temperature that occur have, in general, an amplitude of about 1C, it appears that the thermojunctions should be protected against the sun's radiation which, with a fluctuating wind, may cause a spurious temperature fluctuation of perhaps one tenth that of the true fluctuation.

Accordingly, when recordings are made in strong sunshine, a small shield is held so that, though it is far enough away to offer no obstacle to the air flow, its shadow falls on the thermojunctions and thus shields them from the sun's radiation.

Measurement of vapor-pressure fluctuations.—An instrument has been developed to measure the fluctuations in vapor pressure of the air passing a fixed point, and is described in another paper.[2] It is sufficient, therefore, to give here only a brief description of the principle of the apparatus. The psychometric element consists of two fine wet-bulb thermocouples, built into an electrical network so designed that it simulates the behavior of Regnault's psychrometric equation for small changes in the variables (vapor pressure, saturation vapor pressure at the wet-bulb temperature, and the wet-bulb depression). The output of the circuit gives a measure of the fluctuations of vapor pressure practically linear over the range of variation that occurs naturally.

The tapered capillary tubes, providing the water supply to the fine wet-bulb thermocouples projecting from their lower openings, are shown in fig. 2. The reference junction is buried in the ground with those of the dry-bulb thermocouples.

Method of recording wind, temperature and humidity fluctuations.—The outputs of the hot wire, dry-bulb thermocouple and humidity circuits are each recorded by means of mirror galvanometers. The lights from the galvanometers (four in all) are focused by means of a horizontal cylindrical lens onto a wooden drum which, frictionally driven by a synchronous electric motor, rotates on a horizontal axis parallel to the lens once in five minutes. Photographic paper is wound around the drum, and the whole apparatus housed in a dark room. There are thus provided synchronous traces giving wind, temperature and humidity structure over a five-minute interval for each run.

[2] To be published in *J. sci. Instrum.* (*Ed. note:* since published, **28,** 86–89, 1951.)

TABLE 3. Amplitude distortion and phase lag introduced into recording by critically damped galvanometer of sinusoidal fluctuations of varying periods (expressed as ratio of free period of galvanometer on open circuit)

Ratio of fluctuation period to free period of galvanometer	Ratio of recorded to true amplitude of fluctuation	Phase lag (deg)
8.0	0.98	16
4.0	0.94	29
2.0	0.80	51
1.0	0.50	90
0.5	0.20	114

It is necessary to consider the distortion, both of amplitude and phase, of the fluctuations of the various quantities, introduced by this method of recording. Each galvanometer is critically damped, and thus spurious fluctuations are avoided. The varying resistance of the hot wires, when measuring wind in the open, does not affect materially the damping of their galvanometers, which are sensitive enough to require a series resistance of some thousands of ohms, as shown in the circuit diagrams. The shunt resistances of the hot-wire galvanometers, because of these high series resistors, are little more than their critical damping resistance. Table 3 shows the amplitude and phase distortion introduced into the recording of fluctuations (assumed sinusoidal) of various periods, expressed in terms of the free period of the galvanometer on open circuit, by a critically damped galvanometer.

The table shows that the distortion is serious for fluctuations of the same period as the galvanometer, and that an instrument with a free period something less than one quarter that of the fluctuations is necessary for adequate recording. It will, however, become evident, when the method of analyzing the records is described later, that phase distortion is not material when the individual quantities suffer equal distortion of this sort. As remarked previously, a lower limit to the period of fluctuations that contribute materially to the flux of heat, *etc.*, in the lower atmosphere cannot be set arbitrarily. The galvanometers used in the present investigation have been the lowest-period type available, each with a free period of about two seconds. How far galvanometers of this period are able to give a true record of the fluctuations responsible for heat transport is examined in an appendix to this paper.

3. The observations

Observations have been made at an open, level grassland site, near Melbourne, by mounting the apparatus shown in fig. 1 on a theodolite stand. Thus, the measurements refer to a level about $4\frac{1}{2}$ ft above the ground. Up to the present, observations have been confined to occasions of clear days and clear nights. It is proposed later to extend the observations to

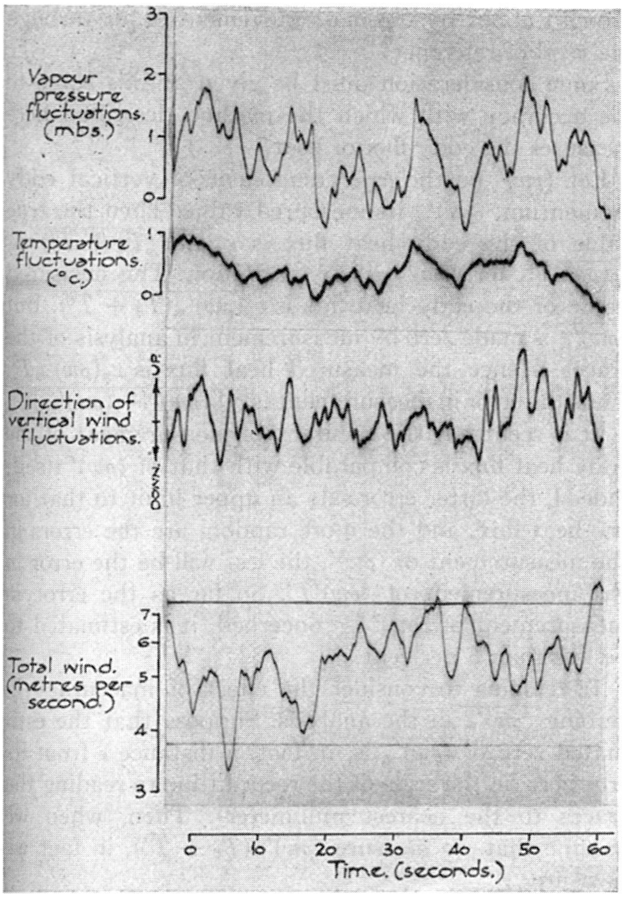

FIG. 7. Traces of fluctuations of vapor pressure, temperature, angle of wind and total wind at 4½ ft on open, level grassland site near Melbourne during clear weather at 1435 EST 21 February 1950.

greater heights and to different types of weather, *e.g.*, occasions of advective warming and cooling of air.

Simultaneous measurements of temperature and humidity are made at two levels (2 and ¼ m above the ground) and near the apparatus. These quantities are measured with electrically-driven Assman psychrometers. It is thus possible to relate the flux of heat and water vapor through the level concerned to the gradients of temperature and humidity, respectively. The instrument is levelled by reference to two spirit levels built into its base; this is done to provide free movement of the vane. The period of a run is five minutes.

Two examples of the traces so obtained are shown in figs. 7 and 8, the former referring to a clear day and the latter to a clear night.

Because fluctuations are, generally, more pronounced during the day than during the night, it has been found necessary to use different sensitivities in the recording of temperature and vapor pressures, and the scales of the fluctuations are shown in the figures.

4. Analysis of the traces

Synchronous points on the traces referring to air flow, its inclination, temperature and vapor pressure,

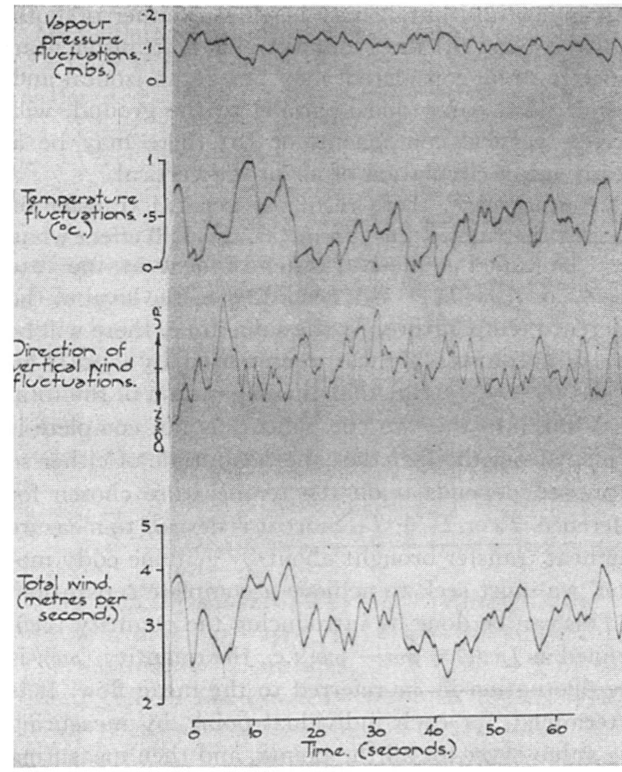

FIG. 8. Traces, similar to those in fig. 7, during clear weather at 2100 EST 5 May 1950.

respectively, are read off the photographic record at 300 points equally spaced over the five-minute run, *i.e.*, at intervals of 1 sec. An accurately ruled transparent grid is used for this purpose. By reference to the calibration curves, the total air flow and the sine of its angle of inclination are determined for each of the points. Multiplication of these quantities yields the value of ρw at each point, *i.e.*, at each instant of time.

Heat flux.—Let us refer again to the expression for the heat flux, $c_p\overline{\rho w T}$ per unit area. As stated earlier, this may be written

$$c_p\overline{\rho w(T_0 + T')},$$

where T_0 is some arbitrary temperature level. Alternatively, another temperature level $T_0 + T_1$ may be chosen, and the heat flux is

$$c_p\overline{\rho w(T_0 + T_1 + T' - T_1)}.$$

Evidently

$$c_p\overline{\rho w(T_0 + T')} = c_p\overline{\rho w(T_0 + T_1 + T' - T_1)},$$

i.e.,

$$c_p[\overline{\rho w}T_0 + \overline{\rho w T'}] = c_p[\overline{\rho w}(T_0 + T_1) + \overline{\rho w(T' - T_1)}], \quad (3)$$

T_0 and T_1 being constant.

The term $\overline{\rho w}$ represents the net rate at which *air mass* is being transferred vertically at the point in question. A net vertical mass transfer, *i.e.*, a mean flow that is not horizontal, may be due to either or both of two causes: (a) the ground itself, the transfer near which is being considered, may not be horizontal and the air flow, being quasi-parallel to the ground, will have a vertical component, or (b) there may be a steady mean circulation of air in the vertical.

In either case, there results a vertical transfer of mass at the rate $\overline{\rho w}$ and, from (3), this will effect what may be called a mass transfer of heat at the rate $c_p\overline{\rho w}T_0$ or $c_p\overline{\rho w}(T_0 + T_1)$, according to the level of the reference temperature. At the same time, there will be an eddy transfer of heat represented by $c_p\overline{\rho w T'}$ or $c_p\overline{[\rho w(T' - T_1)]}$. But that this separation of the total heat flux into the two components is not complete is evident from the fact that the magnitude of either so expressed depends upon the temperature chosen for reference, T_0 or $T_0 + T_1$. Since it is desired to measure the heat transfer brought about by genuine eddy motion, we must seek to achieve a complete separation.

This can be done by introducing the quantity $(\rho w)'$ defined as $(\rho w)' = \rho w - \overline{\rho w}$, *i.e.*, the quantity $(\rho w)'$ is the fluctuation in ρw referred to the mean flow. It is determined, for each individual point, by measuring the value of ρw for all the points, and then measuring the departure of ρw for the point from this mean. Clearly,

$$\overline{(\rho w)'} = \overline{\rho w} - \overline{\overline{\rho w}}. \tag{4}$$

The expression for the heat flux due to fluctuations about the mean flow, the true eddy flux, may thus be written as

$$c_p\overline{(\rho w)'T} = c_p\overline{(\rho w)'T'} = c_p\overline{(\rho w)'(T' - T_1)},$$

in consequence of (4).

In eliminating the mass transfer of heat in this way, we have at the same time provided justification for referring the temperature fluctuations to any convenient level, for the eddy flux is represented either by $c_p\overline{(\rho w)'T'}$ or $c_p\overline{(\rho w)'(T' - T_1)}$.

A point-by-point multiplication of the quantity $(\rho w)'$ by the synchronous value of T' is made, and the mean product, multiplied by c_p, yields the quantity $c_p\overline{(\rho w)'T}$, the heat flux due to eddy motion.

It will be recognized from the foregoing that the determination of $\overline{\rho w}$, and the measurement of fluctuations of ρw from this mean, *is an essential and unavoidable step in the analysis if a distinction is to be made between mass and eddy transfer of heat. It cannot be circumvented by any method of combining directly temperature and wind fluctuations, which would provide a heat flux composed inseparably of the two parts, eddy and mass transfer.*

It is a particular merit of the hot-wire method of measuring air flow that, in providing a simple product of the velocity and density of the air, it makes easy the elimination from the analysis of the heat transfer brought about by the mass movement of air through the level of reference.

Some consideration must be given at this stage to the accuracy with which the method described here measures the eddy flux of heat.

Let $(\rho w)'$ be the *true* component of vertical eddy momentum, $(\rho w)'_m$ its measured value. Then the true value of the eddy heat flux is $c_p\overline{(\rho w)'(T_0 + T')}$ or $c_p\overline{(\rho w)'T'}$, for $\overline{(\rho w)'} = 0$ by definition. The measured value of the eddy heat flux is $c_p\overline{(\rho w)'_m(T_0 + T')}$. But $\overline{(\rho w)'_m}$ is made zero by measurement in analysis of the traces. Hence the measured heat flux is $c_p\overline{(\rho w)'_m T'}$. Thus the error in measurement is $c_p\overline{[(\rho w)' - (\rho w)'_m]T'}$.

It is seen that the accuracy of measurement of the eddy heat flux is comparable with that of $(\rho w)'$ itself; indeed, the latter error sets an upper limit to that for the heat flux, and the more random are the errors in the measurement of $(\rho w)'$, the less will be the error in the measurement of $\overline{(\rho w)'T'}$. So far as the error in measurement of $(\rho w)'$ is concerned, it is estimated to be less than 5 per cent.

It remains to consider the effect of inaccuracy in zeroing $\overline{(\rho w)'_m}$ in the analysis. Suppose that the estimated zero of $\overline{(\rho w)'_m}$ is, in fact, a distance ϵ from its true zero on the scale of the record (due to reading the traces to the nearest millimeter). Then, when we assume that we measure $\overline{(\rho w)'_m(T_0 + T')}$, in fact we measure

$$\begin{aligned}\overline{[(\rho w)'_m - \epsilon](T_0 + T')} &= \overline{(\rho w)'_m}T_0 + \overline{(\rho w)'_m T'} - \epsilon(T_0 + \overline{T'}) \\ &= \epsilon T_0 + \overline{(\rho w)'_m T'} - \epsilon(T_0 + \overline{T'})\end{aligned}$$

[for we have now postulated $\overline{(\rho w)'_m} = \epsilon$, not 0],

$$= \overline{(\rho w)'_m T'} - \epsilon\overline{T'}.$$

Hence, there is introduced an error $\epsilon\overline{T'}$ into the eddy heat flux measurement, due to the inability to zero $\overline{(\rho w)'_m}$ exactly. But if 300 points are used to determine $\overline{(\rho w)'_m}$, and there is a random error of measurement of 1 mm in each point, the value of ϵ is $(1/300)^{\frac{1}{2}}$ mm, *i.e.*, about 1/170 cm. Therefore, since $(\rho w)'_m$ and T' are represented on the traces by lengths of the order of centimeters, the error so introduced into the measurement of the eddy flux will be much less than 1 per cent. This, of course, is aside from the error introduced by inaccuracy in the measurement of instantaneous $(\rho w)'$ discussed above.

It has been felt necessary to present the above discussion at some length because there is a certain subtlety in the argument which can only be clarified by a step-by-step treatment.

Water-vapor flux.—If the vapor pressure is e and temperature T, the density of water vapor ρ_w is given by $\rho_w = e/R_wT$, where R_w is the gas constant for water vapor. As in the case of heat flux, the vapor pressure in the air passing the reference point may be regarded as $e = e_0 + e'$, e_0 being the vapor pressure at

TABLE 4. Measurements of heat and water-vapor eddy flux (positive values indicate upward transfer).

Date	Time (EST)	Cloud amount	Cloud type	Wind speed at 2 m (m/sec)	Wind direction at 2 m	Dry-bulb temp. (deg F)		Lapse rate (× dry adiabatic)	Wet-bulb temp. (deg F)		Vapor pressure (mb)		Humidity gradient (g/cm⁴)	Heat flux (cal cm⁻² sec⁻¹)	Water-vapor flux (g cm⁻² sec⁻¹)	K_H (cm²/sec)	K_w (cm²/sec)
						¼ m	2 m		¼ m	2 m	¼ m	2 m					
21 Feb. 1950	1435	1/10	Cu	5	SSW	74.8	69.8	159	64.3	61.4	16.69	15.47	4.4×10^{-9}	13.8×10^{-4}	1.36×10^{-6}	5×10^{2}	3.1×10^{2}
5 May 1950	2100	0	—	3	NNW	49.7	51.9	−70	48.5	49.8	11.11	11.41	-1.1×10^{-9}	-3.4×10^{-4}	0.05×10^{-6}	2.65×10^{2}	-0.5×10^{2}

some arbitrary level of reference, and e' the fluctuation above that level.

Then the flux of water vapor through the level considered is $\overline{w\rho_w}$, and

$$\overline{w\rho_w} = \overline{w(e_0 + e')/R_w(T_0 + T')}$$
$$= \overline{\rho w(e_0 + e')/R_w\rho(T_0 + T')}.$$

It is necessary to consider the range of variation of the quantities in this expression. ρw, in general, varies from positive to negative, and e varies by a considerable fraction (about 10 per cent) of its average value. On the other hand, $T_0 + T' = p/R\rho$, where p is the atmospheric pressure and R the gas constant for air, is virtually constant. Hence, the expression for the vapor flux may be written as

$$\overline{w\rho_w} = (R/R_w p)\{\overline{[\overline{\rho w} + (\rho w)'][e_0 + e']}\}$$
$$= (0.622/p)[\overline{\rho w}(e_0 + e') + \overline{(\rho w)'}e_0 + \overline{(\rho w)'e'}].$$

In this expression, $\overline{\rho w}(e_0 + e')$ represents the mass transfer of water vapor, and $\overline{(\rho w)'e'}$ an eddy transfer. The middle term vanishes, since $\overline{(\rho w)'} = 0$ by definition. Hence, the quantity we wish to measure, the eddy transfer, is represented by $(0.622/p)\overline{(\rho w)'e'}$.

In the above, no account has been taken of virtual-temperature effects in the expression $\rho(T_0 + T')$, for such influence in the present considerations is quite insignificant. Nor, throughout this investigation, is any account taken of the influence of varying water-vapor content on the behavior of the hot-wire anemometer, for the same reason.

5. Results

In table 4 are presented measurements of the eddy flux of heat and water vapor on the two occasions to which the traces in figs. 7 and 8 refer. The second occasion (5 May) has been included not because it is typical, but because it illustrates a rather anomalous behavior in the flux of water vapor which has been observed on several clear nights. It will be seen that on this occasion there was an *upward* eddy flux of water vapor, small it is true, even though the gradient of absolute humidity was directed downwards. Furthermore, at the time of observation the grass was heavily covered with dew, and this was also the case on the other nights referred to. These observations confirm some surprising results that have been observed in the direct measurement of evaporation from pots sunk level with the ground which, during some clear nights, lose weight even though they are dew covered at the end of the night. It may be that dew, which is commonly regarded as deposition of water vapor from the air is, on occasion, entirely transpiration, some of which may be evaporated into the air. Confirmation and explanation of these results await further work.

The values of the fluxes shown in table 4 are not necessarily typical, but (except for the flux of water vapor on 5 May) they are in accord with general concepts of their respective values.

It has been argued (Priestley and Swinbank, 1947) that the coefficients of eddy transfer, as customarily defined, should vary from one transferred property to another, if the considered property influences the motion. It is therefore of some interest to compare the transfer coefficient for heat with that for water vapor, which can influence the motion only slightly, on the occasions listed in table 4. They have been estimated, for heat, from the equation

$$\text{Heat flux} = K_H c_p \rho (\partial T/\partial z + \Gamma),$$

in the usual notation, and for water vapor from the equation

$$\text{Water-vapor flux} = K_w \rho \; \partial q/\partial z,$$

where q is the specific humidity. The values of the coefficients so calculated are shown in table 4.

Many more observations are required before definite conclusions can be drawn, but the discrepancy in the values of the coefficients listed here is interesting.

6. Further work

The method described in this paper seems to be capable of providing reliable measurements of the vertical transfer of heat and water vapor brought about by the movement of eddies in the lower atmosphere. Furthermore, it can be used for the measurement of these quantities over any sort of surface. The results quoted readily show, by comparison against incoming solar energy or outgoing long-wave radiation, that the values so measured are considerable, and that this mechanism of transfer must be of major importance in the transport of heat and water vapor in these regions.

It must be said that the present method of analysis, point by point, is laborious, and further work is being concentrated on its mechanization. The only obstacle remaining in the way of this is the separation of the calibration curves shown in fig. 6. If the response of this hot-wire circuit can be made to depend only on the angle of inclination of the wind and not on its speed, the whole analysis can be made mechanical. There has also been designed a simple addition to the instrument which, by providing a continuous record of the instantaneous azimuth of the wind, will provide measurements, in exactly the same way as for heat and water vapor, of the eddy transfer of momentum.

When this has been achieved, a program of investigation into heat, water-vapor and momentum flux at various heights, and in a variety of synoptic situations, will be carried through.

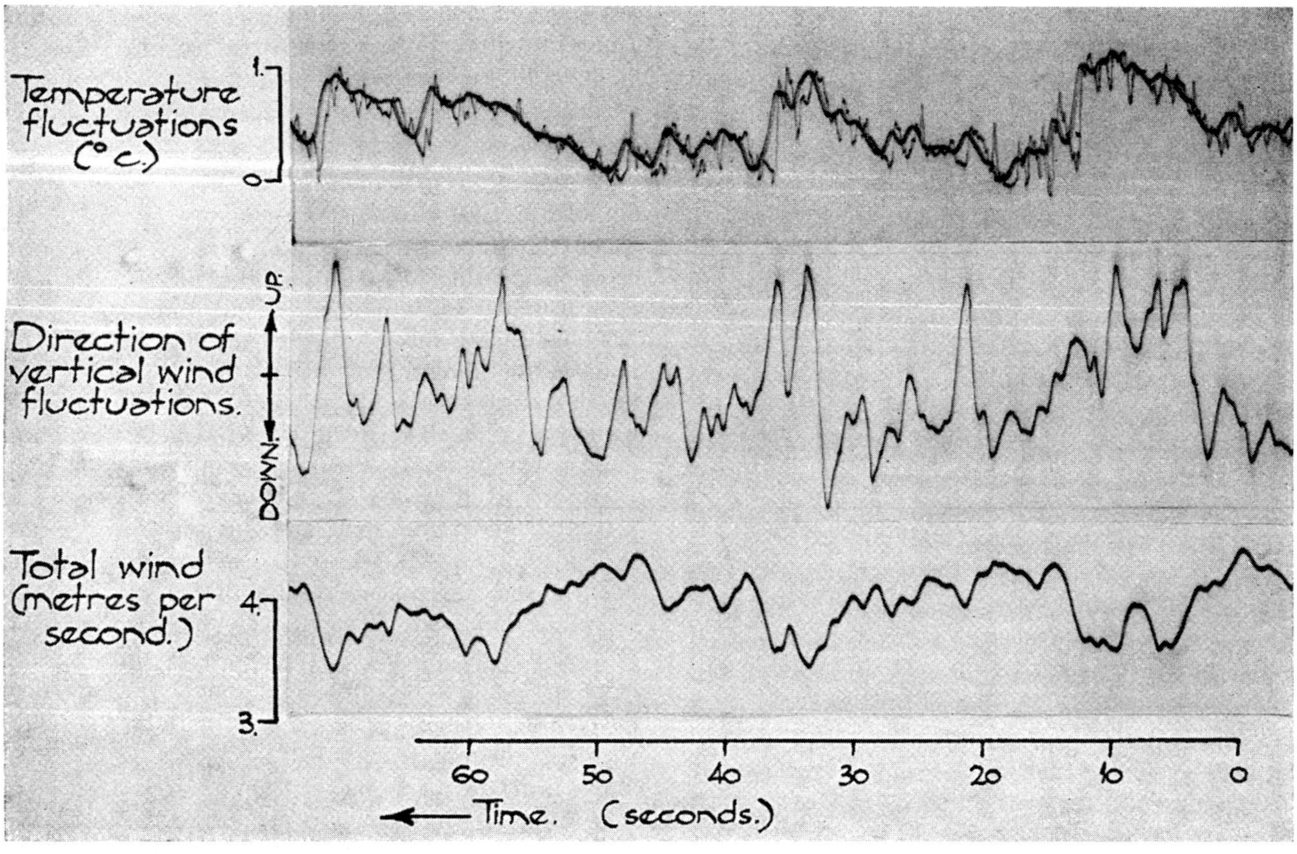

FIG. 9. Superimposed traces of temperature fluctuations measured with short and long period galvanometer, and of angle of wind and total wind at same site during clear weather at 1500 EST 26 July 1950.

APPENDIX

It has been stressed earlier (section 3) that a lower limit to the dimensions of the eddies that contribute materially to the flux of the various properties cannot be set arbitrarily, and that the eddies must be examined quantitatively before the necessary speed of response of the recording apparatus can be decided.

The ideal way to do this would be to obtain traces of the various properties, as described above, using galvanometers all of a certain free period, and, simultaneously and at the same place, another set of traces using galvanometers of much shorter period. A comparison of the fluxes of heat and water vapor calculated from the two sets of data could then be made, and certain conclusions drawn. If the respective fluxes showed no material difference, it could be said that the longer period galvanometers were adequate for the purpose at hand. But if there were differences, this need not be true of either the longer or shorter period instruments, and further comparison against yet faster galvanometers would be necessary.

The galvanometers providing the traces shown in figs. 7 and 8 were all of two-seconds period on open circuit. Instruments of shorter period than this are not readily available, and it was not possible to carry out the comparison described above. Therefore a different comparison has been made as follows: one galvanometer of free period 0.2 sec has been used to provide a record of the temperature fluctuations at a point, and simultaneously and at the same place (*i.e.*, at about one-inch distance) a similar set of thermocouples has provided a record of the temperature fluctuations on a two-second galvanometer.

The circuit resistances of the two galvanometers were so arranged that they were critically damped, of the same voltage sensitivity, and with the same zero on the recording paper. As before, records of the total wind and its inclination were made synchronously, using the same galvanometers as before. Traces so obtained on a clear day (26 July 1950) are shown in fig. 9.

The temperature traces are rather striking in that they reveal that, though the shorter period galvanometer shows detail of fine structure suppressed by the other, this detail consists entirely of short-period fluctuations of small amplitude. The fluctuations of large amplitude are of much longer period, and these the slower galvanometer appears to record faithfully. Some difference in phase lag can be detected in the two traces.

The heat flux has been calculated from each temperature trace in the manner described previously. The values that result are, for the short-period galvanometer, 13×10^{-4} cal cm^{-2} sec^{-1}, and for the other 14×10^{-4} cal cm^{-2} sec^{-1}. The difference between these two values is perhaps within the error of this method of measuring heat flux and, particularly since the shorter period galvanometer trace yields the slightly smaller value, probably not significant.

This test, though not conclusive, affords good support for regarding apparatus able to record fluctuations of a few-seconds period as sufficiently fast for measuring the flux of heat, and of other properties, by eddies in the atmosphere near the ground.

Acknowledgment.—Acknowledgment is gratefully made to Mr. L. F. G. Simmons, of the National Physical Laboratory, Teddington, England, who initiated me into the technique of hot-wire anemometry, and who suggested a method of measuring the vertical wind of which that used is a modification.

REFERENCES

Boussinesq, J., 1901: Physique Mathematique—Sur la pouvoir refroidissant d'un courant liquide ou gazeux. *C. R. Acad. Sci. (Paris)*, **133**, 257.

King, L. V., 1914: On the convection of heat from small cylinders in a stream of fluid: determination of the convection constants of small platinum wires with application to hot wire anemometry. *Phil. Trans. roy. Soc. London*, A, **214**, 373.

Ower, E., 1949: *The measurement of air flow.* London, Chapman and Hall, pp. 240–270.

Priestley, C. H. B., and W. C. Swinbank, 1947: Vertical transport of heat by turbulence in the atmosphere. *Proc. roy. Soc. London*, A, **189**, 543.

Measurements of evaporation and heat transfer in the lower atmosphere by an automatic eddy-correlation technique

By A. J. DYER
*C.S.I.R.O., Division of Meteorological Physics, Aspendale, Victoria, Australia**

(Manuscript received 18 October 1960)

SUMMARY

Following the development of an automatic instrument for the determination of evaporation and sensible-heat transfer by the eddy-correlation technique, a large number of measurements of these eddy fluxes have been made over level pasture-land. The observations, taken at a height of 4 m, are in excellent agreement with the energy balance. Individual measurements, of 5-min duration, are distributed about a strict energy balance with a standard deviation of 31 per cent, due mainly to horizontal gradients of wind, temperature, and humidity below 4 m height. Over a period of several hours these gradients average out to negligible values.

The short response time of the sensing equipment ($\simeq$ 0·3 sec) ensures that, for conditions of moderate and high instability, significant deficiencies in eddy-flux measurements occur only with winds exceeding 10 m sec^{-1}.

A quantity ϵ, the 'evaporation fraction,' i.e., the ratio of evaporation to evaporation plus sensible-heat transfer, is found to increase markedly after rain, and to gradually decrease during the succeeding dry period. This increase can be quantitatively related to the amount of rain. The average value of ϵ during February-March 1960 was 0·56, whilst a value of 0·54 can be computed from annual totals of precipitation and net day-time radiation income.

1. INTRODUCTION

The need for a method to measure the vertical transfer of water vapour and heat in the lowest layers of the atmosphere has long been recognized. A number of workers have attempted to relate the magnitudes of these fluxes to other meteorological parameters using varying degrees of empiricism, and have often employed formulae of doubtful validity.

In the case of water-vapour transfer, or evaporation, the lysimeter technique has provided useful results. This, however, is a somewhat cumbersome method, and is not readily made portable. The value of the aerodynamic method was established by Pasquill (1949), and more recent work has substantiated this in conditions of near neutrality (House, Rider and Tugwell 1960).

The eddy-correlation technique for measuring turbulent transfer, a method first applied by Swinbank (1951), merits extensive exploitation as it relies only on basic physical principles. This has not previously been practicable chiefly because of the laborious nature of the measurements, but this difficulty has now been overcome by the development of an automatic instrument, specially designed to employ this method.

A large number of measurements of evaporation and heat transfer were made at Edithvale, Victoria during 1959-1960. Profile measurements were made simultaneously with the above, and an analysis of these results will be published separately.

2. PRINCIPLE AND INSTRUMENTATION

It can readily be shown (e.g., Swinbank 1951; Priestley 1959), that the average vertical transfer of water vapour E (expressed in latent heat units) and sensible heat H, across unit area of a horizontal plane can be written in the form,

$$E = L\,\overline{(\rho w)'\,q'}, \qquad (1)$$

and

$$H = C_p\,\overline{(\rho w)'\,T'}, \qquad (2)$$

* Present affiliation, Dept. Agric. Engineering, University of California, U.S.A.

where L is the latent heat of vaporization of water, and C_p the specific heat of air at constant pressure; $(\rho w)'$, q' and T' represent the departure of the quantities ρw, q and T from their respective mean values; ρ is the air density, w the vertical wind component, q the specific humidity and T the air temperature. The bar denotes an average over the period of observation. In the derivation of Eqs. (1) and (2) it is assumed that $\overline{\rho w}$ is zero. This is true for a horizontally uniform surface if the period of observation is sufficiently long.

Swinbank (1955) has made use of this technique by recording the fluctuations of ρw, q and T, and subsequently obtaining the fluxes with the aid of a mechanical computer (Taylor and Webb 1955). This work established the practicability of the method, and has provided a most informative body of data. At that stage of instrumental development there remained, however, two main short-comings. Firstly, the limited speed of response of the recording equipment caused the fluxes to be underestimated; and secondly, the analysis and computation of the data was extremely laborious.

To eliminate these difficulties and to provide an instrument suitable for the measurement of evaporation by workers in allied fields, special equipment has been developed to give continuous electrical signals proportional to the relevant quantities, and to multiply and integrate automatically.

A general description of the complete instrument has previously been published (Taylor and Dyer 1958) and detailed information has been given for the anemometer by Taylor (1956), for the integrators by Dyer (1958), for the fine-wire resistance thermometers by McIlroy (1955) and for the electronic analogue of the saturated vapour-pressure curve by Taylor (1958). Pl. X, Fig. 1 (a), (b) and (c) shows the sensing heads and console.

In addition to the above, but not relevant to the present enquiry, the instrument also contained an anemometer (Dyer 1960) to measure the horizontal wind component u, so permitting measurements of momentum flux.

It is not possible to know the mean values of ρw, q and T in advance, so that signals proportional to $\rho w'$, q' and T' cannot be obtained. The use of adjustable working means, α, β and γ obviates this difficulty and the fluxes can then be calculated from the exact relationships,

$$E = L\ \overline{(\rho w)'\ q'} = L\ [\overline{(\rho w - \alpha)\ (q - \beta)} - \overline{(\rho w - \alpha)} \cdot \overline{(q - \beta)}] \quad . \quad . \quad (3)$$

$$H = C_p \overline{(\rho w)'\ T'} = C_p [\overline{(\rho w - \alpha)\ (T - \gamma)} - \overline{(\rho w - \alpha)} \cdot \overline{(T - \gamma)}] \quad . \quad . \quad (4)$$

Hence to determine E, or H, one analogue multiplier and three integrators are required.

The naturally occurring $\overline{\rho w}$ is very close to zero and thus no provision for α is necessary in practice.

The specific humidity term $q - \beta$, is obtained by mixing signals from two bridges containing fine-wire wet- and dry-bulb resistance thermometers. Adjustment of the working means β and γ is provided by the balancing control of the dry-bulb bridge. The above mixing is carried out according to a modified form of the psychrometric equation :

$$q - \beta = q_s\,(T_w) - \frac{C_p}{L}\,(T - T_w) - q_s\,(\delta) + \frac{C_p}{L}\,(\gamma - \delta) \quad . \quad . \quad (5)$$

where δ is a fixed working mean of the wet-bulb bridge and $q_s\,(T_w)$ and $q_s\,(\delta)$ represent the saturation specific humidities at temperatures T_w and δ respectively. A slight modification of the electronic analogue described by Taylor (1958) permits the simulation of the non-linear term $q_s\,(T_w) - q_s\,(\delta) + C_p/L\,(T_w - \delta)$. In the normal mode of operation, that of determining E, the output of the instrument is

$$\begin{aligned} E &= L\ \overline{(\rho w)'\ q'} \\ &= L\ \overline{(\rho w)' \left[q_s'\,(T_w) + \frac{C_p}{L}\, T_w' \right]} - C_p\, \overline{(\rho w)'\ T'} \quad . \quad . \quad (6) \end{aligned}$$

The second term on the right-hand side of Eq. (6) is $-H$, hence the first term is $E + H$.

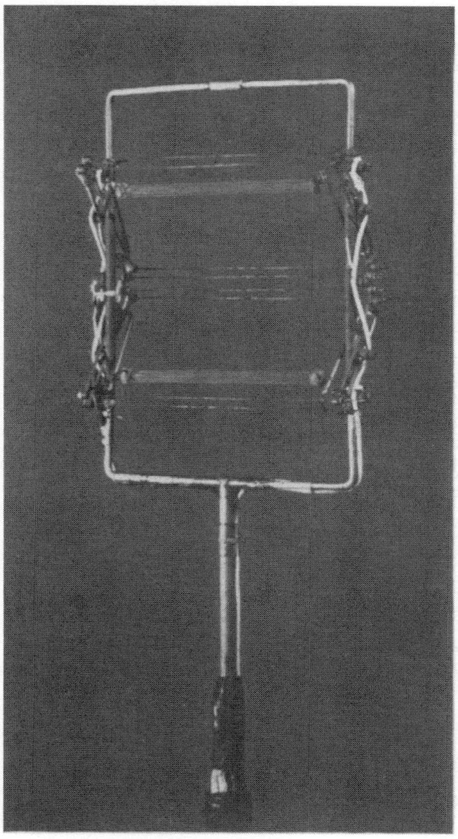

(a) Anemometer.

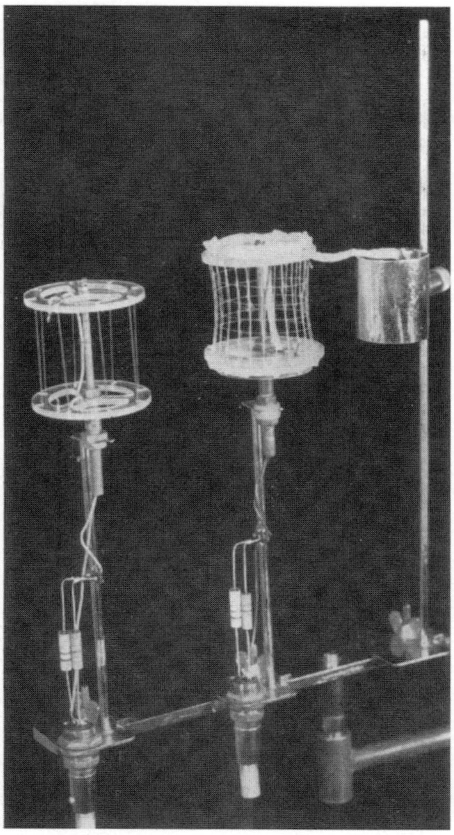

(b) Wet- and dry-bulb hygrometer.

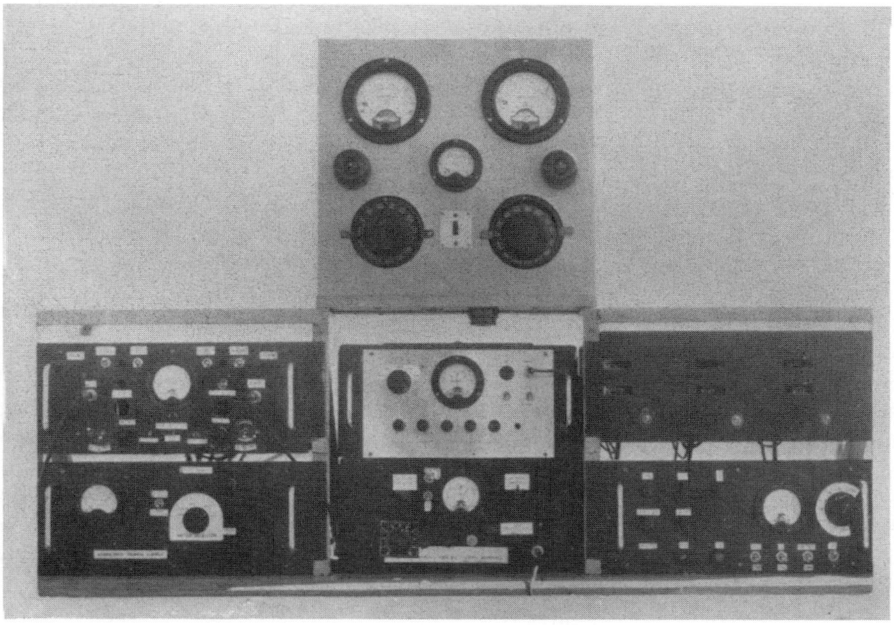

(c) Electric console.

Figure 1. Sensing elements and associated equipment.

Vol. **87** Plate X.

To face page 402

The expression $q_s' (T_w) + C_p T_w'/L$ can be readily shown to represent fluctuations of equivalent temperature, and so the quantity $E + H$ may be conveniently referred to as *equivalent heat.*

By replacing the T element with a fixed resistor (i.e., making $T' = 0$), $E + H$ is determined directly; and similarly, by suppressing the T_w circuit and so removing the first term, the output of the instrument is made equal to $- H$.

With the present equipment only one of E, H, $E + H$, can be determined in a single measurement, but the addition of a second multiplier and two integrators would permit the simultaneous measurement of two of them.

3. Verification of instrument performance

The basic test of the method is that over a reasonably uniform site it should satisfy the equation for energy balance at the ground surface,

$$E + H = R - G \qquad (7)$$

where R is the net radiation income and G the heat flux into the ground. In the use of Eq. (9) all quantities will be expressed in mw cm^{-2}.

A large number of tests, each of 5-min duration, were carried out during 1959-1960 at the C.S.I.R.O. field station at Edithvale, Victoria, Australia. Previous experience indicates that at the measuring heights used in this experiment a 5-min period is adequate (Swinbank 1955; Deacon 1955) and this was subsequently confirmed during the present programme. At the Edithvale site, level pasture-land bearing coarse grass in patchy clumps with occasional stalks up to 70 cm height extends in all directions for at least one mile.

R was measured with a ventilated net radiometer, which was periodically checked against standard silver-disc and Ångström pyrheliometers. Only trivial changes in calibration were found to occur. The output of the net radiometer was continuously recorded and a mean value obtained for each 5-min run.

G was determined by means of laboratory-calibrated flux-plates similar to those described by Deacon (1950). The outputs of seven such flux-plates installed 1 cm below the ground surface were combined to obtain a representative average which was also recorded.

No special attempt was made to allow for the change of heat storage in the shallow layer of soil above the flux plates. The tests were carried out more or less continuously over a period of several hours on each occasion, and the average rate-of-change of heat storage over such a period is easily shown to be negligible.

In the initial testing period the equivalent heat, $E + H$, was determined directly by the eddy correlation instrument, and compared with $R - G$. The ratio, $(E + H)/(R - G)$, called here the 'recovery factor,' and expressed as a percentage, is the fraction of the available energy which is 'recovered'; and so provides a measure of instrument performance. A perfect heat balance would obviously correspond to a recovery factor of 100 per cent.

For the later tests, E and H were measured alternately and then suitably combined to calculate the recovery factor. Except for periods of maintenance, a series of 5-min runs extending over several hours was made on each occasion, with an off-time between runs of either one or two minutes.

The measuring height was initially 1·5 m, and this was later increased to 3·5-4 m when it became evident that the fluxes were being underestimated because of the higher frequencies present in the eddy structure at the lower level. The main body of the results was obtained at either 3·5 m or 4 m, and results from lower levels are used only in connexion with response time considerations.

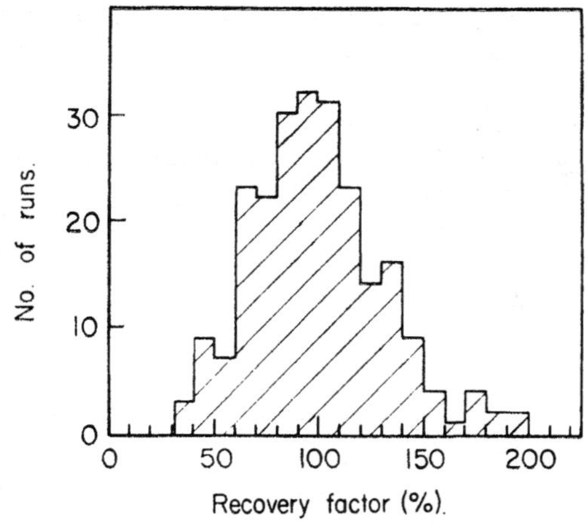

Figure 2. Distribution of recovery factor per cent for equivalent heat ($E + H$) measurements.

(*a*) $E + H$ *measured directly*

During April and October 1959, 311 runs of 5-min each were made at a measuring height of 3·5-4 m in which $E + H$, R and G were determined. The recovery factor for each run was calculated. Individual values of $R - G$ ranged from 10 mw cm^{-2} to 60 mw cm^{-2}.

The overall distribution of the recovery factor is shown in Fig. 2. The mean value is 99 $\pm$ 2 per cent, in excellent agreement with the energy balance. The standard deviation of the distribution is 31 $\pm$ 1 per cent.

A more detailed presentation of the data appears in Table 1. The variation of the mean daily recovery factor is in accord with the distribution of Fig. 2. A similar degree of scatter appears in the detailed daily records, a number of which are plotted in Fig. 3.

TABLE 1. MEAN DAILY RECOVERY FACTOR AT A NUMBER OF MEASURING HEIGHTS AND WIND SPEEDS ($E + H$ MEASURED DIRECTLY)

Date (1959)	Height (metres)	$\overline{R - G}$ (mw/cm^2)	No. of runs	Mean recovery factor (per cent)	u (m sec^{-1})
3 April	1·5	20·6	10	77 ± 10	–
23 April	1·5	27·0	17	86 ± 7	2·6
28 April	1·5	21·7	18	83 ± 7	2·3
24 April	2·5	28·0	19	86 ± 7	2·8
7 April	3·5	26·5	35	104 ± 5	5·0
8 April	3·5	21·6	9	94 ± 6	2·5
9 April	3·5	29·2	24	97 ± 6	4·5
10 April	3·5	21·4	17	102 ± 7	5·3
17 April	3·5	21·2	28	100 ± 6	3·5
20 April	3·5	21·0	9	92 ± 10	5·0
21 April	3·5	17·0	14	97 ± 7	4·0
5 October	4·0	21·8	11	95 ± 8	–
14 October	4·0	30·0	20	95 ± 6	4·7
15 October	4·0	34·8	19	108 ± 6	4·0
16 October	4·0	21·8	24	107 ± 6	4·2
19 October	4·0	36·0	12	87 ± 8	–
21 October	4·0	34·2	17	109 ± 7	7·6
22 October	4·0	16·9	25	102 ± 6	5·3
23 October	4·0	38·1	6	95 ± 12	–
27 October	4·0	15·3	14	93 ± 7	–
28 October	4·0	32·4	27	95 ± 5	–

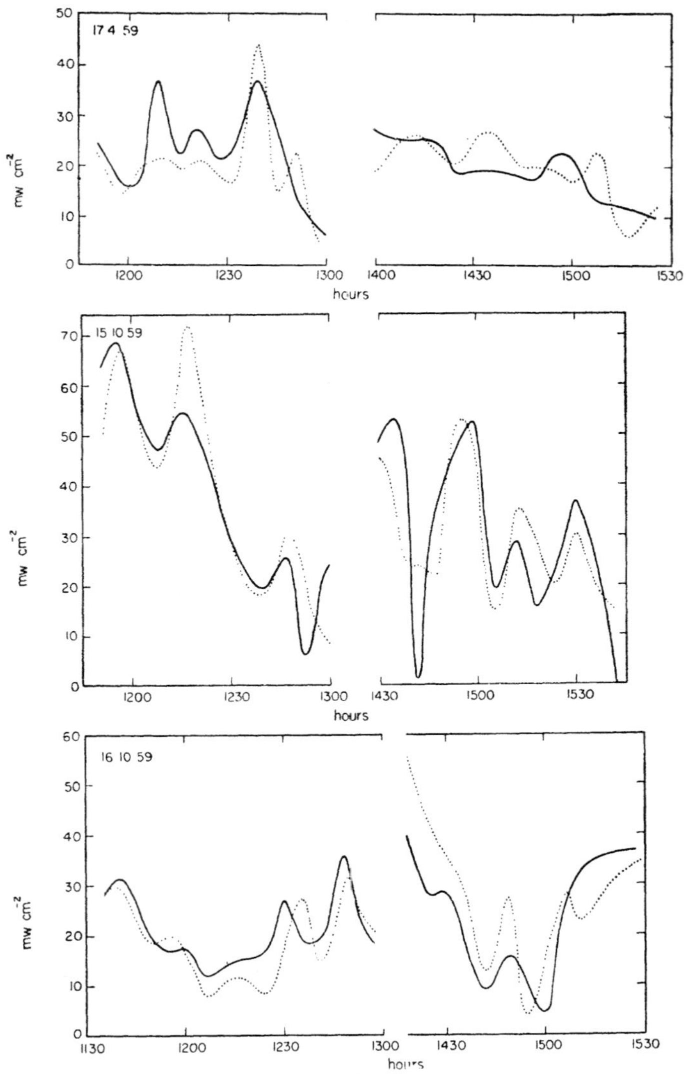

Figure 3. Typical comparison of $E + H$ and $R - G$. $(E + H)$ measured directly.
——— $R - G$; · · · · · $E + H$.

(*b*) *E, H. Measured separately*

Commencing in January 1960, and extending through to April 1960, 417 5-min observations were made at 4 m height.

To compute the energy balance, these results were grouped into units. Each unit run contains two measurements each of E and H taken alternately, and is thus equivalent to two 5-min runs of $E + H$ measured directly. (A measurement of stress or momentum transfer was also included in each unit, but these results are not relevant to the present study).

The recovery factor was calculated for each unit run and the distribution plotted (Fig. 4). The mean value is 96 $\pm$ 2 per cent, again giving a satisfactory energy balance. The standard deviation, 22 $\pm$ 1 per cent, agrees well with the value of $31/\sqrt{2}$ per cent suggested by the results of the previous section.

The variation of the mean daily recovery factor (Table 2) is similar to that in Table 1. A few low values can be associated with greater wind speeds not generally encountered in the previous study. Omitting these, the overall mean recovery factor becomes 100 $\pm$ 2 per cent.

Some typical examples of continuous series of runs are shown in Fig. 5.

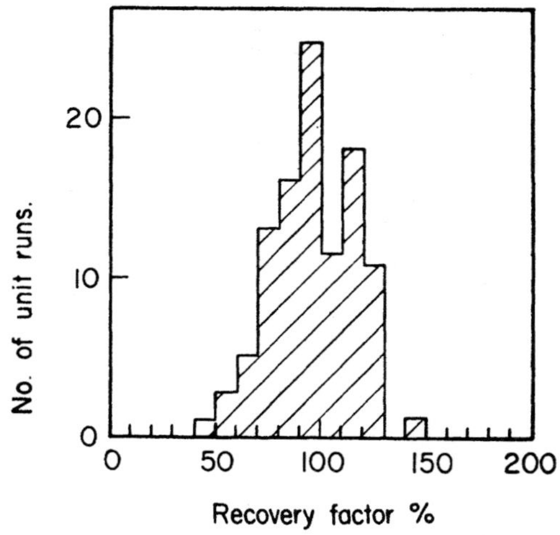

Figure 4. Distribution of recovery factor with E, H measured separately.

TABLE 2. MEAN DAILY RECOVERY FACTOR AT A MEASURING HEIGHT OF 4 m (E AND H MEASURED SEPARATELY)

Date (1960)	No. of 5-min runs	$\overline{E+H}$	$\overline{R+G}$	Recovery factor (per cent)	$\bar{u}$ (m sec^{-1})	$\epsilon = \frac{\overline{E}}{\overline{E}+\overline{H}}$
4 January	11	37·7	39·0	97 ± 13	4·99	0·28
5 January	20	45·0	50·2	90 ± 10	7·28	0·47
6 January	9	61·0	57·0	107 ± 14	4·27	0·53
25 January	8	24·0	23·8	101 ± 14	3·62	0·61
27 January	14	38·3	40·5	94 ± 12	3·24	0·59
29 January	22	43·2	44·7	97 ± 10	4·50	0·60
2 February	8	42·0	44·0	95 ± 14	5·02	0·46
3 February	20	36·7	42·7	85 ± 10	5·62	0·62
12 February	17	34·0	33·8	100 ± 10	5·03	0·52
19 February	7	33·6	32·5	103 ± 15	3·86	0·64
22 February	12	26·1	27·1	96 ± 14	6·28	0·60
26 February	32	33·8	41·8	82 ± 7	6·86	0·49
29 February	7	23·3	27·7	84 ± 15	4·04	0·67
1 March	20	35·8	37·6	96 ± 10	4·04	0·57
2 March	33	39·4	40·0	99 ± 7	4·32	0·64
4 March	22	26·7	23·6	108 ± 9	3·84	0·55
8 March	4	30·1	26·5	111 ± 20	1·96	0·48
9 March	16	27·0	29·2	92 ± 10	5·30	0·50
10 March	13	30·5	30·2	101 ± 14	5·00	0·72
15 March	12	30·2	33·2	91 ± 14	4·98	0·55
17 March	9	36·0	35·2	102 ± 15	4·53	0·49
18 March	23	29·5	29·5	100 ± 9	3·89	0·47
22 March	14	34·7	42·0	83 ± 12	5·51	0·47
23 March	23	14·3	14·1	102 ± 9	3·57	0·45
24 March	14	30·2	30·0	101 ± 12	4·29	0·64
28 March	7	23·2	22·0	105 ± 15	4·12	0·52
1 April	10	14·8	14·2	104 ± 15	4·49	0·32
11 April	17	13·5	13·2	102 ± 10	–	0·55

The 'evaporation fraction' $\epsilon = \overline{E}/(\overline{E} + \overline{H})$, listed in Table 2, can be equated also to $\overline{E}/(\overline{R} - \overline{G})$. ϵ thus expresses the fraction of the available energy used for evaporation and emerges later as a most useful parameter.

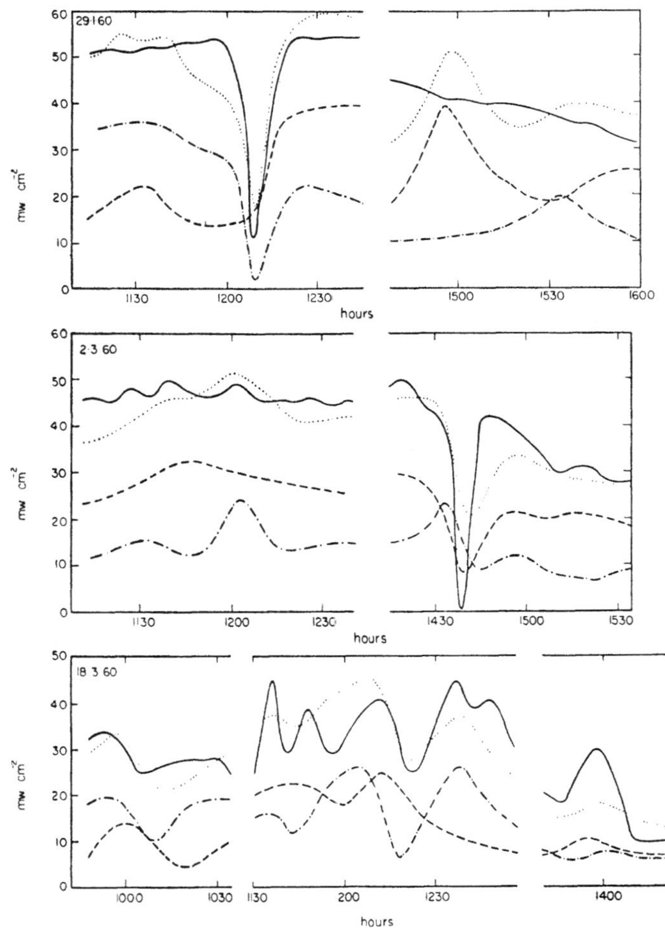

Figure 5. Typical comparisons of $E + H$, $R - G$, with E, H measured separately.
—— $R - G$; ····· $E + H$; ·—·—· H; - - - - E.

(c) *The response-time effect*

It is essential to the eddy-flux technique that the sensing elements should have a sufficiently fast response to record the smallest eddies contributing to the correlation. Both the anemometer and the wet-bulb resistance thermometer have a response time* of approximately 0·3 sec (Taylor 1958; McIlroy 1955).

Deacon (1959) and McIlroy (1960) have presented analyses of response-time limitations, and the various effects of horizontal windspeed $\bar{u}$, response time τ, and measuring height z, have been incorporated by plotting the recovery factor as a function of the dimensionless parameter $\tau\,\bar{u}/z$. Their results have been summarized by the dotted lines of Fig. 6 for the limiting atmospheric conditions of neutrality and high instability.

A comparison with the present results can be made by plotting the data of Tables 1 and 2 in the above manner, and then grouping and averaging in convenient classes of $\tau\,\bar{u}/z$. Most of the data reported here were obtained in conditions of moderate instability, and it is seen from Fig. 6 that these lie within the limiting curves found by Deacon and McIlroy. The low value of $\tau\,\bar{u}/z = 0{\cdot}15$ is not statistically significant but may be due to sluggish operation of the wind-alignment vane at very low wind speeds.

* The term 'response time' is used here to denote the time required for the fraction $1 - e^{-1}$ of a step-function change to be recorded.

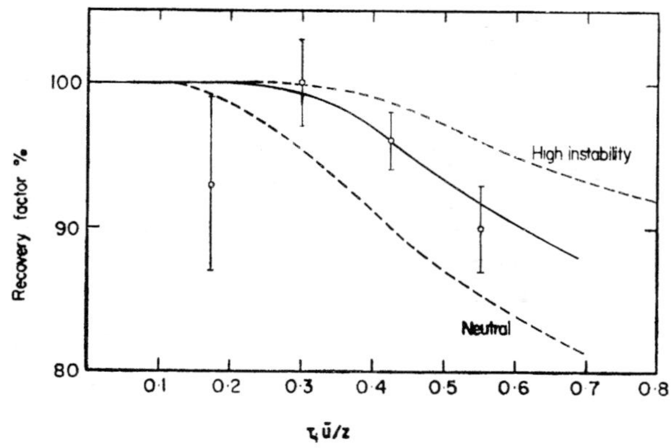

Figure 6. Recovery factor as a function of the parameter $\tau\, u/z$.

From a curve of this type an assessment of the instrumental requirements for a particular situation can be made. For example, with the present equipment operating at a measuring height of 4 m, the recovery factor would fall to 90 per cent at a windspeed of 8·4 m sec^{-1} under conditions of moderate instability.

4. Physical significance of recovery factor distribution

A careful examination of random errors in the radiation, ground-flux and eddy-correlation equipment indicated that these, at the most, would account for about one-third of the observed scatter of the recovery factor for 5-min runs. Of these, errors in the eddy flux equipment were the most significant.

Fluctuating errors of a few per cent would arise in steady conditions from the neglect of heat-storage changes in the 1 cm of soil above the flux plates, and with extremely variable insolation these could increase to as much as 40 per cent for a 5-min period.

The latter situation was, however, comparatively rare, and a significant contribution to the recovery factor distribution must therefore be sought in meteorological effects as well as instrumental.

To assess likely sources of error in determining the turbulent heat flux, Swinbank (1955) made a formal analysis of all terms entering into a detailed energy-balance equation, and concluded that in conditions of steady radiation a maximum error of 20 per cent could occur for a 5-min run as a result of the non-vanishing of the horizontal gradients $\partial\overline{T}/\partial x$ and $\partial\overline{\rho u}/\partial x$. (From the equation of continuity $\overline{\rho w} = 0$ if $\partial\overline{\rho u}/\partial x = 0$). In the case of evaporation, the gradients $\partial\overline{q}/\partial x$ and $\partial\overline{\rho u}/\partial x$, could likewise cause a maximum error of about 20 per cent.

On this basis a maximum discrepancy of 20 per cent might be expected in the simple energy-balance equation. This value must be increased in the present instance for three reasons :

(1) The height of observation (4 m) was greater than that considered by Swinbank (1 m).

(2) Measurements were made under all conditions of radiation.

(3) The observed horizontal gradients leading to the above error estimates were both space and time averages. (The accurate measurement of a horizontal gradient at a point is, of course, extremely difficult, if not impossible).

This interpretation indicates that for a measuring height of 4 m, discrepancies in excess of 20 per cent might arise more readily than Swinbank's analysis would suggest, and with some contribution from the neglect of heat storage a standard deviation of about the observed magnitude (31 per cent) might therefore be anticipated.

The quantities $\partial\overline{\rho u}/\partial x$, $\partial\overline{T}/\partial x$, and $\partial\overline{q}/\partial x$ will, during the course of a day, be variable in sign, and with reasonably uniform surfaces may be expected to average out to negligible values.

Evidently the atmosphere achieves horizontal uniformity by a process of continual adjustment on a scale which becomes apparent if short running times are used. For longer periods, the naturally imposed averaging ensures that the simple energy balance of Eq. (7) is valid.

It is relevant to assess the standard error of a flux measurement by the eddy-correlation method for different periods of observation (Table 3), making use of the recovery-factor distributions reported earlier. In practice, the accuracy will be somewhat better than this because of the neglect of heat-storage changes in the experimental data.

TABLE 3. Standard error of measurement of average eddy-flux at 4 m height for different periods of observation

Period of observation	Standard error (per cent)
5 min	31
10 min	22
30 min	13
1 hr	9
5 hr	4
10 hr	3

Over a period of one day, the average flux can be determined with a standard error of a few per cent. Where short-period results are required, these can be obtained at a somewhat reduced accuracy. In both cases the quality of the measurements would be better in steady-radiation conditions.

This assessment of the accuracy of the method by-passes any consideration of the response-time effect, which as discussed earlier, imposes a minor limitation at high wind speeds only.

5. The relation between G and H

Priestley (1959, p. 97) has discussed the theoretical aspects of sharing between two media of the heat supplied by radiation at the interface, and favours a relation of the form

$$G = \frac{\rho' c' \sqrt{\kappa}}{\rho C_p \sqrt{K}} . H \qquad . \quad . \quad . \quad . \qquad (8)$$

where ρ, ρ' refers to density; C_p, c' to specific heat, and K, κ to diffusivity. The prime denotes the underlying medium. The values of K and κ appropriate here are the bulk values at some distance from the surface.

Analysing the measurements of G and H made by Swinbank (1955), Priestley suggests that a value of K of 10^5 cm^2 sec^{-1} by day provides a suitable agreement between experiment and theory. The value of $\rho' c' \sqrt{\kappa} = 0{\cdot}034$ used for Swinbank's results would not differ greatly from that appropriate to the present data.

Plotting average values of G in ranges of H (Fig. 7), the present results suggest a value of K of 10^5 cm^2 sec^{-1}, for the near-neutral atmosphere, increasing to 10^6 cm^2 sec^{-1} for the very unstable case. Priestley (1959, p. 100) quotes corresponding K values of 10^5 cm^2 sec^{-1} increasing to 10^7 cm^2 sec^{-1}.

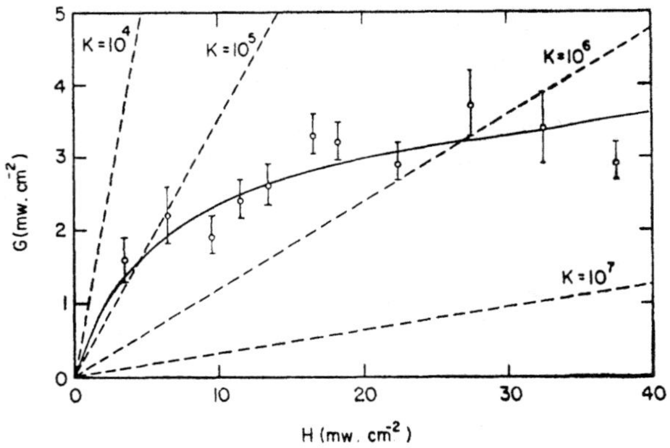

Figure 7. The relation between G and H.

The present results are obviously not universally applicable, and the detail of the $G - H$ relation would be dependent on such things as the type of soil and its moisture content, and the type of crop cover. The theoretical interpretation, nevertheless, would appear to provide a useful guide to the problem of heat sharing, particularly where variations in one parameter are under consideration.

6. Chronological association of precipitation and evaporation

It might be expected that the evaporation fraction ϵ will be chiefly determined by the availability of water for the evaporation process. For example, with rain following a dry period, ϵ would presumably increase and, in the absence of further rainfall, decline again gradually to its initial value. From day to day ϵ would obviously be a more conservative quantity than amount of evaporated water, since the latter would be more closely related to net radiation income, a highly variable quantity.

The data of Table 2 bear on this question, and in Fig. 8 ϵ is plotted in chronological sequence. For the period 20 February to 31 March there is good continuity of measurement, and some degree of correspondence with the expected behaviour of ϵ will be readily observed. The variation of ϵ outside this period cannot be so accurately delineated, but can be reasonably inferred.

Assuming that ϵ, as determined, can be applied to the whole day from sunrise to sunset, the daily totals of evaporation can be computed from known day-time totals of net radiation. G has been taken equal to $0{\cdot}08\,R$, the overall value found in the present experiment. Night-time evaporation (or condensation) is ignored.

The total evaporation thereby found for March is 2·46 in., whereas the total precipitation is only 0·33 in. Much of the water evaporated during this period must, therefore, have come from ground storage of earlier precipitation. Soil moisture profiles taken at the beginning and end of March showed little change down to a depth of 3 ft, suggesting that greater depths were involved. The grass roots are known to extend below this level.

With this method of computing daily evaporation, it is clear that areas under the curve of Fig. 8 can be expressed as amount of evaporated water, and the quantitative association of the peaks of the ϵ-time curve with the immediately preceding rainfall can be examined.

The precipitation data of Fig. 8 have been conveniently grouped (A, B, etc.) and associated areas accurately computed. The broken lines were then fitted to demarcate these areas under the appropriate peaks of the ϵ-time curve and, by trial, it was found possible to do this most readily with lines suggesting an approximately exponential decay of ϵ following a single rainfall. This interpretation of the ϵ-time curve is plausible in general physical terms, and is supported by the consistency of the pattern. The minor

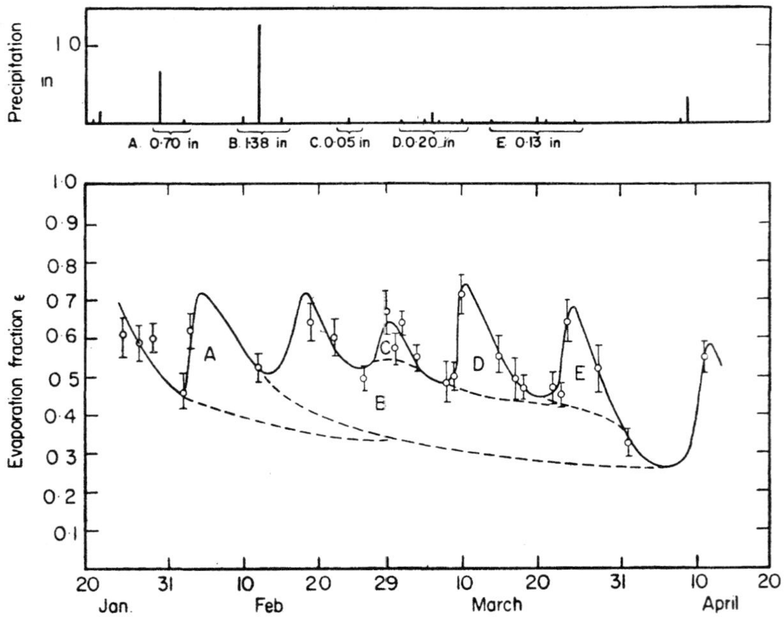

Figure 8. Evaporation fraction ε as a function of time.

discrepancy of group C is probably caused by an error in sampling, typical of such small amounts of rainfall. The detail of this analysis is clearly applicable to the present site alone, and an extension to other areas would require a knowledge of the effects of different types of soils and crops.

It is impracticable to attempt to correlate the fraction of the evaporation regarded as consequent on precipitation several months earlier. However, the long-period effects can be examined to a limited degree by noting that the average value of ϵ during February and March 1960 was 0·56.

Neglecting run-off, a long term (annual) value ϵ_A can be computed from the relationship : annual evaporation $= \epsilon_A \Sigma (R - G) =$ annual precipitation where, as before, R and G refer only to daylight hours.

The annual precipitation at Edithvale, Victoria is 26·1 in., and the annual value ΣR is 91 w hr cm^{-2}, i.e., equivalent to 52·2 in. Allowing for G as previously, we find $\epsilon_A = 0{\cdot}54$.

During the winter months ϵ would presumably increase because of the greater availability of water, but these higher values of ϵ would enter with very little weight into the determination of ϵ_A because of the much smaller radiation totals with which they are associated. Close agreement of ϵ_A with the experimental value found for late summer and early autumn might therefore be expected and is, in fact, obtained.

An interesting feature of Fig. 8, is that on three occasions (2 February, 8-9 March, 22-23 March) low values of ϵ persisted for up to three days following appreciable rainfall, and the maximum value of ϵ was not achieved until the fourth day. It has been suggested to the author that this can be ascribed to the accretion of suberin by the shallow roots during the dry period, and their delayed re-activation after rain. The slow diffusion of moisture down to the more active root systems is an alternative explanation.

7. Conclusions

The eddy-correlation technique has been demonstrated to be a practicable and accurate method for the measurement of evaporation and sensible-heat transfer over a reasonably uniform surface.

In conditions of near-neutrality, the accuracy of the technique is not limited by response-time effects over the range of wind speed normally encountered. With increasing atmospheric instability the useful wind-speed range is considerably extended and, unlike the aerodynamic method, optimum performance is obtained at maximum instability.

With a measuring height of 4 m as used in the present experiment, minor surface irregularities of up to several tens of centimetres are clearly of no consequence, and if sufficient fetch is available, greater irregularities could be readily accommodated by increasing the height and period of observation.

The usefulness of the eddy-correlation method to studies of the type described in the latter part of this paper is self-evident, and with this in mind, the development of portable equipment employing transistor circuitry has been undertaken. This latter instrument, called the 'Evapotron,' has performed satisfactorily in laboratory tests, and field trials will commence shortly.

Acknowledgments

Many members of this Division have made significant contributions to the study of flux measurement by the eddy-correlation technique, and the author has derived much benefit from his association with them. A large part of the instrumentation was carried out by Mr. R. R. McGregor. Miss S-A. Yeo and Miss Y. Mullins assisted with the operation of the eddy-flux instrument, and Mr. J. Stevenson was responsible for many of the radiation and ground-flux measurements. The successful prosecution of the project depended very much on the expert construction of the sensing elements by Mr. G. Harris, and on their maintenance and calibration by Mr. G. Grauze.

References

Deacon, E. L.	1950	*Quart. J. R. Met. Soc.*, **76**, p. 479.
	1955	*C.S.I.R.O. Div. of Met. Phys.*, Tech. Pap. No. 2.
	1959	*Advances in Geophysics*, **6**, p. 211.
Dyer, A. J.	1958	*J. Sci. Instr.*, **35**, p. 240.
	1960	*Ibid.*, **37**, p. 166.
House, G. J., Rider, N. E. and Tugwell, C. P.	1960	*Quart. J. R. Met. Soc.*, **86**, p. 215.
McIlroy, I. C.	1955	*Austr. J. Agric. Res.*, **6**, p. 196.
	1960	*C.S.I.R.O., Div. of Met. Phys.*, Tech. Pap. No. 11.
Pasquill, F.	1949	*Proc. Roy. Soc.* A., **198**, p. 116.
Priestley, C. H. B.	1959	*Turbulent transfer in the lower atmosphere*, Chicago Univ. Press.
Swinbank, W. C.	1951	*J. Met.*, **8**, p. 135.
	1955	*C.S.I.R.O., Div. of Met. Phys.*, Tech. Pap. No. 2.
Taylor, R. J. and Webb, E. K.	1955	*Ibid.*, Tech. Pap. No. 6.
Taylor, R. J. and Dyer, A. J.	1958	*Nature*, **181**, p. 408.
Taylor, R. J.	1958	*J. Sci. Instr.*, **35**, p. 47.
	1956	*Ibid.*, **33**, p. 436.

A NOTE ON ESTIMATING THE EFFECT OF A LIMITED FETCH ON MICROMETEOROLOGICAL EVAPORATION MEASUREMENTS

(Research Note)

J. H. C. GASH

Institute of Hydrology, Wallingford, Oxon, UK

(Received in final form 20 November, 1985)

Abstract. A formula for the effective fetch of micrometeorological evaporation measurements is derived by application of diffusion theory, using Calder's approximation of a uniform wind field and neutral atmospheric stability. This simplification allows estimation of the likely sampling error, which would result from an upwind step-change in evaporation rate, without change in roughness.

1. Introduction

The limitation of a finite fetch ensures that any micrometeorological evaporation measurement will have a sampling error. This error is a contribution to the measured flux made by water vapour which has diffused from vegetation upwind of the surface under study. With the simplifying assumption of no change in roughness at the upwind edge of the surface under study, this sampling error can be quantified by the application of diffusion theory. This approach was first used by Sutton (1934), and later by Dyer (1963) and Pasquill (1972). The analysis given here follows Dyer's, but instead of using Philip's (1959) diffusion equation, the approximate solution suggested by Calder (1952) is used. This solution replaces the power functions used to describe the wind and eddy diffusivity profiles, with the simpler, albeit less realistic, approximation of a uniform wind field. The analytical approach made possible by this approximation leads to a simple formula, which expresses the sampling error in terms of the measurement height, the surface roughness parameters, and the change in evaporation rate at the edge of the surface under study.

2. Theory

2.1. Diffusion from Evaporating Vegetation

An area of transpiring vegetation may be considered as a continuum of line sources of water vapour, each occupying an infinitesimal strip of width δx. The equivalent areal flux of evaporation, E, is given by $Q/\delta x$, where Q is the source strength of line source per unit length. For a sensor mounted at a height z_m above the zero plane displacement, water vapour diffusing from a distance x upwind will on average be sensed at

Boundary-Layer Meteorology **35** (1986) 409–413.

concentration $\rho_v(x, z_m)$. This concentration can be calculated from the basic diffusion equation

$$u \frac{\partial \rho_v(x, z_m)}{\partial z_m} = K_{z_m}(\partial \rho_v(x, z_m)/\partial z_m) \tag{1}$$

where u is the time-averaged horizontal wind velocity in the direction of the mean wind, and K_{z_m} the appropriate eddy diffusivity at height z_m. Applying Calder's (1952) solution for diffusion from an infinite cross-wind line source of passive particles gives

$$\rho_v(x, z_m) = \frac{Q}{ku_* x} e^{-Uz_m/ku_* x} \tag{2}$$

where k is von Karman's constant, u_* is friction velocity, and U is the uniform windspeed, assumed to apply over the whole neutrally buoyant boundary layer. Substituting $E = Q/\delta x$ into Equation (2), then letting δx approach zero and integrating gives

$$\rho_v(z_m) = \frac{E}{ku_*} \int_0^{x_L} \frac{1}{x} e^{-Uz_m/ku_*} \, dx \tag{3}$$

where $\rho_v(z_m)$ is the absolute humidity sensed at height z_m, which is a result of evaporation occurring between the sensor and a distance x_L away.

The vertical gradient of humidity at the height z_m is then given by

$$\frac{\partial \rho_v(z_m)}{\partial z_m} = \frac{E}{ku_*} \frac{\partial}{\partial z_m} \int_0^{x_L} \frac{1}{x} e^{-Uz_m/ku_* x} \, dx \, . \tag{4}$$

Since the derivative of the integral is mathematically the same as the integral of the derivative, we can rewrite Equation (4) as

$$\frac{\partial \rho_v(z_m)}{\partial z_m} = \frac{E}{ku_*} \int_0^{x_L} \frac{1}{x} \frac{\partial}{\partial z_m} e^{-Uz_m/ku_* x} \, dx \, . \tag{5}$$

Differentiating with respect to z_m and then integrating with respect to x gives

$$\frac{\partial \rho_v(z_m)}{\partial z_m} = -\frac{E}{ku_* z_m} e^{-Uz_m/ku_* x_L} \, . \tag{6}$$

As x_L approaches infinity, the exponent in Equation (6) approaches zero and

$$\frac{\partial \rho_v(z_m)}{\partial z_m} = -\frac{E}{ku_* z_m} \tag{7}$$

which is the flux-gradient equation for evaporation, with the eddy diffusivity for water vapour taken as $ku_* z_m$.

2.2. The Effective Fetch

Consider humidity measured at a point with an upwind fetch approaching infinity and a gradient given by Equation (7). A percentage of that gradient will be the result of molecules diffusing from within a distance x_F away from the point of measurement. Let us term that distance the F percent effective fetch. It will be given by the ratio of Equations (6) and (7), which on taking logarithms and rearranging gives

$$x_F = -\frac{Uz_m}{ku_*} 1/\ln(\mathrm{F}/100). \tag{8}$$

Before Equation (8) can be evaluated, a value is needed for the uniform windspeed, U, and the friction velocity, u_*. When comparing his equation with experimental results, Calder selected what he considered to be a typical value of U for the site (Porton Down) where the data were obtained. He chose a value of 5 m s^{-1} at a height of 2 m. As z_0 was known, he was able to calculate u_* from the logarithmic wind profile

$$u = \frac{u_*}{k} \ln(z/z_0). \tag{9}$$

In Equation (8), U and u_* are not required separately, only the ratio U/u_* need be known. For the present analysis, in order that the average velocity of a molecule diffusing from a source to the sensor is the same for the assumed uniform velocity field as it would be for the logarithmic profile, U is assumed to be the average wind velocity given by that profile between the surface and z_m, i.e.

$$U = \int_{z_0}^{z_m} u \, \mathrm{d}z \Big/ \int_{z_0}^{z_m} \mathrm{d}z. \tag{10}$$

Substituting for u from Equation (9) and integrating gives

$$U/u_* = (\ln(z_m/z_0) - 1 + z_0/z_m)k. \tag{11}$$

Substituting from Equation (11) into Equation (8) gives

$$x_F = -\frac{z_m(\ln(z_m/z_0) - 1 + z_0/z_m)}{k^2 \ln(F/100)}. \tag{12}$$

This allows calculation of x_F, the effective fetch, from a simple function of roughness length and measurement height.

2.3. The Effect of a Step Change in Evaporation

Consider a site which comprises a distance upwind of the sensors of length x_F; this is surface A, from which the rate of evaporation is E per unit area. Beyond x_F is surface

B, which stretches towards infinity and from which the rate of evaporation is CE per unit area. Then neglecting the constraint of making an energy balance, we assume neutral stability over both surfaces, which also have the same roughness. The ratio of the observed gradient, to that which would be observed from an infinite fetch of surface A can be calculated as follows.

The observed absolute humidity gradient is given by integrating Equation (5) between the limits zero to x_F, and then again between x_F and x_L to give

$$\frac{\partial \rho_v(z_m)}{\partial z_m} = -\frac{E}{ku_* z_m} \{[e^{-Uz_m/ku_* x}]_0^{x_F} + C[e^{-Uz_m/ku_* x}]_{x_F}^{x_L}\} . \tag{13}$$

Evaluating with x_L approaching infinity gives

$$\frac{\partial \rho_v(z_m)}{\partial z_m} = -\frac{E}{ku_* z_m} (C + (1 - C)\, e^{-Uz_m/ku_* x_F}) . \tag{14}$$

The ratio of this gradient to the gradient which would be observed from an infinite fetch of surface A (Equation (7)), is the fractional error which results from the limited fetch of surface A. In terms of evaporation the ratio of Equations (14) and (7) gives

$$\frac{E_{\text{meas}}}{E} = C + (1 - C) e^{-z_m(\ln(z_m/z_0) - 1 + z_0/z_m)/k^2 x_F} \tag{15}$$

whereas E_{meas} is the measured evaporation and U/u_* has been substituted from Equation (11).

3. Comparison with Dyer's Predictions

In his more rigorous treatment, Dyer used empirical power functions to describe the vertical variation of windspeed and eddy diffusivity. He set (after converting to SI units)

$$u = 1.88\,(z/0.01)^{1/7}$$

and

$$K = 3.20 \times 10^{-3}(z/0.01)^{6/7} .$$

Since in terms of the present model the eddy diffusivity is $ku_* z$ and the wind profile is given by Equation (9), if the two sets of equations are considered equivalent at the measurement height, then they can be solved to give implied values of z_0 for any of Dyer's values of z_m. These are given in Table I together with Dyer's calculation of the 90% effective fetch and the consequent fetch : height ratio. The predictions of Equation (12) are also given for those values of z_m and z_0. The predictions differ over a range from -17 to $+19\%$.

4. Concluding Remarks

If the arguments which lead from Equation (3) to Equation (11) are accepted, then the errors in the predictions will be the result of the approximations implicit in the use of Equation (3). In particular, the assumption of a constant windspeed and the choice of its value are open to question. It is difficult to assess the effect of these assumptions except by comparing the predictions with those made by more rigorous but less easily applicable methods. In that respect, the agreement between the present model and the results of Dyer are encouraging and suggest that the approximation of a uniform windspeed leads to predictions which are not likely to be greatly in error. For the practical purpose of estimating the upwind sample of a micrometeorological evaporation measurement and its likely sampling error, the present method is probably adequate. Table I suggests that the simplifying assumptions lead to errors of about 20% or less.

TABLE I

$z_m(m)$	$z_0(m)$	Dyer		Equation (12)	
		Fetch (m)	Fetch:height ratio	Fetch (m)	Fetch:height ratio
0.5	0.0244	70	140	58	116
1	0.0252	170	170	153	153
2	0.0225	420	210	395	198
5	0.0147	1350	270	1364	273
10	0.0082	3300	330	3449	345
20	0.0035	8100	405	8656	433
50	0.0006	26500	530	28982	560

The 90% effective fetch and the consequent fetch:height ratio as given by Dyer (1963) and as given by Equation (12) for the same measurement height and roughness length.

References

Calder, K. L.: 1952, *Some Recent British Work on the Problem of Diffusion in the Lower Atmosphere*, Proc. U.S. Tech. Conf. Air Poll., McGraw-Hill, New York,pp. 787–792.

Dyer, A. J.: 1963, 'The Adjustment of Profiles and Eddy Fluxes', *Quart. J. R. Meteorol. Soc.* **89**, 276–280.

Pasquill, F.: 1972, 'Some Aspects of Boundary-Layer Description', *Quart. J. R. Meteorol. Soc.* **98**, 469–494.

Philip, J. R.: 1959, 'The Theory of Advection', *J. Meteorol.* **16**, 535–547.

Sutton, O. G.: 1934, 'Wind Structure and Evaporation in a Turbulent Atmosphere', *Proc. Roy. Soc.* **A146**, 701–722.

9 Scintillometer measurements of sensible heat flux over heterogeneous surfaces

J-P. LAGOUARDE[1], K.J. McANENEY[2] and A.E. GREEN[2]
[1]INRA Bioclimatologie, Villenave d'Ornon, France
[2]Hort Research, Kerikeri Research Centre, Kerikeri, New Zealand

INTRODUCTION

The idea of using optical scintillations to infer the surface heat flux should not surprise anyone who has watched the air shimmering above a hot surface in summer. The meteorological variable of interest here is C_N^2 which denotes the strength of refractive turbulence. In turbulence theory, it is common to describe the spatial variability of atmospheric variables by their relevant structure function, in this case the dimensionless $D_N(r)$. For separation distances r in the inertial sub-range of scales about a point x,

$$D_N(r) = <[N(x) - N(x+r)]^2> = C_N^2 r^{2/3} \qquad l_0 << r << L_0$$

where N is refractive index, the angle brackets indicate a time or ensemble average, and l_0 and L_0 are the inner and outer scales respectively. The inner scale marks the transition from the inertial to viscous range of eddy sizes, whereas L_0 describes the size of the dominant inhomogeneities. In the near surface atmosphere l_0 is typically around 1 cm and L_0 is about half the measurement height (Weseley and Derzko, 1975). At visible and near infrared wavelengths, fluctuations in N are due mainly to turbulent variations in air temperature and thus are intimately related to the sensible heat flux. It is this liaison that we seek to exploit in order to quantify sensible heat through measurement of optical scintillations.

Early efforts to derive turbulence information from scintillation measurements of C_N^2 have been reviewed by Green *et al.* (1994) and McAneney *et al.* (1995). These studies also show how, by complementing C_N^2 with a measure of mean windspeed U, both sensible heat H and the momentum fluxes u_* may be derived by a simple iterative procedure. This approach has worked remarkably well above uniform surfaces whose aerodynamic characteristics (surface roughness and the zero plane displacement) are well-described as simple rule-of-thumb fractions of the vegetation height.

Scaling up in hydrology using remote sensing. Edited by J.B. Stewart, E.T. Engman, R.A. Feddes and Y. Kerr

Scintillometers using a point source of electromagnetic radiation saturate under high turbulence, a feature which limits the practical pathlength (L) of visible or near infrared laser scintillometers to around 100 m for near surface measurements. This study used a large aperture instrument (LAS) designed at the NOAA Wave Propagation Laboratory, Boulder, Colorado (Ochs and Cartwright, 1980) with the aim of maintaining linearity between C_N^2 and signal variance over a much larger range of refractive turbulence. McAneney *et al.* (1995) note its operational advantages of short averaging times to reduce uncertainty in turbulence statistics, insensitivity to the effects of flow distortion by obstacles near the transmitter or receiver, and the possibility of housing the instruments behind glass windows in order to achieve long-term monitoring of surface fluxes. However it is the promise of spatially-averaged fluxes at scales of the order of several kilometres that is our primary consideration here as we evaluate possibilities for spatial integration of surface fluxes. In particular the method offers a way of directly validating remote sensing estimates of surface fluxes at comparable spatial scales of resolution.

PHYSICAL PRINCIPLES

The LAS transmitter consists of an incoherent source of near-infrared radiation positioned at the focus of a concave mirror (diameter D=15 cm). Inhomogeneities in refractive index caused by eddy motions along the path distort the wavefront of the transmitted beam and the resulting variations in signal intensity are recorded at a receiver comprising an identical mirror and a photodiode detector. Further details can be found in McAneney *et al.* (1995) and the theoretical basis is given by Clifford *et al.* (1974) and Wang *et al.* (1978).

Turbulent distortions to a wavefront by eddy motions in the optical path may be regarded as a collection of diverging and converging lenses. For a point source of light under conditions of weakly-refracting turbulence, it follows from geometrical optics that the smallest disturbances (of the scale of l_0) are most effective in causing intensity fluctuations at the receiver (Tatarskii, 1993). The incoherent source used in the LAS can be considered as the sum of a large number of point sources; the diverging wavefronts from these will be distorted by eddies of scale l_0. However since $D >> F > l_0$, the effects of fine-scale turbulence are smoothed out over the receiver aperture: F is the first Fresnel zone ($F = (\lambda L)^{1/2}$, λ being the optical wavelength) and gives the scale of diffractional spreading. Thus for the LAS, the most effective eddies are those with a length similar to the aperture diameter and which by focussing or defocussing the entire beam can create a large change in the irradiance at the receiver.

Clifford *et al.* (1974) extended the weak turbulence theory of Tatarskii (1961) to show that the variance of the natural logarithm of the intensity incident at the receiver is

$$\sigma^2_{LnI} = 16\pi^2 k^2 L \int du \int dK K \phi_N(K) \sin^2(K^2 L u(1-u)/2k)[2J_1(x_1)2J_1(x_2)/(x_1 x_2)]^2 \qquad (2)$$

where the integration over u goes between 0 and 1, and the second over K, between 0 and infinity; u is the normalised path distance from the transmitter; $k = 2\pi/\lambda$ is the optical wavenumber; $x_1 = KD_R u/2$ and $x_2 = KD_T(1-u)/2$, where D_R and D_T are the receiver and transmitter apertures; K is the 3-dimensional spatial wavenumber; J_1 is a Bessel function of the first kind of order one; and Φ_N describes the turbulent medium in terms of its Fourier components K

$$\Phi_N(K) = 0.033 C_N^2 K^{-11/3} \qquad (L_o)^{-1} << K << (l_o)^{-1}$$

Assuming equal transmitter and receiver apertures and on the condition that

$$D/F > (0.124 k^{7/6} L^{11/6} <C_N^2>)^{3/5}$$

Wang *et al.* (1978) evaluate Equation 2 to obtain:

$$\sigma^2_{LnI} = 0.892 <C_N^2> D^{-7/3} L^3$$

where

$$<C_N^2> = \int C_N^2(u) W(u) du \tag{3}$$

Equation 3 is the usual calibration equation for LAS, with the weighting function W(u) calculated by numerical integration of Equation 2.

The experimental configurations of most interest in this study were chosen expressly to have C_N^2 varying along the optical path. For a two component surface, we have

$$< C_N^2> = (C_N^2)_1 W_1 + (C_N^2)_2 W_2 \tag{4}$$

where weighting functions W_1 and W_2 follow from Equations 2 and 3 and the subscripts refer to two surfaces. The main object of this paper was to test Equations 2, 3 and 4.

Figure 1 shows the normalised path-weighting function for the case of equal receiver and transmitting apertures. In the case of different sized apertures, Wang *et al.* (1978) give the following expression:

$$<C_N^2> = C\sigma^2_{LnI} (D_R/D_T)^{7/3} L^{-3} \tag{5}$$

where C is a calibration constant that is a function of the ratio D_R/D_T. Figure 1 shows how unequal receiver and transmitter diameters allow some coarse spatial filtering by shifting the position of maximum weighting towards the smaller aperture end.

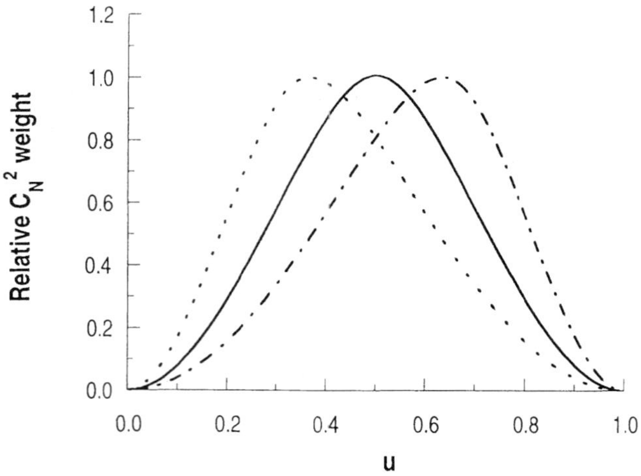

Figure 1 Path-weighting functions for equal apertures (0.15 m) of transmitter and receiver (solid line), $D_T/D_R = 2$ (-·-·-) and $D_T/D_R = 0.5$ (···). All calculations were done for L = 450 m.

FETCH CONSIDERATIONS

In addition to the spatial weighting along the optical path, it is also necessary to consider the source area contributing to the refractive turbulence responsible for scintillation. These turbulent eddies are representative of upwind sources and are advected down or across the optical path. Above a horizontally homogeneous surface this creates no problem but, in the case of changing terrain the situation is more complicated, with Brunet *et al.* (1994) finding that a transition between dry and irrigated vegetation introduced a step change in surface fluxes. Downwind of the transition the air flow gradually adjusts to the new boundary conditions. Calculations of the upwind source area affecting eddy correlation measurements have been proposed by Gash (1986) and Schuepp *et al.* (1990), amongst others, but the question remains as to whether the 'footprint' of a scintillometer is any different. In this respect we recall that a scintillometer is sensitive only to eddies with length scales comparable to the beam diameter. For these smaller eddies, the adjustment to changing surface conditions occurs rapidly but the greater response time or 'memory' of larger eddies means that the low frequency parts of the spectrum may never be completely in equilibrium with the underlying surface (Panofsky and Dutton, 1984). Given the restricted bandwidth of the scintillometer, and the iterative approach to calculation assuming Monin-Obukhov similarity, the flux estimations are based upon the presumption that the turbulence is fully adjusted to the underlying surface. Recalling the Brunet *et al.* (1994) result, we contend that under conditions of local advection, the scintillometer should be relatively insensitive to flux divergence between the surface and the measurement height and thus it provides a more locally representative measurement than eddy correlation measurements at the same height.

EXPERIMENTAL

The experimental site was in the middle of a large (4 km by 5 km) clearing in the forest of *Les Landes*, roughly 25 km south of Bordeaux. We chose large circular fields of maize of radius 350 and 600 m irrigated by central pivot and bordered by fallow fields of tall fescue. The experiment took place in three phases: the first was conducted over a single uniform field, the second dealt with averaging fluxes from two adjacent surfaces of roughly similar aerodynamic characteristics, while the third explored the influence of several factors pertaining to the theoretical weighting curves (Equations 2 - 5). The different configurations are shown in Figure 2 and the pertinent experimental details are summarised in Table 1.

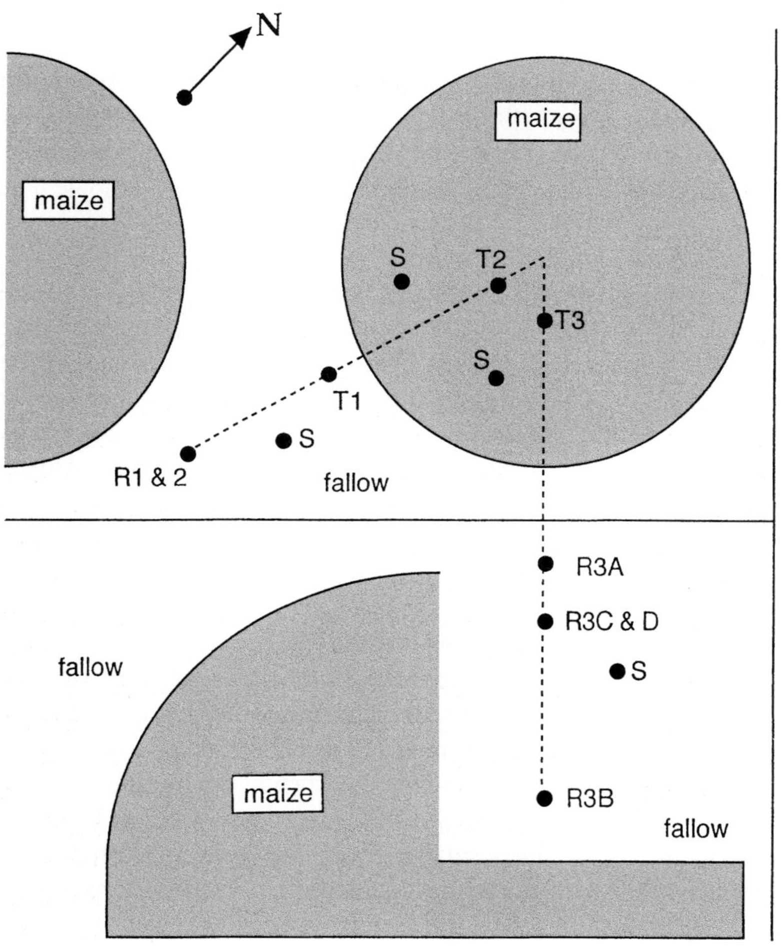

Figure 2 Schematic plan showing experimental layout of the various configurations. T1, 2 and 3 and R1, 2 and 3A, B and C refer to the transmitter and receiver positions summarised in Table 1. S indicates positions of the eddy correlation and micro-meteorological stations.

Table 1 Experimental configuration

Configuration	Surface	Pathlength (m)	D_{Trans} (m)	D_{Rec} (m)	$W_1(W_2)$
1	Fallow	260	0.15	0.15	
2	50% Fallow 50% Maize	600	0.15	0.15	0.50 (0.50)
3A	33% Fallow 67% Maize	450	0.15	0.15	0.84 (0.16)
3B	67% Fallow 33% Maize	870	0.15	0.15	0.21 (0.79)
3C	50% Fallow 50% Maize	600	0.15	0.075	0.69 (0.31)
3D	50% Fallow 50% Maize	600	0.075	0.15	0.35 (0.66)

Phase 1

On 31 May, 1994, a single day of measurements were made over a fallow field with a transmitter-receiver distance of 260 m and light easterly winds. The fescue sward was in its second year of fallow and comprised reasonably dense clumps of foliage 30 cm high occupying some 70% of the soil surface. Above 30 cm, stalks constituted a lower density canopy up to a height of 0.8 to 1.1 m. The scintillometer was mounted at a height z =2.5 m at the top of a stayed micrometeorological mast.

Phase 2

Six days of measurement were undertaken between the 5th and 13th of June over a path crossing 300 m of fallow and 300 m of irrigated maize. Rows of maize were orientated NE-SW and spaced at 80 cm. During this measurement period, the maize grew rapidly from 45 to 80 cm high and by the final date of measurement had obscured almost all the inter-row soil surface. The scintillometer was again mounted at 2.5 m. Irrigation began on June 12 providing a good contrast in soil water availability for the last day of measurement (day of year 164). Prior to this date, occasional heavy rains mitigated expected contrasts in plant water status.

Phase 3

From July 27 to August 3, the trajectory crossed two very different surfaces. By this time the maize had attained a total height of 3.3 m and the scintillation measurement height z was raised to 5.6 m. The fescue field was in its first year of fallow with foliage to a maximum height of 30 cm and bare soil estimated to occupy some 30 to 40% of the surface. Several different experimental configurations were tested over trajectories across 273 m of maize, variable portions of dryland fallow ranging between 150 m and 570 m, and a small (27 m) interconnecting region comprising a partially water-filled ditch and pasture maintained in a well-watered condition by spray drift. In the analysis that follows, the well-watered pasture and ditch were not distinguished from the maize which was assumed to extend 300 m.

In addition to the scintillation measurements, sensible heat flux was measured independently over each surface by eddy correlation using a one-dimensional sonic anemometer and fine thermocouple (Campbell Scientific Inc., Utah, USA). Eddy correlation instruments were sited after considerations of fetch as calculated by Schuepp *et al.* (1990). Net radiation, soil heat flux and other supporting measurements, such as wet and dry-bulb air temperatures and mean windspeed, were also measured at these reference flux positions. During phases 1 and 2, air temperatures and windspeed were measured at two heights to enable backup flux estimations to be made by a simplified aerodynamic method (Panofsky and Dutton 1984). In phase 3, these variables were only monitored at a single height: windspeed at the height of the scintillometer path and estimates of surface roughness (z_0) and zero plane displacement (d_0) are necessary prerequisites to estimating the sensible heat and momentum fluxes by the procedure in Appendix A.

All data were recorded and processed on-line using portable data loggers (Campbell Scientific Inc., Utah) as described by Green *et al.* (1994) and McAneney *et al.* (1995). On occasions rapid (20 Hz) sampling of the vertical windspeed signal from the sonic anemometer was undertaken to allow estimates of surface shear stress (u_*) by the spectral dissipation method (see Appendix B).

RESULTS AND DISCUSSION

Phase 1

This first phase of measurements accorded results in good agreement with the experience already obtained over uniform surfaces in New Zealand (data not shown). Comparison of the eddy correlation measurements of H with estimates from the simplified aerodynamic method established d_0 as 0.42 m for the tall fescue. Using this figure in conjunction with spectral measurements and the scintillometer-derived estimate of the Richardson number (ξ), we determined z_0 to be 0.10 m. This procedure was repeated at the completion of Phase 2 with identical results.

Phase 2

During these measurements, the maize grew rapidly, with attendant changes in its aerodynamic characteristics. The zero plane displacement was considered to be 0.65 of the measured crop heights giving values that varied from 35 cm on day 156 to 50 cm on day 164. These figures were then used to calculate z_0 from spectral measurements carried out on days 156 (z_0 = 5 cm), 161 (z_0 = 7 cm) and 164 (z_0 = 4 cm), as described in Appendix B. Values for other days were obtained by interpolating from these figures: the decrease observed on day 164 is consistent with the eventual closure of the inter-row space and the attainment of a relatively smooth aerodynamic surface. Wind directions were variable, but mostly between 30° and 90° to the LAS axis. On day 158 the wind blew directly down the rows of maize.

With the optical path crossing equal portions of maize and grass, weighting functions W_1 and W_2 should each have been 0.5. Figure 3 plots the scintillometer measure of $\langle C_N^2 \rangle$ against the arithmetic average of the C_N^2 for each field calculated from the measured values of H and U and using Equations A1 to A5. The accord is satisfactory with a slope of 1.11 (± 0.029) and a residual standard deviation of 0.09

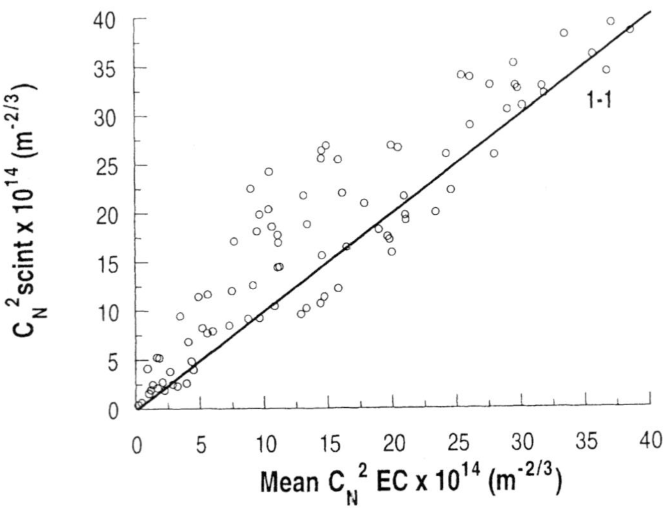

Figure 3 Comparison of scintillometer measurements of $<C_N^2>$ against the mean of the micro meteorological estimates for each field

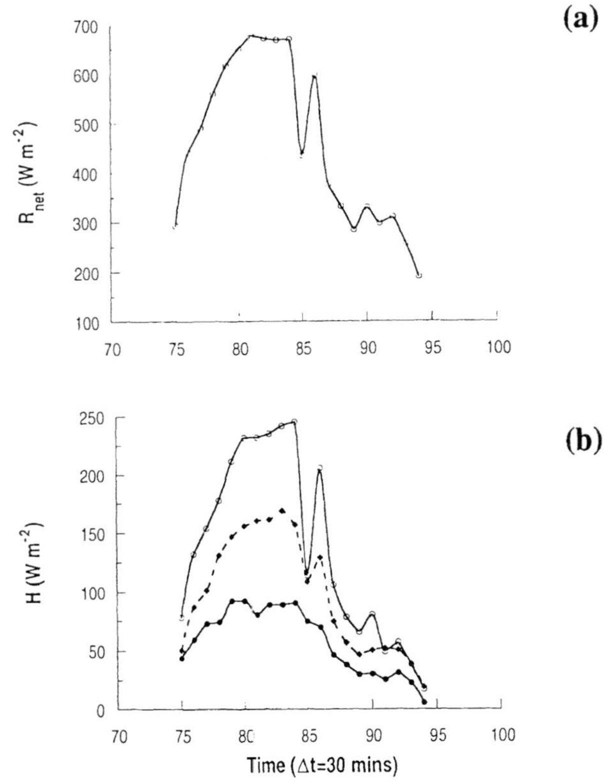

Figure 4 Time series for day 164 of (a) net radiation, and (b) sensible heat derived from the scintillometer (solid diamonds) and direct measurement by eddy correlation in the maize (solid circles) and fallow (open circles) fields. All data points represent 30-minute flux measurements.

x 10^{-14} ($m^{-2/3}$). Eliminating day 158, a day when the wind blew parallel to the rows of maize and when a lower z_0 may well have been more appropriate than the interpolated value, improves the slope to 1.06 (±0.020). This suggests that much of the scatter between days is the result of an imperfect knowledge of z_0 and d_0 rather than because of inaccuracies in the directly measured value of $<C_N^2>$ by scintillation.

The only day with a significant between-field contrast in sensible heat flux was day 164 following overnight irrigation of the maize. On this day, despite having very different canopy architectures, the net aerodynamic characteristics — $\ln((z-d_0)/z_0)$ and hence U — were very similar for the two fields. Thus by means of irrigation, we had effectively imposed a large contrast in water availability upon a more-or-less uniformly rough surface. As can be seen in Figure 4, the scintillometer provides a plausible average of the eddy correlation measurements of sensible heat and all measurements faithfully follow the contemporaneous behaviour of net radiation.

Collecting all available data, Figure 5 illustrates very satisfactory agreement with regression analysis giving

$$H_{scint} = 1.12\ (\pm\ 0.020)H(\text{weighted EC})$$

with a residual standard deviation of 26.7 W m^{-2}, a figure not too different from the value we might expect if comparing two eddy correlation systems at different locations in the same field or at different heights in a constant flux layer. Excluding day 158 improves the slope to 1.03. More importantly, data from day 164 are all scattered about the 1 : 1 line with little bias.

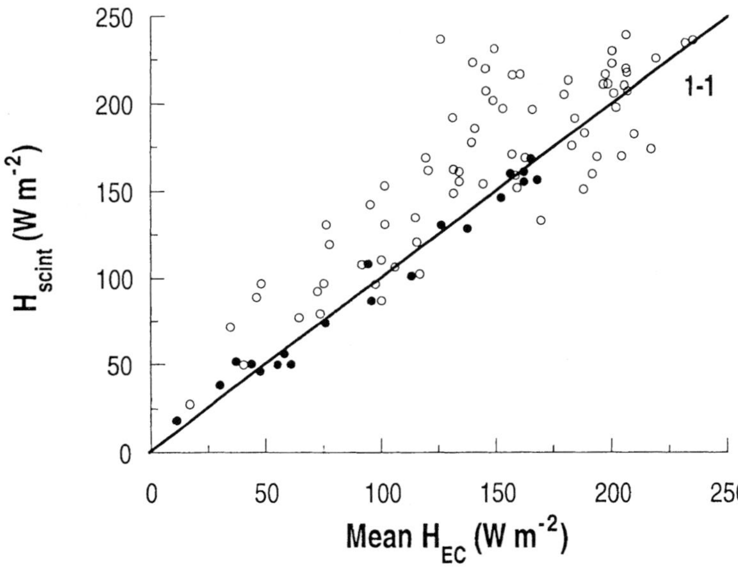

Figure 5 Comparison of sensible heat flux derived from scintillometery against the mean of the eddy correlation values. Day 164 is represented by the solid circles.

Phase 3 Our purpose here was to test Equations 2 - 5 by varying proportions of the trajectory traversing surfaces which differed significantly in both surface roughness and surface energy budget. Two days were also devoted to testing the influence of varying the ratio of the receiver and transmitter apertures. Winds were from the north-east, perpendicular to the path, so that local advection down the optical path was not of itself a concern. Nevertheless sensible heat fluxes were sometimes downwards over the maize. This was most common in the late afternoon on the second day after irrigation, when a combination of large surface roughness, high saturation deficits, bountiful soil water and a falling radiation load predisposed the crop towards such stable conditions (McAneney & Itier 1996). Over the fallow field, daytime conditions were always unstable. The changing sign of the sensible heat along the path poses problems when attempting to calculate a path-averaged heat flux from the scintillometer; we have stopped short of attempting this analysis. The previous success of the scintillometer in Phases 1 and 2 is taken to indicate that the instrument itself responds faithfully to $<C_N^2>$, with a symmetric weighting function when transmitter and receiver apertures are equal. The relevant theoretical weighting functions for the different experimental configurations are listed in Table 1.

The results are summarised in Figure 6. Regression using all the data gives

$$C_N^2 \text{ (weighted EC)} = 1.02(\pm 0.031) C_N^2 \text{ (scint)}$$

with a residual standard deviation of 1.263 x 10^{-14} ($m^{-2/3}$) and a statistically non-significant intercept. While this global result is gratifying, there is a tendency for the weighted C_N^2, as determined by 'back-calculation' from the eddy correlation

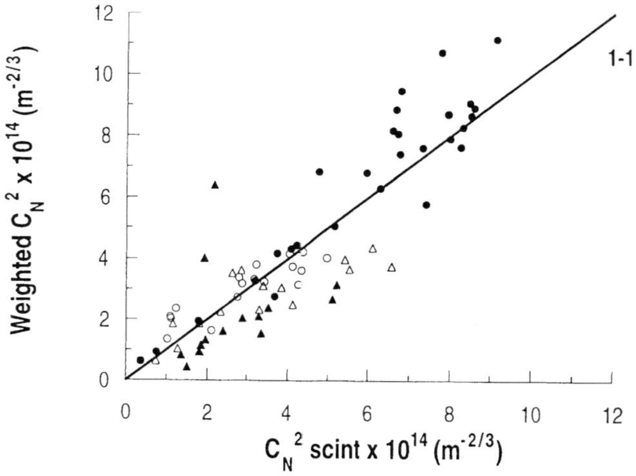

Figure 6 Weighted average of the micro-meteorological estimates of $<C_N^2>$ versus from the scintillometer. Open circles represent configuration 3A (450 m), solid circles 3B (870 = m), and open and closed triangles 3C and 3D.

measurements, to underestimate the scintillometer response when aperture diameters were unequal. The reverse seems true for the longest pathlength configuration. Two factors may explain those apparent deviations from theory. In the former case the apertures were stopped down with a plastic cap having an off-centred circular hole of diameter 7.5 cm. The hole was off-centred in order to minimise blockage by either the central source or the detector mounting assemblies (~2.5 cm diameter) but obviously did not eliminate the effect entirely. The effective D (Equation 3) is somewhat smaller than 7.5 cm because of this, and/or because of an unequal radiance pattern of the light emitting diode (Kohseik, 1987). For the long pathlength case, there is some small uncertainty as to the true path length — 850 m or 900 m — and the authors have negotiated a middle path: the shorter distance would bring the two independent sets of measurements almost perfectly into line. There is an important lesson here and we have since adopted the precaution of a surveyor's range-finder to determine the pathlength accurately.

CONCLUDING REMARKS

The collective results of tests conducted using different path lengths and aperture diameters above surfaces of varying aerodynamic properties and water budgets leads us to conclude that the scintillometer indeed provides a sensitive measure of the path-weighted C_N^2 as suggested by theory. We stress the importance of knowing path-length accurately, as the signal variance varies as the cube of this distance. When the underlying surfaces have similar roughness, a path-averaged sensible heat flux can be extracted without difficulty. This is more problematic should the direction of the sensible heat flux change sign along the path because the scintillometer cannot distinguish between refractive turbulence created under stable or unstable conditions. These problems, plus those created by large contrasts in surface roughness, will be scrutinised in subsequent studies.

ACKNOWLEDGMENTS

The authors gratefully acknowledge the help of Alistair Hall (Hort Research, Palmerston North) in matters theoretical and heretical.

REFERENCES

Brunet, Y., Itier, B., McAneney, K.J. and Lagouarde, J.P. 1994. Downward evolution of scalar fluxes and surface resistance under conditions of local advection. Part II: measurements over barley. *Agric. For. Meteorol.*, **71**,227-245.

Clifford, S.F., Ochs, G.R. and Lawrence, R.S. 1974. Saturation of optical scintillation by strong turbulence. *J. Opt. Soc. Am.*, **64**, 148-154.

Fairall, C.W., Edson, J.B., Larsen, S.E. and Mestayer, P.G. 1990. Inertial dissipation air-sea flux measurements: a prototype system using realtime spectral computations. *J. Atmos. and Ocean Techn.*, **7**, 425-453.

Gash, J.H.C. 1986. A note on estimating the effect of limited fetch on micrometeorological evaporation measurements. *Boundary-layer Meteorol.*, **35**, 409-413.

Green, A.E., McAneney, K.J., Astill, M.S. 1994. Surface layer scintillation measurements of daytime heat and momentum fluxes. *Boundary-layer Meteorol.*, **68,** 357-373.

Kaimal, J.H.C., Wyngaard, J.C., Izumi, Y. and Coté, O.R. 1972. Spectral characteristics of surface layer turbulence. *Quart. J. Roy. Met. Soc.*, **98.** 563-589.

Kohseik, W. 1987. A 15 cm aperture LED scintillomer for C_N^2 and crosswind measurements. KNMI Scientific Report WR 87-3, KNMI, de Bilt, The Netherlands.

McAneney, K.J., Green, A.E. and Astill, M.S. 1995. Large-aperture scintillometry: the homogeneous case. *Agric. For. Met.*, **76**,149-162.

McAneney, K.J. and Itier, B. 1996. Operational limits to the Priestley and Taylor formula. *Irrig. Sci.* In press.

Ochs, G.R. and Cartwright, W.D. 1980. *Optical system model IV for space-averaged wind and C_N^2 measurements.* NOAA Tech. Memo. ERL WPL-52, NOAA Environmental Research Laboratories, Bounder CO.

Panofsky, H.A and Dutton, J.A. 1984.*Atmospheric Turbulence: Models and Methods for Engineering Applications.* John Wiley & Sons, Inc. NY, 397pp.

Schuepp, P.H., Leclerc, M.Y., Macpherson, J.I. and Desjardins, R.L. 1990. Footprint predictions of scalar fluxes from analytical solutions of the diffusion equation. *Boundary-layer Meteorol.*, **50**, 355-373.

Tatarskii, V.I. 1961. *Wave propagation in turbulent medium.* McGraw-Hill, New York.

Tatarskii, V.I. 1993. Review of scintillation phenomena, pp. 2-15. In: Wave Propagation in Random Media (Scintillation) V.I. Tatarskii, A. Ishimaru and V.U. Zavorotny (eds.) Co published by SPIE - The International Soc. for Optical Engineering and Institute of Physics Publishing.

Wang, Ting-i, Ochs, G.R. and Clifford, S.F. 1978. A saturation-resistant optical scintillometer to measure C_N^2. *J. Opt. Soc. Am.*, **68**, 334-338.

Weseley, M.L. and Derrzko, Z.I. 1975. Atmospheric turbulence parameters from visual resolution. *Appl. Optics,* 14 847-853.

Weseley, M.L. 1976. The combined effect of temperature and humidity fluctuations on refractive index. *J. Appl. Meteorol.* 15 43-49.

APPENDIX A

In what follows we describe the estimation of H and u* from optical measurements of C_N^2 and U. Exactly the same equations can be used to calculate C_N^2 from U and H. Both forms of calculations were used for analyses discussed in the text.

For scintillation at visible or near-infrared wavelengths, the temperature structure parameter C_T^2 is related to C_N^2 by:

$$C_T^2 = C_N^2(T^2/\gamma P)^2(1+0.03/\beta)^2 \quad \text{(A1)}$$

where T is the average air temperature (K), P the atmospheric pressure (Pa), the refractive index parameter γ is 7.9×10^{-7} K Pa^{-1}, and β is the Bowen ratio (Weseley, 1976).

Since the sensible heat and momentum fluxes together determine atmospheric stability, which in turn influences turbulent transport, an iterative procedure is necessary to calculate fluxes from C_T^2 and U. Following Panofsky and Dutton (1984), we begin by defining a temperature scale:

$$T_* = -H/\rho c_p u_* \tag{A2}$$

where ρ is the density of air and c_p is the heat capacity at constant pressure. C_T^2 is written in dimensionless form and assumed to be a universal function of ξ (= z/L_{mo}, where L_{mo} is the Monin-Obukhov length)

$$C_T^2/T_*^2 z^{-2/3} = f(\xi)$$

where

$$f(\xi) = 4.9(1 + 7|\xi|)^{-2/3} \qquad \xi \leq 0$$

or

$$f(\xi) = 4.9(1+2.5\xi) \qquad \xi > 0 \tag{A3}$$

and

$$\xi = z/L_{mo} = (z-d_o)g\kappa T_*(1+0.07/\beta)/u_*^2T \tag{A4}$$

where von Karman's constant κ is taken as 0.4 and g is the acceleration due to gravity.

The vertical profile of mean wind velocity is given by

$$u_* = \kappa U/[\ln((z - d_o)/z_o) - ¥_m(\xi)] \tag{A5}$$

where $¥_m$ the appropriate integrated stability correction. For uniform vegetative surfaces, z_o and d_o can be taken as simple fractions of the vegetation height (h_c):

$$d_o = 0.65h_c \qquad \text{and} \qquad z_o \approx 0.13h_c \tag{A6}$$

A5 then allows u_* to be determined given a measurement of U, estimates of z_o and d_o and an initial guess for ξ. C_T^2 is calculated from the measured optical signal (A1) and an assumed initial value for β. T* follows from (A3) and H from (A2). This leads to a second approximation of ξ from (A4). The procedure is repeated until the estimated and calculated values for ξ converge. For stable conditions, the Bowen ratio correction in A4 was ignored to ensure convergence. The calculations adapt easily to a computer-based spreadsheet.

APPENDIX B

Here we describe the spectral dissipation method for determining u_*. After Panofsky and Dutton (1984), spectral densities of the vertical windspeed $S_w(f)$ in the inertial range of frequencies (f) are written as

$$fS_w(f)/u_*^2 = 0.2\phi_\varepsilon(\xi)^{2/3}\kappa^{-2/3}(f(z-d_o)/U)^{-2/3} \quad \text{(B1)}$$

where for unstable air

$$\phi_\varepsilon(\xi) = 1 - \xi$$

and $S_w(f)$ has units of velocity squared per unit frequency; all other units have been defined previously.

The $S_w(f)$ data were then plotted against frequency on a log-log graph, a –5/3 line fitted by eye through the data and $S_w(f)$ determined at the point where f = 2U. (Kaimal *et al.* (1972) and Fairhall *et al.* (1990) expect inertial subrange behaviour provided that f exceeds 2U/z.) Shear stress follows after substituting for $S_w(f)$ in B1 and assuming a value for d_o based on plant height (see Equation A6). Finally, Equation A5 was solved to give a value of z_o to consistent with the above choice of d_o. Clearly z_o and d_o are not independent but this is not a drawback for this particular application. The iterative calculations described in Appendix A provided the estimate of β.

C Atmospheric Control on the Evaporation Process – Commentary

It is self-evident that atmospheric conditions provide an important control on the rate at which water leaves the land surface and enters the atmosphere as water vapour: moist surfaces dry more quickly on sunny, windy days when the air is dry. In this section we select benchmark papers that trace key steps in evolving the understanding of these meteorological controls on the evaporation process when it is assumed that water is "readily available" at the evaporating surface, i.e. when the details of the surface, including whether it is open water, bare soil, or vegetation-covered, are considered to be of secondary importance. The first paper introduces the important concept that evaporation into the atmosphere is a diffusive process, while the second is significant because it introduced the concept of a meteorologically determined potential rate of evaporation from natural surfaces. The fact that evaporation needs energy and is part of a surface energy balance becomes explicit in paper C3, while in C4 it was noticed that in practice, if the horizontal advection of energy is not strong, radiation and surface temperature are often the most important controls on evaporation rate. The final paper is important because, by introducing the two-way interaction between the land surface and the overlying atmosphere, it explained why early ideas on evaporation, including the concept of a potential rate, had proved so robust.

C1 The fundamental equation of evaporation

Dalton, J. (1802) On evaporation.

Understanding of the role of evaporation as a component of the hydrological cycle has been developing since ancient times (Brutsaert, 1982), with the first quantitative measurements of evaporation generally being attributed Edmund Halley in the 17th Century (see Dooge, 1974; Brutsaert, 2005). However, it was John Dalton, a scientist living in the north of England between 1766 and 1844, who was the first to understand the fundamentals of the process of evaporation. Dalton realised that water vapour existed in the atmosphere as a separate entity and that the amount of water vapour the atmosphere could hold (i.e. the saturated vapour pressure) increased with temperature. He measured this temperature dependence and then devised an experiment to quantify the factors controlling evaporation. Evaporation from small canisters of water was measured under different conditions of temperature and ventilation. The experiment led him to describe evaporation as a process driven by a difference in humidity, with a constant of proportionality as an unspecified function of wind speed. In Dalton's words: "*The evaporating force must be universally equal to that of the temperature of the water, diminished by that already in the atmosphere*". Expressing this as an equation would give evaporation, E, as:

$$E = K\,(e_s - e_a)$$

where K is a windspeed-dependent coefficient, e_s is the saturated vapour pressure at the water temperature, and e_a is the vapour pressure of the air above. This basic equation eventually (Brutsaert, 1982) led to the formulation adopted by Penman (see C3) and still forms the basis of modern physically-based models of evaporation.

Oliver & Oliver (2003) and Oliver (2005) give excellent accounts of John Dalton's life and career, which covered a wide range of scientific interests, including his best known contribution to physical chemistry through Dalton's Law of Partial Pressures.

C2 The concept of potential evaporation

Thornthwaite, C. W. (1948) An approach toward a rational classification of climate.

Thornwaite recognised that precipitation data alone could not give a meaningful classification of climate; we also need to know whether the precipitation is greater or less than the water needed for evaporation. Only when this balance between precipitation and evaporation is known can the water available as a hydrological resource be estimated. This awareness led Thornthwaite to develop the concept of *potential evaporation* or *potential evapotranspiration* as he prefers: the evaporation that "*would be possible under ideal conditions of soil moisture and vegetation*". The paper anticipates and addresses many of the criticisms that were subsequently levelled against this concept: the difference between actual and potential evapotranspiration, their complementary relationship and the need to include biological control in estimating actual evapotranspiration.

Thornwaite's formula for estimating potential evapotranspiration is based on temperature and is empirical. Nevertheless it is derived with an understanding of the underlying process and the interrelationships between the factors controlling evaporation. Importantly, it rests on the biological observation that plant growth is strongly correlated to both temperature and transpiration.

As the title implies, the objective of the paper was to derive a methodology for classifying climate and much of the paper is given over to that. While this concern might appear archaic in the context of current research, the final paragraphs of the paper are a reminder that the need to understand the links between climate and vegetation is still very much with us. Predicting the impacts of climate change on vegetation distribution, and the subsequent feedbacks with climate, is very much a current priority for climate modellers. This is still a relevant and thought-provoking paper.

C3 The combination equation

Penman, H. L. (1948) Natural evaporation from open water, bare soil and grass.

Howard Penman was a physicist working at the Rothamsted agricultural research station in Britain. The Penman formula, which is derived in this paper, is sometimes cited (e.g. by Rodda, 2006) as the single most important contribution made to hydrology by a British scientist. There is no doubt that it has been a major influence on a generation of hydrologists.

Penman's formula combines the standard Energy Budget–Bowen Ratio equation with the mass transfer equation (see C1), but introduces the simplification of linearising the slope of the saturation vapour pressure *versus* temperature curve (Δ), over the range between the surface temperature and the air temperature at screen height. Through this simple approximation, Penman could implicitly estimate the surface temperature and the surface vapour pressure from their values measured at screen height, and the value of Δ, which is a known function of temperature. The evaporation can then be estimated from the variables measured at a standard agrometeorological observing station.

The measurements which Penman had available to derive the constants for his equation were far from ideal, consisting basically of an evaporation tank and lysimeters with bare soil and short grass. Nevertheless, the data allowed factors to be derived which modified the estimate of open water evaporation to give the potential evaporation from short grass, well-supplied with water. The Penman formula has been extensively adopted by water professionals, providing a practical tool to both water resource and irrigation engineers. For some 20 years it formed the basis of the FAO (UN Food and Agriculture Organization) standard method for estimating evaporation from irrigated crops (Doorenbos & Pruitt, 1975) and is still widely used today. A whole literature developed in refining the equation and deriving crop factors to modify the "Penman evaporation" for different crops, at different stages of growth. Thom & Oliver (1977) examined the implicit assumptions in Penman's treatment of the aerodynamic resistance to vapour transport, and the effect of the lack of a soil heat flux term on the seasonal distribution of evaporation. They give a well-argued analysis of why, despite it containing some empiricism, the equation has proved to be so remarkably robust.

C4 The approach to equilibrium evaporation

Priestley, C. H. B. & Taylor, R. J. (1972) On the assessment of surface heat flux and evaporation using large scale parameters.

Priestley and Taylor were motivated by a desire to provide simple daily estimates of surface fluxes appropriate for large-scale dynamical computations. They argued that, under these large-scale conditions, atmospheric advection effects are least important and that the evaporation should be related to the rate which would, in theory, occur when the atmosphere is in equilibrium with a large evaporating surface. The most relevant *necessary* parameters are then surface net radiation, R_n, and the rate of change of saturated specific humidity with temperature, s, a function of air temperature. Priestley and Taylor made, in the words of the paper itself, the "*first attempt to probe, in a practical way, the extent to which the same .... parameters might also prove* <u>*sufficient*</u> *for the purpose in hand*", i.e. estimating the evaporation from real landscapes.

We therefore select Priestley and Taylor's paper as a benchmark, not only because it provided new insight into the feedback between the land surface and the atmosphere, but also because it produced a practical evaporation formula, which despite its simplicity, has proved to be remarkably useful. Where water is not limiting, their equation gives evaporation, λE_{PT} as:

$$\lambda E_{PT} = 1.26 \frac{s}{s+\gamma} R_n$$

where $\gamma = C_p/\lambda$, the ratio of the specific heat of air and the latent heat of vaporisation. The second *advection term* in the Penman formula (see C3) is thus replaced by a fixed fraction of the first *radiation term*. The fraction (0.26) was derived empirically using data from five oceanic/lake studies and four land surface studies where water was not limiting. This minimalistic simplification is inherently attractive, but also practically useful because there is then no requirement for measurements of atmospheric humidity or windspeed. For this reason, the equation became popular with hydrologists and irrigation engineers as a simple way to estimate potential evaporation and potential evapotranspiration. Subsequent modelling experiments, which are described in the commentary on the next paper, helped to explain how the Priestley-Taylor equation (and also the Penman formula) could be used to estimate evaporation when applied in humid conditions, in situations with limited local advection, and for vegetation with surface resistances typical of those for agricultural crops. Later, as remotely sensed data became more widely used, the applicability of the Priestley-Taylor equation was further enhanced because surface radiation and surface temperature are more amenable to estimation from remotely sensed data than are other climate variables.

C5 Evaporation into the convective boundary layer

de Bruin, H. A. R. (1983) A model of the Priestley-Taylor parameter, α.

In the 1970s and 1980s, studies made using comparatively simple coupled models led to much improved understanding of the interaction between the overlying atmospheric boundary layer and evaporating land surfaces with uniform land cover. Important papers in this process include those by McNaughton (1976), de Bruin (1989), McNaughton & Spriggs (1986, 1989) and Huntingford & Monteith (1998). We select the 1983 paper by Henk de Bruin as the benchmark representing this field because of the insight it provides into the process of evaporation: it was the first to explain why the concept of potential evapotranspiration, and particularly the Priestley-Taylor equation, can successfully estimate the rate of evaporation for a range of atmospheric and surface conditions.

De Bruin's paper identifies the fundamental reason behind the success of these equations and the role of the dynamics of the convective boundary layer (CBL). During the day, water is evaporating into a CBL that is partially, but not totally, contained by a stable inversion aloft. The CBL underlies a free atmosphere that is usually drier and (in terms of potential temperature) cooler than the CBL. Air is entrained, from the drier, cooler free atmosphere into the mixed layer at a rate that depends on the strength of the temperature gradient through the inversion, but also on the surface sensible heat flux (strictly, buoyancy flux). The mixed layer grows and the warmer, drier air entrained at the top of the mixed layer is moistened by surface evaporation. As a result, the specific humidity of the mixed layer is fairly constant and changes only slowly through the day. However, the temperature of the mixed layer, and therefore the vapour pressure deficit, rises allowing evaporation to exceed equilibrium evaporation rate, depending on the surface resistance of the underlying surface. In practice, de Bruin's analysis showed that, for a range of surface resistances typical of short agricultural crops and for start-of-day atmospheric profiles found in the humid mid-latitudes, the daily average evaporation rate can be approximated by the estimates given by both the Penman formula (paper C3) and the Priestley-Taylor equation (paper C4).

References

Brutsaert, W. H. (1982) *Evaporation into the Atmosphere*. Reidel, Dordrecht, The Netherlands.

Brutsaert, W. H. (2005) *Hydrology: An Introduction*. Cambridge University Press, Cambridge, UK.

De Bruin, H. A. R. (1989) Physical aspects of the planetary boundary layer. In: *Estimation of Areal Evaporation* (ed. by T. A. Black, D. L. Spittlehouse, M. D. Novak & D. T. Price), 117–132. IAHS Publ. 177. IAHS Press, Wallingford, UK.

Dooge, J. C. I. (1974) The development of hydrological concepts in Britain and Ireland between 1674 and 1874. *Hydrol. Sci. Bull.* **19**, 279–302.

Doorenbos, J. & Pruitt, W. O. (1975) *Guidelines for Predicting Crop Water Requirements*. Irrigation and Drainage Paper 24, FAO, Rome, Italy.

Huntingford, C. & Monteith, J. L. (1998) The behaviour of a mixed-layer model of the convective boundary layer coupled to a big leaf model of surface energy partitioning. *Boundary-Layer Met.* **88**, 87–101.

McNaughton, K. G. (1976) Evaporation and advection, I: Evaporation from extensive homogeneous surfaces. *Quart. J. Royal Met. Soc.* **102**, 181–191.

McNaughton, K. G. & Spriggs, T. W. (1986) A mixed-layer model of regional evaporation. *Boundary-Layer Met.* **34**, 243–262.

McNaughton, K. G. & Spriggs, T. W. (1989) An evaluation of the Priestley-Taylor equation. In: *Estimation of Areal Evaporation* (ed. by T. A. Black, D. L. Spittlehouse, M. D. Novak & D. T. Price), 89–104. IAHS Publ. 177. IAHS Press, Wallingford, UK.

Oliver, H. (2005) John Dalton, weatherman and father of modern hydrology. *Manchester Memoirs, Manchester Literary and Philosophical Society* **142**, 107–113.

Oliver, H. & Oliver, S. (2003) Meteorologist's profile – John Dalton. *Weather* **58**, 206–211.

Rodda, J. C. (2006) On the British contribution to international hydrology – an historical perspective. *Hydrol. Sci. J.* **51**(6), 1177–1193.

Thom, A. S. & Oliver, H. R. (1977) On Penman's equation for estimating evaporation. *Quart. J. Royal Met. Soc.* **103**, 345–357.

ESSAY III.

On Evaporation.

When a liquid is exposed to the air, it becomes gradually dissipated in it: the process by which this effect is produced, we call *evaporation.*

Paper C1 is reprinted from *Mem. Proc. Lit. Phil. Soc. Manchester*, 5, part 2, 574–594. This paper by John Dalton is reproduced by kind permission of the Manchester Literary and Philosophical Society.

Many philosophers concur in the theory of chemical solution: atmospheric air, it is said, has an affinity for water; it is a menstruum in which water is soluble to a certain degree. It is allowed notwithstanding by all, that each liquid is convertible into an elastic vapour in vacuo, which can subsist independently in any temperature; but as the utmost forces of these vapours are inferior to the pressure of the atmosphere in ordinary temperatures, they are supposed to be incapable of existing in it in the same way as they do in a torricellian vacuum: hence the notion of affinity is induced.—According to this theory of evaporation, atmospheric air (and every other species of air for aught that appears) dissolves water, alkohol, ether, acids, and even metals. Water below 212° is chemically combined with the gases; above 212° it assumes a new form, and becomes a distinct elastic fluid, called *steam:* whether water first chemically combined with air, and then heated above 212°, is detached from the air or remains with it, the advocates of the theory have not determined.—This theory has always been considered as complex and attended with difficulties; so much that M. Pictet and others have rejected it, and adopted that which admits of distinct elastic vapours in the atmosphere at all temperatures, uncombined with either of the principal constituent gases; as being much more simple and easy of explication than the other; though they do not remove the grand objection to it, arising from atmospheric pressure. It has however been made to appear in these essays, I presume, that the objection to it from pressure, is itself founded upon an ungrounded hypothesis.

Leaving the theory of evaporation for the present, we shall proceed to the experiments.

The following positions have been established by others, and need therefore only to be mentioned here.

1. Some fluids evaporate much more quickly than others.

2. The quantity evaporated is in direct proportion to the surface exposed, all other circumstances alike.

3. An increase of temperature in the liquid is attended with an increase of evaporation, not directly proportionable.

4. Evaporation is greater where there is a stream of air than where the air is stagnant.

5. Evaporation from water is greater the less

the humidity previously existing in the atmosphere, all other circumstances the same.

The objects in view in this essay, are,

1. To determine the precise effect that a variation of temperature has upon the quantity evaporated.

2. To determine the ratio of evaporability of different fluids.

3. To find a rule by which the quantity and effect of previous humidity in the air may be ascertained.

4. From these and other facts to obtain a true theory of evaporation.

On the Evaporation of Water at 212°.

I took a small cylindrical vessel of tin, its diameter 3¼ and depth 2½ inches; and having fixed three pieces of wire to equidistant points of the circumference, they were fastened together at the top and the extremities bent into a hook, by which the vessel might be suspended from the end of a balance, &c. This done, the vessel was nearly filled with water, which was then made to boil over a small red fire in different circumstances: it was held in the hand and removed nearer to or further from the fire, so as to be kept just at the point of ebullition. In this state the vessel and water were weighed true to a grain, and the instant of time noted by a watch; then kept as above at 212° for ten minutes or more and again weighed: and the loss of water by evaporation, per minute, was thus ascertained. The experiments were repeated several times in the same as well as in different circumstances; and the results in no instance differed materially when obtained in the same circumstances.

The least evaporation per minute was 30 grains: this was when the fire, or lamp, was in the middle of a room, the doors and windows shut, and the air calm.

The next degree was 35 grains per minute or thereabouts: this was when the evaporating vessel was over a small fire in the usual fire-place; there being a moderate draught of air, and the room close.

A brisker fire, causing a stronger current of air up the chimney, gave from 35 to 40 grains per minute.

When the windows of the room were open, and a strong wind prevailed, the draught over the fire was proportionally increased, and the evaporation was from 40 to 45 grains per minute.

The extremes that have thus been noticed are 30 and 45 grains per minute: but were the experiment tried in the open air in high winds, I am inclined to believe from a comparison of the observations, that an evaporation of 50, 55 or even 60 grains per minute might be observed.

On the Evaporation of Water below 212°.

I have frequently tried the evaporation at all the temperatures below 212°: it would be tedious to enter into detail of all the experiments, but shall give the results at some remarkable points. In all the high temperatures I used the vessel above mentioned, keeping a thermometer in it, by which I could secure a constant heat, or at least keep it oscillating within narrow limits.

The evaporation from water of 180° was from 18 to 22 grains per minute, according to circumstances; or about $\frac{1}{2}$ of that at 212°.

At 164° it was about $\frac{1}{3}$ of the quantity at the boiling temperature; or from 10 to 16 grains per minute.

At 152° it was only $\frac{1}{4}$ of that at boiling; or from 8 to 12 grains, according to circumstances.

The temperature of 144° afford $\frac{1}{5}$ of the effect at boiling; 138° gave $\frac{1}{6}$; &c.

Having previously to these experiments determined the force of aqueous vapour at all the temperatures under 212°, I was naturally led to examine whether the quantity of water evaporated in a given time bore any proportion to the force of vapour of the same temperature, and was agreeably surprised to find that they exactly corresponded in every part of the thermometric scale: thus the forces of vapour at 212°, 180°, 164°, 152°, 144° and 138° are equal to 30, 15, 10, $7\frac{1}{2}$, 6 and 5 inches of mercury respectively, and the grains of water evaporated per minute in those temperatures were 30, 15, 10, $7\frac{1}{2}$, 6 and 5 also; or numbers proportional to these. Indeed it should be so from the established law of mechanics, that all effects are proportional to the causes producing them. The atmosphere, it should seem, obstructs the diffusion of vapour, which would otherwise be almost instantaneous, as in vacuo; but this obstruction is overcome in proportion to the force of the vapour. The obstruction however cannot arise from the *weight* of the atmosphere, as has till now been supposed; for then it would effectually prevent any vapour from arising under 212°: but it is caused by the *vis inertiæ* of the particles of air; and is similar to that which a stream of water meets with in descending amongst pebbles.

The theory of evaporation being thus manifested from experiments in high temperatures, I found that if it was to be verified by experiments in low temperatures, regard must be had to the force of vapour actually existing in the atmosphere at the time. For instance, if water of 59° were the subject, the force of vapour of that temperature is $\frac{1}{60}$ of the force at 212°, and one might expect the quantity of evaporation $\frac{1}{60}$ also; but if it should happen, as it sometimes does in summer, that an aqueous atmosphere to that amount does already exist, the evaporation, instead of being $\frac{1}{60}$ of that from boiling water, would be nothing at all. On the other hand, if the aqueous atmosphere were less than that, suppose $\frac{1}{2}$ of it, corresponding to 39° of heat, then the effective evaporating force would be $\frac{1}{120}$ of that from boiling water; in short, the evaporating force must be universally equal to that of the temperature of the water, diminished by that already existing in the atmosphere. In order to find the force of the aqueous atmosphere I usually take a tall cylindrical glass jar, dry on the outside, and fill it with cold spring water fresh from the well; if dew be immediately formed on the outside, I pour the water out, let it stand a while to increase in heat, dry the outside of the glass well with a linen cloth, and then pour the water in again; this operation is to be continued till dew ceases to be formed, and then the temperature of the water must be observed; and opposite to it in the table (page 559) will be found the force of vapour in the atmosphere. This must be done in the open air, or at a window; because the air within is generally more humid than that without. Spring water is generally about 50°, and will mostly answer the purpose the three hottest months in the year: in other seasons an artificial cold mixture is required.—The accuracy of the result obtained this way I think scarcely needs to be insisted upon. Glass, and all other hard, smooth substances I have tried, when cooled to a degree below what the surrounding aqueous vapour can support, cause it to be condensed on their surfaces into water. The degree of cold is usually from 1 to 10 below the mean heat of the 24 hours; in summer I have often observed the point as high as 58° or 59°, corresponding to ½ an inch of mercury in force, and once or twice have seen it at 62°: in changeable and windy weather it is liable to considerable fluctuation; but this is not the place to enlarge upon it.

For the purpose of observing the evaporation in atmospheric temperatures I got two light tin vessels, the one 6 inches in diameter and $\frac{1}{2}$ inch deep, the other 8 inches diameter and $\frac{3}{4}$ inch deep; and made to be suspended from a balance, like the former one. When any experiment designed as a test of the theory was made, a quantity of water was put into one of these (generally the 6 inch one, which I preferred) the whole was weighed to a grain; then it was placed in an open window or other exposed situation for 10 or 15 minutes, and again weighed to ascertain the loss by evaporation; at the same time the temperature of the water was observed, the force of the aqueous atmosphere ascertained as above, and the strength of the current of air noticed. From a great variety of experiments made both in the winter and summer, and when the evaporating force was strong and weak, I have found the results entirely conformable with the above theory. The same quantity is evaporated with the same evaporating force thus determined, whatever be the temperature of the air, as near as can be judged; but with the same evaporating force, a strong wind will double the effect produced in a still atmosphere. Thus, if the aqueous atmosphere be correspondent to 40° of temperature and the air be 60°, the evaporation is the same as if the aqueous atmosphere were at 60° of temperature and the air 72°; and in a calm air the evaporation from a vessel of 6 inches in diameter in such circumstances would be about .9 of a grain per minute, and about 1. 8 grains per minute in a very strong wind; the different intermediate quantities being regulated solely by the force of the wind.

The following table exhibits the ratios and quantity of water evaporated in each temperature, derived from the preceding theory, and confirmed by experiments, as far as they have been extended. The first column expresses the temperature; the second, the corresponding force of vapour taken from the preceding table; the other three columns give the number of grains of water that would be evaporated from a surface of 6 inches in diameter in the respective temperatures, on the supposition of there being previously no aqueous vapour in the atmosphere. These columns present the extremes and the mean of evaporation, likely to be noticed, or nearly such: for, the first is calculated upon the supposition of 35 grains loss per minute from the vessel of 3½ inches in diameter; the second, 45 and the third 55 grains per minute.

TABLE

Shewing the force of vapour, and the full evaporating force of every degree of temperature from 20° to 85°, expressed in grains of water that would be raised per minute from a vessel of six inches in diameter, supposing there were no vapour already in the atmosphere.

Temperature.	Force of Vap. inch.	Evaporating Force in Grains.		
212°	30	120	154	189
20°	.129	.52	.67	.82
21	.134	.54	.69	.85
22	.139	.56	.71	.88
23	.144	.58	.73	.91
24	.150	.60	.77	.94
25	.156	.62	.79	.97
26	.162	.65	.82	1. 02
27	.168	.67	.86	1. 05
28	.174	.70	.90	1. 10
29	.180	.72	.93	1. 13
30	.186	.74	.95	1. 17
31	.193	.77	.99	1. 21
32	.200	.80	1. 03	1. 26
33	.207	.83	1. 07	1. 30

TABLE CONTINUED.

Temperature. 212°	Force of Vap. inch. 30	Evaporating Force in Grains. 120	154	189
34°	.214	.86	1. 11	1. 35
35	.221	.80	1. 14	1. 39
36	.229	.92	1. 18	1. 45
37	.237	.95	1. 22	1. 49
38	.245	.98	1. 26	1. 54
39	.254	1. 02	1. 31	1. 60
40	.263	1. 05	1. 35	1. 65
41	.273	1. 09	1. 40	1. 71
42	.283	1. 13	1. 45	1. 78
43	.294	1. 18	1. 51	1. 85
44	.305	1. 22	1. 57	1. 92
45	.316	1. 26	1. 62	1. 99
46	.327	1. 31	1. 68	2. 06
47	.339	1. 36	1. 75	2. 13
48	.351	1. 40	1. 80	2. 20
49	.363	1. 45	1. 86	2. 28
50	.375	1. 50	1. 92	2. 36
51	.388	1. 55	1. 99	2. 44
52	.401	1. 60	2. 06	2. 51
53	.415	1. 66	2. 13	2. 61
54	.429	1. 71	2. 20	2. 69
55	.443	1. 77	2. 28	2. 78
56	.458	1. 83	2. 35	2. 88
57	.474	1. 90	2. 43	2. 98
58	.490	1. 96	2. 52	3. 08
59	.507	2. 03	2. 61	3. 19
60	.524	2. 10	2. 70	3. 30
61	.542	2. 17	2. 79	3. 41

TABLE CONTINUED.

Temperature. 212°	Force of Vap. inch. 30	Evaporating Force in Grains. 120	150	182
62°	.560	2. 24	2. 88	3. 52
63	.578	2. 31	2. 97	3. 63
64	.597	2. 39	3. 07	3. 76
65	.616	2. 46	3. 16	3. 87
66	.635	2. 54	3. 27	3. 99
67	.655	2. 62	3. 37	4. 12
68	.676	2. 70	3. 47	4. 24
69	.698	2. 79	3. 59	4. 38
70	.721	2. 88	3. 70	4. 53
71	.745	2. 98	3. 83	4. 68
72	.770	3. 08	3. 96	4. 84
73	.796	3. 18	4. 09	5. 00
74	.823	3. 29	4. 23	5. 17
75	.851	3. 40	4. 37	5. 34
76	.880	3. 52	4. 52	5. 53
77	.910	3. 65	4. 68	5. 72
78	.940	3. 76	4. 83	5. 91
79	.971	3. 88	4. 99	6. 10
80	1. 00	4. 00	5. 14	6. 29
81	1. 04	4. 16	5. 35	6. 54
82	1. 07	4. 28	5. 50	6. 73
83	1. 10	4. 40	5. 66	6. 91
84	1. 14	4. 56	5. 86	7. 17
85	1. 17	4. 68	6. 07	7. 46

The use of this table will appear from the following problems:

PROBLEM I.

Having given the temperature at which the aqueous atmosphere begins to be condensed into water, and the temperature of the air, to find the quantity of water that would be evaporated in a minute from a vessel of 6 inches diameter.

Solution. Subtract the grains opposite to the lower temperature from those opposite to the higher one, in the first, second or third column of grains, according to the strength of the wind, and the remainder will be the quantity evaporated in a minute, under those circumstances, nearly.

Example. Let the point of condensation be 52°, the temperature of the air 65°, with a moderate breeze.

The number opposite 52° in the second column of grains is 2. 06, and that opposite 65° is 3. 16; the difference, 1. 1 grain, is the evaporation per minute.

PROBLEM II.

Having given the quantity evaporated in a minute, found by experiment, and the temperature of the air, to find the force of the aqueous atmosphere, and the point of condensation.

Solution. Subtract the observed evaporation from that opposite the given temperature in the table; and look above for the number nearest to the remainder in the same column of evaporation, opposite to which will be found the force of the aqueous atmosphere, and the point at which it begins to be condensed.

Example. Finding the evaporation from a vessel of 6 inches in diameter to be 1. 7 grain per minute with a brisk wind, air 62°; what is the weight of the aqueous atmosphere, and the temperature at which it begins to be condensed into water?

The number opposite 62° in the third column of grains is 3. 52, being the whole evaporating force at that temperature in a perfectly dry atmosphere; from which take 1. 7 grains, the real evaporating force observed, and the remainder, 1. 82, corresponds, as per table, to the force .294 inches of mercury, the weight of vapour, and to 43° of temperature*.

* It may be proper to remind the reader that all the experiments on evaporation are understood to be made in the open air, or in a window with a current inward; also it may be observed the evaporation in a close room is much less and is besides irregular, being greater proportionably from a less surface, evidently from the stagnation of the air.

Evaporation of Spirits, Ether, &c.

If the law of evaporation above given apply to water in every part of the scale of heat, no reasonable doubt can be entertained respecting its application to other liquids. I have notwithstanding made several experiments on others, the results of which are conformable to the same law. Some of them follow:—

1. Spirit of wine.—Evaporated from a surface of 4 inches in diameter, 54 grains in 25 minutes: air 53°; aqueous atmosphere at 49°, and beginning to rain with a moderate breeze. It would proportionally have been 121 grains from a vessel of 6 inches in diameter. This gives nearly 5 grains per minute. The same spirit boiled at or near 180°.

Now from the *data*, water of 83° is equivalent in force to spirits of 53°: and it may be seen that the evaporating force of water of 83° is nearly 5 in the first and second columns of grains of the table. It seems probable that the aqueous atmosphere does not diminish the evaporation of spirits as it does that of water.

2. Ether. 1. Put a phial containing ether, and a small tin vessel of 1¾ inch diameter into a scale and balanced them exactly: then poured the ether into the evaporating vessel and put

the phial into the scale again; took out 40 grains from the opposite scale, and waited till the equilibrium was restored: this was in 8 minutes 6 seconds. The air was 50°, and the ether at first 50°; but it rapidly sunk, as was found by dipping a very small bulbed thermometer into it, to 28°. In a window with a moderate breeze.

——— 2 and 3. Repeated the experiment in the same circumstances, except the evaporating vessel, which was now porcelain, and 2½ inches diameter. Lost 40 grains in 3 minutes. Thermometer sunk from 50 to 30°. The two experiments made this way did not differ above one or two grains.

These results reduced to a vessel of 3¼ inches in diameter give

1st. Experiment, loss 17 grains per minute;
2 & 3 ——— — 22½ ———

The reason why the result in the first experiment was something less than in the other two, was evidently owing to the circumstance of its longer duration, by which the ether was the greater part of the time in a low temperature, and consequently evaporated less.—The ether used boiled at 102°. At 50° it was therefore in the capacity of water at 160°. But water at 160°, at most loses only 17 or 18 grains per minute, and less

20° below that temperature. At first view therefore it should seem that ether evaporates quicker than the general law assigns.—But it must be allowed that the temperature of the *air* has some effect upon evaporation, though it has certainly very little. Now ether in the above experiments is acted upon by a current of air of an equal or higher temperature than itself; but water of 160° is usually acted upon by air 100° lower than itself, which is every moment precipitating the vapour formed, and thus obstructing its circulation. This appears to be a sufficient cause for the small difference observed.

With respect to mercury, sulphuric acid, muriate of lime, &c. there can be no doubt but they experience a real evaporation like those above; but it must be very small in proportion as their boiling points are high. And it would be difficult to make experiments upon such of these as have an affinity for aqueous vapour; because their acquisition from the aqueous atmosphere would far exceed their loss by evaporation.

Since writing the above essay, opportunities have occurred to ascertain whether the evaporation from ice is conformable to the same law as that from water. Every one, who has tried the experiment, admits the fact that ice is evaporable.—I have lately made several observations on this subject, the results of which, as far as they go, support the conclusion that the general law of evaporation continues the same below the point of congelation as above it. All the experiments were made in the tin vessel above described of 6 inches in diameter; a quantity of water was suffered to freeze in it, so as to form a circular cake of ice; the vessel and ice were then weighed together, and exposed in the open air for a certain time, after which being again weighed, the loss was found; the force of the aqueous atmosphere was sometimes determined during the experiment by a mixture of pounded ice and salt, in the manner already described.

	Grs.	H.	Grs.		Wind.	Air.
Nov. 5. In the night lost	110 in	9	; or, .20	per m.	N. E. brisk.	28° to 31°
—— at 10 A. M. —	25 in	$1\frac{1}{4}$	; or, .33	——	N. E. mod.	32°
—29. at 1 P. M. —	24 in	$1\frac{3}{4}$	; or, .23	——	calm.	31°
P. M. —	84 in	$9\frac{1}{4}$	; or, .15	——	———	30°
—30. in the night —	94 in	9	; or, .17	——	N. E. mod.	31°
Dec. 19. P. M. —	75 in	8	; or, .16	——	N. E. calm.	26°—28°
In the night —	33 in	11	; or, .05	——	calm.	29°
—20. A. M. —	21 in	2	; or, .175	——	W. mod.	31°

Some of these being made in the night, and of long duration, neither the temperature of the air, nor the force of the aqueous atmosphere could be fairly determined: the second experiment was made under every favourable circumstance, and the aqueous atmosphere found at 22°. By problem 2, at page 588 it would have been determined at $21\frac{1}{2}$°, using the second column of grains in the table.

AN APPROACH TOWARD A RATIONAL CLASSIFICATION OF CLIMATE

C. W. THORNTHWAITE

[With separate map, Pl. I, facing p. 94]

THE direction that the modern study of climate has taken has been dictated largely by the development of meteorological instruments, the establishment of meteorological observatories, and the collection of weather data. The catalogue of climatic elements consists of those that are customarily measured and usually includes temperature, precipitation, atmospheric humidity and pressure, and wind velocity. Increasingly, climatic studies have tended to become statistical analyses of the observations of individual elements. Because of this, climatology has been regarded in some quarters as nothing more than statistical meteorology.

THE ROLE OF EVAPORATION AND TRANSPIRATION

But the sum of the climatic elements that have been under observation does not equal climate. One element conspicuously missing from the list is evaporation. The combined evaporation from the soil surface and transpiration from plants, called "evapotranspiration," represents the transport of water from the earth back to the atmosphere, the reverse of precipitation. The rain gauge measures precipitation within acceptable limits of accuracy. We know reasonably well how rainfall varies from one place to another over the inhabited parts of the earth and also how it varies through the year and from one year to another. On the other hand, no instrument has yet been perfected to measure the water movement from the earth to the atmosphere, and consequently we know next to nothing about the distribution of evapotranspiration in space or time.

We cannot tell whether a climate is moist or dry by knowing the precipitation alone. We must know whether precipitation is greater or less than the water needed for evaporation and transpiration. Precipitation and evapotranspiration are equally important climatic factors. Since precipitation and evapotranspiration are due to different meteorological causes, they are not often the same either in amount or in distribution through the year. In some places more rain falls month after month than evaporates or than the vegetation uses. The surplus moves through the ground and over it to form

➤ DR. THORNTHWAITE, consulting climatologist, has spent the growing seasons of 1946 and 1947 in experiments "fitting crops to weather" (see p. 4).

streams and rivers and flows back to the sea. In other places, month after month, there is less water in the soil than the vegetation would use if it were available. There is no excess of precipitation and no runoff, except locally where the soil is impervious and cannot absorb the rain on the rare occasions when it falls. Consequently, there are no permanent rivers, and there is no drainage to the ocean. In still other areas the rainfall is deficient in one season and excessive in another, so that a period of drought is followed by one with runoff. The march of precipitation through the year almost never coincides with the changing demands for water.

Where precipitation is in excess of water need, the climate is moist. Where the water deficiency is large in comparison with the need, the climate is dry. Where precipitation and water need are equal or nearly equal, the climate is neither humid nor arid.

Potential Evapotranspiration as a Climatic Factor

The vegetation of the desert is sparse and uses little water because water is deficient. If more water were available, the vegetation would be less sparse and would use more water. There is a distinction, then, between the amount of water that actually transpires and evaporates and that which would transpire and evaporate if it were available. When water supply increases, as in a desert irrigation project, evapotranspiration rises to a maximum that depends only on the climate. This we may call "potential evapotranspiration," as distinct from actual evapotranspiration.

We know very little about either actual evapotranspiration or potential evapotranspiration. We shall be able to measure actual evapotranspiration as soon as existing methods are perfected. But to determine potential evapotranspiration is very difficult. Since it does not represent actual transfer of water to the atmosphere but rather the transfer that would be possible under ideal conditions of soil moisture and vegetation, it usually cannot be measured directly but must be determined experimentally. Like actual evapotranspiration, potential evapotranspiration is clearly a climatic element of great importance. By comparing it with precipitation we can obtain a rational definition of the moisture factor.

Precipitation is a strictly physical process, which meteorologists have investigated in much detail. Evapotranspiration is likewise a physical process, yet since it is subject to biological control, it must be studied by methods unfamiliar to the meteorologist. Information concerning evapotranspiration has come chiefly not from the meteorologist but from the biologist. For this reason it is necessary to make use of the literature and apply the methods of plant physiology. Nevertheless, evapotranspiration represents the return flow of water to the atmosphere and is thus an important meteorological process.

The only method so far developed that measures the actual evapotranspiration from a field or any other natural surface without disturbing the vegetation cover in any way is the so-called "vapor transfer" method. Water vapor when it enters the atmosphere from the ground or from plants is carried upward by the moving air. It is carried upward in small eddies or bodies of air that are replaced by drier eddies from above. Although we cannot see water vapor, we can measure it in the air. We find that when evaporation is taking place the amount of moisture is greatest in the air near the ground and decreases with distance above it. If we determine the rate at which the air near the ground is mixing with that above it and at the same time measure the difference in water-vapor content at the two levels, we can determine both the rate and the amount of evapotranspiration. Furthermore, we can determine equally well the amount of water condensed as dew.[1]

This method is not easy either to understand or to use. It is hard to use because it requires physical measurements more precise than are usually made. Furthermore, the coefficient of turbulent transfer of air is not a constant. It varies from time to time and from place to place. It even varies with height at a given time and place. In spite of these difficulties the method can be perfected and will answer many important questions for climatology and biology.[2]

Scientists have tried in various ways to determine the amount of water used by plants. One of the earliest attempts was to remove leaves or branches from a plant, let them dry for a brief time, and weigh them to see how much water they had lost. Another method is to place plants in sealed containers and measure the moisture that accumulates in the confined air. Experimenters have grown thousands of individual plants in pots, weighing them periodically to determine the evapotranspiration losses. These methods are highly artificial, and generalization from them sometimes gives fantastic

[1] See the note by John Leighly: New Occasions and New Duties for Climatology, *Geogr. Rev.*, Vol. 29, 1939, pp. 682–683.

[2] C. W. Thornthwaite and Benjamin Holzman: Measurement of Evaporation from Land and Water Surfaces, *U. S. Dept. of Agric. Tech. Bull. No. 817*, 1942; C. W. Thornthwaite: The Measurement of Evaporation and Transpiration from Natural Surfaces, *Proc. Hydrology Conference, State College, Pa., June 30–July 2, 1941* (*Pennsylvania State College School of Engineering Tech. Bull. No. 27*), 1942, pp. 185–197; *idem:* Atmospheric Turbulence and the Measurement of Evaporation, *Proc. Second Hydraulics Conference, June 1–4, 1942* (*Univ. of Iowa Studies in Engineering, Bull. 27*), Iowa City, 1943, pp. 280–288.

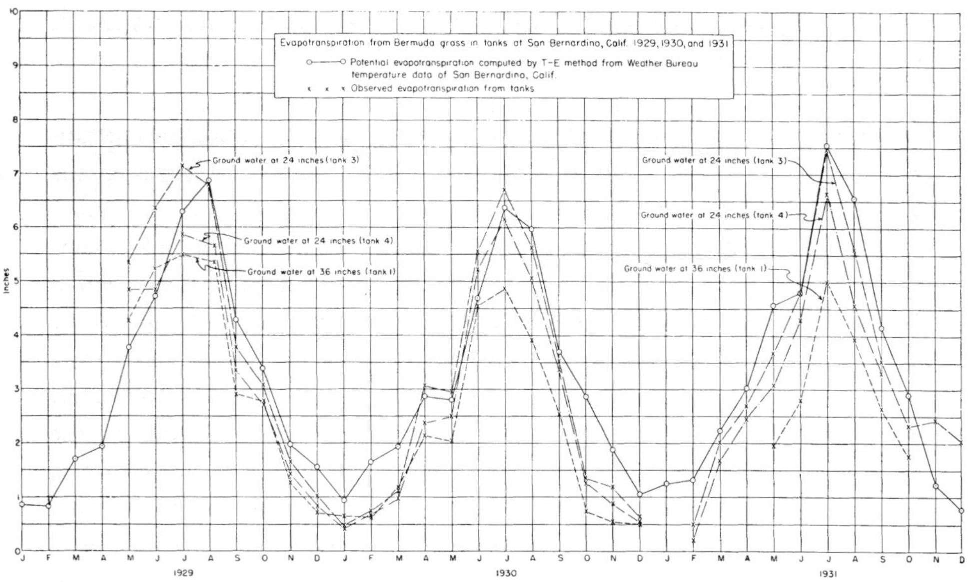

FIG. 1

results. For example, in a German study transpiration from an oak woodland was computed as being more than eight times the actual rainfall.[3]

There are other, less artificial methods of determining both water use and water need. In some irrigated areas rainfall, water applied by irrigation, and water outflow are all measured. The fraction of the water applied that does not run off is the evapotranspiration. Irrigation engineers have determined the evapotranspiration from plants growing in sunken tanks filled to ground level with soil in which water tables are maintained at different predetermined depths beneath the soil surface.[4] Lee found that the annual use of water by salt grass at Independence, Calif., ranged from 13.4 inches with the water table at 5 feet below the ground surface to 48.8 inches with it at 1.5 feet below. Debler reported a range in annual use of water by salt grass at Los Griegos, N. Mex., from 10.1 inches with the water table at 37 inches to 48.4 inches with it at 5 inches. Young and Blaney observed a range at Santa Ana, Calif., from 13.4 inches with free water 4 feet below the surface to 42.8 inches with it 1 foot below. Comparable figures for Bermuda grass at San Bernardino, Calif., were 28.19 inches with the water

[3] B. G. Ivanov: Evaporation under Natural Conditions: Methods of Study and Results Attained (Hydrometeorological Publications), Moscow, 1939. (Translated from the Russian by Headquarters, Army Air Forces Weather Division.)

[4] A. A. Young and H. F. Blaney: Use of Water by Native Vegetation, *California Dept. of Public Works, Division of Water Resources Bull. No. 50*, 1942.

table at 3 feet and 34.37 inches with it at 2 feet. The observations from San Bernardino are shown in Figure 1.

Lowry and Johnson[5] published data on annual water use in 12 irrigated valleys in the western United States and one in the Dominican Republic. Since water is supplied by irrigation, water use in these valleys approximates potential evapotranspiration. It ranges from 18 inches in a mountain valley in Colorado to 58 inches in the Barahona district in the Dominican Republic.

Although the various methods of determining potential evapotranspiration have many faults and the determinations are scattered and few, we get from them an idea of how much water is transpired and evaporated and how much would be if it were available. We find that the rate of evapotranspiration depends on four things: climate, soil-moisture supply, plant cover, and land management. Of these the first two prove to be by far the most important.

Some scientists have believed that transpiration serves no useful purpose for the plant. We now understand that transpiration effectively prevents the plant surfaces that are exposed to sunlight from being overheated. Most plants require sunlight for growth. The energy of the sun combines water and carbon dioxide in the leaves into food, which is later carried to all parts of the plant and used in growth. This process, which is called "photosynthesis," is most efficient when the leaf temperatures are between 85° and 90° F. But a leaf exposed to direct sunlight would quickly become much hotter if the energy of the sun were not disposed of in some way. The surface of dry ground may reach a temperature of 200° F.; temperatures higher than 160° F. have been measured one-fourth of an inch below the ground surface. The plant is admirably designed to dissipate heat, the leaves being like the fins of a radiator, and some of the excess heat is conducted into the adjacent air and carried away in turbulence bodies. In this way the air is heated. But some of the excess heat energy is utilized in transpiration, to change water from a liquid into a vapor. Most of the heat of evaporation must come from the plant. Thus, the greater the intensity of sunshine, the greater will be the tendency to overheating, and the larger will be the transpiration of a plant exposed to it, if water is available for the process. Transpiration is a heat regulator, preventing temperature excesses in both plant and air. Dew formation at night is the reverse of this process and tends to prevent low temperature extremes, since the heat released goes

[5] R. L. Lowry, Jr., and A. F. Johnson: Consumptive Use of Water for Agriculture, *Trans. Amer. Soc. of Civil Engineers*, Vol. 107, 1942, pp. 1243–1266 (Paper No. 2158); discussion, pp. 1267–1302.

mainly to the plant. Both transpiration and growth are related to temperature in the same way.

Atmospheric elements whose influence on transpiration has been studied include solar radiation, air temperature, wind, and atmospheric humidity. These factors are all interrelated. Although solar radiation is the basic factor, there seems to be a closer parallelism between air temperature and transpiration. The temperature of the transpiring part is most closely related to the rate of transpiration.

Transpiration and growth are both affected in the same way by variations in soil moisture. Both increase with increase of available water in the root zone of the soil, to an optimum. Above the optimum both are less, presumably because of poor aeration of the soil, which results in a lack of oxygen to supply the roots and an excess of carbon dioxide.[6] On the other hand, as water in the soil increases above the optimum for growth, direct evaporation from the soil surface also continues to increase.

We do not yet know how much we may increase or decrease transpiration by varying the type of plants or by modifying the plant cover. Since transpiration regulates leaf temperature, and since most plants reach their optimum growth at about the same temperature, we probably cannot change it very much except by reducing the density of the plant cover and thus wasting a part of the solar energy. If all the vegetation is removed from a field, there will be no transpiration. But as long as the root zone of the soil is well supplied with water, the amount of water transpired from a completely covered area will depend more on the amount of solar energy received by the surface and the resultant temperature than on the kind of plants.

Since potential evapotranspiration is an important climatic element, we need to know its distribution over the earth and how it varies through the year and from one year to another. Actual determinations are so few that it would be impossible to make a map of any area by means of them. At the present time the only alternative is to discover a relation between potential evapotranspiration and other climatic factors for which there are abundant data.

Temperature and Growth

Many studies have been made of temperature and growth. Some investigators have measured the elongation of shoots, stems, and roots grow-

[6] A. J. Loustalot: Influence of Soil-Moisture Conditions on Apparent Photosynthesis and Transpiration of Pecan Leaves, *Journ. of Agric. Research,* Vol. 71, 1945, pp. 519–532; G. W. Schneider and N. F. Childers: Influence of Soil Moisture on Photosynthesis, Respiration and Transpiration of Apple Leaves, *Plant Physiology,* Vol. 16, 1941, pp. 565–583.

ing under various controlled temperatures. Others have determined increase in width of leaves and increase in dry weight of the plant. Rate of development of various insects under controlled temperature has been determined. There is always an optimum temperature for growth, a temperature at which the growth rate is highest. At lower or higher temperatures growth is slower. The temperature at which growth is most rapid seems to vary somewhat with the material under study and with the length of exposure but is always near 30° C. (86° F.). In his investigation of maize seedlings Lehenbauer[7] obtained most rapid growth for a 3-hour period at 29° C. and for 9–12 hours at 32° C. He says that for periods of 3 hours or longer the greatest rates of growth occur within the temperature range 29° to 32°. Wadley[8] found that the rate of development of the green bug was most rapid at 30° C. Similarly, there are minimum and maximum temperatures beyond which growth does not occur. These limits vary likewise, but the minimum is near 0° C. and the maximum somewhere above 40° C.

About half a century ago Van't Hoff[9] propounded the principle that the velocity of a chemical reaction doubles or trebles with each rise in temperature of 10° C. This is an exponential law of the form

$$v = ca^t, \qquad (1)$$

in which a has the value 1.0718 when velocity, v, doubles with a 10° rise in temperature, and the value 1.1161 when v trebles. The Van't Hoff law has been applied by biologists to physiological processes. The customary procedure is to determine temperature coefficients from actual growth measurements. The temperature coefficient is the quotient of two growth rates that are separated from each other by a 10° interval of temperature, and by the Van't Hoff rule it is expected to range between 2 and 3. In reality, the coefficient varies much more widely. It exceeds 10.0 in the temperature range between 0° C. and 10° C. and falls steadily to values less than 1.0, above the optimum temperature at which growth and temperature are inversely related.

Lehenbauer[10] published the following table of temperature coefficients for 12-hour growth rates of shoots of maize seedlings, for various 10° ranges of temperature:

---

[7] P. A. Lehenbauer: Growth of Maize Seedlings in Relation to Temperature, *Physiological Researches*, Vol. 1, 1914, pp. 247–288.

[8] F. M. Wadley: Development-Temperature Correlation in the Green Bug, Toxoptera Graminum, *Journ. of Agric. Research*, Vol. 53, 1936, pp. 259–266.

[9] J. H. van't Hoff: Études de dynamique chimique, Amsterdam, 1884.

[10] *Op. cit.*, p. 281.

Temperature, °C.	12–22	13–23	15–25	18–28	20–30	21–31	22–32	25–35	32–42	33–43
Growth rate, 0.01 mm.	9–59	10–64	20–75	28–98	45–108	53–109	59–111	75–86	111–11	101–6
Coefficient	9.56	6.40	3.75	3.50	2.40	2.06	1.88	1.15	0.09	0.06

The temperature coefficient must have values of ∞ at the temperature minimum, 1.0 at the temperature optimum, and 0 at the temperature maximum. Only in the temperature range 20° to 30°, where the temperature coefficient is between 2 and 3, is the Van't Hoff law valid. Since the temperature coefficient drops continuously, the coefficient a in equation (1) must also drop. The Van't Hoff $a = 1.0718$ is valid at only one point on the curve and is no more significant than any other value of a. The fact that a drops steadily and falls below 1.00 when the optimum temperature is reached is far more significant than any average value of a (or of the temperature coefficient).

There is clearly some growth-inhibiting factor that is directly proportionate to temperature. It is only because of the accident that the temperatures which plants ordinarily experience under natural conditions are below the optimum, that growth is popularly assumed to vary directly with temperature. At temperatures above 30° C. the growth-inhibiting factor becomes greater than the growth-stimulating factor and the growth rate falls with rising temperature. Since the exponential equation $v = ca^t$ requires a continuously increasing growth rate with rising temperature, it does not truly express the relation between temperature and growth.

What the growth-inhibiting factor is, is uncertain. It may reasonably be the operation of water deficiency in the plant tissue. The amount of water in a plant is a balance between that absorbed by the roots and that transpired from the leaves. As the transpiration rate rises with increasing temperature, it will presently exceed the rate of water absorption. Then a suction pressure, or tension, develops in the plant, acting as a brake on transpiration and on growth. The moisture balance in the plant is upset, transpiration is reduced, and growth is retarded. A more satisfactory equation of growth, therefore, is the following:

$$v = a \frac{bce^{ct}}{(e^{ct} + b)^2}, \tag{2}$$

in which a, b, and c are constants and e is the base of the Napierian system of logarithms. In this equation the numerator represents the growth-stimulating factor and the denominator the growth-inhibiting factor. The optimum temperature for growth is where numerator and denominator are equal.

When we fit the equation to a series of Lehenbauer's observations on the growth of maize seedlings, we have the following coefficients: $a = 1764.9$,

$b = 1118.8$, $c = .24$. The equation gives v as a percentage of the optimum growth rate. The Lehenbauer data and the growth curve derived from this equation are plotted in Figure 2. Some of the computed values of v are as follows:

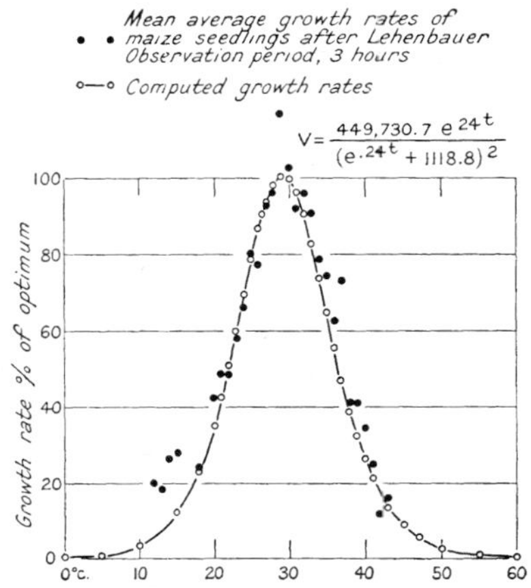

FIG. 2

°C.	v.
0	0.36
5	1.19
10	3.88
15	12.33
20	35.52
25	78.30
28	98.27
30	99.69
32	90.30
35	64.56
40	26.32
45	8.76
50	2.72
55	0.83
60	0.25

DETERMINATION OF POTENTIAL EVAPOTRANSPIRATION

Unfortunately, there are no comparable controlled experiments on the relation of transpiration to temperature. The most reliable measurements of transpiration and evaporation together are for long periods of time. When only monthly or annual totals of evapotranspiration are available, the temperature relation cannot be determined precisely.

The determinations have shown that potential evapotranspiration is high in the southern part of the United States and low in the northern part and that it varies greatly from winter to summer. From observations like those of Figure 1, it has been found that when adjustments are made for variation in day length, there is a close relation between mean monthly temperature and potential evapotranspiration. Study of all available data has resulted in a formula that permits the computation of potential evapotranspiration of a place if its latitude is known and if temperature records are available. The formula is given, and its use described, in the appendix.

Figure 3 shows the distribution of average annual potential evapotranspiration in the United States. It is based on normals of some 3500 Weather Bureau stations, revised to 1930. The average annual water need in the United States ranges from less than 18 inches in the high mountains of the West to more than 60 inches in three isolated areas in the deserts of Arizona and southern California. It is less than 21 inches along the Canadian border of the eastern United States and more than 48 inches in Florida and southern

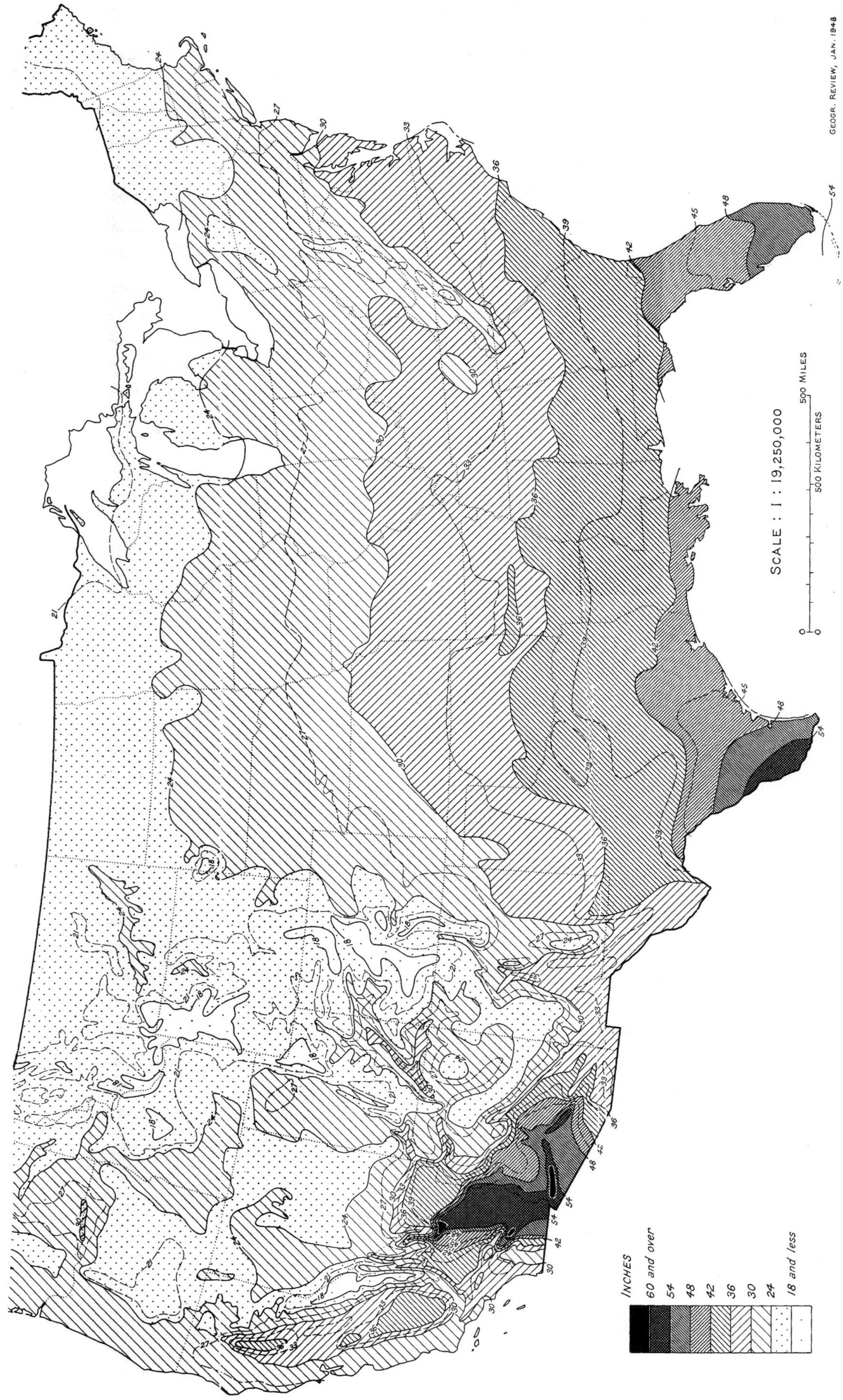
INCHES
60 and over
54
48
42
36
30
24
18 and less
SCALE : 1 : 19,250,000
0 500 MILES
0 500 KILOMETERS
GEOGR. REVIEW, JAN. 1948

FIG. 3

Texas. Although potential evapotranspiration and precipitation are independent climatic elements, in arid regions potential evapotranspiration is increased because of the higher daytime temperatures due to the absence of clouds and rain and because of the small actual evapotranspiration. The high values in the Colorado and Gila Deserts and in the lower Rio Grande Valley are examples. In the arid section of the Columbia River Valley between Washington and Oregon the potential evapotranspiration is more than 30 inches, whereas it is only about 21 inches in the same latitude in the eastern United States.

The march of potential evapotranspiration through the year follows a uniform pattern in most of the country. It is negligible in the winter months as far south as the Gulf Coastal Plain and is only 2 inches a month in southern Florida. It rises to a maximum in July that ranges from 5 inches along the Canadian border to 7 inches on the Gulf coast. In some mountain areas and along the Pacific coast it does not reach 5 inches in any month.

The march of precipitation is highly variable from one region to another. In much of the United States more than half the rain falls in the growing season. In the Pacific Coast States the distribution is reversed. In most places the precipitation is less than the need during a part of the year. In times of excess rainfall water is stored in the soil. The part of this water that is within reach of roots is used before the plants begin to suffer; therefore drought does not begin immediately when rainfall drops below water need. The amount of water in the root zone available to plants varies with the soil structure and the distribution of roots. It is, accordingly, not a constant. However, except in areas of shallow soil the water storage capacity available to mature plants with fully developed root systems varies around a mean that is the equivalent of about 10 centimeters or 4 inches of rainfall. Curves comparing water need and precipitation at selected stations in the United States are given in Figure 4.

In Brevard, N. C. (Fig. 4-A), less than half an inch of water is needed in each of the winter months. The need rises rapidly during the spring, reaches a high point of more than 5 inches in July, and falls rapidly during the autumn. The total average annual water need is 28.50 inches. The precipitation ranges from less than 3½ inches in November to more than 6½ inches in July and is greater than the need in every month. The total average annual precipitation is 61.06 inches, more than twice the potential evapotranspiration. The surplus of 32.56 inches represents runoff. In Salisbury, N. Y. (Fig. 4-B), the average annual potential evapotranspiration is 21.81 inches and the precipitation is 48.39 inches, more than twice as much. In no month is the precipitation less than the need. Here the surplus that runs off amounts to 26.58 inches.

In Bar Harbor, Maine (Fig. 4-C), the precipitation falls to a minimum of less than 3½

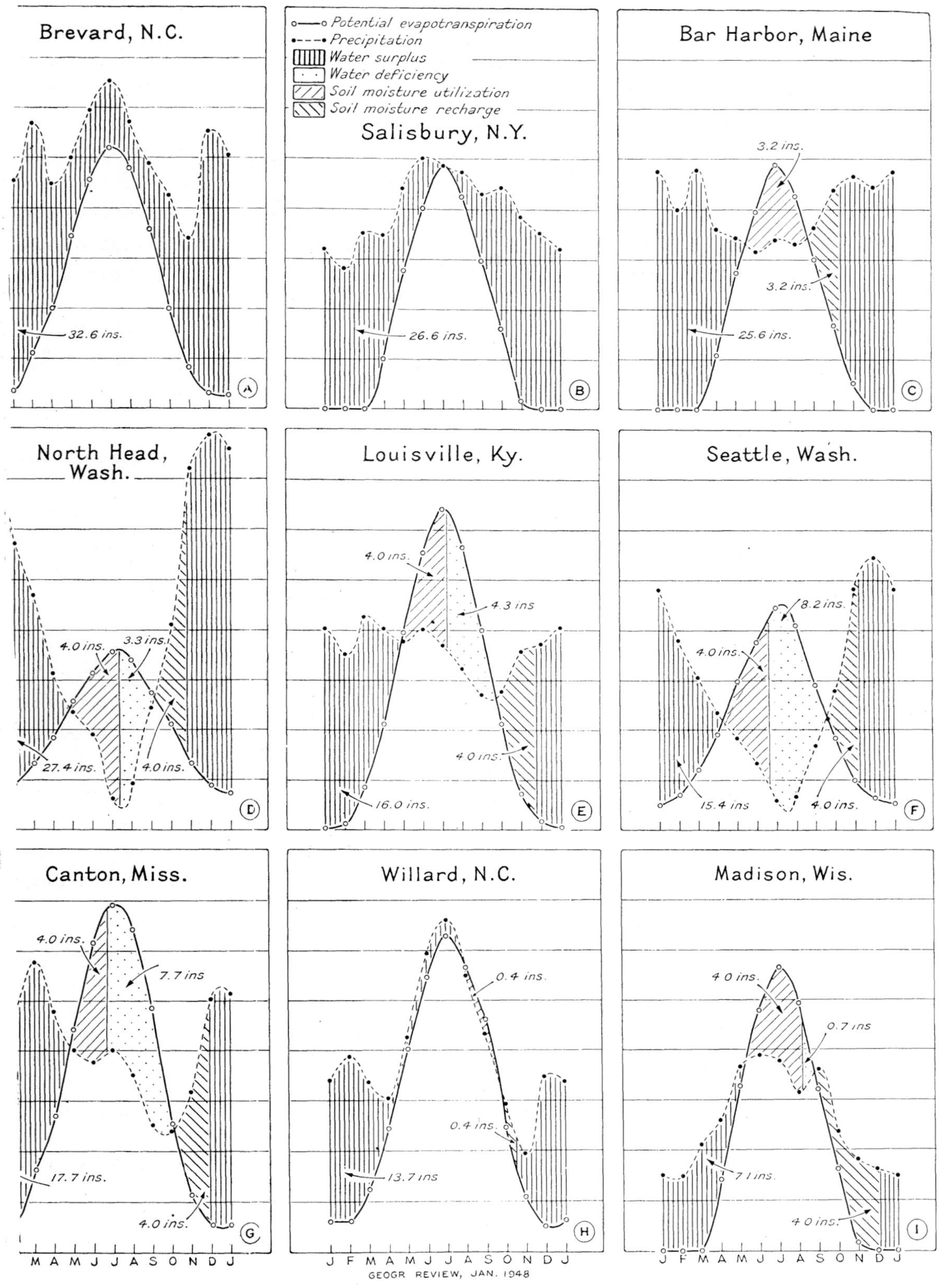

FIG. 4—March of precipitation and potential evapotranspiration at selected stations in the United States.

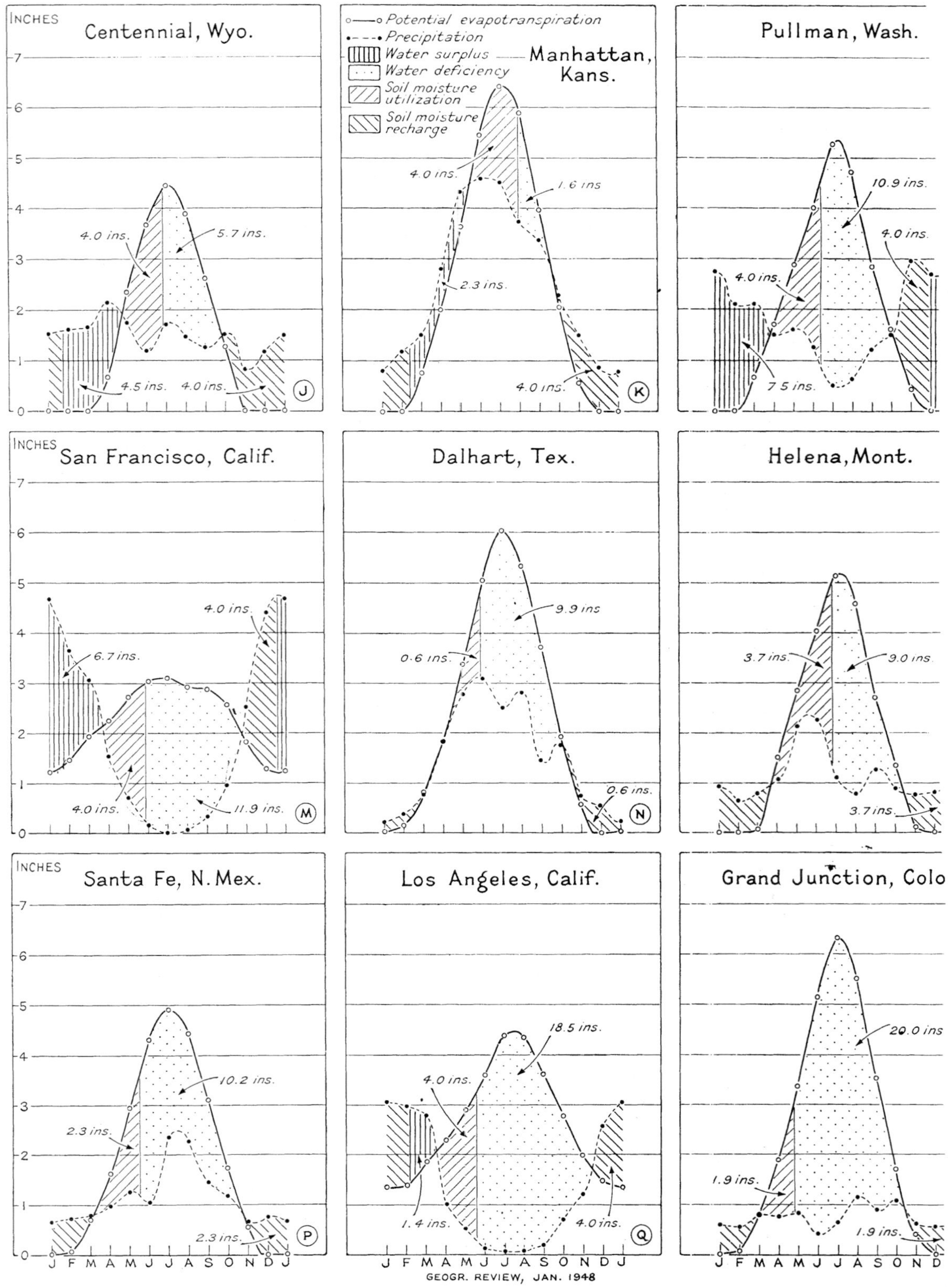
Centennial, Wyo.
INCHES
Potential evapotranspiration
Precipitation
Water surplus
Water deficiency
Soil moisture utilization
Soil moisture recharge
Manhattan, Kans.
Pullman, Wash.
San Francisco, Calif.
Dalhart, Tex.
Helena, Mont.
Santa Fe, N. Mex.
Los Angeles, Calif.
Grand Junction, Colo
J F M A M J J A S O N D J
GEOGR. REVIEW, JAN. 1948

FIG. 4 (cont'd.)

inches in each of the 3 summer months, at exactly the time when potential evapotranspiration rises above 4 inches. Current rainfall is less than need, but the difference is made up from water in storage in the soil. In North Head, Wash. (Fig. 4-D), the precipitation exhibits marked seasonal variation: it is more than 7 inches a month in November, December, and January and less than 1 inch in July and August. Although water need is not large in any month, the July maximum being only 3.6 inches, it is in excess of rainfall during 5 months, from May through September. There is too much rain in winter and too little in summer. In late September, as the days grow cooler and shorter, water need falls below precipitation. For a while the rainfall not immediately needed goes to replace stored soil moisture that has been used up. From then on, the surplus water is lost so far as plant growth is concerned. It raises ground-water levels and produces surface and subsurface runoff. But it is of no benefit to plants and adds nothing to atmospheric moisture. In spring, as the days lengthen and become warmer, plant growth accelerates, and both transpiration and evaporation increase rapidly. By midspring water need exceeds precipitation. Thereafter, until midsummer, the excess demands for water are satisfied from the soil-moisture reserves. When these reserves are exhausted, the vegetation must rely solely on current rainfall. And current rainfall is not enough. Water use falls below water need; the plants suffer, and growth is retarded.

In Louisville, Ky. (Fig. 4-E), rainfall is fairly uniform through the year—between 3.5 and 4 inches a month except in autumn, when it falls below 3 inches. Since potential evapotranspiration is not uniform, however, there is a surplus of water in winter and a deficiency in summer, just as in North Head (Fig. 4-D) and Seattle (Fig. 4-F).

Although annual precipitation in Canton, Miss. (Fig. 4-G), is nearly 50 inches and summer rainfall averages 3.75 inches, the summer potential evapotranspiration is so large that there is a water deficiency of nearly 8.0 inches, almost the same as in Seattle.

In some places monthly precipitation and water need are nearly equal. For example, in Manhattan, Kans. (Fig. 4-K), winter rainfall is only a little greater than potential evapotranspiration and summer rainfall is only a little less. Water surplus and water deficiency are both small. In Madison, Wis. (Fig. 4-I), there is a maximum of rainfall in summer, but even so rainfall does not equal need; there is a small water deficiency in summer and a considerable water surplus in winter and spring. In Willard, N. C. (Fig. 4-H), potential evapotranspiration and precipitation are nearly the same in all the growing-season months. Rainfall is much greater than need in winter, however, and there is a water surplus of nearly 14 inches. In Dalhart, Tex. (Fig. 4-N), on the other hand, rainfall is equal to water need in winter but falls far short in summer, and there is a water deficiency of 10 inches.

In Grand Junction, Colo. (Fig. 4-R), rainfall is slight throughout the year. Water supply satisfies less than one-third of water need, and deficiency is very large. The climate is arid.

The curves of San Francisco, Calif. (Fig. 4-M), and Centennial, Wyo. (Fig. 4-J), show that in both places precipitation fails about equally to provide the water needed in June, July, and August. Water deficiencies, in inches, for June, July, and August are as follows: San Francisco 2.87, 3.07, 2.87; Centennial 2.48, 2.75, 2.44. But there is a vast difference in the relative aridity of the two places. Whereas July rainfall in Centennial provides about 1.5 inches of a needed 4.5 inches, in San Francisco there is no rain at all in July. In one place the water deficiency of July amounts to 62 per cent of the need, in the other to 100 per cent. The aridity of a place in a given period of time depends not on the numerical amount of the water deficiency but rather on the relation of this deficiency to the water need.

TABLE I—COMPARATIVE MOISTURE DATA OF SEATTLE, WASH., AND MANHATTAN, KANS.

In centimeters

ITEM	J	F	M	A	M	J	J	A	S	O	N	D	Y
					Seattle, Wash.								
Potential evap.	1.3	1.8	3.1	4.9	7.6	9.6	11.4	10.5	7.4	4.7	2.5	1.6	66.4
Precipitation	12.3	9.7	7.8	6.0	4.7	3.4	1.5	1.7	4.3	7.1	12.3	13.9	84.7
Storage change	0	0	0	0	−2.9	−6.2	−0.9	0	0	2.4	7.6	0	
Storage	10.0	10.0	10.0	10.0	7.1	0.9	0	0	0	2.4	10.0	10.0	
Actual evap.	1.3	1.8	3.1	4.9	7.6	9.6	2.4	1.7	4.3	4.7	2.5	1.6	45.5
Water deficiency	0	0	0	0	0	0	9.0	8.8	3.1	0	0	0	20.9
Water surplus	11.0	7.9	4.7	1.1	0	0	0	0	0	0	2.2	12.3	39.2
Runoff*	8.9	8.4	6.5	3.8	1.9	1.0	0.5	0.2	0.1	0.1	1.1	6.7	39.2
Moisture ratio	8.47	4.38	1.52	0.22	−.38	−.65	−.87	−.84	−.42	0.54	3.92	7.68	
					Manhattan, Kans.								
Potential evap.	0.0	0.0	1.9	5.1	9.3	13.9	16.4	15.0	10.1	5.2	1.4	0.0	78.3
Precipitation	2.0	3.0	3.8	7.1	11.1	11.7	11.5	9.5	8.6	5.8	3.8	2.2	80.1
Storage change	2.0	2.8	0	0	0	−2.2	−4.9	−2.9	0	0.6	2.4	2.2	
Storage	7.2	10.0	10.0	10.0	10.0	7.8	2.9	0	0	0.6	3.0	5.2	
Actual evap.	0.0	0.0	1.9	5.1	9.3	13.9	16.4	12.4	8.6	5.2	1.4	0.0	74.2
Water deficiency	0.0	0.0	0.0	0.0	0.0	0.0	0.0	2.6	1.5	0.0	0.0	0.0	4.1
Water surplus	0.0	0.2	1.9	2.0	1.7	0.0	0.0	0.0	0.0	0.0	0.0	0.0	5.8
Runoff*	0.0	0.1	1.0	1.5	1.6	0.8	0.4	0.2	0.1	0.1	0.0	0.0	5.8
Moisture ratio	∞	∞	1.00	0.39	0.19	−.16	−.30	−.37	−.15	0.11	1.71	∞	

*Assuming that 50 per cent of the water available for runoff in any month is held over until the following month. In watersheds of less than 100 square miles the percentage is probably smaller.

A moisture ratio that expresses the relative humidity or aridity of a month may be obtained by dividing the difference between precipitation and potential evapotranspiration by potential evapotranspiration,

$$\frac{p-e}{e} \text{ or } \frac{p}{e} - 1. \qquad (3)$$

Positive values of the ratio mean that the precipitation is excessive, negative values that it is deficient. A ratio of zero means that water supply is equal to water need.

WATER SURPLUS AND WATER DEFICIENCY

Numerical values of water surplus and water deficiency may be obtained as a simple bookkeeping procedure, precipitation being treated as income and potential evapotranspiration as outgo, and soil moisture as a reserve that may be drawn upon as long as it lasts. Sample computations for Seattle, Wash., and Manhattan, Kans., are given in Table I.

The distribution of average annual water deficiency in the eastern United States is shown in Figure 5. Deficiencies rise sharply to the west in the Great Plains States and are in excess of 10 inches in most of Texas. Deficiencies

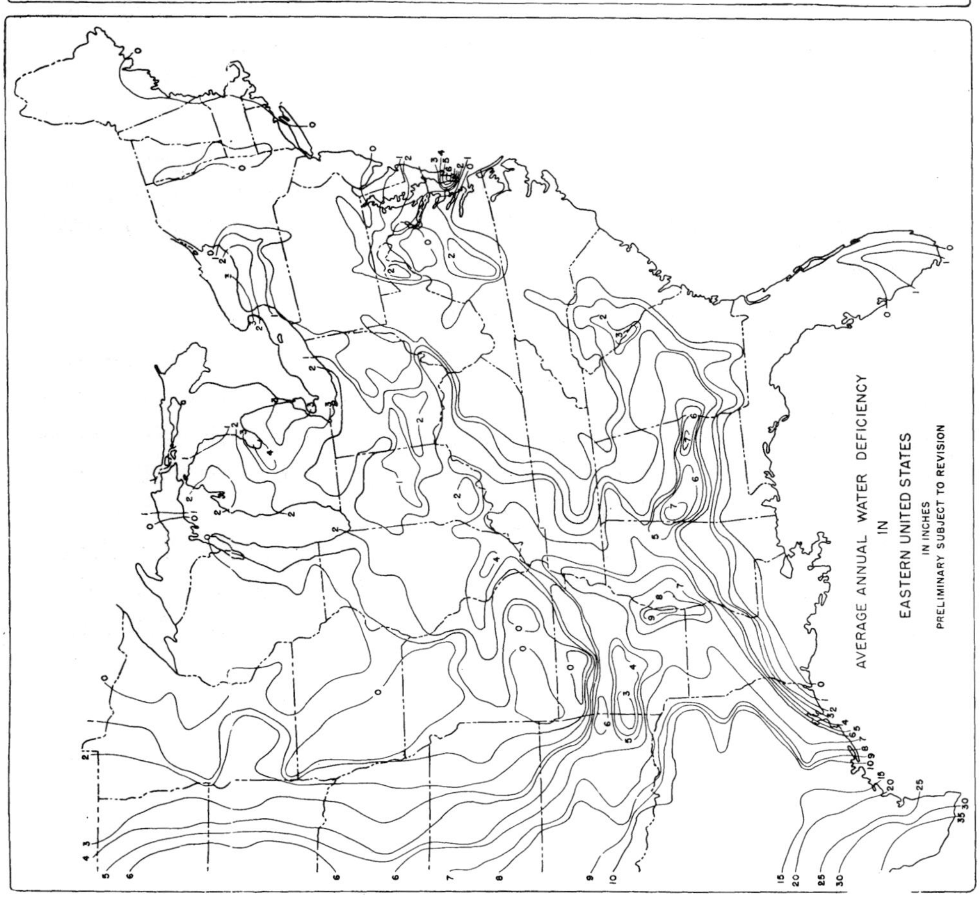
AVERAGE ANNUAL WATER DEFICIENCY
IN
EASTERN UNITED STATES
IN INCHES
PRELIMINARY SUBJECT TO REVISION

Fig. 5

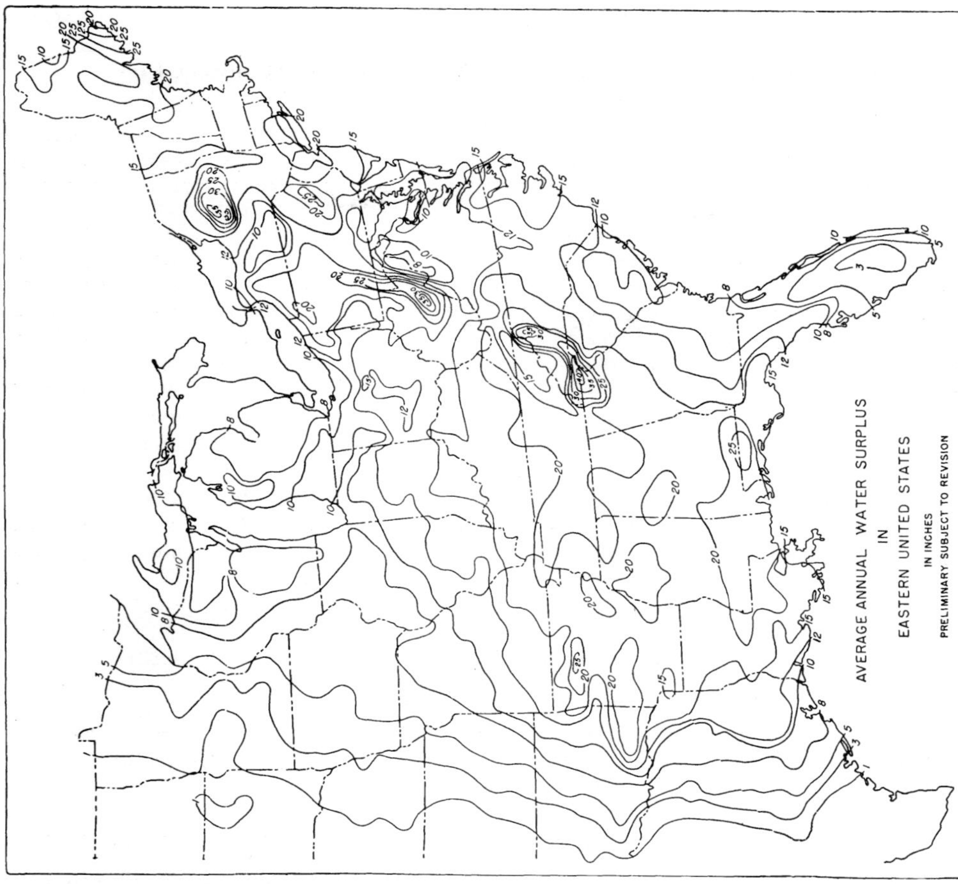
AVERAGE ANNUAL WATER SURPLUS
IN
EASTERN UNITED STATES
IN INCHES
PRELIMINARY SUBJECT TO REVISION

Fig. 6

TABLE II—COMPARATIVE DATA ON IRRIGATED VALLEYS IN THE WESTERN UNITED STATES

VALLEY	WEATHER STATION	OBSERVED WATER USE *cm.*	POTENTIAL EVAPOTRANS. *cm.*	DIFF. *cm.*	DIFF. %
New Fork	Pinedale, Wyo.	46.6	45.4	−1.2	2.6
Michigan and Illinois	Walden, Colo.	45.7	46.9	1.2	2.6
Southwest Area, San Luis	San Luis, Colo.	53.0	51.8	−1.2	2.4
West Tule Lake	Tule Lake, Calif.	67.7	57.9	−9.8	14.5
Garland Div., Shoshone	Lovell, Wyo.	61.6	59.3	−2.3	3.7
North Platte	Torrington, Wyo.	60.4	60.7	0.3	0.5
Mason Creek and Boise	Caldwell, Idaho	66.1	66.5	0.4	0.6
Uncompahgre	Delta, Colo.	68.3	67.5	−0.8	1.2
Mesilla	Agr. College, N. M.	83.2	84.0	0.8	0.9
Greenfields Div., Sun River	Great Falls, Mont.	60.3	60.2	−0.1	0.2
Pecos River near Carlsbad	Carlsbad, N. M.	89.6	91.7	2.1	2.3
Lower Rio Grande	Del Rio, Tex.	115.8	117.0	1.2	1.0
Barahona, Dominican Repub.	Barahona	147.3	146.3	−1.0	0.7

are surprisingly large in various areas of high precipitation in the Southern States—more than 6 inches in the Alabama Black Belt and in the lower Mississippi Valley.

The distribution of average annual water surplus in the eastern United States is shown in Figure 6. It is more than 10 inches in most of the East and Southeast but falls rapidly westward and is less than 1 inch in the Great Plains States. In various centers in the Appalachians the water surplus is more than 35 inches.

RELIABILITY OF COMPUTED POTENTIAL EVAPOTRANSPIRATION

We can test the accuracy of the computed values of potential evapotranspiration in a number of ways. In Table II above the observations of annual water use in irrigated valleys published by Lowry and Johnson[11] are compared with computations of annual potential evapotranspiration. Although the computations refer to points of weather observation and the observations to irrigation projects differing in area, in only one place, West Tule Lake, Calif., is there a significant difference between computed and observed potential evapotranspiration.

There are no comparable direct observations of potential evapotranspiration in the eastern United States. There are, however, measurements of runoff from a great many streams for many years, which may be compared with computed values of water surplus. Observed mean annual runoff in 124 minor watersheds in the Tennessee Valley compared with water surplus

[11] *Op. cit.*, pp. 1253 and 1301.

computed from data of temperature and precipitation at Weather Bureau stations[12] shows a close correspondence, especially noteworthy because the data of the Tennessee Valley Authority refer to areas ranging in size from a few square miles to nearly two thousand and are for a uniform period from 1920 to 1942, whereas the computations were made from averages of temperature and precipitation for periods of various lengths ending in 1930 and refer to the points of weather observation. Computed water surplus and measured runoff are equally close together in Ontario and Mexico, where potential evapotranspiration is respectively much smaller and much larger. The National Resources Board in its report of December 1, 1934, gives a map of average annual runoff in the United States, which may be compared with the map of water surplus, Figure 6. The correspondence is generally close, though there are discrepancies, some of which are due to admitted errors in the original runoff map.

From these and other tests it appears that the computed values of potential evapotranspiration are of the right order of magnitude throughout most of the United States. They are, nevertheless, only approximate. More exact determinations must await further study and the development of a more rational equation. Whether or not the formula can be used without modification to determine potential evapotranspiration in equatorial and polar regions is uncertain. This question also requires further study.

ESSENTIALS OF A CLIMATIC CLASSIFICATION

When the various climatic elements, such as temperature and precipitation, are plotted on maps, the values grade regularly except where gradients steepen on mountain slopes or along seacoasts. Nothing distinguishes any one value in the series from the others; hence divisions of the scales are arbitrary and frequently owe their origin to arithmetical or cartographical convenience. Divisions appearing on maps, selected perhaps to conform to a particular color scheme, gradually assume the role of climatic regions; for example, the 10-inch isohyet as the limit of agriculture.

Köppen made a significant advance in climatic classification when he undertook, first to identify the climatic regions and locate their boundaries from studies of the distribution of vegetation, and then to select numerical climatic values for the boundaries. Penck, De Martonne, and Thornthwaite employed the same method but used data of hydrology and soils to supple-

[12] See Figure 3 (p. 691) in H. G. Wilm, C. W. Thornthwaite, and others: Report of the Committee on Transpiration and Evaporation, 1943–44, *Amer. Geophys. Union Trans. of 1944*, Part 5, Washington, 1945, pp. 683–693.

ment those of vegetation in locating the boundaries of climatic regions. Although this method represents a step forward in that the climatic boundaries are not arbitrary, it is still empirical.

In order to achieve a rational quantitative classification of climate, definite and distinctive break points must be discovered in the climatic series themselves. No such break points exist in the data either of precipitation or of potential evapotranspiration. Both run in continuous series from very low values to very large ones. But when they are taken together, there are some distinctive points, and we have the beginnings of a rational classification.

The primary climatic factors relate to moisture and heat. We are interested in whether a climate is moist or dry and warm or cold. We also need to know whether there is seasonal variation—whether the climate is moist in one season and dry in another.

The Moisture Factor

We cannot tell whether a climate is moist or dry by knowing precipitation alone; we must know whether precipitation is greater or less than potential evapotranspiration. Ignorance of the magnitude of water need has led to the development of a number of moisture indices, which attempt to evaluate the effectiveness of precipitation. It is recognized that water utilization or water loss, that is, evaporation and transpiration, must be related to precipitation somehow. Accordingly, a few of the indices make use of computed or measured values of evaporation usually from free water surfaces, lakes and reservoirs, or from evaporation pans.

In 1905 Transeau[13] produced an index of precipitation effectiveness, by using the quotient of total annual measured precipitation and annual evaporation as computed by Russell[14] for a year for about 150 stations, and made a map of the precipitation-evaporation ratio in the eastern United States. In 1933 Isozaki[15] computed ratios for 99 stations in Japan, using average annual precipitation and annual evaporation from pans, and made a moisture map of the Japanese Empire. Trumble,[16] making use of measurements of evaporation from pans in South Australia, derived an empirical equation relating

[13] E. N. Transeau: Forest Centers of Eastern America, *Amer. Naturalist*, Vol. 39, 1905, pp. 875–889.

[14] Thomas Russell: Depth of Evaporation in the United States, *Monthly Weather Rev.*, Vol. 16, 1888, pp. 235–239.

[15] Masaru Isozaki: Thornthwaite's New Classification of Climate and Its Application to the Climate of Japan, *Journ. of Geogr.*, Tokyo Geographical Society, Vol. 45, 1933, pp. 234–245 (in Japanese).

[16] H. C. Trumble: The Climatic Control of Agriculture in South Australia, *Trans. and Proc. Royal Soc. of South Australia*, Vol. 61, 1937, pp. 41–62; *idem:* Climatic Factors in Relation to the Agricultural Regions of Southern Australia, *Trans. Royal Soc. of South Australia*, Vol. 63, 1939, pp. 36–43.

monthly evaporation to vapor-pressure deficit, which he computed from temperature values. He determined empirically that water need of a month exceeds water supply when computed evaporation is more than three times precipitation. His map of southern Australia shows the number of months of the year when precipitation is less than water need. Wilson[17] produced a moisture index by dividing evaporation from a Livingston porous cup atmometer by annual precipitation and made a moisture map of Ohio.

Numerous investigators, recognizing that temperature is the major control of evaporation, have substituted temperature for evaporation in their moisture indices. The Lang rain factor, $I = P/T$, indicates that the effectiveness varies directly with precipitation and inversely with temperature. De Martonne's index of aridity, $I = P/(T + 10)$, is a slight refinement of Lang's. Köppen's three formulae for delimiting the dry climates, $I = 8P/(5T + 120)$, $I = 2P/(T + 33)$, and $I = P/(T + 7)$, presented in 1918, 1923, and 1928 respectively, are similar to those of Lang and De Martonne. All use annual values of precipitation and temperature given in the metric system.

In 1931 Thornthwaite[18] utilized evaporation data from the continental United States and derived an empirical equation by which the precipitation-evaporation ratio could be determined from monthly values of precipitation and temperature. He used this moisture index in making a climatic map of North America and a later one of the earth.[19]

In 1936 Ångström[20] suggested a modification of De Martonne's index of aridity. He found that the index of aridity was proportionate to duration of precipitation, which was in turn directly proportionate to amount of precipitation and inversely proportionate to an exponential function of temperature. His humidity coefficient is $I = P/(1.07^t)$, in which the denominator of the fraction doubles with each rise of 10° C. in temperature, in accordance with Van't Hoff's law. Ångström published maps of northwestern Europe showing the humidity coefficient in January and July. In 1939 Church and Gueffroy[21] used the Ångström formula to make similar maps of the United States. In 1946 Setzer[22] utilized Van't Hoff's law in the

[17] J. D. Wilson and J. R. Savage: An Evaporation Survey of Ohio, *Ohio Agric. Exper. Sta. Bull.* 564, 1936.

[18] C. W. Thornthwaite: The Climates of North America According to a New Classification, *Geogr. Rev.*, Vol. 21, 1931, pp. 633–655.

[19] *Idem:* The Climates of the Earth, *Geogr. Rev.*, Vol. 23, 1933, pp. 433–440.

[20] Anders Ångström: A Coefficient of Humidity of General Applicability, *Geografiska Annaler*, Vol. 18, 1936, pp. 245–254.

[21] P. E. Church and E. M. Gueffroy: A New Coefficient of Humidity and Its Application to the United States, *Geogr. Rev.*, Vol. 29, 1939, pp. 665–667.

[22] José Setzer: A New Formula for Precipitation Effectiveness, *Geogr. Rev.*, Vol. 36, 1946, pp. 247–263.

development of a moisture index. Although developed independently, Setzer's index is identical with Ångström's. Setzer published a map of the moisture index in the state of São Paulo, Brazil.

Ångström[23] has ably explained what the investigator wishes to accomplish with moisture indices:

> Especially with geographical, geological, or biological problems in view, [one has] tried to express the humidity or aridity of a climate through some kind of coefficient which is expected to be a measure on conditions of humidity not in the simple physical sense, but involving an indication of the climatic tendency to a surplus or deficiency of water with consequences upon the structure of the ground, the existence of rivers, floods, lakes or deserts etc. It is evident that the humidity of the soil is hereby a prime factor, and as this element is more seldom directly measured it seems justified to try to replace it by some kind of coefficient deduced from the current meteorological data.

A Moisture Index

It is now apparent that the actual evaporation and transpiration from the soil is not what must be compared with precipitation in order to obtain a moisture index, but, rather, the potential evapotranspiration. Where precipitation is exactly the same as potential evapotranspiration all the time and water is available just as needed, there is neither water deficiency nor water excess, and the climate is neither moist nor dry. As water deficiency becomes larger with respect to potential evapotranspiration, the climate becomes arid; as water surplus becomes larger, the climate becomes more humid. Where there is a water surplus and no water deficiency, the relation between water surplus and water need constitutes an index of humidity. Similarly, where there is a water deficiency and no surplus, the ratio between water deficiency and water need constitutes an index of aridity. Expressed as percentages these two indices are:

$$I_h = \frac{100s}{n} \quad \text{and} \quad I_a = \frac{100d}{n}, \tag{4}$$

where I_h and I_a are indices of humidity and aridity respectively, s is water surplus, d is water deficiency, and n is water need. The ultimate in the scale of aridity is where there is no precipitation and the water deficiency is consequently equal to the water need, making I_a equal to 100 per cent. As precipitation and potential evapotranspiration are independent of each other, the index of humidity does not reach a limit where water surplus equals water need, that is, where precipitation is twice potential evapotranspiration, but continues above 100 per cent.

Since water surplus and water deficiency occur at different seasons in

---

[23] *Op. cit.*, p. 245.

most places, both must enter into a moisture index, the one affecting it positively, the other negatively. Although a water surplus in one season cannot prevent a deficiency in another except as moisture may be stored in the soil, to a certain extent one may compensate for the other. Water surplus means seasonal additions to subsoil moisture and ground water. Deeply rooted perennials may make partial use of subsoil moisture and thus minimize the effect of drought. Transpiration proceeds, but at reduced rates. For this reason, a surplus of only 6 inches in one season will counteract a deficiency of 10 inches in another. Thus in an over-all moisture index the humidity index has more weight than the aridity index: the latter has only six-tenths the value of the former. The moisture index is:

$$I_m = I_h - .6I_a \text{ or } I_m = \frac{100s - 60d}{n}. \quad (5)$$

Moist climates have positive values of I_m; dry climates have negative values. Figure 7 shows how climatic types are separated in terms of the moisture index, I_m, and makes clear how they are related to water surplus and water deficiency. The various climatic types together with their limits are as follows:

Climatic Type		Moisture Index
A	Perhumid	100 and above
B4	Humid	80 to 100
B3	Humid	60 to 80
B2	Humid	40 to 60
B1	Humid	20 to 40
C2	Moist subhumid	0 to 20
C1	Dry subhumid	−20 to 0
D	Semiarid	−40 to −20
E	Arid	−60 to −40

The index values −60, 0, and 100 are entirely rational limits of moisture regions. That the others are also may be seen in the nomograms of Figure 7. These definitions of moisture regions appear to be in final form. Further work is needed to improve the means for determining potential evapotranspiration, moisture surplus, and moisture deficiency. This may lead to revision of the location of the moisture regions but will not change the definition of them.

These climatic types are the same, and have the same meaning, as those proposed in an earlier climatic classification.[24] However, whereas the limits in the previous classification were determined empirically by study of vegetation, soils, drainage patterns, and so on, these limits are rational and

[24] Thornthwaite, The Climates of North America (*op. cit.*).

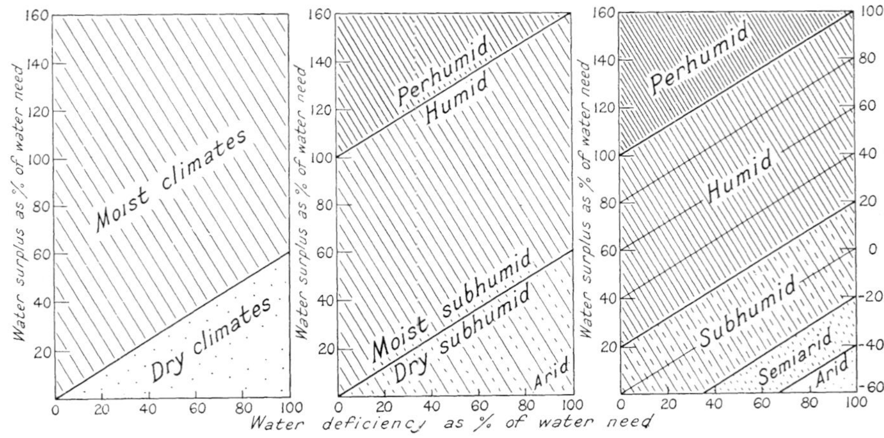

FIG. 7—Delineation of moisture regions on basis of water surplus and water deficiency.

are established solely in terms of the relation between potential evapotranspiration and precipitation. It is possible to convert the present moisture index, I, to the old P–E index by means of the following equation:

$$P\text{–}E = .8I + 48. \qquad (6)$$

The distribution of the moisture regions of the United States is shown in Plate I A. The moist climates of the East and the dry climates of the West are separated by moisture index 0, in a line that extends from western Minnesota southward to the Gulf of Mexico. This is a very important line, since it separates regions of prevailing water surplus from those of water deficiency. Although dry climates predominate in the western United States, there is a belt of moist climates along the Pacific coast as far south as the San Francisco peninsula. Moist climates appear also as islands in the western mountains.

Perhumid climates are not extensive in the United States. They occur along the coasts of Washington, Oregon, and northern California and on the western slopes of the Cascades and the Sierras. There is a single small island in the high Rockies of Colorado, and other islands on high elevations in the Appalachians. A narrow belt along the Maine coast is perhumid. Humid climates are most extensive in the East, but they occur adjacent to perhumid climates on the Pacific coast and on high areas elsewhere in the West. Subhumid climates are most extensive in the Middle West, wide belts of moist subhumid and dry subhumid extending from the Canadian border in Minnesota and North Dakota to the Gulf coast of Texas. Much of the Florida peninsula is moist subhumid. Smaller subhumid areas occur in the West, mostly as belts along the lower mountain slopes. The Great

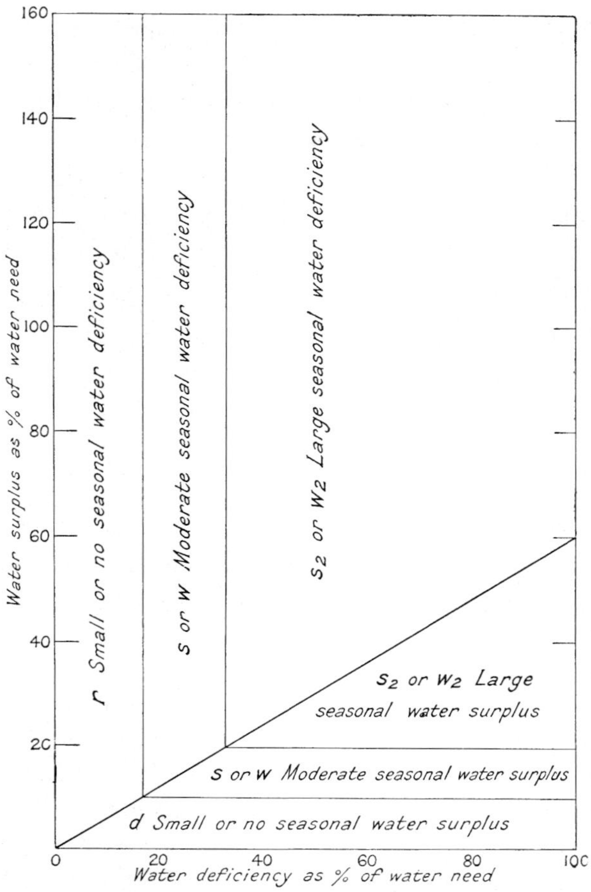

FIG. 8—Nomogram to determine seasonal regime of moisture.

Plains and much of the intermountain region are semiarid. Arid climate is most extensive in the Southwest. There are, however, small arid areas in Washington, Oregon, Idaho, Montana, Wyoming, Utah, and Colorado. Most of the San Joaquin Valley of California is arid.

SEASONAL VARIATION OF EFFECTIVE MOISTURE

It is important to know whether a place is continuously wet or continuously dry, or whether it is wet in one season and dry in another. The moisture index can indicate how humid or how arid a climate is, but it cannot at the same time distinguish climates with seasonal moisture variation from those without it. In the moist climates, if there is a dry season, we need to know how dry it is; in the dry climates, how wet the wet season is if one occurs. In moist climates water deficiency may be large, moderate, small, or nonexistent. In dry climates the same is true of water surplus (see Fig. 8).

By definition, a large water deficiency or surplus is an amount sufficient to make the climate one grade drier or moister than it would be otherwise. For example, in San Francisco the water surplus of 6.68 inches (24.6 per cent of the water need) makes the climate less dry, changing it from semiarid (D) to dry subhumid (C_1). Los Angeles, with a somewhat larger water deficiency and a smaller winter surplus, remains semiarid (D). A moderate water surplus or deficiency will, by definition, change the climate one-half grade.

The symbols s and s_2 indicate respectively moderate and large seasonal variation of moisture, with the drier season occurring in summer; w and w_2 are used similarly where the drier season occurs in winter. Where water need and precipitation march through the year approximately parallel to each other, there will be little or no seasonal variation of moisture. Here the symbols r and d are used, the first to designate the areas with little or no water deficiency in moist climates, and the second little or no water surplus in dry climates. The climatic subdivisions are defined in terms of the humidity and aridity indices as follows:

Moist Climates (A, B, C_2)		Aridity Index
r	little or no water deficiency	0 – 16.7
s	moderate summer water deficiency	16.7 – 33.3
w	moderate winter water deficiency	16.7 – 33.3
s_2	large summer water deficiency	33.3+
w_2	large winter water deficiency	33.3+
Dry Climates (C_1, D, E)		**Humidity Index**
d	little or no water surplus	0 – 10
s	moderate winter water surplus	10 – 20
w	moderate summer water surplus	10 – 20
s_2	large winter water surplus	20+
w_2	large summer water surplus	20+

The symbols s, s_2, w, and w_2 have the same meaning in both moist and dry climates in spite of the fact that they are defined differently. They refer to the season when rainfall is most deficient.

The winter dry, w and w_2, types do not occur in the United States. In fact, these types are much less widely distributed over the earth than would be deduced from a study of the seasonal march of precipitation. Water need is naturally larger in summer than in winter; and in many areas of maximum summer rainfall, water need and precipitation are essentially together through the year, and there is neither summer water surplus nor winter water deficiency. For example, in Nanking, China, the average precipitation of the three summer months is 55.0 centimeters and of the winter months only 12.0; but since water need varies even more widely between winter and summer, the climate is not of the w type. The water surplus occurs in late winter and spring, not in summer.

In Figure 9 representative stations are grouped according to whether the seasonal range of effective moisture is small, moderate, or large. Those in which precipitation most nearly parallels potential evapotranspiration and both water surplus and water deficiency are small are Ames, Iowa (C_2r), Alexandria, Minn. (C_2r), and Grafton, N. Dak. (C_1d).

The distribution in the United States of the climatic subtypes based on

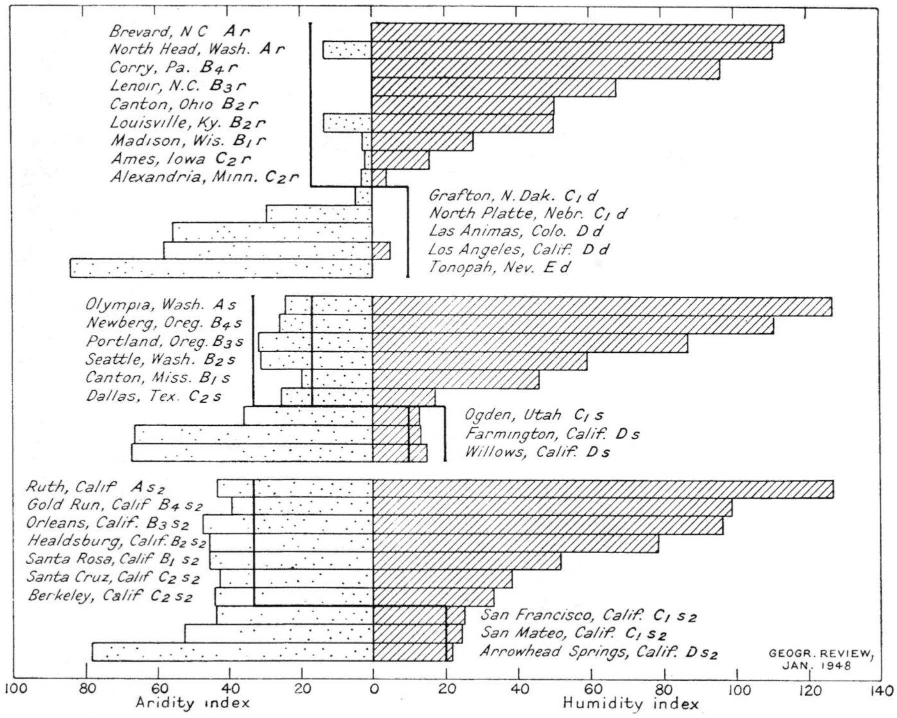

FIG. 9—Seasonal regime of moisture at selected stations.

seasonal variation of effective moisture is shown in Plate I B. Types r and d, representing small seasonal variation, are most extensive. Types s and s_2 prevail in the Pacific-coast region, where precipitation comes mainly in winter; s_2 does not occur except where the march of precipitation is opposite to that of water need and for this reason has its most extensive development in California. Type s may occur as well in regions of prevailing summer rain if summer water need sufficiently exceeds it. There are several areas of s climate in the southern United States and a few isolated stations in the Northeast. Canton, Miss. (B_1s), and Dallas, Tex. (C_2s), are examples. A large s area extends from eastern Texas and central Oklahoma eastward into Arkansas and Louisiana. A smaller area occupies the lower Mississippi Valley, and another crosses Alabama from Georgia to Mississippi along the well-known Black Belt.

The existence of the s type in the southeastern United States is surprising. Summer rainfall is fairly large, and it is not generally realized that water need is larger. Summer drought is serious in much of the eastern United States, even in the r type climates. In the s type summer drought is chronic. Supplementary irrigation becomes an important need.

AN INDEX OF THERMAL EFFICIENCY

Potential evapotranspiration is an index of thermal efficiency. It possesses the virtue of being an expression of day length as well as of temperature. It is not merely a growth index but expresses growth in terms of the water that is needed for growth. Given in the same units as precipitation it relates thermal efficiency to precipitation effectiveness.

In equatorial regions, where mean monthly temperature does not vary appreciably through the year, a mean annual temperature of 23° C. (73.4° F.) is a reasonable boundary between megathermal and mesothermal climates. Away from the equator, where there is a seasonal variation of temperature, mean annual temperature at the boundary is lower, 21.5° C. (70.7° F.), because the reduced growth and water need of winter are more than offset by an accelerated growth and an increased water need of summer. Where the mean temperature of every month is 23°, and if the station is directly on the equator, so that there is no variation in day length, the potential evapotranspiration is 114.0 centimeters (44.88 inches). This may be taken as the index that separates megathermal and mesothermal climates. Other boundaries in the series are in a descending geometric progression (p. 82).

Climatic regions based on temperature efficiency are analogous to those derived from the moisture index and are designated by similar symbols. Microthermal (C′) and mesothermal (B′) climates are subdivided like the subhumid (C) and humid (B) climates.[25] The distribution of the thermal regions in the United States is shown in Plate I C. Mesothermal (B′) climates are most extensive. Microthermal (C′) climates occur along most of the Canadian border and extend southward along the Rocky Mountains in a continuous belt to southern New Mexico. There are three areas of megathermal (A′) climate: southern Florida, southern Texas, and the arid parts of southern California and southwestern Arizona. Only small isolated areas of tundra (D′) occur, in the high Sierras of California, the northern Rockies of Wyoming and Montana, and the central Rockies of Colorado. There are no instrumental records of frost (E′) climate in the United States, and it does not appear on the map.

The megathermal (A′) climates of the United States are different from those of equatorial regions. Summer days are longer and hotter, and winter days are shorter and colder. Temperature and day length work together

[25] Although the symbols are the same as those used in the original classification, the climatic limits differ considerably from the old ones. Consequently, the symbols have different meanings. For example, in the original scheme tundra was E′ and frost climate F′. Mesothermal climates are more extensive than previously shown.

to bring about seasonal variations in potential evapotranspiration or temperature efficiency. The difference is illustrated in Figure 10-A. Whereas at Belém, Brazil, the range is only from 10.3 to 13.1 centimeters, at Barahona, Dominican Republic, it is from 8.2 to 16.2 centimeters, and at Miami, Fla., from 5.2 to 17.8 centimeters. At the mesothermal border the mean tem-

TE Index		Climatic Type	
cm.	*in.*		
		E'	Frost
14.2	5.61		
		D'	Tundra
28.5	11.22		
		C'_1	
42.7	16.83		Microthermal
		C'_2	
57.0	22.44		
		B'_1	
71.2	28.05		
		B'_2	
85.5	33.66		Mesothermal
		B'_3	
99.7	39.27		
		B'_4	
114.0	44.88		
		A'	Megathermal

perature of the coldest month is as low as 50° F. in some stations. Freezing temperatures may occur almost annually, whereas in equatorial stations frosts are entirely unknown. For this reason certain plants popularly associated with tropical climates do not grow in the megathermal climates of the United States. However, plant growth continues, although at reduced rates, throughout the winter.

The mesothermal (B′) climates occupy a wide range of latitudes in the eastern and central United States. Summer temperatures fall off slowly from south to north: the mean July temperature in the lower Rio Grande Valley of Texas is 85° F., and it is above 70° F. in the Yellowstone Valley of Montana. Day length, on the other hand, increases from south to north—about 14 hours in southern Texas and 16 hours in northern Montana. Consequently, average July potential evapotranspiration decreases slowly: it is about 18 centimeters in southern Texas and 15 centimeters in Montana. The fact that winter temperatures and day length both decrease from south to north has less influence because potential evapotranspiration is small even on the southern margin of the climatic zone. However, the growing season becomes shorter, and the northern edge of the zone, where annual potential evapotranspiration is only 50 per cent of the maximum for the climatic type, is reached near the Canadian border. The four subdivisions of the

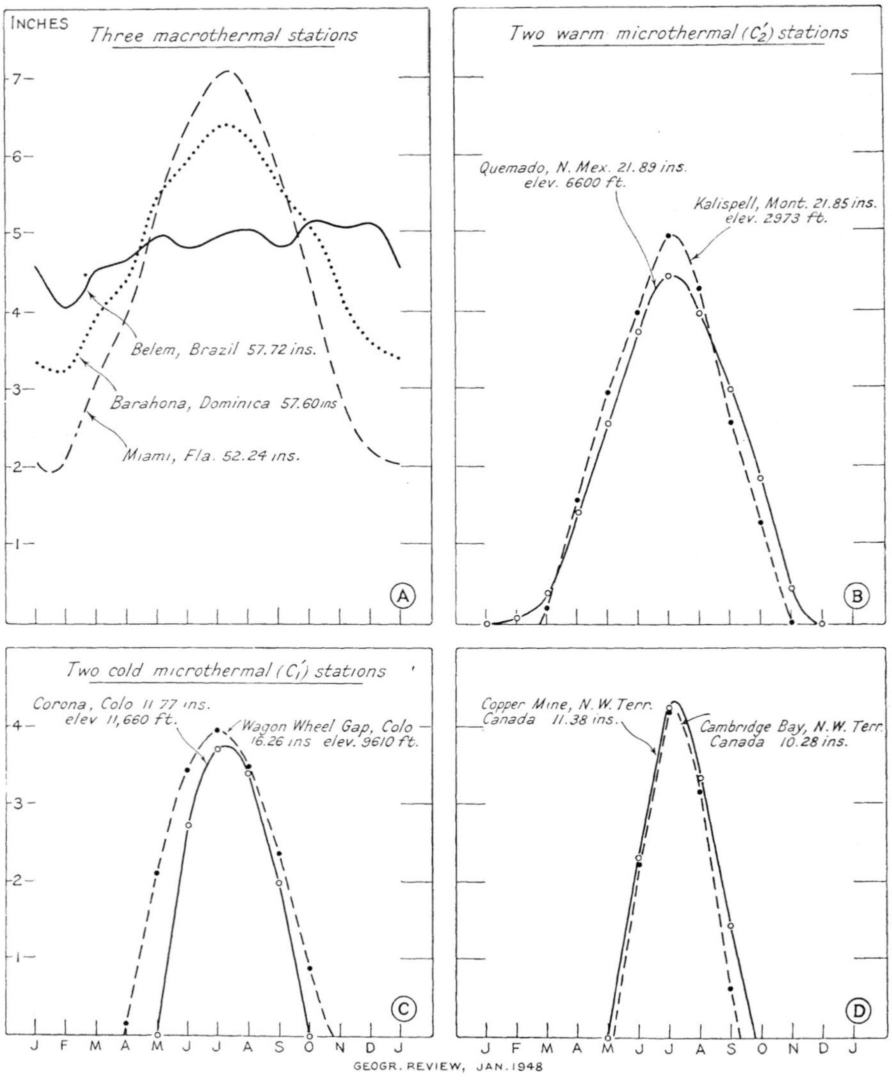

FIG. 10—March of potential evaporation. For *macrothermal* read *megathermal* (a preferable form, suggested by Mr. A. B. Hoen).

mesothermal climates appear to have real validity, but since no appropriate names for them have been found, they are identified only by subscripts 1 to 4 following symbol B′.

Microthermal (C′) climates extend continuously from the Canadian border in Montana to within 200 miles of the Mexican border in New Mexico. They are at high elevations in the southern part of their range and descend to lower elevations in the north. Quemado, N. Mex., and Kalispell, Mont., have the same annual potential evapotranspiration and are both warm microthermal (C'_2); the first is 6600 feet above sea level, the second only 2973 feet. Figure 10-B shows the march of potential evapotranspiration in these two stations. Spring and summer temperatures are considerably lower in Kalispell. The growing season is shorter, but since summer days are longer, higher rates of water use and more rapid growth compensate for shorter time of plant activity.

The colder type of microthermal climates (C'_1) is not areally important in the United States.[26] The two most extensive areas are in Wyoming and Colorado, though smaller areas are found in several western states. Wagon Wheel Gap, Colo., is near the warm limit of the type, and Corona, Colo., is near the cold limit (Fig. 10-C). Although the temperatures of the warmest months are low (Corona mean July temperature, 48.8° F.; Wagon Wheel Gap, 56.2° F.), the potential evapotranspiration is surprisingly large, nearly 10 centimeters (4 inches). Annual precipitation is in excess of water need in both places, but summer rainfall is less than summer potential evapotranspiration. At Wagon Wheel Gap summer water deficiency amounts to 2.2 inches. Short growing season and low summer temperature do not prevent droughts.

No currently reporting Weather Bureau station in the United States has tundra (D') climate. However, it undoubtedly occurs in various isolated high areas, mostly in the West. The summit of Mt. Washington, in New Hampshire, has an annual water need of 12.9 inches, which places it on the tundra margin. Figure 10-D shows the march of potential evapotranspiration in two stations in the far north of Canada. Both are very near the microthermal-tundra boundary; Coppermine is just above the line, and Cambridge Bay just below. Although the mean temperature of July, the warmest month, is only about 50° F. in both places, the water need is astonishingly high, being above 10 centimeters (4 inches). Average annual potential evapotranspiration exceeds precipitation in both places, and summer drought is a characteristic feature of the climate. In southern and eastern Greenland, on the other hand, precipitation greatly exceeds evapotranspiration, and there is a large water surplus, a part of which nourishes the icecap, and a part of which runs off in summer.

In the climate of perpetual frost (E') temperatures need not remain continuously below the freezing point. Although frost effectively prevents vegetation growth, and thus transpiration, nevertheless some water passes into the atmosphere by evaporation or sublimation. At the outer limit of the frost climate potential evapotranspiration is 14.2 centimeters (5.61

[26] This type is equivalent to taiga (D') in the previous classification.

inches). It is evident that even in the frost climate moisture differentiation is important, because if precipitation exceeds potential evapotranspiration, ice will accumulate and a glacier will develop. On the other hand, if precipitation is less than this critical value, ice accumulation is impossible. It is not certain that true frost (E′) climate exists anywhere in the United States. A snow line or mountain glacier does not necessarily indicate a frost climate.

Summer Concentration of Thermal Efficiency

At the equator, where day length is the same throughout the year and where temperature is also uniform, seasonal variation of potential evapotranspiration will be small. With no variation no season can be called summer, and the potential evapotranspiration of any consecutive three months will constitute 25 per cent of the annual total. On the other hand, in the Polar Regions, where the growing season is entirely within the three summer months, the potential evapotranspiration of these months will constitute 100 per cent of the total. Between these extremes, as potential evapotranspiration falls from that characteristic of megathermal (A′) climates to that of frost (E′) climates, the part that is concentrated in summer gradually rises from 25 per cent to 100 per cent. This rise results from an increasing length of midsummer days, and an increase in the length of winter, with increase in latitude.

The astronomical motions of the earth tend to produce a certain fixed relation between the summer index of thermal efficiency and the annual index. This relation has been approximately determined from a series of stations in the interior low plains of North America. The summer concentration appears to be inversely proportionate to the logarithm of the annual index. The relation is described by the equation

$$s = 157.76 - 66.44 \log E, \tag{7}$$

in which s is summer concentration in percentages and E is potential evapotranspiration in inches.

But there are a number of meteorological situations in which the relation is modified or altered and is, in consequence, abnormal. On small oceanic islands and seacoasts dominated by onshore winds in middle and high latitudes summer temperatures are less warm, and winter temperatures less cold, than elsewhere in these latitudes. The summer concentration of thermal efficiency is lower than it should be for the latitude. In mountains there is a more or less uniform reduction of temperature throughout the year. Annual and summer thermal-efficiency indices are both reduced, but not in the same proportion. Here, too, the summer concentration of thermal

Table III—Comparative Moisture Data of Selected Stations

In inches

Station	Water need	Summer need %	Precip-itation	Water surplus	Water def.	Surplus % of need	Def. % of need	Moisture index	Climatic type
A. Brevard, N. C.	28.50	51.1	61.06	32.56	0.00	114.2	0.	114.2	$A\,B'_2rb'_4$
B. Salisbury, N. Y.	21.80	67.2	48.39	26.58	0.00	121.8	0.	121.8	$A\,C'_2rb'_1$
C. Bar Harbor, Maine	22.32	59.1	47.09	25.55	0.00	114.5	0.	114.5	$A\,C'_2rb'_2$
D. North Head, Wash.	24.72	41.1	48.86	27.40	3.27	110.8	13.2	102.9	$A\,B'_1ra'$
E. Louisville, Ky.	31.77	55.5	43.47	15.95	4.28	50.2	13.5	42.2	$B_2B'_2rb'_3$
F. Seattle, Wash.	26.14	46.4	33.35	15.43	8.23	59.0	31.5	40.2	$B_2B'_1sa'$
G. Canton, Miss.	38.70	50.6	49.10	17.68	7.68	45.6	19.8	33.8	$B_1B'_3sb'_4$
H. Willard, N. C.	35.35	49.7	49.05	13.70	0.00	38.8	0.	38.8	$B_1B'_3rb'_4$
I. Madison, Wis.	25.12	61.4	31.54	7.09	0.67	28.2	2.7	26.6	$B_1B'_1rb'_2$
J. Centennial, Wyo.	18.86	63.7	17.68	4.49	5.67	23.8	30.1	5.7	$C_2C'_2sb'_1$
K. Manhattan, Kans.	30.83	57.8	31.50	2.28	1.61	7.4	5.2	4.3	$C_2B'_2rb'_2$
L. Pullman, Wash.	24.13	58.1	20.71	7.48	10.91	31.0	45.3	3.9	$C_2B'_1s_2b'_2$
M. San Francisco, Calif.	27.09	33.3	21.88	6.68	11.89	24.6	43.8	−1.7	$C_1B'_1s_2a'$
N. Dalhart, Tex.	29.02	56.9	19.09	0.00	9.92	0.	34.2	−20.5	$D\,B'_2db'_2$
O. Helena, Mont.	22.32	48.1	13.31	0.00	9.01	0.	40.4	−24.2	$D\,C'_2db'_4$
P. Santa Fe, N. Mex.	24.41	56.1	14.21	0.00	10.19	0.	41.7	−25.1	$D\,B'_1db'_3$
Q. Los Angeles, Calif.	32.01	38.7	14.96	1.42	18.47	4.4	57.7	−30.2	$D\,B'_2da'$
R. Grand Junction, Colo.	28.74	59.2	8.70	0.00	20.04	0.	69.9	−41.8	$E\,B'_2db'_2$
S. Fresno, Calif.	36.22	52.8	9.17	0.00	27.04	0.	74.7	−44.8	$E\,B'_3db'_3$
T. Tonopah, Nev.	25.43	58.7	4.84	0.00	21.38	0.	84.0	−50.4	$E\,B'_1db'_2$

efficiency is abnormally low. On the other hand, in some regions the normal seasonal march of temperature is exaggerated. Summer temperatures are increased by advection of hot tropical air, and winter temperatures are lowered by advection of cold polar air. In these regions the summer concentration is abnormally high.

Potential Evapotranspiration *Ins.*	Potential Evapotranspiration *Cms.*	Temperature-Efficiency Type	Summer Concentration *Percentage*	Summer Concentration *Type*
		A'		a'
44.88	114.0		48.0	
		B'_4		b'_4
39.27	99.7		51.9	
		B'_3		b'_3
33.66	85.5		56.3	
		B'_2		b'_2
28.05	71.2		61.6	
		B'_1		b'_1
22.44	57.0		68.0	
		C'_2		c'_2
16.83	42.7		76.3	
		C'_1		c'_1
11.22	28.5		88.0	
		D'		d'
5.61	14.2			
		E'		

The extent to which the summer-concentration percentage fails to meet the above requirements is a measure of its abnormality. San Francisco, for example, has a potential evapotranspiration of 27.09 inches and a summer concentration of 33.3 per cent. Thus summer concentration is that of full megathermal (a') climate, though the temperature-efficiency type is actually only first mesothermal (B'_1). San Francisco is an example of a so-called "marine" climate.

Summer concentration of thermal efficiency in the United States is shown in Plate I D. The displacement of the various zones from the comparable ones of thermal efficiency (Pl. I C) is a measure of oceanity or continentality.

The abnormality of the summer-concentration percentage deserves special study, but it will be more profitable on a world basis than for the United States alone.

Elements of the Classification

Four symbols used together give a complete description of a climate. The necessary data for classifying the stations in Figure 4 are presented in Table III. Water need, in the first column, is, of course, potential evapotranspiration. The second column gives the percentage that summer potential evapotranspiration is of the annual total. The column labeled "Surplus as percentage of need" gives the humidity index, and that labeled "Deficiency as percentage of need" the index of aridity. The moisture index is obtained by subtracting six-tenths of the latter from the former.

The various subdivisions of mesothermal, microthermal, and humid climatic types do not have individual names but can be referred to only by symbol. Thus we can say first, second, third, or fourth mesothermal. Brevard, N. C. ($AB'_2rb'_4$), is perhumid, second mesothermal, with no season of water deficiency, and a temperature-efficiency regime normal to fourth mesothermal. San Francisco, Calif. ($C_1B'_1s_2a'$), is dry subhumid, first mesothermal, with large winter water surplus, and a temperature-efficiency regime normal to megathermal.

Superficially the present system is similar to its predecessor[27] in that the same factors are employed; namely a moisture factor, a heat factor, and the seasonal variation of the two. Actually, the two systems are fundamentally different. In the earlier classification, climatic types were identified, and boundaries were located, empirically, through study of the distribution of vegetation, soils, drainage features, and so on. In the present classification, climates are defined rationally, and boundaries are determined by the data.

The difference may be illustrated by the change in point of view respecting vegetation. The earlier study adopted Köppen's position that the plant is a meteorological instrument which integrates the various factors of climate and which, with experience, can be "read" like a thermometer or a rain gauge. In the present study, vegetation is regarded as a physical mechanism by means of which water is transported from the soil to the atmosphere; it is the machinery of evaporation as the cloud is the machinery of precipitation.

Climatic boundaries are determined rationally by comparing precipitation and evapotranspiration. The subdivisions of the older classification were justly criticized as being vegetation regions climatically determined. The present climatic regions are not open to this criticism, since they come from a study of the climatic data themselves and not from a study of vegetation.

This classification can be improved. A first step will be to develop better means of determining potential evapotranspiration. Additional observations are needed, particularly in the tropics and in high latitudes. With new data available, the present formula can be revised, or perhaps a new and more rational formula can be devised. A truly rational method of delimiting the temperature-efficiency regions has not yet been developed. A relation be-

---

[27] Thornthwaite, The Climates of North America (*op. cit.*).

tween the heat factor and the moisture factor may exist that will provide the rational basis. So far it has not been discovered.

There is an encouraging prospect that this climatic classification which is developed independently of other geographical factors such as vegetation, soils, and land use, may provide the key to their geographical distribution.

Soil-forming processes are related to water surpluses and water deficiencies; so are hydrological regimes and drainage patterns. The problem of the origin of the prairies may come near solution in an analysis of annual frequency of water deficiency and water surplus. Furthermore, much can be learned regarding soil productivity and best land use through a study of magnitude and frequency of water deficiency. Finally, we have a better understanding than ever before of the qualities of a climate when we are able to compare the potential evapotranspiration through the year with the precipitation.

APPENDIX I

THE DETERMINATION OF POTENTIAL EVAPOTRANSPIRATION

The relation between mean monthly temperature and potential evapotranspiration adjusted to a standard month of 30 days, each having 12 hours of possible sunshine, in four selected areas is shown in Figure 11. In each area the relationship within the range of temperature involved is well expressed by an equation of the form

$$e = ct^a, \qquad (8)$$

in which e is monthly evapotranspiration in centimeters and t is mean monthly temperature in °C. From these observations it is seen that there is no simple relationship between monthly evapotranspiration and monthly temperature. The coefficients c and a vary from one place to another. Thus an equation having coefficients derived from observations made in a warm climate does not yield correct values of potential evapotranspiration for an area having a cold climate, and vice versa. In Figure 12 the lines showing the relationship between temperature and evapotranspiration in several areas tend to converge where potential evapotranspiration is 13.5 centimeters and temperature is 26.5° C. At lower temperatures there is increasing divergence in potential evapotranspiration.

In a general equation constants c and a must be allowed to vary with a factor that is small in cold climates and large in hot climates. Mean annual temperature is not satisfactory because in some places it is affected by below-freezing temperatures. A special equation was developed for the purpose. A monthly index is obtained from the equation $i = (t/5)^{1.514}$. Summation of the 12 monthly values gives an appropriate heat index, I. While this index varies from 0 to 160, the exponent a in the above equation varies from 0 to 4.25. The relation between the two is closely approximated by the expression

$$a = 0.000000675I^3 - 0.0000771I^2 + 0.01792I + 0.49239. \qquad (9)$$

The coefficient c in equation (8) above varies inversely with I. From these relations a general equation for potential evapotranspiration was obtained. It is

$$e = 1.6\,(10t/I)^a, \tag{10}$$

in which a has the value given in equation (9).

The formula gives unadjusted rates of potential evapotranspiration. Since the number of

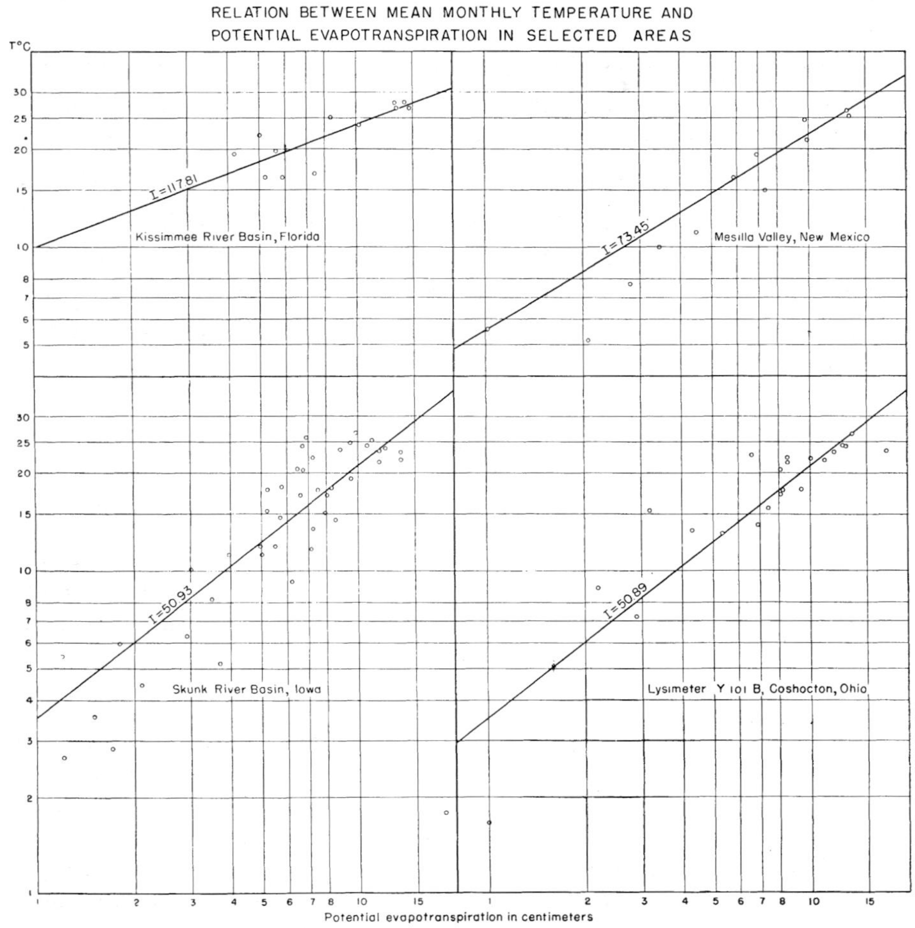

FIG. 11

days in a month ranges from 28 to 31 (nearly 11 per cent) and the number of hours in the day between sunrise and sunset, when evapotranspiration principally takes place, varies with the season and with latitude, it becomes necessary to reduce or increase the unadjusted rates by a factor that varies with the month and with latitude.

This mathematical development is far from satisfactory. It is empirical, and the general equation does not accord with the newly developed law of growth. Furthermore, the equation is completely lacking in mathematical elegance. It is very complicated and without nomograms and tables as computing aids would be quite unworkable. The chief obstacle at present to the development of a rational equation is the lack of understanding of why potential evapotranspiration corresponding to a given temperature is not the same everywhere. Not until this matter is understood will a rational method be possible.

In spite of the lack of a theoretical foundation, this method makes it possible to arrive at values of potential evapotranspiration that are approximately correct and that give an entirely new approach to the problems of climatic classification.

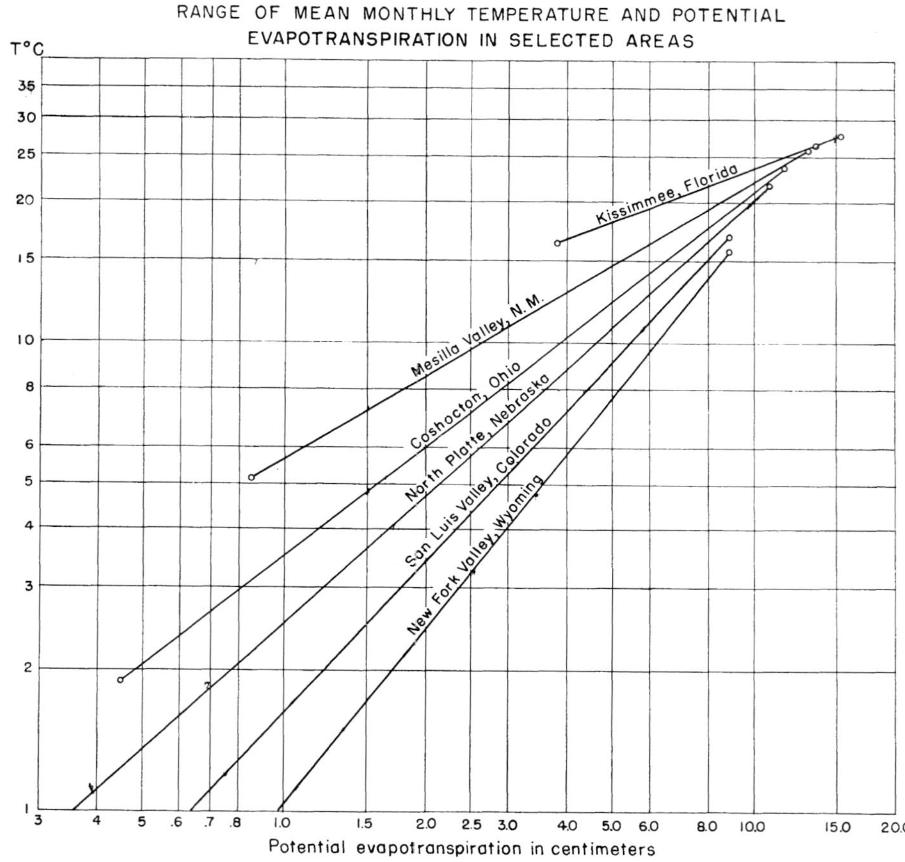

FIG. 12

In order to determine potential evapotranspiration, mean monthly values of temperature must be available, and the latitude of the station must be known. Three steps are involved in the computation, and all three are accomplished by means of a nomogram and tables.

The first step is to obtain the heat index, *I*. Table IV gives monthly values of *i* corresponding to monthly mean temperatures. Summation of the 12 monthly values gives the index, *I*.

The next step is to determine unadjusted values of potential evapotranspiration from the nomogram in Figure 13.[28] Since there is a linear relation between the logarithm of temperature and the logarithm of unadjusted potential evapotranspiration, straight lines on the nomogram define the relationship. All lines pass through the point of convergence at

[28] A table has been prepared for this step, but it is too long to publish here. It will be placed on file at the American Geographical Society.

TABLE IV

T°C	.0	.1	.2	.3	.4	.5	.6	.7	.8	.9
0			.01	.01	.02	.03	.04	.05	.06	.07
1	.09	.10	.12	.13	.15	.16	.18	.20	.21	.23
2	.25	.27	.29	.31	.33	.35	.37	.39	.42	.44
3	.46	.48	.51	.53	.56	.58	.61	.63	.66	.69
4	.71	.74	.77	.80	.82	.85	.88	.91	.94	.97
5	1.00	1.03	1.06	1.09	1.12	1.16	1.19	1.22	1.25	1.29
6	1.32	1.35	1.39	1.42	1.45	1.49	1.52	1.56	1.59	1.63
7	1.66	1.70	1.74	1.77	1.81	1.85	1.89	1.92	1.96	2.00
8	2.04	2.08	2.12	2.15	2.19	2.23	2.27	2.31	2.35	2.39
9	2.44	2.48	2.52	2.56	2.60	2.64	2.69	2.73	2.77	2.81
10	2.86	2.90	2.94	2.99	3.03	3.08	3.12	3.16	3.21	3.25
11	3.30	3.34	3.39	3.44	3.48	3.53	3.58	3.62	3.67	3.72
12	3.76	3.81	3.86	3.91	3.96	4.00	4.05	4.10	4.15	4.20
13	4.25	4.30	4.35	4.40	4.45	4.50	4.55	4.60	4.65	4.70
14	4.75	4.81	4.86	4.91	4.96	5.01	5.07	5.12	5.17	5.22
15	5.28	5.33	5.38	5.44	5.49	5.55	5.60	5.65	5.71	5.76
16	5.82	5.87	5.93	5.98	6.04	6.10	6.15	6.21	6.26	6.32
17	6.38	6.44	6.49	6.55	6.61	6.66	6.72	6.78	6.84	6.90
18	6.95	7.01	7.07	7.13	7.19	7.25	7.31	7.37	7.43	7.49
19	7.55	7.61	7.67	7.73	7 79	7.85	7.91	7.97	8.03	8.10
20	8.16	8.22	8.28	8.34	8.41	8.47	8.53	8.59	8.66	8.72
21	8.78	8.85	8.91	8.97	9.04	9.10	9.17	9.23	9.29	9.36
22	9.42	9.49	9.55	9.62	9.68	9.75	9.82	9.88	9.95	10.01
23	10.08	10.15	10.21	10.28	10.35	10.41	10.48	10.55	10.62	10.68
24	10.75	10.82	10.89	10.95	11.02	11.09	11.16	11.23	11.30	11.37
25	11.44	11.50	11.57	11.64	11.71	11.78	11.85	11.92	11.99	12.06
26	12.13	12.21	12.28	12.35	12.42	12.49	12.56	12.63	12.70	12.78
27	12.85	12.92	12.99	13.07	13.14	13.21	13.28	13.36	13.43	13.50
28	13.58	13.65	13.72	13.80	13.87	13.94	14.02	14.09	14.17	14.24
29	14.32	14.39	14.47	14.54	14.62	14.69	14.77	14.84	14.92	14.99
30	15.07	15.15	15.22	15.30	15.38	15.45	15.53	15.61	15.68	15.76
31	15.84	15.92	15.99	16.07	16.15	16.23	16.30	16.38	16.46	16.54
32	16.62	16.70	16.78	16.85	16.93	17.01	17.09	17.17	17.25	17.33
33	17.41	17.49	17.57	17.65	17.73	17.81	17.89	17.97	18.05	18.13
34	18.22	18.30	18.38	18.46	18.54	18.62	18.70	18.79	18.87	18.95
35	19.03	19.11	19.20	19.28	19.36	19.45	19.53	19.61	19.69	19.78
36	19.86	19.95	20.03	20.11	20.20	20.28	20.36	20.45	20.53	20.62
37	20.70	20.79	20.87	20.96	21.04	21.13	21.21	21.30	21.38	21.47
38	21.56	21.64	21.73	21.81	21.90	21.99	22.07	22.16	22.25	22.33
39	22.42	22.51	22.59	22.68	22.77	22.86	22.95	23.03	23.12	23.21
40	23.30									

t = 26.5° C. and PE = 13.5 cm. The slope of the line is determined by the heat index of the station. For example, the heat index of Brevard, N. C., is 56.0, and the line ruled on the nomogram represents the relationship between potential evapotranspiration and temperature at that place. At a mean temperature of 10° C. (50° F.) the unadjusted potential evapotranspiration is 3.6 centimeters.[29] Knowing the index (I) of the station, one sets a straightedge in

[29] For accurate computation it is desirable to make a nomogram on a larger scale on logarithm paper that is commercially available. Keuffel and Esser No. 359–112L (logarithmic, 2 x 3 cycles) is satisfactory. The point of convergence and the heat-index scale are easily added. A transparent rule may be used.

TABLE V—MEAN POSSIBLE DURATION OF SUNLIGHT IN THE NORTHERN AND SOUTHERN HEMISPHERES EXPRESSED IN UNITS OF 30 DAYS OF 12 HOURS EACH

N. LAT.	J	F	M	A	M	J	J	A	S	O	N	D
0	1.04	.94	1.04	1.01	1.04	1.01	1.04	1.04	1.01	1.04	1.01	1.04
5	1.02	.93	1.03	1.02	1.06	1.03	1.06	1.05	1.01	1.03	.99	1.02
10	1.00	.91	1.03	1.03	1.08	1.06	1.08	1.07	1.02	1.02	.98	.99
15	.97	.91	1.03	1.04	1.11	1.08	1.12	1.08	1.02	1.01	.95	.97
20	.95	.90	1.03	1.05	1.13	1.11	1.14	1.11	1.02	1.00	.93	.94
25	.93	.89	1.03	1.06	1.15	1.14	1.17	1.12	1.02	.99	.91	.91
26	.92	.88	1.03	1.06	1.15	1.15	1.17	1.12	1.02	.99	.91	.91
27	.92	.88	1.03	1.07	1.16	1.15	1.18	1.13	1.02	.99	.90	.90
28	.91	.88	1.03	1.07	1.16	1.16	1.18	1.13	1.02	.98	.90	.90
29	.91	.87	1.03	1.07	1.17	1.16	1.19	1.13	1.03	.98	.90	.89
30	.90	.87	1.03	1.08	1.18	1.17	1.20	1.14	1.03	.98	.89	.88
31	.90	.87	1.03	1.08	1.18	1.18	1.20	1.14	1.03	.98	.89	.88
32	.89	.86	1.03	1.08	1.19	1.19	1.21	1.15	1.03	.98	.88	.87
33	.88	.86	1.03	1.09	1.19	1.20	1.22	1.15	1.03	.97	.88	.86
34	.88	.85	1.03	1.09	1.20	1.20	1.22	1.16	1.03	.97	.87	.86
35	.87	.85	1.03	1.09	1.21	1.21	1.23	1.16	1.03	.97	.86	.85
36	.87	.85	1.03	1.10	1.21	1.22	1.24	1.16	1.03	.97	.86	.84
37	.86	.84	1.03	1.10	1.22	1.23	1.25	1.17	1.03	.97	.85	.83
38	.85	.84	1.03	1.10	1.23	1.24	1.25	1.17	1.04	.96	.84	.83
39	.85	.84	1.03	1.11	1.23	1.24	1.26	1.18	1.04	.96	.84	.82
40	.84	.83	1.03	1.11	1.24	1.25	1.27	1.18	1.04	.96	.83	.81
41	.83	.83	1.03	1.11	1.25	1.26	1.27	1.19	1.04	.96	.82	.80
42	.82	.83	1.03	1.12	1.26	1.27	1.28	1.19	1.04	.95	.82	.79
43	.81	.82	1.02	1.12	1.26	1.28	1.29	1.20	1.04	.95	.81	.77
44	.81	.82	1.02	1.13	1.27	1.29	1.30	1.20	1.04	.95	.80	.76
45	.80	.81	1.02	1.13	1.28	1.29	1.31	1.21	1.04	.94	.79	.75
46	.79	.81	1.02	1.13	1.29	1.31	1.32	1.22	1.04	.94	.79	.74
47	.77	.80	1.02	1.14	1.30	1.32	1.33	1.22	1.04	.93	.78	.73
48	.76	.80	1.02	1.14	1.31	1.33	1.34	1.23	1.05	.93	.77	.72
49	.75	.79	1.02	1.14	1.32	1.34	1.35	1.24	1.05	.93	.76	.71
50	.74	.78	1.02	1.15	1.33	1.36	1.37	1.25	1.06	.92	.76	.70
S. LAT.												
5	1.06	.95	1.04	1.00	1.02	.99	1.02	1.03	1.00	1.05	1.03	1.06
10	1.08	.97	1.05	.99	1.01	.96	1.00	1.01	1.00	1.06	1.05	1.10
15	1.12	.98	1.05	.98	.98	.94	.97	1.00	1.00	1.07	1.07	1.12
20	1.14	1.00	1.05	.97	.96	.91	.95	.99	1.00	1.08	1.09	1.15
25	1.17	1.01	1.05	.96	.94	.88	.93	.98	1.00	1.10	1.11	1.18
30	1.20	1.03	1.06	.95	.92	.85	.90	.96	1.00	1.12	1.14	1.21
35	1.23	1.04	1.06	.94	.89	.82	.87	.94	1.00	1.13	1.17	1.25
40	1.27	1.06	1.07	.93	.86	.78	.84	.92	1.00	1.15	1.20	1.29
42	1.28	1.07	1.07	.92	.85	.76	.82	.92	1.00	1.16	1.22	1.31
44	1.30	1.08	1.07	.92	.83	.74	.81	.91	.99	1.17	1.23	1.33
46	1.32	1.10	1.07	.91	.82	.72	.79	.90	.99	1.17	1.25	1.35
48	1.34	1.11	1.08	.90	.80	.70	.76	.89	.99	1.18	1.27	1.37
50	1.37	1.12	1.08	.89	.77	.67	.74	.88	.99	1.19	1.29	1.41

the appropriate position on the nomogram and reads potential evapotranspiration corresponding to the given mean temperature of the month. The nomogram is used only when tempera-

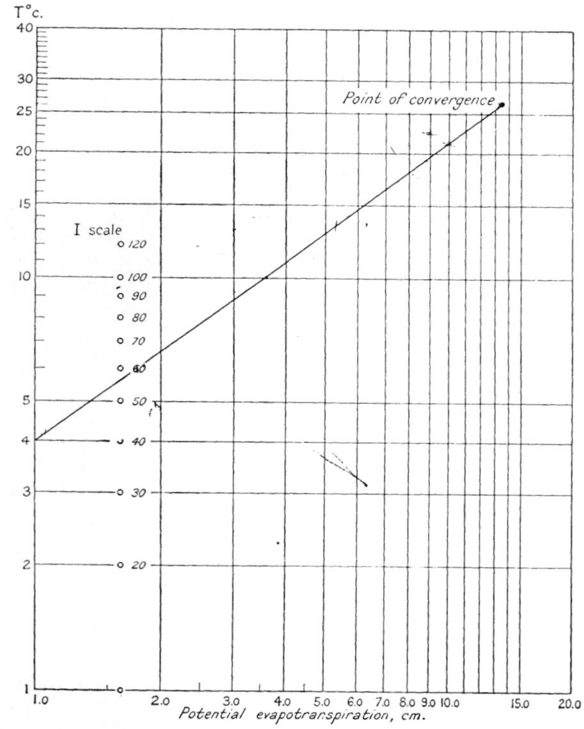

T.°C.	PE
26.5	13.50
27.0	13.95
27.5	14.37
28.0	14.78
28.5	15.17
29.0	15.54
29.5	15.89
30.0	16.21
30.5	16.52
31.0	16.80
31.5	17.07
32.0	17.31
32.5	17.53
33.0	17.72
33.5	17.90
34.0	18.05
34.5	18.18
35.0	18.29
35.5	18.37
36.0	18.43
36.5	18.47
37.0	18.49
37.5	18.50
38.0	18.50

FIG. 13—Nomogram for determining potential evapotranspiration and table for use at higher temperatures.

ture is 26.5° C. or less. (The accompanying table gives potential evapotranspiration corresponding to the higher temperatures). Twelve values are obtained for the 12 months. These are unadjusted values for months of 30 days of 12 hours each.

Finally, these values of potential evapotranspiration are adjusted for day and month length. Table V gives correction factors by which the unadjusted potential evapotranspiration of each month must be multiplied. The corrections must be the appropriate ones for the latitude of the station. The correction factor for 50° N. latitude will be used for all stations farther north. The same limitation applies in the Southern Hemisphere. Potential evapotranspiration is in centimeters. Summation of the 12 monthly values gives annual potential evapotranspiration in centimeters. The computations for Brevard, N. C., follow:

	J	F	M	A	M	J	J	A	S	O	N	D	Y
t.°C.	0.9	1.1	4.7	9.2	13.3	17.3	18.8	18.4	16.1	10.3	4.7	1.2	9.7
I	0.07	0.10	0.91	2.52	4.40	6.55	7.43	7.19	5.87	2.99	0.91	0.12	39.06
Unadj. PE	0.2	0.4	2.0	4.2	6.3	8.4	9.2	9.0	7.2	4.7	1.9	0.4	
Adj. PE	0.2	0.3	2.1	4.6	7.6	10.2	11.4	10.4	7.4	4.6	1.6	0.3	60.7

AVERAGE ANNUAL THERMAL EFFICIENCY

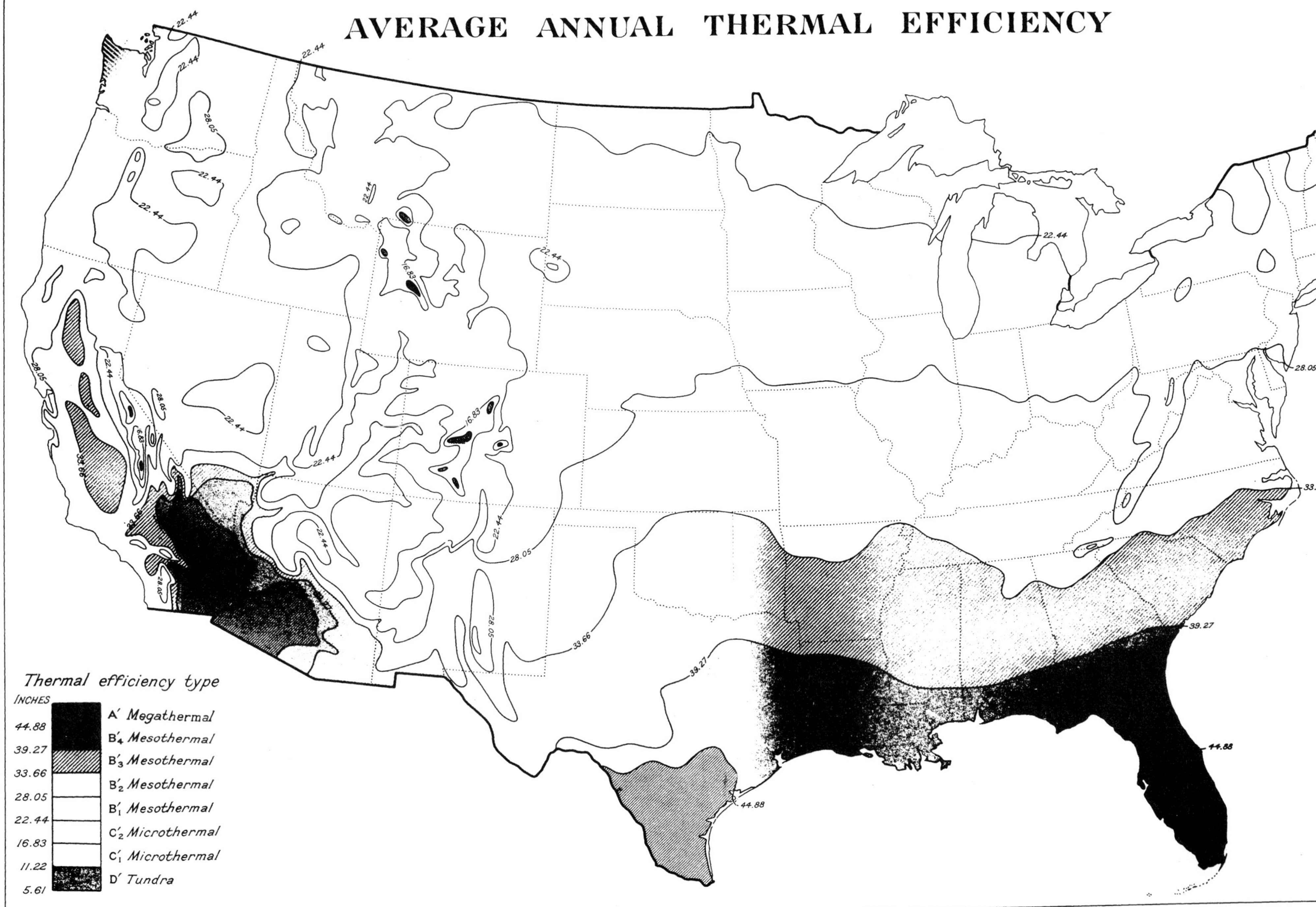

SEASONAL VARIATION OF EFFECTIVE MOISTURE

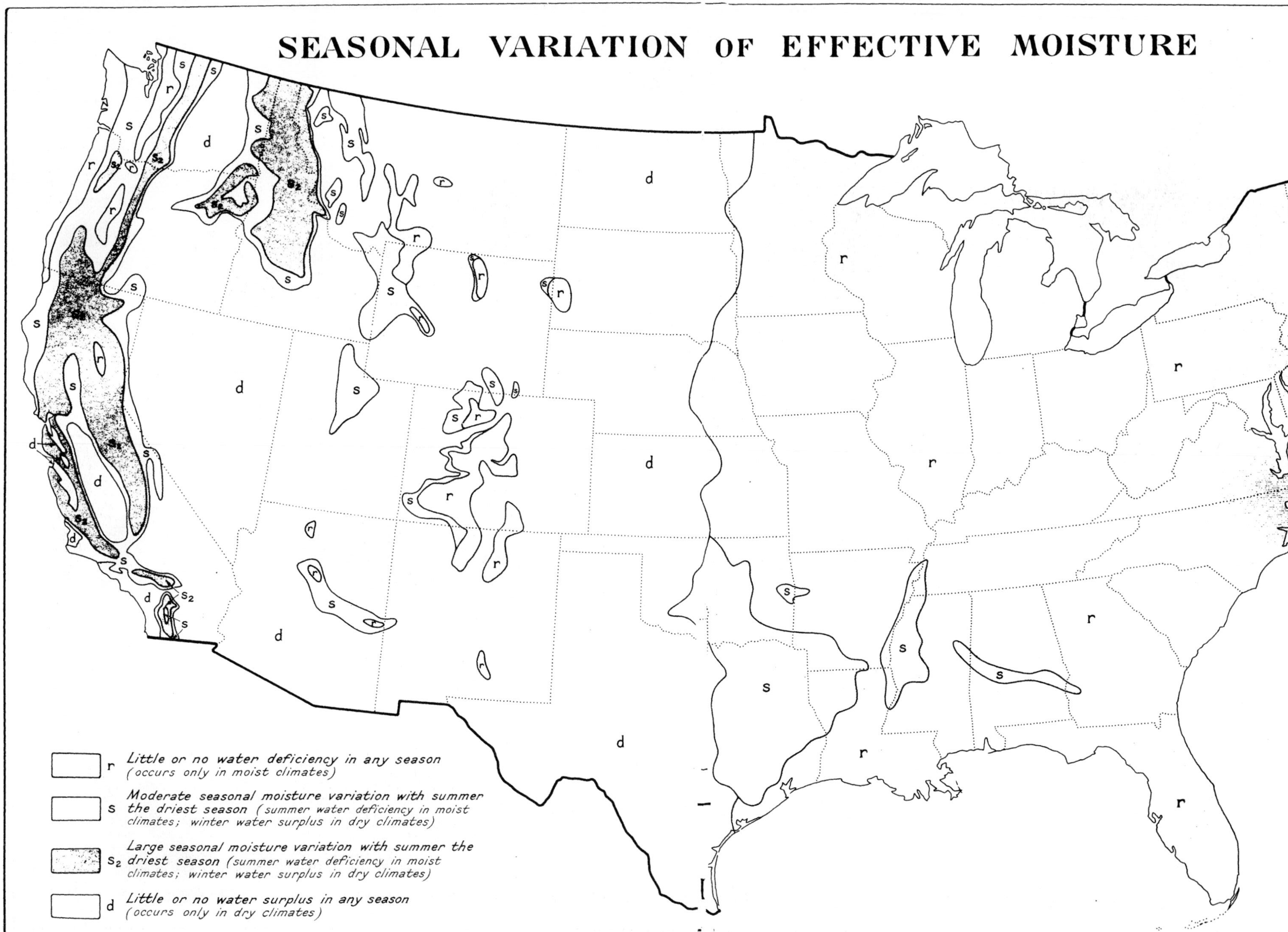

The American Geographical Society of New York

MOISTURE REGIONS IN THE UNITED STATES

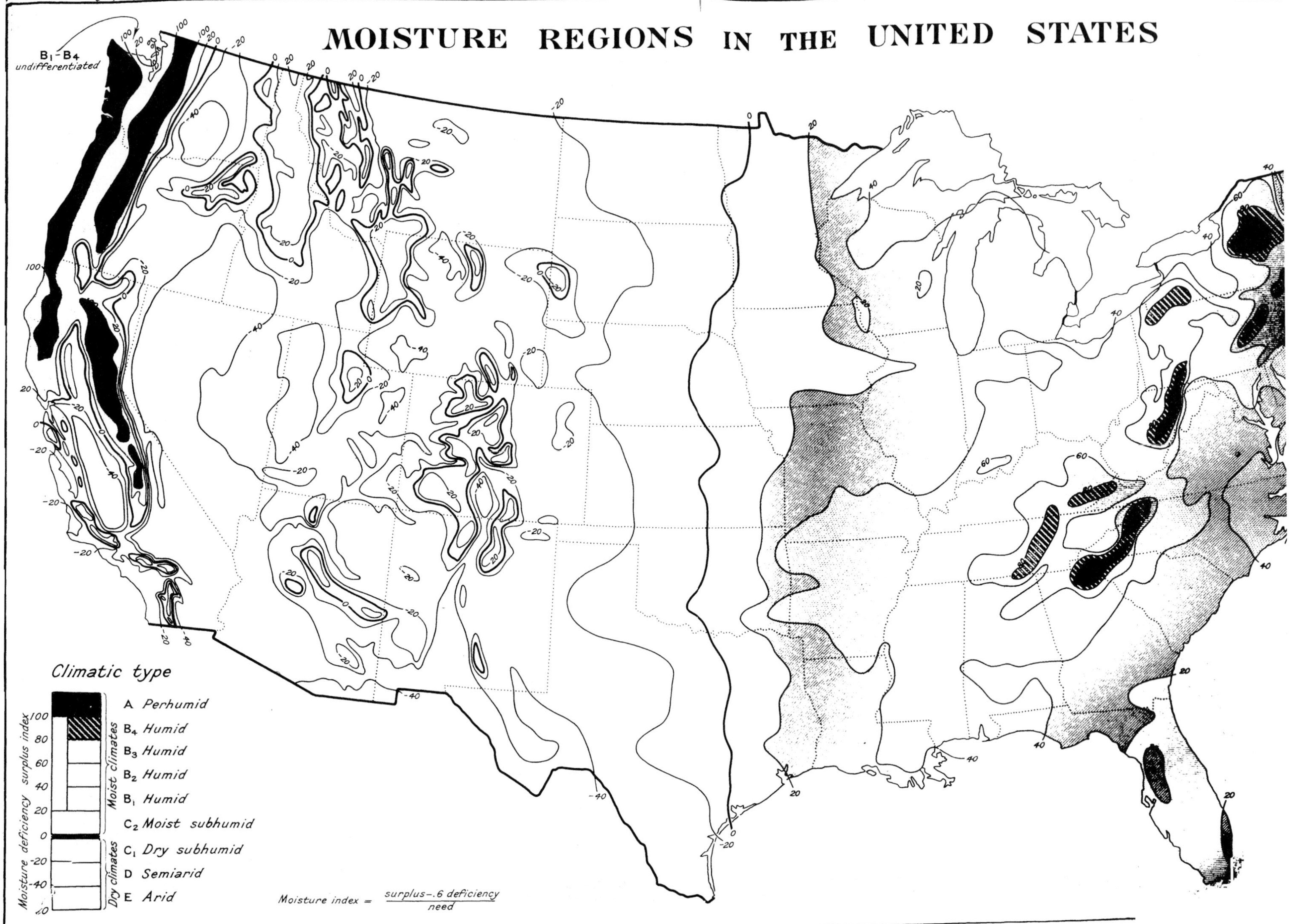

SUMMER CONCENTRATION OF THERMAL EFFICIENCY

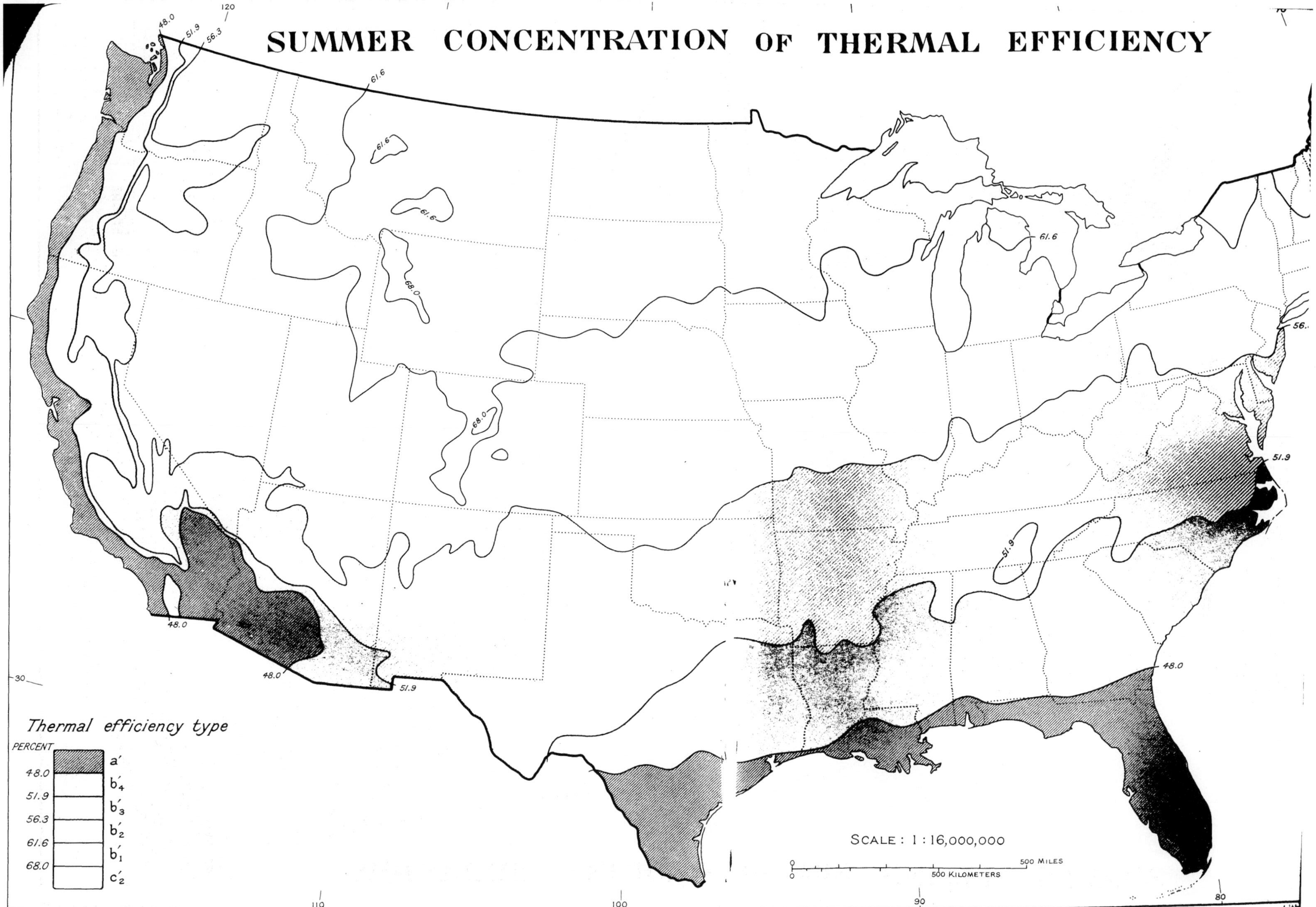

Natural evaporation from open water, bare soil and grass

By H. L. Penman

Physics Department, Rothamsted Experimental Station, Harpenden, Herts.

(*Communicated by B. A. Keen, F.R.S.—Received* 9 *October* 1947)

[Plate 3]

Two theoretical approaches to evaporation from saturated surfaces are outlined, the first being on an aerodynamic basis in which evaporation is regarded as due to turbulent transport of vapour by a process of eddy diffusion, and the second being on an energy basis in which evaporation is regarded as one of the ways of degrading incoming radiation. Neither approach is new, but a combination is suggested that eliminates the parameter measured with most difficulty—surface temperature—and provides for the first time an opportunity to make theoretical estimates of evaporation rates from standard meteorological data, estimates that can be retrospective.

Experimental work to test these theories shows that the aerodynamic approach is not adequate and an empirical expression, previously obtained in America, is a better description of evaporation from open water. The energy balance is found to be quite successful. Evaporation rates from wet bare soil and from turf with an adequate supply of water are obtained as fractions of that from open water, the fraction for turf showing a seasonal change attributed to the annual cycle of length of daylight. Finally, the experimental results are applied to data published elsewhere and it is shown that a satisfactory account can be given of open water evaporation at four widely spaced sites in America and Europe, the results for bare soil receive a reasonable check in India, and application of the results for turf shows good agreement with estimates of evaporation from catchment areas in the British Isles.

List of symbols used

x, y, z	Co-ordinate axes downwind, acrosswind and vertical.
u_z	Mean horizontal wind velocity in x direction measured at height z; usually in miles/day.
T_s, T_a, T_d	Temperature of surface, air and dewpoint; usually °F.
e_s, e_a, e_d	Saturation vapour pressure at above temperature; usually in mm. Hg.
h	Relative humidity $= e_d/e_a$.
Δ	de_a/dT_a.
ϕ	$(e_s-e_a)/(e_s-e_d)$.
α	Constant defining hydrolapse.
β	Bowen's ratio $= \gamma(T_s-T_a)/(e_s-e_d)$.

γ	Constant of wet and dry bulb hygrometer equation; in °F and mm. Hg, $\gamma = 0{\cdot}27$.
E_0, E_B, E_T	Evaporation rate from open water, bare soil and turf; usually in mm./day.
E_a	Value of E_0 obtained by putting $e_s = e_a$ in sink strength formula.
R_C	Short-wave radiation from sun and sky; usually in evaporation equivalent of mm./day.
R_A	Angot value of R_c for a completely transparent atmosphere.
r	Radiation reflexion coefficient. (Also used for runoff without any possibility of confusion.)
$m/10$	Fraction of sky covered by cloud.
n/N	Ratio of actual/possible hours of sunshine.
H	Net radiant energy available at surface.
K, S, C	Parts of H used in convective transfer to air, storage in water, conduction to surround.
5B, 16T	Depths to water-table (in.) and type of cover (bare, turf).
λ	Specific yield of soil.
R, D	Rainfall, drainage.
B	Beaufort wind force.

1. Introduction

Three kinds of surface are important in the return of rain to the atmosphere. For extended areas of land, they are, in order of importance: vegetation, on which plant leaves act as transpiring surfaces; bare or fallow soil, from which water evaporates at, or just below, the soil-air interface; and open water, from which evaporation takes place directly. Although the last may be of predominant importance locally, e.g. in attempting to assess the water balance of lakes and reservoirs, the chief justification of the great attention given to it (see § 2, below) is found in the opportunity it presents of providing a reproducible surface of known properties. Because of this, it is convenient to approach the problems of the dependence of evaporation from bare and cropped soil on weather conditions through a study of evaporation from open water, seeking an absolute relation between weather elements and open water evaporation, and comparative relations between losses from the soil and losses from open water exposed to the same weather.

Evaporation from bare soil involves complex soil factors as well as atmospheric conditions: transpiration studies add to these further important physical and biological features, for a plant's root system can draw on moisture throughout a considerable depth of soil, its aerial parts permit vapour transfer throughout a considerable thickness of air, and its photo-sensitive stomatal mechanism restricts this transfer, in general, to the hours of daylight. A complete survey of evaporation from bare soil and of transpiration from crops should take account of all relevant factors, but the present account will be largely restricted to consideration of the early stages that would arise after thorough wetting of the soil by rain or irrigation, when soil type, crop type and root range are of little importance.

2. The estimation of evaporation from weather data

Two requirements must be met to permit continued evaporation. There must be a supply of energy to provide the latent heat of vaporization, and there must be some mechanism for removing the vapour, i.e. there must be a sink for vapour. Analytical attacks on the problem start from one of these two points and it is convenient to consider the latter first as it has been the more popular.

(a) *Sink strength*

(i) *Empirical equations*

Until recent years the approach was empirical, a hundred years' work since Dalton having produced little improvement in the form of equation he gave. In essentials it is

$$E = (e_s - e_d)\, f(u), \tag{1}$$

where E is the evaporation in unit time, e_s is the vapour pressure at the evaporating surface, e_d is the vapour pressure in the atmosphere above, and $f(u)$ is a function of the horizontal wind velocity. For water, e_s is known if the surface temperature is known. Of the many empirical formulae cast into this form, one due to Rohwer (1931) summarizes results of very intensive work at Fort Collins, Colorado, at 5000 ft. above sea-level. Other things being equal, Rohwer found a small variation of evaporation rates with atmospheric pressure, and reduced to conditions at sea-level, his equation for the daily rate from an open water surface 3 ft. square is

$$E = 0{\cdot}40(e_s - e_d)\,(1 + 0{\cdot}27u_0)\ \text{mm./day}, \tag{2}$$

where vapour pressures are in mm. mercury, and wind speed at ground level is in m.p.h. Examining the effect of size of surface on evaporation rates, over a period of 485 days, he compared the observed values of evaporation from a large surface 86 ft. diameter with the estimates based on (2), and found the mean value of observed/estimated to be 0·77. There is some bias here, however, for the average wind speed over the whole period was only 1·50 m.p.h., and examination of the individual daily records shows that on the rare occasions of a wind speed in excess of 3 m.p.h. the correction factor is nearly unity. The ground wind velocity, u_0, is an extrapolated value estimated from a number of readings at various heights, and if from Rohwer's u, z curve we interpolate at 2 m., the relation becomes

$$E = 0{\cdot}40(e_s - e_d)\,(1 + 0{\cdot}17u_2)\ \text{mm./day}, \tag{3}$$

and except at very low wind speeds might be expected to apply to large open water surfaces.

(ii) *Aerodynamic equations*

As an alternative to this empirical treatment an aerodynamical approach has been made in recent years. A simple treatment (Penman 1940) showed that the right order of magnitude could be obtained by assuming that the main resistance to the evaporation current is provided by a thin layer of air (*c.* 1 to 3 mm. thick) next to the surface, in which air movement is essentially non-turbulent, and vapour movement across which is by a process of molecular diffusion. The more formal treatment, having wider implications than the solution of evaporation problems, has considered the turbulent mixing and transport of the vapour outside this sublaminar boundary layer, and it attempts to take into account the dependence of evaporation rates per unit area upon size and shape of the test area as well as upon weather elements. An account of this work is given by Brunt (1939) up to and including the work of O. G. Sutton (1934). Extension of Sutton's work by W. G. L. Sutton (1943) and Pasquill (1943) has given an expression for the *total* evaporation from a rectangular strip of length x_0 downwind and width y_0:

$$E(x_0, y_0) = C(e_s - e'_d)\, u_2^{0\cdot 76} x_0^{0\cdot 88} y_0, \tag{4}$$

where C is a constant related to the absolute temperature, e'_d is the vapour pressure of the air at a height great enough to be unaffected by the evaporation, and u_2 is the wind velocity at $z = 2$ m. Although e'_d is unobservable, it has been possible to use the same general theory to express the shape of the hydrolapse, and to set $(e_s - e_d) = \alpha(e_s - e'_d)$ where e_d is the measured value at screen height, and α ($\doteqdot 0\cdot 52$) is almost independent of u and x_0. Differentiating (4), the rate of evaporation *at* x_0 is obtained, and, substituting numerical values appropriate to zero temperature gradient, it becomes

$$E = 0\cdot 11(e_s - e_d)\, u_2^{0\cdot 76} x_0^{-0\cdot 12} \text{ mm./day}. \tag{5}$$

In the open it is impossible to fix the position of the leading edge, but arbitrarily putting $x_0 = 1\cdot 6 \times 10^6$ cm. (10 miles), the evaporation rate becomes

$$E = 0\cdot 376(e_s - e_d)\, u_2^{0\cdot 76} \text{ mm./day}, \tag{6}$$

where e_s and e_d are in mm. mercury, and u_2 is now in miles/hr. If u_2 is measured in miles/day—a practical convenience—the rate is

$$E = 0\cdot 033(e_s - e_d)\, u_2^{0\cdot 76} \text{ mm./day}. \tag{6a}$$

Notes. (1) The assumption of zero temperature gradient involves the identity of e_s and e_a, where e_a is the saturation vapour pressure at the mean air temperature; $e_s - e_d$ then becomes $e_a(1-h)$.

(2) A tenfold increase in x_0 will decrease E in the ratio $(1/10)^{0\cdot 12}$, i.e. to $1/1\cdot 3$, and the constant in (6*a*) will become 0·025.

(*b*) *Energy balance*

Certain simplifying assumptions are needed; where they are known to be reasonable, reliable estimates of evaporation are possible. Using as the unit of energy the amount required to evaporate 1/10 g. of water at air temperature (59 cal.) it is possible to build up the following expression for the heat budget, H, taking into account the incoming short-wave radiation from sun and sky, and the long-wave exchanges between earth and sky (Brunt 1939; equation 15, p. 136; equation 25, p. 144):

$$H = R_C(1 - r - \mu) - \sigma T_a^4(0\cdot 56 - 0\cdot 092\sqrt{e_d})\,(1 - 0\cdot 09m), \tag{7}$$

where R_C is the measured short-wave radiation/cm.^{2}/day,

r is the reflexion coefficient for the surface,

μ is the fraction of R_C used in photosynthesis,

σT_a^4 is the theoretical black-body radiation at T_a °K,

e_d is in mm. mercury,

and $m/10$ is the fraction of sky covered with cloud.

Using the convenient symbols of Cummings & Richardson (1927), the heat budget is used in evaporation, E, heating of the air, K, heating of the test material, S, and heating of the surroundings of the test material, C, i.e.

$$H = E + K + S + C. \tag{8}$$

Over a period of several days, and frequently over a single day, the change in the stored heat, S, is negligible compared with other changes and the same may be true of the heat conducted through the walls of the test material container. Thus, (8) can often be safely reduced to

$$H = E + K. \tag{9}$$

The transport of vapour and the transport of heat by eddy diffusion are, in essentials, controlled by the same mechanism, and apart from the differences in the molecular constants, the one is expected to be governed by $(e_s - e_d)$ where the other is governed by $(T_s - T_a)$. To a very good approximation, therefore, it is possible to write down the ratio of K/E in the form

$$K/E = \beta = \gamma(T_s - T_a)/(e_s - e_d), \tag{10}$$

where β, the ratio symbolized by Bowen (1926) as R, has the value -1 in the standard wet and dry bulb hygrometer equation, and γ is the standard constant of this equation. In °F and mm. Hg, $\gamma = 0{\cdot}27$.

Thus $$H = E(1+\beta), \quad E = H/(1+\beta). \tag{11}$$

Of the terms on the right-hand side of (7), the radiation term will rarely be directly measurable, but for periods of the order of a month or more it can be estimated from duration of sunshine. Angot has given tables of the total radiation to be expected if the atmosphere were perfectly transparent (Brunt (1939), p. 112), and there appears to be a general correlation between R_C/R_A and n/N in the form $R_C/R_A = a + bn/N$, where n/N is the ratio actual/possible hours of sunshine. For Virginia, U.S.A., Kimball (1914) finds $a = 0{\cdot}22$, $b = 0{\cdot}54$; for Canberra, Australia, Prescott (1940) finds $a = 0{\cdot}25$, $b = 0{\cdot}54$. At Rothamsted, monthly values over the period 1931–40 lead to $a = 0{\cdot}18$, $b = 0{\cdot}55$, with a suggestion of a seasonal variation. Using these latter constants we have

$$R_C = R_A(0{\cdot}18 + 0{\cdot}55n/N). \tag{12}$$

In terms of the maximum to be expected (R_N; for $n = N$), equation (12) becomes

$$R_C = R_N(0{\cdot}25 + 0{\cdot}75n/N), \tag{12a}$$

agreeing with the form given by Angstrom for Stockholm (Brunt (1939), p. 127).

The value of μ is very small (*c.* 0·005) and can be neglected. The value of r will vary with season and type of surface. For water its annual mean will be about 0·06 in the British Isles; for bare soil, about 0·10; and for turf might be about 0·20 (Geiger (1927) quoting Angstrom). Note that r and β will be the only effective factors in (7) discriminating between the different types of surface.

The terms expressing the net flow of radiation to and from a cloudless sky are due to Brunt and are based on the mean values of the constants obtained in six correlations of the energy flow with mean air temperature. Sverdrup (1945) gives a diagram indicating values of the same order but with slightly greater values of $\sigma T^4 f(e_d)$. The uncertainty here, however, is negligible compared with that arising from the cloudiness term. It is obvious that cloud control of long-wave radiation must depend upon cloud type, and as a provisional expedient to make some allowance for this it is proposed to set $m/10 = 1 - n/N$. Equation (7) therefore reduces to

$$H = E(1+\beta) = (1-r)\,R_A(0{\cdot}18 + 0{\cdot}55n/N) - \sigma T_a^4(0{\cdot}56 - 0{\cdot}092\sqrt{e_d})\,(0{\cdot}10 + 0{\cdot}90n/N) \tag{13}$$

where $R_A(0{\cdot}18 + 0{\cdot}55n/N)$ is to be replaced by R_C when this is known. The parameters represented on the right-hand side of (13) are easily determined; to obtain E it is necessary to obtain β, which involves knowing the surface temperature (equation 10). The sink strength approach also involves this knowledge, and although arrangements can be made to measure it experimentally, it is desirable to eliminate it for the prediction of evaporation or for a survey of evaporation as a climatic element.

(c) *Combination of sink strength and energy balance*

From (1), expected to take the form of (6) or (6a), we have

$$E = (e_s - e_d)\,f(u). \tag{1}$$

Let E_a be the value of E obtained by putting e_a instead of e_s. Then

$$E_a = (e_a - e_d)\,f(u),$$

i.e.
$$E_a/E = 1 - (e_s - e_a)/(e_s - e_d) = 1 - \phi \text{ say.} \tag{14}$$

From (10) and (11);

$$E = H/(1+\beta) = H/[1 + \gamma(T_s - T_a)/(e_s - e_d)].$$

If we set $T_s - T_a = (e_s - e_a)/\Delta$, where Δ is the slope of the $e:T$ curve at $T = T_a$, then

$$H/E = 1 + \gamma(e_s - e_a)/\Delta(e_s - e_d) = 1 + \gamma\phi/\Delta. \tag{15}$$

From (14) and (15);
$$E = (H\Delta + E_a\gamma)/(\Delta + \gamma), \tag{16}$$

$$e_s = (e_a - \phi e_d)/(1 - \phi), \tag{17}$$

i.e. E can be estimated from air conditions only, and, if required, an estimate of surface temperature can be obtained that might be useful outside evaporation studies.

In addition to the constants, readily obtainable from standard sources, the weather parameters needed are mean air temperature, mean dewpoint, mean wind velocity at a standard height and mean duration of sunshine.

3. Rothamsted experiment 1944, 1945

(a) *Experimental site*

Experiments have been carried out in the meteorological enclosure at Rothamsted, situated at about 420 ft. above O.D. in open parkland in the Chiltern Hills. The enclosure includes a brick-lined pit, 8 ft. deep and 20 ft. diameter, around which twelve cylinders were set in the soil in 1924, five of them being filled with a sandy loam from Woburn (Bedfordshire), the soil texture being uniform throughout. The cylinders are 6 ft. deep and 2 ft. 6 in. diameter and are made of cast iron lined with a $\frac{1}{2}$ in. layer of bitumen painted concrete; the bottoms have a slope down to an outlet pipe accessible from the pit. (See figure 1 and figure 7, plate 3.)

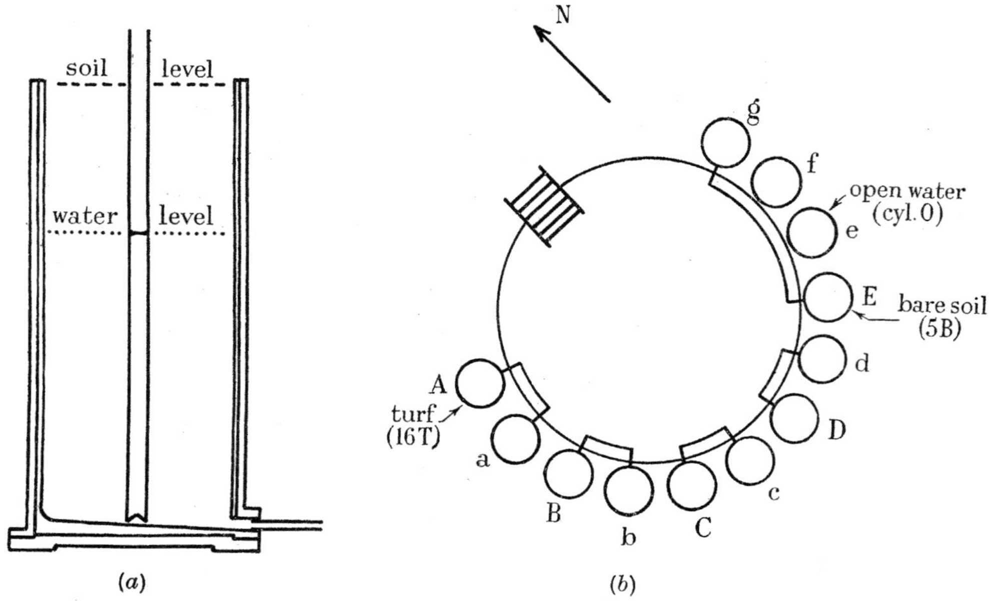

Figure 1. (*a*) Section of cylinder, and (*b*) plan of the pit.

The soil was left to settle and weather for 16 years so that some semblance of natural structure could be attained and by May 1940 the settlement amounted to 6 in. This was made good by a further supply of Woburn soil, experimental work was done in 1941 and 1942 and in the spring of 1944 there was a further slight topping up in preparation for the work now to be described.

FIGURE 7. The experimental site June 1944, looking north-east.

(*Facing p.* 126)

Ten cylinders were joined up in pairs at the outlets, each soil cylinder being connected to an unfilled cylinder referred to below as the 'minor', so forming a set of U tubes. Figure 1 shows the arrangement schematically, with three cylinders labelled (O, 5B and 16T). These are the main ones to be discussed and were the same in both years; changes were made in the others early in 1945. Waterproof covers were provided for the minors to prevent entry of rain and to reduce evaporation losses to negligible amounts. On A and C turves were laid in April 1944 and on D in March 1945; the other soil surfaces were kept bare.

At the outset the minors were filled with water until the soil or turf surfaces were flooded and then water was run out until the water-table had reached a pre-determined depth below the soil surface. The depths and the nature of the surface are given in table 1:

TABLE 1. DEPTHS (IN.) OF WATER-TABLE AND NATURE OF SURFACE

cylinder ...	A	B	C	D	E
1944	16 (T)	16 (B)	10 (T)	10 (B)	5 (B)
1945	16 (T)	24 (B)	24 (T)	36 (T)	5 (B)

Cylinder e was filled to near the brim and the level kept at about 1 in. below. This was the first open water standard, referred to as cylinder O. In the early summer of 1945 a tank of sheet galvanized iron, 2 ft. 6 in. diameter and 2 ft. deep was supplied by the Meteorological Office and was set up at the north end of the enclosure about 50 yd. north of the pit. A hole 1 ft. 9 in. deep was dug into which the tank fitted firmly, and the water-level was kept at or near ground level, so leaving a projecting rim of 3 in. This tank, referred to as tank MO, had the same area as the cylinder but was shallower, had a thinner and more conducting wall material, was completely surrounded by turf-covered soil whereas the other had the pit on one side, and had a higher effective rim.

Ground level round the cylinders had been raised so that soil level was the same inside and out except on one side of cylinder O where the topping up was not complete; for all the cylinders there was a big discontinuity in surface on the pit side and although the pit should have been roofed in, it was not practicable at the time and a major objection to the experimental site had to be accepted as unavoidable. Figure 7, plate 3, shows the exposure to the north-east and the disposition of some of the other components of the enclosure; in the centre, beyond the pit, is the large rain gauge (1/1000 acre) used for rainfall values employed below, and to the right of it are visible two of the bare drain gauges on which earlier Rothamsted work was based. The general exposure here is good, although the presence of the large gauges might affect local eddies with east and north-east winds. The general exposure to north-west was equally good, that to south-west a little worse, and that to south-east poorest of all due to an extended belt of trees, the nearest being about 80 yd. away.

The surround varied during the experiment. The local exterior topping up remained bare for a while, but a crop of weeds soon developed and gave a local cover nearly enough equivalent to the turf of the enclosure. By the summer of 1945 there was a fair amount of grass in this and it was possible to cut it with a mower. Growth was very rapid during 1945 and part of the surround, too rough for a mower, got out of hand and for a while there was a stand of tall grass on the west side which may have had an adverse effect on transpiration from cylinder A (16T). In both years the field to the east was sown with mangolds giving a green cover from June until late autumn, at which time the soil surface was moist and remained moist for the remainder of the winters. The field to north and west carried an oat crop in 1944, about 1 ft. high in May, 3 ft. high throughout July and harvested in early August; the undersown clover then provided a green cover for the remainder of the season, was grazed during 1945, but grew away from the animals and a hay crop was eventually taken from it. To the south there was a pasture, kept short by grazing. Thus, except for short periods, the experimental surfaces could be reasonably described as being in the midst of an extensive area of short vegetation and as long as this transpired at maximal rate, then for so long did the experimental conditions come close to satisfying the basic assumptions from which equation (6) was derived.

(b) *Measurements and calibrations*

In addition to the normal 09.00 observations of a third-order meteorological station, supplementary measurements were made.

(i) *Temperature.* Mercury-in-steel thermographs were set up on cylinders C, D, E and O; that on D was transferred to tank MO in June 1945. The long bulbs were horizontal and were either pressed into the soil or supported in the water so that about half was below the surface and half above the surface; the water bulbs had sheaths of muslin to ensure that they were always wet all over. They were calibrated in place, and in all cases the corrected 'surface' temperature is more truly the mean temperature of the top few mm. of soil or water. To find the daily mean a smooth curve was drawn through the thermogram and a mean of six readings at 4 hr. intervals found; this was corrected from the calibration curve and the corresponding value of the saturation vapour pressure taken as the daily mean value of e_s.

Only one value of the dewpoint was obtained per day and although this was found to be adequate for long period surveys it was not always adequate for individual days, particularly where there was a pronounced change during the day. The Dunstable Branch of the Meteorological Office kindly supplied estimates of the dewpoint at 6 hr. intervals for each day and from these it was possible to see the way in which the dewpoint had changed in a given period of 24 hr. and to weight the Rothamsted values accordingly. Obtaining a reliable daily mean value of the dewpoint remains one of the main experimental problems to be solved.

The mean air temperature is never of first order of importance and it has been sufficient to take the conventional mean of maximum and minimum for this parameter.

(ii) *Wind.* A three-cup anemometer was set up at 2 m. in the south-west corner of the enclosure in 1944, and was moved to the middle of the enclosure early in 1945 so as to be about half-way between the pit and the new MO tank. The scale reading was read once daily, and in view of this it was an obvious convenience to use miles/day as the unit of wind velocity (equation (6*a*); figures 2 and 3). The instrument was calibrated in December 1944 in terms of a similar instrument with smaller cups, with which a calibration curve was supplied. The curve was non-linear at low speeds and calibration of the experimental instrument is consequently uncertain in this region: table 2 shows the result:

TABLE 2. CALIBRATION OF ANEMOMETER (MILES/DAY)

observed	0	30	60	90	120	150	210	270
corrected	38	50	78	106	133	161	215	272

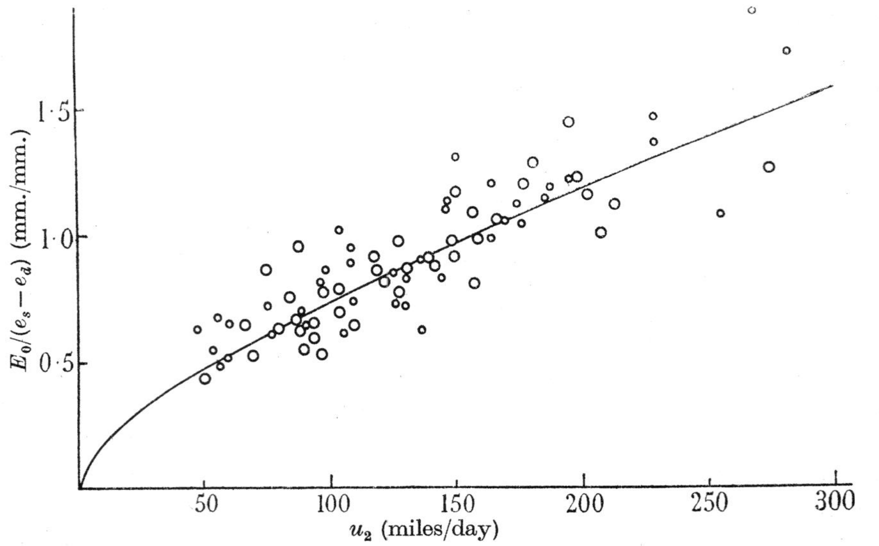

FIGURE 2. Daily evaporation per unit partial pressure difference for open water surface (cylinder O). The curve is: $E_0/(e_s-e_d) = 0{\cdot}033\,u_2^{0{\cdot}68}$, ○ 1944, ◯ 1945.

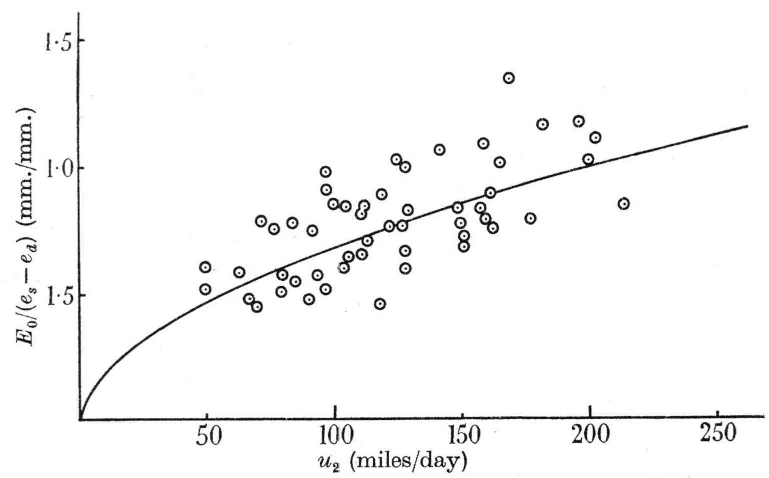

FIGURE 3. Daily evaporation per unit partial pressure difference for open water surface (tank MO). The curve is: $E_0/(e_s-e_d) = 0{\cdot}065\,u_2^{0{\cdot}54}$.

A Dines wind recorder on the laboratory roof, about $\frac{1}{4}$ mile away, was used to estimate direction and variation of wind speed during the day, where necessary.

(iii) *Radiation.* A continuous record of radiation intensity on a horizontal surface was obtained each day, the total area under the trace measured by planimeter and the figure so obtained converted to the equivalent number of mm. of water that the total energy would evaporate at air temperature.

The duration of sunshine was obtained from a Campbell-Stokes recorder.

(iv) *Evaporation.* Daily measurements were made of the depth of the water-levels below arbitrary zeroes.

Cylinders A, B, C and D. A rigid cradle was built into the top of the minor into which could be fitted a framework carrying a screw ending in a sharp pointed dip-stick. The screw carried a scale that moved past a fixed index mark and readings could be made to better than $\frac{1}{5}$ mm. except on very windy days.

Cylinders E and O. Measurements here were somewhat cruder. A solid straight edge was placed across the top of the minor in a marked position and, by means of a guide, a pointed rule was slid down until it just touched the water surface. With care, readings could be reproduced to within about $\frac{1}{5}$ mm., and in the major part of the experiment this was adequate accuracy.

Tank MO. A hook gauge reading to $\frac{1}{100}$ in. was used.

Changes in level are due to evaporation or rainfall, both being excluded from the minors. A fall in level takes place in both arms of the U system when evaporation occurs, so that for a change in minor level of δz, the total evaporation is greater, and may be set equal to $\delta z(1+\lambda)$, where λ is the specific yield of the soil with water-table at z cm. below the surface. If, over a period, the measured rainfall is R, then the total evaporation is given by

$$E = \delta z(1+\lambda)+R. \tag{18}$$

As λ is a function of z, measurements were made to give the values in table 3.

TABLE 3. SPECIFIC YIELD OF SOIL

depth of water-table (in.)	5	10	16	24	36
specific yield, λ (cm./cm.)	0·02	0·04	0·10	0·13	0·13

4. RESULTS FOR INDIVIDUAL DAYS

(a) *Open water: sink strength*

Equation (6*a*) was tested by plotting $E_0/(e_s-e_d)$ against u_2. Figure 2 shows the result for cylinder O with the 1944 and 1945 data distinguished, figure 3 shows the result for tank MO from mid-June 1945, and in figure 4 are given values of $\Sigma E_0/\Sigma(e_s-e_d)$ for wind speed ranges sufficiently wide to include at least four observations in the summations. The data represented in these figures have been selected as follows: (i) rain days have been excluded as there is some uncertainty about the uniformity of rainfall distribution over the site; on such a day the fall in level is made up of evaporation minus rainfall. (ii) For cylinder O, only those days have been used in which $E_0 > 2{\cdot}5$ mm. and $(e_s - e_d) > 2{\cdot}5$ mm.; for tank MO, to increase the number of available results, the limits were lowered to 2·0 mm. The experimental errors tend to be absolute and the uncertainty in the ratio increases greatly as E_0 and Δe become very small.

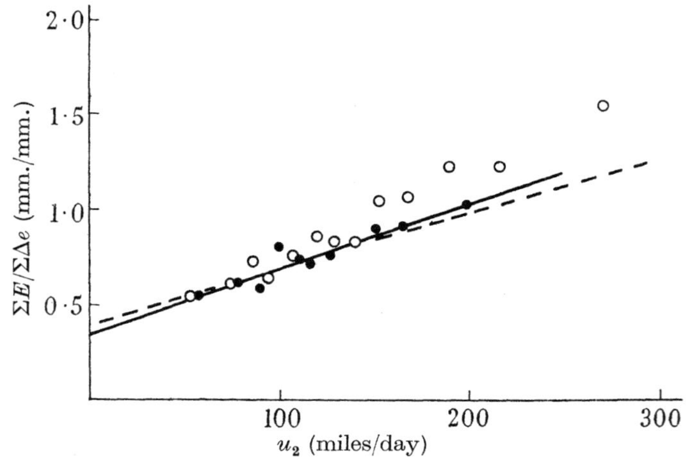

FIGURE 4. Mean daily evaporation per unit partial pressure difference for open water (groups of days having approximately the same average wind speed). The continuous line is: $E_0/(e_s - e_d) = 0{\cdot}35(1 + 9{\cdot}8 \times 10^{-3} u_2)$. ○ cylinder O; —●— tank MO; ---- Rohwer.

The scatter in figures 2 and 3, although not very much worse than that obtained by other workers doing indoor experiments, is considerable. A number of obvious contributory factors have been examined and shown to be of slight effect: dryness of the surround, wind distribution during the day, height of rim and season of the year. The main sources are undoubtedly the meteorological observations themselves; in increasing order of importance they are: dewpoint and surface temperature determinations and wind velocity measurements. Concentrating on the last, it is doubtful whether a measurement at a single height and the assumption of zero velocity at ground level are sufficient to define the wind velocity profile even over a smooth surface; they cannot be expected to take account of the local turbulence introduced by many obstructions and surface irregularities. These will vary with wind direction, and an analysis of the cylinder O results showed that all high values of $E/\Delta e$ at high u_2 were for days with north-east winds, i.e. days in which the local exposure would be most conducive to extra turbulence. It seems, therefore, that in spite of their greater scatter, the tank MO results are probably a better guide to a general law than the cylinder O results, and this is supported by the comparison (figure 4) with Rohwer's results. (Because of the scatter it is probable that the mean curves do not differ significantly.) The mean curves show that there is a linear relationship between $E/(e_s - e_d)$ and u_2, but for comparison with the theoretical form two curves could be fitted: (i) through the overall mean value of $E/\Delta e$ and of u_2, a curve $E/\Delta e = bu_2^{0{\cdot}76}$ (cf. equation $6a$) and, (ii) through the two general means obtained from the groups

9-2

of points lying to the left and right of the overall mean, a curve $E/\Delta e = au_2^n$. The results, in decreasing order of efficiency, are given in table 4.

TABLE 4. VALUES OF $f(u)$ IN $E_0 = (e_s - e_d)\,f(u)$

	cylinder O	tank MO	Rohwer*
best	$0{\cdot}30(1+14{\cdot}2\times10^{-3}u_2)$	$0{\cdot}35(1+9{\cdot}8\times10^{-3}u_2)$	$0{\cdot}40(1+7{\cdot}1\times10^{-3}u_2)$
good	$0{\cdot}033u_2^{0{\cdot}68}$	$0{\cdot}065u_2^{0{\cdot}54}$	—
fair	$0{\cdot}021u_2^{0{\cdot}76}$	$0{\cdot}019u_2^{0{\cdot}76}$	—

The curves drawn in figures 2 and 3 are the 'good' curves, wind velocities being in miles per day.

From the above, it is concluded that: (i) the best form of (1) for practical use is

$$E_0 = 0{\cdot}35(1+9{\cdot}8\times10^{-3}u_2)\,(e_s - e_d)\,\text{mm./day}, \quad u_2 \text{ in m.p.d.}; \tag{19}$$

(ii) for analysis, demanding a curve passing through the origin, the power of the wind velocity is much nearer $\frac{1}{2}$ than $\frac{3}{4}$; (iii) if the form of equation (6*a*) is to be maintained, the constant is to be reduced from 0·033 to about 0·020, a result that may be due to inaccurate assumptions about the distance away of the hypothetical 'leading edge', may be due to the departure from zero temperature gradient, or may be due to an inaccurate value of α in specifying the shape of the hydro-lapse.

(*b*) *Open water: energy balance*

The initial objective in this approach was an application to extended periods in which the assumptions made in reducing (8) to (9) would be reasonable; the main discussion will be of this aspect and appears below, but in view of its success it seemed worth while extending the application to individual days. An estimate of E_{MO}, based on energy, was obtained for most days between mid-June and the end of September 1945 and results are shown in figure 5 and table 5. The latter includes a representative sample of the data of the figure, the choice being made at roughly 6-day intervals with an attempt to give reasonable ranges of wind velocities and sunshine conditions, and affords a comparison of the estimates based on energy balance and sink strength (equation 19) with each other and with the observed values of daily evaporation. The values of $(e_s - e_d)$ range from 1·10 to 7·40 mm., those of σT^4 from 13·6 to 14·8, those of $(0{\cdot}56 - 0{\cdot}092\sqrt{e_d})$ from 0·30 to 0·26, the product of the last two functions tending to be constant, and those of $(0{\cdot}10 + 0{\cdot}90n/N)$ from 0·10 to 0·70. From the last, it will be seen that the most important term in the back radiation is the cloudiness factor—the least certain of any. Although they could be deduced from other columns, values of β are given. The figure is extremely encouraging. With the table it shows that the energy balance estimate is too big in mid-summer but improves later in the year. The change in the reflexion coefficient (here taken as constant throughout) would act in the opposite sense, and apart from

* Wing-Cdr Frost, in a private communication, has stated that observations at Poona, India, which were reduced to the form $E/\Delta e = au^{0{\cdot}38}$ by the experimenters can be equally well fitted by

$$E = 0{\cdot}40(1+7{\cdot}3\times10^{-3}u_2)\,(e_s - e_d)\ \text{mm./day}.$$

This is almost indistinguishable from the Rohwer equation.

any deficiencies in (13) itself it is probable that the main cause of the over-estimate is the warming of the bottom of the tank to a higher temperature than the outside soil so producing a positive value of the factor C in (8). The neglect of C in (9), therefore, leads to an over-estimate of H and hence of E.

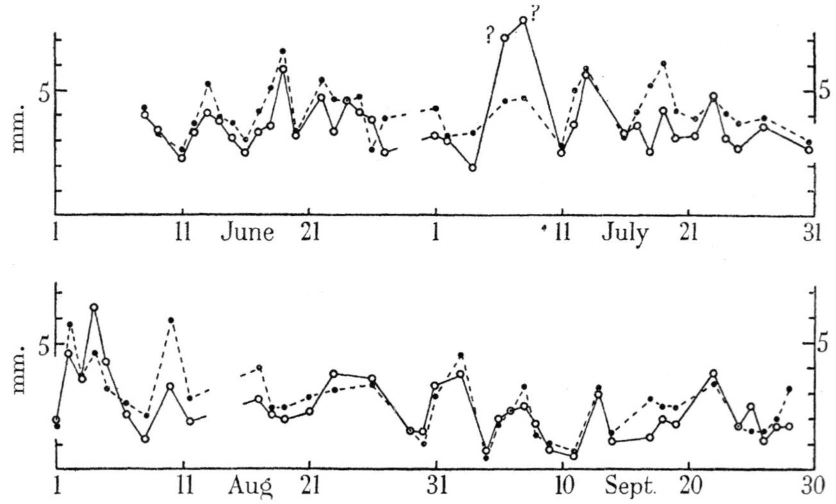

FIGURE 5. Comparison of the observed daily evaporation from open water (tank MO) and estimates based on the energy balance. —○— observed; --●-- energy balance estimate. Year 1945.

TABLE 5. ENERGY BALANCE ESTIMATE FOR TANK MO FOR INDIVIDUAL DAYS, COMPARED WITH THE OBSERVED VALUE AND THE SINK STRENGTH ESTIMATE (EQUATION 19)

date	u_2	$0{\cdot}95R_c$	H		evaporation (mm./day)		
					estimates		observed
1945	(m.p.d.)	(mm./day)	(mm./day)	β	$H/(1+\beta)$	$\Delta ef(u)$	E_{MO}
June 11	149	4·36	3·64	0·37	2·6	2·4	2·3
16	80	5·62	4·52	0·48	3·0	2·7	2·5
22	92	8·28	5·58	0·03	5·4	4·2	4·7
27	118	7·19	4·82	0·35	3·9	4·1	2·5
July 1	197	7·16	5·01	0·16	4·3	4·7	3·2
8	96	8·49	5·98	0·28	4·7	4·8	7·8?
12	50	8·19	5·80	0·15	5·0	3·1	3·6
17	111	6·45	4·78	0·15	4·2	3·2	3·6
23	128	9·09	5·76	0·23	4·7	5·7	4·8
30–31	129	3·72	2·78	−0·07	3·0	2·6	2·7
Aug. 3	67	7·58	4·11	0·12	3·6	4·3	3·6
8	80	3·65	2·67	0·25	2·1	1·4	1·2
17	122	5·65	3·76	−0·07	4·0	2·8	2·8
23	182	5·00	3·43	0·06	3·2	3·2	3·8
26	63	7·23	3·90	0·17	3·4	3·4	3·6
Sept. 2	196	5·40	3·70	−0·18	4·5	3·3	3·8
8	128	1·90	1·53	0·15	1·3	1·4	1·8
9	50	1·43	1·07	0·04	1·0	0·8	0·8
14	133	1·72	1·36	−0·02	1·4	0·9	1·1
22	169	4·52	2·34	−0·31	3·4	2·3	3·8
27	146	3·17	1·63	−0·17	2·0	2·3	1·7

9-3

From the table the sink strength estimates based on the fitted curve appear to be a little better than the energy balance estimates. The correlation coefficients between observed and estimated values are about 0·80 in each case, and when it is remembered that the fitting has reduced the sink strength estimate to about 60 % of its theoretical value it is apparent that on the basis of the original predictions the energy balance estimate is the better. The two estimates agree on 8 July when the observed value was extremely high and was queried at the time of observation. In open country there are several causes of spurious high readings; rabbits and birds appreciate an open pool on a hot day and although the enclosure was refitted with wire netting for this experiment it is impracticable to take measures to ensure 100 % freedom from intrusion. Leaks are usually unidirectional and although a big leak is easily noticed, a small one, particularly if variable, could easily be overlooked. Replication is the only safe control here, and the close agreement of the sink strength results for cylinder O and tank MO is regarded as confirmation of the general water-tightness of both containers.

(c) *Evaporation on individual days* (*other surfaces*)

No detailed analysis has been attempted, for two reasons. With water-tables at some distance below the surface there is always some drying out of the soil above the water-table that does not affect the water-table, so that on rain-free days the movement of the water-table does not represent all the evaporation taking place. On a rain day, no rise in water-table will occur until this accumulated deficit of moisture has been made up and on such days the estimate of evaporation based on water-table movement and rainfall will be excessive.

The second reason is that changes in soil temperature from day to day produce changes in the surface tension of the water held in the soil above the water-table and slight water movements take place accordingly. During dry periods the movements of the water-table in cylinder 24 B were due almost entirely to temperature changes, and evaporation was negligible by comparison.

Over an extended period the effect of temperature changes under the saturated surfaces is negligible; if the period is chosen so that its beginning and end are marked by a rise in water-table due to rain then one can be sure that the soil moisture content above the water-table is the same at beginning and end, and (18) can be applied with confidence. Such periods have been termed 'natural periods' (Penman & Schofield 1941) and are the basis of the ensuing discussion.

5. Evaporation in natural periods

(a) *General*

The main discussion will be confined to the open water surfaces, the bare soil with water-table at 5 in. and the turf with water-table at 16 in., but it will be useful to give a brief account of the performance of other cylinders. The bare soil cylinders with water-tables at 10, 16 and 24 in. failed to satisfy the condition of continuous saturation at the surface; the first showed signs of drying during extended periods of rainfree weather and the other two were almost continuously dry except for a day or two after rain. Under turf, the water-table at 10 in. was probably too high for the proper development of the plant, and both growth and transpiration were slightly less than for cylinder 16T; the 1945 results showed transpiration to be about the same for cylinders 16T and 24T, although the crop yield was greater on the latter; the turf over the deepest water-table (cylinder 36T) failed to establish itself with the same luxuriance as the others and little of any value has yet emerged from the results.

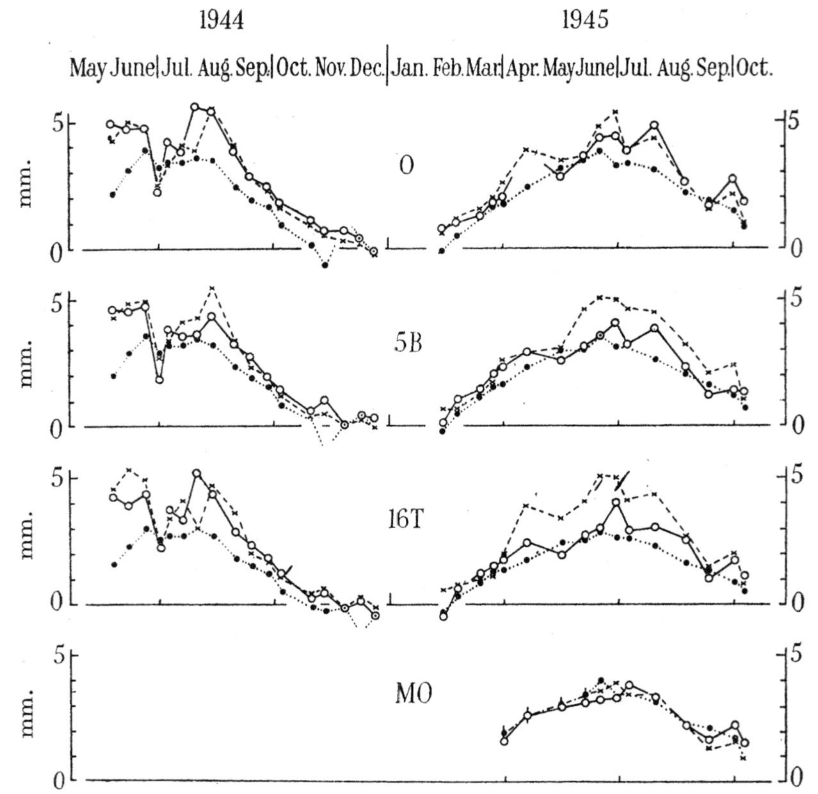

FIGURE 6. Mean daily evaporation in natural periods for surfaces with non-limiting water supply. (Open-water: O and MO. Bare soil with water-table at 5 in.: 5B. Turfed soil with water-table at 16 in.: 16T.) —O— Observed valves. Estimates: – – × – – sink strength, ··●·· energy balance and combined.

The results of 18 months for cylinders O, 5B, 16T and tank MO are summarized in figure 6, and with a few supplementary notes the figure should be almost self-explanatory. Apart from the large winter gap, when snow and ice prevented readings from being obtained, there are one or two minor gaps of a few days when records were unobtainable. The lines drawn have no significance other than helping the eye to follow seasonal changes.

The sink strength estimates are based on the mean wind velocity for the period, and the mean values of T_s and T_d from which e_s and e_d were derived. For cylinders O, 5B and 16T the linear expression derived for cylinder O was used (table 4, column 1); for tank MO the corresponding equation (19) was used.

The radiation estimates are based on (13) and (9), using reflexion coefficients of 0·05, 0·10 and 0·20 for open water, bare soil and turf respectively; for cylinder O and tank MO, therefore, the value of H was the same, but the values of β were not necessarily the same, depending upon the values of the surface temperatures.

As the surface temperature of tank MO was not measured until mid-June 1945, the combined estimate (equation 16) was made for the preceding natural periods.

(b) *Comments and conclusions*

(i) For cylinder O the sink strength estimate is a reasonably good fit throughout, suggesting that in selecting results for figure 2 there has been no undue bias.

(ii) For cylinders 5B and 16T the sink strength estimate is good in 1944 but too big in 1945. It is difficult to interpret the turf results as the thermometer bulb was usually covered by growing grass of varying length; perhaps one ought to be surprised at such an exposure combined with a formula based on open water leading to results so near observation. In the case of the bare soil, a reasonable explanation is available. During the summers, and particularly in 1945, a tough wiry weed established itself on the surface and by its mere physical presence probably slowed up evaporation by decreasing wind speed over the surface. As a result of this a little more heat would be available for warming the surface, i.e. T_s, and hence e_s and (e_s-e_d), would increase, so leading to an increased estimate as a result of a decreased observed evaporation.

(iii) For cylinders O, 5B and 16T the radiation estimates are almost invariably too low. In some of the winter periods the value of β reached very uncertain values near -1 and the derived values of $H/(1+\beta)$ were absurd. These values are not plotted.

(iv) For tank MO both estimates are reasonably good and table 5 and figure 5 above may be regarded as giving the fine structure of some of the results shown in figure 6.

(v) Comparison of the results for cylinder O and tank MO so far quoted seem, superficially, to be in conflict; the same sink strength formula fits both whereas the radiation estimates fit the tank results only. The evaporation from the tank was usually about 25 % less than that from the cylinder (table 6 below), but the surface temperature was correspondingly lower, leading to the near agreement of table 5. It is suggested that the increased evaporation from cylinder O was due to the greater surface temperature, i.e. to an extra supply of energy being available to it that was not available for the tank, and that the exclusion of this additional energy supply from the energy balance for cylinder O led to the noted underestimates of evaporation on this basis. Whereas tank MO had an all-soil surround, O and the other cylinders had a hollow on one side. The air in this would warm up during the day to temperatures much in excess of the soil or water in the cylinders at the same depth, leading to an inflow of heat, i.e. to a negative value of C in (8), at least for midsummer months.

It is reasonable to assume that the effect would be the same order for all cylinders, i.e. that *relative* values of evaporation from the three kinds of saturated surfaces would not be materially affected and that one of the major aims of the experiment has not been upset by the deficiency of the experimental site.

6. Conclusions from results of preceding paragraphs

It is convenient to interpolate a short set of conclusions based on the study of daily and periodic evaporation. Without repeating the reservations already made concerning experimental accuracy and theoretical adequacy, we have:

(*a*) A sink strength formula has been obtained that agrees closely with the results of intensive work by Rohwer, and is substantially the same for two open water surfaces having different environments.

(*b*) An energy balance has led to close agreement with observed values for one of these surfaces in which the conditions most nearly satisfy the basic assumptions made in striking the balance.

(*c*) The discrepancy for the other surface is of a kind that can reasonably be applicable to all the cylinders similarly exposed, so that relative values of evaporation rates may be deduced.

7. Results (continued): relative evaporation rates

Table 6 shows the seasonal variation in evaporation from cylinder O, expressed in *in./day*, and the relative rates for bare and turfed soil, the natural periods being grouped roughly in calendar months. For 1945 the mean daily rates for tank MO are included. The bare soil cylinder was set up for a new experiment in December 1945 involving destruction of the surface.

Natural surfaces that are bare, such as arable land before a crop is established, will rarely remain moist in summer when weeds can grow, and will, under ordinary cultivation, rarely grow weeds in winter when the soil will remain moist. It is, therefore, permissible to select results from the preceding table and to state that the evaporation rate from a freshly wetted bare soil will be about 90% of that from an open water surface exposed to the same weather. This is in agreement with the results of indoor experiments (Penman 1941) and other outdoor work (e.g. White 1932).

It is not so easy to reach a firm decision about the grass surface. Assuming, for simplicity, that the leaf temperature is always the same as that of the open water surface, the ratio E_T/E_0 will depend upon the length of daylight (N hr.), and the difference between the minimum surface temperature and the dewpoint. If this difference is large compared with the diurnal temperature change, the value of E_T/E_0 will be of the order of $N/24$, i.e. will range from about 0·70 in summer, to about 0·30 in winter at Rothamsted. As the excess of minimum over dewpoint temperature decreases, both these extreme ratios increase, and when the difference is zero they will be of the order 0·95 and 0·58 respectively, with a value of about 0·83 at the equinoxes (assuming a sinusoidal variation of diurnal temperature change). When the dewpoint temperature is greater than the surface minimum, conditions are more complicated, as condensation will take place on both kinds of surface, and until the dew is evaporated both will behave as open water surfaces whatever the light conditions. Under these conditions the relative evaporations over a whole day will approach equality, although the absolute amounts will tend to be small.

TABLE 6. SEASONAL VARIATION IN RELATIVE EVAPORATION

period 1944	E_0 (in./day)	E_B/E_0	E_T/E_0	period 1945	E_0 (in./day)	E_B/E_0	E_T/E_0	E_{MO} (in./day)
Feb.	—	—	—	8 Feb.–4 Mar.	0·04	0·78	0·31	—
Mar.	—	—	—	5 Mar.–7 April	0·06	1·10	0·86	0·06
April	—	–	—	8–30 April	0·10?	1·05?	0·92?	0·10
9 May–1 June	0·18	0·91	0·83	4–25 May; 28 May–9 June	0·11	0·89	0·69	0·11
2–28 June	0·19	0·98	0·88	10–22 June; 23 June–2 July	0·16	0·83	0·72	0·12
29 June–4 Aug.	0·16	0·82	0·89	3–12 July; 16 July–8 Aug.	0·17	0·84	0·71	0·14
5 Aug.–2 Sept.	0·19	0·81	0·78	12 Aug.–2 Sept.	0·11	0·87	0·96	0·08
5–30 Sept.	0·10	0·91	0·79	6–15 Sept.; 16–24 Sept.	0·09	0·61	0·61	0·07
1–10 Oct.	0·05	0·65	0·48	25 Sept.–22 Oct.	0·07	0·70	0·64	0·06
20 Oct.–6 Nov.								
7–13 Nov.	0·03	0·54	0·0	1–28 Nov.	0·03	0·7?	0·6	0·02
20–30 Nov.								
1–22 Dec.	0·01	0·8	−0·4	29 Nov.–22 Dec.	0·05	—	−0·0	0·05

The simple initial assumption is not likely to be realized in practice, and both mean surface temperature and its daily amplitude will be important. In winter, when the important temperature differences are likely to be small and unequal there may be more condensation on one surface than another and so negative ratios might be obtained (December 1944 and 1945).

It is clear that a more detailed study of this part of the problem is needed, and at the moment only a limited generalization can be made with the reservation that it may apply to the Rothamsted site and the years 1944–45 only. Using the *totals*, for midsummer periods (May–August inclusive) the ratio is 0·81; for equinoctial periods (March–April, September–October) it is 0·72; for midwinter the results are too few and too erratic to attempt expression of a ratio. We can, however, get an indication of the order of magnitude from another source. At Fleam Dyke, Cambridge, two drain gauges, one bare and the other turfed, are maintained side by side. Over winter months, when the summer drying has been made good in both gauges, E_B is greater than E_T. For the months December–March inclusive, in the years 1939–40 to 1942–43, the totals were:

$$\Sigma E_B = 10{\cdot}5\,\text{in.},\quad \Sigma E_T = 7{\cdot}6\,\text{in.}\quad \text{i.e.}\quad E_T/E_B = 0{\cdot}725$$

(data kindly supplied by Mr Porteous, Engineer-in-charge, Cambridge University and Town Waterworks Co.). Putting $E_B/E_T = 0{\cdot}90$, then $E_T/E_0 = 0{\cdot}65$, and as the March transpiration would be greater than that in November, the midwinter ratio would be smaller. To a satisfactory degree of accuracy it may be rounded off to 0·60.

8. Conclusions from results of seasonal variation in relative evaporation

(*a*) The evaporation rate from continuously wet bare soil is 0·9 times that from an open water surface exposed to the same weather conditions in all seasons.

(*b*) The corresponding relative evaporation rate from turf with a plentiful water supply varies with season of the year. Provisional values of E_T/E_0 for southern England are:

Midwinter (November–February)	0·6
Spring and autumn (March–April, September–October)	0·7
Midsummer (May–August)	0·8
Whole year	0·75

(*c*) The discrepancy between cylinder O and tank MO is greatest in midsummer when the effect of extra heat flow through the walls of cylinder O is likely to be most important.

9. Tests on other data

(*a*) *Data required and methods of using them*

Before the conclusions of § 8 can be applied to soil surfaces it is necessary to estimate the evaporation that would take place from an open water surface exposed to the same weather. To avoid reference back, the requirements are repeated: (i) for the sink strength estimate it is necessary to know the mean surface temperature, the mean dewpoint temperature and the mean wind velocity. Then

$$E_0 = 0{\cdot}35(1+9{\cdot}8\times10^{-3}u_2)\,(e_s-e_d)\ \text{mm./day}; \tag{19}$$

(ii) for the energy balance estimate the requirements are: mean daily short-wave radiation (or mean daily duration of sunshine), mean daily cloudiness (or mean daily duration of sunshine), mean air temperature, mean dewpoint temperature, and mean surface temperature. Then

$$E_0 = H/(1+\beta)\ \text{mm./day}, \tag{11}$$

where
$$\left.\begin{aligned} H &= R_C(1-r)-\sigma T_a^4(0{\cdot}56-0{\cdot}092\sqrt{e_d})\,(0{\cdot}10+0{\cdot}90n/N)\\ \text{and}\quad R_C &\doteqdot R_A(0{\cdot}18+0{\cdot}55n/N);\end{aligned}\right\} \tag{13}$$

(iii) for the combined estimate there must be known, mean air temperature, mean dewpoint temperature, mean wind velocity, and mean daily duration of sunshine. Then

$$E_0 = (H\Delta+0{\cdot}27E_a)/(\Delta+0{\cdot}27)\ \text{mm./day}, \tag{16}$$

where H is given by (13) above, and

$$E_a = 0{\cdot}35(1+9{\cdot}8\times10^{-3}u_2)(e_a-e_d) \quad \text{(cf. (19))}.$$

The values of N and R_A will vary with latitude and season, but are readily obtainable from standard sources. The value of r has a similar double variance but it will probably be sufficiently accurate to use a constant value of 0·05.

For general use, where E_0 has not been directly measured, the third method is most useful, and as the dependence on wind speed is not very critical, a Beaufort wind force can be substituted for u_2 so giving an opportunity of making an evaporation estimate from the data of a weather map. To convert the sink strength formula it is sufficient to use the 'good' expression for cylinder O (table 4) combined with two standard conversion factors:

$$E_0 = 0{\cdot}033u_2^{0{\cdot}68}(e_s-e_d), \quad u_2 \text{ in miles/day};$$

$$u_2 = 0{\cdot}78u_{10};$$

$$u_{10} = 1{\cdot}87B^{\frac{3}{2}}\times 24, \quad u_{10} \text{ in miles/day};$$

i.e.

$$E_0 = 0{\cdot}033[0{\cdot}78\times24\times1{\cdot}87B^{\frac{3}{2}}]^{0{\cdot}68}(e_s-e_d);$$

$$= 0{\cdot}37B^{1{\cdot}02}(e_s-e_d) \text{ mm./day}.$$

The coarseness of the Beaufort estimates of wind force suggests that this may safely be reduced to

$$E_0 = 0{\cdot}37B(e_s-e_d) \text{ mm./day}, \qquad (20)$$

giving

$$E_a = 0{\cdot}37B(e_a-e_d) \text{ mm./day}$$

for use in (16).

Although evaporation data exist for many sites over long periods of time there are not many sets that have sufficient contemporary meteorological data alongside to enable a comparison of observed and predicted evaporation to be made. The few cases discussed below have been chosen to give a fairly wide variety of sites and types of surface.

(b) *Open water*

(i) Fitzgerald (1886). Pans were floated in the middle of an 85 acre reservoir at *Chestnut Hill, Boston, Mass.* Table 7 gives results for a pan 10 ft. diameter and 10 ft. deep, filled to within 3 in., and sunk to within 6 in. of the top. The anemometer was at 30·5 ft. above the surface. Values of E_0, u, T_s, T_a and h were measured on 8 days between June and October 1885. Using (19) the following results are obtained, E_{19} representing the estimates.

Table 7. Daily evaporation at Boston, U.S.A.

day ...	1	2	3	4	5	6	7	8
u_2 (m.p.d.)	223	135	116	150	127	75	51	129
(e_s-e_d) (mm.)	10·4	7·6	7·8	7·7	10·0	8·0	6·3	2·5
E_{19} (mm./day)	11·6	6·2	5·8	6·7	7·8	4·9	3·3	2·0
E_0 (mm./day)	10·7	6·8	6·1	6·6	7·1	7·1	4·1	2·5

(ii) Visentini (1936). A pan was floated in the reservoir at *Molato, Italy*, with an anemometer on the dam. Values of T_s, T_a, u and h are given. In the absence of information it is assumed that the anemometer height was 2 m. and the calibration the same as for the Rothamsted anemometer.

TABLE 8. MEAN DAILY EVAPORATION AT MOLATO, ITALY

1934 ...	Aug.	Sept.	Oct.	Nov.	Dec.
u (m.p.d.)	67	48	44	82	77
$(e_s - e_d)$ (mm.)	8·2	6·85	5·8	3·2	1·95
E_{19} (mm./day)	4·8	3·5	2·9	2·1	1·2
E_0 (mm./day)	5·1	3·9	4·1	2·3	1·3

1935 ...	Mar.	Apr.	May	June	July	Aug.	Sept.
u (m.p.d.)	106	96	99	87	91	70	72
$(e_s - e_d)$ (mm.)	3·0	5·25	3·9	8·2	8·5	6·3	5·4
E_{19} (mm./day)	2·1	3·6	2·7	5·4	5·7	3·9	3·3
E_0 (mm./day)	1·8	2·7	3·0	6·2	7·9	5·7	4·3

(iii) Davydov (1936). Measurements were made on *Sevan Lake, Soviet Armenia* (no details). Values of $(e_s - e_d)$ between lake surface and 10 cm. above, and of wind velocity at 9 m. are given. Results are means for 1927–30:

TABLE 9. MEAN DAILY EVAPORATION AT SEVAN LAKE, SOVIET ARMENIA

	Apr.	May	June	July	Aug.	Sept.	Oct.	Nov.	Dec.
u_2 (m.p.d.)	136	106	123	157	119	144	140	153	183
$(e_s - e_d)$ (mm.)	0·1	0·0	1·5	2·3	3·9	4·7	4·9	3·8	3·3
E_{19} (mm./day)	0·1	0·0	1·2	2·0	3·0	4·0	4·1	3·3	3·2
E_0 (mm./day)	0·4	0·2	1·2	2·1	3·4	4·5	4·1	3·6	2·9

(iv) Ray (1931). Monthly means for a standard U.S. Weather Bureau evaporation pan are given for 1917–30 for *San Juan, Puerto Rico*, together with u_0, T_a, mean saturation deficit and mean hours of sunshine. For speed of computation a 12 hr. day has been assumed for the whole year. As results are given in in./month and $(e_a - e_d)$ in inches of mercury, the same form is kept in table 10:

TABLE 10. MEAN MONTHLY EVAPORATION AT SAN JUAN, PUERTO RICO

	Jan.	Feb.	Mar.	Apr.	May	June
E_a (in./month)	5·35	4·75	7·3	7·2	7·6	7·4
H (in./month)	3·8	5·15	6·1	7·0	7·05	7·7
E_{16} (in./month)	4·25	5·05	6·4	7·05	7·2	7·65
E_0 (in./month)	5·55	5·55	7·6	7·85	7·75	7·45

	July	Aug.	Sept.	Oct.	Nov.	Dec.
E_a (in./month)	8·55	8·15	6·05	5·0	4·8	5·75
H (in./month)	7·1	7·55	6·4	5·3	4·3	3·8
E_{16} (in./month)	7·55	7·7	7·3	5·2	4·45	4·3
E_0 (in./month)	8·15	8·05	6·6	5·85	5·1	5·45

(c) *Bare soil*

There are few records available and information about related weather is even more rare. By the courtesy of the Indian authorities, the rainfall, drainage and other weather records for Pusa have been made available from 1907 to 1934. Pusa lies in the monsoon region of Asia, at latitude 26° N. in the Ganges basin, and in most years it is a fairly reasonable assumption that once the monsoon has broken, bare soil will remain moist for most of the monsoon period.

In 1906 four drain gauges were constructed, each 1/1000 acre in area, without disturbing the soil, two being 3 ft. deep and two 6 ft. deep. One at each depth was kept bare and the other pair cropped, and daily records of rainfall and drainage for months in which at least one gauge ran are available from 1907–34. The crop carried by gauges II (6 ft.) and IV (3 ft.) during the monsoon period was either maize (1907–15) or sunn hemp (1916–34), sowing taking place in June each year. Both are tall plants (8 ft. and 6 ft. high), and, standing well above the turf surround, they would be better ventilated, intercept more radiation and expose a larger transpiring surface to the air than a patch of the same size in the middle of a large field. Transpiration would be abnormally great, and the evidence of the cropped gauges is to be rejected on the ground that the same surface was not typical of its environment.

The records were separated into natural periods in which the difference between rainfall and drainage (or drainage and run-off) can be equated to the evaporation of the period. As drainage continued for several days after rain it was not always possible to decide which was the last rainfall causing drainage, and there are inevitable uncertainties in the estimation of $(R-D)$ per day as a result. As elsewhere, all estimates of evaporation are very much dependent upon the drain gauge receiving the same amount of rain as the rain gauge.

The values of $R-D$ for gauges I and II (6 ft. and 3 ft. bare) usually agreed closely and there was general consistency in the performance of gauges II and IV. A condensed summary for 2 years shows the order of evaporation per day from all four; further analysis will be restricted to the records for gauge I.

TABLE 11. COMPARATIVE EVAPORATION AT PUSA

period	R/day (in.)	$(R-D)$/day (in.) I (6′B)	II (6′C)	III (3′B)	IV (3′C)
1911					
8–18 July	0·34	0·17	0·24	0·15	0·23
19 July–25 Aug.	0·47	0·16	0·37	0·15	0·35
26 Aug.–26 Sept.	0·27	0·12	0·24	0·12	>0·27
27 Sept.–13 Oct.	0·22	0·12	0·11	0·10	no D
1922					
2–26 July	0·72	0·11	0·22	0·08	0·13
27 July–9 Aug.	0·53	0·09	0·21	0·05	?
10–20 Aug.	0·24	0·12	0·20	0·10	?
21 Aug.–23 Sept.	0·21	0·09	0·22	0·06	0·19
29 Sept.–5 Oct.	0·19	0·13	0·19	0·14	>0·19

Weather observations at Pusa (1911–33) were made at 08.00 hr. and included dry and wet bulb temperatures, air maximum and minimum temperatures, anemometer readings at 08.00 and 08.03 hr., cloud amount, and 'instrumental test observations', the last being the actual readings of the three dry thermometers, presumably expected to be equal when read. These readings never agreed and differences were erratic, ranging up to 3° F, the usual order being dry > max. > min.

Assuming that the anemometer ran continuously between 08.03 hr. one day and 08.00 hr. the next, the run-of-the-wind per day was obtained for all days except when the reading passed through an unknown zero; the height is assumed to be 2 m. and the calibration the same as the Rothamsted instrument (table 2 above). From the mean air temperature and the 08.00 hr. value of dewpoint the value of the mean saturation deficit was obtained. From these, values of E_a were obtained.

The determination of H had to be based on a single cloud estimate per day; comment is unnecessary. There seemed little point in evaluating it for all periods in all years and only two, 1911 and 1922, are considered in detail (table 12). Mean wind speeds (as measured) ranged from 33 to 133 miles/day, mean and 08.00 hr. air temperatures from 81 to 86° F, 08.00 hr. humidity from 87 to 92 % and estimated n/N from 0·00 to 0·64.

(*d*) *Cropped soil*

Using the annual summary of the Monthly Weather Report, data for 70 stations in the British Isles have been abstracted for the years 1930–39 and long period means obtained for mean air temperature, mean vapour pressure, mean Beaufort wind force and the mean ratios of actual/possible hours of sunshine. From these, values of E_a and H were obtained for each site and values of E_0 derived, using equation (16). Converted to inches per year, an evaporation map of the British Isles was obtained

TABLE 12. ESTIMATED E_B AND MEASURED $(R-D)$/DAY FOR BARE SOIL: PUSA 1911 AND 1922

period	R/day (in.)	E_a/day (in.)	H/day (in.)	E_{16} (in./day)	$0{\cdot}9E_{16}$ (in.)	$(R-D)$/day (in.)
1911						
8–18 July	0·34	0·08	0·11	0·10	0·09	0·17
19 July–29 Aug.	0·46	0·11	0·16	0·15	0·13	0·16
30 Aug.–8 Sept.	0·17	0·17	0·13	0·14	0·13	0·11
9–26 Sept.	0·32	0·08	0·16	0·14	0·12	0·14
27–30 Sept.	0·18	0·07	0·11	0·10	0·09	0·14
1–13 Oct.	0·22	0·09	0·14	0·13	0·11	0·10
1922						
2–20 July	0·80	0·09	0·12	0·11	0·10	0·10
21–26 July	0·45	0·12	0·12	0·12	0·11	0·15
27 July–9 Aug.	0·54	0·12	0·14	0·13	0·12	0·09
10–15 Aug.	0·12	0·04	0·08	0·07	0·06	0·08
16–20 Aug.	0·38	0·12	0·17	0·16	0·14	0·18
21–26 Aug.	0·16	0·07	0·12	0·11	0·10	0·10
27–30 Aug.	0·25	0·09	0·10	0·10	0·09	0·16
31 Aug.–5 Sept.	0·25	0·08	0·12	0·11	0·10	0·13
6–10 Sept.	0·15	0·09	0·22	0·18	0·17	0·13
11–28 Sept.	0·34	0·09	0·14	0·13	0·11	0·14
29 Sept.–5 Oct.	0·19	0·02	0·19	0·14	0·13	0·14

showing the probable value of annual evaporation from open water in exposed sites. From the conclusions in § 8 above one would expect the corresponding value of the annual evaporation from cropped land to be $\frac{3}{4}$ of E_0 if the crops transpired at maximum rates all the year; in practice the rates will be less than this because of the ripening process in annual vegetation and/or the lack of summer rainfall, particularly in south-east England, but in the table below this conversion factor is applied uniformly. The table shows the observed difference between rainfall and runoff for certain catchment areas (Lloyd 1940, 1941, 1942) (these are rather monotonous), the observed difference multiplied by $\frac{4}{3}$, which should be the expected corresponding open water evaporation, and the estimated value of E_0 based on annual mean values of weather elements for stations somewhere near, if not in, the catchment area.

TABLE 13. EVAPORATION FROM CATCHMENT AREAS

catchment	period	mean rainfall-runoff (in./year)	$4/3(R-r)$	estimated E_0 for nearby sites (in./year)
Lea	1928–36	19·2	26	Greenwich, 25 Rothamsted, 20
Thames	1928–36	18·7	25	Kew, 26 Oxford, 24
Severn	1928–36	18·8	25	Ross-on-Wye, 24 Cheltenham, 21 Shrewsbury, 21
Vrnwy	1932–38	19·1	26	Shrewsbury, 21 Sealand, 24 Llandudno, 23
Rivington	1932–38	17·4	23	Stonyhurst, 21 Hutton, 18
Spey	1936	10·3	14	Dalwhinnie, 17

Detailed examination of Rothamsted data has shown that the sum of the twelve monthly estimates exceeds the annual estimate by about 10 %, due to the extra weight which should be given to summer evaporation. A similar increase is to be expected for other sites and should be borne in mind in reading table 13.

10. GENERAL CONCLUSION

A detailed discussion of the data presented in § 9 (*b*), (*c*) and (*d*) would be intolerably long, and much of it would be concerned with the adequacy of the observations rather than the adequacy of the equations on which the estimates are based. The general impression is satisfactory for all three types of surface, and the wide range of climatic regions employed indicates that something of universal significance has been obtained in the results of §§ 3 to 8, although there must inevitably be something of the time and place at which the experimental work was done included in the equations.

Two aspects of evaporation have been under review. There is that of the physicist and mathematician seeking facts to fit a formula; sufficient has emerged to show possible sources of weakness in the theoretical treatment of both sink strength and energy balance and to show the relative importance of the quantities that must be measured to obtain adequate accuracy in experimental trials. There is also that of the 'practical' man—water engineer or meteorologist—seeking a formula to fit the facts. Formulae have been given and where all the necessary measurements can be made, or values forecast, then reliable evaporation rates can be estimated or predicted. There are still empirical aspects of both sink strength and energy balance estimates and until these are removed there must always be some doubt about the possibility of successful translation in space or time of the formulae.

The work described in the preceding pages is an extension of research carried out at Rothamsted Experimental Station before 1941. It was resumed in 1944 at the request of the Meteorological Office, and it is a pleasure to be able to record my appreciation of the assistance given by several branches of the Office in the form of equipment, information and advice. I am particularly indebted for many helpful discussions to Mr C. S. Durst of the Meteorological Office, and to my colleague, Dr R. K. Schofield. To the Director of the Meteorological Office and the Meteorological Research Committee I am grateful for the helpful reception given to the report on which this paper is based and for permission to publish it.

Finally, this work could not have been done at short notice if the basic equipment of the enclosure had not been available: to Dr B. A. Keen, F.R.S., until recently Head of the Physics Department at Rothamsted, especially grateful acknowledgement is due for the foresight that provided the installation with full knowledge that it would be many years before the soil would be fit for experimental work.

References

Bowen, I. S. 1926 *Phys. Rev.* **27**, 779.
Brunt, D. 1939 *Physical and dynamical meteorology.* Camb. Univ. Press.
Cummings, N. W. & Richardson, B. 1927 *Phys. Rev.* **30**, 527.
Davydov, V. K. 1936 *Assoc. Intern. D'Hydrologie Sci. Bull.* no. 22, 142.
Fitzgerald, D. 1886 *Trans. Amer. Soc. Civ. Engrs*, **15**, 581.
Geiger, R. 1927 *Das Klima der bodennahen Luftschicht.* Brunswick: F. Vieweg.
Kimball, H. H. 1914 *Mon. Weath. Rev., Wash.* **42**, 474.
Lloyd, D. 1940 *Quart. J.R. Met. Soc.* **66**, 181.
Lloyd, D. 1941 *Quart. J.R. Met. Soc.* **67**, 33.
Lloyd, D. 1942 *Quart. J.R. Met. Soc.* **68**, 35.
Pasquill, F. 1943 *Proc. Roy. Soc.* A, **182**, 75.
Penman, H. L. 1940 *Quart. J.R. Met. Soc.* **66**, 401.
Penman, H. L. 1941 *J. Agric. Sci.* **31**, 454.
Penman, H. L. & Schofield, R. K. 1941 *J. Agric. Sci.* **31**, 74.
Prescott, J. A. 1940 *Trans. Roy. Soc. S. Aust.* **64**, 114.
Ray, C. L. 1931 *Mon. Weath. Rev., Wash.* **59**, 192.
Rohwer, C. 1931 *U.S. Dept. Agric. Tech. Bull.* 271.
Sutton, O. G. 1934 *Proc. Roy. Soc.* A, **146**, 701.
Sutton, W. G. L. 1943 *Proc. Roy. Soc.* A, **182**, 48.
Sverdrup, H. V. 1945 *Oceanography for meteorologists.* London: George Allen and Unwin.
Visentini, M. 1936 See Davydov. Bull. no. 22, p. 119.
White, W. N. 1932 U.S. Dept. Interior. Water Supply Paper 659.

On the Assessment of Surface Heat Flux and Evaporation Using Large-Scale Parameters

C. H. B. PRIESTLEY and R. J. TAYLOR—*Division of Atmospheric Physics, Commonwealth Scientific and Industrial Research Organization, Aspendale, Victoria, Australia*

ABSTRACT—In an introductory review it is reemphasized that the large-scale parameterization of the surface fluxes of sensible and latent heat is properly expressed in terms of energetic considerations over land while formulas of the bulk aerodynamic type are most suitable over the sea. A general framework is suggested.

Data from a number of saturated land sites and open water sites in the absence of advection suggest a widely applicable formula for the relationship between sensible and latent heat fluxes.

For drying land surfaces, we assume that the evaporation rate is given by the same formula for evaporation multiplied by a factor. This factor is found to remain at unity while an amount of water, varying from one site to another, is evaporated. Following this a linear decrease sets in, reducing the evaporation rate to zero after a further 5 cm of evaporation, the same at several sites examined.

1. INTRODUCTION

In recent years, advances in our knowledge of larger scale dynamics have placed new requirements on our understanding of boundary layer problems with a consequent emphasis on the mutual nature of the interplay between these two types of process. Students of the boundary layer need to review their understanding in the context of this interplay and must abandon some of the idealizations they have been wont to make; students of the larger scale processes, in turn, are coming to think about the boundary layer as an inherent part of the problem.

Two sharply contrasting philosophies appear to be emerging and it may be useful to discuss the symptoms and arrest the trend before there is an undesirable hardening of attitudes. The "separatist" approach aims to treat the boundary layer development as a closed or quasi-closed problem and to use the solution so obtained to provide lower boundary conditions for the "free" atmosphere. This way of thinking is a direct continuation of the way the subject has developed historically. By contrast, the "merged" approach takes the view that a well designed, sufficiently detailed model of the atmosphere will generate its own special layers, such as the tropopause and the boundary layer, and that it is undesirable to set up a separate computational system for part of the troposphere to serve as input to the remainder.

Clearly, the most important situations for the larger scale dynamics are the changing ones where the time-dependent and advective terms are significant. These are the very cases when the governing conditions of the boundary layers (geostrophic wind, vertical stability, etc.) are subject to changes determined by the atmosphere above and thus are the ones when the separatist approach is least applicable. We must recognize that there is a danger that boundary layer treatments will be consolidated into frameworks that are inherently ill suited for handling the evolutionary aspects of the problems.

This is not to deny that steady-state, one-dimensional treatments still have an important role to play as an essential preliminary to the solution of the more important and complicated evolutionary problems. Again, the separatist approach may have positive advantage in special conditions; for example, when the top of the boundary layer is well defined, as by a sharp inversion, and when conditions in it, though varying to a degree significant for the larger scales, are doing so as a result of purely internal processes.

With whatever vertical flux (heat, vapor, momentum, angular momentum) we are concerned, there are two main facets to the problem: determination of the surface value of the flux and consideration of its variation with height. In principle, the merged approach might look on the first as a boundary condition and the second as something that the large-scale model itself, given sufficient detail, will provide; but to state the problem thus, at the present stage of knowledge and model formulation, is to state it too simply. Beginning quite close to the surface, the vertical flux will be subject to the process of progressive hand-over up the scale of eddy sizes (Priestley 1967). Of special interest is the stage of transition from the subgrid scale to the grid scale—in modeling terms, transition from a stage where the flux requires parameterization to one in which it presumably does not—and this is likely to be a gradual process. Unless these stages are accurately represented in the model, there is the danger that nature's processes will be seriously distorted.

About a decade ago, one of us (Priestley 1959, chapter 8) attempted a systematic appraisal of how the heat flux, and by implication the evaporation, at the surface would behave in time-dependent situations. It emerged from this discussion that three quite different circumstances need to be distinguished:

1. Heating over the sea with an inversion to confine the depth.
2. Heating over the sea without an inversion.
3. Heating over land.

The dominant controls, and hence the evolutionary pattern, are quite different in the three cases.[1] The implication is clear that differences in controls will require differences in parameterization schemes.

The rate of conduction of heat near the interface between two media is governed by the quantity $\rho c\sqrt{K}$ pertaining to each medium. Here ρ is density, c is specific heat (at constant pressure for air), and K is the thermal diffusivity, molecular or eddy as appropriate. This combination may conveniently be termed the *conductive capacity* and its typical values respectively for open water, the atmosphere, and solid ground generally form a strongly *descending* sequence. Heat provided at the interface tends to be conducted into the two media at rates proportional to their conductive capacities.[2] However, when the thermal balance between the two media is disturbed, as by advection, the heat subsequently passing from one to the other over any not-too-short time is governed by and proportional to the conductive capacity of the lower ranking medium.

Over the oceans, it is probably not unreasonable to look on the fluxes as occasioned by preexisting differences of temperature and humidity between surface and air. Because of the very large conductive capacity of water and because the radiation penetrates into it, the local value of net radiation does not control the evaporation and heat flux into the air. Over the land, it does. For periods of 24 hr or so, the amount of heat taken up or released by the soil is, generally speaking, not a large component of the surface energy balance. Consequently, the available radiant energy exercises a real and immediate physical constraint on the fluxes of vapor and sensible heat over land.

Large-scale dynamical computations, if they are to be extended beyond about 2 days, require as input the rates of evaporation, E, and sensible heat transfer, H, at the earth's surface. We shall consider the minimum requirements to be the daily rates of these quantities, with a spatial resolution comparable with the main network of other observations; that is, to borrow from the Global Atmospheric Research Program (GARP) specifications, on the order of several hundreds of kilometers.

Over land, as indicated above, the sum, $LE+H$ (where L is the latent heat of evaporation of water), is strongly governed by the net radiation, R, at the earth's surface. It is equally clear that the apportionment of energy between LE and H will be governed by the dryness of the ground and, because of the nature of the relationship between saturated vapor pressure and temperature, by the general level of surface temperature, $\overline{T}_0$. Parameterization in terms of the slope of the saturated vapor pressure curve has frequently been suggested [e.g., Priestley (1959), Bryson and Kuhn (1962)]. Full consideration of the surface energy balance, not reproduced here but found in any appropriate reference work, shows that $\overline{T}_0$ is itself strongly governed by R and ground dryness. Thus, knowledge of R and of ground dryness constitutes a *necessary* prerequisite for the estimation of the main space variations in daily LE and H on the scale considered here. This paper might well be described as a first attempt to probe, in a practical way, the extent to which the same two prerequisite parameters might also prove *sufficient* for the purpose in hand.

There will be no room here for detailed discussion as to how one arrives at the two main parameters, R and ground dryness, and the subsidiary parameter, $\overline{T}_0$. Suffice it to emphasize that any valid model must provide them before it can aspire to provide LE and H.

It is probable that the fluxes over land on the space scale considered are determined to first order by the radiation actually falling on the area rather than by advected energy. The radiation received will increase as the square of the gridpoint separation, whereas advective effects will increase more or less linearly because the difference in horizontal fluxes of heat and vapor at the upwind and downwind edges of an area will not continue to increase indefinitely as these edges are moved farther apart. While the argument indicates that the grid-scale effects of advection will generally be minor over land, we would hope eventually to be in a state of knowledge where advected energy values would emerge from the calculations, and allowance could be made for them.

Although we shall be concerned with the problem of estimating the total input of heat and vapor to the atmosphere from a large area, the only observations available as a basis are those from individual sites, subject in some cases to quite apparent small-scale nonuniformity and advection. The steps that have been taken to compensate as far as possible for this are described below, *but this essential difference between our starting material and what we hope to derive from it must be kept constantly in mind.*

2. A GENERAL FRAMEWORK

For investigation of the apportionment between heat flux and evaporation, an appropriate framework is provided by the following consideration.

Swinbank and Dyer (1967) have shown that the profiles of specific humidity, q, and temperature, T, are similar over land. Moreover, both the eddy conductivity, K_H, and the eddy diffusivity, K_W, tend to the neutral value of eddy viscosity, ku_*z, at low levels in the turbulent layer or at small $z/|L|$ (Dyer 1967) within the limits of experimental error. Here, k is von Kármán's constant,

[1] The case of cooling from below was not included as it was considered less important *dynamically* though its significance in special processes (e.g., advection fog) is obvious.

[2] Note that where eddy-*K*s are used or implied, their most appropriate values may depend also on the time scale of the phenomenon: values used in the present time scale (days) would require reconsideration before use in seasonal or climatic-change applications.

u_* the friction velocity, z the height, and L the Monin-Obukhov stability length parameter.[3] K_H and K_W are then not merely proportional, but are equal. Accordingly, both q and T satisfy the same one-dimensional diffusion equation

$$\frac{\partial q, T}{\partial t}=\frac{\partial}{\partial z}\left(K\frac{\partial q, T}{\partial z}\right). \tag{1}$$

If the surface, denoted by the suffix zero, is saturated, q_0 takes on the saturation value $q_s(T_0)$. Let $\overline{T}$ be some constant value, central to the range of variation, and consider the variable

$$\psi=q-q_s(\overline{T})-\left(\frac{\partial q_s}{\partial T}\right)_{T=\overline{T}}(T-\overline{T}). \tag{2}$$

This variable also satisfies eq (1). Provided that the range of variation is not too large, we may approximate by "linearizing" the q_s,T curve, which implies that $\psi_0=0$ for all time. It had been earlier concluded by one of the authors (Priestley 1959) that the solution would then be $\psi=0$ for all t *and* z and thus

$$\frac{LE}{H}=\frac{L}{c_p}\left(\frac{\partial q_s}{\partial T}\right)_{T=\overline{T}}=\frac{s}{\gamma}, \tag{3}$$

c_p is the specific heat of air at constant pressure, s is defined as $\partial q_s/\partial T$ at the appropriate temperature, and γ is c_p/L.

However, this is not the most general solution, for ψ can increase or decrease with height without conflicting with eq (1) provided that

$$K\frac{\partial \psi}{\partial z} \text{ is independent of } z. \tag{4}$$

This complementary function represents any vertical flow through the system without accumulation, and the solution will not be definitive until some further restrictive condition for ψ is identified. Meanwhile, it is appropriate to examine whether or not eq (3) still represents a principal part of the relationship between H and LE. We shall accordingly analyze our data in terms of the quantity, α, defined by

$$\frac{LE}{LE+H}=\alpha\frac{s}{s+\gamma}. \tag{5}$$

Here, α will be unity for saturated surfaces only in the special case $\psi=0$. Clearly, α is related to the Bowen ratio, β $(=H/LE)$, by the relationship

$$\beta=\frac{1-\alpha\dfrac{s}{s+\gamma}}{\alpha\dfrac{s}{s+\gamma}}. \tag{6}$$

[3] The use of L in this sense as well as to denote latent heat is standard. No confusion is likely to ensue.

The quantity α may also be useful in analyzing data from unsaturated surfaces; in this case, the only a priori expectation would be that α would have a smaller value than for saturated surfaces; indeed, the ratio of the two alphas could be regarded as an index of aridity.

Penman (1948) gave an equation for the rate of evaporation from a saturated soil surface that can be written in the form

$$LE=\frac{s}{s+\gamma}(R-G)+\frac{\gamma}{s+\gamma}h(q_s-q) \tag{7a}$$

where R is the net incoming radiation, G is the heat flux into the ground (so that $R-G=LE+H$), h is a suitably defined transfer coefficient, and q_s is taken at the temperature of the air. In the absence of advection, eq (7a) allows us to put limits that are not entirely intuitive on the possible daily average rate of evaporation. In the first place, it is unlikely that an inversion will prevail so as to make H negative. If H is not negative, then

$$LE<R-G, \qquad \text{i.e., } \alpha<\frac{s+\gamma}{s}. \tag{7b}$$

Secondly, it is unlikely that the saturation deficit term in eq (7a) will become negative and thus produce condensation. Absence of condensation requires that

$$LE>\frac{s}{s+\gamma}(R-G), \qquad \text{i.e., } \alpha>1. \tag{7c}$$

The term $s/(s+\gamma)$ varies from 0.56 at a surface temperature of 10°C to 0.82 at 35°C. This temperature range embraces all the observations treated. When the surface is known to be saturated, the conditions (7b, 7c) can be used as minimum criteria for the absence of advective effects.

3. EVAPORATION AND HEATING OVER THE LAND (SATURATED SURFACE)

a. CSIRO Lysimeters

The Commonwealth Scientific and Industrial Research Organization (CSIRO) lysimeter installation at Aspendale, Victoria, Australia, has been described by McIlroy and Angus (1963). It consists essentially of 12 weighed containers of 1.6-m diameter and 1-m depth. Provision is made for measuring the drainage from each container so that, with due allowance for rainfall, the evaporation can be measured. Only evaporation from pots carrying a pasture crop similar to that on the immediately surrounding area will be considered. The lysimeter pots are irrigated when necessary so that only occasions following heavy rain, when the difference between them and the surrounding countryside is at a minimum, can be used. The records were searched to find days conforming to the following conditions:

1. No rain and no irrigation.
2. At least 0.80 in. of rain in the previous 1, 2, or 3 days (a quite arbitrary definition of "heavy rain").

3. Availability of ancillary data, including 24-hr net radiation, since heat flux is not measured directly at this site, but α can be defined in a form equivalent to eq (5) as

$$\alpha=\frac{LE}{\frac{s}{s+\gamma}(R-G)}. \tag{8}$$

Altogether, 36 days fulfilling these conditions were found and, of those, 14 also satisfied conditions (7b, 7c). Using eq (8), we calculated a mean α of 1.34 ± 0.05 for these 14 days.[4] Of the 22 days not satisfiying conditions (7b, 7c), 21 did not satisfy condition (7b) whereas only one failed to satisfy condition (7c). This would indicate that, in these general weather conditions at least, the site is one where there is convergence of energy by advection so that the value 1.34 for α might be expected to be too high.

In this calculation, 24-hr totals of evaporation and net radiation were used, heat flux into the ground was neglected, and s was calculated at instrument shelter temperature because surface temperatures were not available.

b. University of Wisconsin Lysimeter

This installation has been described by Black et al. (1968) and the results of a series of observations with it on a snap bean crop will be presented by Black et al. in a future publication. Copies of the original data on which the latter paper is based have kindly been made available to us. They include evaporation, rainfall, net radiation, surface temperature, and vapor pressure at 1.4-m height for a period of 54 days in July–September 1968. On 23 of these days, condition (7b) was not fulfilled and these are excluded. Condition (7c) was not satisfied on a number of occasions, but judgment on these must be deferred until the degree of saturation of the surface has been considered later in the paper. The days in question are, therefore, not included here.

These observations permit us to evaluate α not only in the form given in eq (8) but also in the equivalent

$$\alpha=\frac{L\Delta q}{\frac{s}{s+\gamma}(L\Delta q+c_p\Delta T)} \tag{9}$$

where Δ implies a difference over a given height interval. Equations (9) and (8) are equivalent because K_H and K_W are equal. Values of α in both forms [assuming surface saturation in eq (9)] were calculated and plotted against the cumulative total of evaporation minus precipitation, P, since the beginning of the series. For negative values of $\int(E-P)dt$ greater than 4 cm of water, the two forms [eq (8) and (9)] agree well and this, then, has been taken as the criterion of surface saturation. There are 23 values of α giving a mean of 1.30 ± 0.03.

[4] This $\pm$ notation will be used to denote the standard error of the mean.

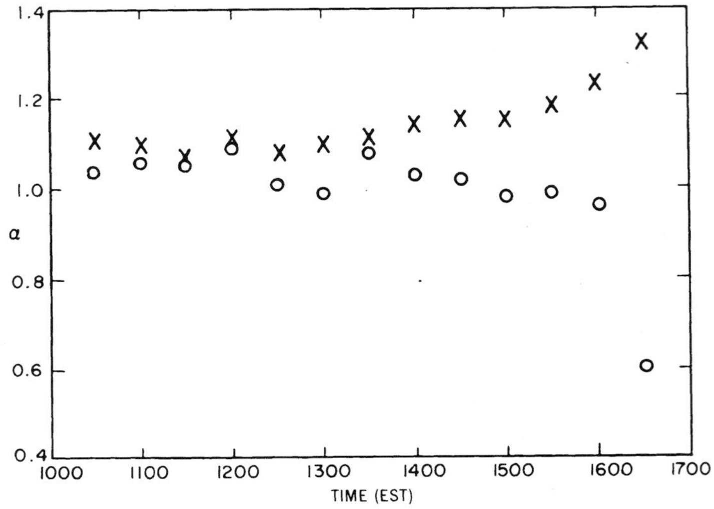

FIGURE 1.—Fluxatron observations of Mar. 17, 1970; $X=L\Delta q/\{[s/(s+\gamma)]\times(L\Delta q+c_p\Delta T)\}$ and $O=LE/\{[s/(s+\gamma)](R-G)\}$.

c. Fluxatron Measurements

An interesting series of observations was obtained in March of 1970 (Dyer and Hicks 1970) during a micrometeorological expedition to Gurley in northern New South Wales. Heavy rain (approx. 1.5 in.) fell during the evening of March 16 and observations on the following day can be taken, with some confidence, as referring to a saturated surface of recently plowed bare soil.

That part of the data relevant here comprises 13 half-hourly averages of R, G, and H (the last by eddy correlation technique) accompanied by soil surface temperature and temperature and humidity at 1.5 m. The evaporation rate can be deduced from the surface energy balance.

Both eq (8) and (9) were used to calculate α and the results are given in figure 1. The good agreement between the two estimates of α from 1030 to 1330 LST confirms that the surface was indeed saturated then. The fall-off in true evaporation rate, and the overestimation of it due to the assumption of a saturated surface, as the surface dries out, are clearly shown. The mean α for the period from 1030 to 1330 LST is 1.08 ± 0.01.

This value of α differs significantly from those above, which are similar to others to be discussed below. The explanation offered is that the occasion was one of cold advection; the soil heat flux was, throughout the day, directed upward.

d. Wangara Data

An expedition, code-named "Wangara," was made in the winter of 1967 to the neighborhood of Hay in southern New South Wales mainly to investigate mesoscale momentum transfer, but a number of observations were also made that are relevant here. The site provided an extensive area of very good horizontal uniformity, The observations to be used here comprise net radiation.

heat flux into the ground, and dry-bulb temperatures at 1, 2, and 4 m together with deduced values of the shearing stress. All details are given by Clarke et al. (1971). Values of heat flux were calculated in lapse conditions according to the flux-gradient relation suggested by Dyer (1967) and in neutral and inversion conditions by assuming that the transfer coefficients for heat and momentum were equal. Evaporation was then determined by energy balance. Only over one period of 6 days immediately following a fall of 1.5 cm of rain can we be reasonably confident that the surface acted as a saturated one, and this period gives $\alpha=1.33\pm0.21$. The comparatively low precision of the determination is no doubt connected with the fact that the fluxes being measured were very small.

4. EVAPORATION AND HEATING OVER THE OCEANS

a. Heating Beneath an Inversion

When heating occurs over the sea with a well-defined inversion to limit its effects, the conditions are favorable for the application of the separatist approach. An effective, practical, and, to a large extent, definitive treatment of this problem was proposed by Burke (1945), who specified the heat input at the bottom by the bulk aerodynamic formula

$$H_0=-c_p\rho Cu(T-T_0) \tag{10}$$

with a similar formula for moisture input by evaporation. (T and the wind speed, u, are referred to a convenient near-surface standard level.) He assumed that the heat and vapor so supplied to the air were redistributed so as to preserve an adiabatic lapse rate (dry with constant specific humidity up to condensation level and saturated above). He thereby determined the evolutionary behavior of temperature and humidity and the change in height of the mixed (boundary) layer with time and distance due to entrainment. Any change in height due to dynamical influences would presumably be additive.

Equivalently, in terms of friction velocity u_*,

$$H_0=-c_p\rho C_*u_*(T-T_0). \tag{11}$$

C and C_* are heat transfer coefficients that depend on the reference height selected for T and u, and, if this height is not low enough, on the stability.

Charnock and Ellison (1967) elaborated on this approach, replacing the assumption of an adiabatic profile by the similarity profile

$$k(T-T_0)=\frac{H}{\rho c_p u_*}\left[f\left(\frac{z}{L}\right)-\ln\frac{z_T}{L}\right] \tag{12}$$

with a similar formula for humidity. Here z_T is the thermal equivalent of the roughness length, L is the Monin-Obukhov stability length parameter, and f is the universal function for unstable conditions that has, in effect, been established by experiment (Dyer 1967) at least up to $z/|L|=3$. It is a matter for trial whether this more elaborate treatment gives results sufficiently different form Burke's to justify the increased complexity.

Large-scale application, requiring a relationship between the transfer and large-scale variables, may imply some formal preference for eq (11) over (10). More importantly, our reference level for T and u is either so high that variations of C and C_* with stability must be specified or low enough to avoid this difficulty. In the latter case a sufficiently low level must be included in the numerical model.

Because of the arrangement of terms, it has sometimes passed unnoticed that eq (12) is a specific form of eq (11), giving the explicit dependence of C_* on reference height and stability:

$$C_*=k\left[f\left(\frac{z}{L}\right)-\ln\frac{z_T}{L}\right]^{-1} \text{exponent.} \tag{13}$$

The introduction of similarity arguments may prove of greater practical importance in the determination of H_0 than in problems of the redistribution of heat so gained and of the rise of the top of the heated layer. For these last two problems, Burke's treatment may remain adequate.

b. Heating Without an Inversion

When there is no inversion (below the tropopause) to confine the heating, the rate of adjustment of T and T_0 is slower so that large heating rates can be sustained over much longer trajectories. For numerical application, the difficult part of the problem is to know how the medium- and large-scale cumuli will redistribute the heat in the vertical and thereby control the evolution of T.

Even in the presence of turbulence and convection, the transfer of heat is limited by the conductive capacities of the two media (sec. 1). One of the authors (Priestley 1959) was able to obtain interesting evolutionary results and contrasts showing that the sustained heating was not only far greater over water than over land or ice, but that its pattern of time or trajectory dependence had cyclogenetic implications in the one case and anticyclogenetic in the other. Such results should have first-order verisimilitude, but the model was obviously crude, a constant K being used to avoid the pretence of finer understanding. Apart from the use of a transfer-type expression such as eq (10), it is clear that this is not essentially a boundary layer problem nor one of turbulence in the usual sense, and that we must look to a better knowledge of the dynamics of convection to improve our models.

c. Parameterization Needs for Energy Fluxes Over the Sea

In either of the two foregoing cases, the heat transfer at the bottom will be given by a formula such as eq (10), (11), or (12) so that the large-scale parameters must include sea-surface temperature and wind, temperature and humidity at the lowest layers of the model, and some measure of stability if this lowest level is more than a few meters above the surface. The dependence of C and C_* on stability can be specified in exact terms only for surface layers up to some tens of meters thick; for the deeper layers, say 100–1000 m, reliable data are only now beginning to emerge (e.g., Clarke 1970).

In subsection 4a, the remaining large-scale parameter is the depth of the mixed layer. This completes the formulation, assuming that no energy (or momentum) penetrates beyond the inversion and that the interaction is purely one of height change, thermally and dynamically produced. The separatist approach, then, has some clear advantages, and Burke (1945) and Charnock and Ellison (1967) have set out the full framework of equations in some detail. In subsection 4b, this advantage disappears because the remaining large-scale parameter is some property of the free atmosphere chosen suitably to indicate its power to transfer heat by convection and because the heating may remain large over sufficient length of trajectory to involve feedback of variability into the large-scale parameters such as u.

5. APPLICATION OF ADVECTION-FREE OCEANIC DATA

a. The Derivation of α

Although the current local value of net radiation over the sea need not dominate the total exchange, one would expect the ratio LE/H, and hence α as defined by eq (5), *in advection-free situations* to be determined by processes similar to those discussed in sections 2 and 3. We shall accordingly analyze several sets of high-quality temperature and humidity measurements over water, not only as being relevant to the advection-free value of the ratio over water, but also as adding weight to the determination of this ratio over *any* saturated surface. The known equality of K_H and K_W over land may be tentatively extended to water surfaces and α may again be expressed by eq (9).

b. Indian Ocean Data (CSIRO)

Deacon and Stevenson (1968) have reported observations made on two Indian Ocean cruises numbered Dm 1/62 and G 4/62 of 30 and 27 days, respectively. Use will be made here only of the surface temperatures and temperatures and humidities at 3 m in the air. The average surface temperature and the average air-sea temperature and humidity differences were calculated for each day, and each such average was treated as one observation in what follows. Values of α from these data are, according to eq (9),

1.26 ± 0.01 for Dm 1/62

and

1.30 ± 0.01 for G 4/62.

A number of days showed temperature inversions. Comparatively common, too, were days of very large lapse rates with temperature differences (air-sea) as great as $-6°$C. These facts suggest that certain days must have been affected by the advection of energy and, in the hope of minimizing such effects, "selected days" (defined as those with a temperature difference, in the lapse sense, lying between 0° and 1°C) were analyzed separately. The resulting values of α are

1.25 ± 0.01 for Dm 1/62 (19 days)

and

1.31 ± 0.01 for G 4/62 (15 days).

Thus, there is no difference on either cruise between the totality of days and those specially selected. This would suggest that the observations are about equally affected by advection of energy toward and away from the areas concerned.

c. Indian Ocean (University of Washington)

Paulson (1967) described observations made on 16 days of cruising on the Indian Ocean in February–March 1964. From the maps that he gives, it appears that 9 days (February 22–March 2, omitting February 24 when no observations were made) were spent satisfactorily far from land and only these days have been used. The observations comprised temperature and humidity at six heights (that were not always the same) ranging from 114 to 815 cm. Each day's observations were averaged separately; the resulting average values of c_pT were plotted against those of Lq, and a line was fitted by eye to find the best value of $c_p\Delta T/L\Delta q$. In every case, the points did fit a straight line very closely (as required by the equality of K_H and K_W). Surface temperatures being unavailable, s was found for the air temperature at the lowest level of measurement. The resulting mean value of α is 1.20 ± 0.03.

d. Lake Eucumbene

We also include in this section the results of one set of experiments on a large water storage. Webb (1960) in a study of evaporation from Lake Eucumbene gives, among other things, water surface temperature and temperature and humidity at a height of 4 m. The data are given as averages over 3 hr, and for each such period Webb has also given, from his records of wind speed and direction, a statement of the adequacy of the over-water fetch. Only the best cases (fetch of ½ mi or more) have been used and, so as not to limit unnecessarily the number of days available for analysis, only the three 3-hr periods from 0900 to 1800 LST have been taken into account. As will be noted later, ½ mi of fetch over water may be inadequate for an equilibrium state to be set up; nevertheless, the mean α obtained from the 26 available days was 1.25 ± 0.03.

e. Atlantic Ocean

When this paper was in an advanced state of preparation, the observations of Hoeber (1970*a*) came to our attention. These are values of H and LE derived from gradient measurements (making due allowance for thermal stratification) taken over a period of some 2½ weeks in the Atlantic Ocean at the Equator near longitude 30° W. We have used the data as presented in his figure 5 that show the mean diurnal course of these quantities and have calculated the mean values of H and LE. The mean water temperature at 20-cm depth is 25.6°C (Hoeber 1970*b*) and s has been evaluated at this temperature; the value of α so obtained [5] is 1.30 ± 0.02. Although this value was not originally included in the calculation of the mean α in section 6, it does not in fact disturb it.

6. THE VALUE OF α

There is some evidence of warm advection having affected the lysimeter results and cold advection the fluxatron measurements so it seems that the best estimate of α is the overall mean (land and water) of 1.26. This remains the same if the Lake Eucumbene observations are rejected. If the fluxatron results are rejected as being too discordant with the others, the mean is 1.28. This uncertainty in α is hardly important compared with the natural variability of the components E and H. In the following discussion, the evaporation from a horizontally uniform saturated surface (i.e., the potential evaporation, PE) will be given, in energy units, by

$$PE = 1.26 \frac{s}{s+\gamma} (R-G). \tag{14}$$

If α is 1.26, it follows from eq (6) that the Bowen ratio

$$\beta = \frac{H}{LE} = \left(1 - 1.26 \frac{s}{s+\gamma}\right) \Big/ 1.26 \frac{s}{s+\gamma} \tag{15}$$

and is thus a function of the surface temperature. The mean observed value of H/LE from each set of data is compared with that derived from eq (15) in figure 2, indicating that the variation with temperature is correctly accounted for by the inclusion of the factor $s/(s+\gamma)$ in eq (14). The disparity in the case of the fluxatron observations has already been discussed. The observations were also examined for any possible wind speed dependence. In one case (CSIRO Indian Ocean observations, selected days), there was a statistically significant correlation, but the slope of the regression line was so small that the total variation in α thus predicated was negligible in relation to the general uncertainty in its best value.

In concluding this section, we reiterate that the values of α obtained here are intended primarily for apportioning the net radiation, R, over substantial saturated land areas. They may also be used, with caution, for assigning Bowen ratios (but not apportioning R) in advection-free situations over water. Their inapplicability in the more general context over water is indicated by reference to the climatic atlas of Budyko (1955) where considerable oceanic areas, well removed from land, have monthly heat fluxes directed downward.

[5] The standard error was evaluated from mean 24-hr values of H and LE derived from the table given by Hoeber (1969).

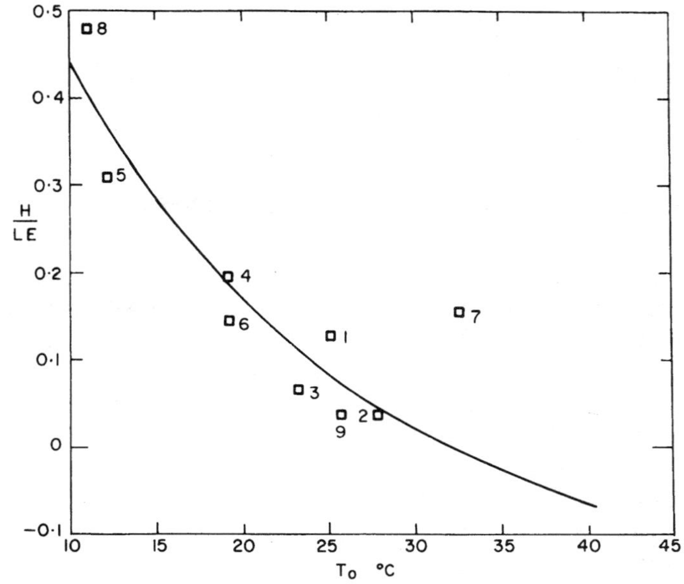

FIGURE 2.—Ratio of H to LE plotted against temperature. Values for (1) Indian Ocean (University of Washington), (2) Indian Ocean (CSIRO Dm 1/62), (3) Indian Ocean (CSIRO G 4/62), (4) Lake Eucumbene, (5) CSIRO lysimeter, (6) University of Wisconsin lysimeter, (7) fluxatron, (8) Wangara, and (9) Atlantic Ocean are plotted.

7. UNSATURATED LAND SURFACES

a. Statement of the Problem

When evaporation is to be related to the potential evaporation rate [eq (14)], some measure of soil wetness is obviously needed. This is not to exclude the possibility that other soil and crop parameters may also be necessary. As has been foreshadowed by its use as an independent variable in subsection 3b, the measure chosen is the accumulated actual (not potential) evaporation minus precipitation as this offers hope of evaluation from the rainfall and radiation records. If the ratio of actual to potential evaporation is a linear function of the accumulated actual evaporation, then it can be shown that this ratio is a negative exponential function of accumulated potential evaporation. This latter relationship is clearly the basis on which the "soil moisture retention tables" of Thornthwaite and Mather (1957) have been calculated.

Data from five different sources have been analyzed to examine the evaporation rate in the drying-out phase

b. CSIRO Lysimeters

Observations from CSIRO lysimeters were found in the records for two periods following heavy rain when no

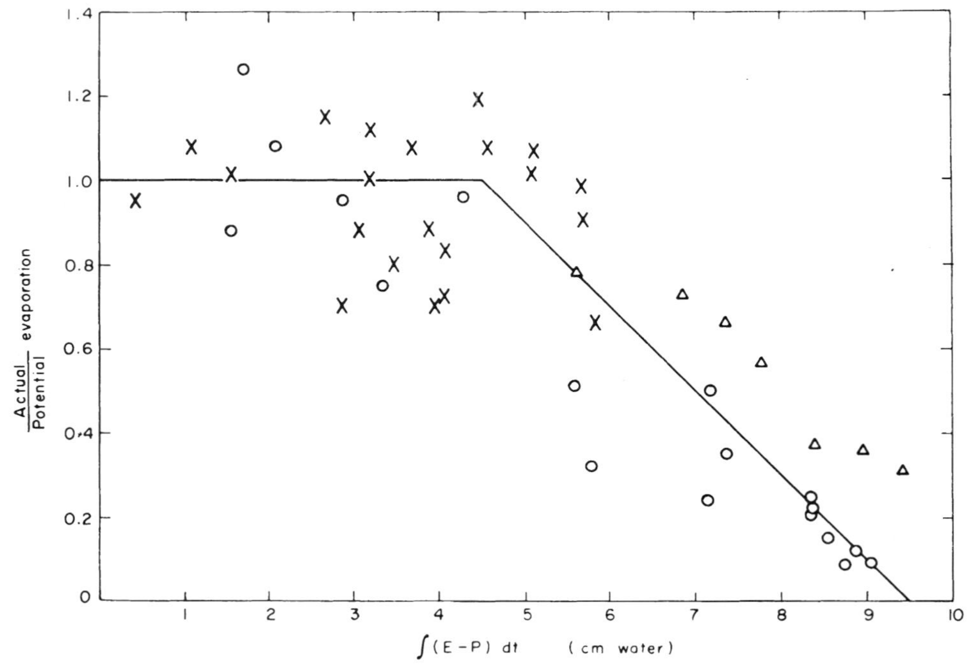

FIGURE 3.—The drying phase as shown by (X) CSIRO lysimeters, (O) three crops at Katherine, and (Δ) Great Plains.

irrigation was applied and good ancillary data existed (although radiation values are absent for some days). They are Feb. 11–25, 1963, and Dec. 2–22, 1968. Other such "drying periods" were also found, but they did not follow heavy rain, and the measurements were clearly much affected by advection; they are not considered here.

In figure 3, these observations are shown as the ratio, actual:potential evaporation rates, plotted against $\int(E-P)dt$.

c. Three Crops at Katherine, Northern Territory, Australia

Slatyer (1956) reports the results of observations on three crops (cotton, peanuts, and sorghum) at Katherine over 5 weeks of dry weather following heavy rain. The measurements comprise soil moisture [by gypsum block down to 64 in. (1.63 m) with auger sampling down to 8 in. (0.20 m)] and pan evaporation rates. Evaporation from the crops was calculated from the soil moisture and normalized by Slatyer with respect to pan evaporation raised to the power 0.75 (Prescott 1949, 1951).

To make use of these observations, we must assume the following:

1. This system of normalization does, at least approximately, compensate for variations in net radiation, wind speed, etc.
2. The first observation in each crop is at the potential rate—that is, each crop provides a point (not plotted) as (0.0, 1.0) in figure 3.
3. The change in soil moisture adequately measures $\int(E-P)dt$.

The results are shown in figure 3.

d. Great Plains

Lettau and Davidson (1957) report the results of observations at O'Neill, Nebr., including evaporation rate, net radiation, heat flux into the soil, and soil moisture down to 40 cm (deeper on some days, but a constant depth is needed here). Values of the ordinate in figure 3 can thus be assessed in absolute terms, but the abscissa can only be estimated, relative to an unknown zero, from changes in the water content in the top 40 cm of soil. To locate these observations in figure 3, we drew the line by eye to fit the observations discussed in subsections 7b and 7c and placed the first observation on it. The other points were then positioned in accordance with their location relative to the first observation.

e. U.S. Water Conservation Laboratory, Phoenix, Arizona, Lysimeters

Van Bavel (1967) reports a series of observations that are relevant to the present study. A field (8000 m^2 area) was irrigated by flooding on May 28, 1964, and observations proceeded from May 29 to June 28, at which time the soil water potential had fallen to about −15 bars. The evaporation rate from weighed lysimeters is available together with net radiation and temperature at an unstated height in the air. Soil heat flux was not measured but is stated to be small; in any case, 24-hr totals of evaporation and net radiation are used; therefore, effect of soil heat flux would probably be negligible.

With such a small field, irrigated and in an arid climate, effects of advection must be expected to be considerable and some allowance must be made for them. When the ratio of actual to potential evaporation [the latter from

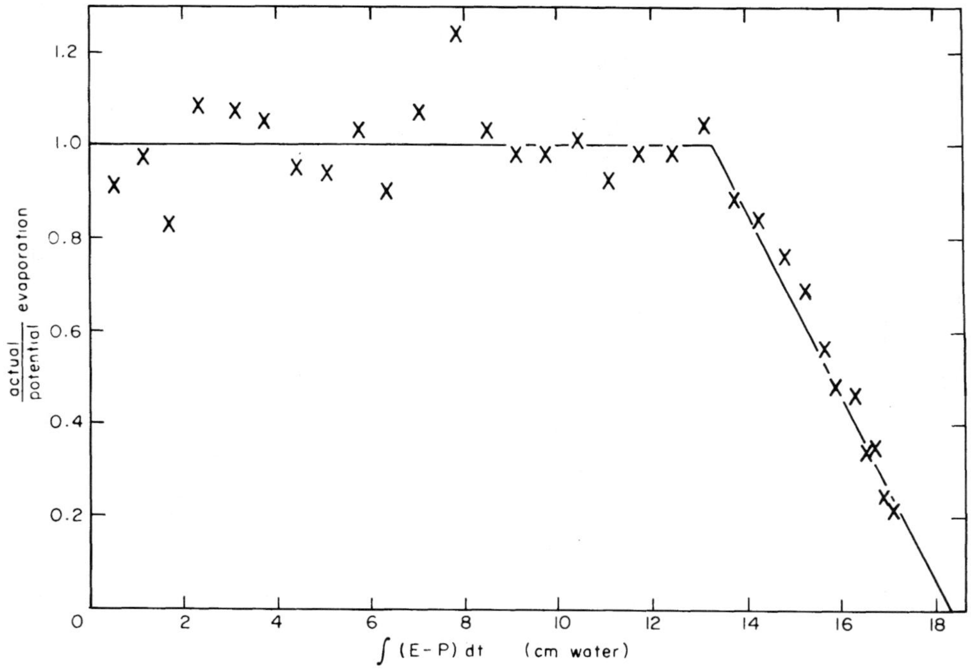

FIGURE 4.—The drying phase as shown by U.S. Water Conservation Laboratory lysimeters.

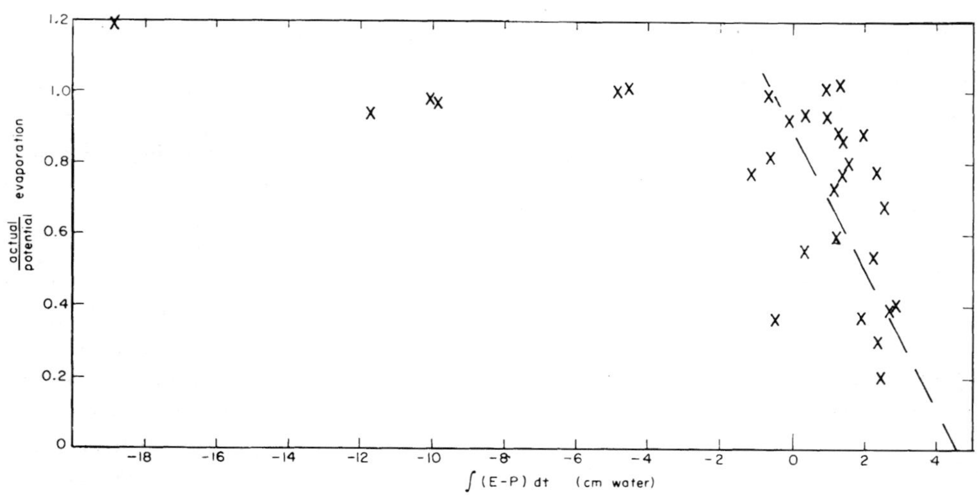

FIGURE 5.—The drying phase as shown by the University of Wisconsin lysimeter. The broken line has been drawn with the same slope as the descending lines of figures 3 and 4.

eq (14)] was plotted against time, it stayed reasonably constant until June 17 after which a pronounced downward trend developed. The average value of the ratio during this quasi-constant period was 1.35. Since the general meteorological conditions did not appear to vary greatly, it has been assumed that the effect of advection was to increase all evaporation rates by 35 percent and that they should be divided by 1.35 to approximate those that would have been measured in a horizontally homogeneous situation. The results are shown in figure 4. Comparing figures 3 and 4, we see that a good deal more water was evaporated in the latter case before the evaporation departed from the potential rate, but the equality of the slopes of the descending lines should be noted.

f. University of Wisconsin Lysimeters

The data referred to in subsection 3b include a number of days when the surface was clearly not saturated. The results from all observations, excluding only those when LE was greater than R, are shown in figure 5. In this graph, the origin of the abscissa refers to the day when the series of observations started and is thus quite arbitrary in terms of water content because the observations were not begun, as in other cases, immediately after heavy rain or irrigation. The broken line is drawn to have the same slope as the descending lines of figures 3 and 4, and, within the limits imposed by the considerable degree of scatter, the observations can be said to be not inconsistent with it.

g. Summary of Drying-Out Results

As the soil dries out, observations over a considerable variety of crops are consistent with the view that the ratio of actual to potential evaporation rate falls off linearly with increasing $\int(E-P)dt$ and approaches zero when the potential evaporation rate is about 5 cm more than it was when the surface first ceased to behave as a saturated one. The very large degree of scatter in figures 3 and 5 is not really as important as would appear at first sight—the observations have been normalized with respect to radiation and, with the removal of the major cause of variability, the relative effect of others appears magnified.

An important point on which the present analysis can throw no light is knowledge of when evaporation rate first begins to fall below the potential. The CSIRO lysimeter and Katherine crop results would put this point at about $\int(E-P)dt=4.5$ cm so that the soil would be completely dry when 9.5 cm of water had evaporated. [It is interesting to note that Slatyer (1966) has said that the Katherine site, after being thoroughly dried out, can absorb up to about 4 in. (10.2 cm) of rain before runoff begins.] The U.S. Water Conservation Laboratory would put the "downturn" point at about 13 cm and the results from the University of Wisconsin lysimeter indicate that this site can behave as a saturated surface while at least 18 cm of water are evaporated. On the other hand, the bare soil surface of figure 1 shows a decreasing evaporation rate (that was continued on the following days, though not as markedly so as at 1630 LST on Mar. 17, 1970) after only about 0.2 cm of water had evaporated; observers on the spot (Hicks 1970) did, in fact, notice that a visibly dry crust was forming on the soil at this stage.

It is clear that the differences in downturn point evidenced in figures 3, 4, and 5 should be examined with reference to soil, crop, and other factors, but we believe this requires a study in soil-plant water movement for which conclusions should be based on data from a wider distribution of sites than we have been able to assemble.

8. CONCLUDING DISCUSSION

In the context of GARP, one of our basic aims must be the specification of heat flux and evaporation over the land surfaces of the globe, and what has been written indicates that an energy approach to the problem is both physically realistic and operationally practical.

However, the apportionment of net radiation between heat flux and evaporation requires a knowledge of the distribution of R itself. This is not the place to discuss how one determines actual or predicted maps of R, but there is a need to make certain general points. The first, and principal, is that the mapping of R (and particularly its predicted value) will be highly sensitive to the knowledge and predictability of cloud amount and type. Here again, then, is not only a feedback process between the boundary layer and the free atmosphere, but also another example of how our prospective ability to link boundary layer considerations into larger scale processes is limited essentially by an ignorance in *the latter area;* that is, of convection, etc., in this case. Moreover, so severe does this limitation appear to be that it would effectively limit our immediate hopes in the practical mapping of R to the first-order variations in space and time. Hence, the objective has been defined as the mapping of the main variations in the daily totals of R on a space-scale comparable with that of several hundred kilometers already specified for GARP. Although the scales are comparable, they are not identical, for the energy-input mapping must be governed by consideration of the major variations in land form, land use, and hydrology.

In principle, R is calculable from the more elaborate dynamical models, though with what accuracy (bearing in mind particularly its dependence on cloud amount) remains to be seen. R can also be measured directly.[6] A world network of observations needs to be established:

1. For its own sake.
2. To provide the material for later calculations if these are to eventuate operationally.
3. To encourage every meteorologist to accept net radiation into his daily thinking as one of the basic synoptic variables as would be entirely appropriate in a modernized scientific approach to the subject.

The problem of sampling and representativeness of net radiation is not different in kind from those encountered with the accepted basic elements near the surface (temperature, wind, rainfall, etc.) and the typical accuracy of a good net radiometer, about 5 percent for daily totals, is more than adequate.

Given R, by whatever means, potential evaporation can be estimated from eq (14). It is hoped that other workers will further test the tentative conclusion that α is about 1.26 for saturated surfaces using any good quality advection-free data that has not been treated here. Meanwhile, it seems appropriate to begin to look for possible explanations; that is, what physical considerations might impose the additional constraint on ψ that the analysis of section 2 shows to be involved.

The ratio of molecular diffusivity of water vapor to thermal diffusivity in air has been reviewed by Montgomery (1947) and estimated as about 1.19, more or less independently of temperature (Deacon and Webb 1962). Thus, it does not appear that details of the laminar sublayer can fully explain an α of 1.26, but the possibility that these play some part in the process must not be overlooked.

Some earlier concepts, at least of the lower layers of the marine atmosphere, envisaged these layers as conditioned by the systematic transport of vapor from its source at the

[6] Whether R is best measured directly on a full network basis or derived from an easier network measurement, such as global radiation by conversion factors obtained from extensive intercomparisons, is a matter for others to discuss; note that such conversion factors will depend on albedo, season, and other variables.

surface to a condensation sink in the atmosphere above. This might suggest a mechanism whereby the height, temperature, etc., of the sink exercised control on the general level of q and hence of ψ in the boundary layer. We do not favor this type of explanation, preferring to regard any such sink as an "effect" rather than a "cause." Moreover, many of the data here were obtained under cloudless conditions; also $\alpha > 1$ implies a negative ψ (given $K_H = K_W$) whereas condensation implies a positive ψ in consequence of the curvature of the q_s, T relationship.

It is relevant here to refer to two earlier studies. Linacre (1964) examined the relationship between the temperatures of freely evaporating leaves in bright sunshine and of the air and reached the conclusion that leaves are hotter than the air up to about 33°C and, above that, they are cooler. Priestley (1966) examined this result in terms of the average daily maximum temperature for each month reported by island observing stations and by land stations after periods of heavy rain. His conclusion was that, in the radiation climates that actually exist in nature, air temperatures over a well-watered surface do not rise above 90°–93°F (32°–34°C). Now eq (15) implies that H becomes negative if $s/(s + \gamma)$ exceeds 1/1.26. This occurs at 32°C (see also fig. 2). The agreement is striking, although its implications may not yet be entirely clear.

Allowances will have to be made for variations in vegetation and soil moisture retention properties when assessing the fluxes from unsaturated surfaces, but this is not entirely a new requirement; knowledge of the vegetation type and amount is implied in the requirement for the estimation of R. Much of the information on water-holding capacity has no doubt already been gathered for agricultural and land-use studies and merely needs to be properly collated to suit it for introduction into meteorology.

It has been made clear that allowance for energy exchange over land in the context of GARP, or in the operational exercises expected to follow from GARP, will require the availability of maps of the appropriate "fixed" boundary conditions; that is, land use and vegetation, albedo (Posey and Clapp 1964), etc. Anticipating the discussion of momentum transfer, a similar need will exist for maps of surface roughness or low-level drag coefficient. Finally, whereas the utilization of the framework discussed here will necessarily involve the prediction of ground moisture, a program of extensive observation will serve not only to improve the technique of prediction but also to provide a continuous updating of initial conditions for the calculations. It is important, therefore, to lend every encouragement to the development of techniques for remote sensing of ground moisture.

ACKNOWLEDGMENTS

We are grateful to T. A. Black for making available to us the observations discussed in subsections 3b and 6f and to H. Hoeber for supplying the mean water temperature for his Atlantic Ocean observations.

REFERENCES

Black, T. A., Thurtell, G. W., and Tanner, C. B., "Hydraulic Load Cell Lysimeter, Construction, Calibration, and Tests," *Soil Science Society of America Proceedings,* Vol. 32, No. 5, Madison, Wis., Sept.–Oct. **1968**, pp. 623–629.

Bryson, Reid A., and Kuhn, Peter M., "Some Regional Heat Budget Values for Northern Canada," *Geographic Bulletin* No. 17, Geographical Branch, Department of Mines and Technical Surveys, Ottawa, Canada, May **1962**, pp. 57–65.

Budyko, M. I. (Editor), "Atlas Teplovago Balansa" (Atlas of Thermal Balance), Glavnaia Geofizcheskaia Observatoriia, Leningrad, U.S.S.R., **1955**, 41 pp.

Burke, C. L., "Transformation of Polar Continental Air to Polar Maritime Air," *Journal of Meteorology,* Vol. 2, No. 2, June **1945**, pp. 94–112.

Charnock, H., and Ellison, T. H., "The Boundary Layer in Relation to Large-Scale Motions of the Atmosphere and Ocean," *Report of the Study Conference on Global Atmospheric Research Programme (GARP), Stockholm, Sweden, June 28–July 11, 1967,* ICSU/IUGG Committee on Atmospheric Sciences, COSPAR, World Meteorological Organization, Geneva, Switzerland, Appendix III, **1967**, pp. 1–16.

Clarke, R. H., "Observational Studies in the Atmospheric Boundary Layer," *Quarterly Journal of the Royal Meteorological Society,* Vol. 96, No. 407, London, England, Jan. **1970**, pp. 91–114.

Clarke, R. H., with Dyer, A. J., Brook, R. R., Reid, D. G., and Troup, A. J., "The Wangara Experiment: Boundary Layer Data," Commonwealth Scientific and Industrial Research Organization, Division of Meteorological Physics, *Technical Paper* No. 19, Aspendale, Australia, **1971**, 352 pp.

Deacon, E. L., and Stevenson, J., "Radiation and Associated Observations Made on Indian Ocean Cruises: Pt. 1. Cruises During 1962," Commonwealth Scientific and Industrial Research Organization, Division of Meteorological Physics, *Technical Paper* No. 16, Aspendale, Australia, **1968**, 22 pp.

Deacon, E. L., and Webb, E. K., "Small-Scale Interactions," *The Sea: Ideas and Observations on Progress in the Study of the Seas: Vol. 1. Physical Oceanography,* Interscience Publishers, New York, N.Y., **1962**, 864 pp. (see pp. 43–87).

Dyer, A. J., "The Turbulent Transport of Heat and Water Vapour in an Unstable Atmosphere," *Quarterly Journal of the Royal Meteorological Society,* Vol. 93, No. 398, London, England, Oct. **1967**, pp. 501–508.

Dyer, A. J., and Hicks, B. B., "Flux-Gradient Relationships in the Constant Flux Layer," *Quarterly Journal of the Royal Meteorological Society,* Vol. 96, No. 410. London, England, Oct. **1970**, pp. 715–721.

Hicks, B. B., CSIRO Division of Atmospheric Physics, Aspendale, Victoria, Australia, **1970** (personal communication).

Hoeber, Heinrich, "Wind-, Temperatur- und Feuchteprofile in der wassernahen Luftschicht über dem äquatorialen Atlantik," (Wind, Temperature and Humidity Profiles in the Surface Air Layer Over the Equatorial Atlantic), *"Meteor" Forschungsergebnisse,* Reihe B, Heft 3, Gebrüder Borntraeger, Berlin, Germany, **1969**, pp. 1–26.

Hoeber, Heinrich, "Measurements in the Atmospheric Surface Layer Above the Tropical Atlantic," *Weather,* Vol. 25, No. 9, London, England, Sept. **1970*a***, pp. 419–424.

Hoeber, Heinrich, Meteorologisches Institut, Universität Hamburg, Germany, **1970*b*** (personal communication).

Lettau, Heinz H., and Davidson, Ben, "Exploring the Atmosphere's First Mile: Vol. 2. Site Description and Data Tabulation," *Proceedings of the Great Plains Turbulence Field Program, O'Neill, Nebraska, August 1–September 8, 1953,* Pergamon Press, New York, N.Y., **1957**, pp. 377–575.

Linacre, E. T., "A note on a Feature of Leaf an d Air Temperatures," *Agricultural Meteorology,* Vol. 1, No. 1, Elsevier Publishing Co., Amsterdam, The Netherlands, Mar. **1964**, pp. 66–72.

McIlroy, I. C., and Angus, D. E., "The Aspendale Multiple Weighed Lysimeter Installation," Commonwealth Scientific and Industrial Research Organization, Division of Meteorological Physics, *Technical Paper* No. 14, Aspendale Australia, **1963**, 29 pp.
Montgomery, Ray B., "Viscosity and Thermal Conductivity of Air and Diffusivity of Water Vapour in Air," *Journal of Meteorology,* Vol. 4, No. 6, Dec. **1947**, pp. 193–196.
Paulson, Clayton A., "Profiles of Wind Speed, Temperature and Humidity Over the Sea," *Scientific Report,* NSF Grant GP-2418, Department of Atmospheric Sciences, University of Washington, Seattle, Oct. **1967**, 128 pp.
Penman, Howard Latimer, "Natural Evaporation From Open Water, Bare Soil and Grass," *Proceedings of the Royal Society of London,* Ser. A, Vol. 193, England, **1948**, pp. 120–145.
Posey, Julian W., and Clapp, Philip F., "Global Distribution of Normal Surface Albedo," *Geofisica Internacional,* Vol. 4, No. 1, Mexico City, D. F., Mexico, **1964**, pp. 33–48.
Prescott, J. A., "A Climatic Index for the Leaching Factor in Soil Formation," *Journal of Soil Science,* Vol. 1, No. 1, Oxford University Press, London, England, Mar. **1949**, pp. 9–19.
Prescott, J. A., "Climatic Expressions and Generalized Climatic Zones in Relation to Soils and Vegetation," *Proceedings of the Specialist Conference in Agriculture: Plant and Animal Nutrition in Relation to Soil and Climatic Factors, Adelaide and Canberra, Australia, August 22–September 15, 1949,* Her Majesty's Stationery Office, London, England, **1951**, pp. 27–32.
Priestley, C. H. B., *Turbulent Transfer in the Lower Atmosphere,* the University of Chicago Press, Ill., **1959**, 130 pp.
Priestley, C. H. B., "The Limitation of Temperature by Evaporation in Hot Climates," *Agricultural Meteorology,* Vol. 3, No. 3/4, Elsevier Publishing Co., Amsterdam, The Netherlands, **1966** pp. 241–246.
Priestley, C. H. B., "Handover in scale of the fluxes of momentum, heat etc. in the atmospheric boundary layer," The Physics of Fluids, Vol. 10, No. 9, Part 2, **1967**, pp S38–S46.
Slatyer, R. O., "Evapotranspiration in Relation to Soil Mositure," *Netherlands Journal of Agricultural Science,* Vol. 4, P. van der Bergh Publishers, Wageningen, The Netherlands, Feb. **1956,** pp. 73–76.
Slatyer, R. O., "Agricultural Climatology of the Katherine Area, N T.," Commonwealth Scientific and Industrial Research Organization, Division of Land Research and Regional Survey, *Technical Paper* No. 13, Canberra, Australia, Feb. **1966,** 39 pp.
Swinbank, W. C., and Dyer, A. J., "An Experimental Study in Micro-Meteorology," *Quarterly Journal of the Royal Meteorological Society,* Vol. 93, No. 398, London, England, Oct. **1967,** pp. 494–500.
Thornthwaite, Charles Warren, and Mather, J. R., "Instructions and Tables for Computing Potential Evapotranspiration and the Water Balance," *Publications in Climatology,* Vol. 10, No. 3, Drexel Institute of Technology, Laboratory of Climatology, Philadelphia, Pa., **1957**, pp. 185–310.
Van Bavel, C. H. M., "Changes in Canopy Resistance to Water Loss From Alfalfa Induced by Soil Water Depletion," *Agricultural Meteorology,* Vol. 4, No. 3, Elsevier Publishing Co., Amsterdam, The Netherlands, May **1967**, pp. 165–176.
Webb, E. K., "An Investigation of the Evaporation From Lake Eucumbene," Commonwealth Scientific and Industrial Research Organization, Division of Meteorological Physics, *Technical Paper* No. 10, Aspendale, Australia, **1960**, 75 pp.

[*Received January 8, 1971; revised June 14, 1971*]

A Model for the Priestley-Taylor Parameter α

H. A. R. DE BRUIN

Royal Netherlands Meteorological Institute, De Bilt, the Netherlands

(Manuscript received 13 May 1982, in final form 17 January 1983)

ABSTRACT

The behavior of the Priestley-Taylor parameter α is investigated using a simplified atmospheric boundary-layer model coupled to the Penman-Monteith equation. It is found that under conditions typical for a sunny summer day in the mid-latitudes, α is primarily determined by the surface resistance r_s ($\alpha \approx 1.3$ when $r_s = 0$, $\alpha \approx 1$ when $r_s \approx 60$ s m^{-1}, and $\alpha \approx 0.6$ when $r_s \approx 250$ s m^{-1}). This is in good agreement with experimental values reported in literature. The model is an extension of the models by McNaughton (1976) and Perrier (1980).

1. Introduction

This paper deals with the parameter α, introduced by Priestley and Taylor (1972) and defined by

$$\lambda E = \alpha \frac{s}{s+\gamma}(Q^* - G), \tag{1}$$

where E is the evaporation, Q^* the net radiation, G the soil heat flux density, λ the latent heat of vaporization, $\gamma \equiv c_p/\lambda$ the psychrometric constant, c_p the specific heat of air at constant pressure, and s the slope of the saturation specific humidity–temperature curve at air temperature.

Priestley (1959) and Slatyer and McIlroy (1961) showed that when saturated air passes over a wet surface λE will become equal to $\lambda E_{EQ} \equiv s/(s + \gamma) \times (Q^* - G)$ and consequently $\alpha \rightarrow 1$. Slatyer and McIlroy (1961) argued that when unsaturated air is in contact with a wet surface over a long fetch it will tend to become saturated and thus then α tends to become 1. Therefore they referred to the quantity E_{EQ} as the *equilibrium* evaporation.

It is noted that when $\alpha = 1$ the Bowen ratio $\beta = \gamma/s$. This follows directly from (1) and the energy balance equation, i.e.,

$$Q^* - G = H + \lambda E, \tag{2}$$

where H is the sensible heat flux density and $\beta \equiv H/\lambda E$.

Air above extensive wet surfaces, such as the oceans, is seldom saturated. This explains why for wet surfaces α is often found to be 20–30% greater than unity (Priestley and Taylor, 1972; Brutsaert, 1982).

More recently McNaughton (1976) and, independently, Perrier (1980) presented a more general model for α with which they showed that α tends to become unity also in the case where unsaturated air passes over even an unsaturated surface. Their basic assumptions were that the atmospheric boundary layer (ABL) has a constant height and is a closed system, i.e., at its top there is no exchange of heat and water vapor with the air above. Several data sets, mostly referring to "moderately dry" conditions, confirm $\alpha \approx 1$ (e.g., McNaughton, 1976; Perrier, 1980; Brutsaert, 1982). However, the McNaughton-Perrier model (hereafter called MP model) does not explain why for wet surfaces usually $\alpha > 1$, or why under dry conditions $\alpha < 1$. Monteith (1981) suggested that an explanation for this can be the fact that in the MP model for entrainment of dry air at the top of the ABL is ignored.

In this paper an extension of the MP model is given in which the ABL has a variable height and entrainment of dry air at the top of the ABL is permitted. It will be shown how α depends on the surface resistance r_s and other factors.

Our model is a simplified ABL model coupled to the Penman-Monteith equation for estimating the surfact fluxes of heat and water vapor.

2. The McNaughton-Perrier model

In order to introduce the reader to the problem first a brief description of the MP model, in the form presented by Perrier (1980), is given. Assuming the ABL is a closed box of constant height he showed that the dew point deficit $Y = T_a - T_d$ (T_a = the air temperature, T_d = the dew point) obeys the differential equation

$$s\frac{dY}{dt} = \frac{sH - \gamma(\lambda E)}{\rho c_p h}. \tag{3}$$

Combining (3) with the Penman-Monteith equation for H and λE (see Section 3) leads to

$$\frac{dY}{dt} + \frac{Y}{\tau} = \frac{Y_e}{\tau}, \tag{4}$$

in which the time constant τ is given by

$$\tau = [r_a + \gamma r_s/(s + \gamma)]h \tag{5}$$

and Y_e by

$$Y_e = \frac{\gamma}{s + \gamma}\left(\frac{Q^* - G}{\rho c_p}\right)r_s. \tag{6}$$

In (5) and (6) r_a and r_s are the aerodynamic and surface resistance respectively.

Now the case is considered where at $t = t_0$ Y_e changes suddenly from $(Y_e)_1$ to $(Y_e)_2$. This can be due either to a sudden change in r_s [the case considered by McNaughton (1976)] or to a sudden change in $(Q^* - G)$ as considered by Perrier (1980). Then it follows from (4) that Y tends to $(Y_e)_2$. The time scale for this process is τ: when $t \approx 3\tau + t_0$, $Y \approx (Y_e)_2$ and consequently $dY/dt \to 0$. Then it is seen from (3) that $\beta \to \gamma/s$ or $\alpha \to 1$.

It is noted that McNaughton (1976) followed a somewhat different approach to arrive at the same result, but his basic assumptions were the same.

The main limitations of the MP model is that the ABL is assumed to be a closed box of constant height. In the case considered above heat is continuously supplied at the surface. In a closed box model this can lead to unrealistically high ABL temperatures. Since s, and thus β, depends on temperature, this reflects directly on the surface fluxes and therefore also on the predicted α. In our extension of the MP model these problems will be avoided.

3. The present model

a. General

In our approach we will couple an existing ABL model to the Penman-Monteith equation, giving a more physically realistic model. For our study this approach has one disadvantage, namely the full set of governing equations are too complicated to allow a direct comparison with the original MP model. To allow comparisons to be made we confined ourselves to those conditions under which the flux of water vapor at the top of the ABL is proportional to the surface evaporation. It appears, under these conditions, that a model is then obtained, which is, in its form, similar to the original MP model. Furthermore, we will restrict ourselves to weather conditions typical for a summer day in temperate latitudes.

Notwithstanding these restrictions our model still contains the most important physical elements, e.g., the ABL height is variable and there is entrainment at the top of the ABL.

b. The boundary layer sub-model

To describe the relevant boundary layer parameters we will use the recent papers by Driedonks (1982a,b) concerning the dynamics of a well-mixed ABL. On a clear day the profiles of the conservative quantities θ (potential temperature) and q (specific humidity) have the approximate forms shown in Fig. 1. Here θ is defined with respect to the surface air pressure instead of 1000 mb generally used. Thus θ is the air temperature at $z = 0$. Between z_L and h, q and θ reach constant values (of q_m and θ_m, respectively) as a result of turbulent mixing. At $z = h$ the ABL is capped by an inversion: for $z > h$ there is stable air characterized by $\gamma_\theta = d\theta/dz$ and $\gamma_q = dq/dz$. The transition layer between the well-mixed and the air aloft is usually small, so that the profiles can be approximated as shown in Fig. 1. At $z = h$ they have a step change characterized by $\Delta\theta$ and Δq.

In the surface layer (constant flux layer) the gradients of θ and q are steep: θ and q increase rapidly during daytime, especially close to the ground, from θ_m and q_m at $z = z_L$ to their surface values θ_s and q_s.

Usually the thickness of the surface layer is small compared to h, so that its heat capacity can be neglected. Furthermore we neglect the effects of horizontal advection, large-scale vertical motions and divergence of radiation.

1) PARAMETERS OF ENTRAINMENT OF DRY AND WARM AIR AT $z = h$

When the depth of the boundary layer is increasing stable air with a higher potential temperature is drawn into it. As a result there is a downward heat flux at $z = h$. For practical applications good results are obtained when this flux is taken to be proportional to the flux at the surface (Tennekes, 1973; Driedonks, 1982b). Together with the fact that 1) θ is constant in the boundary layer and 2) the depths of the tran-

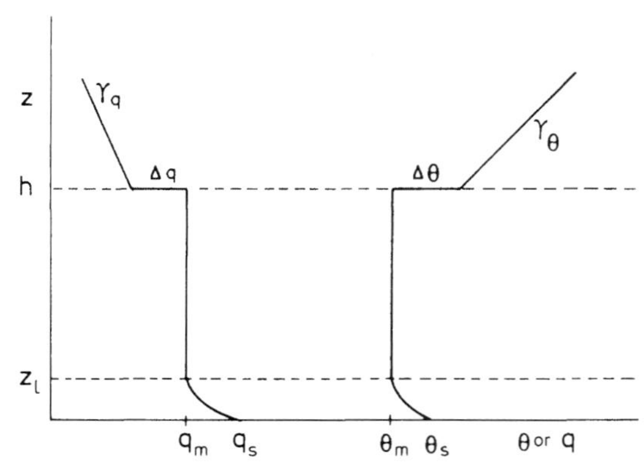

FIG. 1. Profiles of potential temperature (θ) and specific humidity (q) in a well-mixed atmospheric boundary layer.

sition layers at the surface and the inversion are negligibly small, this leads to

$$\frac{d\theta_m}{dt} = \frac{aH}{\rho c_p h}, \tag{7}$$

where a is a constant equal to 1.2 (Driedonks, 1982b).

The evolution of q_m is described by a set of equations in dh/dt, Δq, γ_q and $d(\Delta q)/dt$ (Driedonks, 1982b). It is our purpose to investigate the sensitivity of the Priestley-Taylor parameter α to changes in the entrainment rate of dry air at the top of the ABL. We don't intend to apply our model to actual situations, e.g., for weather forecasts. Moreover, we want to make a direct comparison between the MP model and ours. Therefore to simplify the set of equations describing the entrainment of water vapor, we will confine ourselves to those situations in which the flux of water vapor at the top of the ABL is proportional to the surface evaporation E. In that case we arrive at a rate equation for q_m similar to that for θ_m:

$$\frac{dq_m}{dt} = \frac{bE}{\rho h}. \tag{8}$$

In (8) b is the parameter describing the rate of entrainment of dry air at the top of the ABL. Through this parameter we are able to vary this rate in our calculation: when $b = 1$ there is no flux of moisture at $z = h$, while when $b = 0$ the humidity increase due to the surface evaporation is just balanced by the humidity decrease caused by entrainment of dry air. Also b can be negative. In that case the entrainment of dry air exceeds the surface evaporation. Use of (8) has the advantage that we are able to derive from it a simple differential equation for δq, which makes a direct comparison with the original MP model possible. Of course, the disadvantage is that (8) is not generally applicable. However, it will be seen later that α appears to be rather insensitive to the value of b. Therefore the use of (8) causes less restriction than one would expect at first sight.

Data on q_m are scarce. Driedonks (1981) published observations of q_m at Cabauw for 8 days. When variations of q_m of 15% or less are ignored, q_m was about constant on 4–5 days, implying that b was about 0 in those cases. Furthermore on one day q_m increased clearly, while on two others q_m tended to decrease. From this information we estimate that usually b varies between $-0.5 \leqslant b \leqslant 0.5$.

It will be shown in Section 3c that the surface fluxes depend on the specific humidity deficit δq at the top of the surface layer (at $z = z_L$) defined by $\delta q = q_w(\theta_m, p_L) - q_m$, where q_w is the saturation specific humidity at $\theta = \theta_m$ and air pressure $p = p_L$ ($=p$ at z_L). We will assume that p_L is constant with time. Usually, the surface layer is so thin that the variation of p in this layer can be neglected. Then we can omit the dependence of q_w on p in the following.

We now approximate $q_w(\theta_m)$ by

$$q_w(\theta_m) = q_w(\theta_r) + s_r(\theta_m - \theta_r), \tag{9}$$

where θ_r is a reference potential temperature which is the average of θ_m on a particular day, and s_r the derivative of q_w at $\theta = \theta_r$. Eqs. (7)–(9) lead to

$$\frac{d(\delta q)}{dt} = \frac{as_rH - b\gamma\lambda E}{\rho c_p h}. \tag{10}$$

2) THE EVOLUTION OF θ_m AND h

Under the conditions we are considering, i.e., a sunny summer day h and θ_m are given by Driedonks (1982a)

$$h(t) = \{2(2a - 1)[I(t) - \delta_0]\gamma_\theta^{-1}\}^{1/2} \tag{11}$$

and

$$\theta_m(t) = \theta_{00} + \gamma_\theta\left(\frac{a}{2a - 1}\right)h(t), \tag{12}$$

in which

$$I(t) = \int_{t_0}^{t} \frac{H}{\rho c_p} dt', \tag{13}$$

where t_0 is the time of effective sunrise, i.e., the time at which H becomes positive. The quantities δ_0 and θ_{00} are determined by the initial values of $\Delta\theta$, θ_m and h (denoted with an index 0): $\theta_{00} = \theta_{m0} + \Delta\theta_0 - \gamma_\theta h_0$ and $\delta_0 = h_0\Delta\theta_0 - \frac{1}{2}\gamma_\theta h_0^2$.

Eqs. (11) and (12) require that $I(t) > \delta_0$ and $h \gtrsim 1.2h_0$. In most cases these conditions are satisfied a few hours after t_0. Therefore we will start our calculation at $t_0^* = t_0 + 2$ h so that we do not run into these problems. Eqs. (11) and (12) require also that γ_θ is constant with time.

c. The surface fluxes

1) THE PENMAN-MONTEITH EQUATION FOR E AND H

The surface fluxes H and E are governed by the Penman-Monteith equation (Monteith, 1981)

$$\lambda E = \frac{s(Q^* - G) + (\rho c_p/r_a)\delta q}{s + \gamma^*} \tag{14}$$

and

$$H = \frac{\gamma^*(Q^* - G) - (\rho c_p/r_a)\delta q}{s + \gamma^*}, \tag{15}$$

where $\gamma^* = \gamma(1 + r_s r_a^{-1})$. It is assumed that r_s does not depend on weather variables and that it is determined only by surface conditions.

In our study we will apply (14) and (15) to the entire surface layer, i.e., the level at which δq must be specified is $z = z_L$. Then the aerodynamic resistance is

$$r_a = \int_{z_{0h}}^{z_L} \frac{dz}{K_h}, \tag{16}$$

in which z_{0h} is the roughness length and K_h the turbulent exchange coefficient for sensible heat and water vapor.

Thom and Oliver (1977) showed that r_a is mainly determined by wind speed. It is noted that, since we consider the entire surface layer, r_a in our case is somewhat greater than the values reported in the literature, which usually refer to a reference height of 2 m. We estimate in our case for short vegetation that r_a has typical values of $r_a \approx 25$ s m^{-1} at high wind speeds and $r_a \approx 75$ s m^{-1} at $u \approx 2$ m s^{-1}. The values for taller vegetation will be lower.

2) NET RADIATION AND SOIL HEAT FLUX

Because vegetation insulated the surface efficiently from solar radiation the soil heat flux density G of a cropped surface is usually small compared to the net radiation Q^*. For practical calculations it is permitted to take it as a small fraction of Q^*. For a grass cover at Cabauw during daytime De Bruin and Holtslag (1982) found

$$G = 0.1Q^*. \tag{17}$$

This empirical expression will be used in our calculations, i.e., we will confine ourselves to surfaces covered with a vegetation completely shading the ground.

Since only cloudless days are considered, the net radiation can be written as

$$Q^* = (1 - A)K^{\downarrow} + c\sigma\theta_m^{\,6} - \sigma\theta_m^{\,4} - 4\sigma\theta_m^{\,3}(\theta_s - \theta_m), \tag{18}$$

where A is the albedo, $K^{\downarrow}$ the global radiation, $c = 9.4 \times 10^{-6}$ K^{-2} and σ the Stefan-Boltzmann constant. Note that θ_m and θ_s are expressed in K. In (18) we used the empirical expressions of Swinbank (1963) for the incoming long-wave radiation. The last term of (18) is a relatively small correction term which can be evaluated making use of (15), (17) and

$$H = \rho c_p(\theta_s - \theta_m)r_a^{-1}. \tag{19}$$

d. Differential equation for δq

Substitution of (14) and (15) into (10) leads to

$$\frac{d(\delta q)}{dt} + \frac{\delta q}{\tau} = \frac{(\delta q)_e}{\tau}, \tag{20}$$

which, in this form, is similar to the MP equation (4). Now the time constant τ is given by

$$\tau = r_a h \frac{s + \gamma^*}{as_r + b\gamma} \tag{21}$$

and the quantity $(\delta q)_e$ by

$$(\delta q)_e = \frac{(Q^* - G)r_a(s_r a\gamma^* - b\gamma s)}{\rho c_p(as_r + b\gamma)}. \tag{22}$$

In the case of a closed ABL, $a = b = 1$. Because $\delta q \approx sY$ and $s \approx s_r$, it is seen that then (20) reduces to the MP equation (4). However in our model h is variable. Under the conditions we are considering h will grow rapidly in the morning. Since τ is proportional to h, τ has the same behavior.

The quantity of $(\delta q)_e$ represents the external forcing. When r_s and r_a are constant, $(\delta q)_e$ will approximately behave as a sine with its maximum at about local noon, because it is proportional to $(Q^* - G)$.

4. Comparison with the McNaughton-Perrier model

In order to get an impression of the differences between the MP model and the present model we will consider here the case treated before in Section 2 where $(\delta q)_e$ undergoes a step change at $t = t_0$. Then, according to (20), δq will approach the new value of $(\delta q)_e$, while $d(\delta q)/dt$ will become zero. As a result $\beta \rightarrow (b\gamma)/(as_r)$. This follows from (10). We have seen in Section 2 that the closed ABL model of MP leads to $\beta \rightarrow (\gamma/s)$. We found that $a \approx 1.2$, whereas b lies between about -0.5 and 0.5. So our model yields for the case considered that β varies between about $-0.4\gamma/s$ and $+0.4\gamma/s$ (taking $s \approx s_r$).

This example illustrates that our extension of the MP model can have significant effects. The example used here, however, is not realistic. This can be seen as follows. When we take $s = s_r \approx 10^{-3}$ K^{-1}, $r_a = 50$ s m^{-1}, $h = 500$ m, $b = 0$ and $a = 1.2$, the time constant τ is at least 7 h ($r_s \geq 0$). Due to the properties of Eq. (20), δq will reach a new equilibrium state in about $t = 3\tau$. This implies that this is the case after at least 20 h. In reality, therefore, an equilibrium never will be reached. In the next section we will treat more realistic examples.

5. Results from model calculations

In this section we will apply our model under conditions typical for a clear summer day in the Netherlands. We will calculate the Priestley-Taylor parameter α for different values of 1) the parameter b which determines the entrainment of dry are at the top of the boundary layer, 2) the surface resistance r_s and 3) the aerodynamic resistance r_a. As initial conditions we used $t_0^* = 0700$ GMT, $h = 180$ m, $\theta_{m0} = 18.5$°C, $\delta q = 5$ g kg^{-1} and $\gamma_\theta = 5 \times 10^{-3}$ K m^{-1}. We also used the observed solar radiation data (½ h values). In Fig. 2 the diurnal variation of the calculated α is shown for different values of r_s. The aerodynamic resistance r_a is taken 50 s m^{-1} and $b = 0$ (i.e., q_m is a constant). It is seen that α has a diurnal variation, but during the midday hours, when the fluxes are largest, α is fairly constant. Also the curves referring to different values of r_s are very similar. Therefore, they can be

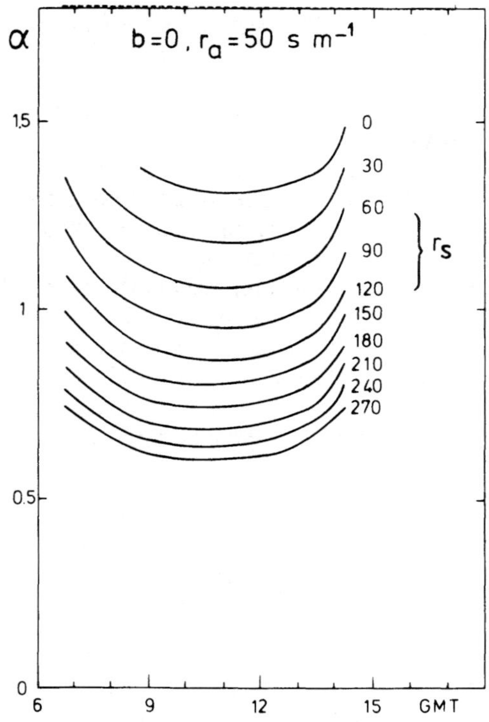

FIG. 2. Daytime variation of the computed α for different values of r_s; $b = 0$ and $r_a = 50$ s m^{-1}.

characterized by one point. For this we chose α_{11}, which is the value of α at 1100 GMT.

In Fig. 3 α_{11} is shown as a function of r_s for different values of b, while r_a is taken constant at $r_a = 50$ s m^{-1}. The main features are as follows:

1) α_{11} varies between about 1.2 and 1.4 at $r_s = 0$. This is in good agreement with the values reported in literature for well-watered surfaces (Priestley-Taylor, 1972; Monteith, 1981; Brutseart, 1982).

2) At $r_s = 270$ s m^{-1} and $\alpha_{11} \approx 0.6$, while in the region between $r_s = 30$ and $r_s = 120$ s m^{-1}, α_{11} varies between 0.85 and 1.25. These figures agree well with those found experimentally by De Bruin and Holtslag (1982).

3) The dependence of α_{11} on b is relatively weak and is insignificant for $r_s > 150$ s m^{-1}. Only for small values of r_s are variations in b of any importance.

4) When $r_s \approx 60$ s m^{-1}, which is a typical summer value for grass in the temperate climates (Thom and Oliver, 1977), α_{11} in about 1. This again is in good agreement with experimental data (McNaughton, 1976; Perrier, 1980; Brutsaert, 1982).

In Fig. 4 the dependence of the calculated α_{11} on r_a and r_s is shown when $b = 0$. We can conclude that for $r_s < 120$ s m^{-1} α_{11} depends only slightly on r_a, but for $r_s > 120$ s m^{-1} α is more sensitive to variations of r_a. Generally, r_a increases with decreasing wind speed. Thus the model predicts that when the surface is dry α decreases with increasing wind speed.

Usually, $r_s < 120$ s m^{-1} so that in most cases the wind speed has only a minor influence on α and thus on the surface fluxes. This agrees with the findings of De Bruin and Holtslag (1982).

It is interesting to note that the α values as predicted by the model for very rough surfaces (low r_a values) tend to be relatively low for $r_s > 60$ s m^{-1} and to be relatively high for low r_s. This is in agreement

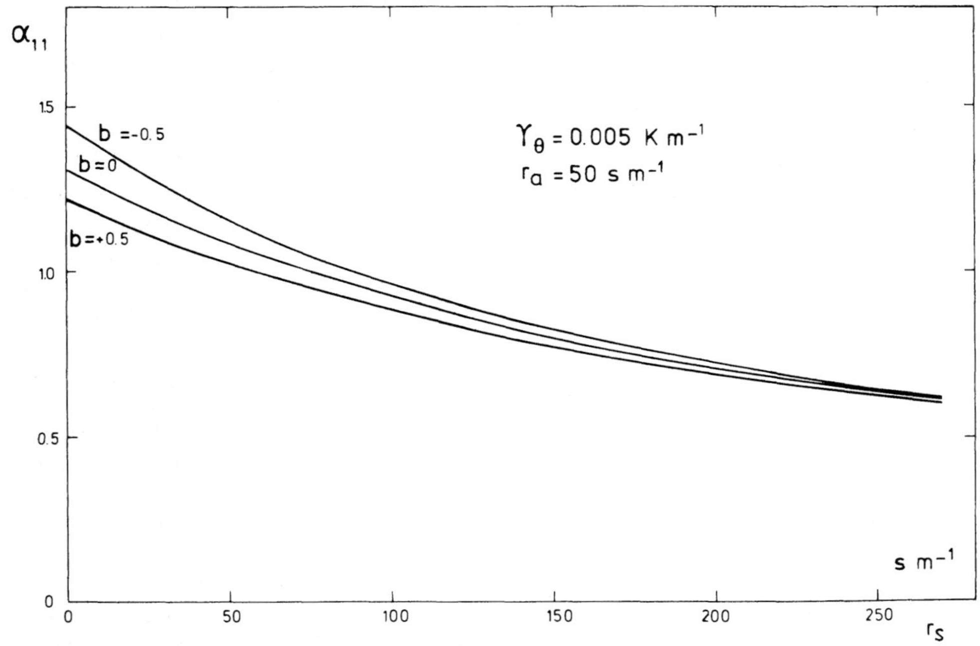

FIG. 3. Calculated α_{11} as a function of r_s for different values of b; $r_a = 50$ s m^{-1}.

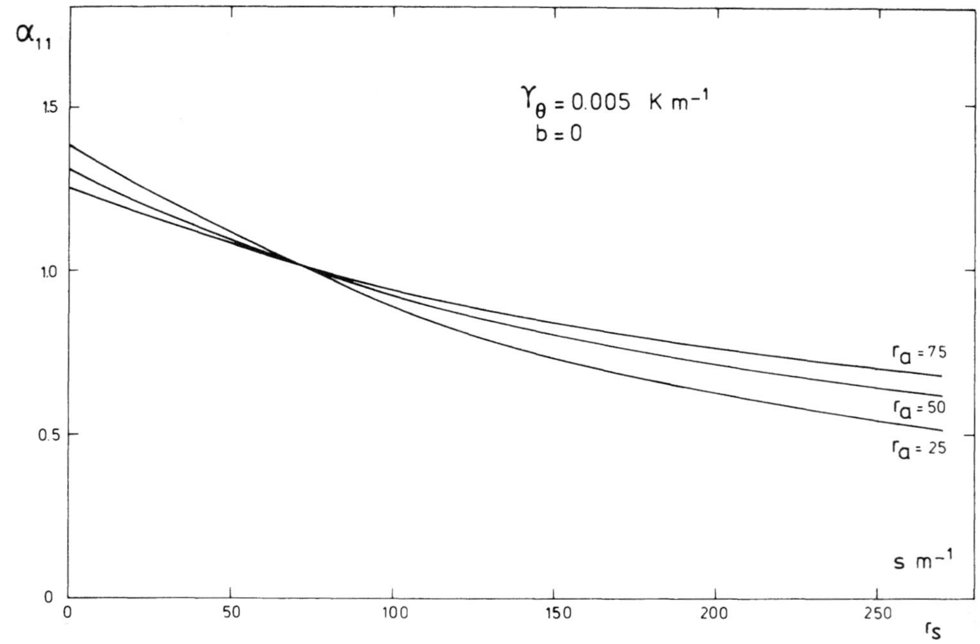

FIG. 4. Calculated α_{11} as a function of r_s for different values of r_a; $b = 0$.

with the findings of Shuttleworth and Calder (1979). For a pine forest in Norfolk, England, they obtained an α of about 0.7 when the canopy was dry—corresponding to a r_s of $\sim$120 s m^{-1} (Gash and Stewart 1975) and an $\alpha \approx 1.5$ when the canopy was wet ($r_s \approx 0$). For another forest in Wales they even found that $\alpha \approx 10$ when $r_s = 0$, but in this case advection must have played a dominant role.

The analysis given above reveals that on clear days α is primarily determined by r_s. Therefore it is interesting to investigate whether observations of α and r_s behave as predicted by the model. For this we analyzed the set of micrometeorological data collected above short grass in the summer of 1977 at Cabauw (the Netherlands) described by De Bruin and Holtslag (1982). In Fig. 5 the observed values of α_{11} are plotted

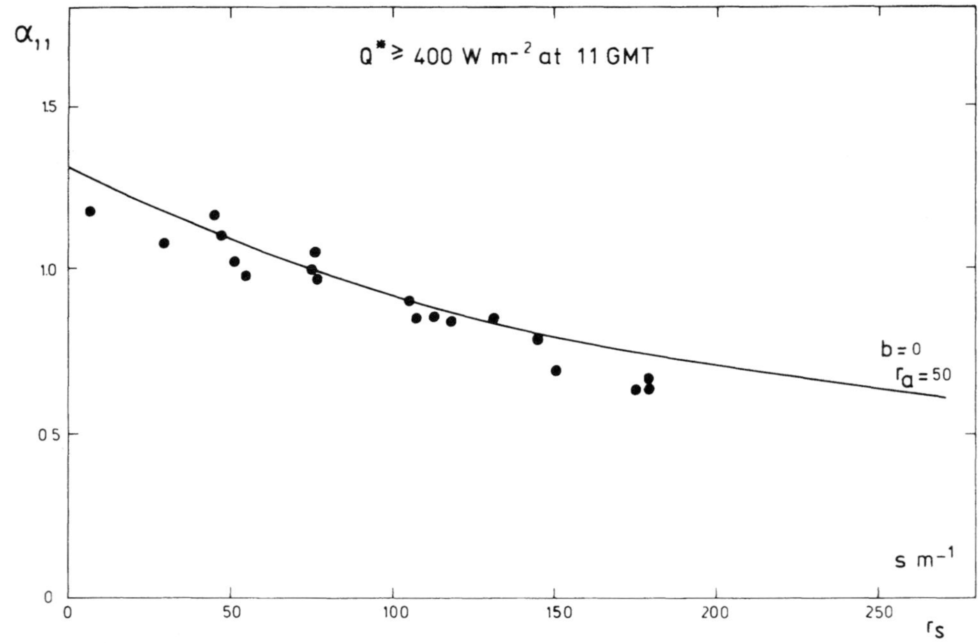

FIG. 5. Comparison between observed and computed values of α_{11} vs. r_s; observations at Cabauw in 1977, $b = 0$, $r_a = 50$ s m^{-1} and $Q^* \geq 400$ W m^{-2}.

against the corresponding values or r_s. The data depicted in Fig. 5 refer to hours around 1100 GMT during which the average net radiation was at least 400 W m^{-2}. Then it is likely that the conditions were similar to those we considered. In Fig. 5 the curve is also drawn representing the relation between α_{11} and r_s as calculated with our model taking $b = 0$ and $r_a = 50$ s m^{-1}, using the initial weather data mentioned above. From Fig. 5 we may conclude that the results of our model are encouraging: there is a good fit between the dependence of α on r_s and the model prediction.

6. Conclusions and discussion

In this study a model for the Priestley-Taylor parameter α is presented which is an atmospheric boundary layer (ABL) model coupled to the Penman-Monteith equation. It is an extension of the model published earlier by McNaughton (1976) and Perrier (1980). These authors assumed the ABL was a closed box of constant height. In our approach the ABL height is variable and depends on the heat input at the surface. Also the entrainment of, usually, warmer and drier air at the top of the ABL is taken into account.

It is shown that our approach differs significantly from the MP model. It appears that the calculated values of α agree well with those reported in literature (e.g., Priestley and Taylor, 1972; Monteith, 1981; Brutsaert, 1982). For example, an α of about 1.3 is found for well-watered surfaces, while $\alpha \approx 1$ under moderately dry conditions ($r_s \approx 60$ s m^{-1}). Also it is found that, although α is also a function of wind speed, surface roughness and the entrainment of dry air at the top of the ABL, it depends primarily on the surface resistance r_s. This result supports the use of Priestley-Taylor like parameterization of the surface heat fluxes, for example, in weather forecast models as proposed for instance by De Bruin and Holtslag (1982) and applied by Reiff *et al.* (1983).

In order to make our model comparable with the original MP model we simplified the equations describing the entrainment of dry air at the top of the ABL. In this way a simple differential equation for the specific humidity could be obtained. This restriction implies that the model is less generally applicable. Because it is our purpose to study the behavior of α under different conditions rather than to built a complete model; e.g., for weather forecasts, these limitations are of minor importance for this study. Moreover it appears that α is rather insensitive to variations in the parameter b [see Eq. (8)] that accounts in our model for the entrainment of dry air at the top of the ABL. This implies that our approach leads to more relatistic results than one would have thought at first sight.

It should be noted that the above results apply to the conditions for which we did our calculation, i.e., for vegetation completely shading the ground, and (initial) weather conditions, which are typical for a clear summer day in the temperate climates.

Acknowledgments. A part of this work has been done during a stay at the Institut National de Récherche Agronomique, Versailles (France). The final draft of this paper was written while the author worked at the Institute of Hydrology, Wallingford (United Kingdom). The helpful discussions with Alain Perrier are acknowledged. Also the author wishes to thank Bert Wartena, Henk Tennekes and John Stewart for their valuable comments.

REFERENCES

Brutsaert, W. H., 1982: *Evaporation into the Atmosphere.* Reidel, 299 pp.

De Bruin, H. A. R., and A. A. M. Holtslag, 1982: A simple parameterization of the surface fluxes of sensible and latent heat during daytime compared with the Penman-Monteith concept. *J. Appl. Meteor.*, 21, 1610–1621.

Driedonks, A. G. M., 1981: *Dynamics of the Well-Mixed Atmospheric Boundary Layer.* Doctoral dissertation, KNMI, DeBilt, Holland [Sci. Rep. WR 81-2], 189 pp.

——, 1982a: Sensitivity analysis of the equations for a convective mixed layer. *Bound.-Layer Meteor.*, **22,** 475–480.

——, 1982b: Models and observations of the growth of the atmospheric boundary layer, *Bound.-Layer Meteor.*, **23,** 283–306.

Gash, J. H. C., and J. B. Stewart, 1975: The average surface resistance of a pine forest derived from Bowen ratio measurements. *Bound.-Layer Meteor.*, **8,** 453–464.

McNaughton, K. G., 1976: Evaporation and advection I: Evaporation from extensive homogeneous surfaces. *Quart. J. Roy. Meteor. Soc.*, **102,** 181–191.

Monteith, J. L., 1981: Evaporation and surface temperature. *Quart. J. Roy. Meteor. Soc.*, **107,** 1–27.

Perrier, A., 1980: Étude micro-climatique des relations entre les propriétés de surface et les caractéristique de l'air: application aux échanges régionaux. *Météor. Environ.*, EVRY (France).

Priestley, C. H. B., 1959: *Turbulent Transfer in the Lower Atmosphere.* University of Chicago Press, 130 pp.

——, and R. J. Taylor, 1972: On the assessment of surface heat flux and evaporation using large-scale parameters. *Mon. Wea. Rev.*, **100,** 81–92.

Reiff, J., A. P. van Ulden, G. Cats, D. Blaauboer and H. A. R. De Bruin, 1983: An air-mass transformation model for short-range weather forecasting. Part I: The model. *Submitted to J. Climate Appl. Meteor.*

Shuttleworth, W. J., and I. R. Calder, 1979: Has the Priestley equation any relevance to forest evaporation? *J. Appl. Meteor.*, **18,** 639–676.

Slatyer, R. O., and I. C. McIlroy, 1961: *Practical Microclimatology.* CRIRO, Melbourne, Australia, 310 pp.

Swinbank, W. C., 1963: Long-wave radiation from clear skies. *Quart. J. Roy. Meteor. Soc.*, **89,** 339–348.

Tennekes, H., 1973: A model for the dynamics of the inversion above a convective boundary layer. *J. Atmos. Sci.*, **30,** 558–567.

Thom, A. S., and H. R. Oliver, 1977: On Penman's equation for estimating regional evaporation. *Quart. J. Roy. Meteor. Soc.*, **103,** 345–357.

D Vegetation Controls on Evaporation – Commentary

The papers in Section C show how the concept of potential evaporation was developed. This concept is attractive and useful: attractive because it allows evaporation to be calculated only from routinely measured meteorological variables, and useful because in humid climates, much of the time, it provides a worthwhile estimate of the actual evaporation. However, this is not always the case, and equating actual with potential evaporation can give incorrect estimates and be misleading – vegetation does not always act like moist green blotting paper. The papers in this section explore the development of the science necessary to deal with the situations when evaporation may not necessarily be at the potential rate.

The first set of papers charts the realization that the high rate of evaporation from wet forest canopies is the principal reason for differences between the evaporative water loss from forested and grass catchments. The observation that net radiation is not the only source of energy, and the subsequent modelling of the process in terms of the other driving meteorological variables, leads to the understanding of how forest evaporation will vary with rainfall. The second set of papers then deals with the situations when plants exercise stomatal control to restrict their rate of transpiration. These papers show how the theory has developed to scale up this effect from leaf to canopy, and to estimate transpiration from formulae that account for vegetation's need to allow carbon dioxide to enter leaves through their stomata for photosynthesis but to conserve water by restricting transpiration loss through these same stomata.

INTERCEPTION

D1 The effect of upland afforestation

Law, F. (1958) Measurements of rainfall, interception and evaporation losses in a plantation of Sitka spruce trees.

In the 1950s, there was a policy of planting coniferous forests over upland Britain. This policy was largely driven by economic and social aims; specifically, decreasing the reliance on imported timber and creating rural employment. The effects of this afforestation on water resources was not a major consideration, although in Britain, as elsewhere in the world, the benefits of forests in protecting catchments were recognised. The possible impact on evaporation, and thus runoff, was not seriously considered until Frank Law published this paper.

Law was an engineer responsible for managing a reservoir in a high rainfall area of northern England. Concerned that the reservoir consistently failed to deliver its designed yield, he initiated a small process study. A natural lysimeter was created in a forest plot using concrete walls sunk into an impervious clay sub-soil. The measurements of runoff from the forest lysimeter were compared with the drainage from a grass-covered lysimeter or "percolation gauge". Law's suspicion that the forest was responsible for the reduced yield was confirmed when results revealed that the forest plot delivered a third less runoff than the grass.

The results were not accepted immediately: Law's plot was small and the high rate of evaporation could have been the result of an edge effect. Moreover, the conclusions from paired catchment experiments in other countries were equivocal and the Penman formula, which was gaining acceptance at that time, implied that the evaporation should be controlled mainly by meteorology, not vegetation cover. Nevertheless, the experiment was hugely influential, eventually leading to the establishment of the UK Institute of Hydrology and an extensive programme of process and catchment scale research into the effects of forests on water resources. Ultimately, the controversy was solved through process studies of rainfall interception and transpiration, as discussed in the following papers.

D2 Evaporation as the driver of interception loss

Horton, R. E. (1919) Rainfall interception.

R. E. Horton was a giant in the history of hydrology; he was the first to identify and formulate many of the basic processes which make up the hydrological cycle. This is not the only volume of the *Benchmark Papers in Hydrology Series* to contain one of his papers; Beven (2006), for example, reproduces two of Horton's papers and gives more details about his life and work. Although Horton's work on rainfall interception is poorly acknowledged, his paper that we have selected is nevertheless also a classic. This paper was way ahead of its time and there is little, if anything, which should be rewritten today. Critically, Horton recognised that the interception loss was driven by evaporation, and that the influence of tree species was felt largely through the storage capacity which evaporates at the end of a storm. This division of the evaporation into that during rainfall and that which occurs after rainfall ceases is the basis of Gash's analytical model (see D6), which is probably the most widely applied physically-based model of rainfall interception today. Thus, indirectly at least, Horton's paper has been hugely influential.

Horton correctly predicted that interception loss would vary with rainfall climate: the number of storms and their intensity and duration. Indeed, he identified rainfall duration as a key variable controlling evaporation. However, recognising that rainfall duration measurements were relatively rare, he went on to express interception in terms of the more commonly measured rainfall amount. He made explicit the assumption, based on his measurements, that the two were linearly related, and that interception loss could be predicted from a relationship between gross rainfall and throughfall. This approach proved to be of great practical use although, paradoxically, its widespread adoption resulted in Horton's insight into the process being apparently lost for some 60 years! During those intervening years, studies were motivated more by looking at how the interception loss varied with tree species than with the rainfall climate in which they were growing.

D3 Intercepted rainfall and the surface energy balance

Rutter, A. J. (1967) An analysis of evaporation from a stand of Scots Pine.

In 1965, an International Symposium on Forest Hydrology was held at Pennsylvania State University. It was attended by many of the leading researchers in the subject and the proceedings volume, edited by W. E. Sopper and H. W. Lull, was a key reference book for young scientists entering the field at that time. We consider the paper presented by A. J. Rutter to be the most influential among those presented. It described how a detailed leaf-scale process study could be scaled up to provide new, quantitative understanding of how evaporation from forests was different from that from short vegetation. Working in a pine forest, Rutter made measurements of the evaporation rate from incised branches in both wet and dry conditions. He found that "*the rate of evaporation of intercepted water is, on average, about 4 times as great as the transpiration rate in the same environmental conditions*". Extending this result, he calculated the balance between the transpiration and interception components of the annual evaporation and demonstrated how these changed between winter and summer.

Rutter found that winter evaporation exceeded the estimated net radiation by a factor of four. Critically important, Rutter also used small thermocouples to measure the difference between leaf and air temperature. These measurements showed that, under wet conditions, the air temperature was greater than the leaf temperature, implying an input of sensible heat and allowing him to speculate that wet canopy evaporation was being supported by cooling the surrounding air, i.e. by advected energy.

D4 Modelling interception as an evaporation-driven process

Rutter, A. J., Kershaw, K. A., Robins, P. C. & Morton, A. J. (1971) A predictive model of rainfall interception in forests. I. Derivation of the model from observations in a stand of Corsican pine.

Building on the understanding and ideas in the previous paper (D3), Jack Rutter and his team at Imperial College, London, revolutionised the way in which the process of interception was modelled. The Rutter model described in this paper exploited the computer technology, which was becoming available at that time, to calculate a running mass balance of the depth of water on the forest canopy. At the heart of the model is the estimate of evaporation made using the Penman-Monteith equation (see D8) used in predictive mode: because the canopy is saturated, the surface resistance is set to zero in this equation. The model is driven not just by rainfall, but by meteorological data – net radiation, temperature, humidity and wind speed – because the process of interception is explicitly driven by evaporation rate. Because the Penman-Monteith equation is derived from the physics of the surface energy balance, the empiricism in the model is thus reduced to specifying physically meaningful parameters such as those describing the surface roughness and canopy capacity. This makes the model more generally applicable.

In subsequent papers, Rutter *et al.* applied the model to different types of forest (1975) and investigated the sensitivities of the predicted evaporation to the model parameters and driving variables (1977). A whole family of models based on the Rutter model now exist. The state of the art is currently given by the version proposed by Valente *et al.* (1997), who incorporated improved boundary conditions and a model framework for use in sparse canopies; the model reverts to the original Rutter model in closed-canopy conditions.

D5 Measurement of wet canopy evaporation

Stewart, J. B. (1977) Evaporation from the wet canopy of a pine forest.

Rutter's measurements (see D3) on pine needles showed that, under wet conditions, the canopy was at a lower temperature than the air. This implied that energy for evaporation was being supplied to the canopy by cooling the surrounding air. Rutter's paper is carefully worded and the discussion at the end of the paper reveals that it was not necessarily clear where this energy was coming from. Perhaps this was a local edge effect associated with the limited size of the stand of trees Rutter studied? Thus, it was not known whether a downward flux of sensible heat could occur for long periods and over extensive areas of forest. This paper by John Stewart answered that question by presenting the first micrometeorological observations of wet canopy evaporation from an extensive forest stand. The paper demonstrates that in wet conditions evaporation frequently exceeds the net radiation available to support it. Stewart's measurements were made close to the centre of Thetford Forest, an extensive pine forest some 30 km across, in eastern England. Consequently, the fetch over uniform forest was long enough for the atmospheric boundary layer to have come to equilibrium with the surface, leading to the conclusion that large-scale advection must be capable of providing the energy.

The knowledge that, at least in the maritime climate of western Europe, forests could draw on advected energy to evaporate intercepted rainfall provided an important input to the debate on whether forest used more water than the grass. It gave credence to the emerging realization that the evaporation of intercepted rainfall was likely to be the key factor in determining the difference between the water use of these two vegetation types.

D6 Expressing the essence of the process in a simple formula

Gash, J. H. C. (1979) An analytical model of rainfall interception in forests.

This paper builds both on the insights in Horton's paper (D2) and the numerical water budgeting and application of the Penman-Monteith equation used in the Rutter model (D4). It takes a storm-based approach to derive a comparatively simple formula which explicitly separates the components of interception loss. The model makes the point that, in addition to the tree structure parameters of the Rutter model, the duration of rainfall is critical in controlling the overall magnitude of the evaporation.

By making average rainfall rate an explicit parameter, the model allows rainfall amount to be used as a surrogate for rainfall duration. Tropical climates with high intensity, short duration storms (e.g. Lloyd *et al.*, 1988) are thus dealt with equally effectively as the long-duration synoptic rainfall events found in temperate climates (e.g. Gash *et al.*, 1980). With an empirical value of the average evaporation rate during rainfall, $\overline{E}$, and the (optional) assumption of only one storm per day, the model can be applied to predict the evaporation of intercepted rainfall from daily rainfall data, removing the need for the rarely-available above-canopy meteorological data required by the Rutter model. Values of $\overline{E}$ have subsequently proved to be fairly conservative, with similar rates being observed in temperate and tropical conditions.

As with the Rutter model, the one-dimensional description used in the original Gash model fails for sparse forest canopies. Gash *et al.* (1995) proposed a revised version with sensible boundary conditions which has a sounder treatment of these canopies. This version was, in turn, further modified by Valente *et al.* (1997) to give an improved treatment of the energy conservation. Later, van Dijk *et al.* (2001a,b) also derived a useful improvement in which the canopy store is treated as a variable, allowing the model to be used for fast growing crops.

D7 Explaining the difference between evaporation from forest and grass

Shuttleworth, W. J. & Calder, I. R. (1979) Has the Priestley-Taylor equation any relevance to forest evaporation?

This paper was the first to directly challenge the then current practice of estimating forest evaporation with formulae such as the Penman (C3) and Priestley-Taylor (C4), which take no account of the different controls on the evaporation rates of intercepted and transpired water. Shuttleworth and Calder used the insight gained from the process studies into forest evaporation carried out in Britain during the 1970s, which had shown that control of transpiration is dominated by a large surface resistance which is larger than for grassland, while wet canopy evaporation is controlled by the aerodynamic resistance, which is much lower than for grassland.

Contrasting the results from the high-rainfall, upland forest on Plynlimon in Wales with those from Thetford Forest in the low-rainfall, east of England, the paper demonstrates that for forest, the Priestley-Taylor equation fails to estimate the evaporation in either wet or dry conditions. The paper thus uses observational data to confirm that forests do not evaporate at the potential rate and that separate representation and modelling of the transpiration and interception contributions to total evaporation are essential. This is the approach now adopted in all state-of-the-art vegetation–atmosphere transfer models, in both meteorology and hydrology.

Because the interception term is rainfall dependent, the actual evaporation from a forest will depend on the nature and duration of precipitation. Consequently, as this paper makes clear, the difference between the evaporation from a forested and grassed catchment will vary geographically as the balance between the interception and transpiration terms changes. The paper also provides an explicit physical explanation of why forest evaporation is so variable.

TRANSPIRATION

D8 The big leaf model

Monteith, J. L. (1965) Evaporation and environment.

The concept of leaf stomatal resistance was formulated during the 1950s (see Commentary to B1), but it is generally accepted that this paper by John Monteith took the important extra step of proposing a bulk *surface resistance* which represented the net effect of the whole canopy acting together. The essential assumption is that all the radiation capture and the partitioning of energy into latent and sensible heat can be described as if it occurred at a single level, the effective source-sink height. Once the assumption is made that energy partition is made, in effect, by a single *big leaf* at an effective source-sink height, the derivation of the equation becomes straightforward. The known relationship between saturated vapour pressure and temperature is used to eliminate the average surface temperature of the canopy, with this relationship being approximated in the derivation by a straight line between the surface and air temperatures.

The so-called big leaf concept was not universally well-received at the time Monteith proposed it, particularly by the advocates of multi-layer computer modelling. Most famously, the respected theoretician J. R. Philip (as recorded by Smiles, 2005) "*felt that the simplification of canopy exchange processes inherent in Monteith's one-dimensional 'big leaf' model was unrealistic and misleading*". Philip's forcefulness in attacking the big leaf concept led to a rejoinder in Monteith's *Principles of Environmental Physics* (Monteith, 1973) that not only did the experimental evidence support the use of crop or surface resistance as an index of the physiological control of water loss by a crop canopy, but also that no more appropriate index had yet been devised "*despite attacks based on armchair speculation divorced from field experience*". This conflict between single and multilayer modelling was later resolved by Shuttleworth (1976), who demonstrated that it was possible to write a multi-layer model in analytically continuous form and create a one-dimensional description in which evaporation is described by a combination equation similar to the Penman-Monteith equation.

Although Monteith intended that bulk surface resistance be used as a diagnostic of plant behaviour, it has now become the most widely used predictive evaporation model for closed canopies, with calculations made from meteorological data for wet canopies with surface resistance set to zero, and for dry canopies using an empirical sub-model of surface resistance. A specific form of the Penman-Monteith equation, in which the aerodynamic properties and surface resistances of the crop are specified, is now recommended by FAO (UN

Food and Agriculture Organization) for calculating reference crop evaporation (Allen *et al.*, 1988). Sometimes this specific form is wrongly referred to by practitioners as being the Penman-Monteith equation, but this equation is, of course, much more generally applicable.

D9 Stomatal response

Jarvis, P. G. (1976) The interpretation of the variations in leaf water potential and stomatal conductance found in canopies in the field.

The availability of porometers (see B1) allowed plant physiologists to make field measurements of how stomata and leaf water potential varied in response to different environmental conditions. This paper by Paul Jarvis brought together the knowledge gained from this work by expressing it in an analytical framework. The description assumes that a maximum *stomatal conductance* (the reciprocal of resistance), which may never occur in nature, is modified by a multiplicative combination of factors which are considered to act independently. Normalized functions were derived representing the controls resulting from changes in photon flux density, temperature, leaf water potential, vapour pressure deficit and ambient carbon dioxide concentration. The resulting set of equations introduced a means of representing environmental controls on stomatal conductance, and provided a foundation for a practical method of estimating plant controls on transpiration.

At the time this paper was published, micro-meteorologists were using measurements of evaporation to derive the bulk surface resistance (or its reciprocal surface conductance) of canopies. These surface conductances were then used predictively, with evaporation modelled by the Penman-Monteith equation (D8). Initially the conductance was merely assigned a typical value, or a prescribed daily trend, sometimes with an allowance for the effect of soil moisture limitation (e.g. Gash & Stewart, 1977; Dolman *et al.*, 1990). However, Jarvis's paper provided the basis for a more realistic model which captured the response of plant stomata to the environment in which the vegetation is growing. Stewart (1988) adapted Jarvis's formulation to give the *Jarvis-Stewart model*, which describes the conductance in terms of routinely measured variables by substituting solar radiation for photon flux density, soil moisture deficit for leaf water potential, and neglecting carbon dioxide concentration. Stewart used multivariate regression to derive a complete set of parameters, an approach subsequently adopted by others (e.g. Dolman *et al.*, 1990; Wright *et al.*, 1995, 1996). Because global climate models reproduce the variables driving the Jarvis-Stewart model, it could also be adopted for estimating evaporation within in these global-scale models, and it is still widely used within meteorological and hydrological models.

D10 Introducing photosynthesis

Ball, J. T., Woodrow, I. E. & Berry, J. A. (1986) A model predicting stomatal conductance and its contribution to the control of photosynthesis under different environmental conditions.

This paper built on the increased understanding of plant biochemistry to introduce a new approach to modelling stomatal control, which is based on the primary function of stomata: allowing carbon dioxide to enter the stomatal cavities. In the model, the assimilation of carbon is viewed as the controlling factor. This approach is less empirical than Jarvis's approach (D9) and therefore more transferable from one plant species to the next. The only dependency is on the plant's photosynthetic function, i.e. whether it has either C_3 or C_4 photosynthetic chemistry. A key difference between this formula for stomatal conductance and the Jarvis-Stewart formula is that stress factors (apart from that for humidity) are no longer combined as a product; rather, one factor is considered to be the dominant limitation on carbon assimilation and hence on stomatal conductance.

This alternative means of describing the behaviour of stomata was soon introduced into global climate models (e.g. Sellers *et al.*, 1996) as a means of reducing the range of vegetation parameters needed. This has sometimes been referred to as the "greening" of the representation of the land surface used in global climate models. In such models, the form of the Ball-Berry equation for stomatal conductance adopted (e.g. Sellers *et al.*, 1996) is:

$$g_s = m(A_n/C_s)P_lF_e + g_{\min}$$

where $g_{\min}$ is a prescribed minimum stomatal conductance; m is a slope parameter (~9 for C_3 plants); A_n is the net carbon assimilation; C_n is the partial pressure of carbon dioxide; P_l is atmospheric pressure adjacent to the leaf; and F_e is a humidity dependent stress factor. In the original Ball-Berry equation, F_e was numerically equal to the relative humidity, but in some land surface models (e.g. Dickinson *et al.*, 1998) it is assumed to be a function of vapour-pressure deficit. In this formula, the simplest estimate of A_n is that given (Farquhar & Sharkey, 1982) as:

$$A_n = \min(J_c, J_e, J_s)$$

where J_c, J_e, J_s are functions expressing the assimilation rates when limited by the Rubisco enzyme, light or transport capacity, respectively, for C3 and C4 plants; see Collatz *et al.* (1991, 1992), Sellers *et al.* (1996) and Cox (2001).

The need for global climate models to predict both the evaporation and carbon dioxide flux has meant that formulations based on the Ball-Berry approach are now the preferred option in all global models in which the land surface is coupled to describe atmosphere–biosphere feedback.

D11 Sparse crops

Shuttleworth, W. J. & Wallace, J. S. (1985) Evaporation from sparse crops – an energy combination theory.

This paper provided the missing framework for describing evaporation from sparse vegetation canopies. The equation takes a one-dimensional approach and treats the canopy as being semi-transparent. The heat fluxes from the leaves and the soil are then assumed to blend in the mean canopy airflow at the height of the effective canopy source. The *Shuttleworth-Wallace equation* reduces to the Penman-Monteith equation at the asymptotic limits of complete canopy cover and bare soil.

The paper provided working assumptions for the parameters in the equations for the purpose of illustration, but their precise definition remained speculative until later papers addressed this. The Shuttleworth-Wallace equation has been used extensively, proving to be particularly useful in describing growing agricultural crops and row crops (e.g. Lafleur & Rose, 1990; Farahani & Ahuja, 1996; Zhou *et al.*, 2006).The basic theoretical framework of Shuttleworth-Wallace has been extended to improve the description of the aerodynamic resistances (Shuttleworth & Gurney, 1990), include multi-species canopies and the exchange of CO_2 (e.g. Wallace, 1996; Wallace & Verhoef, 2000), and incorporate two canopies evaporating in parallel (Dolman, 1993; Huntingford *et al.*, 1995).

References

Allen, R. G., Pereira, L. S., Raes, D. & Smith, M. (1998) *Crop Evapotranspiration – Guidelines for Computing Crop Water Requirements*. FAO Irrigation and Drainage Paper 56. FAO, Rome, Italy.

Beven, K. J. (2006) *Streamflow Generation Processes*. Benchmark Papers in Hydrology, vol. 1. IAHS Press, Wallingford, UK.

Collatz, J. G., Ball, J. T. Grivet, C. & Berry, J. A. (1991) Physiological and environmental regulation of stomatal conductance, photosynthesis, and transpiration: a model that includes a laminar boundary layer. *Agric. Forest Met.* **54**, 107–136.

Collatz, J. G., Ribas-Carbo, M. & Berry, J. A. (1992) Coupled photosynthesis–stomatal conductance model for leaves of C_4 plants. *Australian J. Plant Physiology* **19**, 519–538.

Cox, P. M. (2001) Description of the TRIFFID global dynamical vegetation model. *Hadley Centre Technical Note 24*. Hadley Centre, Bracknell, UK.

Dickinson, R. E., Shaikh, M., Bryant, R. & Graumlich, L. (1998) Interactive canopies for a climate model. *J. Climate* **28**, 2823–2836.

Dolman, A. J. (1993) A multiple-source land surface energy balance model for use in general circulation models. *Agric. Forest Met.* **65**, 21–45.

Dolman, A. J., Gash, J. H. C., Roberts, J. & Shuttleworth, W. J. (1990) Stomatal and surface conductance of tropical rainforest. *Agric. Forest Met.* **54**, 303–318.

Farahani, H. J. & Ahuja, L. R. (1996) Evapotranspiration modeling of partial canopy/residue-covered fields. *Trans. ASAE* **39**, 2051–2064.

Farquhar, G. D. & Sharkey, T. D. (1982) Stomatal conductance and photosynthesis. *Annual Rev. Plant Physiology* **33**, 317–345.

Gash, J. H. C. & Stewart, J. B. (1977) The evaporation from Thetford Forest in 1975. *J. Hydrol.* **35**, 385–396.

Gash, J. H. C., Wright, I. R. & Lloyd, C. R. (1980) Comparative estimates of interception loss from three coniferous forests in Great Britain. *J. Hydrol.* **48**, 89–105.

Gash, J. H. C., Lloyd, C. R. & Lachaud, G. (1995) Estimating sparse forest rainfall interception with an analytical model. *J. Hydrol.* **170**, 79–86.

Huntingford, C., Allen, S. J. & Harding, R. J. (1995) An intercomparison of single and dual-source vegetation–atmosphere transfer models applied to transpiration from Sahelian savannah. *Boundary-Layer Met.* **74**, 397–418.

Lafleur, P. M. & Rose, W. R. (1990) Application of an energy combination model for evaporation from sparse canopies. *Agric. Forest Met.* **49**, 135–153.

Lloyd, C. R., Gash, J. H. C., Shuttleworth, W. J. & Marques, A. de O. (1988) The measurement and modelling of rainfall interception by Amazonian rain forest. *Agric. Forest Met.* **43**, 277–294.

Monteith, J. L. (1973) *Principles of Environmental Physics*. Edward Arnold, London, UK.

Rutter, A. J. & Morton, A. J. (1977) A predictive model of rainfall interception in forests. III. Sensitivity of the model to stand parameters and meteorological variables. *J. Appl. Ecol.* **14**, 567–588.

Rutter, A. J., Morton, A. J. & Robins, P. C. (1975) A predictive model of rainfall interception in forests. II. Generalization of the model and comparison with observations in some coniferous and hardwood stands. *J. Appl. Ecol.* **12**, 367–380.

Sellers, P. J., Randall, D. A., Collatz, G. J., Berry, J., Field, C., Dazlich, D. A., Zhang, C. & Bounoua, L. (1996) A revised land surface parameterization (SiB2) for atmospheric GCMs. 1, Model formulation. J. *Climate* **9**, 676–705.

Shuttleworth, W. J. (1976) A one-dimensional theoretical description of the vegetation-atmosphere interaction. *Boundary-Layer Met.* **10**, 271-301.

Shuttleworth, W. J. & Gurney, R. J. (1990) The theoretical relationship between foliage temperature and canopy resistance in sparse crops, *Quart. J. Royal Met. Soc.* **116**, 497–519.

Smiles, D. (2005) John Robert Philip. *Biographical Memoirs of Fellows of the Royal Society* **51**, 327–345.

Stewart, J. B. (1988) Modelling surface conductance of pine forest. *Agric. Forest Met.* **43**, 19-35.

Valente, F., David, J. S. & Gash, J. H. C. (1997) Modelling interception loss for two sparse eucalypt and pine forests in central Portugal using reformulated Rutter and Gash analytical models. *J. Hydrol.* **190**, 141–162.

van Dijk, A. I. J. M. & Bruijnzeel, L. A. (2001a) Modelling rainfall interception by vegetation of variable density using an adapted analytical model. Part 1. Model description. *J. Hydrol.* **247**, 230–238.

van Dijk, A. I. J. M. & Bruijnzeel, L. A. (2001b) Modelling rainfall interception by vegetation of variable density using an adapted analytical model. Part 2. Model validation for a tropical upland mixed cropping system. *J. Hydrol.* **247**, 230–238.

Wallace, J. S. (1996) Evaporation and radiation interception by neighbouring species. *Quart. J. Royal Met. Soc.* **123**, 1885–1905.

Wallace, J. S. & Verhoef, A. (2000) Modelling interactions in mixed-plant communities: light, water and carbon dioxide. In: *Leaf Development and Canopy Growth* (ed. by B. Marshall & J. A. Roberts), 204–250. Sheffield Academic Press, Sheffield, UK.

Wright, I. R., Manzi, A. O. & da Rocha, H. R. (1995) Canopy surface conductance of Amazonian pasture: model application and calibration for canopy climate. *Agric. Forest Met.* **75**, 51–70.

Wright, I. R., Gash. J. H. C., da Rocha, H. R. & Roberts, J. M. (1996) Modelling surface conductance for Amazonian pasture and forest. In: *Amazonian Deforestation and Climate* (ed. by J. H. C. Gash, C. A. Nobre, J. M. Roberts & R. L. Victoria), 438–458. Wiley, Chichester, UK.

Zhou, M. C., Ishidaira, H., Hapuarachchi, H. P., Magome, J., Klein, A. S. & Takeuchi, K. (2006) Estimating potential evaporation using Shuttleworth-Wallace model and NOAA-AVHRR NDVI data to feed a distributed hydrological model of the Mekong River basin. *J. Hydrol.* **327**, 151–173.

MEASUREMENT OF RAINFALL, INTERCEPTION AND EVAPORATION LOSSES IN A PLANTATION OF SITKA SPRUCE TREES

FRANK LAW

Engineer to the Fylde Water Board, Blackpool, Lancashire, England

ABSTRACT

A forest lysimeter has been formed by the construction of a concrete wall around one-ninth of an acre (0.045 ha) within a dense plantation of sitka spruce trees. It is thought that the underlying clay is so impermeable that no leakage of water occurs, and the runoff is collected in a tank and measured by meter.

Precipitation is measured by three gauges on poles near the top of the canopy and by gauges at ground level in adjacent open areas. The amount of water reaching ground level is determined by ten rain gauges distributed over the forest floor and by troughs which girdle some of the tree boles and collect stem flow.

During the dry period from 4 July 1955 to 8 July 1956 the precipitation on the lysimeter amounted to almost 39 inches (only 70% of the long-period average); 24 inches of this reached the forest floor and the runoff was equivalent to less than 11 inches of rain. The gross loss was therefore 28 inches, compared with 16.6 inches over the adjacent water supply cathment and 15.5 inches on a grass-covered gauge nearby.

The techniques used for measuring precipitation on the canopy and the amount penetrating same and reaching ground are described and the results analysed.

INTRODUCTION

At present, British water supply undertakings are being encouraged to afforest their water catchments and extensive areas have or are being planted with trees. Compared with other countries, relatively little research on the water balance of woodlands has been carried out in Great Britain. Some of the results obtained in other regions have indicated that a tree cover may reduce the yield of water more than a short herbaceous vegetation but the magnitude of this effect, if it occurs at all under British conditions, is largely unknown. Consequently, at the end of 1954, the author, with the permission of the Fylde Water Board, began an investigation of the water balance in a plantation of sitka spruce, *Picea sitchesis.* The results obtained in this plantation are being tentatively compared with those from small grasscovered percolation gauges and with the water catchment as a whole, which at present is mainly grassland or moorland. This problem of the effects of afforestation on water yield is important to the Fylde Water Board since ten years previously it had leased over 2,000 acres (i.e. 22%) of its Hodder catchment to the Forestry Commission. The area was mainly rough grazing but has now been almost completely planted with conifers which can be expected to produce a dense forest cover within the next 20 years.

The purpose of this paper is to describe the techniques used in examining the water balance in the spruce plantation, to analyse some of the results and to suggest the need for similar experiments in larger forest plots.

LOCATION

The Hodder catchment is situated in the Yorkshire Pennine hills about 20 miles to the north east of Preston. Between 1929 and 1932 a number of small woodland

LYSIMETER AT STOCKS RESERVOIR

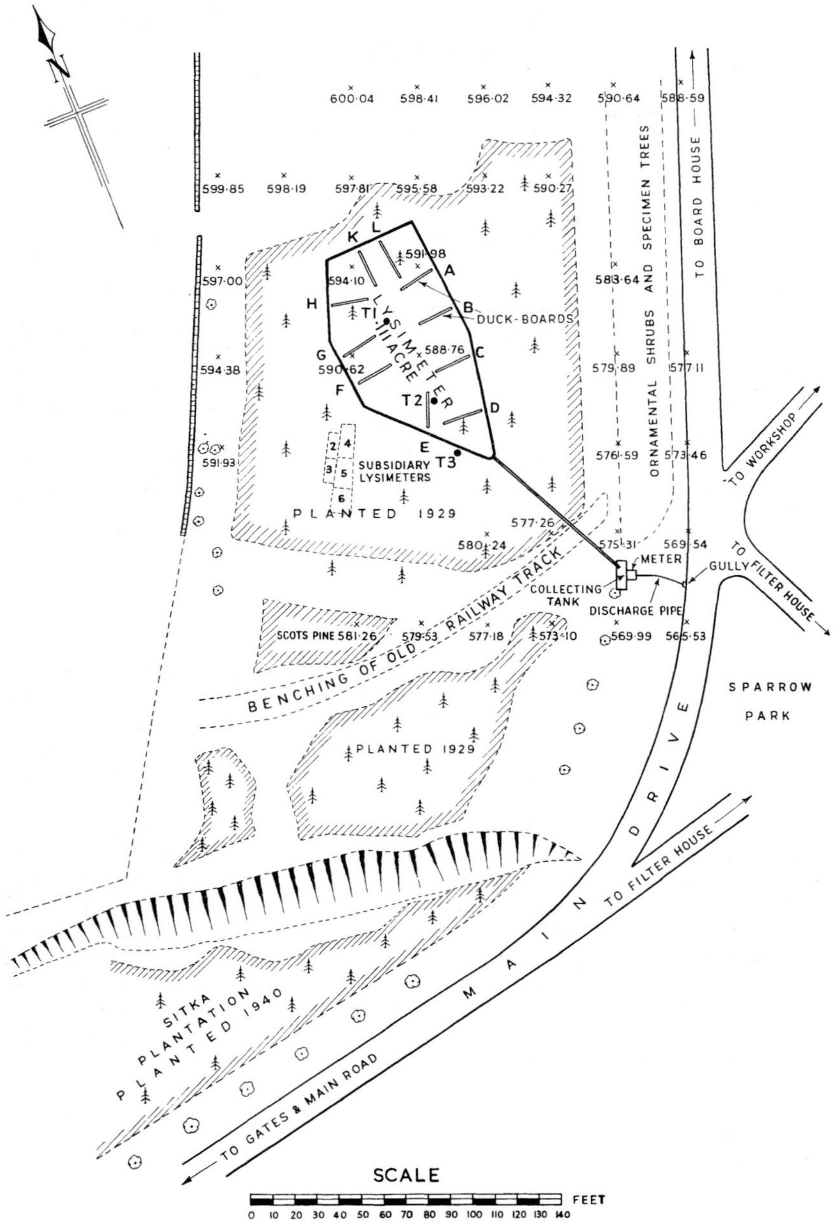

Fig. 1

398

blocks were planted for amenity purposes near the Stocks reservoir at Hodder and one of these plantations was selected for study. The plantation is located on a gentle south-facing slope at an elevation of about 600 feet above the sea and has a total area of 0.6 acre (0.24 ha). The plantation is in a relatively exposed situation and the average rainfall recorded at a climatological station close-by is 57 inches.

The experimental area of 0.111 acres (0.045 ha) is located more or less in the centre of the forest plot. (See Fig. 1.)

Treatment of Site

The sitka spruce was planted in 1929 using normal sylvicultural techniques. The last thinning was in 1952 when 60 trees were removed from the experimental area to leave 95 trees. The remaining trees ranged in height from 26 to 33 feet when the experiments commenced in 1955 and had an average trunk diameter of 6·5 inches at 4 feet above ground level. The current annual height increment is about 2 feet. When the plantation was thinned all the tree boles except the marginal ones were pruned. The canopy is alive down the whole length of the edge trees and forms an effective barrier to the wind, tending to isolate the central experimental area from the surrounding grassland. The wind-speed within the plot is approximately 40% of that at the Climatological Station 200 yards to the north-east.

The experimental plot was enclosed by a concrete cut-off wall approx. 9 inches wide and sunk down to a depth of about 2 feet in the soil and into the underlying clay. The wall gradually narrows above to a width of 6 inches at a height of 6 inches above ground level. The plot is roughly diamond shaped and has a fall of 1 in 10 along the longer diagonal to the lowest corner where there is an outlet through which water can flow. Field tiles were laid on the inside of the cut-off wall and surrounded with gravel. Three stone drains were intercepted during the work, and, where necessary, the intercepted drains were joined up by field drains on the outside of the wall. The top-soil is only about 12 inches deep. It is thought that the clay sub-soil is impervious and that the concrete wall effectively isolates the experimental area from sub-surface flow of water so that the experimental area functions as a lysimeter.

Observations

Complete records have been taken in the plot since July 1955, but partial data are available from March 1955. The precipitation falling upon the plot has been measured by rain gauges outside the experimental area and by rain gauges erected on poles so that they are above the forest canopy. The amount of incident rainfall penetrating to the forest floor has been determined with rain gauges beneath the trees and stem garlands to collect the water flowing down the tree trunks. Run off through the outlet at the lower corner of the lysimeter is determined by means of a flow meter. No measurements of soil moisture have been taken but the observation periods have been selected so that at their beginnings and ends the soil was probably near « field capacity ».

Measurement of rainfall upon the plot

Several rain gauges are maintained by the Board in the area around the lysimeter and three of these were selected as basic gauges for the experiment, namely:

399

	Catch in the calendar year 1956	Deviation from mean of three
No. 11 situated approx. 250 yds. to the North	54.5 ins.	+ 0.9%
No. 18 situated approx. 400 yds. to the S. W.	53.1 ins.	— 1.6%
No. 21 D situated approx. 200 yds. to the N. E.	54.4 ins.	+ 0.7%
Mean for 1956	54.0 ins.	

All are at sites inspected and approved by the Meteorological Office but only No. 21 D has a turf-wall surround.

The mean of these three gauges has been taken as the rainfall outside the plantation, but as the plantation is comparatively small it was thought that wind-eddies may cause some of the rain to be blown over the top. Consequently, in mid-June 1955, three rain gauges were fixed at the tops of poles within the plantation canopy, the water therefrom being led down tubes fastened to the poles to containers at the base.

For the first nine months of the experiment these tree-top gauges consisted of the top portions of Meteorological Office, Mark II rain gauges, i.e. in external appearance they consisted of a cylinder 5 inches diameter by 7 inches deep.

The first, T1, was on a pole only 21 feet 6 inches high because it was situated in a clear space (approximately 5 feet diameter at the base of the canopy and wider above) but the other two were on higher poles, T2 being 29 feet high and T3 27 feet 6 inches high.

Over this first period from June 1955 to early March 1956, the catch of the tree-top gauges averaged 88% of the three outside gauges, whilst between themselves T2 was consistently lowest and T3 the highest, the ratio T1/T2 being 1.02 and T3/T2 1.05.

Dr. H. R. Mill (1900) had found that the rainfall caught by a plain gauge at 10 feet above ground level was 90 to 91% of that at the standard level of 1 foot above the ground, and at 25 feet above ground level about 75% to 88%, but when the experiment commenced it was not known what was the relative exposure of rain gauges situated at 20 to 30 feet above ground level within a plantation canopy. In March 1956, therefore, a replica of the tree-top gauges was fixed at 10 feet height at an experimental rain gauge site nearby (known as No. 10 site). At the same time T3 was fitted with a type of nipher shield, and a similarly shielded gauge was set up at No. 10 site.

The ratio between the catch of T3 and the mean of T1 and T2 had previously been 1.04, but after the change to nipher shield the ratio jumped to 1.16, so it appeared that the shield caused an increase of about 12% in the catch of T3. Over the same period the shielded gauge 10 feet high at No. 10 site (No. 10E) caught 8% more than did No. 10F which was the replica of T1 and T2, but 10 feet high in the open. Consequently, it appeared that the exposure of the tree-top gauges was 50% greater than the equivalent of 10 feet high above the ground in the open, and that the corrections to be made should be 50% greater than those found at No. 10 site — namely 3.3% deficiency on 10E and 12.0% deficiency on 10F, relative to the basic turf wall gauge No. 10.

Uplifting the catch of T3 over the relevant period (43.2 inches) by 1.5 times 3.3% gave 45.4 inches, whilst uplifting T1 and T2 (36.8 inches) by 1.5 times 12.0% × 88/82 (to allow for the more serious adjustment) gave 43.9 inches as compared with 44.9 inches for the average of the three outside gauges for the same period. The factor 88/82 is derived from (100-12) divided by (100 — 1.5 × 12).

It appeared, therefore, at the end of December 1956, that the rainfall on the plantation was *not* appreciably reduced from the average rainfall outside the plantation.

Since August 1956, however, it had been noticed that the ratio between the tree-top gauges and the outside gauges had been lowering. This may be due to growth of the trees (which as previously stated is estimated a 2 feet per annum), and on the 7th January 1957, T1 was lifted 5 feet to test this theory.

From 7th January to 17th February 1957 the catches were:

Mean of 3 outside gauges	8.23 inches
Tree-top gauge T1	7.56 inches
Tree-top gauge T2	6.38 inches
Tree-top gauge T3	7.33 inches.

The ratio T3/T2 remained approximately as before, at 1.15, but T1/T2 jumped from 1.02 to 1.19.

The height of T1 was still less than T2 and T3, so on 18th February 1957 it was lifted a further 5 feet, to determine whether there could be an optimum ratio of catch in a tree-top gauge. At the time of writing insufficient information has been obtained to make any reliable observations thereon, but it would seem that in the early stages of the experiment T1 and T2 were being deprived of rain by nearby branches of the trees, whilst T3 was giving somewhat similar results because it was more exposed to the prevailing winds than were the other two. It seems possible that the slender tops of the trees cause a considerable amount of turbulence and that this turbulence is so great that a rain gauge makes little extra difference — consequently the necessity for shielding of the gauge may not be as great as was thought earlier in the experiment.

It does seem definite, however, that the «tree-top» gauges should be *above* the canopy.

In collaboration with Dr. J. D. Ovington of the Nature Conservancy Organisation, three further tree-top gauges have now been installed. These have a different nipher shield without a horizontal «brim», because it has been found at the No. 10 site that elevated shielded gauges with «brims» catch excessive amounts when high winds coincide with «drizzly» rain — and the more complicated the nipher shield the greater the excess catch.

The poles carrying these three additional gauges are adjustable for 1 foot increments of height, so as to facilitate experiments within and just above the canopy. As a check three similarly shielded gauges are being erected at No. 10 site, with the rims 18 inches, 10 feet and 30 feet respectively, above ground level. Reliable conclusions from these last additions to the equipment cannot be expected for some considerable time, but it may be possible to give some data as an addendum to this paper when published.

In the early stages of the experiment, as a temporary measure, the rain falling on the canopy was estimated as the mean of the six gauges — three outside and three tree-top. This definitely appears to give a low estimate, but has been retained for this paper.

Measurement of rainfall under the canopy

At ground level within the lysimeter ten Meteorological Office, Mark II rain gauges are being used to measure the water dripping from the trees. These ten gauges have been moved about in random manner from time to time in order to overcome the variation in the catches depending upon whether or not they are receiving drips from the branches of the trees.

At first the movement was made at the earliest convenient time after one inch of rain had been measured outside and the gauges were placed randomly by the observer, but latterly this movement has been more frequent (approximating to every half inch of rain outside the plantation) and the spacing of the gauges has been to a more definite plan, but still with attempted randomness.

To prevent damage to the plantation floor by trampling in wet weather, ten wooden duckboards (caillebotis), each 20 feet long were laid down in December 1955, and, since then, each of the ten gauges has been placed alongside its duckboard at one of 36 alternative points marked off on the duckboard, 18 on one side and 18 on the other. The line of each duckboard was selected so that at some points the gauges would be near to a trunk, and at other points remote from the trunks. In order to ensure that the placing of the gauges is random, the point at which the gauge is placed is selected out of the 36 alternatives by two throws of a dice. It was thought that this arrangement would give a more truly random pattern of rain gauges and would obviate bias on the part of the observer, but results have not confirmed this as will be seen from Table 1.

In 1955 the coefficient of variation of the catches in the ten gauges in any one pattern varied from 16% to 63% with an average of 33%. When the patterns were aggregated the coefficients of variation dropped to 9.8% for the first 11 patterns and 9.4% for the subsequent 13 patterns, whilst for the total 24 patterns the coefficient of variation was reduced to 6.4%. It will be seen that these are approximately in inverse proportion to the square root of the number of patterns aggregated — i.e. what would be expected by statistical reasoning if the patterns were random.

On the other hand in 1956, with the gauges set out randomly alongside the 10 duckboards, whilst the coefficients of variation for individual patterns still ranged from 20% to 65%, with an average of 38%, the coefficients for aggregates of patterns have been much higher than would be expected from the inverse square root ratio — indeed after 54 patterns the coefficient of variation was 11.8%, whereas it would have been only 5.4% for true randomness. In other words there has been more bias with the duckboards than previously; but though there is this obvious bias and variation between the gauges, as the duckboards are fairly uniformly spread over the lysimeter area, it is considered the average may be more reliable than the standard error of the mean would suggest. Incidentally, it would appear that duckboards **B** and H are under denser foliage than the others and that the interception is, therefore, greater.

Since mid-March 1957 the method of siting the gauges has been amended by making a further throw of the dice and placing the gauge concerned at a distance in feet from the duckboard equal to this third throw of the dice. The result will be that the gauges will sample an area about 13 feet wide by 20 feet long, and the 10 gauges will therefore, be sampling a total of 290 square yards out of the 537 square yards within the lysimeter wall. If circumstances permit, the results of this amendment will be given in an addendum when the paper is published.

To demonstrate the variability between the catches of the gauges in any one pattern those for the period 9 July - 26 August 1956 (a particularly wet period) are set out in Table 2.

Measurement of Water Flowing down the Tree Stems

The amount of water flowing down the tree stems has been measured by fixing garlands round the trunks of some of the trees at chest height, and leading the flow therefrom into containers alongside the tree. At the start of the experiment three of the trees were thus garlanded, but in view of the variation in the amount of water

TABLE 1

	No. of Patterns	Catches by Plantation Floor Gauges — Inches											Coeff: of Variation %
		A	B	C	D	E	F	G	H	K	L	Mean	
16 Mar. - 3 July 1955	11	7.39	9.00	6.61	8.79	8.67	9.01	8.54	8.69	7.13	7.97	8.18	9.8
4 July - 31 Dec. 1955	13	11.13	12.78	12.18	11.84	11.93	10.58	10.61	9.03	11.15	10.87	11.22	9.4
Sub-total 1955	24	18.52	21.78	18.79	20.63	20.60	19.59	19.15	17.72	18.38	18.84	19.40	6.4
1 Jan. - 8 July 1956	22	11.99	9.68	10.62	10.83	11.04	10.51	9.24	7.80	8.99	11.64	10.23	12.6
9 July - 26 Aug. 1956	12	8.67	7.64	10.83	10.75	10.91	8.84	13.5-	8.73	10.03	8.88	9.88	17.1
27 Aug. - 31 Dec. 1956	20	10.11	8.55	10.89	11.38	12.49	7.83	12.05	8.86	8.69	11.51	10.24	16.2
Sub-total 1956 (Duckboards)	54	30.77	25.87	32.34	32.96	34.44	27.18	34.79	25.39	27.71	32.03	30.35	11.8
Grand Total	78	49.29	47.65	51.13	53.59	55.04	46.77	53.94	43.11	46.09	50.87	49.75	7.7

TABLE 2

Plantation Floor Gauge Catches

1956		A	B	C	D	E	F	G	H	K	L	Mean	Coeff. of Variation
													%
July	9 - 15	.04	.64	.32	.60	.46	.30	.22	.36	.03	.20	.32	65
	16 - 23	.12	.21	.24	.09	.15	.06	.27	.06	.06	.12	.14	56
	24 - 27	.60	.39	1.03	1.90	1.45	.77	•1.59	.82	1.35	.91	1.08	44
	28 - 29	.44	.73	.83	.67	.71	.36	1.02	.33	.89	.55	.65	35
	30 - Aug. 1	1.19	1.73	1.16	1.29	2.16	1.34	3.40	.88	2.26	.94	1.63	48
August	2 - 6	.67	.55	.93	1.11	.59	1.41	1.00	.72	.58	.50	.81	37
	7 - 12	1.06	.74	1.35	1.29	1.48	1.45	1.26	1.50	.94	1.51	1.26	21
	13 - 15	1.08	.36	.67	.98	1.33	.56	1.85	1.23	.53	1.13	.97	47
	16	.36	.53	1.01	.27	.45	.52	.35	.52	.40	.59	.50	51
	17 - 19	.70	.80	1.50	.83	1.15	1.21	.78	.70	.96	.97	.96	27
	20 - 24	1.87	.79	1.58	1.46	.55	.72	1.44	1.15	1.58	1.14	1.23	35
	25 - 26	.54	.17	.21	.26	.43	.14	.32	.46	.45	.32	.33	42
Totals for Period 9 July - 26 Aug. (12 «patterns»)		8.67	7.64	10.83	10.75	10.91	8.84	13.50	8.73	10.03	8.88	9.88	17.1

collected from tree to tree the number of trees sampled was increased to five from 2nd February 1956, the results being shown in Table 3. The ratios between the water collected for the individual trees is very great, but the reasons for these differences are not apparent. The amount of water flowing down the stem is, however, small compared with the total amount of precipitation. In the 18 months during which observations have been made, stem flow averaged only 7% of the rain falling on the trees. It is hoped that it will be possible to make more measurements of water-flow down the tree stems in the future.

THROUGHFALL AND INTERCEPTION

The sum of the drip from the trees and the flow down the stems is known as «Thr ughfall» and the difference between the estimated rainfall over the canopy and the throughfall is termed «Interception».

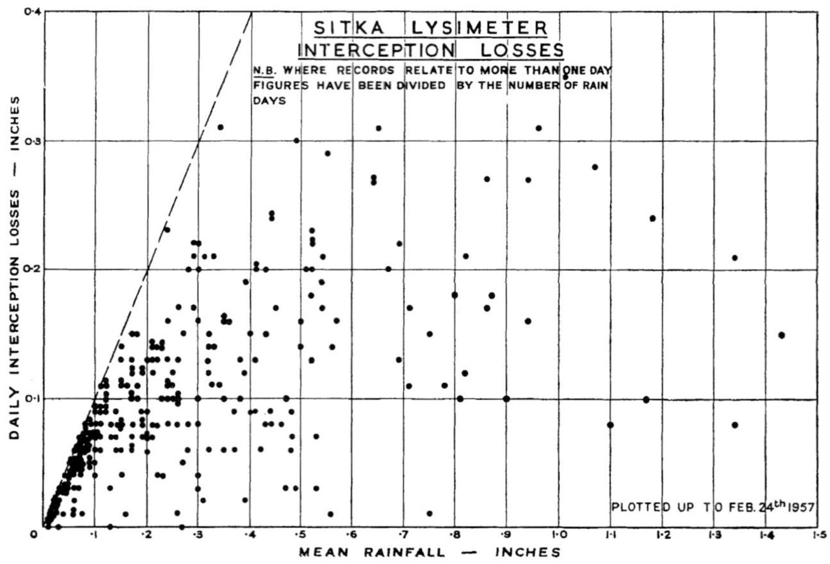

Fig. 2

In Fig. 2 the daily interceptions have been plotted against estimated rainfalls on the canopy. It will be seen that it is a scatter diagram with no definite trend.

On the other hand if the values are grouped into rainfalls on the canopy of 0-.1 inch, .11 to .20 inch . . . . . .51 to .60 inch and .61 inch and upwards for the summer and winter 6-months periods we obtain the results shown in Fig. 3, with a solid curve through the aggregate means. It will be seen that, on the average, the amount intercepted is not a constant ratio of the rainfall — it is greater with light rains of short duration than with heavy or prolonged rains. With falls of 0.2 inches per day the interception averages 50%, whilst for 1.0 inch per day it is only 19%.

It is impossible to say at this stage what significance there is in the departures of the seasonal means from the mean line, except that the interceptions in a dry summer seem to be the highest — which is the worst from a water supply point of view.

TABLE 3
Comparison of Catches of Tree Bole Gauges

	Tree Bole Gauge Catches-inches					Means		Estimated Tree-top Rainfall ins.	Mean of 3 tree-bole as percentage of rainfall
	B1	B2	B3	B4	B5	of 3 B1, B2, B3	of 5 B1-B5		
1955									
July 4 - Dec. 31	2.06	1.03	1.31	—	—	1.47	—	19.73	7.5
1956									
Jan. 1 - Feb. 1	.64	.37	.32	—	—	.44	—	5.45	8.1
Feb. 2 - July 8	1.07	.58	.66	.95	.18	.77	.69	13.58	5.7
July 9 - Aug. 26	1.13	.79	.69	1.23	.44	.87	.86	14.76	5.9
Aug. 27 - Dec. 31	1.95	.81	1.18	2.04	.37	1.31	1.27	16.68	7.8
Totals 2 Feb. - 31 Dec.	4.15	2.18	2.53	4.22	.99	2.95	2.82	45.02	6.3

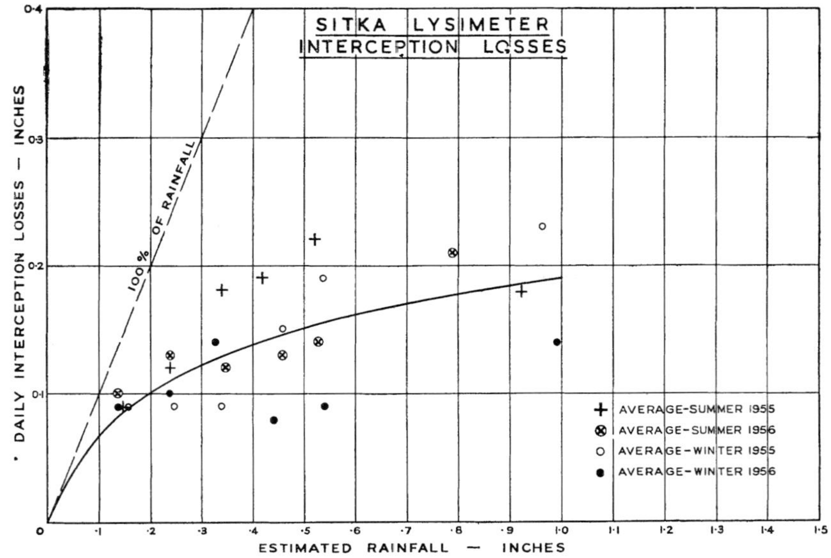

Fig. 3

On the other hand, part (if not the whole) of this apparent effect may be due to increasing shelter of the tree-top gauges as the trees became taller. This would lower the average value for the six gauges relative to the true rainfall on the plantation, but further investigation of this matter will have to await more accurate determination of the rainfall upon the canopy.

It is, however, noteworthy that the interceptions in the winter of 1955 were as high as those of the summer of 1956, despite the fact that theoretically the drying potential is vastly greater in summer than in winter.

Measurement of Runoff

From the lowest corner of the plot wall a six inch pipe has been taken to a 1,200 gallon(5, 450 litre) storage tank from which the drainwater is lead through a 2 ins. diameter semi-positive waterworks meter. Between the storage tank and the meter there is a strainer to prevent the conifer needles from choking the meter, and, on the downstream side of the meter the pipe is arranged so that the meter does not come into operation until the tank is approximately one-third full, so that the water cannot dribble through the meter. Each morning prior to reading the meter the observer lowers this flexible pipe so that the tank is fully drained off as quickly as possible. A similar meter has been fixed to the overflow from the tank, but the overflow has never yet come into operation during the course of the experiment.

The weekly runoffs from the lysimeter are shown in Fig. 4, along with the corresponding runoff from the catchment as a whole and from the grasscovered gauge A21. The weekly rainfall on rain gauge 21D is also shown to a smaller scale, but it should be noted that both runoffs and rainfalls are not strictly comparable because the general rainfall on the catchment is greater than that of gauge 21D, which in turn is greater than the estimated rainfall on the top of the plantation canopy.

407

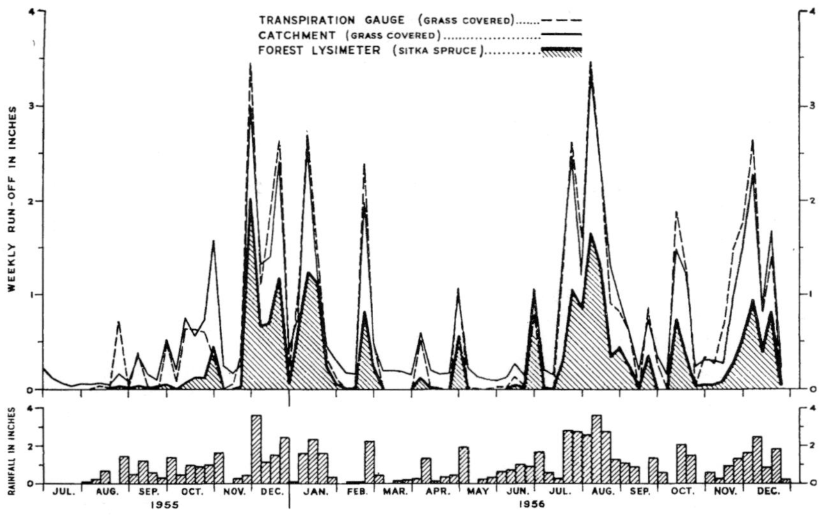

Fig. 4

It will be seen that the transpiration gauge and catchment runoffs are quite similar but the lysimeter runoffs are far less.

Preliminary Results

Preliminary results have already been published (Law 1956) and these are set out in Table 4, together with the corresponding results for the calendar year 1956. The first 12 months period was quite dry, with rainfall amounting to only 70% of the Meteorological Office estimate of the long-period average, and this has been divided into three portions, terminal dates being times when the soil moisture was thought to be at «field capacity»:

24 weeks	4 July to 18 December 1955
12 weeks	19 December to 11 March 1956, and
17 weeks	2 March to 8 July.

The calendar year 1956 commenced dry, but July and August were wet—the latter month particularly so— and the year's rainfall was approximately **95**% of the long period average.

In addition to the data for the Sitka plantation lysimeter there are also in Table 4 corresponding figures for evaporation from the evaporation tank and percolation gauges in the Climatological Station 200 yards to the north-east of the lysimeter, and the rainfall, runoff and losses over the catchment as a whole.

Noteworthy points in this table are:

(*a*) During the dry period from 4 July 1955 to 8 July 1956 the precipitation on the lysimeter amounted (in round figures) to 39 inches, but of this only 24 inches reached the forest floor and the runoff was less than 11 inches. The gross loss was, therefore, 28 inches, compared with 16.6 inches over the catchment area and

TABLE 4

Stocks Reservoir, Slaidburn, Yorkshire (W. R.)

Summary of data in successive periods from 4 July 1955 to 8 July 1956 and Calendar Year 1956

	12 months' Period with approximately 70% of average annual rainfall				Calendar year 1956 with 95% of average annual rainfall
	4 July to 18 Dec.	19 Dec. to 11 Mar.	12 Mar. to 8 July	TOTAL 4 July/55 to 8 July/56	
No. of Weeks	24	12	17	53	52
	ins.	ins.	ins.	ins.	ins.
Sitka Plantation Lysimeter					
(1) Mean of 3 Outside Gauges	17.29	12.68	10.96	40.93	53.96
(2) Mean of 3 Tree-top Gauges	15.35	11.06	10.18	36.59	46.99
(3) Estimated Mean Rainfall	16.32	11.87	10.57	38.76	50.47
(4) Estimated Interception	5.87	4.25	4.51	14.63	16.78
(5) Estimated Throughfall	10.45	7.62	6.06	24.13	33.69
(6) Lysimeter Runoff	3.55	6.30	0.91	10.76	16.14
(7) Gross Losses	12.77	5.57	9.66	28.00	34.33
(8) Net Losses without Interception	6.90	1.32	5.15	13.37	17.55
Experimental Results in Climatological Station					
(9) Evaporation from 6 feet Square Tank	8.98	(1.00)	8.40	18.38	16.97
(10) Losses from Grass-covered Percolation Gauge A21	7.45	(0.22)	7.86	15.53	15.70
(11) Losses from Grass-covered Percolation Gauge (kept irrigated) (Composite figures based upon Gauges B21, C21 and E21)	10.52	(1.05)	8.83	20.40	17.07
(12) Calculated evaporation from Water Surface by Penman's 1954 «Rome» formula	9.93	0.29	10.38	20.60	17.48
(13) Runoff from Percolation Gauge A21	10.11	12.83	2.61	25.55	39.34
Catchment Area of 9,253 Acres					
(14) Estimated General Rainfall	18.66	14.49	11.66	44.81	60.15
(15) Runoff	10.85	12.36	5.02	28.23	40.72
(16) Estimated losses	7.81	2.13	6.64	16.58	19.43

NOTE: The results in brackets are doubtfull, possibly due to melting snow and other causes.

15.5 inches on the grass-covered percolation gauge A21. In other words the losses in the lysimeter were 11 or 12 inches higher.

(*b*) The nett losses in the lysimeter (line 8) in the warmer periods are less than elsewhere (lines 9, 10, 11 and 16) — so it would appear that for considerable periods the trees were short of moisture, or on certain days the trees were not transpiring because the available energy was being used up in the evaporation of intercepted rainfall. In the calendar year 1956, however, the nett losses are greater than those of the percolation gauge A21 and it would seem that at 8 July 1956 there was still a soil moisture deficiency in the lysimeter, but that at no time during the year was there a deficiency great enough to restrict the transpiration rate, as occurred apparently in 1955.

(*c*) During the winter period 19 December 1955 to 11 March 1956 the nett loss in the lysimeter was only 1.32 inches, so it would not seem that there was much leakage, if any at all, through the clay base. Losses from the catchment and from the experimental gauges were of the same order, although there are certain unexplainable differences in the latter, probably due to melting snow, etc., but possibly due to the fact that gauges A21 and B21 have sloping surfaces facing into the prevailing wind (and A21 more steep than B21 at that time) and the rainfall on same not being measured absolutely accurately. This difficulty of measuring rainfall on a sloping surface has not yet been overcome and additional percolation gauges C21 and E21 have been formed with level surfaces.

Effect upon the Yield of the Catchment

The yield of a catchment is normally estimated in Great Britain as the runoff in a year having a rainfall of 80% of the average. In the case of the Stocks Reservoir catchment, where the experiments have been carried out, the average annual rainfall is estimated to be 62.9 inches and the nominal yield is therefore (80% of 62.9 inches) minus (17.5 inches evaporation losses) = 32.8 inches. An increase of 11 inches in the losses would therefore reduce the yield by one third. The nett effect is worse, however, because usually between 20% and 33% has to be given to the stream as compensation water, and the effective loss for supply purposes, resulting from 11 inches increase of the losses, is 42% to 50%, depending on the ratio of compensation water.

Watertightness of the Lysimeter Base

It is thought that the clay sub-soil of the lysimeter is impervious but there can be no certainty. As a check two subsidiary lysimeters L2 and L3, have been constructed to the south-west — L2 contains two tree stumps, and L3 one tree stump and one ill-shaped tree which is to be cut down. These subsidiary lysimeters have been surrounded by concrete walls similar to that of the main lysimeter and the intention is to cover them with plastic sheeting and then irrigate with measured quantities daily. If the differences between the irrigation and the runoff are significant, then the same may apply to the main lysimeter.

Futur Work

Alongside L2 and L3, three other subsidiary lysimeters have been formed, L4, L5 and L6, but each of these contains two good trees, and it is hoped, by suitable covers of plastic sheeting and irrigation, to estimate —

(*a*) what transpiration can take place if the trees are adequately provided with moisture;

(*b*) to what extent transpiration is reduced when intercepted moisture is being evaporated.

Consideration is also being given to the practicability of gauging one of the reservoir feeders, in the drainage area of which a considerable proportion of the planting by the Forestry Commission has taken place. This would, however, involve considerable expense and it would be another fifteen years before worthwhile results would be available.

The author hopes that in the near future the Nature Conservancy, the Forestry Commission and other interested bodies will commence similar experiments in larger forests, because it would seem that the main possibility of error in the present experiments is that they have been carried out in a small plantation, and in a large forest the evaporation of moisture hanging on the trees may not be so rapid.

REFERENCES

LAW, F. (1956) «The effect of Afforestation upon the Yield of Water Catchment Areas», British Association for the Advancement of Science, Sheffield, 1956, printed in *Journal of the British Waterworks Association*. November, 1956. pp. 489-494.

MILL, H. R. (1900) «The Development of Rainfall Measurement in the last Forty Years», in a publication entitled British Rainfall, 1900.

MONTHLY WEATHER REVIEW

CHARLES F. BROOKS, Editor.

VOL. 47, No. 9.
W. B. No. 695.

SEPTEMBER, 1919.

CLOSED Nov. 4, 1919
ISSUED Dec. 16, 1919

RAINFALL INTERCEPTION.

By ROBERT E. HORTON, Consulting Engineer.

[Dated: Albany, N. Y., Sept. 3, 1919.]

SYNOPSIS.—1. Rainfall interception represents a loss of precipitation which would otherwise be available to the soil.

2. The loss takes place through evaporative processes, but may, for convenience, be subdivided into *(a)* interception storage, and *(b)* evaporation during rain.

3. The amount of interception loss is primarily a function of the storage capacity of the plant surface, the duration of precipitation, and the evaporation rate during precipitation.

4. Since there is generally a fairly close correlation between shower duration and amount of precipitation, estimates of interception loss can, for practical purposes, be expressed in terms of precipitation amount per shower.

5. The interception storage loss for trees varies from 0.02 to 0.07 inch per shower, and approaches these values for well-developed crops.

6. The interception storage loss for trees in woods is greater, but the evaporation loss during rain is less than for trees in the open.

7. The percentage of total precipitation loss is greater in light than in heavy showers, ranging from nearly 100 per cent where the total rainfall does not exceed the interception storage capacity to about 25 per cent as an average constant rate for most trees in heavy rains of long duration.

8. Light showers are much more frequent than heavy ones, and the interception loss for a given precipitation in a month or season varies largely, according to the rainfall distribution.

9. Expressing the interception loss in terms of depth on the horizontal projected area shadowed by the vegetation, the loss per shower of a given amount is very nearly the same for various broad-leaved trees during the summer season.

10. The amount of water reaching the ground by running down the trunks of trees may amount to a relatively large volume when measured in gallons for a smooth bark tree in a long heavy rain. It is, however, a relatively small percentage, commonly 1 to 5 per cent, of the total precipitation. The percentage increases from zero in light showers to a maximum constant percentage in heavy showers of long duration.

11. Different interceptometers under the same tree will give fairly consistent results, if so placed that they do not receive direct rainfall, and if they stand under a complete leaf cover of average density.

12. So far as the experimental data go, there is little evidence of watershed effect or dripping of water from the periphery of the crown to a greater extent than through the crown itself.

13. The interception loss from needle-leaved trees, such as pines and hemlocks, is greater both as regards interception storage and evaporation during rain than from broad-leaved trees.

14. The average duration of showers of a given intensity is greatest in winter and the colder summer months, and least in midsummer or thunder-storm months, whereas the evaporation rate is greatest in midsummer and least in the colder months. As a result of the opposite effects of these two factors affecting interception loss, the average loss per shower of a given intensity seems to be nearly constant throughout the different months of the summer period, May to October, inclusive.

15. Data are insufficient for a final determination of the relative losses from trees in winter and in summer. Apparently the winter and summer losses for a given monthly precipitation for needle-leaved trees are about equal, whereas for deciduous, broad-leaved trees the winter interception loss appears to be about 50 per cent as great when the trees are defoliated as during the growing season.

16. Interception loss from full-grown field crops approaches in value that from trees, but owing to the short time during which crops stand on the ground in a fully developed stage of growth, the total annual interception loss from cropped areas is very much smaller than from wooded areas.

17. The average interception loss from 11 trees, excluding peripheral interceptometers and excluding hickory, for which the results are defective, during the summer of 1918 was 40 per cent of the precipitation.

147217—19——1

INTRODUCTION.

A large amount of data has been accumulated on this subject. There does not appear, however, to have been any thorough and complete analysis of all the available data, and it is unfortunate that not even a reasonably complete digest of the experimental observations is available in English. Furthermore, the processes involved do not seem to have been carefully analyzed, and, as a result, many of the experimental data are not in a form permitting interpretation of the results to the best advantage.

The subject is one on which it is somewhat difficult to experiment in a satisfactory manner, and it is not surprising that the conclusions hitherto drawn by different authorities are sometimes at variance, and many of the data are seemingly discordant.

TABLE No. 1.—*Summary of rainfall interception data for forests.*

Wood.	Per cent loss.			Station, duration, authority, etc.
	Winter.	Summer.	Year.	
(1)	(2)	(3)	(4)	(5)
Mixed			25	16 German stations(ref. 1), p. 106,
Do			16	3 Swiss stations (ref. 1) 16-year average, p. 106.
(?)			16	Nancy, Bellefontaine, 11 years.
Evergreens	20.90	[1] 26	[2] 23	German stations (ref. 1), p. 107.
Deciduous	19.60	[1] 35	[2] 26	Do.
General average	21.40	[1] 30	[2] 25	
Larch			15	Swiss stations, 12 years (ref. 1), p. 131.
Spruce			23	
Beech			10	
Do			24	Prussian stations (ref. 1), p. 131.
Spruce			27.7	Do.
Do			39	Raphael Zon (ref. 10), p. 230.
Broad leaved			13	Raphael Zon.
Beech			15	Raphael Zon, from Ney; crown loss after deducting trunk run-off (ref. 10), p. 230.
Pine			20	
Spruce			33.33	
Blue beech	5.85		8.48	Mathieu at Nancy, 11 years (ref. 10, p. 231).
Beech, 20-year			2	Bühler, 2 to 3 years (ref. 10, p. 230).
Beech, 50-year			27	
Beech, 60-year			23	
Beech, 90-year			17	

a Forest influences.
b Final report National Waterways Commission.
[1] May–Sept. 5/12 year.
[2] Approximate, deduced by proportion from columns (2) and (3).

Table No. 1 contains a digest of the results of different experiments, and of the conclusions of different authorities therefrom.

In comparison with some of the European results, the following statement by H. S. Graves is pertinent: (Monthly Weather Review, Dec., 1914, **42:** 671).

Many and exact measurements have demonstrated that a forest cover intercepts from 15 to 80 per cent of the precipitation, according to the species of trees, density of the stand, age of the forest, and other factors. Thus pine forests of the North intercept only about 20 per cent, spruce about 40 per cent, and fir nearly 60 per cent of the total precipitation that falls in the open, the amount that runs off along the trunks in some species is very small, less than 1 per cent, in others, beech for instance, it is 5 per cent.

Harrington in Forest Influences, (**1**) says:

"It seems that the deciduous trees withhold more of the precipitation through the entire year than do the evergreens."

Zon (**10**) states:

"As a result of a great number of investigations it may be assumed that coniferous forests intercept more precipitation than broad-leaved forests."

Imbeaux gives the opinion that the interception loss is 50 per cent in coniferous, and 20 to 30 in deciduous woods. (Essai-Programme of Hydrology.)

PHYSICS OF RAINFALL INTERCEPTION.

It is a matter of common observation that the percentage of precipitation reaching the ground in forest or on fields with growing crops is very small in the earliest stages of a rain, increasing as the duration of the storm increases, the total amount reaching the ground being small for short light showers, and increasing for severe prolonged storms. General observations also lead to the following conclusions:

When rain begins, drops striking leaves are mostly retained, spreading over the leaf surfaces in a thin layer or collecting in drops or blotches at points, edges, or on ridges or in depressions of the leaf surface. Only a meager spattered fall reaches the ground, until the leaf surfaces have retained a certain volume of water, dependent on the position of the leaf surface, whether horizontal or inclined, on the form of the leaf, and on the surface tension relations between the water and the leaf surface, on the wind velocity, the intensity of the rainfall, and the size and impact of the falling drops. When the maximum surface storage capacity for a given leaf is reached, added water striking the leaf causes one after another of the drops to accumulate on the leaf edges at the lower points. Each drop grows in size (the air being still) until the weight of the drop overbalances the surface tension between the drop and the leaf film, when it falls, perhaps to the ground, perhaps to a lower leaf hitherto more sheltered. These drops may also be shaken off by wind or by impact of rain on the leaf. The leaf system temporarily stores the precipitation, transforming the original rain drops usually into larger drops. In the meantime the films and drops on the leaves are freely exposed to evaporation.

It is evident that the amount of interception in a given shower comprises two elements. The first may be called interception storage. If the shower continues, and its volume is sufficient, the leaves and branches will reach a state where no more water can be stored on their surfaces. Thereafter, if there is no wind, the rain would drop off as fast as it fell, were it not for the fact that even during rain there is a considerable evaporation loss from the enormous wet surface exposed by the tree and its foliage. As long as this evaporation loss continues and after the interception storage is filled, the amount of rain reaching the ground is measured by the difference between the rate of rainfall and the evaporation loss. When the rain ceases the interception storage still remains on the tree and is subsequently lost by evaporation. If there is wind accompanying the rain, then, owing to motion of the leaves and branches, it is probable that the maximum interception storage capacity for the given tree is materially reduced as compared with still air conditions. Furthermore, in such a case, after the rain has ceased, a part of the interception storage remaining on the tree may be shaken off by the wind, and the storage loss in such a case is measured only by the portion of the interception storage which is lost by evaporation and is not shaken off the tree after the rain has ceased. One effect of wind is, therefore, to reduce materially the interception storage. As regards evaporation loss during rain, the effect of wind is, of course, to increase it materially.

The difference between interception losses with and without wind is illustrated by the accompanying figure 1. If there is no wind, and the rain falls gently, it is nearly all intercepted until the interception storage capacity is reached—thereafter in the absence of wind, evaporation proceeds slowly, the remainder of the precipitation dripping off the leaves, generally in large drops, and reaching the ground. For a sharp shower with wind, the interception storage is filled only to a limited extent, drops being temporarily stored on the leaves and then shaken off. The evaporation rate may, however, be materially increased, so that while the depth of interception during the earlier part of the storm is likely to be less than for a storm without wind, the total interception depth for a long-continued storm with wind may be the greater of the two, owing to increased evaporation.

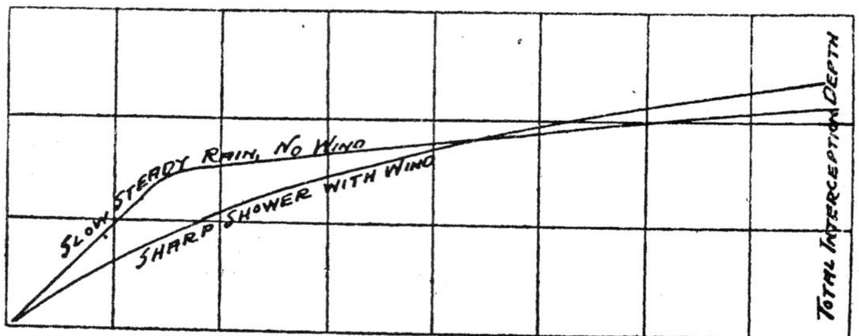

FIG. 1.—The effect of wind and character of shower on interception loss.

The maximum interception storage can be approximately determined from records taken for short, light showers, during which nearly all the rainfall was intercepted.

The fundamental storage equation—inflow equals outflow plus gain, or, minus loss of storage—applies to this process, but after the leaf storage is saturated, leaves are freely exposed to evaporation, the inflow rate minus evaporation rate equals outflow rate.

For the storm as a whole, the following relation holds: Inflow minus total evaporation during storm equals total outflow plus leaf storage at end of storm.

It will be seen that during the greater portion of a long rain capable of producing a severe storm, run-off rate equals precipitation rate minus evaporation rate. In general, for a storm sufficient to saturate the leaf storage.

Total interception equals leaf storage capacity plus evaporation loss during the storm.

Owing to the great extent of leaf surface, the evaporation loss from leaf surfaces is much greater than from the projected area shaded by the tree, but is likely to be relatively small per unit of exposed surface compared with the evaporation rate in fair weather, owing to the smaller saturation deficit common during rain, and the approximate equality between leaf-surface and air temperature.

The superficial storage capacity of a plant is approximately constant at a given stage of growth or leaf development. By storage capacity is meant the depth of water on the projected area covered by the plant which can be stored or detained on the plant surface in still air.

If T = duration of the storm in hours.

E_r = evaporation rate in inches depth per hour during the storm.

K_1 = ratio of the evaporation surface to the projected area.

S_j = interception storage capacity in inches depth on the projected area.

P = precipitation rate per unit of time.

Then the total interception loss is—

$$J = S_j + K_1 E_r T \quad (1)$$

and the percentage loss—

$$\frac{J}{PT} = \frac{S_j + K_1 E_r T}{PT} \quad (2)$$

This formula indicates that the percentage loss decreases as the duration of the storm increases; and furthermore, since the numerator is independent of the rain intensity, the percentage loss decreases as the intensity of the storm increases.

Following this reasoning, we should expect the percentage of rainfall intercepted to be less for heavy rain than for light rains. Zon states this to be a fact (*a*) (**10**) but he does not give any data in support of his conclusions. Most of the data hitherto published in the form in which presented are quite doscordnat as regards the relation of rainfall intensity to amount of interception, but this is quite certainly due in part to the presentation of results in monthly or seasonal totals regardless of rainfall distribution.

THE CHARACTER OF INTERCEPTION STORAGE ON DIFFERENT PLANTS.

Observation and sketches were made of the amount of water accumulated on different plant surfaces after a rainfall on July 12, 1915, of 0.12 inch at night, with no wind. Sketches of typical leaves, showing the mode of water storage or accumulation thereon, are contained in the accompanying fig. 2.

Leaves of different plants vary greatly in the manner in which rain falling on them is retained. Many leaves become wetted over their entire upper surface with a thin film of water which is not shown in the sketches. There does not seem to be any regular rule as to this, as leaves which appear bright and waxy as well as others having dull surfaces both become wetted in some cases, whereas in other cases, according to the configuration of the surface of the leaf, water accumulates on both classes of leaves only in drops or blotches. Of course, water tends to accumulate in capillary spaces of all forms. In some cases where the entire leaf surface becomes wetted the film thickens in the depressions along the lines of the veins. More generally the entire leaf surface does not become wetted, and in such cases the water which accumulates in drops on the leaf surfaces is mostly concentrated on the plateaus or ridges between the lines of veins. Apparently the majority of leaves do not become appreciably wetted on the underside, excepting where drops and blotches overflow from the edges.

The possibilities of interception storage are revealed by observations by the author on July 12, 1915, of more than 100 water drops retained per leaf on leaves of horse-chestnut, oak, and aspen, in addition to blotches and films, and 100 or more drops on single stems of rye.

During the defoliated season the author has observed large drops about 1 inch apart clinging after a cold rain to the underside of every twig or horizontal branch of a maple. In warm weather this water runs off more easily and the interception storage is then largely on leaf surfaces.

An approximate estimate of the interception storage can be arrived at by counting the number of drops per unit of plant surface, estimating their diameter and volume.

Volumes of small spheres per million

Diameter.	Volume per million.
In inches.	*Cubic inches.*
1/32	15.98
1/16	127.83
3/32	450.40
1/8	1,022.60
5/32	2,085.00
3/16	3,451.40
7/32	5,721.20
1/4	8,181.20

For example, a crop of rye containing 3,000 stalks per acre, with storage equal to one hundred and twenty $\frac{1}{8}$-inch drops per stalk, would contain, exclusive of water in the heads, 213.1 cubic feet of interception storage per acre. This is equivalent to a depth of 0.047 inch on the surface.

Again, a tree having one-half million leaves, with an average of twenty $\frac{1}{8}$-inch diameter drops per leaf, would contain 5.92 cubic feet of interception storage. If the crown diameter was 40 feet, the projected area being 1,256 square feet, the interception storage would be equivalent to 0.0564 inch.

EXPERIMENTAL DATA OF INTERCEPTION.

In order to determine the numerical factors for calculating interception losses, an effort was first made to utilize existing experimental data. For this purpose, the interception records for the Adlisberg and Haidenhaus forest meteorological stations were analyzed for the years 1889 and 1890. The recorded precipitations on each day when rain fell were grouped together according to the amount of precipitation at the station in the open, averages for each group were taken both for the station in open and for stations in the forest.

TABLES 2, *A*, *B*, *C*.—*Analysis of Adlisberg records precipitation arranged by daily amounts caught by gage in open, in inches.*

Range mm. per day.	Average inches per day.	Beech (1).	Beech (2).	Beech (3).	Average beeches (2) and (3)
(1)	(2)	(3)	(4)	(5)	(6)
B—BEECHES, SUMMER PERCENTAGES CAUGHT.					
0–15	0.10	[1] 92.80	77.30	79.60	78.45
5–10	.30	95.51	78.59	83.91	81.25
10–20	.60	93.41	80.50	87.50	84.00
Over 20	.90	86.20	80.44	85.30	[2] 82.37
A—BEECHES, WINTER PERCENTAGES.					
0–5	0.10	97.30	72.05	80.43	71.24
5–10	.30	100.00	64.15	68.27	66.21
10–20	.60	70.63	42.25	51.88	47.06
Over 20	.90	87.51	....	....	....
C—FIRS, PERCENTAGES CAUGHT.					
		Winter.	*Summer.*		
0–5	0.10	42.95	52.50	....	....
5–10	.30	44.65	68.90	....	....
10–20	.60	54.76	81.10	....	....
Over 20	.90	75.74	79.80	....	....

[1] For full two-year period. [2] Number of observations small.

Winter and summer records were separated, the six months period, November to April, inclusive, being counted as winter. Tables No. 2, A, B, and C contain a summary of these studies. The results for fir trees, Table 2 C, show a fairly consistent increase in the percentage of precipitation reaching the ground as the rainfall in inches per day increases. This is true in the case of fir for both winter and summer conditions. In the case of beech trees, one station was maintained throughout the whole period, and two additional stations during the year 1890.

Station No. 1 for beech, which covers the entire period, shows in the case of many storms a measured precipitation greater than that in the open. The average percentage of rainfall reaching the ground for station No. 1 under beech is indicated in column (2) of Tables 2 B and 2 A, respectively. It will be noted that the percentage reaching the ground under beech, station No. 1, is much larger than the percentage reaching the ground under beeches Nos. 2 and 3, and this is indicated by columns (3) and (4) of the same tables.

If station No. 1, for beech, is included, the average for the three stations does not show any consistent relation between rainfall intensity and amount of interception loss. Excluding station No. 1, the record for beeches shows a fairly consistent but small increase in the percentage of rainfall reaching the ground, with increased rainfall rate during the summer season, as indicated by column (5) of Table 2 B. The results fail, however, to show a consistent increase; and show, in fact, a consistent apparent decrease in percentages for the mean of beech at stations Nos. 2 and 3, for winter conditions. In

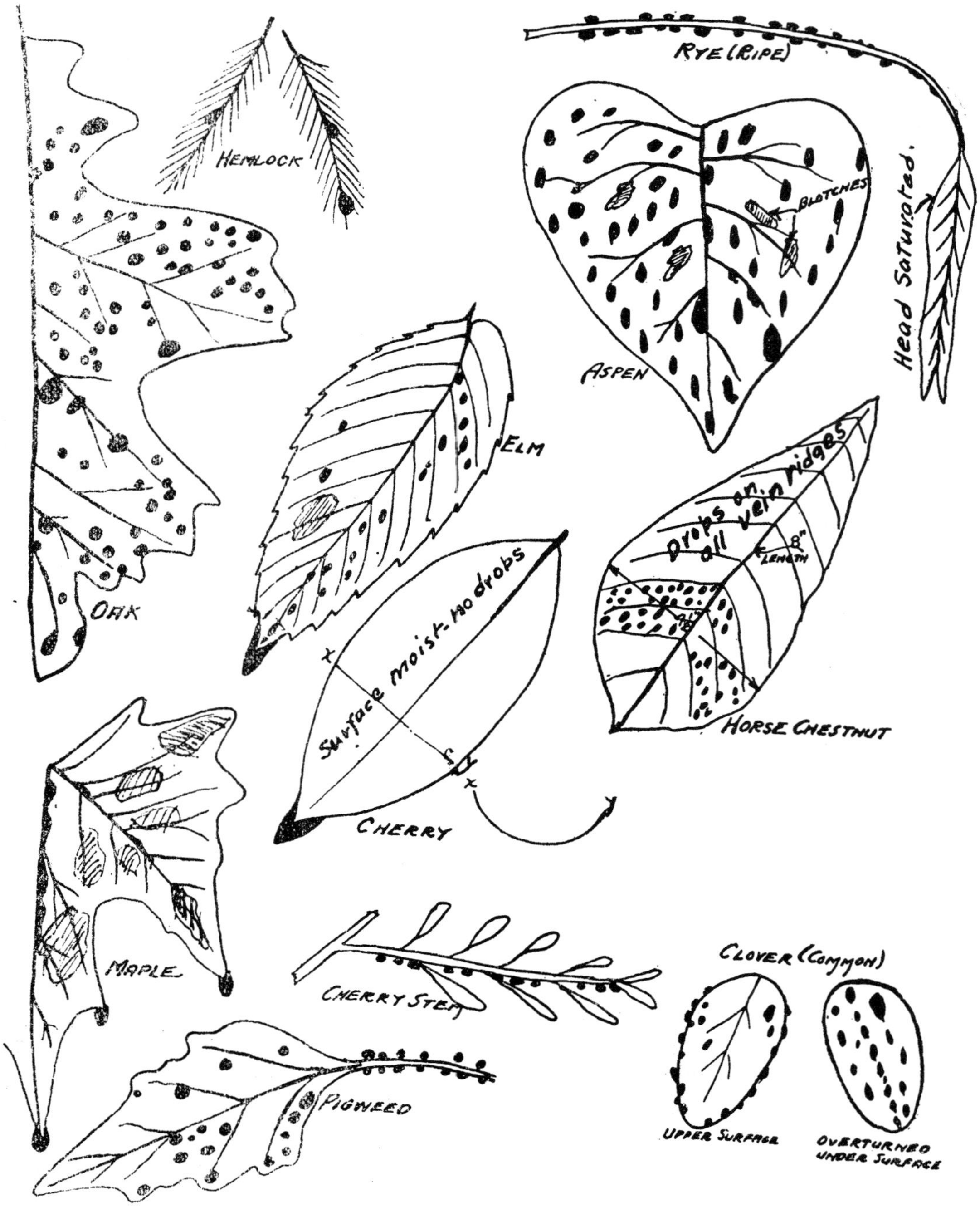

FIG. 2.—Interception storage on various plants.

general it appears that the *a priori* conclusion is fairly confirmed by these experiments as well as could be expected, taking into account the evident large experimental errors existing.

Interception by beech crowns of different ages (Bühler).

	Age of stand, years.			
	20	50	60	80
Proportion reaching ground	0.98	0.73	0.77	0.83
Proportion intercepted	.02	.27	.23	.17

Taking the Adlisberg records by months, arranged in order of magnitude by rainfall rates, we find, for the summer, the following:

TABLE NO. 3.—*Monthly precipitation caught under trees, per cent of that in the open, Adlisberg.*

Inches.	Days.	Rate.	Per cent caught.				
			Fir.	Beech (1).	Beech (2).	Beech (3).	Mean, Beech (2) and (3).
(1)	(2)	(3)	(4)	(5)	(6)	(7)	(8)
2.23	20	0.111	57.2	90.7	73.4	78.5	75.9
4.33	20	.217	77.2	107.2	77.5	89.4	83.4
4.21	17	.240	72.6	94.0	75.4	83.0	79.2
5.08	14	.360	95.2	87.0	82.6	85.8	84.2
1.47	4	.370	90.2	96.0	82.6	97.6	90.2
9.42	19	.480	78.6	94.4	74.9	80.0	77.5

This table shows a general tendency to increase in percentage of rainfall caught by gage in forest as rainfall rate increases. The same is true at Haidenhaus, as indicated by Table No. 4.

The Adlisberg and Haidenhaus records both show an apparent decrease in percentage caught in the forest in winter as the rate per storm increases.

TABLE NO. 4.—*Haidenhaus interception data; analysis of monthly records on basis of average rainfall rate per storm day.*

Precipitation in open.			Precipitation in forest, per cent of open.	
Inches.	Days.	Inches per day.	Deciduous.	Evergreen.
(1)	(2)	(3)	(4)	(5)
WINTER.				
0.18	3	0.06	102.2	22.2
1.16	11	.105	86.8	56.2
1.33	3	.11	157.0	78.0
2.62	22	.12	73.4	26.3
3.02	17	.17	80.2	53.7
SUMMER.				
2.65	18	0.146	61.7	42.8
4.73	21	.225	70.2	45.6
1.55	6	.26	70.3	57.6
4.89	15	.326	70.0	56.0
5.36	13	.41	71.0	62.8
9.30	20	.46	77.2	58.2

THE AUTHOR'S EXPERIMENTS ON INTERCEPTION.

The data of interception thus far reviewed are in the form of annual, monthly, or, at best, daily averages. Published data of rainfall interception in individual showers are meager. The following observations were made by Seckendorff during a continuous downpour of rain, which lasted from the morning of June 12th to the night of June 14th. The total precipitation was 52.6 millimeters (2.07 inches).[1]

Interception of rainfall by trees.

Tree.	Precipitation in inches.	Percentage reaching ground.	
		Not including water running down tree trunk.	Including water running down tree trunk.
Beech	2.07	54.0	61.6
Oak	2.07	62.5	68.9
Maple	2.07	65.2	69.4
Pine	2.07	30.6	31.6

In order to provide data for analysis on the basis of individual showers the experiments described below were carried out.

TABLE 5.—*Summary of Ebermayer's experiments on rainfall interception.*

Station.	Forest.	Precipitation in open, inches per month.		Percentage caught in forest.		Loss per cent.		Precipitation in forest.	
		Winter.	Summer.	Winter.	Summer.	Winter.	Summer.	Winter.	Summer.
(1)	(2)	(3)	(4)	(5)	(6)	(7)	(8)	(9)	(10)
Seeshaupt	40-year-old fir	2.55	3.28	68.6	76.2	41.4	23.8	1.75	2.50
Duschelberg	40-year-old pine	6.79	2.73	73.6	72.5	26.4	27.5	5.00	1.98
Ebrach	50-year-old pine	2.86	1.61	90.6	69.6	9.4	30.4	2.59	1.12
Altenfurth	Wild pine	2.34	1.83	75.4	68.9	24.6	31.1	1.765	1.26
Mean, evergreens		3.64	2.36	77.05	71.8	22.95	28.2	2.78	1.72
Johanneskreutz	60-year old beech	3.49	3.16	72.5	79.0	27.5	21.0	2.53	2.50
Rohrbrunn	60-year-old beech	4.04	2.96	80.0	82.4	20.0	17.6	3.23	2.44
Mean, beech		3.76	3.06	76.25	80.7	23.75	19.3	2.88	2.47

[1] Lueger, Wasserversorgung der Städte.

During the period July to November, inclusive, 1917, and April to October, inclusive, 1918, interceptometers were maintained under trees of various kinds at the hydrologic laboratory of the author near Albany, N. Y. The accompanying map, fig. 3, shows the positions of the different trees and the accompanying Table No. 6 gives the size of each tree and other details. The rain gages used as interceptometers were galvanized iron pans, each 17 inches in diameter and 5 inches deep. A $\frac{1}{2}$-inch pipe nipple was secured in the bottom of each pan near the side, the pan was supported at a height of about 1 foot above ground, and the nipple was inserted in the neck of a 1-gallon glass bottle. Under the major portions of the crowns of the trees there was complete leaf cover, but varying in thickness or density. The interceptometers were placed as nearly as possible under complete leaf-cover of average thickness.

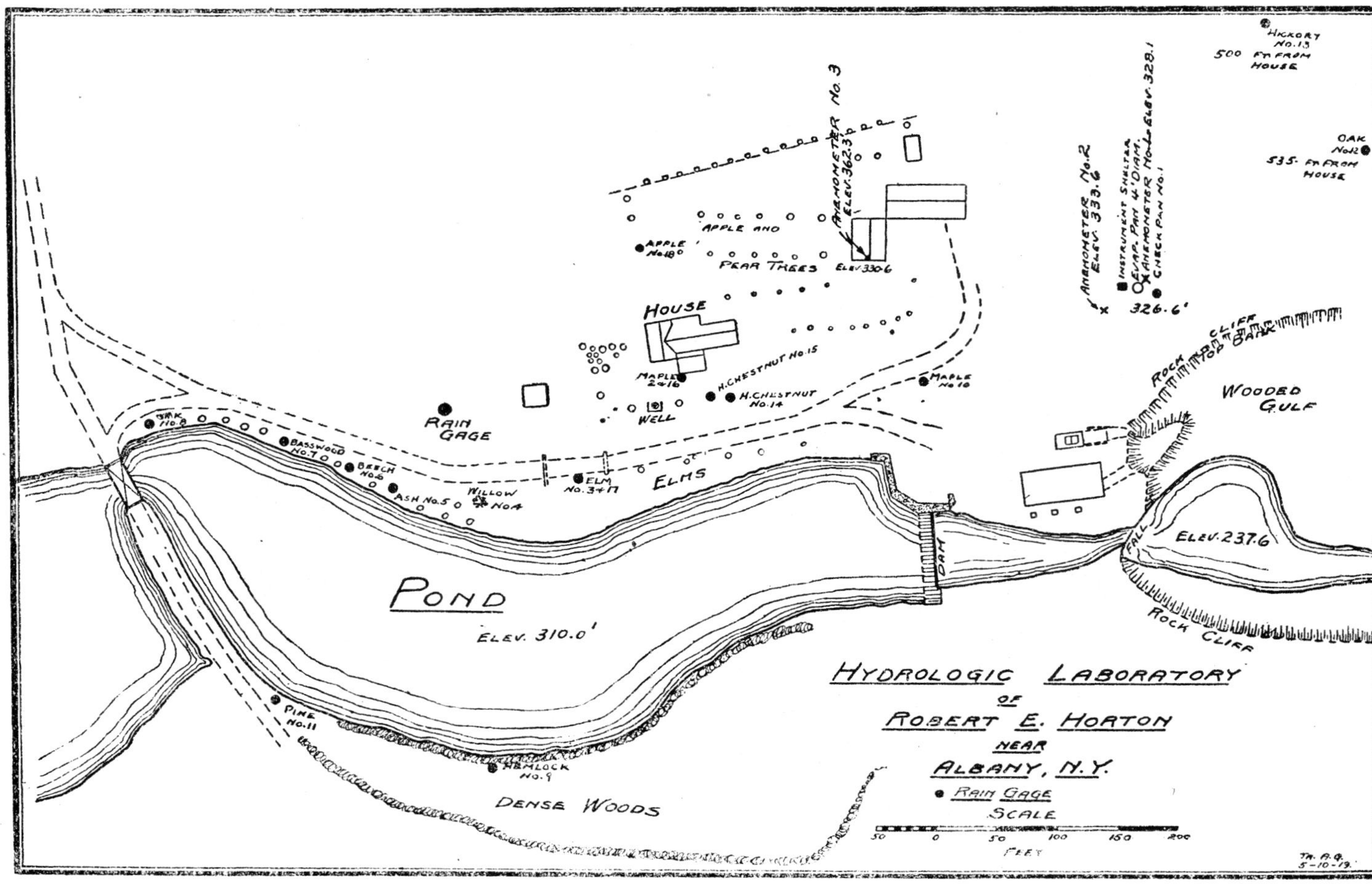

FIG. 3.—Map of author's hydrologic laboratory near Albany, N. Y.

Duplicate interceptometers were placed under maple, horse-chestnut, and elm trees, one in each case being near the trunk or about midway between the trunk and periphery of the tree, the other being just within the periphery. The peripheral interceptometers probably received direct rainfall rather than drippings from the tree, especially in the case of the elm, as the branches of this tree were 15 feet or more above ground, and the tree was in an exposed position, with the interceptometer on the south side, so that rains from the south or southwest falling at an angle with the wind could not be prevented from entering the gage directly.

In order to compare the catch by the interceptometers with that from an ordinary rain gage, a check interceptometer was maintained in the instrument inclosure alongside of a Friez tipping-bucket rain gage. This interceptometer gave readings practically identical in all cases with those obtained from the Friez rain gage.

A triangle of 3 rain gages was used, the rain gages being in the positions indicated on the map (fig. 3). There were nearly always slight differences in the amounts of rain caught by these gages, and in reducing the interceptometer results the rainfall on the tree crown has been taken as equal to that indicated by the mean of the three gages, although it is possible that a somewhat more accurate result might have been obtained by applying the results of a given rain gage to the records from trees to which the given rain gage stood nearest. The interceptometers were read in each instance as soon as practicable after the rain ceased, usually within an hour or two at most. Measurements were not, however, taken for each temporary cessation of rainfall. If, for example, two showers occurred separated by a rainless interval of not to exceed one hour, the rainfall for both showers was included in a single measurement.

The accompanying photographs, figures 4 to 9, inclusive, show several of the interceptometers and the trees in conjunction with them.

During the early part of 1917 the depths in the interceptometer bottles were measured with a rain-gage stick, the bottles having been previously calibrated by weighing. Later the water caught in the bottles has been measured in a calibrated can. The number of canfuls and the fraction of a canful, measured with a rain-gage stick, were recorded in each instance. The can was carefully calibrated for different depths by weighing on an accurate torsion balance. This method of measuring the water

FIG. 4.—Evaporation station and check pan.

FIG. 5.—Interceptometer under hemlocks.

FIG. 6.—Interceptometer under willow shrubs.

FIG. 7.—Interceptometer under elm.

FIG. 8.—Interceptometer under elm, near view.

FIG. 9.—Interceptometer under ash, beech, and basswood.

caught in the tree-trunk tubs has been used throughout the work.

TABLE No. 6.—*Data for interceptometers.*

Pan No.	Tree.	Diameter of trunk.	Height.	Interceptometer to trunk C to C (A).	Interceptometer to periphery (B).[1]	Projected area.	Depth on area of 2 quarts =115.5 cubic inches.
		Feet.	*Feet.*	*Feet.*	*Feet.*	*Sq. ft.*	
1	Test gage						
2	Maple (house)	1.00		1½ sw	14	620	0.00130
3	Elm	1.95	55	8 se	13	1,240	.00065
4	Willow			1½ s	2	64	.01250
5	Ash	1.95		4½ n	18	1,320	.00061
6	Beech	.70		4½ ne	{ 7, 14 }	200	.00401
7	Basswood	.75		8 n	{ 0, 7 }	240	.00335
8	Oak	2.90	30	5 nw	14	625	.00128
9	Hemlock	1.40	42	5 se	9	350	.00229
10	Maple	.60	47	4 s	6½	285	.00282
11	Pine	2.10	65	4 n	13	1,035	.00078
12	Oak	2.40	65	12 sw	13	1,440	.00056
13	Hickory	1.45	50	6½ sw	11	950	.00085
14	Horse chestnut	1.90		13 ne	2½	780	.00103
15	....do	1.40		3 nw	12	435	.00185
16	Maple (house)	1.00		10.5 sw	5	620	.00130
17	Elm	1.95	55	12 s	9	1,240	.00065
18	Apple	1.85	30	5½ ne	13	845	.00095

[1] Where two values are given the greater one is to outside edge of foliage. Smaller one is to edge of foliage draining to trunk.

WATER RETURNED BY TREE TRUNKS.

In this study the data cited thus far relate to interception by tree crowns alone. In many instances, some portion of the intercepted water runs down the tree branches and trunk and so ultimately reaches the ground. The total interception loss by trees was determined by Riegler at Nancy by the use of gages of the same area as the tree crown, arrangements being made to include in the catch of the gage the part of the rainfall which flowed down the trunk. Riegler's experiments cited by Harrington have been reduced to percentages, and are presented in Table No. 7. These experiments also indicate the small differences in interception by various kinds of broad-leaved trees. The portion of the intercepted rainfall which reaches the ground by way of the tree trunk is apparently much smaller for evergreen than for broad-leaved trees. Zon states that the percentage of the total rainfall passing down the tree trunk varies from 0.7 of 1 per cent to 3 per cent for evergreens, and may be as high as 15 per cent for broad-leaved trees.

TABLE No. 7.—*Riegler's experiments on interception by trees.*

[Rain falling on tree crown=100 per cent.]

	Rain falling through crown on soil.	Rain running off trunk.	Total per cent reaching soil.	Loss per cent.
(1)	(2)	(3)	(4)	(5)
Beech	65.4	12.8	78.2	21.8
Oak	73.6	5.7	79.3	20.7
Maple	71.5	6.0	77.5	22.5
Spruce	39.8	1.4	41.2	58.8

NOTE.—Rain gage equal in area to tree crown. Results for spruce void, because part of rain ran off tips of outward inclined branches and was not caught by gage.—Harrington, Forest Influences, p. 133.

In order to determine the amount of water running down the trunks of trees, in the author's experiments, a small lead trough was constructed around each tree trunk, as shown on fig. 10. The troughs were made of lead flashing about one-sixteenth inch thick, cut into strips 2½ inches wide. The strip was first rolled while straight into approximately the form shown in the cross-section B, figure 10. The straight trough was filled with sand to preserve its form, and was wound around a prepared portion of the tree trunk, usually about 2 feet above ground. The tree trunk was prepared by removing rough scales and smoothing down the bark, care being taken not to make any deep incisions which would injure the tree. The trough was first tacked to the tree on the side opposite the pan, about midlength of the trough, then each end was carefully wound around to the opposite side in such a manner as to give the trough a slight inclination. The wider side of the trough which rested against the tree was nailed at various points with small nails, and the edge of the lead caulked as tightly as possible into all crevices and irregularities in the trunk.

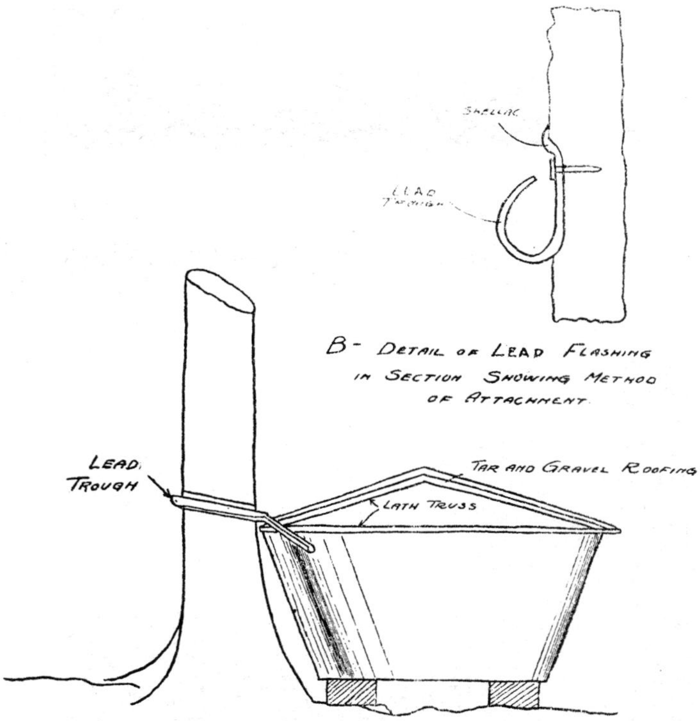

FIG. 10.—Typical tree-truck interceptometer with section of tub and cover.

The sand was then removed from the trough, the shorter end was bent over and into the longer projecting end of the trough, and the longer end was bent down so as to convey the water into the catch pan.

In 1917 melted paraffin was used to secure a water-tight joint between the upper edge of the lead and the bark of the tree. This would work well for a while, but afterwards would scale off, requiring frequent renewal to prevent leakage.

During 1918 several coats of thick shellac were used instead, with better results. The outside edge of the trough was bent over so as to leave an opening about one-fourth of an inch wide so as to prevent direct rainfall entering the trough in any considerable quantity. Covered 5-gallon galvanized pails were first used to catch the run-off, but it was found that for nearly all trees these would overflow, in a rain of a half inch or more, and large galvanized iron wash tubs were substituted, these tubs having a capacity of about 35 gallons each. Even then in some cases the tubs under certain trees would overflow during very heavy rains.

To prevent direct rainfall entering the tubs, and also to reduce evaporation, covers were made by constructing a light truss work of lath on which heavy tar and graveled roofing paper was tacked, thus providing a light, strong roof with inclination sufficient to carry the rain off readily, the roofing paper being lapped over the lead trough where the latter passed over the edge of the tub. The covers are held in position by weights and wires.

Early in the investigation it was found that there were surprisingly large variations in the amount of water caught in the tubs under different trees, and it soon became evident that smooth bark trees carry relatively large quantities of water down their trunks, while shag dark color, that from pine and hemlock being nearly as black as molasses.

The volume of water caught from the trunk of a large smooth bark tree in a heavy shower was often 20 to 30 gallons. How this occurs is esily understood when one considers than a film 0.01 inch thick flowing down the trunk of a tree 3 feet in circumference at 10 feet per minute amounts to 216 cubic inches per hour. As the results subsequently given show, the water running down the tree trunks, when reduced to equivalent depth on the projected area of the tree crown, is relatively slight, as appears from Tables Nos. 8 and 9, which show the amounts of precipitation caught by the tree trunk interceptometers.

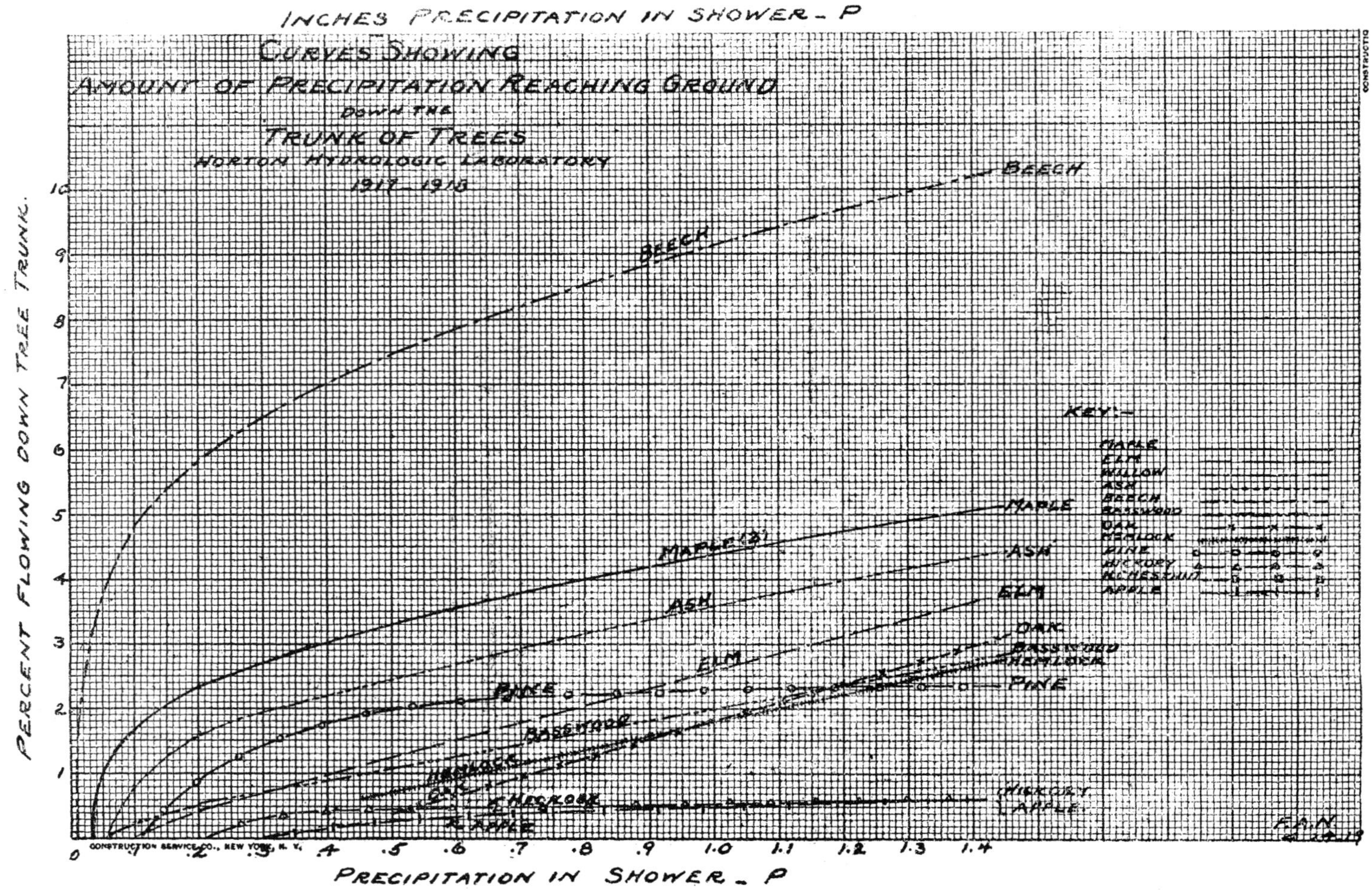

FIG. 11.—Summary of curves showing the amount of precipitation reaching ground down the tree trunks.

bark hickories, oaks, pines, and hemlocks permit but little water to pass down the trunks of the trees.

Since the trees stood in the open or in hedges they were more exposed than similar trees in a dense forest. In the case of a rain with driving wind, striking the exposed side of the tree, it is probable that a not inconsiderable portion of the water running down the trunks resulted from rain which struck the trunks directly, whereas, in the case of trees in dense forests this would not occur to any great extent. Close observation of the trees, and comparison of the results in rains which descended vertically and in strong winds indicate that this condition, while marked in some rains, does not usually prevail, and it may be safely assumed that nearly all the water running down the trunks would also have run down the trunks of trees in the forest. In nearly all cases, the water running down the tree trunks was of exceedingly

TABLE 8.—*Summary of amount of water in inches of precipitation flowing down trunk of tree for storms of various magnitudes.*

Number showers.	Kind of tree.	Precipitation (inches.)					
		0–0.05	0.05–0.10	0.10–0.30	0.30–0.60	0.60–1.0	1.0–2.0
2	Maple	T.	0.001	0.005	0.014	0.038	0.078
10	.....do	T.	.001	.004	.010	.020	.053
3	Elm	T.	T.	.001	.004	.014	.042
17	.....do	....	....	....	....	....	....
4	Willow	....	....	....	....	....	....
5	Ash	T.	T.	.003	.010	.017	.052
6	Beech	.001	.003	.011	.028	.064	.127
7	Basswood	T.	T.	.001	.003	.008	.032
8	Oak	....	....	....	....	....	....
12	.....do	T.	T.	T.	.001	.005	.033
9	Hemlock	T.	T.	.001	.002	.007	.030
11	Pine	T.	T.	.001	.008	.019	.027
13	Hickory	T.	T.	T.	.002	.004	.007
14	Horse-chestnut	....	....	....	....	....	....
15	.....do	....	....	....	....	....	....
18	Apple	T.	T.	T.	.001	.003	.007

TABLE No. 9.—*Summary of amount of water in per cent of total precipitation per shower flowing down trunk of tree for storms of various magnitudes.*

Number showers.	Kind of tree.	Precipitation in shower (inches).					
		0-0.05	0.05-0.10	0.10-0.30	0.30-0.60	0.60-1.0	1.0-2.0
2	Maple	T.	1.5	2.7	3.3	5.2	6.1
10	.....do	T.	1.5	2.2	2.4	2.5	3.8
	Mean	T.	1.5	2.4	2.8	3.8	5.0
3	Elm	T.	T.	0.5	1.0	1.9	3.3
5	Ash	T.	T.	1.6	2.4	2.5	4.1
6	Beech	3.2	4.5	6.0	6.7	8.5	9.9
7	Basswood	T.	T.	0.6	0.7	1.3	2.5
12	Oak	T.	T.	T.	0.2	0.7	2.6
9	Hemlock	T.	T.	0.6	0.5	0.9	2.4
11	Pine	T.	T.	0.6	1.9	2.5	2.2
13	Hickory	T.	T.	T.	0.5	0.5	0.6
18	Apple	T.	T.	T.	0.2	0.4	0.6

Smoothed curves of these results are given on figure 11. In general the rough bark trunks conduct the least water to the ground, and in the case of apple, shag bark hickory, oak and hemlock very little rain runs down the trunks in showers of less than 0.2 to 0.3 inch. The curves are generally of parabolic form, showing a smaller percentage running down the tree trunks in light than in heavy showers.

TOTAL INTERCEPTION LOSS AS A PERCENTAGE OF THE PRECIPITATION PER SHOWER.

In working up the experimental records, data were tabulated in groups, each group including showers for which the precipitation fell between assigned limits. The group means for the different showers and interceptometers are contained in Table No. 10.

TABLE 10.—*Analysis of 1917-1918 interceptometer records.*

[Horton Hydrologic Laboratory, Voorheesville, N. Y.]

No.	Tree. Kind of tree.	Precipitation, 0-0.05. Number of showers.	Amount of precipitation.	Duration of rain.	Loss.	Precipitation, 0.05-0.10. Number of showers.	Amount of precipitation.	Duration of rain.	Loss.	Precipitation, 0.10-0.30. Number of showers.	Amount of precipitation.	Duration of rain.	Loss.	Precipitation, 0.30-0.60. Number of showers.	Amount of precipitation.	Duration of rain.	Loss.	Precipitation, 0.60-1.0. Number of showers.	Amount of precipitation.	Duration of rain.	Loss.	Precipitation, 1.0-2.0. Number of showers.	Amount of precipitation.	Duration of rain.	Loss.
			In.	Hrs.	In.		In.	Hrs.	In.		In.	Hrs.	In.		In.	Hrs.	In.		In.	Hrs.	In.		In.	Hrs.	In.
2	Maple	12	0.031	0.71	0.026	18	0.067	1.18	0.012	31	0.184	2.18	0.063	17	0.418	3.63	0.123	5	0.724	6.31	0.170	2	1.278	8.70	0.262
10	.....do	5	.034	1.13	.027	12	.065	1.27	.014	20	.185	2.40	.093	12	.416	4.11	.188	3	.806	8.06	.421	2	1.410	5.12	.504
16	.....do.[1]	3	.013	1.15	.018	12	.065	1.27	.035	17	.187	2.33	.057	12	.416	4.11	[2]—.016	5	.725	6.31	.095	1	1.257	10.91	.735
	Mean	......	.036	1.00	.024	......	.066	1.24	.040	......	.185	2.30	.071	......	.417	3.95	.098	......	.752	6.89	.230	......	1.315	8.24	.500
3	Elm	12	.031	0.71	.025	18	.067	1.18	.053	31	.184	2.18	.083	17	.418	3.63	.139	5	.724	6.31	.189	1	1.257	10.91	.187
17	.....do.[1]	2	.013	0.24	.018	12	.065	1.27	.039	18	.192	2.28	.066	11	.417	4.43	.070	4	.755	6.71	.074	......	......	......	......
	Mean	......	.037	0.48	.022	......	.066	1.22	.046	......	.188	2.23	.075	......	.418	4.03	.104	......	.740	6.51	.131	......	1.257	10.91	.187
4	Willow[3]	12	.031	0.71	.024	18	.067	1.18	.046	31	.184	2.18	.087	17	.418	3.63	.172	5	.724	6.31	.367	2	1.278	8.70	.564
5	Ash	12	.031	0.71	.011	18	.067	1.18	.028	31	.184	2.18	.049	17	.418	3.63	[2].105	4	.694	5.57	.262	2	1.278	8.70	.268
6	Beech	12	.031	0.71	.023	18	.067	1.18	.035	31	.184	2.18	.061	17	.418	3.63	.130	4	.755	7.21	.121	2	1.278	8.70	.316
7	Basswood[3]	9	.027	0.52	.022	6	.070	0.98	.035	13	.173	2.07	[2].016	5	.424	2.51	[2]—.005	1	.603	4.68	.215	1	1.257	10.91	[2]—.028
8	Oak[3]	9	.027	0.52	0.25	6	.070	0.98	.053	13	.173	2.07	.073	5	.42	2.51	.177	1	.603	4.68	.319	1	1.257	10.91	.340
12	.....do	5	.034	1.13	[2].002	12	.065	1.27	.042	20	.185	2.40	.051	12	.416	4.11	.050	4	.755	7.22	.121	1	1.257	10.91	.396
	Mean	......	.030	0.82	.014	......	.068	1.12	.048	......	.179	2.23	......	......	.420	3.31	.113	......	.679	5.95	.220	......	1.257	10.91	.368
9	Hemlock	11	.032	0.67	.022	18	.067	1.18	.056	28	.180	2.22	.088	16	.417	3.79	.105	4	.755	7.21	.188	1	1.257	10.91	.489
11	Pine	11	.032	0.67	.023	18	.067	1.18	.055	28	.180	2.22	.089	16	.417	3.79	.151	4	.755	7.22	.114	1	1.257	10.91	.026
13	Hickory	5	.034	1.13	[2].003	11	.066	1.32	.048	20	.185	2.40	[2].041	12	.416	4.11	[2]—.010	4	.755	7.22	[2].059	1	1.257	10.91	[2]—.125
14	Horsechestnut[1,3]	5	.034	1.13	.022	12	.065	1.27	.043	20	.185	2.40	[2].058	12	.416	4.11	.111	5	.725	6.31	.083	1	1.300	6.50	.345
15	.....do.[3]	5	.034	1.13	.024	12	.065	1.27	.041	20	.185	2.40	.088	12	.416	4.11	.160	5	.725	6.31	.270	1	1.300	6.50	.493
	Mean	......	.034	1.13	.023	......	.065	1.27	.042	......	.185	2.40	.073	......	.416	4.11	.135	......	.725	6.31	.176	......	1.300	6.50	.419
18	Apple	5	.034	1.13	.018	12	.065	1.27	.041	20	.185	2.40	.076	12	.416	4.10	.106	4	.755	7.22	.176	1	1.257	10.91	.349
	Mean of all	......	.033	0.83	.020	......	.066	1.20	.043	......	.183	2.26	.067	......	.418	3.76	.102	......	.726	6.52	.191	......	1.277	9.44	.320
	Mean of all except Nos. 4, 7, 8, 14, and 15	......	.034	0.84	.018	......	.066	1.23	.043	......	.185	2.28	.068	......	.417	3.92	.093	......	.746	6.88	.166	......	1.277	9.78	.310
	Mean of all except Nos. 14, 16, and 17	......	.032	0.83	.020	......	.067	1.19	.044	......	.182	2.25	.068	......	.417	3.66	.111	......	.724	6.54	.214	......	1.277	9.55	.289

[1] Near periphery of tree cover.
[2] Mean is low due to few large negative results.
[3] Trees for which trunk water was not collected and for which correction has been made.

Table No. 11 shows the same results, in condensed form, and reduced to the basis of percentages of the total precipitation per shower which was lost or intercepted, corrected for rain running down the tree trunks. The average loss ranges from 70 per cent of the total in very slight showers, to about 24 per cent in heavy, long-continued rains

TABLE No. 11.—*Summary of interceptometer records showing amount of water lost to ground by tree interception.*

Tree.		Percentage of precipitation lost.					
No.	Kind of tree.	0-0.05	0.05-0.10	0.10-0.30	0.30-0.60	0.60-1.0	1.0-2.0
2	Maple	83.8	62.7	34.2	29.4	23.5	20.5
10	....do	79.4	67.7	50.3	45.2	52.6	35.7
16	....do.[1]	41.8	54.0	30.5	([2])	13.1	58.5
	Mean Nos. 2 and 10	81.6	65.2	42.2	37.3	38.0	28.1
3	Elms	80.6	79.1	45.1	33.3	26.1	14.9
17	....do.[1]	41.8	60.0	34.4	16.8	9.8	....
	Mean	61.2	69.6	39.8	25.0	18.0	14.9
4	Willow[3]	77.4	68.7	47.3	41.1	50.6	44.1
5	Ash	35.5	41.8	26.6	[2] 25.1	37.7	21.0
6	Beech	74.2	52.2	33.1	31.1	16.0	24.7
7	Basswood[3]	81.5	50.0	([2])	([2])	35.6	([2])
8	Oak[3]	92.5	75.7	42.2	41.8	52.8	27.1
12	....do.[3]	([2])	64.7	27.6	12.0	16.0	31.5
	Mean[3]	....	70.2	34.9	26.9	34.4	29.8
9	Hemlock	68.7	83.6	48.9	25.2	24.9	38.9
11	Pine	71.9	82.1	49.5	36.2	15.1	2.1
13	Hickory	([2])	72.7	[2] 22.2	([2])	[2] 7.8	([2])
14	Horse chestnut[1 3]	64.7	66.2	[2] 31.4	26.7	11.4	26.5
15	....do.[3]	70.5	63.1	47.6	38.5	37.2	37.9
	Mean[3]	67.6	64.6	39.5	32.6	24.3	32.2
18	Apple	52.9	63.1	41.1	25.5	23.3	27.8
	Mean of all	70.5	65.1	38.2	39.5	26.7	29.4
	Mean of all except 4, 7, 8, 14, 15, 16, 17	68.4	67.0	37.6	29.6	24.6	24.3

[1] Near periphery of tree cover.
[2] Percentage too small: Too many negative results involved in one mean.
[3] Not corrected for water running down the trunk; all others are corrected.

Figure 12 shows the percentage loss for all trees, on a comparative basis. In general, with regard to all the curves, the plotted points are very consistent for smaller amounts of rainfall where the number of observations included in a group mean was relatively large. There were several showers of less than 0.05 inch rain, in which nothing was caught in the interceptometers, and which showers were not included in making out the group means. The effect of including these would be to make the percentage interception for rainfall amounts of less than 0.05 inch somewhat larger. The low loss for heavy showers in the case of the hickory and elm, especially the former, is undoubtedly due to the exposure of the trees and the height of the crown above ground. As a result of these conditions, the interceptometers received direct precipitation in some showers, accompanied by wind blowing from the same side of the trees as that on which the interceptometer was placed. The curves are in general hyperbolic in form, and could be expressed by formulae of the type

$$\text{Per cent loss} = a + \frac{b}{P}$$

in which a and b are constants, and P is the amount of precipitation per shower. Here a represents interception storage depth, and $a+b$ is the ordinate of the asymptotic line, which the curve approaches as the amount of precipitation increases indefinitely. In other words, the interception loss approaches a constant percentage of the total precipitation in very heavy rains.

Figure 13 is an average of all the curves, expressed in a similar manner.

INTERCEPTION DEPTH ON PROJECTED TREE-CROWN AREA PER SHOWER.

In figure 14, the results are shown graphically in terms of the amount of precipitation loss over the projected area of the crown of the tree.

Figure 15 is the mean of the curves shown on figure 14. This is a straight line, except near the origin, and would be apparently a straight line throughout but for the fact that in very light showers, less than sufficient to satisfy the interception storage, some of the rain may be shaken off the trees by the wind.

The amount of interception storage for each tree is approximately determined by extending the line or curve on figure 14 to the zero precipitation line, in the direction determined by the portion of the curve plotted for rains exceeding the interception storage. For rains less in amount than the average interception storage, the interception loss would be in general 100 per cent, if there were no wind, but may be less, if a portion of the storage remaining on the tree at the end of the shower is shaken off by the wind.

The accompanying formulæ show the average interception, corrected for water running down the trunks of the trees, expressed in terms of precipitation per shower. In the linear formulæ the first constant represents the interception storage and the second the limiting minimum proportion of the rain lost as the amount per shower increases.

Summer season interception per shower by various trees.

$J=0.015+0.23\ P_s$ .... Ash.
$J=0.04+0.18\ P_s$ .... Apple.
$J=0.04+0.20\ P_s$ .... Horse-chestnut.
$J=0.02+0.23\ P_s$ .... Beech.[1]
$J=0.03+0.22\ P_s$ .... Oak (mean).
$J=0.03+0.23\ P_s$ .... Maple.
$J=0.23\ P_s^{1/2}$ .... Elm (No. 3).[2]
$J=0.13\ P_s^{1/2}$ .... Basswood.
$J=0.20\ P_s^{1/2}$ .... Hemlock and pine.
$J=0.02+0.40\ P_s$ .... Willow shrubs.[3]

The formulas apply only when P is greater than the constant, otherwise $J=P$, nearly.

In deriving these formulæ, results obtained by peripheral interceptometers have been disregarded for reasons elsewhere stated. It will be noted by reference to figure 12 that the interception loss for oak, maple, ash, beech, and horse-chestnut are very nearly the same, and an average formula for these trees may be used. In the case of all except the hemlock, pine, and elm trees, the interception curves either in terms of precipitation per shower, or shower duration are straight lines. In the case of the hemlock, pine, and elm, the curves both in terms of amount and duration are parabolic in form. In the case of the pine and hemlock, the bark and leaves seem to absorb a relatively large amount of precipitation as hygroscopic moisture.

The relatively large loss by interception from willows, which in this case were shrubs of 8 to 10 feet in height, is notable. This loss greatly exceeds that for other trees, except the basswood, even after correction for water running down the trunks, which correction in the case of the willow shrubs was estimated from the data for other trees.

The following Table No. 12 shows the mean precipitation in each shower and the mean loss per shower for each kind of tree.

The inaccuracy of using average percentages in calculating interception losses in individual cases is illustrated by the following example. The average precipitation per shower during the 78 experiments on pine was 0.22 inch, and the average interception loss from the

[1] Tree crown partly defoliated.
[2] Apparently deficient for storage, because of high crown and direct catch of rain.
[3] Probably excessive, apparently much water runs down stem.

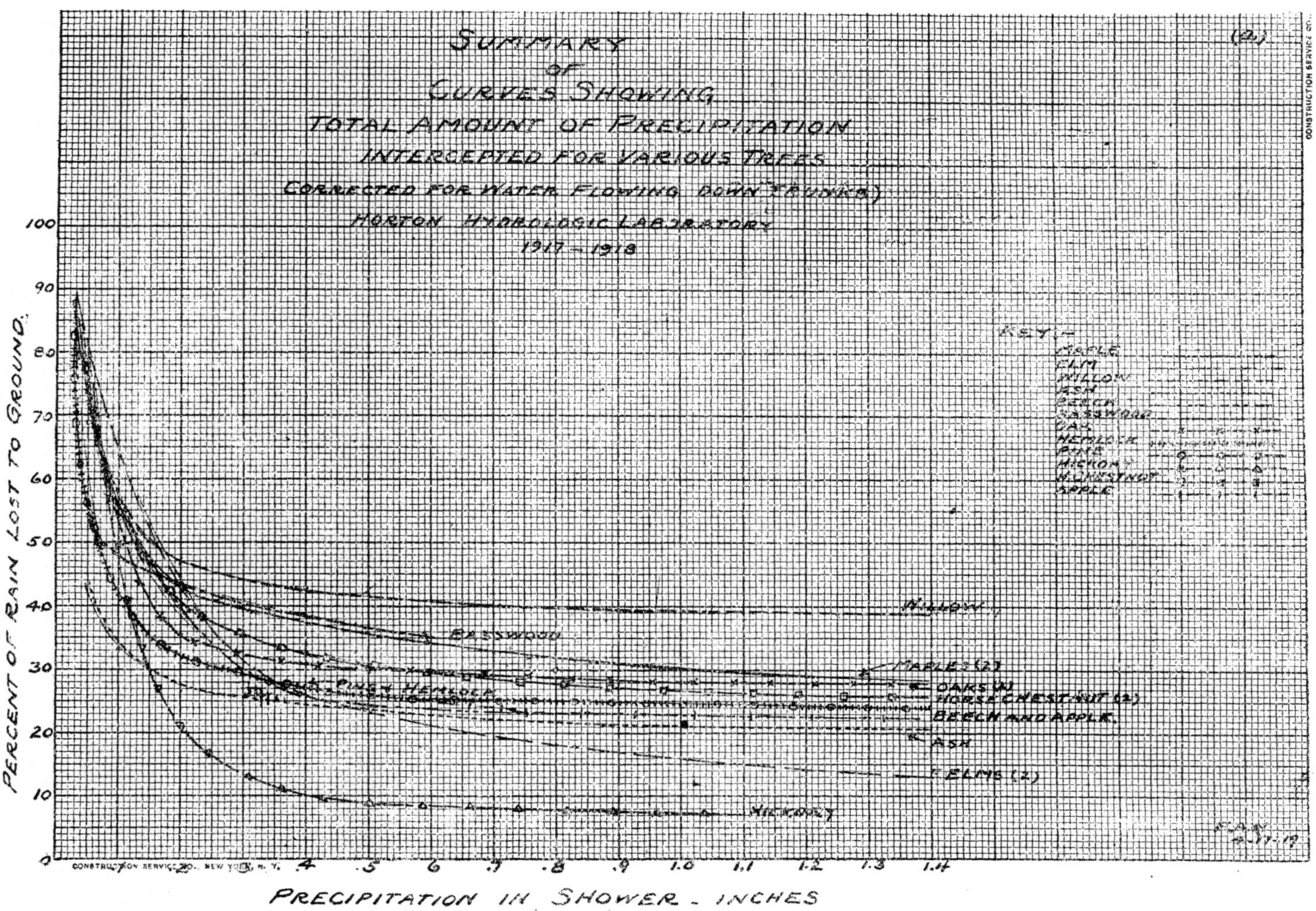

FIG. 12.—Summary of curves showing total amount of precipitation intercepted for various trees. (Corrected for water flowing down trunks.)

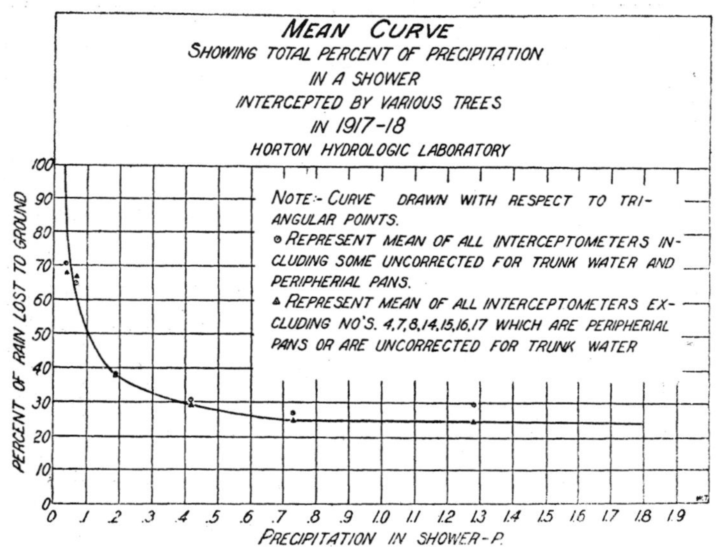

FIG. 13.—Mean curve showing total per cent of precipitation in a shower intercepted by various trees in 1917–18.

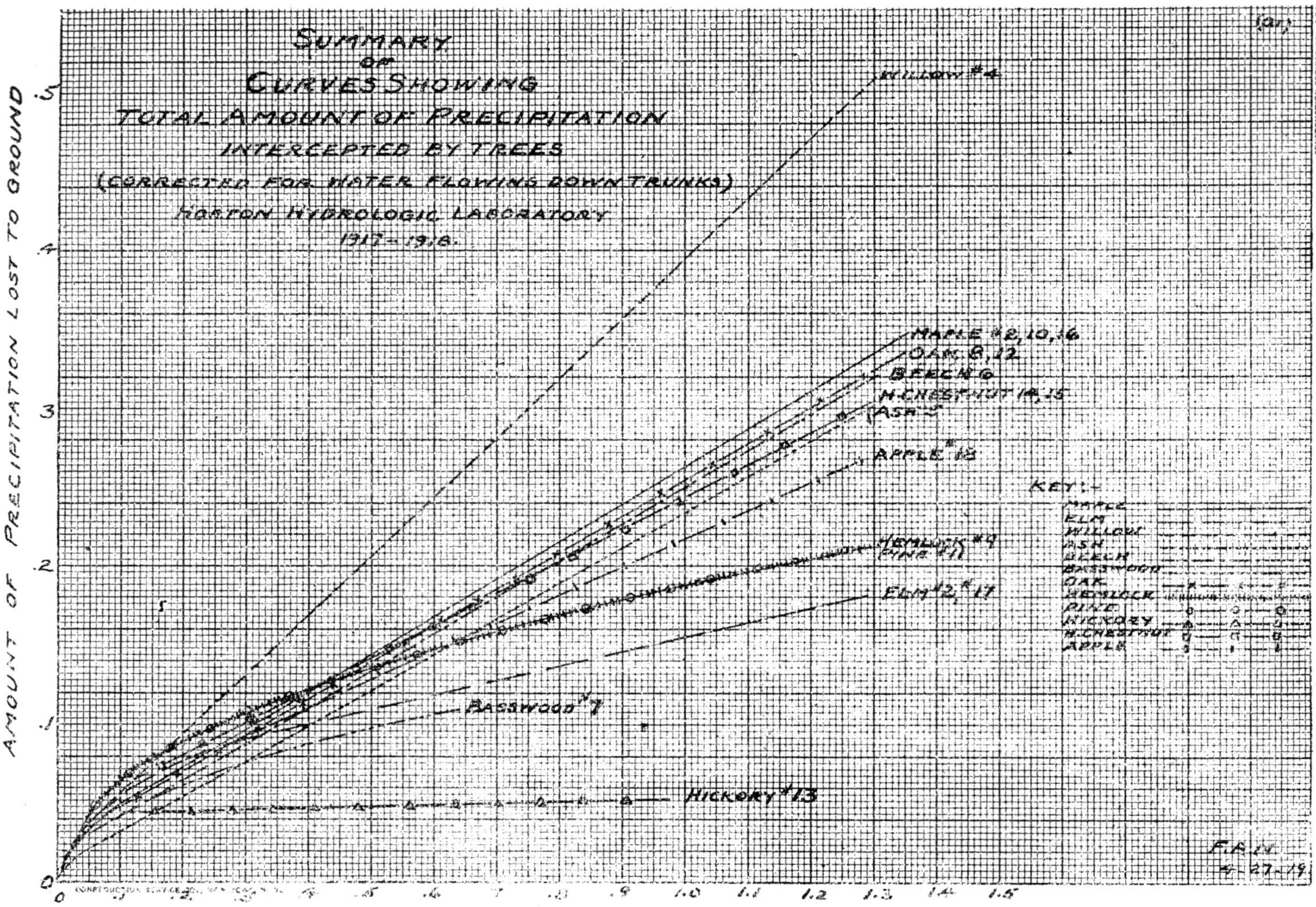

FIG. 14.—Summary of curves showing total amount of precipitation **intercepted by trees.** **(Corrected for** water flowing down trunks.)

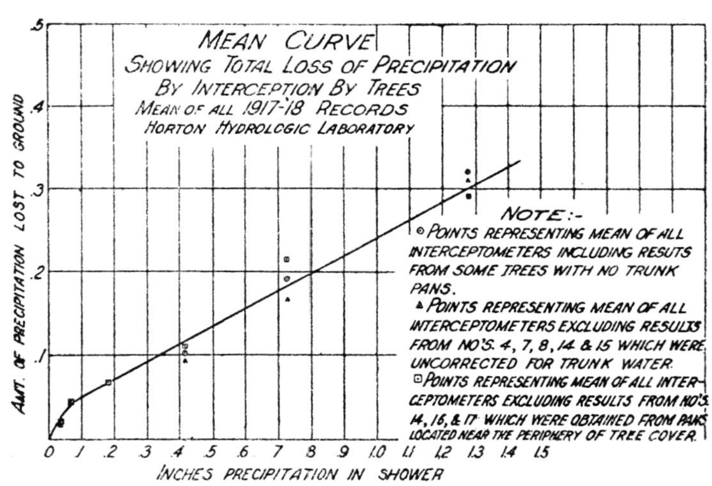

FIG. 15.—Mean curve showing total loss of precipitation by interception by **trees.**
(Mean of all 1917-18 records.)

pine tree was 0.085 inch per shower, or 38.7 per cent. Consider a month with 5 inches precipitation—based on the average percentage there would be a total loss of 1.935 inches. If, for example, the precipitation during the given month consisted of 50 showers of 0.10 inch, the loss would be 3.15 inches, while if the rain had fallen in 5 showers of 1 inch each the loss would have been only 0.915 inch.

INTERCEPTION LOSS IN TERMS OF SHOWER DURATION.

Figure 16 contains curves showing the precipitation loss expressed in terms of duration of the shower in hours.

It should be noted that the data were not tabulated directly in terms of shower duration. The method of deriving the group means sometimes includes showers of short duration and high intensity in the same group

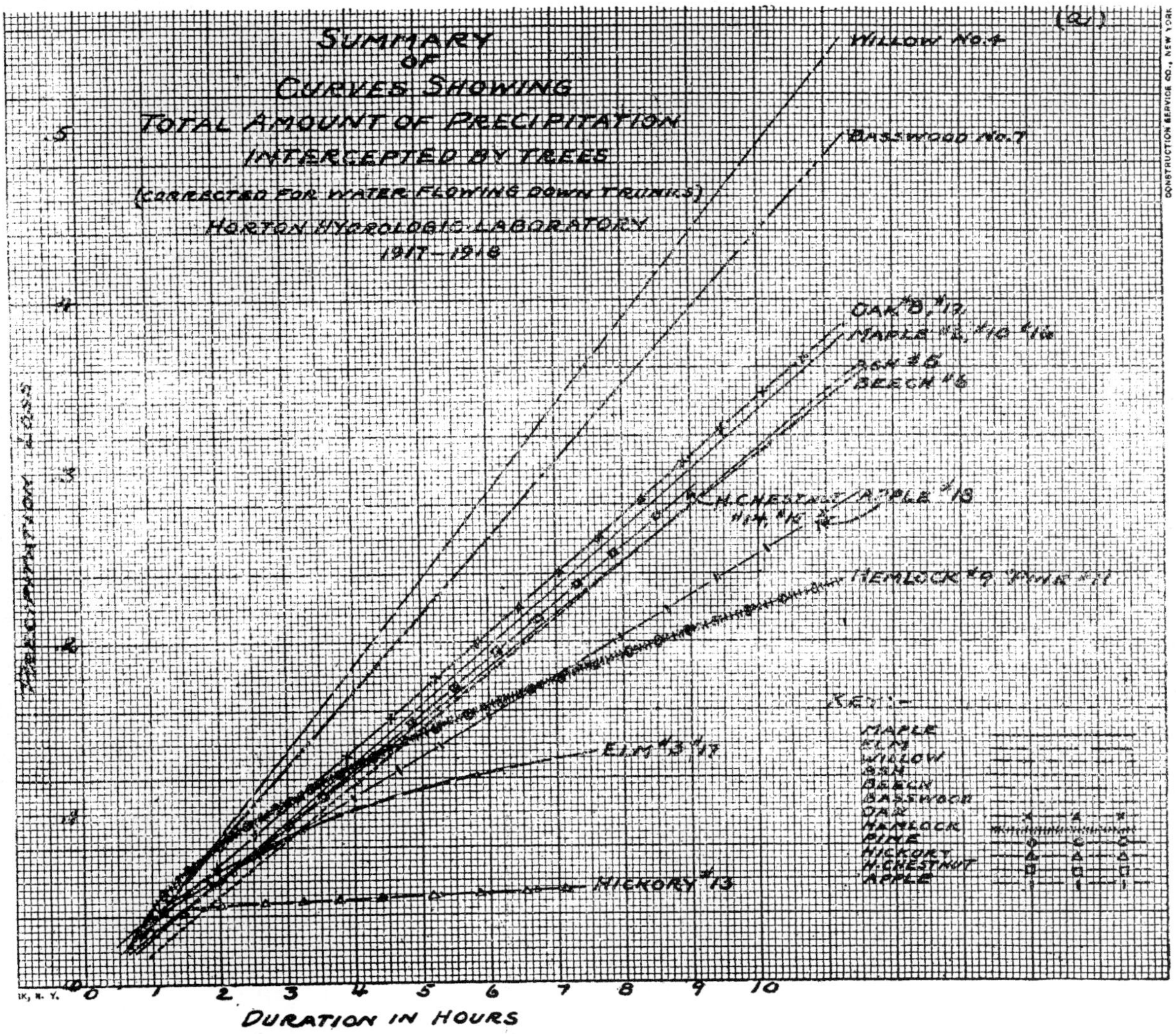

FIG. 16.—Summary of curves showing total amount of precipitation intercepted by trees. (Corrected for water flowing down trunks.)

TABLE 12.—*Average results, 1917–18, interceptometer records.*

Tree.		Number of showers.	Precipitation (inches).		Loss (inches).		Notes.
No.	Kind of tree.		Total.	Mean.	Total.	Mean.	
2	Maple	85	17.152	.200	6.454	.076	Near trunk.
10	.....do	54	14.880	.290	7.059	.131	Under group of small trees with undergrowth.
16	.....do	50	13.962	.279	2.461	.049	Same as Z but near edge.[1]
3	Elm	84	19.265	.229	7.325	.087	
17	.....do	47	11.929	.254	2.758	.059	
4	Willow	85	20.564	.242	9.700	.114	([1]).
5	Ash	84	19.720	.235	5.524	.066	
6	Beech	84	19.922	.237	5.435	.065	
7	Basswood	35	6.892	.197	.778	.022	([1]).
8	Oak	35	6.892	.197	3.036	.087	1917 only.[1]
12	.....do	54	13.919	.258	3.014	.056	1918 only.
9	Hemlock	78	17.547	.225	6.635	.085	
11	Pine	78	17.547	.225	6.633	.085	
13	Hickory	53	13.865	.262	.994	.019	Do.
14	Horse chestnut	55	14.567	.265	3.878	.071	([1]).
15	.....do	55	14.567	.265	6.135	.112	([1]).
18	Apple	54	13.919	.258	4.427	.082	1918 only.

[1] No trunk interceptometer used.

with showers of very much longer duration but lower intensity. It is probable that if the data had been tabulated, and the means taken for groups of showers of similar duration, regardless of the amount of precipitation, a somewhat closer correlation between showers and interception loss would have resulted.

During July to October, 1917, there were 42 rainfall days and 54 showers of .01 inch or more, an average of about 1.3 showers per rainfall day.

As illustrating the importance of rainfall duration in relation to interception loss, consider a month with 2 inches precipitation in 10 showers of 2 hours' duration each, then for oak the interception loss would be $10 \times .07 = 0.70$ inch, while if there had been 4 showers of 2 hours' duration each the loss would have been only 0.28 inch, with the same amount of precipitation for the month.

During April to October, 1918, there were 70 rainfall days, with 130 showers of 0.01 inch or more, an average of 1.88 showers per rainfall day.

If we consider the interception in each shower as at least equal to the interception storage, then as an approximation, 1.5 showers per rainfall day may be considered as a fair average, as this will represent the approximate number of showers per day with 0.01 inch or more of rain, the interception in showers of less than 0.01 inch being about sufficient to make up the deficiency in interception storage for showers of 0.02 to 0.03 inch.

There is in general a fairly close relation between the amount of precipitation per shower and duration of the shower, at the station where these records were kept. This relation for the years 1917 and 1918, as shown by Fig. 17, is represented by the equation $h_s = 7.4\ P_s^{0.70}$ where h_s is the average duration of a shower in hours, and P_s is the average of the amount of precipitation per shower in inches. In this study it was considered that

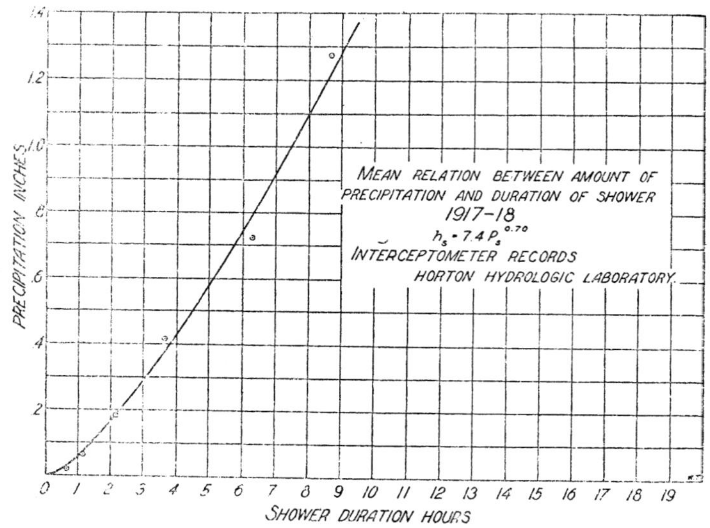

FIG. 17.—Mean relation between amount of precipitation and duration of shower, 1917–18.

any period of one hour or more in which there did not fall 0.01 inch measured rainfall would terminate an antecedent shower.

For practical purposes, it will probably often be found more convenient to utilize interception results or formulæ expressed in terms of amount of precipitation rather than in terms of shower duration, although the latter method of expressing results appears to be the more logical.

RELATION OF INTERCEPTION TO EVAPORATION.

Since interception losses are in reality evaporation losses, it might naturally be expected that there would be a fairly close relation between the relative amount of interception losses in different months of the year, and the relative evaporation in the same months. The data available previous to the experiments of the author do not show any consistent relation of this kind.

TABLE NO. 13.—*Monthly distribution of interception losses, Adlisburg Switzerland, 1889–90.*

Month.	Number rainfall days.	Precipitation in the open.	Total loss (inches).		Loss per rainfall day (inches).	
			Under beech.	Under fir.	Beech.	Fir.
(1)	(2)	(3)	(4)	(5)	(6)	(7)
January	17	2.68	....	0.93	....	0.055
February	26	3.71	....	1.84	....	.071
March	25	3.29	0.30	2.09	0.012	.084
April	32	6.00	.73	3.10	.023	.097
May	15	3.39	.043	1.75	.003	.117
June	43	9.85	....	4.52	....	.105
July	35	9.33	1.18	2.39	.034	.068
August	40	14.90	.75	3.07	.019	.077
September	16	5.29	.28	.77	.017	.048
October	34	9.90	1.38	1.74	.041	.051
November	36	4.72	.30	1.86	.008	.052
December	15	1.34	....	.64	....	.043

TABLE NO. 14.—*Monthly distribution of interception loss, Haidenhaus, Switzerland, 1890.*

Month.	Mean air temperature (° F.).	Number rainfall days.	Precipitation in the open.	Total loss (inches).		Loss per rainfall day (inches).	
				Under beech.	Under fir.	Beech.	Fir.
(1)	(2)	(3)	(4)	(5)	(6)	(7)	(8)
January	23.2	15	3.36	0.76	1.53	0.051	0.102
February	24.8	5	.20	.008	.16	.0016	.032
March	29.9	11	1.16	.15	.51	.014	.046
April	39.4	19	3.02	.60	1.40	.032	.074
May	53.96	15	4.86	1.45	2.13	.097	.142
June	59.9	19	2.65	1.02	1.51	.054	.079
July	59.9	21	4.72	1.40	2.57	.067	.122
August	57.6	21	9.30	2.13	3.90	.101	.186
September	49.1	6	1.55	.46	.66	.077	.110
October	42.3	15	5.36	1.56	2.00	.104	.133
November	33.8	23	2.51	.70	1.92	.030	.084
December	27.7	4	.33	....	.075	....	.019

This fact is exemplified by Tables Nos. 13 and 14, showing the monthly interception losses at Adlisberg and Haidenhaus. It will be noted that while the interception loss is generally greater in the summer months, May to October, inclusive, than in the winter, the monthly results do not show any very consistent relation to the average evaporation curve, which, as well known, increases from May until about August 1, and then decreases.

TABLE No. 15.—*Monthly summary, rainfall caught in interception pans, inches, Horton hydrologic laboratory.*

Month.	Evaporation (inches).	Precipitation.			Check pan No. 1.	Maple (house), No. 2.		Elm, No. 3.		Willow, No. 4.		Ash,[1] No. 5.		Beech, No. 6.		Basswood, No. 7.		Oak, No. 8.	
		No. 1, T. B.	No. 2, Erwin's.	No. 3, field.		Pan.	Trunk.	Pan.	Trunk.	Pan.	Trunk.	Pan.	Trunk.	Pan.	Trunk.	Pan.	Trunk.	Pan.	Trunk.
(1)	(2)	(3)	(4)	(5)	(6)	(7)	(8)	(9)	(10)	(11)	(12)	(13)	(14)	(15)	(16)	(17)	(18)	(19)	(20)
1917.																			
August	5.41	2.45	2.36	2.19	2.423	1.430	0.044	1.132	0.006	1.008	......	1.842	0.024	1.602	0.107	2.156	0.015	1.243	......
September	3.72	1.54	1.38	1.51	1.282	.665	.035	.671	.012	.588	......	1.143	.022	1.046	.079	1.528	.013	.954	......
October	1.61	3.66	3.46	4.10	2.748	2.528	.245	2.999	.093	2.329	......	3.180	.157	2.658	.387	3.029	.074	2.344	......
1918.																			
May	5.280	2.86	2.73	2.92	2.616	1.866	.118	1.818	.029	1.772	......	2.164	.094	1.666	.266	......	......	......	......
June	5.056	2.15	2.08	2.30	2.093	1.288	.066	1.272	.017	1.022	......	1.262	.050	1.206	.165	......	......	......	......
July	5.931	1.56	1.63	1.72	1.587	1.060	.034	.857	.007	.820	......	1.241	.032	.996	.075	......	......	......	......
August [2]	5.921	1.65	1.55	1.62	1.577	1.038	.035	1.049	.010	.800	......	1.013	.042	.906	.104	......	......	......	......
September	2.749	4.33	4.27	4.65	4.015	2.598	.199	1.608	.061	2.875	......	1.673	.063	2.440	.287	......	......	......	......
October	2.212	1.68	1.67	1.78	1.647	1.225	.112	1.107	.118	1.272	......	1.072	.045	.920	.140	......	......	......	......
Mean for 1918	4.525	2.37	2.32	2.50	2.256	1.513	.094	1.325	.040	1.427	......	1.404	.054	1.356	.173	......	......	......	......
1919.																			
April	2.486	2.26	2.19	2.43	......	......	......	......	......	......	......	......	......	......	......	......	......	......	......
May	4.514	4.41	4.41	4.84	......	......	......	......	......	......	......	......	......	......	......	......	......	......	......
June	5.289	1.68	1.61	1.74	......	......	......	......	......	......	......	......	......	......	......	......	......	......	......
July	6.275	3.58	3.48	3.70	......	......	......	......	......	......	......	......	......	......	......	......	......	......	......

Month.	Hemlock, No. 9.		Maple (bank), No. 10.		White pine, No. 11.		Oak (east), No. 12.		Hickory, No. 13.		Horse chestnut, No. 14.		Horse chestnut (west), No. 15.		Mapel (house), No. 16.		Elm, No. 17.		Apple, No. 18.	
	Pan.	Trunk.	Pan.	Trunk.	Pan.	Trunk.	Pan.	Trunk.	Pan.	Trunk.	Pan.	Trunk.	Pan.	Trunk.	Pan.	Trunk.	Pan.	Trunk.	Pan.	Trunk.
(1)	(21)	(22)	(23)	(24)	(25)	(26)	(27)	(28)	(29)	(30)	(31)	(32)	(33)	(34)	(35)	(36)	(37)	(38)	(39)	(40)
1917.																				
August	1.424	......	......	......	1.119	0.003	......	......	......	......	......	......	......	......	......	......	......	......	......	......
September	1.037	......	......	......	1.050	.003	......	......	......	......	......	......	......	......	......	......	......	......	......	......
October	2.655	0.062	1.807	0.136	3.197	.051	3.177	0.037	3.978	0.014	3.526	......	3.134	......	......	......	3.530	0.093	2.754	0.023
1918.																				
May	1.970	.022	1.471	.067	1.414	.048	2.387	.009	2.703	.009	2.041	......	1.761	......	2.423	......	2.095	......	1.870	.006
June	1.512	.006	.978	.039	1.560	.047	1.794	.005	2.146	.006	1.515	......	1.174	......	1.852	......	1.424	......	1.562	.004
July	.725	.004	.718	.019	1.059	.004	1.104	.001	1.344	.005	1.206	......	.769	......	1.060	......	1.125	......	1.062	.000
August [2]	1.211	.006	1.052	.025	.999	.039	1.475	.017	1.826	.014	1.328	......	.896	......	1.485	......	1.337	......	1.332	.012
September	1.461	.013	2.749	.196	1.542	.064	1.968	.036	2.674	.022	3.060	......	2.312	......	2.609	......	1.916	......	1.922	.031
October	.810	.000	.890	.060	.670	.022	1.317	.000	1.473	.005	1.048	......	1.109	......	1.454	......	1.270	......	1.118	.002
Mean for 1918	1.281	.008	1.377	.068	1.207	.037	1.674	.011	2.028	.010	1.700	......	1.337	......	1.714	......	1.528	......	1.478	.009

[1] About one-half leaves stripped by caterpillars in 1917.

[2] Not monthly total; 2 days (25 and 29) excluded.

TABLE No. 16.—*Monthly summary, rainfall losses by interception. May–October, inclusive, 1918, Horton hydrologic laboratory.*

	Pan number.	Monthly losses (inches).								Mean loss in per cent of rainfall.
		May.	June.	July.	August.[1]	September.	October.	Total.	Mean.	
(1)	(2)	(3)	(4)	(5)	(6)	(7)	(8)	(9)	(10)	(11)
Evaporation	....	5.280	5.056	5.931	5.921	2.749	2.212	27.149	4.525	......
Precipitation	....	2.837	2.177	1.637	1.610	4.412	1.700	14.373	2.396	......
Check pan	1	2.616	2.093	1.587	1.577	4.015	1.647	13.535	2.256	......
Maple (house)[2]	2	.853	.823	.543	.537	1.615	.363	4.734	.789	33.0
Elm	3	.990	.888	.773	.551	2.743	.475	6.420	1.070	44.7
Willow [3]	4	1.065	1.155	.817	.810	1.537	.428	5.812	.969	40.4
Ash	5	.579	.865	.364	.555	2.676	.583	5.622	.937	39.1
Beech	6	.905	.806	.566	.600	1.685	.640	5.202	.867	36.2
Hemlock	9	.845	.659	.908	.393	2.938	.890	6.633	1.106	46.2
Maple (bank)[4]	10	1.299	1.160	.900	.533	1.467	.750	6.109	1.018	42.5
White pine	11	1.375	.570	.574	.572	2.806	1.008	6.905	1.151	48.1
Oak (east)	12	.441	.378	.532	.118	2.408	.383	4.260	.710	29.6
Hickory	13	.125	.025	.288	......	1.716	.222	2.376	.475	19.8
Horse chestnut [3 5]	14	.796	.662	.431	.282	1.352	.652	4.175	.696	29.1
Horse chestnut (west)[3 6]	15	1.076	1.003	.868	.714	2.100	.591	6.352	1.059	44.2
Maple (house)[3 5]	16	.414	.325	.577	.125	2.403	.246	4.090	.682	28.5
Elm [3 5]	17	.742	.753	.512	.273	2.496	.430	5.206	.868	36.2
Apple	18	.961	.611	.575	.264	2.459	.580	5.450	.908	37.9

[1] Two days (25 and 29) excluded; values are not monthly totals.
[2] Near trunk.
[3] Without trunk interception.
[4] Under group of small trees with undergrowth.
[5] Near edge.
[6] Near trunk with undergrowth.

NOTE.—2 and 16 and 3 and 17 under same tree.

Table 15 shows the results of the author's experiments, expressed in terms of the monthly amount of precipitation caught in each interceptometer, and Table No. 16 shows the monthly interception losses. Here again there is no apparent relation between the monthly evaporation and the amount of interception, in fact if such relation exists, it would evidently require long statistical records in order that it might be revealed by data analyzed on the monthly basis.

As a further test to reveal whether the experiments indicate the existence of a close correlation between interception loss and evaporation loss, the data for showers of 0.1 to 0.3 inch precipitation were analyzed, as shown in Table No. 17. This method of analysis should eliminate some of the uncertainties of the presentation of data in the form given in Tables Nos. 13, 14, and 16, inasmuch as only results for showers of about equal intensity are compared in Table No. 17. Here, however, there is, as before, no consistent relation between evaporation rate and interception loss. Apparently this relation, although it probably exists, is not very marked. One of the principal reasons is apparently the fact that rainfall duration is generally much greater in the autumn, winter, and spring than during the midsummer months. Interception loss is proportional to evaporation rate and rainfall duration. It so happens that during the months when the evaporation is the greatest the rainfall duration is the least, and vice versa. As a result, the interception losses are more nearly constant than the evaporation rate.

Until more detailed studies have been made, it appears that for practical purposes interception formulæ deduced for summer conditions may be applied throughout any of the months May to October, inclusive, without sensible error.

As regards watershed effect, somewhat better results might be obtained by placing peripheral interceptometers immediately underneath the marginal leaves of the tree, instead of on the ground.

In order to test further the effect of different exposures of interceptometers under the same tree crown, a series of six interceptometers was placed under each of the three tree crowns, that is, interceptometers were placed in geometrical order, without reference to the character of the foliage above them. They were placed at angular spaces of 60 degrees around the center, and at distances alternating 10 and 5 feet, beginning 10 feet to the north of the center. The center used was the center of the tree, except in the case of the hemlocks, where it was the center of a group of three trees, about 18, 8, and 12 inches in diameter, respectively, standing nearly in line, the larger trees 15 feet part, and the smaller tree between them.

Results obtained from these groups of interceptometers in four showers, and the averages, are shown in the accompanying Table No. 18, in column No. 3, of which is also shown the character of the cover over each interceptometer. Some of these interceptometers were partially outside of the projected area underneath the tree crown. Others had only thin foilage over them. Those marked "Thick" agree generally for the same tree. These were all underneath portions of the tree crown where there was nearly a complete roof of foliage, or representing average conditions for well-developed growth, subject to the natural requirement that the leaves must have a suitable exposure to the light. The interceptometers marked "Thick" in the subjoined table represent conditions substantially the same as those used in the two years' experimental series. At the same time, Table No. 18 shows that the thicker and denser the foliage, the smaller in general is the amount of rain caught by a gauge underneath the tree.

TABLE No. 17.—*Monthly means of rainfall interception for showers of 0.10 to 0.30 inch, Horton hydrologic laboratory, 1917–18.*

Month.	Mean precipitation per shower.	Mean shower duration (hours).	Evaporation (inches).[1]	Net loss, inches per month.															
				Maple.			Ash No. 5	Beech No. 6	Willow No. 4.[2]	Oak Nos. 8 and 12.	Hemlock No. 9.	Pine No. 11.	Elm.		Horse chestnut.		Apple No. 18.	Mean loss.
				No. 2.	No. 10.	No. 16.							No. 3.	No. 17.	No. 14.	No. 15.		
(1)	(2)	(3)	(4)	(5)	(6)	(7)	(8)	(9)	(10)	(11)	(12)	(13)	(14)	(15)	(16)	(17)	(18)	(19)
1917.																		
July	0.215	1.83	[3] 5.370	0.061	....	....	0.023	0.061	0.068	0.107	....	....	0.103	....	....	....	....	....
August	.173	1.93	[3] 6.260	.039	....	....	.041	0.42	.092	.065	0.083	0.080	.075	....	....	....	....	....
September	.133	.83	[3] 3.820	.063	....	....	.045	.027	.066	.028	.009	.048	.037	....	....	....	....	....
October	.126	3.47	[3] 1.705	.039	0.083	....	.059	.085	.076	{.068 .044}	.080	.085	.054	....	0.076	0.064	0.076	....
1918.																		
May	.202	[4] 2.29	5.220	.073	.087	0.034	.045	.065	.077	.041	.071	.089	.074	0.062	.068	.084	.090	0.069
June	[4] .267	....	5.057	.126	.184	.130	.143	.158	.171	.166	.196	.146	.212	.156	.146	.156	.130	[4] .158
July	.138	.52	5.928	.111	.103	.099	.081	.053	.110	.083	.120	.105	.083	.079	.024	.122	.077	.090
August	.183	1.58	6.323	.064	.108	.081	.027	.044	.087	.041	.082	.076	.114	.074	.041	.081	.055	070
September	.117	7.08	2.744	.107	.117	.102	.046	.099	.087	.031	.117	.117	.077	.077	.102	.077	.087	.088
October	.179	2.89	2.190	.042	.062	.027	.055	.065	.079	.031	.097	.104	.046	.036	.041	.074	.058	.056

[1] 4-foot U. S. W. B. standard pan.
[2] Shrubs, no record of trunks.
[3] A few days missing during the month. Totals increased proportionally.
[4] Record of one storm only.

COMPARATIVE CATCH OF DIFFERENT INTERCEPTOMETERS UNDER THE SAME TREE.

In most cases the interceptometer was placed either within about 4 feet of the trunk of the tree, or in the case of the larger trees, about midway from the trunk to the periphery. In the case of the hemlock, maple, and horse-chestnut trees, an additional interceptometer was placed just within the periphery of the tree, with a view to determining the extent of watershed effect afforded by the tree crown. The branches of the tree above the peripheral interceptometer were at heights of about 10 feet in the case of the horse chestnut and maple and 20 feet in the case of the elm. It was early discovered that the peripheral interceptometers placed in this way would not give reliable results. In many cases, especially when the rain fell at an angle, or came from the same side of the tree on which the interceptometer was placed, the interceptometer would catch direct rain.

The peripheral interceptometers caught about 70 per cent of the true rain, as compared with about 60 per cent for the others. The peripheral interceptometers apparently caught the direct rain in about one-half the showers, which would account for the increased catch.

The results, as far as they go, do not indicate any considerable watershed effect, and it appears that results obtained from interceptometers placed about midway between the tree trunk and the periphery of the crown give results which may be accepted as fairly representing the average interception underneath the entire projected area of the tree crown.

TABLE No. 18.—*Comparative catch of interceptometers under the same trees.*

Date, June (1919)	26	27	20	27	Average.
Shower began	2:00 p. m.	9:30 p. m.	9:06 a. m.	8:40 a. m.	
Shower ended	3:15 p. m.	11:30 p. m.	10:30 a. m.	4:30 p. m.	
Duration, hours	1.250	2.000	1.400	7.830	3.120
Precipitation, inches	.040	.170	.250	.290	.187
Precipitation, inches per hour	.032	.085	.179	.037	.083

Position.	Distance (ft.).	Character of cover.	Amount caught, inches.				
(1)	(2)	(3)	(4)	(5)	(6)	(7)	(8)
		HORSE-CHESTNUT.					
N.	10	Medium	0.017	0.122	0.218	0.242	0.1498
N. 60° E.	5	Thin (edge)	.025	.178	.210	.266	.1697
S. 60° E.	10	....do	.008	.090	.162	.162	.1055
S.	5	Thick	T.	.098	.090	.186	.0935
S. 60° W.	10	Very thick	T.	.074	.146	.138	.0895
N. 60° W.	15	Thick	T.	.058	.074	.130	.0655
		Average	.0083	.1033	.1500	.1873	.1122
		APPLE.					
N.	10	Medium	0	0.058	0.170	0.146	0.0935
N. 60° E.	5	Thick	T.	.082	.122	.146	.0875
S. 60° E.	10	Edge thin	.033	.178	.130	.258	.1497
S.	5	Thick	.008	.082	.066	.130	.0715
S. 60° W.	10	Edge open	.042	.194	.258	.258	.1880
N. 60° W.	5	Thick	0	.082	.202	.186	.1175
		Average	.0138	.1127	.1580	.1873	.1178
		HEMLOCK.					
N.	10	Edge high, med.	T.	0.042	0.122	0.154	0.0795
N. 60° E.	5	Thick	.008	.098	.130	.154	.0975
S. 60° E.	10	Thin, underbrush.	T.	.090	.130	.186	.1015
S.	5	Thick	.008	.138	.130	.154	.1075
S. 60° W.	10	Edge high	.008	.114	.315	.186	.1558
N. 60° W.	5	Very thick	.008	.082	.122	.082	.0735
		Average	.0053	.0940	.1582	.1527	.1025

RELATIVE INTERCEPTION IN FOREST AND OPEN.

Since the trees in all cases in the author's experiments, excepting, perhaps, the hemlock, were more exposed than trees in a dense forest, it is possible that the results do not accurately represent the rainfall interception which would take place within the body of the forest. Information regarding the interception by scattered trees and shrubs such as cover large areas and from orchards and hedges is, however, useful and important, and furthermore, the difference in the results here obtained from those which would have been obtained had the interceptometers been placed in a dense forest can be inferred to some extent at least.

So far as can be judged from the limited data available, the storage loss in an active wind is reduced to from one-third to one-half of the amounts occurring when the air is still. The evaporation losses are quite certainly increased by wind action to at least an equal degree and probably more. There is, of course, some wind effect even in the denser forests, especially near the tree tops.

Barring differences in evaporation rate due to difference in humidity and temperature within the forest and in the open, which differences are at a minimum during rain, it appears that the interception loss from trees in the open is probably less in very light showers and more in long rains than from trees within a forest. In a forest, however, the entire interception loss does not take place on the trees. If there is an undergrowth, the water dropping from the trees may be intercepted by shrubs, herbs, or grasses underneath the trees.

147217—19——2

WINTER AND SUMMER INTERCEPTION COMPARED.

Ebermayer's experiments (Table No. 5) show greater interception losses for the same precipitation in the winter than in the summer, although the difference in the case of broad-leaved trees is relative slight. The Swiss experiments (Tables Nos. 13 and 14), on the other hand, show considerably greater losses during the summer months than in the winter months, especially in the case of the beeches.

J. C. Alter (MONTHLY WEATHER REVIEW, May, 1911, **39**: 760) states:

"It has been observed in well forested regions that a downfall of 4 or 5 inches of snow may be almost entirely supported by the branches of the evergreen trees, even when deposited in a high wind, provided the snow was sufficiently moist when it fell. In such cases a subsequent freeze may attach it firmly to the branches. Since nearly all heavy snows come during only moderately cold weather, and often actually begin as rain, the amount of moisture that clings to the trees to be subsequently evaporated is very great. It has been variously estimated at from 50 to 80 per cent of the fall of snow, under the varying conditions that exist over the forest."

Anyone who has stood beneath a tree to escape a driving rain would be pretty certain to conclude that the amount of interception is greater in summer than in winter for broad-leaved trees.

The small difference indicated by some forest experiments between the percentage of interception in winter and summer by deciduous trees was noted by Harrington (**1**). He says:

"Admitting (though there are some reasons for doubt) that the rainfall is actually the same over a wood and a place outside but near, this small action of the foliage as compared with the branches and twigs requires explanation, and, whatever the explanation may be, it must apply to deciduous trees, as evergreens show no difference in these months. No satisfactory explanation occurs to me."

The apparently anomalous results of the experiments are, however, capable of explanation, and are probably due to a combination of several causes.

1. The winter precipitation falls largely as snow and not as rain. The storage capacity of the trunks, branches, and twigs of deciduous trees for either moist snow or for rain or sleet, falling under conditions such that the precipitation freezes to the tree surfaces, is undoubtedly very much greater per unit of surface than the storage capacity of leaves or branches of trees for summer rain, so situated as to be substantially protected from the wind. Unlike rain, intercepted snow does not run down the trunks of the trees.

2. The angle to the vertical at which snow approaches the earth is, as a rule, very much greater than the inclination of rain. If we view the projection of a forest on a horizontal plane, there will be seen in most cases a considerable percentage of open spaces. Viewed from above, at any angle to the vertical, the percentage of open space visibly decreases as the angle increases, becoming in most cases zero for angles to the vertical as great as those at which snow ordinarily falls. Thus it appears that the effective interception surface is likely to be considerably greater for precipitation falling as snow than for precipitation falling as rain, owing to the greater inclination of the former.

3. At many of the forest meteorological stations the average precipitation rate is less, and the duration greater, in winter than in summer, so that, the percentage of inter-

ception loss being greater for light than for heavy storms, and increasing with the storm's duration, one would naturally expect in these conditions that the average percentage loss in winter, other things being equal, would be relatively greater than in summer. In the case of rain falling during cold weather, the increased surface tension may tend to concentrate interception storage into drops, reducing the film area and evaporation loss.

In the general formula

$$J = S_J + KE_r T_s$$

the storage S_J may be much greater in winter than in summer, E_r is less, but T_s is greater, for winter than for summer conditions.

4. The heaviest interception of snow by needle-leaved trees, such as spruce and fir, is more likely to slide off than is the interception in lighter snow storms. A comparison of measured precipitation and run-off in the winter season for northern streams often shows the apparent water losses remarkably small, so small as to be apparently incompatible with interception losses in winter equal to those in summer.

The average seasonal results at Haidenhaus and Adlisberg in terms of interception loss per rainfall day, as derived from tables Nos. 13 and 14, are as follows:

Mean interception loss per rainfall day.

	May-October, inclusive.	November-April, inclusive.
	Inches.	*Inches.*
Haidenhaus:		
Beech	0.083	0.026
Fir	.129	.060
Adlisberg:		
Beech	.023	.014
Fir	.078	.067

These data show the summer interception loss in all cases to be greater than that for the winter, and for the Haidenhaus station the summer loss for both beech and fir is more than double the winter loss.

It is significant that at both stations the excess of summer over winter losses is greater in the case of beech than in the case of the fir trees.

It seems impossible, without further experimental data, to determine definitely the winter interception losses. Probably the best that can be done is to assume for the present that in the case of evergreen trees, under average conditions, the winter and summer losses by interception are about equal. As regards broad-leaved deciduous trees, it is the author's opinion that the winter interception losses for average conditions in northern United States are probably about 50 per cent of the summer interception losses. It appears likely that under some special conditions, as, for example, where the summer precipitation is concentrated in short, heavy showers, and the winter precipitation occurs in numerous light showers of long duration, the winter interception losses from such trees may approach equality with or even exceed the summer interception losses.

Interception of snow in forests in Russia has been determined by the Imperial Agronomic Institute of Moscow by measurements carried on for five years. Numerous snow samples were taken, weighed, and reduced to equivalent water depths in forests of different kinds, as shown in Table 19. It will be noted that the water equivalent of snow in the older and denser forests ranged from 40 to 80 per cent of the average water equivalent of snow on the ground in small clearings and young plantations, for which the measured depth is probably very near the true actual depth of snow falling over the forests. The depth in cultivated fields has no particular bearing, as it was affected largely both by drifting and melting.

TABLE 19.—*Russian experiments on the interceptive influence of forests on snowfall* (*Zon.*).[1]

	Number of areas examined.	Number of measurements of snow depth.	Number of snow samples weighed.	Average thickness of snow.	Water equivalent of snow depth.
(1)	(2)	(3)	(4)	(5)	(6)
				Inches.	*Inches.*
1. Young plantations (2 to 4 years old) and small clearings within the forest	20	259	7	21.9	5.1
2. Birch forest (35 to 75 years old)	11	377	27	22.2	5.0
3. Oak forest (25 to 90 years old)	2	63	3	23.5	5.6
4. Pine forests	32	887	56	15.5	3.1
Young (25 to 35 years old)	25	662	43	15.2	3.1
Old (60 to 90 years old)	7	225	13	16.4	3.2
5. Spruce forest (25 to 35 years old)	21	460	29	9.7	2.1
6. Pine forest with admixture of birch (65 to 75 years old)	4	6	3	20.0	4.4
Pine forest with admixture of larch (25 to 35 years old)	3	74	2	15.2	3.1
Pine forest with admixture of spruce (35 years old)	5	157	9	12,9	2.9
7. Spruce forest with admixture of larch	3	57	2	14.1	3.1
8. Cultivated field	1	332	8	13.0	3.1

[1] Final report of the National Waterways Commission. . p. 241.

SECONDARY INTERCEPTION.

As regards forest and brush land, the herbaceous vegetation, if any, should be included in estimating the extent and condition of cover. In the European forest experiments, from which most of our data of interception by forests are derived, it appears that the rain gages were not as a rule placed underneath a growth of underbrush, and that what may be termed secondary and tertiary interception—i. e., water caught and retained by underbrush beneath the main forest growth, and that caught and retained by herbaceous vegetation beneath the underbrush, have not been taken into account in the experimental data.

In the case of orchards, for example, either where crops are grown between and under the trees, or where the soil is sodded, the total interception loss is the sum of the losses for partial cover by trees and for complete cover by sod or crops, the latter is in part secondary.

Table 20 contains illustrations of the increase in interception by undergrowth. The experimental results for pan No. 4 under willow brush, figure 16, indicate that the interception loss from dense shrubs may equal that from mature trees in light showers, and is one-half to two-thirds as great in heavier rains. The total interception is somewhat less than the sum of the interception losses for the different classes of vegetation, since the lower layer or layers of vegetation receive only the part of the total precipitation not intercepted by the higher vegetation.

Using the type of formula

$$J = a + bP$$

and letting J_1, J_2, J_3 represent respectively the interception loss for the upper (trees), middle (underbrush), and lower (herbs), layers of vegetation, and J the total, then

$$J_1 = a_1 + b_1 P$$
$$J_2 = a_2 + b_2 (P - J)$$
$$J_3 = a_3 + b_3 (P - J - J_1)$$
$$J = c_1 J + c_2 J_2 + c_3 J_3$$

where c_1, c_2, and c_3 are projection factors, or proportions of the total ground area which would be shaded by a vertical sun over the different classes of vegetation.

Allowing for reduced density of cover, and in the light of the results of forest experiments on comparative evaporation rate in woods and open, the coefficients b_2 and b_3 may fairly be taken at one-third to one-half, or say 40 per cent of their values for open exposures.

The importance of the lower vegetation is illustrated by the following calculation, for a maple forest with a dense growth of underbrush such as sometimes occurs.

Using the following values of the constants:

Layer.	a	b	c
Maple trees	0.03	0.23	90
Underbrush	.02	.115	70
Herbs	.01	.05	40

Then for $P=0.1$, $J_1=0.053$, and $P-J=0.047$; $J_2=.0254$; $P-J_1-J_2=0.0216$; $J_3=.011$, and $J=0.0477+0.0178+0.0044=0.0699$.

Here secondary interception increases the total loss about 50 per cent.

RAINFALL INTERCEPTION BY CROPS AND OTHER HERBACEOUS VEGETATION.

The rainfall loss by interception from growing crops and vegetation other than forests has not hitherto been experimentally determined. For crops like wheat, corn in drills, grass, peas, millet, etc., which, when approaching maturity quite fully shadow the ground, it appears certain that the interception percentage approaches in value that for broad-leaved forests. Other crops, like corn, potatoes, tobacco, cotton or beans, planted in hills, do not as a rule completely shadow the ground at any stage of their growth.

Experiments were made by the author on two dates in 1915 with a view to determining the relative interception under trees under different conditions, and under various other kinds of vegetable cover from the same precipitation. The results are contained in the accompanying Table No. 20. It will be noted that in all cases the percentage loss by interception was larger on July 2 from a rainfall of 0.27 inch than that on July 1 from a rainfall of 1.82 inches.

TABLE NO. 20.—*Comparative interception for different vegetation.*

	July 1, 1915.			July 2, 1915.[1]		
	Inches caught.	Interception.		Inches caught.	Interception.	
		Inches.	Per cent.		Inches.	Per cent.
United States Weather Bureau rain gage in open	1.82	0	0	0.27	0	0
North side 18-inch diameter horse-chestnut	1.30	0.52	29	0.08	0.19	70
South side 18-inch diameter horse-chestnut	1.20	0.62	34	0.04	0.23	85
Horse-chestnut and rosebush	0.70	1.12	61.5	0.054	0.216	80
Under 10-inch diameter elm [2]	2.10	−0.28	......	0.122	0.148	55
Elm, 10 inches diameter, with undergrowth	0.70	1.12	61.5	0.100	0.17	63
Soft maple, 6 inches diameter	1.40	0.42	23.1	0.047	0.223	83
Maple brush	1.60	0.22	12.1	......	......	......
Rye	1.70	0.12	6.6	0.122	0.158	59
	1.40	0.42	23.1	0.119	0.151	56
Red clover, a few spears over gage	1.60	0.22	12.1	0.23	0.040	15

[1] Probably interception is somewhat too large, owing to evaporation loss before measurement.
[2] Water dripped into gage from end of overhanging branch.

The interception by rye was about one-half that by mature horse-chestnut trees in a heavy shower, and about three-fourths in a light shower. Red clover intercepted 20 to 40 per cent as much.

Interception by herbaceous vegetation appears to be more largely a matter of storage than in the case of interception by trees. The extent and nature of interception storage for some plants is illustrated in figure 2.

In the absence of more experimental data, the interception storage for various crops has been estimated from observations similar to those shown in figure 2, and the evaporation coefficients have been assumed about in proportion to the plant surface or density of cover as compared with trees.

According to Zon (**10**), Ney estimated the ratio of foliage area to the area covered by forests and crops as follows:

	One side.	Both sides.
Middle-aged beech forest	8.4	16.8
Cereals	7.4	14.8
Clover	5.6	11.2
Grass	4.8	9.6

These figures indicate that the density of interception cover for grass and cereals is five-eighths to seven-eighths that for beech forests.

The above figures are for leaf surfaces only, not including stems and trunks. The entire leaf surfaces of a plant are not usually fully wetted, but the deficiency is partly made up by the moist surfaces of stems and trunks.

In the case of rapidly growing crops, as corn and grains, the interception evidently varies with the stage of growth. In general the density of cover and the projection factors each increase about in proportion to the height of the plant for field crops which do not completely shadow the ground.

In view of the need of experimental data on the interception losses from field crops, the following methods of experimentation are suggested:

In the case of good-sized plants, such as corn or potatoes, one or more plants may be grown in a potometer of suitable size, the surface of the potometer being covered with a thin rubber sheeting, secured tightly around the stem or stems of the plants, to prevent rain entering the potometer. When exposed, the potometer should stand in a pan of considerably larger size, placed alongside of another similar large pan containing no potometer. The difference in the amount of rain caught in the two large pans, reduced to equivalent depth on the projected area of the plant, will represent the total interception loss by the plant.

In the case of grasses or small grains, this method can not be applied, but the interception loss can be approximately determined as follows:

The grass or grain is grown in duplicate potometers. The transpiration loss from both potometers is determined by weighing the potometers at the beginning and end of a given time interval before a rain. During rain one of the potometers is exposed to the rain, and the other is exposed to similar air conditions, but protected from rain. At the end of the rain, after the interception storage has evaporated from the plants in the exposed potometer, both are again weighed, and a second test of the relative transpiration rates is made. The weights of the exposed potometers will give the ratio of the transpiration loss during the rain to the transpiration loss preceding and following the rain. Applying this correction ratio to the measured transpiration loss for the exposed potometer preceding and following rain will give the approximate transpiration loss for the exposed potometer during the rain.

Since the soil of the exposed potometer is uncovered, it will catch the entire rain less the interception, and the interception loss from the exposed potometer will be approximately equal to the measured precipitation in a rain gage, minus the gain in weight during the rain of the exposed potometer reduced to inches depth on the surface, plus the transpiration loss similarly expressed.

Pending the acquisition of more experimental data, tentative formulæ for interception by crops have been developed, as described in a subsequent paragraph.

WORKING FORMULAS FOR INTERCEPTION LOSS.

In view of the fact that the author's experiments on trees represented mainly conditions in hedges or open, and that in some cases the experiments apparently did not show true average conditions, and for trees with high crowns did not indicate the full interception storage which ordinarily occurs, experimental formulae have been somewhat revised for practical working purposes.

Additional formulæ have been prepared to represent conditions in woods, and for field crops. These formulae are given below.

In the case of field crops, the interception loss has been assumed proportional to the height of the plants at the date for which the calculations were made. A column has been added, showing the average projected area shaded by the plants. This is in most cases, except grasses and drilled grains, a function of the height of the plant. The formulæ represent interception loss on the projected area.

In the case of dense woods, the projected area may closely approach but seldom quite equals 100 per cent of the total area. For thin woods, such, for example, as scrub or jack pine lands, the projected area is commonly 33 to 66 per cent of the total. In scattered groves or brush pastures, it may have any value from zero to 100 per cent.

To obtain the mean interception depth over the total wooded or cropped area, the calculated interception loss is to be multiplied by the projection factor.

The formulas given for woods differ from those for trees of the same kind in the open, in the use of a larger constant for interception storage and a smaller evaporation factor. For average showers, the resulting loss is about the same for a given tree in the woods as in the open, but for heavy long continued rains, the formulas for woods give smaller results.

In the case of many crops, such as corn, cotton, grass, or tobacco, the density of the interception cover increases nearly in proportion to the height of the plant. For this reason the factor h is necessary in the formulas for crops.

In the case of trees, while it is true that Riegler's experiments show somewhat greater interception loss for large mature beeches than for younger trees, yet the difference is by no means as great as in the case of cultivated crops. As regards trees, especially in the woods, the effect of growth, in many instances, is to elevate the entire crown of the tree to a greater height. This may be accomplished by the lower and more shaded branches dying as the upper branches continue to grow. As a result, the density of cover or its thickness in a given vertical line increases less rapidly than the height of the tree.

Again, where the stand is very close, the crowns of adjacent trees may overlap. As a result, however, of requirement for light and air, the density of cover of overlapping crowns is usually no greater than the average density of cover under a single tree, although the projection factor would naturally be greater the thicker the stand of the trees. No attempt has been made to allow for variation in density of cover, but an allowance for this may be made, based on judgment and included in the projection factor.

In practical calculations, additional formulas will be needed for various classes of vegetation, such as sugar cane, rice, cranberries, heather, swamp elder beds, sagebrush, chaparral, cattail flags, and various truck crops where grown extensively, such as sugar beets, onions, or celery, as well as for additional kinds of trees.

The close agreement in amount of interception loss by the different kinds of trees, on the one hand, and in the apparent interception loss by different classes of crops of similar nature, as indicated by observations of the extent of cover and interception storage, leads to the suggestion that for practical purposes in calculations of interception losses, various kinds of trees or crops can be combined in a single group, and the same formula used for all plants of a given group.

WORKING FORMULAS FOR PRIMARY INTERCEPTION LOSS PER SHOWER ON PROJECTED AREAS OF TREES AND PLANTS.

Orchards	$J=0.04+0.18\ P_s$
Chestnut, hedge and open	$J=0.04+0.20\ P_s$
Chestnut, in woods	$J=0.06+0.15\ P_s$
Ash, hedges and open	$J=0.015+0.23\ P_s$
Ash, in woods	$J=0.02+0.18\ P_s$
Beech, hedges and open	$J=0.03+0.23\ P_s$
Beech, woods	$J=0.04+0.18\ P_s$
Oak, hedges and open	$J=0.03+0.22\ P_s$
Oak, woods	$J=0.05+0.18\ P_s$
Maple, hedges and open	$J=0.03+0.23\ P_s$
Maple, woods	$J=0.04+0.18\ P_s$
Willow shrubs	$J=0.02+0.40\ P_s$
Elm, hedges and open	$J=0.03+0.23\ P_s^{\frac{1}{2}}$
Elm, woods	$J=0.04+0.18\ P_s^{\frac{1}{2}}$
Basswood, hedges and open	$J=0.03+0.13\ P_s^{\frac{1}{2}}$
Basswood, woods	$J=0.05+0.10\ P_s^{\frac{1}{2}}$
Hemlock and pine, hedges and open	$J=0.03+0.20\ P_s^{\frac{1}{2}}$
Hemlock and pine, woods	$J=0.05+0.20\ P_s^{\frac{1}{2}}$

WORKING FORMULAS FOR PRIMARY INTERCEPTION LOSS PER SHOWER ON GRASSES AND FIELD CROPS.

	Per cent.[1]
Clover and meadow grass, $J=(0.005+0.08\ P_s)h$	1.00
Forage, alfalfa, vetch, millet, etc., $J=(0.01+0.10\ P_s)h$	1.00
Small grains, rye, wheat, barley, $J=(0.005+0.05\ P_s)h$	1.00
Beans, potatoes, cabbage, and other small hilled crops, $J=(0.02+0.15\ P_s)h$	$\frac{1}{4}\ h$.
Tobacco, $J=(0.01+0.08\ P_s)h$	$\frac{1}{5}\ h$.
Cotton, $J=(0.015+0.10\ P_s)h$	$\frac{1}{3}\ h$.
Buckwheat, $J=(0.01+0.12\ P_s)h$	1.00
Corn, planted in hills or rows, $J=(0.005+0.005\ P_s)h$	0.1 h.
Fodder corn, sorghum, Kaffir corn, etc., sowed in drills, $J=(0.007+0.006\ P_s)h$	1.00

Average height of plants in feet=h.

CALCULATION OF INTERCEPTION LOSSES.

It is evident that interception losses, which may amount to one-third or more of the precipitation, should not be disregarded in estimating run-off or yield of drainage basins. Heretofore they have been usually included in general water losses and not separately estimated. More accurate results can often be obtained by direct estimation of the interception losses.

This may be accomplished well enough for some purposes by the use of percentage factors. Greater accuracy will usually be obtained by the use of interception formulas, taking into account the rainfall distribution. For such calculations the data needed are: The monthly precipitation, the number of rainfall days, the average number of showers per rainfall day, and the character, pro-

[1] Ordinary projected area of total.

jection factor, and density of cover or stage of development of the vegetation.

As an example, the following calculations have been made of the interception losses on the Seneca River drainage basin above Seneca Falls, N. Y., for the summer of 1915. The cultural conditions were determined by inspection and counting over sample areas or belts crossing the drainage basin. Areas devoted to roads and villages have been mainly included with grasslands, while an allowance for garden areas has been made in connection with truck crops.

During the months when the crop is generally harvested the interception loss may be taken as the mean of the amount for two conditions: (1) For mature crops, (2) for their residual stubble. Since grain stubble is usually seeded, or contains weeds, it affords a condition as regards water losses nearly identical to grass of equal height, and may be so considered. The estimation of interception losses in a given locality requires some knowledge of farm practice and the rotation of crops and usual dates of seeding and harvest, in order that fair allowance may be made for the portion of fallow, or newly plowed ground, and for other conditions dependent on farm practice.

The method of calculation of interception losses based on the average shower intensity does not take into account rainfall distribution to the same extent as a calculation based on individual rainfall amounts per shower or day. The labor involved in such calculations is, however, usually prohibitive. Light rains occur much more frequently than heavy ones, and occasion relatively greater interception losses. It follows that the use of the average monthly shower intensity in calculating interception is likely to lead to results slightly too small, in the majority of cases. If desired, a correction factor can be applied, based on the statistical law of distribution of showers of different amounts.

Table No. 21 shows the calculated interception losses for Seneca River Basin during the summer months. It will be noted that while the interception depth on the projected area, for full-grown crops, approaches in value that for trees, yet the average loss per unit area from crop land is much less than from wooded lands—(1) because the projection factor for many crops, especially when young, is smaller than for woods; (2) crops are at approximately their full stage of development, as a rule, for only one to three months per year. During the remainder of the growing season the loss from the small plants, or from the stubble subsequent to harvesting, is greatly reduced. The latter is the cause of marked decrease in the total interception loss for September, as compared with August, in Table No. 21. In the case of trees, the interception capacity remains nearly constant throughout the summer season.

If stubble or fallow land is allowed to grow up to weeds, it may increase the interception loss materially. The interception loss by some plants and weeds is greatly augmented when the plants are in blossom, as in the case of red clover. Wild carrot, which may grow up in a meadow after haying, often has 10 or more blossoms per square foot. Each flowering head is a sponge like structure, which persistently retains a teaspoonful or more of water after every rain. Weeds generally exert a pernicious influence in desiccating the ground through interception as well as in other ways.

In conclusion, credit is due to Mr. James Erwin, for patient, careful work in taking the large number of readings involved in the author's experiments, and to Dr. Floyd A. Nagler, and Mr. Geo. E. Cook, C. E., for the reduction of the several thousand observations involved in this study.

TABLE NO. 21.—*Example of calculated interception losses over cropped area, Seneca River drainage basin above Seneca Falls, N. Y., 1914.*

	May.	June.	July.	Aug.	Sept.	Oct.	Total.
(1)	(2)	(3)	(4)	(5)	(6)	(7)	(8)
P= precipitation on basin [1]	5.110	2.920	2.200	6.320	1.5700	1.920	20.040
d_p= rainfall days [2]	14.000	11.000	12.500	14.000	16.0000	10.000	77.500
$P_d = P/d_p$	.365	.265	.183	.452	.1000	.192	.259
S_n= number of showers [3]	21.000	16.500	18.800	21.000	24.0000	15.000	116.300
P_s= precipitation per shower	.243	.177	.117	.301	.0654	.128	.172
CROP.							
Meadow:							
Per cent of area	22.000	22.000	22.000	17.000	17.0000	17.000	......
Height = h	1.000	2.000	[4].500	[4].500	[4].5000	[4].500	......
$J=(0.005+0.08\ P_s)h$	.024	.038	.007	.015	.0050	.008	[5] 1.818
J^1	.111	.138	.029	.054	.0200	.020	.372
Pasture:							
Per cent of area	25.000	25.000	25.000	[6]18.000	[6]18.0000	[6]18.000	......
Height = h	.500	.500	.500	.500	.5000	.500	......
$J=(0.005+0.08\ P_s)h$	.012	.010	.007	.015	.0050	.008	[5] 1.104
J^1	.063	.041	.033	.057	.0220	.022	.238
Wheat, rye, barley:							
Per cent of area	7.100	7.100	7.100	[4,6]5.000	[4,6]4.0000	[4,6]4.000	......
Height = h	1.000	3.000	4.500	.500	.5000	.500	......
$J=(0.005+0.05\ P_s)h$	.017	.041	.049	.010	.0040	.006	[5] 2.352
J^1	.025	.048	.065	.010	.0040	.004	.156
Oats:							
Per cent of area	8.800	8.800	8.800	6.000	4.0000	4.000	......
Height = h	.500	1.000	2.000	[4,6].500	[4,6].5000	[4,6].500	......
$J=(0.007+0.07\ P_s)h$	.012	.019	.030	.014	.0060	.008	[5] 1.687
J^1	.022	.028	.050	.018	.0060	.005	.129
Potatoes, beans, cabbage, etc.:							
Per cent of area	10.000	10.000	10.000	10.000	8.0000	2.000	......
Height = h	......	.400	.800	1.000	1.0000	1.000	......
$J_1=(0.02+0.15\ P_s)\frac{h^2}{4}$	......	.002	.006	.016	.0070	.010	[5].800
J^1	......	.003	.011	.034	.0130	.003	.064
Corn:							
Per cent of area	4.400	4.400	4.400	4.400	4.4000	4.400	......
Height = h	[6]0	0.700	2.500	5.500	6.5000	0	......
$J_1=(0.005+0.005\ P_s)\frac{h^2}{10}$	......	0	.003	.020	.0220	......	[5] 1.003
J^1	......	0	.002	.019	.0230	0	.044
Buckwheat:							
Per cent of area	......	......	2.700	2.700	2.7000	2.700	......
Height = h	......	......	.500	1.500	2.5000	.500	......
$J=(0.007+0.07P_s)h$	......	......	.008	.042	.0300	.008	[5] 1.872
J^1	......	......	.004	.024	.0190	.003	.050
Orchard and vineyeard [7]	5.000	5.000	5.000	5.000	5.0000	5.000	......
Trees:							
$J_1=(0.04+0.18\ P_s)\ 0.33$	.028	.024	.020	.031	.0170	.021	......
J^1	.029	.020	.019	.033	.0200	.016	.137
Grass:							
Height = h	1.000	2.000	.500	.500	.5000	.500	......
$J_1=(0.005+0.08\ P_s)\ h\times\frac{5}{6}$	.200	.032	.006	.013	.0040	.007	......
J^1	.021	.026	.006	.014	.0050	.005	......
Total of trees and grass J_1=	.048	.056	.026	.044	.0210	.028	[5] 4.269
Do. J^1	.050	.046	.024	.046	.0250	.021	.212
Woods (mixed hardwood):							
Per cent of area [8,9]	10.000	10.000	10.000	10.000	10.0000	10.000	......
$J=0.04+0.18\ P_s$	.084	.072	.061	.094	.0520	.063	[5] 8.268
J^1	.176	.119	.115	.197	.1250	.094	.826
Roads and bare surfaces	5.000	5.000	5.000	5.000	5.0000	5.000	......
No loss	......	......	......	......	......	......	0
Total intercepted loss	.476	.443	.352	.492	.2770	.188	2.228

[1] Weighted mean.
[2] Mean of Shortsville and Wedgewood.
[3] Estimated at 1.5 showers per rainfall day.
[4] Stubble.
[5] Total for 100 per cent area and all showers.
[6] Remainder new plowed.
[7] Estimated at 33 per cent of cover for trees, 2 rods apart, 20 feet diameter crowns. Grass interception under trees taken at 50 per cent of that in open or $\frac{5}{6}$ of value for open meadow.
[8] Including wooded swamps.
[9] Estimated at 85 per cent cover and 50 per cent added for secondary interception.

NOTE.—Where crop does not afford complete cover, the projection factor is included in the formula, and the interception depth on the entire cropped area is designated J_1.

REFERENCES ON INTERCEPTION.

1. Forest Influences, Bul. No. 7. U. S. Forest Service, 1893.
2. **Ebermayer,** Influence of Forests on Air and Soil. (Ger.) April, 1873.
3. Ebermayer's Experiments, Results translated by R. E. Horton, *Michigan Engineer,* 1909.
4. **Alter, J. C.**—Where the Snow lies in Summer. MONTHLY WEATHER REVIEW, May, 1911, **39:** 758-760. Shows photo of trees bearing six inches of moist snow, and illustrates the great snow interception by conifers.
5. **Horton, R. E.**—Deforestation, Drainage & Tillage. *Michigan Engineer,* 1906.
6. **Moore, Willis L.**—The Influence of Forests on Climate and on Floods. Washington, 1910.
7. **Buhler, Dr. A.**—Swiss Central Bureau for Forest Research Investigation. Proceedings, vol. 1, Zurich, 1891. (Ger.) Contains detailed results of Adlisberg and other Swiss stations.
8. **Chittenden, H. M.**—Forests, Reservoirs, and Stream Flow. *Trans. Am. Soc. C. E.,* vol. 57, 1909, pp. 244-546.
9. **Ney.**—Forests and Springs. * * * Tubingen, 1893. (Ger.)
10. **Zon, Raphael.**—Forests and Water in the light of Scientific Investigation. Final Report, National Waterways Commission, Washington, 1912, pp. 205-73. This paper contains a very full bibliography (pp. 274-302) of forest meteorology, where additional references to interception will be found.

AN ANALYSIS OF EVAPORATION FROM A STAND OF SCOTS PINE

A. J. RUTTER

Botany Department, Imperial College, London, England

ABSTRACT

Evaporation from a forest of *Pinus sylvestris* in southeast England was estimated from two sets of observations. The first consisted of direct determinations of interception, transpiration, and evaporation from the soil surface in forty-four sample periods each of 24 hr, during 3 years; the second, of observations on changes in soil moisture when the soil was drier than field capacity, during 6 years. Resistances to vapor diffusion and transport were determined, namely the internal resistance of the leaf in relation to stomatal behavior, and the resistance between the leaf and the air above the canopy in relation to wind-speed. These resistances, suitably transformed, were incorporated in a formula due to Penman which was then found to give a satisfactory fit with observed evaporation. The rate of evaporation of intercepted water was many times as great as the transpiration rate in similar conditions and the theoretical treatment accounted for this. The significance of interception in total evaporation from this forest is briefly discussed.

THE purpose of this paper is to examine some observations on evaporation in a forest of *Pinus sylvestris* in southeast England and to determine to what extent they can be accounted for quantitatively in terms of physical properties of the environment and of the trees.

SITE, AND SOIL MOISTURE CONDITIONS

The observations were made in a plantation of *P. sylvestris* of area about 5 acres (10 ha), which was surrounded by other plantations of the same species and differing only a little in height and age, so that there were fairly uniform conditions 0.5 to 1.0 mile in all directions around the experimental plot. The trees were planted in 1941 and when the observations were begun in 1957 their average height was 19 ft (5.7 m). The number of trees per acre was about 1700 (3600 per ha) and the annual height increment was about 1.5 ft (45 cm). The stand had the advantage that the canopy was still within fairly easy reach of the ground while its weight, about 9000 kg/ha, was almost as great as that recorded for much older stands (Ovington, 1957).

The deepest roots extended to about 6 ft (180 cm), and 37.0 cm of available water could be stored within this depth of soil (Rutter and Fourt, 1965). The effects of drying of the soil on transpiration were investigated in an experiment which is not yet published, but which showed that only in the very dry summer of 1959 was transpiration of the plantation reduced below that of a plot which was established within it and on which the soil was maintained near field capacity by irrigation. Even then the reduction was by no more than 20 percent and occurred only in August and September. It is therefore assumed that, except in these two months, the values of transpiration to be discussed in this paper are independent of variation in soil moisture, i.e. they represent potential transpiration.

ESTIMATES OF EVAPORATION FROM OBSERVATIONS ON SOIL WATER

By a method which has been fully described and discussed (Rutter, 1964), estimates of evaporation were obtained from observations on soil moisture in each of the 6 years 1957 to 1962.

Paper D3 is reprinted from pages 403–416 of *International Symposium on Forest Hydrology* (ed. by W. E. Sopper & H. W. Lull).

The method begins with the equation

$$E = P \pm \Delta S + C - R$$

where E = evaporation, P = precipitation, ΔS = change in storage in the root zone, C = rise by capillarity from the water table to the root zone, and R = runoff or drainage to the water table.

By careful observations with tensiometers a period was recognized each year which extended from the last occasion in spring when the soil was at field capacity and drainage had just ceased, to the first occasion in winter when field capacity was re-established and drainage was resumed. During this period there was neither runoff nor drainage to the water table and between its beginning and end it was assumed that there was no net change in storage. The total of evaporation for the period was therefore

$$E = P + C.$$

This period, of soil-water deficit, began each year in May and extended to some time between October and March according to conditions. The estimates of evaporation were therefore for long periods and included a large part of the total annual evaporation. They were compared with estimates of evaporation from open water (E_o) for the same periods, made by Penman's (1956) formula. The component P varied in different years from 40 to 53 cm, which was equivalent to 1.0 to 1.2 E_o with a mean of 1.1 E_o. C could not be neglected since on this site the water table was near to the roots in the first half of the period of soil-water deficit. It was estimated by the method of Gardner and Fireman (1958) to be between 4.5 and 11.0 cm in different years, and to be equivalent to an additional 0.1 to 0.2 E_o. The mean of 4 years was 8 cm. However, the calculations were made from observations which were not entirely suitable and there remained some uncertainty about the magnitude of capillary rise. Investigations were then extended to further sites (Rutter and Fourt, 1965) in which the water tables were much deeper in the soil. From 3 years of observations on four sites including that on which the original work was done it was concluded that on average the potential evaporation from the plantations was about 1.2 E_o during the period of soil-water deficit. By comparing the original site with an adjacent one on which no water table could be found within 3 m of the deepest roots, it was concluded that capillary rise on the former had been over-estimated and that it was nearer 2 cm than 8 cm per annum.

In the 3 years 1960–2 in which the four afforested sites were studied, the same methods were applied to an adjacent grass-covered site. Here, transpiration was apparently restricted at times by drought and shallow rooting, but the observations were consistent with the assumption that the *potential* transpiration of the grass was 0.7–0.8 E_o in the summer months. The potential transpiration rate of the pine plantations was therefore about 50 percent greater than that of grass. It cannot, however, be assumed that the factor 1.2 E_o would have universal validity wherever *P. sylvestris* might be grown and an attempt has been made to find an explanation for this result in terms which would have general application. While the observation of the length of the period of soil-water deficit gives, in favorable circumstances, good estimates of the total evaporation over a long period of time, the very length of the period means that little can be learnt by this method about the relation of evaporation to environmental conditions. For this purpose the following observations were of more value.

OBSERVATIONS ON SAMPLE DAYS

On 44 days in the 3 years 1957 to 1959, about once in 2 or 3 weeks in summer, and once a month in winter, direct observations on evaporation were made for periods of 24 hr. The days were chosen without regard to weather conditions and included both rainy and dry days. Full details of some of the methods are still to be published and will be only briefly outlined.

Transpiration was measured by the loss in weight of detached branches or leaves in the first 10 min after removal from the tree. Between weighings the detached parts were hung in approximately the positions from which they had been taken. It was established experimentally that during this period there was little or no reduction of transpiration caused by excision. A carefully designed sampling program took account of the diurnal variation and also distinguished the difference between three layers of the canopy and between the north and south sides of trees. Estimates of transpiration per unit leaf weight were combined with canopy weights per unit ground area which were determined from the product of trees per unit area, branches per tree, and weights of foliage per branch, in a sampling program which allowed the seasonal variation in canopy weight to be followed.

The methods used to measure interception have already been described (Rutter, 1963), and the rate of evaporation of intercepted water was also followed by weighing detached branches or leaves when the canopy was wet.

Evaporation from the soil surface was measured by weighing small lysimeters which contained samples of the uppermost 10 cm of soil.

All these methods are obviously open to criticism, but their accuracy was tested by comparing estimates of evaporation obtained from observations on the length of the period of soil-water deficit, with estimates for the same period obtained from the direct observations on sample days. This comparison is shown in Table 1. The close agreement between the means for 3 years is partly due to the arbitrary choice of a figure for capillary rise, which may be underestimated by a few centimeters, but it is clear that the sample days gave reasonably accurate estimates of evaporation.

COMPARISON OF EVAPORATION ON SAMPLE DAYS WITH E_o

This comparison is made in Fig. 1, where each double histogram is the mean of 4 sample days, except for rainless days in March, April and July which are based on 3

TABLE 1

A Comparison of Estimates of Evaporation Made by Two Methods

Period of water deficit	1957 28 May to 23 Dec.	1958 16 May to 25 Oct.	1959-60 8 May to 22 Feb.	Mean
No. of days	210	163	290	
Precipitation, cm	46.6 ± 3.7	40.6 ± 1.9	53.1 ± 2.2	
Assumed capillary rise, cm	2.0	2.0	2.0	
Total, cm	48.6	42.6	55.1	48.8
No. of sample days	11	9	12	
Estimate from sample days, cm	40.1 ± 3.8	45.5	60.7	48.7

Standard errors are shown where replicated sampling allowed their calculation. The standard errors of precipitation arise because of the variation in the estimate of the period of water deficit obtained from four replicate sets of tensiometers. The periods shown at the head of the columns are means of these four.

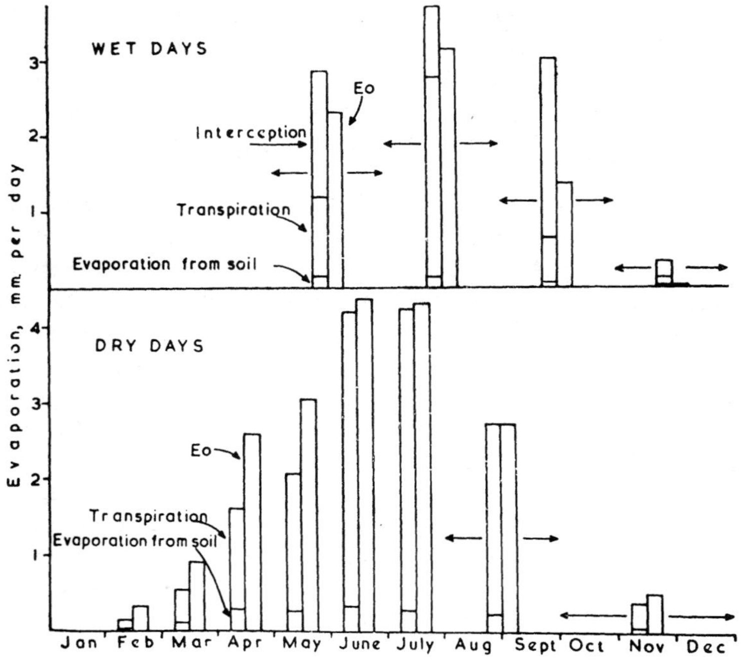

FIG. 1. A comparison of measured evaporation on sample days with calculated evaporation from an open water surface (E_o). Each double histogram is the mean of 3 or 4 days. Horizontal arrows show the period to which the histograms relate, where this exceeds 1 month.

days. It will be seen that on rainless days the observed evaporation was considerably less than E_o from February to May (no observations were made in January) and that the absolute difference is large in April and May. Observed evaporation was approximately equal to E_o from June to December, on rainless days. On days when rain fell the observed evaporation, apparently enhanced by interception, considerably exceeded E_o irrespective of the time of year. It has already been shown (Rutter, 1963) that in this plantation the saturated canopy retains 1.6 to 1.8 mm water, that this can be evaporated in a few hours in summer, and that even in winter it is almost always evaporated within 24 hr. In all the sample days the interception shown is the quantity which was evaporated within the 24 hr period. The individual days on which the histograms are based show fairly consistently the same relations as are shown by their means in Fig. 1. It is clear that the ratio of E:E_o = 1.2 is a generalization which may be adequate on this site for long periods but that it varies seasonally and also with the incidence of wet periods.

RESISTANCES TO WATER VAPOR TRANSPORT

The rate of evaporation from a wet surface can be expressed by

$$E = (c_s - c)\frac{1}{r} \text{ g cm}^{-2}\text{ sec}^{-1} \qquad (1)$$

where c_s and c are the vapor concentrations in g/cm^3 at the surface and in the air above it respectively, and r is the resistance to vapor diffusion or transport, in sec/cm, between surface and the plane at which c is measured.

In moving away from the cells within the leaf, water vapor has to pass a number of resistances which may be grouped as follows:

r_i, the internal resistance of the leaf, mainly though not entirely due to the stomates;

r_s, the resistance of the layer of still air in contact with the surface of the leaf; and

r_t, the resistance of the turbulent air between the outer edge of the surface layer and the plane at which air humidity is measured.

r_s and r_t are both reduced by wind and will be grouped as r_a, the air resistance. For a transpiring leaf, equation (1) may be rewritten

$$E_T = (c_s - c)\frac{1}{r_i + r_a}\ \mathrm{g\,cm^{-2}\,sec^{-1}} \qquad (2)$$

where c_s is now the concentration of saturated air corresponding to the temperature of the leaf.

Similarly the evaporation of intercepted water from a wet leaf surface may be written

$$E_I = (c_s - c)\frac{1}{r_a}\ \mathrm{g\,cm^{-2}\,sec^{-1}} \qquad (3)$$

For all sample days in 1959 observations on the rates of transpiration and the evaporation of intercepted water were combined with measurements which would enable the resistances in equations (2) and (3) to be calculated. The difference between the temperature of the leaf and that of the air circulating in the canopy was measured by thermocouples in which one junction was in contact with a leaf and the other was in the air at a distance of about 15 cm. (In practice the first junction was held between the two leaves of a fascicle which were lightly bound around it.) Four such double junctions were placed at random in the canopy and were moved to new positions each sample day. The temperature and the humidity of the air circulating in the canopy was measured with a whirling psychrometer. It would have been better if the reference point of measurement had been above, rather than within, the canopy, but the analysis which was to be made was not fully foreseen at the time. Fortunately however, observations made at 2-hourly intervals during the sample days on temperature and humidity both within and above the canopy, enabled the resistance in this part of the vapor transport path to be estimated.

The result of all these observations was that the mean resistance to the diffusion of water vapour from wet leaves into the air circulating in the canopy was found to be 1.24 sec/cm, with a range from 0.6 to 3.4 sec/cm, and the corresponding mean for transpiring leaves with dry surfaces was 9.1 sec/cm, with a range from 4.0 to 22.0 sec/cm. The sources of variation will be examined later. This implied that the internal resistance averaged about 8 sec/cm. (Note that the leaf area on which evaporation rate was based included both sides of the leaf and was calculated from length and breadth on the assumption that the leaves are half-cylinders.) The internal resistances seemed high compared with other published results. Monteith (1965), for instance, quotes values for seven species, estimated from transpiration rates by a number of workers, which lie between 0.8 and 3.3 sec/cm. A rough check was therefore made by measuring the stomates in a few sections and counting their number per unit area. The formula of Penman and Schofield (1951) was used to calculate the resistance of the stomatal array. Estimates ranged from 7.6 sec/cm when 60 percent of the stomates were open to 22.4 sec/cm when 20 percent were open. No great accuracy can be claimed for the observations but they confirmed the order of magnitude of the estimates obtained from transpiration rates.

The large difference between the resistances associated with wet and dry leaves suggested an explanation for the higher rates of evaporation relative to E_o on wet days which are shown in Fig. 1, but further analysis was necessary to see whether the observed evaporation rates were consistent with the resistances and the prevailing environmental conditions.

INCORPORATION OF RESISTANCES IN THE PENMAN FORMULA

The usefulness of equations (2) and (3) is limited because the surface temperature of vegetation is not easily recorded. An outstanding feature of the Penman formula is that it combines an equation for vapor transport with an equation describing the energy relations of transpiration. Both these equations require a knowledge of surface temperature, but by combining them surface temperature is eliminated. Penman (1952) showed how resistances to vapor transport could be included in his formula to give the expression

$$E_T = \frac{\Delta H_T + \gamma E_a}{\Delta + \gamma(1/SD)} \text{ mm/day,} \quad (4)$$

Δ is the slope of the saturation vapor pressure curve at mean air temperature and γ is the constant of the wet- and dry-bulb psychrometer equation. H_T is the net radiation at the vegetated surface, divided by the latent heat of vaporization of water, and

$$E_a = 0.35(e_d - e_a)\cdot f(u) \text{ mm/day} \quad (5)$$

where $(e_d - e_a)$ is the vapor pressure deficit of the air in millimeters of mercury and $f(u)$ is a function of wind-speed;

$$S = \frac{R_a}{R_a + R_i}$$

where R_a and R_i are the air and leaf resistances (sec/cm) above unit ground area, and D is a factor which is designed to account for the restriction of transpiration by the closure of the stomates at night. Its formula need not be reproduced here.

Following Penman (1952) it is possible to show that $f(u) = 2.6/R_a$ whence substituting in equation (5),

$$E_a = 0.91(e_d - e_a)\frac{1}{R_a} \text{ mm/day.} \quad (6)$$

Note that the factor 0.91 accounts for the difference in units between equation (6) and equations (1), (2) and (3).

The derivation and assumptions of equation (4) have been further discussed by Businger (1956), Tanner and Pelton (1960), Slatyer and McIlroy (1961), and Monteith (1965).

In order to incorporate the measured resistances (r) into equations (4) and (6) it must first be noted that they are resistances associated with unit leaf area, and that R_a in equation (6) is the air resistance over unit ground area. Clearly the ratio $R_a/(R_a + R_i)$ is dimensionless, but for convenience R_i also has been calculated from r_i. The successive leaves above unit area of ground can be regarded as resistances in parallel and values of R can be obtained from r by dividing the latter by the ratio of leaf area to ground area.

It will be remembered that r_a was measured in the first place between the leaves and the air circulating in the canopy. The corresponding component of R_a is plotted against

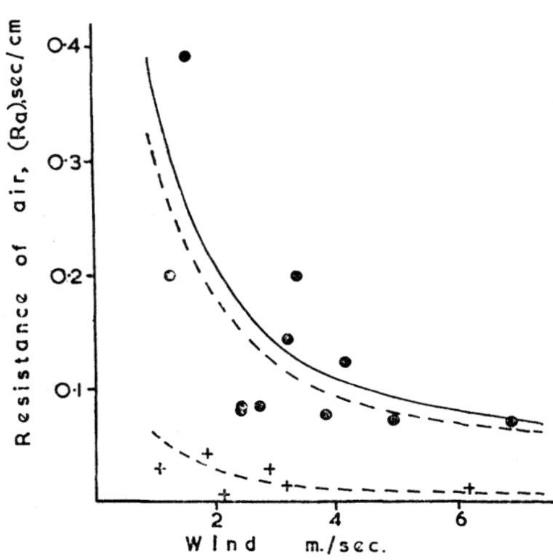

FIG. 2. The relation between wind-speed at 2 m height at the reference station and the resistance to vapor transport (R_a) between the canopy and the air above it. Circles and upper dotted line: resistance between leaves and air within the canopy. Crosses and lower dotted line: resistance between air within canopy and air above canopy. Continuous line: total air resistance.

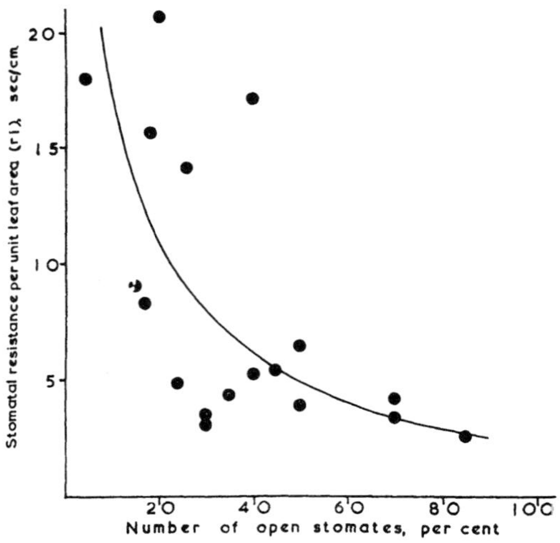

FIG. 3. The relation between internal resistance of the leaves (r_i) and the proportion of open stomates, as judged by infiltration with 9:1 xylene:ethyl alcohol.

windspeed at the reference station, as dark circles in Fig. 2, and a curve is fitted by eye. Determination of the small component of R_a between the air in the canopy and the air above it are plotted as crosses in Fig. 2. These values are not sufficiently accurate to detect the effect of wind and the curve drawn through them simply assumes that their product with wind-speed is a constant. The two components are added to give the uppermost curve representing the total R_a between the leaves and the air about 6 ft above the canopy. This curve conforms closely to the formula

$$R_a = 0.37 w_2^{-0.84}$$

where w_2 is wind-speed, m/sec, at 2 m height at the reference station.

Alternatively, the wind function $f(u) = 2.6/R_a$ in equation (5) is fairly closely represented by $2+0.09u_2$ where u_2 is the wind speed in miles per day, at 2 m height at the reference station.

In the case of observed values of r_i+r_a obtained from transpiring leaves, the expected value of r_a was calculated from the wind-speed, and subtracted to give r_i. The values of r_i were very variable but in about half the examples the stomates had been examined by infiltration with a solution of gentian violet in a 1:9 mixture of ethyl alcohol and xylene and the proportion of stomates open to this mixture had been estimated. It was found that much of the variation in internal resistance was related to stomatal behavior. The internal resistance has been plotted against the percentage of open stomates in Fig. 3 and a curve fitted by eye. The further calculation of the seasonal fluctuation of R_i is shown in Fig. 4. On various occasions throughout 3 years the stomates had been examined by this infiltration technique and the lowermost part of Fig. 4 shows

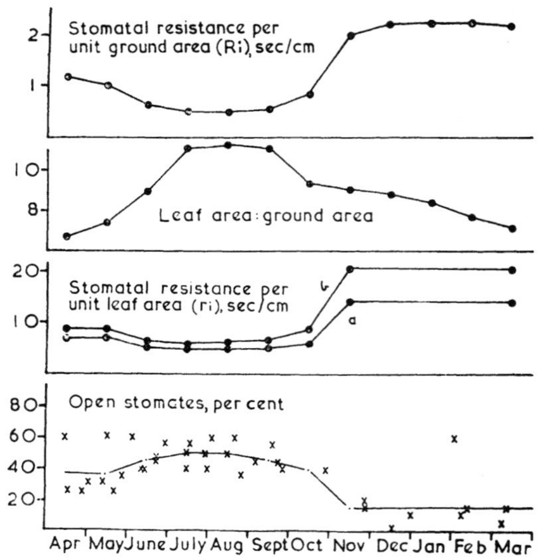

FIG. 4. Stages in the calculation of internal resistance (R_i) per unit ground area. Percentages of stomates open to infiltration are data from 3 years' observations and each point is a mean value for the period between 3 hr after sunrise and 3 hr before sunset. Combining these with Fig. 3 gives the annual cycle (a) of internal resistance per unit leaf area (r_i) during the same daily period. This is corrected to represent in (b) the means for the whole period sunrise to sunset; (b) is divided by (leaf area: ground area) + 1 to give stomatal resistance per unit ground area (R_i). The quantity +1 is a modification designed to include evaporation from the soil with transpiration.

an annual cycle of stomatal behavior in which a line joins the means for each month. Above this is shown (a) the corresponding values of r_i for each month derived from Fig. 3. The stomates were observed only in the period from 3 hr after sunrise to 3 hr before sunset. Since transpiration was negligible at night it could be assumed that the stomatal conductance rose from zero at sunrise and fell to zero again at sunset and it could be calculated that the mean internal resistances in line (a) were underestimated by amounts varying from 25 percent in the summer months to 55 percent in mid-winter because of the exclusion of 3 hr at the beginning and end of the day. The mean internal resistances were therefore correspondingly increased in line (b). In the next section of Fig. 4 is shown the seasonal variation of the ratio leaf area (both sides): ground area, based on 3 years' observation.

Some allowance had to be made for evaporation from the soil surface and since it was found that evaporation per unit ground surface was approximately equal to transpiration per unit leaf area, the ground was empirically treated as an extra unit leaf area and the mean internal resistances in line (b) were divided by 1+ (leaf area: ground area) to give the values of R_i in the uppermost part of the figure. For the purpose of further calculations the values of R_i were rounded off to

April and May	1.10 sec/cm
June	0.65 ,,
July–September	0.55 ,,
October	0.95 ,,
November–March	2.20 ,,

Finally, values of D, the daylength factor, were calculated by the formula in Penman (1952) for each of the sample days. They were usually between 0.95 and 1.0 in June, July and August, and declined to about 0.7 to 0.75 in December to February.

SOURCES OF METEOROLOGICAL DATA FOR THE CALCULATION OF POTENTIAL EVAPORATION

H_T in equation (4) was calculated following Penman (1956), employing hours of sunshine as the measure of incoming radiation, and assuming that the pine canopy reflected 15 percent of the incoming radiation. A reflection factor of 14 percent is quoted by Thornthwaite (1954) from Angstrom (1925). Observations of hours of sunshine, and air temperature and humidity were obtained from a meteorological station about 7 miles (12 km) from the forest. The anemometer at this station proved to be unreliable and wind records had to be obtained from a more distant station. As has already been explained a relation between R_a and wind at this station was established (Fig. 2). Because R_a is small, the term γE_a is 6 to 9 times as large as ΔH_T in the numerator of equation (4). The vapor pressure in E_a must therefore be determined accurately and the mean of four 6-hourly observations of e_a was used.

Since evaporation from the pine plantation exceeds that from a grass cover it might be expected to alter the temperature and humidity microclimate, in which case meteorological observations above the canopy would be preferable to those obtained at a standard meteorological station above short grass. In one year, 1959, observations of temperature and humidity were made at 2-hr intervals above the canopy on sample days for comparison with the records from the reference station.

POTENTIAL EVAPORATION ON THE SAMPLE DAYS

Values of E_T were now calculated for rainless sample days using equations (4) and (6), and the effects of incorporating the stomatal and daylength factors were investigated in three stages. First, mean annual values of both R_i and D were used. Then R_i was allowed its seasonal variation while D was held at its mean value. Finally both R_i and D

were allowed to vary. The regressions of observed evaporation (E_A) on calculated evaporation (E_T) and their errors, were as follows:

CASE 1. $R_i = -1.35$, $D = 0.87$.
$E_A = -0.34 + 1.61\ E_T$ mm/day.
Standard error of divergences of E_A from line $= \pm 0.50$ mm/day.

CASE 2. R_i varying, $D = 0.87$.
$E_A = -0.01 + 0.90\ E_T$ mm/day. (7)
S.E. $= \pm 0.40$ mm/day.

CASE 3. R_i and D varying.
$E_A = 0.14 + 0.83\ E_T$ mm/day.
S.E. $= \pm 0.40$ mm/day.

It will be seen that allowing R_i to vary while holding D at 0.87 resulted in a line which passed through the origin, had a slope very near to 1, and about which the error was appreciably reduced compared with the line when R_i was held constant at its mean value. But allowing D to vary gave a line with a slope further from 1, which no longer passed exactly through the origin, and which produced no further reduction in the error variation. The method of estimating E_A embodied in Case 2 and equation (7) was therefore adopted in further work.

In the case of the sample days on which rain fell, the appropriate formula for the evaporation of intercepted water is,

$$E_I = \frac{\Delta H_T + \gamma E_a}{\Delta + \gamma}\ \text{mm/day.} \qquad (8)$$

All symbols have the same meaning as in equations (4) and (6).

The known interception was divided by E_I to give the time needed for the evaporation of intercepted water. The remaining time was then multiplied by E_T to give the expected transpiration plus evaporation from the soil, E_T' mm. The regression of E_A' (i.e. observed transpiration plus soil evaporation) on E_T' was calculated, allowing R_i to vary and holding D constant as before. It was

$$E_A' = -0.02 + 0.95E_T'\ \text{mm},$$

with a standard error of ± 1.2 mm. The close agreement with equation (7) demonstrates the reliability of equation (8) for determining the time needed to evaporate intercepted water and hence the time remaining for transpiration. In passing it may be noted that for the same environmental conditions E_I may be anything from 2 to 10 times as great as E_T with a mean of about 4 times as great.

The close correspondence between observed and calculated evaporation on the sample days is further illustrated by Fig. 5.

E_T and E_T' exceeded E_A and E_A' on 44 sample days in 3 years by about 11 percent and this might be because the meteorological data used were not measured above the canopy. In 1959, when observations above the canopy were made on the sample days, air temperature was on average 0.7°C less than at the reference station and the average vapor pressure deficit was 2.25 mm Hg compared with 2.45 mm Hg at the reference station. E_T calculated at the reference station overestimated E_A by 6.5 percent in this particular year, and when calculated from meteorological observations above the canopy it underestimated by 1.5 percent. This suggests that meteorological observations above the canopy are preferable but the differences are so small that they may as well be due to the difference of 7 miles between the two locations as to the difference in vegetation.

CALCULATION OF EVAPORATION DURING THE PERIODS OF SOIL-WATER DEFICIT

Estimates of total evaporation obtained from observations on the length of the periods of soil-water deficit in 6 years provide a further test for the evaporation formula. It will be remembered that there is no reason to suppose that transpiration was reduced below its potential rate during these periods. Potential evaporation was calculated for each month by the formula

$$E_T' + I = 0.9E_T(n - I/E_I) + I\ \text{mm} \qquad (9)$$

where n is the number of days in the month

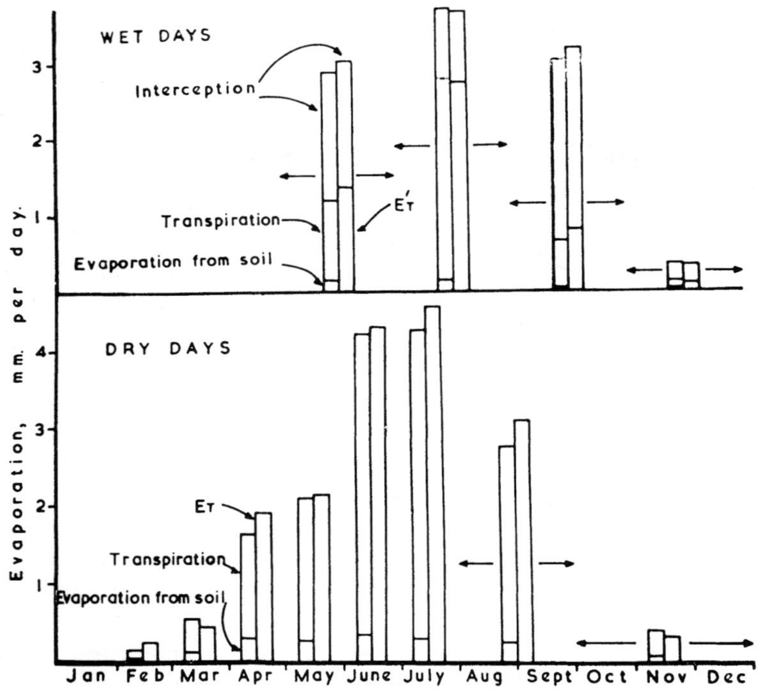

FIG. 5. A comparison of measured evaporation with calculated evaporation, on sample days. In the upper part of the figure observed interception is shown in each half of the double histograms E'_T and E_T are the calculated values of the sum of transpiration and evaporation from the soil. on wet and dry days respectively. Each double histogram is the mean of 3 or 4 days. Horizontal arrows show the period to which the histograms relate when this exceeds 1 month.

and I is the total interception in the month. For April 1957 to March 1960, I was known. For April 1960 to March 1963 it was estimated from the daily totals of precipitation and a graph relating interception to the precipitation falling in 24 hr (fig. 3 in Rutter, 1963).

It will be remembered that evaporation in the period of soil-water deficit is the sum of precipitation and capillary rise. Table 2 shows, for 6 years, precipitation during the period of soil-water deficit, and the sum of $E'_T + I$ (equation (9)). $E'_T + I$ should exceed precipitation by an amount equal to capillary rise, and does in fact exceed it in each period, on average by 4.1 cm. Capillary rise was calculated by Gardner and Fireman's (1958) method to be 8.0, 9.0, 11.0 and 4.5 cm in the years 1957 to 1960 respectively, with an average of 8.0 cm, and was estimated in 1960, 1961 and 1962 from comparison with a plot with a deep water table, to average 2.2 cm per year. The average of $E'_T + I - P$ lies between these two estimates of mean capillary rise, and calculation has therefore given a satisfactory agreement with observation within the limits fixed by the uncertainty about capillary rise. Only in 1959 does the calculation give a rather unsatisfactory fit. E'_T for 1959 has been purposely reduced below its calculated level, by 10 percent in August and by 16 percent in September because an experiment had suggested that transpiration was reduced by these amounts following drying of the soil. It is possible, however, that in this unusually dry summer, transpiration was reduced by rather more than was indicated by the experiment referred to, and that this accounts for the rather large discrepancy between P and $E'_T + I$.

TABLE 2

A Comparison of Estimates of Evaporation made from Observations on the Length of the Period of Soil-Water Deficit and from the Potential Evaporation Formula

Period of soil-water deficit:	1957 28 May to 23 Dec.	1958 16 May to 25 Oct.	1959-60 8 May to 22 Feb.	1960 6 May to 10 Dec.	1961-2 1 June to 20 Jan.	1962-3 13 May to 15 Mar.
Precipitation* (P), cm	46.6	40.6	53.1	50.0	49.2	47.7
E'_T + Interception (I), cm	51.9	43.0	66.0	51.1	48.8	49.8
$E'_T + I - P$, cm	5.3	2.4	13.1	1.1	0.4	2.1

*For the period of soil-water deficit, Evaporation = Precipitation + Capillary rise.

On the other hand it should be noted that the direct estimate of capillary rise by Gardner and Fireman's method was particularly high in this year.

DISCUSSION

It seems necessary first of all to ask whether any two derived quantities which have been compared may be subject to the same errors or alternatively have been determined by completely independent methods. The rates of transpiration and evaporation of intercepted water per unit area of forest on the sample days may be subject to errors in sampling the variation in time and space and to errors in determining the weight of the canopy per unit area of forest. The measurements of transpiration may also have been affected by excision of the branches or leaves. Since the resistances R_i and R_a are calculated from vapor pressure gradients divided by the evaporation rates per unit area of forest they are subject to all the errors contained in the evaporation rates. Errors in R_a will cause errors proportionately almost as large in the numerators of E_T and E_I (equations (4) and (8)), and errors in R_i will affect the denominator of E_T. While it can be said that equations (4) and (8) provide satisfactory generalizations of the relation of the observations to environmental factors, it must be recognized that the equations will contain errors proportionately similar to any which exist in the observations.

However, when calculations of $E'_T + I$ (equation (9)) are compared with estimates of evaporation made from observations on the length of the period of soil-water deficit, independent estimates are being compared. In fact the hypothesis examined in Table 2 is that, for the period of soil-water deficit,

$$E'_T + I = P + C$$

or that

$$E'_T + P - T = P + C$$

and hence

$$E'_T = T + C$$

where T is the equivalent depth of water which falls through the canopy and runs down the trunks. T was determined by gages (Rutter, 1963) during a period defined by observations in the soil, and C was also estimated from observations in the soil. E_T, on the other hand, is calculated from meteorological conditions and characteristics of the canopy, and its accuracy is therefore tested, in Table 2, by completely independent observations.

The main point to be made by this paper is that when all the quantities (including R_a, R_i and D) in Penman's 1952 formula are determined, a satisfactory fit with observation is obtained for pine forest, a type of vegetation which differs markedly in structure and, it is now found in evaporation, from grass and short crops. This conclusion is qualified in two respects. The first is that the theoretical expression has to be multiplied by a factor of 0.9 to make it fit observation and the second is that it includes a factor 0.87 for the symbol D which is assumed to represent daylength effects but which has not been convincingly shown to do so. It has been shown that the factor 0.9 could represent the effect of using meteorological observations made 7 miles away over grass, instead of over the pine canopy investigated. So far as D is concerned, the internal resistances employed represent the mean between sunrise and sunset and it is reasonable to include a factor to allow for stomatal closure and very high resistance at night. But allowing this factor to vary seasonally, as it does, gives an overestimate of evaporation in summer and an underestimate in winter, compared with the observations. This could easily be due to some seasonally varying bias in the calculation either of mean daily meteorological conditions or of mean daily internal resistance, which acts against the seasonal variation of D.

Many of the results obtained in the course of this work offer subjects for comparison with other relevant papers, especially those of Ryhiner and Rijtema (1963) and Monteith (1965), but these comparisons require the presentation of further data and are considered to be outside the scope of the paper except for points which will be made briefly.

Significance of Interception

Both from theory and in practice it has been found that the rate of evaporation of intercepted water is, on average, about 4 times as great as the transpiration rate in the same environmental conditions. It might therefore be expected that evaporation would be increased in wet weather when interception is greater. In Fig. 6 interception and total evaporation $(E_T' + I)$ are plotted against rainfall for the periods April to September and October to March, for each of 6 years.

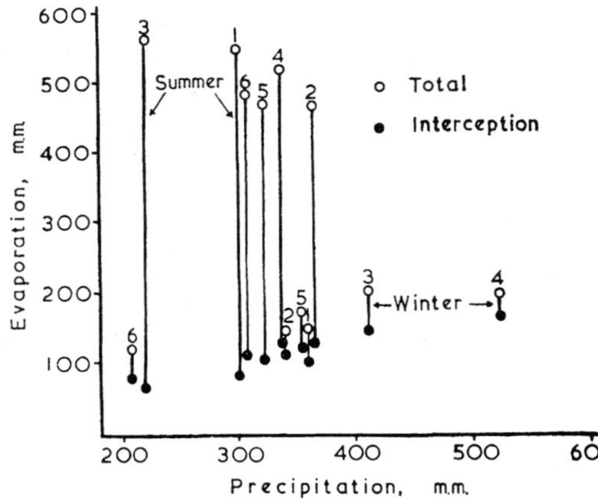

FIG. 6. The relation of interception and total evaporation to rainfall, in the periods April to September (summer) and October to March (winter), for 6 years. Closed circles: interceptions. Open circles: total evaporation. The lines joining them are calculated E_T. Long lines: April to September. Short lines: October to March. Numbers 1 to 6 are 1957 to 1962 respectively, reckoning the year from April to March in the following year.

It will be seen that in the winter period, when interception accounts for most of the evaporation, evaporation tends to increase with rainfall. But in the summer months when transpiration is far more important, as rainfall and interception increase, E_T' and the total of evaporation tend to decrease, presumably because rainy seasons are also cloudy ones, and solar radiation is less. When the results are summed for complete years, there is no discernible relation between rainfall and evaporation.

Energy Relations

The calculated, and observed, rates of evaporation of intercepted water are of the order of 1 to 3 mm/day in winter and 7 to 10 mm/day in summer and they exceed the equivalent intensity of net radiation considerably. It must therefore be concluded that the additional energy required is obtained from the air. There is other evidence that this is so, for wet leaves were found to be 0.3° to 1.0°C cooler, according to conditions of radiation and vapor pressure deficit, than the air around them whereas in the same range of environmental conditions transpiring leaves were only 0 to 0.1°C cooler than the air. Moreover, the air above the wet canopy was on average 1°C cooler than air at the reference meteorological station, whereas the air above the dry canopy was only 0.5°C cooler.

In summer the temperature of the transpiring canopy was 0 to 0.2°C higher by day than that of the air moving through it. Calculated net radiation, converted to the equivalent evaporation rate, barely exceeded the sum of interception and calculated transpiration, during the months April to September. Very few observations were made in winter on the temperature of the transpiring, as distinct from the wet, canopy, but the sum of interception and calculated transpiration was about 4 times as great as calculated net radiation for the months October to March, which is equivalent to an excess over net radiation of 17 percent for the whole year. Although the plot was within a forest with a diameter of about 2 miles, it appeared therefore to receive appreciable advected energy in winter.

Instability of the Formula

A practical advantage of Penman's formula for the potential transpiration of grass

$$E_T = \frac{\Delta H + \gamma E_a}{\Delta + \gamma} \text{ mm/day,}$$

where $E_a = 0.35(e_d - e_a)(1 + 0.01u_2)$
and u_2 = wind-speed at 2 m height in miles/day,

is that in temperate climates in summer ΔH is 3 to 5 times γE_a and the whole expression is fairly insensitive to inaccuracies in E_a. But in the formula

$$E_T = \frac{\Delta H + \gamma E_a}{\Delta + \gamma(1/SD)} \quad (4)$$

where E_a is now 0.91 $(e_d - e_a)(1/R_a)$ as in equation (6), the second part of the numerator is about 6 times as large as ΔH in summer, at least in the pine forest where R_a is small, and the second part of the denominator is similarly larger than Δ. Equation (4) is therefore very sensitive to inaccuracies in either R_a or R_i and this reduces its practical use since it is not satisfactory to adopt assumed values. It is difficult to suggest some other more practical expression, for it is unlikely that a standard factor combined with E_o (e.g. 1.2 E_o) would have validity over a range of climates, and net radiation alone does not give a very good estimate of evaporation.

Determination of Resistances to Vapor Transport

The methods of determining R_a and R_i by starting from the leaf surface are cumbersome, though they could have been improved if the use to which they would ultimately be put had been foreseen.

Monteith (1963) has described a method of estimating both the internal and external resistances from micrometeorological profiles and he has used this method (Monteith, 1965) to calculate the resistances of pine forest near Munich in June and July from the data of Baumgartner (1956). At a mean wind-speed of 2 m/sec the air resistance was 0.06 sec/cm and the internal resistance was 0.62 sec/cm.

With a wind-speed of 2 m/sec above the canopy, which corresponds to about 3 m/sec at the reference station in Fig. 2, the air resistance in the present work was 0.14 sec/cm. The internal resistance in June and July averaged about 0.60 sec/cm. The correspondence of internal resistances in the two cases is striking, although the present work suggests that there is an important seasonal variation which must be measured. Monteith's calculations, like the present observations, show that the air resistance is low, but the difference between the two cases is important for if the value of 0.06 sec/cm were applied to the present results it would result in a very large over-estimate of evaporation.

LITERATURE CITED

ANGSTROM, A. (1925) The albedo of various surfaces of ground. *Geog. Ann. Stockholm* **7**, 323-34.

BAUMGARTNER, A. (1956) Untersuchungen über den Wärme- und Wasserhaushalt eines jungen Waldes. *Ber. deutsch. Wetterd., Bad Kissingen* **28**, 2-53.

BUSINGER, J. A. (1956) Some remarks on Penman's equation for the evapotranspiration. *Neth. Jour. Agr. Sci.* **4**, 77-80.

GARDNER, W. R. and M. FIREMAN (1958) Laboratory studies of evaporation from soil columns in the presence of a water table. *Soil Sci.* **85**, 244-9.

MONTEITH, J. L. (1963) Gas exchange in plant communities. In: Evans, L. T., *Environmental Control of Plant Growth*, pp. 95-110. Academic Press, New York and London.

MONTEITH, J. L. (1965) Evaporation and environment in the state and movement of water in living organisms. *Symp. Soc. Expt. Biol.* **19**, 205-37.

OVINGTON, J. D. (1957) Dry matter production by *Pinus sylvestris* L. *Ann. Bot. Lond.* N.S. **21**, 287-314.

PENMAN, H. L. (1952) The physical bases of irrigation control. *Proc. 13th Interntl. Hort. Congr., London* **2**, 913-24.

PENMAN, H. L. (1956) Evaporation—an introductory survey. *Neth. Jour. Agr. Sci.* **4**, 9-29.

PENMAN, H. L. and R. K. SCHOFIELD (1951) Some physical aspects of assimilation and transpiration. *Symp. Soc. Expt. Biol.* **5**, 115-29.

RUTTER, A. J. (1963) Studies on the water relations of *Pinus sylvestris* in plantation conditions. I. Measurements of rainfall and interception. *Jour. Ecol.* **51**, 191-203.

RUTTER, A. J. (1964) *Ibid.* II. The annual cycle of soil moisture change and derived estimates of evaporation. *Jour. Appl. Ecol.* **1**, 29-44.

RUTTER, A. J. and D. F. FOURT (1965) *Ibid.* III. A comparison of soil-water changes and estimates of total evaporation on four afforested and one grass-covered site. *Jour. Appl. Ecol* **2**, 197-209.

RYHINER, A. H. and P. E. RIJTEMA (1963) Berekening van de gemiddelde groote van de verdamping van naldhout (*Pinus nigra, austriaca*) volgens de methode van McIlroy. *Inst. v. Cultuurtech. en Waterhuishouding*, Nota 213 (unpublished).

SLATYER, R. O. and I. C. MCILROY (1961) *Practica Microclimatology*. UNESCO, Paris.

TANNER, C. B. and W. L. PELTON (1960) Potential evapotranspiration estimates by the approximate energy balance method of Penman. *Jour. Geophys. Res.* **63**, 3391-3413.

THORNTHWAITE, C. W. (1954) A re-examination of the concept and measurement of potential transpiration. In Mather, J. R., *The Measurement of Potential Evapotranspiration. Publications in Climatology*, Seabrook, N.J. **7**, 200-9.

DISCUSSION

WAGGONER: Monteith has devised a micrometeorological means of estimating the stomatal resistance and this he did in a field of barley. At the same time I estimated the stomatal resistance of single leaves in that barley field by means of a porometer and the surprising thing was that, although the leaf area index of this barley field was as high as ten, the ratio between the stomatal resistance of single leaves estimated by the porometer and the stomatal resistance of the whole canopy was not ten, but was more like four. Therefore, I wonder, if your 10 percent overestimate could not be explained by merely using a smaller ratio. Would dividing the single needle stomatal resistance by a factor of 5 instead of 10 bring your observations well into line or perhaps over correct them?

RUTTER: It would result in a very large underestimate of the observed evaporation rates. The stomatal resistance which I found per unit leaf area was high compared with any published figures and so I tried to check it by actual measurements of stomatal size and number, and by using the formula which Penman and Schofield developed. I got a good agreement through this calculation. The stomata are rather few in number and narrow, and the pores are rather long, and this accounts for the high resistance. The other check which I have is that Monteith, using the method to which you have referred, has calculated the stomatal resistance to a pine stand from Dr. Baumgartner's data and the figure which he gets is very close to mine. Mine were obtained through a year and Baumgartner's data were for June to July only.

WAGGONER: Do you mean that the stomatal resistances that Monteith calculated from Baumgartner's data were approximately 0.2 cm and yours for a single leaf was 2.0 cm?

RUTTER: We have to make an adjustment for units since Monteith uses the unit sec/cm. That 2 cm has to be divided by the leaf area index, and when I have done that and changed to Monteith's units I obtain almost exactly the same figure as Monteith.

BAUMGARTNER: To calculate the transpiration of a stand we need a full profile for integrating purposes. At what level in the canopy were the resistances measured?

RUTTER: The transpiration rates from which the resistances were calculated were always measured in two layers of the canopy. The third layer contained no more than 10 percent of the leaves. I measured the leaf temperatures separately in the two layers, and the transpiration rates. The main temperature difference is between the actual leaf surface and the air circulating within the canopy. The temperature and humidity differences between the air circulating in the canopy and the air above it are very small and I did not measure air temperature and humidity separately in the two layers of the canopy. The resistances which have been used in the formula are the means of those obtained in the two main layers of the canopy.

Agricultural Meteorology – Elsevier Publishing Company, Amsterdam – Printed in The Netherlands

A PREDICTIVE MODEL OF RAINFALL INTERCEPTION IN FORESTS, 1. DERIVATION OF THE MODEL FROM OBSERVATIONS IN A PLANTATION OF CORSICAN PINE

A. J. RUTTER, K. A. KERSHAW[1], P. C. ROBINS AND A. J. MORTON
Department of Botany, Imperial College, London (Great Britain)
(Received September 25, 1970)

ABSTRACT

RUTTER, A. J., KERSHAW, K. A., ROBINS, P. C. and MORTON, A. J., 1971. A predictive model of rainfall interception in forests, 1. Derivation of the model from observations in a plantation of Corsican pine. *Agric. Meteorol.*, 9: 367–384.

A model for the prediction of interception loss in pine canopies is described. Central to the model is the storage of water on the canopy. This store is added to by intercepted rainfall and depleted by evaporation and drainage. Since the rates of evaporation and drainage vary with the amount of water on the canopy, the model is constructed to calculate a running balance in time of rainfall, throughfall, evaporation and changes in canopy storage. Evaporation from the wet canopy constitutes the interception loss.

Methods of deriving the relevant parameters of the canopy are discussed. In addition, standard meteorological data, recorded if possible at hourly intervals, are required.

Calculations made by the model agreed satisfactorily with observed monthly interception losses in a stand of Corsican pine, for a period of 18 months.

INTRODUCTION

Of the rain falling annually on a forest, the proportion that is intercepted at leaf and stem surfaces and evaporated from them (the interception loss) is commonly between 20 and 40% in conifers and between 10 and 20% in hardwoods (ZINKE, 1967). Since the total annual evaporation (including transpiration) must normally be less than the annual rainfall, it is clear that evaporation of intercepted water is a considerable proportion of total evaporation.

BURGY and POMEROY (1958), in an experiment with lysimeters in a field of rye-grass, concluded that the rate of evaporation of water intercepted by the grass did not differ substantially from the potential transpiration rate of grass in the same climatic conditions. On the other hand WAGGONER et al. (1969) found that the rate of evaporation from a wetted corn crop was at least twice that from the unwetted crop, and RUTTER (1967) found the rate of evaporation of water intercepted by pine forest to be several times as great as the transpiration rate in the

[1] Present address: Department of Biology, McMaster University, Hamilton, Ontario, Canada.

same conditions. Rutter showed that this was because the aerodynamic resistance to transport of water vapour away from the canopy surfaces was very small compared with the internal resistance to vapour diffusion on the transpiration pathway. It has also been shown (RUTTER, 1968; SCEICZ et al., 1969) that low external resistance is an expected consequence of the high aerodynamic roughness of forests.

In order to estimate total evaporation from forest, RIJTEMA (1965) and RUTTER (1967) have employed formulae that combine the transpiration rate and the rate of evaporation of intercepted water, assuming that the two processes do not take place simultaneously from the same leaf surface. A disadvantage of these formulae is that they require that the amount of water, intercepted in a given period and available for evaporation, should be independently measured unless there is some reliable way of predicting it. A simple method of prediction is through the use of established relations of interception loss to rainfall, either as percentages or regressions. In some cases, e.g., the hardwood forests of the eastern United States, regressions of annual net rainfall on annual gross rainfall have been found to be remarkably constant (HELVEY and PATRIC, 1965). However, as ZINKE (1967) and others have pointed out, interception losses, both as absolute amounts and as proportions of gross rainfall are likely to vary with both evaporation climate and intensity and duration of rainfall. It cannot be expected that interception by a given species will be a constant proportion of gross rainfall either in regions with different climates or in all seasons of the year. This paper attempts a more fundamental analysis of the interception process.

DATA

The data used were obtained in a plantation of *Pinus nigra* in southeast England. The plantation was 40 years old and was last thinned 7 years before the observations, to leave 600 trees/ha with a mean height of about 20 m. The observations and methods, fully described by ROBINS (1969), were of two kinds:

(*1*) For about 8 months, gross rainfall and throughfall (collected in a long, calibrated trough) were recorded on chart recorders which allowed resolution of good quality traces into 5-min periods. Mean canopy temperature, and the temperature and humidity above the canopy were recorded on a similar time scale. Wind speed above the canopy was obtained as hourly runs. A random selection of storms from this period was analysed and used to construct a model of the interception process.

(*2*) Records of gross rainfall and throughfall were obtained at approximately daily intervals for 18 months. Since records of canopy temperature were not made throughout this period the basic model was adapted to make use of estimated net radiation, and was then tested against the whole of these data. Stem-flow was measured but was infrequent and negligible, and throughfall and net rainfall were therefore assumed identical (for definitions of terms see ZINKE, 1967). An extension of the model to stands with stem-flow has still to be attempted.

SUMMARY OF THE ANALYSIS

The canopy was regarded as having a capacity to store water on its surface, which is filled by rainfall and emptied by evaporation and drainage.

See Appendix for the list of symbols.

Components of the water balance

Let R, T and E be rates of rainfall, throughfall and evaporation respectively, and ΣR, ΣT and ΣE be totals of these components in a given time (i.e., rate × time). Then the interception loss in a storm, i.e., the water intercepted and evaporated between the time when rain begins to fall on a dry canopy, and the time when the canopy is finally dry again, is:

$$I = \Sigma E = \Sigma R - \Sigma T \qquad (1)$$

Let the proportion of rain which falls through the canopy without striking a surface (i.e., without interception) be p. Then the water balance of the canopy for any period within a storm may be written:

$$(1 - p)\Sigma R = \Sigma D + \Sigma E \pm \Delta C \qquad (2)$$

where ΣD is the amount of water draining or dripping from the canopy (D is a rate), and ΔC is the change in the amount of water, C, stored on the canopy. Note that:

$$\Sigma T = p\Sigma R + \Sigma D \qquad (3)$$

It is assumed that there is a minimum quantity of water, S, required to wet all the canopy surfaces. This corresponds to the storage capacity of ZINKE (1967) following HORTON (1919), or canopy saturation value of LEYTON et al. (1967). It is comparable to field capacity of a soil; as useful in practice and as difficult to define precisely. C may have values greater or less than S. All the above quantities are expressed as mm depth of water.

Evaporation rate

It is assumed that there is a potential evaporation rate E_p which obtains when all canopy surfaces are wet, i.e., when $C > S$. When $C \leqslant S$ it is assumed that:

$$E = E_p \cdot C/S \qquad (4)$$

This assumption is admittedly crude since it can be observed that at the beginning of a storm evaporation takes place from a canopy wet mainly at the top, while in the last stages of drying it takes place from a canopy wet only at the bottom.

A given value of C/S may therefore represent the same quantity of water exposed in very different ways to evaporation.

Assuming that as the amount of water on the canopy approaches zero there is no drainage of water from it, then eq.4 may be re-written:

$$-\frac{\mathrm{d}C}{\mathrm{d}t} = E_p \cdot \frac{C}{S} \tag{4a}$$

which leads to the unrealistic conclusion that C becomes zero only after infinite time. However, if values appropriate to conditions after cessation of rain are given to the symbols in eq.6, approximate mean rates of E_p are 0.8 and 0.4 mm/h for summer and winter respectively. Substituting these for E_p in eq.4a and putting $S = 1$ mm it is predicted that C will decline to 1% of S in about 6 and 12 hours respectively, times which correspond approximately with visual observations of times required for complete drying of the canopy. The small amount of water calculated as remaining on the canopy to infinite time has no quantitative significance and it will be shown that eq.4 has given a satisfactory fit to observations.

When mean surface temperature of the canopy was measured, potential evaporation was calculated from:

$$E_p = \frac{\rho c(e_s(T) - e_d)}{\lambda \gamma r_a} \tag{5}$$

When surface temperature was not known, the formula used was:

$$E_p = \frac{sR_n + \rho c(e_a(T) - e_d)/r_a}{\lambda (s + \gamma)} \tag{6}$$

For derivation and discussion of eq.5, 6 and 8 see MONTEITH (1965) and SLATYER (1967, pp.47–51, 62–64).

Drainage rate

The analysis of drainage from the canopy will be described below. It is sufficient at this stage to say that ln D is linearly related to C by a drainage coefficient b.

Operation of the model

The total interception loss is the integral of evaporation rate in time. While $C < S$, evaporation rate varies with the amount of water C, on the canopy (eq.4) and C in turn is determined by the balance between rates of rainfall, drainage and evaporation (eq.2). Given certain parameters of the canopy, and the variation in time of rainfall and climatic conditions affecting evaporation, the model constructs a running balance in time of C, E and T. The parameters of the canopy are α (the reflection coefficient for short-wave radiation, affecting R_n), r_a, b, p, and S. Since the right hand or aerodynamic component in the numerator of E_p (eq.6) is usually several times larger than sR_n, α need not be known very precisely and has been assumed to be 0.15. Methods of obtaining the other parameters of the canopy will now be described.

DERIVATIONS

Aerodynamic resistance, r_a

ROBINS (1969) described a method of deriving r_a from the observations referred to on p.368 under (*1*). He found an inverse linear relation between r_a and wind-speed 2 m above the canopy:

$$r_a = \frac{f}{u} \tag{7}$$

and the value of f was 9.2.

Given wind profiles above the forest, r_a may be calculated (MONTEITH, 1965) from:

$$r_a = \frac{1}{k^2 u} \left\{ \ln \left(\frac{z-d}{z_0} \right) \right\}^2 \tag{8}$$

Where, as will be usual, wind profiles have not been measured, d and z_0 may be approximately estimated from the height of the trees. Simplifying from the results of KUNG (1961) and SCEICZ et al. (1969), the following relations are suggested for moderate wind-speeds:

$$d = 0.75 \text{ h}$$

$$z_0 = 0.1 \text{ h}$$

For the stand in which Robins worked, the use of these empirical relations for $h = 20$ m, and $z = 22$ m, gives:

$$\frac{1}{k^2} \left\{ \ln \left(\frac{z-d}{z_0} \right) \right\}^2 = 9.8$$

The correspondence with the value of f estimated by Robins by a quite independent method, is satisfactory but more evidence is needed before it could be accepted that r_a can be sufficiently accurately estimated from windspeed and stand height.

"*Free*" *throughfall coefficient* (p) *and canopy storage capacity* (S)

Fig.1 (from ROBINS, 1969) shows the relation of throughfall to rainfall for periods of approximately 24 h. The figure has features common to many published data of interception. From the origin, throughfall increases at first in a fairly constant ratio to rainfall. Between rainfall 1 and 1.5 mm the upper envelope of the

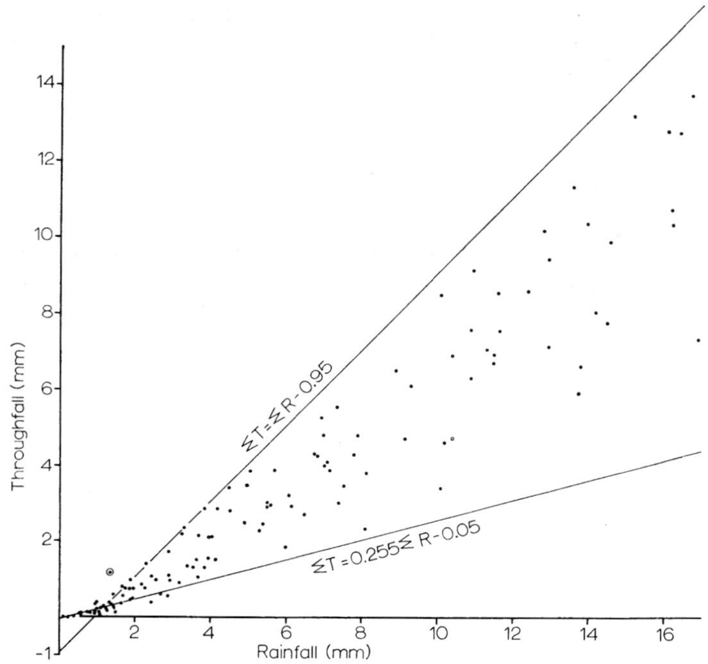

Fig.1. The relation of throughfall to rainfall, for approximate 24 hour periods, in a stand of Corsican pine. The ringed point was neglected.

observations steepens, and the values of throughfall for a given value of rainfall show considerable scatter. It is assumed first that the inflection of the upper envelope represents the attainment of canopy saturation, and that to the left of this point the data may be represented by $\Sigma T = p \Sigma R$. Following LEYTON et al. (1967) it is further assumed that the vertical scatter of points to the right of the inflection is mainly due to variation in evaporation from the wet canopy, and that the upper envelope of these points represents the relation of throughfall to rainfall when evaporation is minimal. Ideally this envelope would be expected to be a line of unit slope:

$$\Sigma T = \Sigma R - S$$

from which S is obtained as the negative intercept on the throughfall axis. This treatment is subjective both in recognition of the inflection corresponding to the point of canopy saturation, and in fitting the upper envelope to the scattered points, but it is not difficult to use if the points are numerous enough. From Fig.1, S is estimated to be between 0.9 and 1 mm. A regression fitted to data for rainfall less than 1 mm gives a value of p of about 0.25. Note that if there were no errors of measurement, throughfall observations may exceed $\Sigma R - S$ when $\Sigma R < S/(1-p)$, in this case when $\Sigma R < 1.3$ mm.

Another approach to the estimation of S will now be described. ZINKE (1967) defined storage capacity as "the depth of water on the projected area covered by the plant which can be stored or detained on the plant in still air". After rain has ceased, and in the absence of wind, the time at which C falls to S should be recognisable by the cessation of drainage from the canopy. In practice, conditions are seldom windless, and since in addition the drainage rate declines exponentially, the time when drainage ceases cannot be recognised precisely. Arbitrarily therefore S was redefined as the value of C when the drainage rate falls to 0.002 mm/min. Below this rate, drainage was irregularly distributed in time (drops shaken off by gusts of wind), difficult to measure accurately, and usually much less as a rate than calculated evaporation (see Fig.3 and 4.) A method was devised for estimating S, so defined, from the data of a given storm.

In the period from the beginning of rain until the time when drainage after rain finally declines to 0.002 mm/min:

$$\Sigma R - \Sigma T = \Sigma E + S \qquad (9)$$

Note that (from eq.4) ΣE is dependent on the value of S. A computer programme (SIM 2) was devised to construct in 5-min steps a running balance of observed rainfall and throughfall, calculated evaporation and, by difference, C. It began with an assumed value of S and by successive approximations derived a value of S which satisfied eq.9 at the critical end time when $D = 0.002$ mm/min.

TABLE I

CHARACTERISTICS OF FOUR STORMS USED IN DEVELOPMENT OF THE MODEL, AND ESTIMATES OF STORAGE CAPACITY AND DRAINAGE COEFFICIENT DERIVED FROM THEM

	19 June	*29 July*	*27 Oct.*	*31 Oct./1 Nov.*
Period analysed	14h35–21h00	13h55–23h00	07h05–12h05	22h45–06h00
Duration of rain (h)	3.2	4.8	3.7	4.9
Mean intensity of rain (mm/h)	3.9	1.6	2.8	2.2
Mean windspeed (m/sec)	1.3	2.8	3.3	2.3
Mean evaporation rate (estimated) (mm/h)	0.13	0.24	0.18	0.03
Storage capacity (S), mm	1.02	1.01	1.13	1.06
Drainage coefficient (b)	4.6	3.5	3.0	3.9
Residual drainage (mm)	0.09	0.08	0.04	0.09

The calculation was made with data from four storms. These storms were a random selection from the total data, and Table I shows that they covered a fair range of seasonal and climatic variation. The values of S obtained varied from 1.01 to 1.13 mm, with a mean of 1.05. The small discrepancy between this and the value obtained from Fig.1 can be approximately accounted for by the residual drainage (Table I) at rates less than 0.002 mm/min. The close agreement between two independent methods of estimating S gives confidence in the assumption of eq.4 which is inherent in the second method.

Drainage coefficient, b

SIM 2 gave estimates of C at 5-min intervals throughout the storms. It is emphasized that these were based on actual observations of R and T, and on calculations only in respect of E which is a quite small component of the water balance.

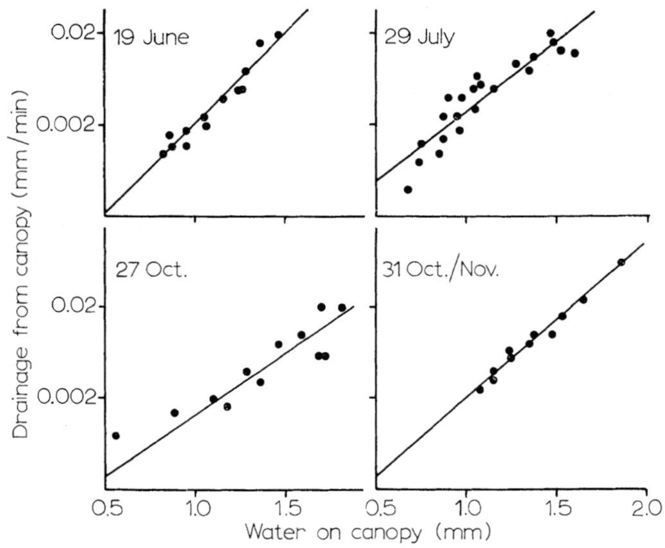

Fig.2. The relation of the observed rate of drainage (logarithmic scale) from the canopy in rainless periods, to the calculated amount of water on the canopy.

In Fig.2, D is plotted on a logarithmic scale against values of C, calculated by SIM 2, when the canopy is draining after rain. The relation between ln D and C is approximately linear, i.e.:

$$D = \exp(a + bC) \tag{10}$$

and b has values between 3.0 and 4.6. It was found that this variation had negligible effect on predictions from the final model and a mean value of 3.7 was taken in subsequent calculations. By definition, when $C = S$, $D = 0.002$ mm/min and therefore $a = \ln(0.002) - bS$.

It is uncertain whether the curve may be extrapolated far below $C = S$, or whether D should fall to zero when C is about 0.9 mm. However, below this level of C, values of D on the extrapolated line are so small that their difference from zero is unimportant.

Equations for predicting throughfall

The drainage rates in Fig.2 are plotted against values of C calculated as means of the values at the beginning and end of 5-min periods, and in all but the highest points C was not significantly changed by drainage within the period. But for higher values, deliberately excluded from the diagram, C is significantly changed during a 5-min period and D within the period must be obtained by integration.

Re-writing eq.10 as:

$$-\frac{dC}{dt} = K \exp(bC)$$

integrating dC from time t_0 to t, and neglecting evaporation, it can be shown that:

$$C_t = \frac{\ln [tbK + \exp(-bC_0)]}{-b} \quad (11)$$

whence $\Sigma D = C_0 - C_t$.

For predicting drainage during rainfall it was assumed that eq.10 still holds. The rate of change of C has now to include the rate at which rain is added to canopy surfaces. Still neglecting evaporation:

$$-\frac{dC}{dt} = K \exp(bC) - Q \quad (12)$$

where $Q = (1-p)R$. Integrating dC from t_0 to t gives:

$$bC_t = \ln Q + b(C_0 + tQ) - \ln [K\exp(b(C_0 + tQ)) - K\exp(bC_0) + Q] \quad (13)$$

whence $\Sigma T = \Sigma R - C_t + C_0$.

Eq.11 and 13 neglect E. When $C \geqslant S$, and E is therefore independent of C, then Q can be put equal to $(1-p)R - E_p$ in eq.12. When $C < S$ and $-dC/dt = K \exp(bC) - (1-p)R + E_p \cdot C/S$, an approximate step-wise calculation was used to follow the changes in C, instead of a simultaneous integration of drainage and evaporation.

THE MODEL TESTED ON FOUR STORMS (SIM 4)

Throughfall, evaporation and water on the canopy were now calculated by a computer programme for 5-min intervals in the four storms from which mean values of S and b had been estimated. The data used were observations on rainfall, canopy surface temperature, and wind-speed and vapour pressure above the canopy; p was taken from Fig.1 to be 0.25, f to be 9.2, S to be 1.05 mm and b, 3.7. At the time after each storm when D fell to 0.002 mm/min, the balances shown in Table II were struck. The tests of the success of the prediction were first that the calculated water on the canopy at this time should be about 1.05 mm; and second that the sum of this quantity and calculated evaporation should equal the observed difference

TABLE II

A COMPARISON BETWEEN OBSERVED AND CALCULATED INTERCEPTION LOSSES IN THE FOUR STORMS SHOWN IN TABLE I (mm)

	20 June	*29 July*	*27 Oct.*	*31 Oct./1 Nov.*
Observations:				
Rainfall	12.40	7.60	10.10	10.60
Throughfall	10.55	4.34	8.08	9.34
Loss	1.85	3.26	2.02	1.26
Calculations:				
Evaporation	0.84	2.17	0.90	0.20
Water remaining on canopy	1.11	1.09	1.02	1.08
Total	1.95	3.26	1.92	1.28

between rainfall and throughfall, the interception loss. Both these tests gave satisfactory results (see Table II).

A more exacting test is that throughout a storm calculated throughfall should agree with the observations, since the method of calculating throughfall depends on assumptions about the relation of both drainage and evaporation rates to

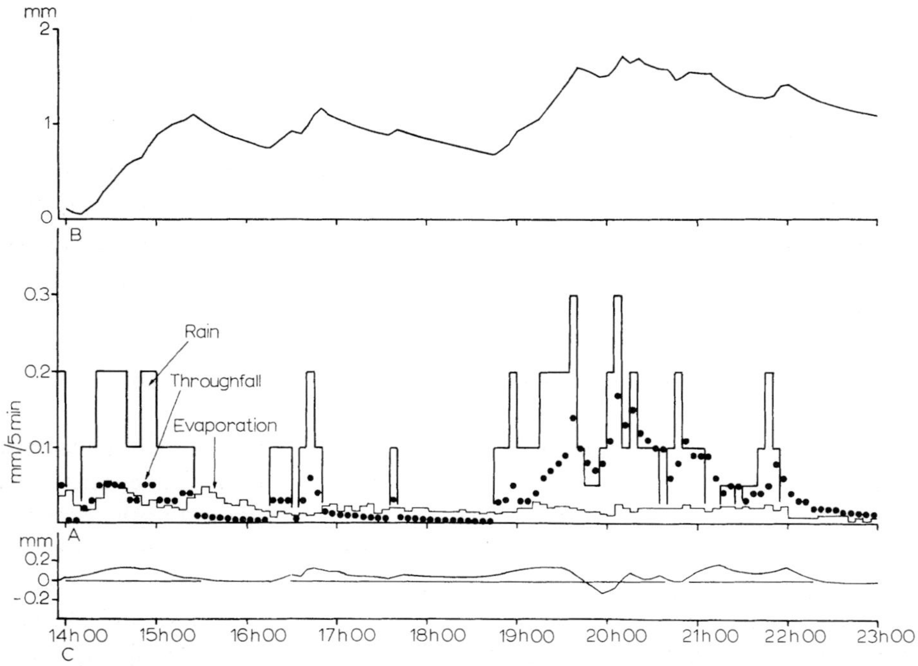

Fig.3. Storm of 29 July (see Tables I and II). A. Observed rainfall, calculated throughfall and evaporation. B. Calculated water on canopy. C. Accumulated difference between observed and calculated throughfall, on same scale as B. Calculations by SIM 4.

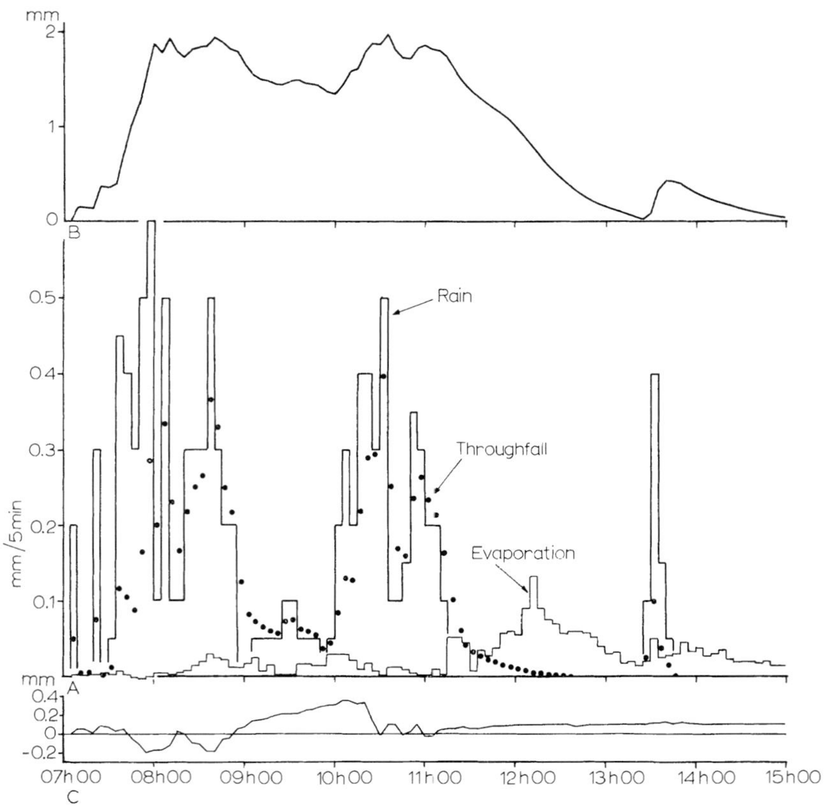

Fig.4. Storm of 27 October (see Tables I and II). Further explanation as Fig.3.

canopy water. Fig.3 and 4 show observed rainfall in the storms of 29 July and 27 October together with the predicted course of throughfall, evaporation and water on the canopy. Section C in each figure shows the accumulating difference between observed and calculated throughfall. This fluctuates during rain but is never greater than the equivalent of about 5 min throughfall and is therefore no more than would be expected from errors of resolving the trace on the throughfall chart into 5-min periods.

During the 9 hours illustrated in Fig.3, C was less than S, and E therefore less than E_p, for 5 h, the canopy sometimes being wetted and sometimes dried. This storm was therefore a good test of the use of eq.4. Near the end of the period in Fig.4, eq.4 predicts that the canopy will be almost dry. The agreement between observed and calculated throughfall, following rain between 13h20 and 13h45, is consistent with this prediction.

Fig.3 (from 19h00) and 4 (from 07h00) both show that throughfall is less than rainfall while the water on the canopy rises. Thereafter rainfall and throughfall are approximately equal, and the water content of the canopy is relatively

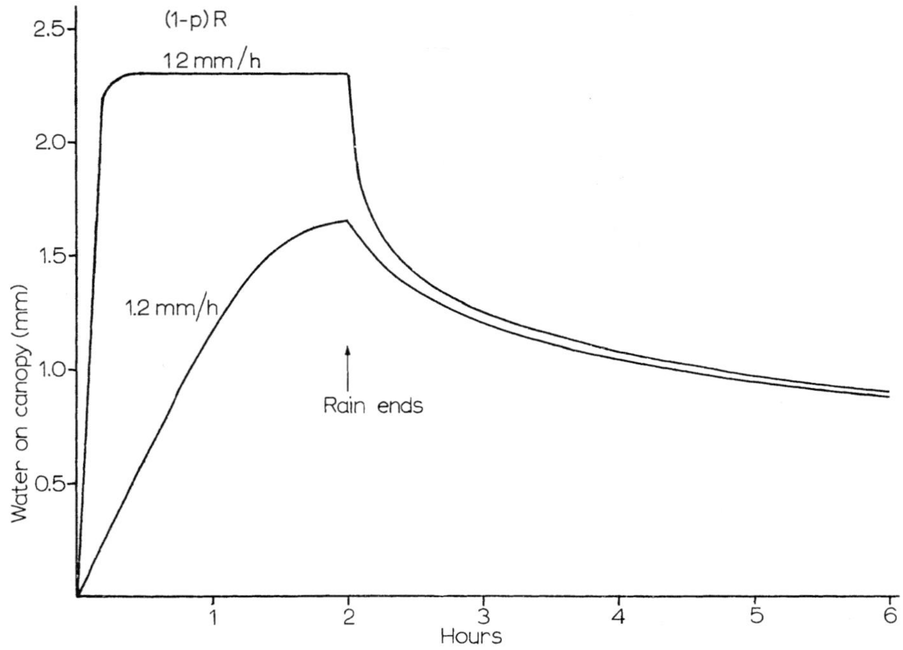

Fig.5. Predicted changes in the water on the canopy during 2 h rain followed by a rainless period.

stable although fluctuating with rainfall intensity. It follows from eq.12 that in constant rainfall intensity C should increase to a steady state whose value is linearly related to ln $[(1-p)R]$. Fig.5 shows the calculated changes in C during 2 h rain for values of $(1-p)R$ equal to 0.02 or 0.2 mm/min. The steady state values of C are 1.6 and 2.3 mm respectively. The right hand side of the figure shows the calculated decline of C from these values when rain stops, and with no evaporation. Canopy water in excess of about 1.5 mm drains off very rapidly so that after 20–30 min drainage, C is almost independent of the value from which the decline began.

THE MODEL TESTED ON 18 MONTH'S RECORDS (SIM 5)

Two objections which could be made to the foregoing tests are first that parameters and relations derived from four storms were used to predict behaviour only in those storms; and second that observations on mean canopy temperature were required and these are not usually available.

The programme was therefore changed by substituting eq.6 for eq.5 in the calculation of E_p; i.e., eliminating surface temperature by introducing net radiation. The model was then tested against 18 months' interception records.

Net radiation was estimated from other meteorological observations by the equations of PENMAN (1956):

$$R_n = R_i(1 - \alpha) - R_b$$

$$R_i = R_s(0.18 + 0.55n/N) \quad (14)$$

$$R_b = \sigma(T^0 + 273)^4 (0.56 - 0.078\, e_d^{\frac{1}{2}})(0.10 + 0.90n/N) \quad (15)$$

From MILANKOVITCH (1930), cited by FRANK and LEE (1966), instantaneous values of extra-terrestrial radiation on a horizontal plane are given by:

$$R_s = \frac{R_o}{v^2} (\sin\theta \cdot \sin\delta + \cos\theta \cdot \cos\delta \cdot \cos\omega t)$$

This expression is integrated to give the extra-terrestrial radiation for any hour of the day, measured from t_o = noon:

$$\Sigma R_s = \frac{R_o}{v^2} \left\{ \sin\theta \cdot \sin\delta \cdot (t_{n+1} - t_n) + \frac{\cos\theta \cdot \cos\delta}{\omega} (\sin\omega t_{n+1} - \sin\omega t_n) \right\}$$

It was assumed that eq.14 and 15 are applicable in any hour of the day and the only difficulty was in obtaining hourly equivalents of n/N. The hourly records available to us included visual estimates of cloud cover, in eights of the whole sky. HARRIS (1969), showed that daily values of n/N were related to daily means of estimated cloud cover m, by:

$$n/N = 0.91 - 0.0054m - 0.0143m^2$$

and it was assumed that the same equation could be used on an hourly basis.

Several meteorological stations in S.E. England keep hourly records, though none is very near the Bramshill site. The climate at Boscombe Down, about 80 km west, seemed most like that of Bramshill. Table III summarises and compares the

TABLE III

COMPARATIVE SUMMARY OF THE CLIMATES AT BOSCOMBE DOWN AND SOUTH FARNBOROUGH DURING THE PERIOD OF THE INVESTIGATION

Date	*Rain, mm*		*Wind, m/sec at 10 m*		*Temperature, °C*		*Vapour pressure deficit at 09h00, mbar*		*Daily hours sunshine*	
	B.D.	*S.F.*	*B.D.*	*S.F.*	*B.D.*	*S.F.*	*B.D.*	*S.F.*	*B.D.*	*S.F.*
June–Aug. 1966	195	225	4.4	4.2	15.2	15.6	3.4	3.4	5.6	6.5
Sept.–Nov. 1966	234	195	4.5	3.9	9.9	10.1	1.6	1.6	3.2	3.2
Dec.1966–Febr.1967	227	187	5.8	5.4	5.2	5.2	0.9	0.8	2.0	2.2
Mar.–May 1967	191	211	5.9	5.7	8.5	8.4	2.1	2.2	5.1	5.4
June–Aug. 1967	126	159	4.0	3.9	15.6	16.2	3.6	4.1	6.6	7.4
Sept.–Nov. 1967	277	234	5.3	5.3	10.1	10.4	1.4	1.6	3.1	3.1
Total	1250	1211								
Mean			5.0	4.7	10.8	11.0	2.2	2.3	4.3	4.6

climates of Boscombe Down and south Farnborough, which is about 8 km from Bramshill, during the 18 months of the interception records. The overall means are very similar. Windspeeds were consistently higher at Boscombe Down but this difference was removed before use by a constant factor. South Farnborough had a little more sunshine in summer when mean temperatures were about 0.5 °C higher. The greatest differences were in June 1966 and July and August 1967 when mean temperatures at south Farnborough were about 1.5 °C higher, partial vapour pressures were similar, but vapour pressure deficits were 15–20% higher because of the higher temperatures in these months.

Hourly records from Boscombe Down were obtained on punched cards and the following parameters transferred to magnetic tape: rainfall and run of the wind during the hour; duration of rain in tenths of an hour; air temperature, relative humidity and cloud cover, at hourly intervals. Temperature and vapour pressure during the hour were calculated as the means of the values at the beginning and the end. Wind was corrected first by the ratio of mean velocities over the whole 18 months at south Farnborough and Boscombe Down; and then by the ratio of measured wind 2 m above the forest to wind at south Farnborough in more limited periods in which this comparison had been made.

Although the duration of rainfall was recorded in tenths of an hour, there was no information on the distribution of these within the hour, and after some

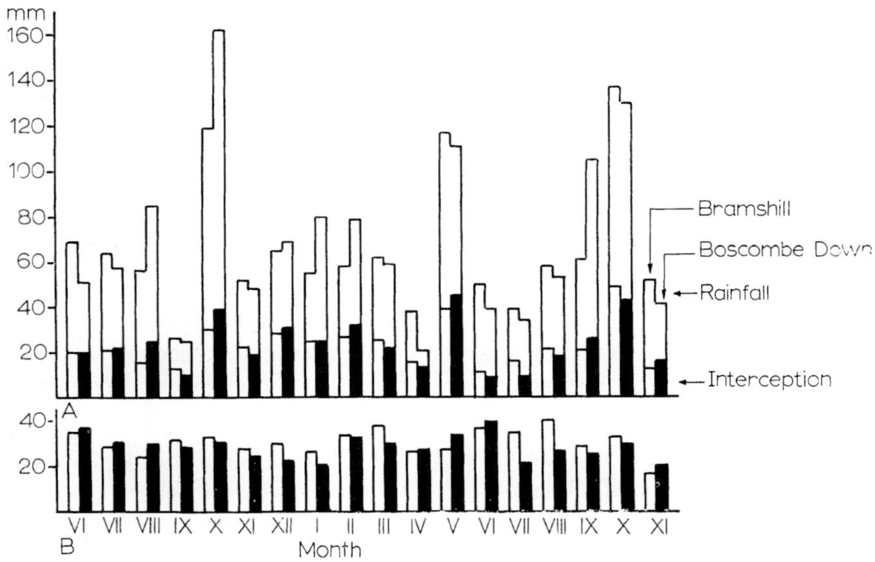

Fig.6. A comparison of observed and predicted interception loss for 18 months. A. Each double histogram shows above, rainfall at Bramshill (left) and Boscombe Down (right); and below, interception loss observed at Bramshill (left) and computed for Boscombe Down (right). B. Interception loss per 100 hours rain observed at Bramshill (left) and computed for Boscombe Down (right).

initial testing it was assumed that rain was evenly distributed throughout the hour. Other conditions were assumed to remain constant within the hour, but the accounting procedure of the programme was made by steps of 0.1 h, since the amount of water on the canopy, and hence the ratio E/E_p can change markedly within an hour.

Total rainfall in the 18-month period was 1,176 mm at Bramshill and 1,250 mm at Boscombe Down. Total interception measured at Bramshill was 415 mm and calculated for Boscombe Down, 427 mm. Fig.6A shows a close correlation between observed and calculated interception loss per month. In 12 out of the 18 months the agreement is within 5 mm. The correlation coefficient is 0.9 ($P < 0.001$).

DISCUSSION

Throughout the 18-months calculations the parameters of the canopy were assumed to remain constant. It is recognised, however, that they may vary. p and S would be expected to vary with the amount and disposition of the foliage, b with the inclination of the leaves and wettability of their surfaces. These three parameters would be expected to vary markedly with season in deciduous woodland and to show some seasonal variation even in an evergreen canopy. The aerodynamic parameters d and z_0 may also vary with the state of the canopy and SCEICZ et al. (1969) have shown that in a pine canopy they varied with windspeed. However, ROBINS' (1969) finding of a linear inverse relation between r_a and windspeed in Corsican pine canopy implies either that d and z_0 did not change with windspeed or that there is some compensating factor not recognised in the combination of eq.5 and 8.

The above comments do not exhaust the possibilities of variation in canopy parameters which can be envisaged. Before attempting to elaborate the model it seems sensible, however, to investigate its sensitivity to such variation, and so discover which parameters must be known accurately and which need be only approximately estimated.

Interception loss has frequently been expressed as a percentage of rainfall. Expressed in this way it would be expected, and frequently has been shown, to be inversely related to rainfall intensity. This relation can be expanded: interception loss, percent of rainfall = interception loss per 100 h rain ÷ rainfall intensity per hour; and interception loss in a given time (storm, week or month) = interception loss per h rain × duration of rainfall.

The second of these two related expressions is perhaps the more useful since it is a product of rate of evaporation and time, although it should be noted that the duration of rain is a biassed expression of the time available for evaporation since the canopy remains wet for some hours after rain ceases. Fig.7 shows that a great part of the interception loss per month is accounted for by duration of rainfall. Fig.6B examines the correlation between interception loss per 100 h rain, observed

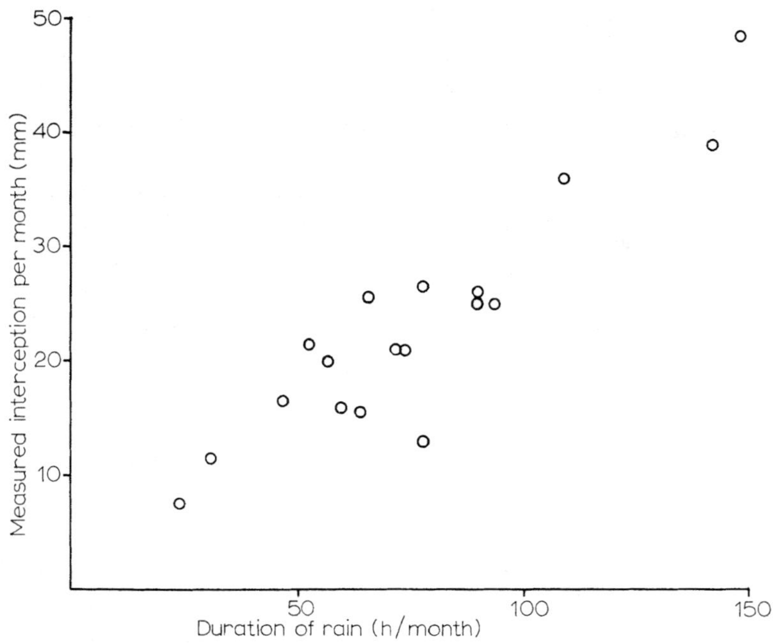

Fig.7. The relation of observed interception loss to number of hours rain, per month, at Bramshill.

at Bramshill and calculated at Boscombe Down. Observed and calculated rates agree fairly well except in July and August 1967. Excluding these months mean interception rates per 100 h rain are 27.2 and 28.5 mm at Bramshill and Boscombe Down respectively; the correlation coefficient for monthly values is 0.65 ($P < 0.01$). However, if the two months are included the correlation coefficient becomes 0.4 and is not significant.

No simple explanation can be given for the discrepancies in these two months. The observations at Bramshill appear to be reliable. Mean monthly temperature and vapour pressure deficit at Bramshill in July and August 1967 exceeded values at Boscombe Down by more than was usual over the whole 18-months period. But this was the case also in June 1966 when there was no discrepancy between observations and calculation. Even though one may point to differences in environment, it still has to be shown that these are sufficient to account for the discrepancies considered. This raises the question of the sensitivity of the model, not only to variation in canopy parameters considered above, but also to variation in environmental factors. These questions will be considered in a further paper.

ACKNOWLEDGEMENTS

We are grateful to the Conservator (S.E.) for permission to work in Forestry

Commission plantations, and to the Meteorological office for the hourly climatic records from Boscombe Down and other data. The work would have been impossible without the generous gift by the Royal Society of towers to give access to the canopy; further equipment was provided by the Natural Environment Research Council. We thank also Dr. J. Barber for mathematical assistance. The observations were made by P.C.R. and included in a thesis awarded the Ph.D. of London University.

REFERENCES

BURGY, R. H. and POMEROY, C. R., 1958. Interception losses in grassy vegetation. *Trans. Am. Geophys. Union*, 39: 1095–1100.

FRANK, E. C. and LEE, R., 1966. Potential solar beam irradiation on slopes; tables for 30° to 50° latitude. *U.S. Dept. Agr., Forest Serv., Res. Paper*, R.M. 18: 116 pp.

HARRIS, G. P., 1969. *A Study of the Ecology of Corticolous Lichens.* Thesis, Univ. London, 239 pp.

HELVEY, J. D. and PATRIC, J. M., 1965. Canopy and litter interception of rainfall by hardwoods of eastern United States. *Water Resources Res.*, 1: 193–206.

HORTON, R. E., 1919. Rainfall interception. *Monthly Weather Rev.*, 47: 603–623.

KUNG, E., 1961. Derivation of roughness parameters from wind profile data above tall vegetation. *Wisc. Univ., Dept. Meteorol., Ann. Rept.*, pp.27–33.

LEYTON, L., REYNOLDS, E. R. C. and THOMPSON, F. B., 1967. Rainfall interception in forest and moorland. In: W. E. SOPPER and H. W. LULL (Editors), *Forest Hydrology* Pergamon, Oxford, pp.163–178.

MILANKOVITCH, M., 1930. Mathematische Klimalehre. *Handbuch Klimatologie*, 1: 1–176.

MONTEITH, J. L., 1965. Evaporation and environment. *Symp. Soc. Exptl. Biol.*, 19: 205–234.

PENMAN, H. L., 1956. Evaporation—an introductory survey. *Neth. J. Agric. Sci.*, 4: 9–29.

RIJTEMA, P. E., 1965. An analysis of actual evapotranspiration. *Agric. Res. Rept., Wageningen*, 659: 107 pp.

ROBINS, P. C., 1969. *Comparative Studies of Evaporation from Pinus nigra and Pseudotsuga menziesii, with Particular Reference to Air and Stomatal Resistances.* Thesis, Univ. London, 151 pp.

RUTTER, A. J., 1967. An analysis of evaporation from a stand of Scots pine. In: W. E. SOPPER and H. W. LULL (Editors), *Forest Hydrology.* Pergamon, Oxford, pp.403–417.

RUTTER, A. J., 1968. Water consumption by forests. In: T. T. KOZLOWSKI (Editor), *Water Deficits and Plant Growth, 2.* Academic Press, London and New York, pp.23–84.

SCEICZ, S., ENDRÖDI, G. and TAJCHMAN, S., 1969. Aerodynamic and surface factors in evaporation. *Water Resources Res.*, 5: 380–394.

SLATYER, R. O., 1967. *Plant-Water Relationships.* Academic Press, London and New York, 366 pp.

WAGGONER, P. E., BEGG, J. E. and TURNER, N. C., 1969. Evaporation of dew. *Agric. Meteorol.*, 6: 227–230.

ZINKE, P. J., 1967. Forest interception studies in the United States. In: W. E. SOPPER and H. W. LULL (Editors), *Forest Hydrology.* Pergamon, Oxford, pp.137–161.

Agric. Meteorol., 9 (1971/1972) 367–384

Evaporation From the Wet Canopy of a Pine Forest

J. B. STEWART[1]

Institute of Hydrology, Wallingford, Oxon, England

Measurements of the energy budget of a pine forest were made on 70 days when the canopy was wet. Out of 245 20-min periods when the canopy was wholly wet, 173 were occasions when the latent heat flux exceeded the net radiation, the additional energy being provided by a downward flux of sensible heat. Under the same level of radiation the average rate of evaporation of intercepted precipitation has been found to be three times the average rate of transpiration, so that the loss by evaporation of intercepted precipitation is only partly compensated by the suppression of transpiration. It is concluded that reliable estimates of the water resources of forested areas require a separate calculation of the interception and transpiration components of the total evaporation loss.

The hydrologist, predicting the water yield of a forested catchment, needs to estimate the total amount of precipitation intercepted by the trees, since like the transpiration, this intercepted precipitation evaporates back into the atmosphere and does not contribute to the runoff. However, transpiration is greatly reduced whenever intercepted precipitation is present on the foliage, since any energy available will evaporate the water residing on the surfaces of the leaves in preference to that within the leaves. Since in a particular environment, evaporation is dependent on the input of energy, it has been suggested that evaporation of intercepted precipitation merely replaces transpiration [*Leyton and Carlisle*, 1959; *Penman*, 1963], and that interception does not need to be considered as an additional loss separate from transpiration. However, using values of interception, given by the difference between gross precipitation and throughfall measured over periods of a week or longer, *Rutter* [1963], *Patric* [1966], *Helvey* [1967], and *Leyton et al.* [1967] have shown that, particularly in winter months, the energy used to evaporate the measured interception exceeded that provided by net radiation during the same periods; in contrast, transpiration from vegetation typically consumes only 80–90% of net radiation [*Aslyng*, 1974]. If these results are generally applicable, they imply that interception exceeds the suppressed transpiration and therefore causes an additional loss from the water resources of the area. *Rutter* [1967] went a stage further and used a modified version of the Penman formula [*Penman*, 1952] with values of the air and leaf resistances derived from measurements of the changes in weight of detached branches with wet and dry leaves, respectively, to show that for the same environmental conditions the rate of evaporation of intercepted water was, on average, about four times as great as the transpiration from the same forest. Rutter put forward the hypothesis that when the rate of evaporation of intercepted precipitation exceeded the input of energy as net radiation, there must be a downward flux of sensible heat from the air. This hypothesis was supported by his observations of wet leaves being cooler than the air temperature. The general availability of this additional energy has been queried by *Hewlett* [1967] and *Pereira* [1967] among others. It has been suggested that it is the result of leading edge advection and therefore that the results are not applicable tc large areas of forest.

There are two uncertainties involved in the evaporation of intercepted precipitation: first, is the rate of evaporation of intercepted precipitation greater than the rate of transpiration under the same environmental conditions? (i.e., does the occurrence of interception cause a reduction in water yield of forested catchments?); and second, can the rate of evaporation exceed the input of energy in the form of net radiation? To try to resolve these uncertainties, *Murphy and Knoerr* [1975] developed a model describing the exchange of energy between a vegetated surface and the atmosphere and used it to calculate the rates of evaporation from wet and dry forest canopies under the same environmental conditions. They found that over a wide range of boundary conditions, defined by the relative humidity, solar radiation, wind speed, and temperature, the ratio of the latent heat flux from a wet canopy to that from a dry canopy lay between $1\frac{1}{2}$ and 2. They also found that for some boundary conditions the latent heat flux from the wet canopy exceeded the net radiation and there was a downward flux of sensible heat. Since this latter result was very dependent on the assumed environmental conditions, they were unable to say whether this would apply in general.

The present paper analyzes measurements made by micrometeorological techniques of the rate of evaporation from the wet canopy of Thetford Forest (the second largest forest in the United Kingdom) over periods of less than 1 hour. The present data provide direct evidence to confirm Rutter's hypothesis and show that there is often a downward flux of sensible heat to maintain rates of evaporation of intercepted precipitation in excess of that resulting from the input of energy from net radiation alone. Besides presenting the average rate of evaporation of intercepted precipitation, this result is also compared to an estimate of the transpiration for the same environmental conditions derived from earlier analyses of data collected over the same forest on dry days, i.e., when only transpiration was occurring. These results provide sufficient information for the calculation of the net interception loss, i.e., the difference between the interception and the suppressed transpiration.

THEORY

The total annual evaporation E from any surface covered by vegetation can be considered in three parts: E_S, the evaporation from the soil; E_T, the evaporation of water transpired from the plants through their leaves (transpiration); and E_I, the evaporation of that part of the gross precipitation intercepted by the vegetation. For a period of 1 month or longer the total precipitation intercepted by the forest canopy and re-

[1] On secondment from the United Kingdom Meteorological Office.

Paper number 7W0412.

turned to the atmosphere by evaporation is the difference between the gross precipitation and the precipitation reaching the soil, i.e.,

$$E_I = I = P_G - P_T - P_S \quad (1)$$

where I is the interception, P_G is the gross precipitation, P_T is the throughfall, and P_S is the stemflow. The energy $\lambda \mathbf{E}_I$ required to evaporate the intercepted precipitation can be computed from measurements of the energy budget made while the canopy is wet. For an extensive uniform area of vegetation the three-dimensional energy budget can be treated as if it were in one dimension and is described by

$$\mathbf{R}_n - \mathbf{G} - \mathbf{S} - \mathbf{P} = \mathbf{H} + \lambda \mathbf{E}_I \quad (2)$$

where $\mathbf{R}_n$ is the net radiation, measured above the canopy at height z, $\mathbf{G}$ is the energy flux into the ground, $\mathbf{S}$ is the total flux of energy into storage between the net radiometer height and ground level, $\mathbf{P}$ is the net rate of energy absorption by photosynthesis and respiration, and $\mathbf{H}$ is the sensible heat flux. By introducing the Bowen ratio β to apportion the sensible and latent heat fluxes, (2) can be rearranged to give

$$\lambda \mathbf{E}_I = \mathbf{A}/(1 + \beta) \quad (3)$$

where $\mathbf{A}$ is the 'available' energy, given by

$$\mathbf{A} = \mathbf{R}_n - \mathbf{G} - \mathbf{S} - \mathbf{P} \quad (4)$$

and by assuming the transfer coefficients for sensible and latent heat to be equal the Bowen ratio is given by

$$\beta = \frac{\mathbf{H}}{\lambda \mathbf{E}} = \frac{c_p \partial\theta/\partial z}{\lambda \partial q/\partial z} \quad (5)$$

where $\partial\theta/\partial z$, $\partial q/\partial z$ are the gradients of potential temperature and specific humidity against height z.

The latent heat flux is also given by the Monteith-Penman equation [*Monteith*, 1965], which can be written in the form

$$\lambda \mathbf{E} = \frac{\Delta' \mathbf{A} + \rho c_p \{q_w(\theta_z) - q_z\}/r_{V,H}}{\Delta' + c_p(1 + r_{ST}/r_{V,H})/\lambda} \quad (6)$$

where

- λ latent heat of vaporization of water;
- Δ' mean slope of the saturated specific humidity curve at the mean of the surface and air temperatures;
- ρ, c_p density, specific heat of dry air, respectively;
- θ_z potential temperature of the air at height z;
- $q_w(\theta_z)$ saturated specific humidity at θ_z;
- q_z specific humidity of the air at height z;
- $q_w(\theta_z) - q_z$ specific humidity deficit;
- $r_{V,H}$ aerodynamic resistance to the transfer of latent and sensible heat (assumed to be equal) between the surface and height z;
- r_{ST} surface resistance.

It has been shown theoretically [*Shuttleworth*, 1975] that r_{ST} is negligible when the canopy is wet, so that the Monteith-Penman equation reduces to

$$\lambda \mathbf{E}_I = \frac{\Delta' \mathbf{A} + \rho c_p \{q_w(\theta_z) - q_z\}/r_{V,H}}{\Delta' + c_p/\lambda} \quad (7)$$

However, even when the canopy is wet, an 'apparent surface resistance' can always be derived by rearranging (6) into the form

$$r_{ST} = [(\Delta'\lambda/c_p)\beta - 1]r_{V,H} + \rho\lambda(1 + \beta)[q_w(\theta_z) - q_z]/\mathbf{A} \quad (8)$$

and can be used to indicate whether the canopy is completely wet.

SITE

Thetford Forest (latitude 52°25′N, longitude 0°40′E, altitude 43 m) was chosen as a suitable site, first because of the flat topography and second because of its size (approximately 200 km² in total). The experimental site is situated near the center of one of the three main blocks of the forest, which has an area of 70 km² and consists of a mixture of Scots and Corsican pine (*Pinus sylvestris* L. and *Pinus nigra* var. *maritima* (Ait.) Melv.) extending continuously for between 2 and 10 km in all directions (except for three sectors of approximately 40° in which there are clearings up to 1 km across at distances of 0.5–1.5 km). Therefore a fully developed boundary layer can be expected to extend to much greater heights than those to which the profile measurements were made [*Pasquill*, 1972]. The elevation of the ground in the area varies by less than 30 m. The site itself is in a stand of Scots pine planted in 1931: by 1972 it had reached an average height of 16 m and had been thinned to a density of about 800 trees per hectare.

ANALYSIS, MEASUREMENTS, AND THEIR ERRORS

The latent heat flux was calculated from measurements of the energy budget, with the Bowen ratio used to apportion the 'available' energy into the sensible and latent heat fluxes. To determine the Bowen ratio, measurements of potential temperature and specific humidity were made above the forest canopy at heights of 17.1, 18.4, 20.1, 21.9, 24.1, 26.8, and 30.2 m above the ground. The temperature of each wet and dry bulb thermometer was recorded once a minute by a computer-controlled automatic data acquisition system, described in detail by *Oliver and Oliver* [1973]. *Gash and Stewart* [1975] describe the construction and operation of the psychrometers together with details of the methods of analysis used to derive the Bowen ratio.

The gradients of specific humidity and potential temperature are very small over the aerodynamically rough forest. Between the measurement heights of 17 and 30 m the differences were typically −0.1 g kg⁻¹ for specific humidity and between +0.05°C and −0.05°C for temperature when the canopy was wet. Most of the error in the measurement of the Bowen ratio is attributable to the errors in the measurement of these small humidity gradients. *McNeil and Shuttleworth* [1975] have investigated the errors in the measurements of the latent heat fluxes derived from profile measurements made over this forest and have shown that there can be large errors in measurements of the Bowen ratio on a particular day, but that over a number of days, random interchanging of the psychrometers greatly reduces these systematic errors.

The measurements of net radiation were made by 4–6 net radiometers of the Funk type [*Funk*, 1959] manufactured by Solar Radiation Instruments. These instruments are nominally calibrated to ±2½% [*Monteith*, 1972]; however, calibrations carried out at two independent laboratories differed systematically by about 5%. The instruments were mounted at a height of 17.8 m on booms extending 5 m from the supporting tower. The resolution of the integrating digital voltmeter used in the automatic data acquisition system to measure the output of the radiometers every 15 s was equivalent to about 0.1 W m^{-2}, with an accuracy of about 0.2 W m^{-2}.

Errors in the measurement of net radiation can occur as a result of poor levelling, large spatial variability, and the presence of raindrops on the polythene domes. It can be shown geometrically that a tilt of 1 degree can cause an error in the measurement of the direct beam of the solar radiation of up to 1% at a solar altitude of 60°, and of up to 10% at a solar altitude of 10°. However, under the predominantly overcast conditions occurring during and shortly after rainfall, which are of interest for this paper, this error should be negligible. To obtain a spatially representative measurement of net radiation, the sensors were mounted at distances of 7–50 m apart. When averaged over 1 hour, the measurements by individual net radiometers generally differed by no more than 5%. A preliminary experimental investigation into the systematic error introduced by the presence of raindrops on the radiometer domes was carried out. The results indicated that the error in the measurement of net radiation was less than 3% under cloudy conditions; the effect of this source of error is therefore comparable to that from other sources. This conclusion is supported by M. H. Unsworth (private communication, 1976), who has made theoretical estimates of this error based on experimental studies and has shown that the error can be expected to be typically 1 W m^{-2} under overcast conditions. (*Peterson et al.* [1973] have shown that the presence of water drops has a negligible systematic effect on measurements of the short-wave component of the net all-wave radiation.)

The other components making up the available energy were measured or estimated in the way described by *Stewart and Thom* [1973]. When the canopy was wet, temperature and humidity changes were small and radiation low, so that the storage terms and the soil heat flux were small, generally less than 10% of the net radiation, so that net radiation is a very close approximation to available energy.

To differentiate between transpiration and the evaporation of intercepted water, six wetness sensors were installed within the forest canopy space. These wetness indicators were similar in principle to those described by *Woodhead* [1969]. Each sensor was an 'artificial leaf,' 15 mm wide by 140 mm long and 2 mm thick, made from electronic circuit board, which consists of parallel copper strips 2 mm wide separated by insulation of similar width. Current flowing across the insulation between the copper strips could be measured whenever water was present on the artificial leaf to complete the circuit. To reduce electrolysis at the copper surface, the circuit was only turned on when a measurement was required. This measurement was made manually, and each sensor interrogated in turn one to four times per hour. To prevent standing water invalidating the results, the artificial leaves were mounted at 45° to the horizontal and the copper strips terminated 10 mm from the lower end. At each measurement level within the canopy (11.1, 13.0, and 15.8 m), two artificial leaves were mounted facing in opposite directions so that at least one would be well exposed to the wind, in much the same way as natural pine needles. Visual and tactile inspection of the canopy indicated that the wetness indicator determined the period of wetness of the canopy successfully to within about 30 min, except when the canopy was wetted by contact with fog.

RESULTS

Between March and October in 1972 and 1973, measurements were made for 708 20-min periods spread over 70 days, when the available energy was greater than 20 W m^{-2} and the leaf wetness sensors showed that the canopy was at least 'partly' wet. (If all the wetness sensors were indicating that water was lying on the artificial leaves, the canopy was described as 'completely' wet: if only some of the wetness sensors were indicating, the canopy was described as 'partly' wet.) For each of these periods the latent heat flux was computed from (3), while the apparent surface resistance was calculated from (8). To simplify the calculations, the average value of 6 s m^{-1} obtained by *Stewart and Thom* [1973] for the aerodynamic resistance was used in (8).

If the canopy is wholly wet, then the apparent surface resistance should be virtually zero [*Shuttleworth,* 1975]. Although any error in the micrometeorological measurements or any difference between the actual and the assumed aerodynamic resistance will result in a departure from zero, the apparent surface resistance will still have a mean value of zero if the errors and differences are randomly distributed. However, if the canopy is not wholly wet, the apparent surface resistance will have a value intermediate between zero and the value of about 100 s m^{-1} appropriate to a dry canopy [*Gash and Stewart,* 1975].

Figure 1 shows the frequency distribution of the apparent surface resistance values for the canopy 'partly' wet and 'completely' wet, respectively. In Figure 1 (top) for the partly wet canopy, there is a very wide distribution of values of apparent surface resistance of which half are greater than 20 s m^{-1}. By contrast in Figure 1 (bottom) when all the wetness sensors indicated surface water, more than four fifths of the apparent surface resistance values are contained within a narrow band (between -20 s m^{-1} and $+20$ s m^{-1}) centered on zero. This implies that when all the wetness sensors are giving a reading, the canopy is probably completely wet. To ensure that only a completely wet canopy was considered, analysis was restricted to cases for which all wetness sensors were indicating the presence of surface water on the artificial leaves and the apparent surface resistance was between -10 s m^{-1} and $+10$ s m^{-1}. (This second restriction had the useful side effect of removing cases where the Bowen ratio measurements were very seriously in error.) In 35 days of data, there were 245 20-min periods which satisfied these criteria. The mean value of the apparent surface resistance for these periods was -1.0 s m^{-1} (indicating that the use of the mean value of 6 s m^{-1} for the aerodynamic resistance was, on average, acceptable).

It was found that for more than two thirds of the 245 20-min periods, there was an inversion of potential temperature over the forest, implying a downward sensible heat flux and a latent heat flux greater than the available energy. The values of the latent heat flux for the individual periods have been plotted against available energy in Figure 2. Also shown in this figure are lines representing the latent heat flux calculated from (7) for various values of specific humidity deficit, using the mean air temperature (for the 245 periods) of 9°C and an aerodynamic resistance of 6 s m^{-1}. While some of the scatter of the points in Figure 2 is certainly due to errors in the measurements of Bowen ratio, the rest of the variation could well be the result of small differences in the specific humidity deficit between periods, i.e., differences of ± 0.3 g kg^{-1} enclose more than 90% of the values. Table 1 presents the average results, which show that the average downward sensible heat flux remained virtually constant for values of available energy between 33 and 147 W m^{-2} and then steadily increased from this negative value as the available energy increased above 147 W m^{-2}

PREVIOUS RESULTS

The results for Thetford Forest are similar to those obtained by *Moore* [1976], who made micrometeorological measurements on about 60 days in the period July–October 1973 over a forest of Radiata pine (*Pinus radiata* D. Don) near Mount

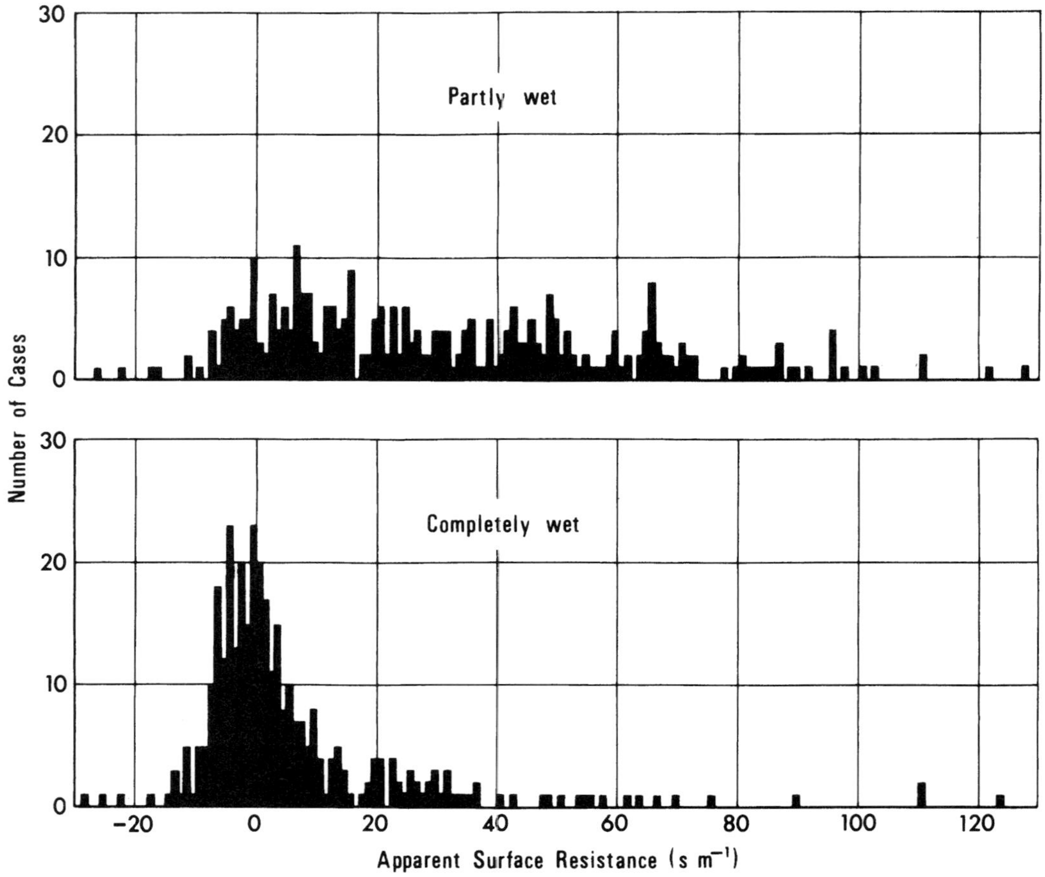

Fig. 1. The frequency distribution of the apparent surface resistance when the canopy was (top) partly wet and (bottom) completely wet as indicated by the wetness sensors.

Gambia in South Australia. On 16 days when the forest canopy was wet, the Bowen ratio remained very close to zero (±0.3), whereas on 16 days when the canopy was dry, the Bowen ratio was between 0.6 and 1.0 for most of the day. In this way, evaporation from the wet forest used up approximately all the net radiation, while the transpiration from the dry forest only took up about 60%.

In contrast, measurements on 18 days in July 1970 over a forest of Douglas Fir (*Pseudotsuga menziesii* (Mirb.) Franco) at Haney, Vancouver, showed that intercepted precipitation evaporated only about 20% more rapidly than transpired water [*McNaughton and Black*, 1973]. During daylight hours on the 2 days when rain occurred, the Bowen ratio was negative only for the first and last hour of each day and on average was +0.3 (T. A. Black, private communication, 1975). Hence even on these wet days, only 80% of the energy available for partitioning into sensible and latent heat was used for evaporation.

DISCUSSION

These results show that often when the forest canopy was wet, additional energy was provided by advection of unsaturated air over the forest. On occasions of prolonged canopy wetness the sensible heat flux continued in a downard direction for many hours: it could not therefore have come merely from the energy stored within and over the forest. Since the distance of the experimental site from the edge of the forest is much greater than that over which leading edge influences occur, the effect cannot be the result of small-scale leading edge advection. Advection must therefore be occurring on a much larger scale, almost certainly because there is such effective turbulent exchange between the forest and the air passing over it, about five times as effective as that for grassland.

Murphy and Knoerr [1975], using a model of the exchange between the canopy of a wet forest and the atmosphere, demonstrated that the extra energy required to maintain the evaporation of intercepted precipitation at a higher rate than that of transpiration was provided by a decreased or even negative sensible heat flux from the canopy and an increase in net radiation as a result of decreased long-wave reradiation

TABLE 1. Mean Values of the Latent Heat Flux and the Available Energy When the Forest Canopy Was Completely Wet (All the Wetness Sensors Were Indicating), the Apparent Surface Resistance Was Between -10 s m^{-1} and $+10$ s m^{-1} and the Available Energy Was Greater than 20 W m^{-2}

Range of **A**, W m^{-2}	Number of 20-min Periods	Mean Value of **A**, W m^{-2}	Mean Value of **H**, W m^{-2}	Mean Value of $\lambda \mathbf{E}_l$, W m^{-2}
21–25	14	23.9	−49.0	72.9
26–32	19	29.9	−53.2	83.1
33–42	23	37.1	−37.6	74.7
43–57	37	51.0	−28.4	79.4
58–77	29	67.7	−28.0	95.7
78–107	39	91.4	−35.7	127.1
108–147	38	125.8	−32.4	158.2
148–207	23	170.7	−18.8	189.5
208–287	13	239.9	12.3	227.6
288–440	9	352.0	59.0	293.0
>441	1	827.0	152.0	675.0
Mean		102.0	−27.1	129.1
Ratio $\lambda \mathbf{E}_l/\mathbf{A}$				1.27

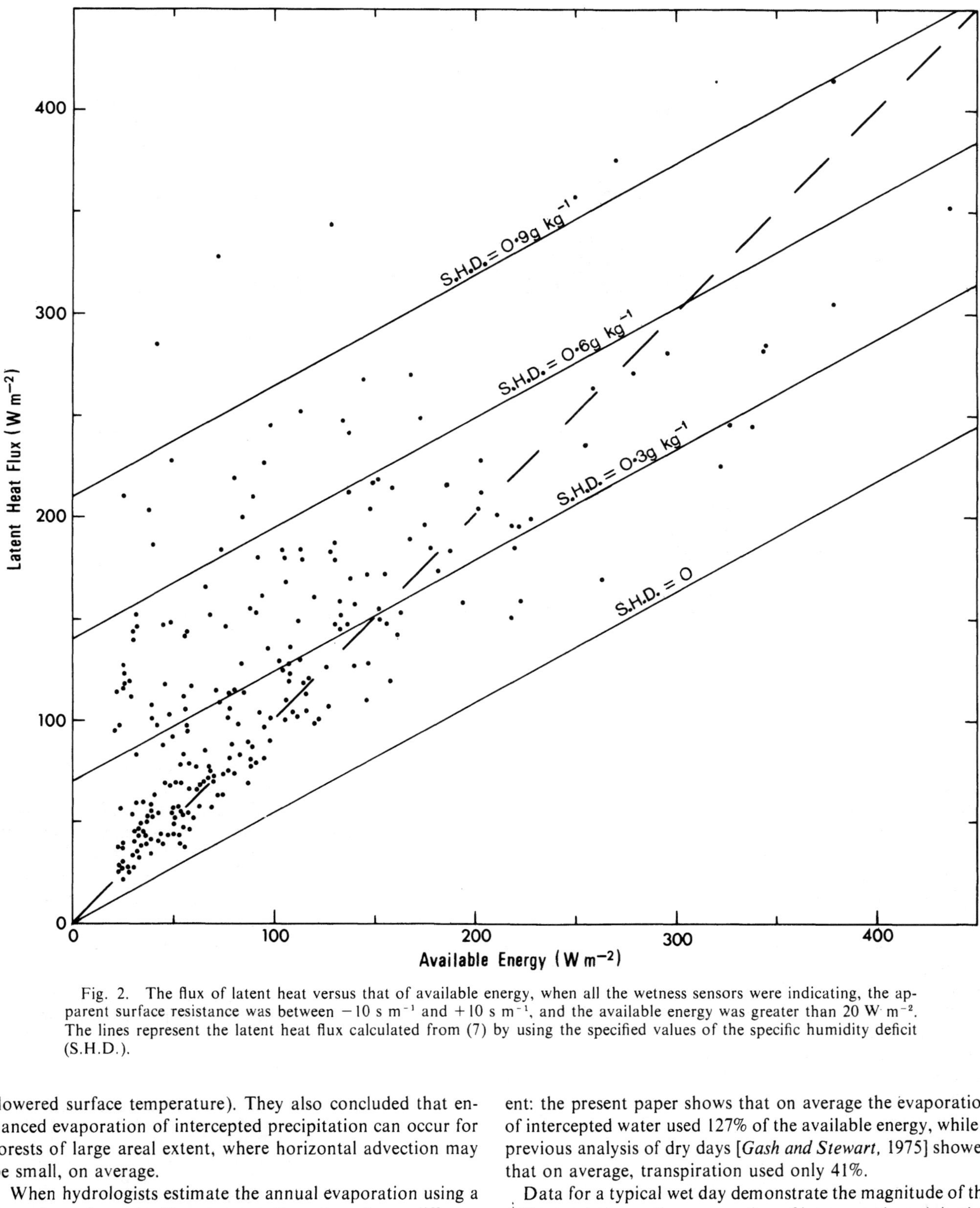

Fig. 2. The flux of latent heat versus that of available energy, when all the wetness sensors were indicating, the apparent surface resistance was between -10 s m^{-1} and $+10$ s m^{-1}, and the available energy was greater than 20 W m^{-2}. The lines represent the latent heat flux calculated from (7) by using the specified values of the specific humidity deficit (S.H.D.).

(lowered surface temperature). They also concluded that enhanced evaporation of intercepted precipitation can occur for forests of large areal extent, where horizontal advection may be small, on average.

When hydrologists estimate the annual evaporation using a formula such as the Penman equation, there is no differentiation between transpiration and evaporation of intercepted precipitation. This practice is supported for short vegetation by the experimental evidence of *McMillan and Burgy* [1960], who found there was no significant difference between the evaporation from lysimeters containing wetted and unwetted vigorously growing rye grass (*Lolium perene* L.). Similar results have also been obtained by *McIlroy and Angus* [1964]. However, the results for Thetford Forest show that the rates of evaporation for wet and dry canopy conditions are very different: the present paper shows that on average the evaporation of intercepted water used 127% of the available energy, while a previous analysis of dry days [*Gash and Stewart,* 1975] showed that on average, transpiration used only 41%.

Data for a typical wet day demonstrate the magnitude of the difference between the evaporation of intercepted precipitation and transpiration. On June 20, 1973, it rained intermittently throughout the daylight hours, with the available energy varying between 21 and 97 W m^{-2}, with a mean value of 50 W m^{-2}. The total available energy for the day was 2.14 MJ m^{-2}, while the measured latent heat loss was 2.62 MJ m^{-2}, which corresponds to 1.05 mm of water. On the assumption that the average rate of conversion of available energy into latent heat found on dry days would have applied in the radiation conditions prevailing on this day had there been no precipitation, the transpiration loss would have been 0.35 mm of water (0.88 MJ m^{-2}), a difference of 0.70 mm.

The net interception loss for a year can be calculated from the measurements of interception reported by *Gash and Stewart* [1977]. Measurements of the precipitation components were made in Thetford Forest throughout 1975 and were used with (1) to give a total interception for the year of 214 mm. The present analysis has shown that on average the rate of evaporation of intercepted precipitation is 3.1 times the rate of transpiration under the same radiation conditions. This result indicates that the presence of intercepted precipitation on the canopy suppressed 69 mm of transpiration, giving a net interception loss of 145 mm for the year. *Gash and Stewart* [1977] also estimated the total evaporation for 1975, i.e., the sum of the interception and transpiration components, and obtained a figure of 566 mm. Hence the net interception loss in this case accounted for 34% of the total evaporation.

SUMMARY AND CONCLUSIONS

When the canopy was wet, the latent heat flux often exceeded the available energy, the additional energy being provided by a downward flux of sensible heat. Since the fetch over forest to the measurement site exceeded 2 km in all directions, the source of the additional energy must be large-scale advection rather than small-scale leading edge advection. One is obliged to draw the conclusion that advected energy comes from areas upwind of the forest where net radiation was not all used for evaporation and where there was therefore an input of sensible heat to the atmosphere. If the air within the boundary layer is moving at 5 m s^{-1}, it will completely pass over even the largest forests in the United Kingdom, Kielder and Thetford forests, which have maximum dimensions of 30 and 15 km, respectively, within 2 hours: advection from nonforested areas is always able to occur unless winds are very light. The total loss of energy as latent heat over this larger area need not exceed the input of energy as net radiation; the disquiet of earlier workers [*Hewlett,* 1967; *Pereira,* 1967] is removed if the problem is treated in three dimensions.

Comparison of the average results for the evaporation from the wet forest canopy with that for transpiration shows that under identical radiation conditions the rate of evaporation of intercepted precipitation from this forest is three times the transpiration rate that occurs when the canopy is dry. The loss by evaporation of intercepted precipitation is therefore only partly compensated by the consequent suppression of transpiration. In contrast, earlier measurements for well-watered grass-covered surfaces showed that there was no detectable difference between the transpiration rate and rate of evaporation of water intercepted by the grass, implying complete compensation. It is therefore concluded that water resources of forests, unlike low vegetation, are affected by the occurrence of interception.

Since the duration of evaporation of intercepted precipitation depends on the duration of canopy wetness, which is related to the frequency of precipitation, the proportion of the annual evaporation from a forest that is due to evaporation of intercepted precipitation will vary greatly even over the United Kingdom. For Thetford Forest and other areas of tall vegetation (where the aerodynamic resistance is small in comparison to the surface resistance), the use of any formula, which does not treat the two components of evaporation separately, will give results with a large systematic error unless by chance a low rate of transpiration balances a high rate of evaporation of intercepted precipitation. For the accurate assessment of the water resources of forested areas, it is essential that hydrologists use appropriate formulas for estimating the evaporation of intercepted precipitation and transpiration separately, before combining the results to give the total water loss by evaporation.

Acknowledgments. The author gratefully acknowledges the assistance of all members of the Institute's hydrometeorological team, who were jointly responsible for the collection and computation of the data described in this paper. This research was supported in part with funds provided by the Department of the Environment under contract DGR 480/35.

REFERENCES

Aslyng, H. C., Evaporation and plant production directly related to global radiation, *Nordic Hydrol., 5,* 247–256, 1974.

Funk, J. P., Improved polythene-shielded net radiometer, *J. Sci. Instrum., 36,* 267–270, 1959.

Gash, J. H. C., and J. B. Stewart, The average surface resistance of a pine forest derived from Bowen ratio measurements, *Boundary-Layer Meteorol., 8,* 453–464, 1975.

Gash, J. H. C., and J. B. Stewart, The evaporation from Thetford Forest during 1975, *J. Hydrol.,* in press, 1977.

Helvey, J. D., Interception by eastern white pine, *Water Resour. Res., 3,* 723–729, 1967.

Hewlett, J. D., Summary of forests and precipitation session, in *International Symposium on Forest Hydrology,* edited by W. E. Sopper and H. W. Lull, pp. 241–243, Pergamon, New York, 1967.

Leyton, L., and A. Carlisle, Measurement and interpretation of interception by forest stands, *Int. Ass. Hydrol. Sci. Hannoversch-Munden, 48,* 111–119, 1959.

Leyton, L., E. R. C. Reynolds, and F. B. Thompson, Rainfall interception in forest and moorland, in *International Symposium on Forest Hydrology,* edited by W. E. Sopper and H. W. Lull, pp. 163–178, Pergamon, New York, 1967.

McIlroy, I. C., and D. E. Angus, Grass, water and soil evaporation at Aspendale, *Agr. Meteorol., 1,* 201–224, 1964.

McMillan, W. D., and R. H. Burgy, Interception loss from grass, *J. Geophys. Res., 65,* 2389–2394, 1960.

McNaughton, K. G., and T. A. Black, A study of evapotranspiration from a Douglas fir forest using the energy balance approach, *Water Resour. Res., 9,* 1579–1590, 1973.

McNeil, D. D., and W. J. Shuttleworth, Comparative measurements of the energy fluxes over a pine forest, *Boundary-Layer Meteorol., 9,* 297–313, 1975.

Monteith, J. L., Evaporation and environment, *Symp. Soc. Exp. Biol., 19,* 206–234, 1965.

Monteith, J. L., Survey of instruments for micrometeorology, *IBP Handbook,* no. 22, Blackwell, Oxford, England, 1972.

Moore, C. J., Eddy flux measurements above a pine forest, *Quart. J. Roy. Meteorol. Soc., 102,* 913–918, 1976.

Murphy, C. E., and K. R. Knoerr, The evaporation of intercepted rainfall from a forest stand: An analysis by simulation, *Water Resour. Res., 11,* 273–280, 1975.

Oliver, S. A., and H. R. Oliver, A computer method for automatic acquisition of wind speed data, *J. Phys. Eng. Sci. Instrum., 6,* 401–403, 1973.

Pasquill, F., Some aspects of boundary-layer description, *Quart. J. Roy. Meteorol. Soc., 98,* 469–494, 1972.

Patric, J. H., Rainfall interception by mature coniferous forests of S. E. Alaska, *J. Soil Water Conserv., 21,* 229–231, 1966.

Penman, H. L., The physical bases of irrigation control, *Proc. 13th Int. Hort. Congr., London, 2,* 913–924, 1952.

Penman, H. L., Vegetation and hydrology, *Tech. Commun. 53,* Commonwealth Bur. of Soils, Harpenden, England, 1963.

Pereira, H. C., Discussion on: Rainfall interception in forest and moorland, in *International Symposium on Forest Hydrology,* edited by W. E. Sopper and H. W. Lull, p. 117, Pergamon, New York, 1967.

Peterson, J. T., E. C. Flowers, and J. H. Rudisill, Dew and frost deposition on pyranometers, *J. Appl. Meteorol., 12,* 1231–1233, 1973.

Rutter, A. J., Studies in the water relations of *Pinus sylvestris* in

plantation conditions, 1, Measurements of rainfall and interception, *J. Ecol.*, *51*, 191–203, 1963.
Rutter, A. J., An analysis of evaporation from a stand of Scots pine, in *International Symposium on Forest Hydrology*, edited by W. E. Sopper and H. W. Lull, pp. 403–417, Pergamon, New York, 1967.
Shuttleworth, W. J., The concept of intrinsic surface resistance: Energy budgets at a partially wet surface, *Boundary-Layer Meteorol.*, *8*, 81–99, 1975.
Stewart, J. B., and A. S. Thom, Energy budgets in pine forest, *Quart. J. Roy. Meteorol. Soc.*, *99*, 154–170, 1973.
Woodhead, T., An investigation into some micro-meteorological aspects of coffee berry disease, *Agr. Meteorol.*, *6*, 195–210, 1969.

(Received May 1976;
revised March 1977;
accepted May 1977.)

(*From the* QUARTERLY JOURNAL OF THE ROYAL METEOROLOGICAL SOCIETY, Vol. 105, No. 443, January 1979)
Quart. J. R. Met. Soc. (1979), **105,** pp. 43–55

551.573:556.133

An analytical model of rainfall interception by forests

By J. H. C. GASH
Institute of Hydrology, Wallingford, Oxfordshire

(Received 22 December 1977; revised 2 June 1978)

SUMMARY

The description of the evaporation of rainfall intercepted by forests in terms of a regression of evaporation loss on incident rainfall is discussed and some of the assumptions implicit in that method are re-examined. The two major factors which control the evaporation of intercepted rainfall are identified. These are: (i) the amount of time that the canopy spends saturated during rainfall and the evaporation rate applicable under these conditions; and (ii) the canopy saturation capacity and the number of times this store is emptied, by drying out after the cessation of rainfall. A model is then constructed which is conceptually similar to the Rutter model, but which replaces that model's numerical approach with an analysis by storm events. The evaporation from a saturated canopy during rainfall is estimated from the Penman–Monteith equation; the evaporation after rain has ceased, the effect of small storms insufficient to saturate the canopy, wetting-up the canopy and evaporation from the trunks are added as separate terms. The model has been tested against data from Thetford Forest in East Anglia, with satisfactory agreement between observation and estimation. It is suggested that the model may be capable of making useful estimates of the evaporation of intercepted rainfall, solely from rainfall measurements.

1. INTRODUCTION

There have been many previous studies of the interception and evaporation of rainfall by forests. The results of these studies have frequently been expressed in the form of empirical regression equations between interception loss and rainfall: that is, as functions of the form

$$I = aP_G + b \qquad (1)$$

where I is the depth of water intercepted and lost by evaporation, the interception loss; P_G, the gross rainfall incident on the canopy; and a and b are regression coefficients. Such an equation can be used either to describe sets of storm data, or, if it is assumed there is only one rainfall event per day, to describe daily interception loss as a function of daily gross rainfall. The reviews of Zinke (1967) and Blake (1975) contain examples of this approach. Helvey and Patric (1965) reviewed results from studies in hardwood forests in the eastern United States and concluded that two regression equations, one for winter and one for summer, were adequate to describe the loss from all hardwood forests in that region. However, the approach has been criticized (e.g. Jackson 1975) for taking no account of such variables as rainfall intensity and duration, and the interval between storms. A further criticism is that the results are empirical and can be extrapolated only to similar forests in areas of climate similar to that where the original data were collected.

In contrast to the regression equation approach, Rutter *et al.* (1971) have constructed a physically based computer model, which uses inputs of rainfall and the meteorological variables controlling evaporation, to calculate a running water balance of a forest canopy of known structure, thereby producing an estimate of the interception loss. The model has been tested against measurements from several forests (Rutter *et al.* 1975) and must be considered the most rigorous method for estimating interception presently available. The model does, however, have practical disadvantages: it requires hourly meteorological data, which are seldom available, and a complex computer programme, which is time consuming to construct and operate.

This paper re-examines some of the assumptions implicit in the regression equation approach. Acknowledging these assumptions, a model is then defined which, while retaining some of the simplicity of the regression equation, includes much of the fundamental physical reasoning explicit in the Rutter model.

2. Theory of regression equations

For any storm large enough to saturate the canopy, Horton (1919) was the first to express interception loss as

$$I = \int_0^t E\,dt + S \qquad (2)$$

where E is the evaporation rate of intercepted water during rainfall; S, the canopy capacity (the amount of water on the canopy when rainfall and throughfall have ceased and the canopy is saturated); t, the duration of rainfall; and evaporation from trunks is neglected. If we now separate evaporation before and after saturation of the canopy:

$$I = \int_0^{t'} E\,dt + \int_{t'}^{t} E\,dt + S \qquad (3)$$

where t' is the time taken for saturation of the canopy to occur.

Defining a mean evaporation rate, $\bar{E}$, from a saturated canopy during rainfall by

$$\bar{E} = (1/(t-t'))\int_{t'}^{t} E\,dt \qquad (4)$$

with rainfall similarly described by a mean rainfall rate:

$$\bar{R} = (1/(t-t'))\int_{t'}^{t} R\,dt,$$

we have

$$P_G - P_G' = \bar{R}(t-t') \qquad (5)$$

where P_G' is the amount of rain necessary to saturate the canopy. Assuming there is no drip from the canopy before saturation, P_G' is also given by

$$(1-p-p_t)P_G' = S + \int_0^{t'} E\,dt \qquad (6)$$

where, following Rutter *et al.* (1971, 1975), p is the free throughfall coefficient, i.e. the proportion of rain which falls through the canopy without striking a surface, and p_t, the proportion which is diverted to the trunks as stemflow.

Substituting Eqs. (4) and (5) into Eq. (3)

$$I = \int_0^{t'} E\,dt + (\bar{E}/\bar{R})(P_G - P_G') + S.$$

Substituting for P_G' from Eq. (6) and rearranging:

$$I = (\bar{E}/\bar{R})P_G + \left(S + \int_0^{t'} E\,dt\right)\{1-(\bar{E}/\bar{R})(1-p-p_t)^{-1}\} \qquad (7)$$

Equating the coefficients of Eqs. (1) and (7), the coefficients of the regression equation can be identified as:

$$a = \bar{E}/\bar{R}, \text{ and } b = \left(S + \int_0^{t'} E\,dt\right)\{1-(\bar{E}/\bar{R})(1-p-p_t)^{-1}\}.$$

Where single values of a and b are derived from a regression involving several storms, their values are considered representative of any individual storm in the set. Thus, in regressing interception loss on rainfall, an implicit assumption is made that $\bar{E}/\bar{R}$ is constant and that the actual deviations of individual points from the line represent only random deviations from that constant value. The fact that in many such regressions interception loss and rainfall are highly correlated, suggests that this assumption is often a good approximation to the truth.

Earlier workers have often equated the coefficient b with the canopy capacity, S – see, for example, the review of Zinke (1967). Rutter (1975) has pointed out that values of S derived in this way can be considered only approximate, as they contain storms where the evaporation is non-zero. The above expression shows that the error introduced by this assumption will also have some dependence on the parameters p, $\bar{E}$ and $\bar{R}$. However, a regression including only storms in zero evaporation conditions, for which $\bar{E}$, and therefore a, would be zero, will yield a value for the coefficient b which is an estimate of the canopy capacity, S. This is effectively the method proposed by Leyton *et al.* (1967).

The use of an equation such as (1), expressed in terms of rainfall, rather than an equation such as (3), expressed in terms of rainfall duration, was proposed by Horton (1919) after examining the relationship between rainfall amount and rainfall duration and finding a near linear relation, i.e. a near constant rainfall rate. He concluded that: ". . . it will probably often be found more convenient to utilize interception results of formulae expressed in terms of amounts of precipitation rather than in terms of shower duration, although the latter method of expressing results appears to be more logical".

Unfortunately, expressing interception results in terms of precipitation introduces empiricism not only into $\bar{E}$, the mean evaporation rate, but also unnecessarily into $\bar{R}$, the mean rainfall rate. Rainfall duration is simple to measure and making it explicit, as in the original equation, would have greatly aided the understanding of the many interception studies undertaken since then.

3. Derivation of the model

(*a*) *Simplifying assumptions*

The object is to create a model of interception loss which is simpler and easier to apply than the Rutter model, but at the same time contains a great deal of the objectivity and physical reasoning behind that model. The primary assumption is that it is possible to represent the real rainfall pattern by a series of discrete storms, separated by sufficiently long intervals for the canopy and trunks to dry. Within this simplifying assumption, the procedure then adopted is to perform an analytical integration of the total evaporation loss of intercepted rainfall, by replacing the actual rate of rainfall and evaporation applicable to each storm by the average rates deduced for all storms. In order to carry out this analytical integration procedure it is necessary to introduce the following simplifying assumptions:

1. The meteorological conditions prevailing during any wetting-up of the canopy are sufficiently similar to those prevailing for the rest of the storms, that mean conditions of rainfall and evaporation, over all storms in the period, can be considered to apply in these transient situations.

2. The logarithmic dependence of the 'drip-rate' on the degree of canopy saturation, observed by Rutter *et al.* (1971), is sufficiently sensitive that there is in effect virtually no drip from the canopy during wetting-up; and the amount of water on the canopy at the end of a storm is quickly reduced (within 20 to 30 minutes) to S, the minimum value necessary for saturation, independent of the initial value when rain ceased.

The effect of these assumptions in the model can best be assessed by its application to real data and by analysing the sensitivity of the results to the various parameters. Their consequence, however, is in effect to separate the dependence of interception loss on 'forest' parameters, such as storage, S, from meteorological variables, such as evaporation rate. By making the dependence of interception loss on these parameters explicit, rather than implicit as they are in a regression equation, it is hoped that results can be more readily extrapolated between forests and climatic regions.

(*b*) *Evaporation from a saturated canopy*

Consider a series of n storms each large enough to saturate the canopy and each separated by a sufficient period for the canopy to dry. If Eq. (3) is applied to each storm the total interception loss is given by

$$\sum_{j=1}^{n} I_j = \sum_{j=1}^{n}\left[\int_0^{t_j'} E\,dt + \int_{t_j'}^{t_j} E\,dt\right] + nS \qquad (8)$$

If $\bar{E}$ is now redefined by

$$\bar{E} = \sum_{j=1}^{n}\int_{t_j'}^{t_j} E\,dt \Big/ \sum_{j=1}^{n}(t_j - t_j'),$$

while similarly $\bar{R}$ is given by

$$\bar{R} = \sum_{j=1}^{n}\int_{t_j'}^{t_j} R\,dt \Big/ \sum_{j=1}^{n}(t_j - t_j'),$$

or

$$\bar{R} = \sum_{j=1}^{n}(P_{Gj} - P'_{Gj}) \Big/ \sum_{j=1}^{n}(t_j - t_j'),$$

equation (8) can be written as

$$\sum_{j=1}^{n} I_j = \sum_{j=1}^{n}\left[\int_0^{t_j'} E\,dt + (\bar{E}/\bar{R})(P_{Gj} - P'_{Gj})\right] + nS \qquad (9)$$

(*c*) *Wetting-up the canopy*

From Eq. (6) the interception loss for the period before saturation is $\int_0^{t'} E\,dt = (1 - p - p_t)P'_G - S$. Substituting this into Eq. (9) for all n storms,

$$\sum_{j=1}^{n} I_j = (1 - p - p_t)\sum_{j=1}^{n} P'_{Gj} + (\bar{E}/\bar{R})\sum_{j=1}^{n}(P_{Gj} - P'_{Gj}) \qquad (10)$$

For small storms insufficient to saturate the canopy completely, $I = (1 - p - p_t)P_G$. Including these (say) m storms, the total interception loss is

$$\sum_{j=1}^{n+m} I_j = (1 - p - p_t)\sum_{j=1}^{n} P'_{Gj} + (\bar{E}/\bar{R})\sum_{j=1}^{n}(P_{Gj} - P'_{Gj}) + (1 - p - p_t)\sum_{j=1}^{m} P_{Gj} \qquad (11)$$

(d) Evaporation from the trunks

Emulating the Rutter model, where the trunk water balance is conceived as being separate from the canopy water balance, the proportion of the rain which is directed to the stem is removed from the incident rainfall before the canopy water balance is considered. The stemflow for each storm with rainfall greater than S_t/p_t is described by stemflow $= p_tP_G - S_t$, S_t being the trunk water store, which must be filled before any stemflow is apparent at the base of the trunks.

Evaporation from trunks is likely to be small compared with evaporation from the canopy (Rutter and Morton (1977) calculate it to be typically 1 to 5% of the above-canopy value during rainfall) and it is neglected here. Interception loss from trunks is then only the amount of water which remains on the trunks after stemflow has ceased. If the assumption is made that the interval between storms is sufficiently long to allow all the water in the trunk store to evaporate, then the interception loss from the trunks is given by

$$\sum_{j=1}^{n+m} I_j = qS_t + p_t \sum_{j=1}^{n+m-q} P_{Gj},$$

when out of the total $(n+m)$ storms there are q storms above the critical rainfall, S_t/p_t, necessary to fill the trunk store.

Equation (11) can now be rewritten as

$$\sum_{j=1}^{n+m} I_j = (1-p-p_t)\sum_{j=1}^{n} P'_{Gj} + (\bar{E}/\bar{R})\sum_{j=1}^{n}(P_{Gj}-P'_{Gj}) + (1-p-p_t)\sum_{j=1}^{m} P_{Gj} + qS_t + p_t \sum_{j=1}^{m+n-q} P_{Gj} \qquad (12)$$

(e) The rainfall necessary to saturate the canopy

Evaporation from partially wet canopies and its description in terms of the Monteith–Penman equation has been the subject of recent discussion (Shuttleworth 1976; Monteith 1977). However, Shuttleworth (1978) has demonstrated that the somewhat arbitrary assumption made by Rutter *et al.* (1971) leads to a theoretically reasonable description of the partially wet canopy. That assumption is that evaporation from the canopy is described by $E = (C/S)E_p$, where E_p is the evaporation which would occur from a totally wet canopy and C is the amount of water on the canopy.

This is applied to conditions on the canopy before saturation is reached, with the further assumption that mean evaporation and rainfall rates also apply. The evaporation during this period is then given by $E = (C/S)\bar{E}$. Assuming there is no water dripping from the canopy before saturation, the rate of change of water on the canopy can be described by $dC/dt = (1-p-p_t)\bar{R} - E$, or $dC/dt = (1-p-p_t)\bar{R} - (\bar{E}/S)C$, which has the solution

$$C = (1-p-p_t)(\bar{R}S/\bar{E})\{1-\exp(-\bar{E}t/S)\},$$

or

$$t = -(S/\bar{E})\ln\{1-(\bar{E}C/\bar{R}S)(1-p-p_t)^{-1}\}.$$

The time taken for saturation to occur is therefore

$$t' = -(S/\bar{E})\ln\{1-(\bar{E}/\bar{R})(1-p-p_t)^{-1}\}.$$

Following the assumption that the mean rainfall rate applies, at the point when the canopy reaches saturation the rainfall is given by $P'_G = \bar{R}t'$; therefore

$$P'_G = (-\bar{R}S/\bar{E})\ln\{1-(\bar{E}/\bar{R})(1-p-p_t)^{-1}\} \quad . \quad . \qquad (13)$$

and Eq. (12) can be rewritten with P'_G considered as a constant for the period:

$$\sum_{j=1}^{n+m} I_j = n(1-p-p_t)P'_G+(\bar{E}/\bar{R})\sum_{j=1}^{n}(P_{Gj}-P'_G)+(1-p-p_t)\sum_{j=1}^{m}P_{Gj}+qS_t+p_t\sum_{j=1}^{m+n-q}P_{Gj} \qquad (14)$$

4. Application of the model to Thetford Forest data

(a) *Site and instrumentation*

Thetford Forest is a pine forest of over 200 km^2 situated in East Anglia. Observations were made in a stand of Scots pine (*Pinus sylvestris* L.) planted in 1931 and now thinned to a density of some 800 trees per hectare.

Details of the instrumentation have been given previously by Gash and Stewart (1977). Interception loss was determined from the difference between gross rainfall, throughfall and stemflow. Throughfall was measured with a random grid of 24 raingauges which were measured and rearranged within an area 30 m × 15 m every two weeks.

Gross rainfall was measured both above the canopy and in an adjacent clearing. Gash and Stewart found good agreement between the two gauges and used the above-canopy gauge in their analysis. That gauge is also used in this analysis. In March 1976 the gauge was changed from a resolution of 0·05 mm per tip to 0·25 mm per tip.

Two automatic weather stations (Strangeways 1972) were mounted on a tower above the forest. Measurements of net radiation, temperature, wet bulb depression and windspeed were made every five minutes and recorded on magnetic tape. Readings from the two stations were later averaged to produce a single set of hourly average values. The measurements were made throughout 1975 and 1976; during that period the weather stations were recalibrated four times.

(b) *The forest structure parameters*

The canopy capacity, S, is defined as the amount of water present on the canopy in conditions of zero evaporation, when throughfall has ceased. S for the Thetford site has been determined to be 0·8 mm (Gash and Morton 1978) with an estimated error of about $\pm 10\%$.

The free throughfall coefficient, p, was estimated to be 0·32 from an analysis of small storms, or 0·26 from a measurement of canopy cover (Gash and Morton).

The stemflow in a mature Scots pine stand is a small component of the water balance. At Thetford the total stemflow for 1975 amounted to less than 2% of the rainfall for that period. Nevertheless, the following equation describing stemflow has been determined (Gash and Morton):

$$\text{stemflow} = 0{\cdot}016(\pm 0{\cdot}003)P_G - 0{\cdot}014(\pm 0{\cdot}035),$$

which implies values of 0·016 and 0·014 mm for p_t and S_t respectively.

(c) *Procedure*

The model, specified by Eq. (12), has been applied to the Thetford data for a total of twenty-one four-week periods. Periods were omitted when either the weather station data were incomplete, or there was snow.

The theory requires that $\bar{E}$ and $\bar{R}$ should be calculated from those hours when the canopy is saturated; however, in practice, to determine whether or not this condition is satisfied requires the drawing up of an hourly water balance, as in the Rutter model. To maintain the simplicity of the model it is therefore necessary to approximate this condition by selecting a rainfall rate, above which it is considered the canopy will always be saturated. Once the canopy is saturated, the rainfall rate necessary to maintain saturation is $E/(1-p-p_t)$, typically 0·3 mm h^{-1}; if the canopy is completely dry the rainfall necessary to achieve saturation in the hour is P'_G, typically 1·3 mm. As a compromise, the threshold rate of 0·5 mm h^{-1} was chosen as representing saturated conditions. It is accepted that this will include some hours of rainfall when the canopy is only partially wet before saturation and will exclude some hours when the canopy is saturated during light rainfall. The following procedure was therefore adopted:

1. For each four-week period, hours with $P_G \geqslant 0{\cdot}5$ mm were isolated and the evaporation calculated using the Monteith–Penman equation as in the Rutter model:

$$\lambda E = (\Delta R_n + \rho c_p D/r_a)(\Delta+\gamma)^{-1},$$

where c_p is the specific heat of air at constant pressure (J kg^{-1}K^{-1}); E, the evaporation rate (kg m^{-2}s^{-1}); r_a, the aerodynamic resistance (s m^{-1}); R_n, the net radiation (W m^{-2}); D, the vapour pressure deficit (mb); γ, the psychrometric constant (mb K^{-1}); Δ, the slope of the saturation vapour pressure curve at air temperature (mb K^{-1}); and ρ, the density of air (kg m^{-3}).

The aerodynamic resistance, as in the Rutter model, was calculated from the expression $r_a = (1/k^2u)\{\ln(z-d)/z_0\}^2$, with $k = 0{\cdot}41$ and u the windspeed at height z (m s^{-1}), and the assumption that $d = 0{\cdot}75h$ and $z_0 = 0{\cdot}10h$, when h is the vegetation height (m) and $z = h+2$ (m).

Under the assumption that hours with $P_G \geqslant 0{\cdot}5$ mm represent saturated canopy conditions, these hourly values of E were then averaged over the four-week period to give an estimate of $\bar{E}$.

2. Similarly, for the same hours P_G was extracted and, making the same assumption, averaged to give $\bar{R}$.

3. These values were then used to calculate P'_G from Eq. (13) for each four-week period, using the following values of the other parameters (Gash and Morton): $S = 0{\cdot}8$ mm, $p = 0{\cdot}32$, $p_t = 0{\cdot}016$.

4. Making the further assumption that there is only one storm per rainday, raindays were divided into those with $P_G \geqslant P'_G$ and those with $P_G < P'_G$. These two sets of daily rainfall amounts were then summed to give $\sum_{j=1}^{n} P_{Gj}$ and $\sum_{j=1}^{m} P_{Gj}$.

5. The number of raindays, n, with $P_G \geqslant P'_G$ was noted.

6. The number of raindays, q, with $P_G \geqslant S_t/p_t$ was noted, and the rainfalls of raindays with $P_G < S_t/p_t$ were summed.

The procedure provided sufficient information for the interception loss to be calculated from Eq. (14).

(*d*) *Results*

The calculated values of $\bar{E}$, $\bar{R}$ and P'_G for each four-week period are given in Fig. 1. The error bars represent standard deviations of the mean hourly values of $\bar{E}$ and $\bar{R}$; the numbers of hours data used are also shown.

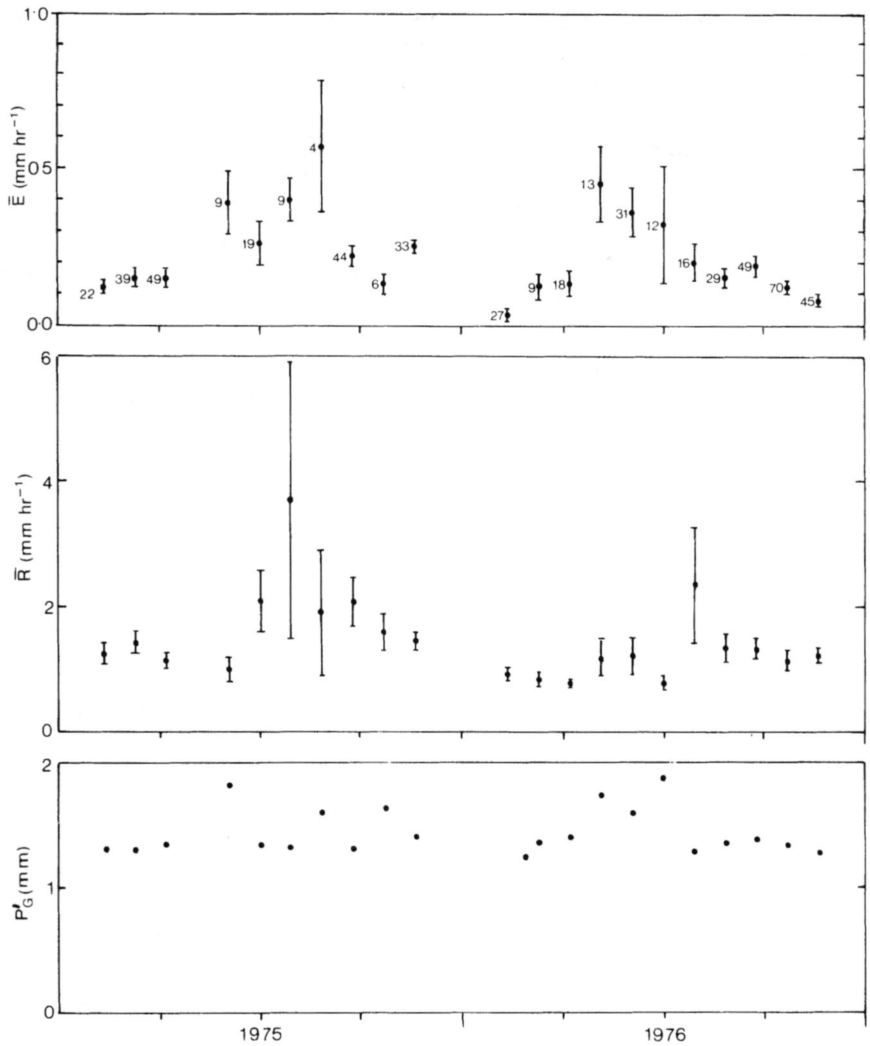

Figure 1. $\bar{E}$, the mean evaporation rate for hours when $P_G \geqslant 0{\cdot}5$ mm, $\bar{R}$, the mean rainfall rate for those hours, and P'_G, the rainfall necessary to saturate the canopy, for four-week periods during 1975 and 1976. The standard deviation of the means and the number of hours data used are shown.

There is no significant correlation between $\bar{E}$ and $\bar{R}$ in either 1975 or 1976, the coefficients of determination, R^2, being 0·178 and 0·008 respectively.

The cumulative totals of observed and estimated interception loss are shown in Fig. 2. The maximum difference between observed and estimated loss is 16 mm. The final totals of observed and estimated loss are 245 and 257 mm respectively; the model therefore overestimates the interception loss by 12 mm, or 5% of the observed loss.

The calculations were made over four-week periods to give a better appreciation of the performance of the model and variability of the estimates. There is, however, no reason why a single calculation cannot be made on the data.

When calculations are performed for the whole period the inputs to the model are: $\bar{E} = 0{\cdot}19\,\mathrm{mm\,h^{-1}}$; $\bar{R} = 1{\cdot}38\,\mathrm{mm\,h^{-1}}$; $\sum_{j=1}^{j=n} P_{Gj} = 822{\cdot}1$ mm; $\sum_{j=1}^{j=m} P_{Gj} = 73{\cdot}7$ mm; and

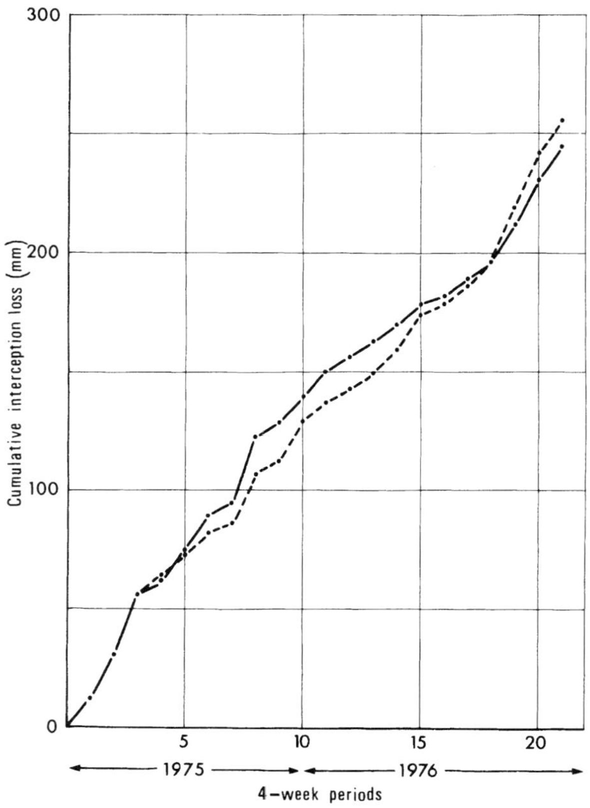

Figure 2. Cumulative totals of measured (continuous line) and estimated (broken line) interception loss, for 21 four-week periods during 1975 and 1976.

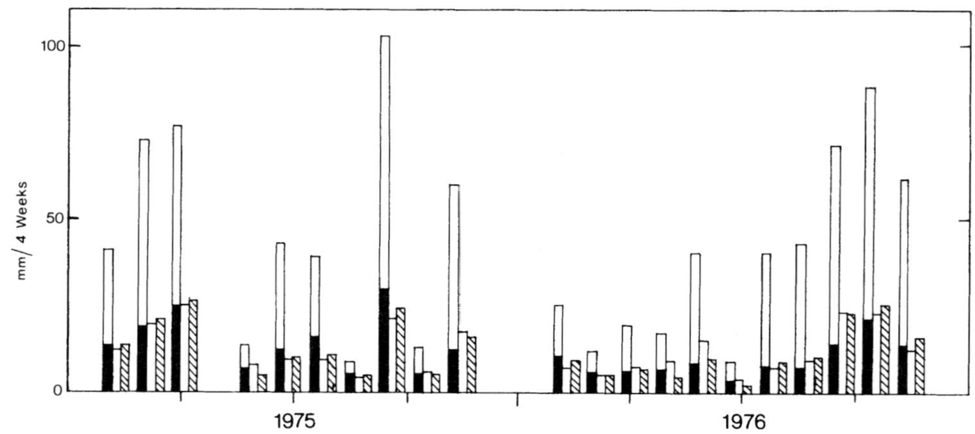

Figure 3. Each block shows from left to right: total rainfall, with measured interception loss in black; estimated interception loss, using the parameters derived for each period (plain); and the estimated interception loss, using the mean parameters derived from the whole data set (hatched).

$\sum_{j=1}^{j=m+n-q} P_{Gj} = 46{\cdot}5\,\text{mm}$; n, m and q being 132, 161 and 157 respectively. This gives $P'_G = 1{\cdot}35\,\text{mm}$.

The interception loss is then made up as shown in Table 1. The rainfall and the observed and estimated interception loss for each period are also shown in Fig. 3, together with the interception loss estimated using the mean parameters derived for the whole period, but applied to each individual period. In both cases a regression of observed on calculated interception loss gave a line whose slope did not differ significantly from unity, and an intercept which did not differ significantly from zero.

TABLE 1. THE COMPONENTS OF THE TOTAL INTERCEPTION LOSS, CALCULATED FOR THE WHOLE PERIOD OF 84 WEEKS OF AVAILABLE DATA DURING 1975 AND 1976

Component of interception loss	Analytical form	Value (mm)
Small storms	$(1-p-p_t)\sum_{j=1}^{m} P_{Gj}$	48·9
Wetting-up the canopy	$n(1-p-p_t)P'_G-nS$	12·5
Evaporation from saturation until rainfall ceases	$(\bar{E}/\bar{R})\sum_{j=1}^{n}(P_{Gj}-P'_G)$	87·4
Evaporation after rainfall ceases	nS	105·6
Evaporation from trunks	$qS_t+p_t\sum_{j=1}^{m+n-q} P_{Gj}$	2·9
Total interception loss		257·3

5. ERROR ANALYSIS

The error in the model prediction due to any variable, X, is estimated as $\delta I = (\partial I/\partial X)\delta X$. If the variables are independent, the resultant errors may then be summed quadratically to give the total error in the estimate.

The percentage errors in the two stemflow terms are large, but the absolute magnitude of evaporation from trunks is, in this case, so small that the errors can be neglected. For the purposes of this analysis the error in the measurement of rainfall is also neglected. The remaining variables are $\bar{E}$, S and $(1-p-p_t)$.

(a) *The error in* $\bar{E}$

The error in $\bar{E}$ will depend on the accuracy of the measurements, in particular that of wet bulb depression, and also on the validity of the assumptions made in determining r_a. These assumptions, which are discussed by Rutter *et al.* (1975), take account of atmospheric stability (Thom *et al.* 1975), and of the effects discussed by Thom (1971, 1972), but essentially estimate the parameters d and z_0 from empirical relationships. $\bar{E}$, the mean evaporation rate, has been calculated under the assumption that the evaporation for hours when $P_G \geqslant 0{\cdot}5\,\text{mm}$ is equivalent to the evaporation for all hours of rainfall onto a saturated canopy. This assumption is obviously an approximation to the truth. However, an independent estimate of $\bar{E}$ is available for the same site in 1972 and 1973, from direct measurements of evaporation rate made using the Bowen ratio technique when the canopy was saturated (Stewart 1977). Periods when the available energy was less than $20\,\text{W}\,\text{m}^{-2}$ ($0{\cdot}03\,\text{mm}\,\text{h}^{-1}$) were omitted from the analysis due to large percentage errors in the measurement technique at low energy levels; but a total of 245 twenty-minute periods gave a mean evaporation rate of 129·1 W m^{-2}, equivalent to 0·19 mm h^{-1}. This agreement with the estimate of $\bar{E}$ reported here suggests that the assumptions made in the derivation of $\bar{E}$ are not seriously in error. $\bar{E}$ is relatively insensitive to changes in the arbitrarily selected threshold of $P_G \geqslant 0{\cdot}5$ mm h^{-1}. Changing the threshold to 0·25 and to 1·0 mm h^{-1} changed $\bar{E}$ from 0·19 mm h^{-1} to 0·21 and 0·17 mm h^{-1}, respectively. An error of ±15% was therefore arbitrarily assigned to $\bar{E}$ and the consequent error in the interception loss, given by $\delta I = (\partial I/\partial \bar{E})\delta \bar{E}$, was ±14·9 mm, or ±6% of the total loss.

(*b*) *The error in S*

The canopy capacity is assumed to have been determined to within ±10% (Gash and Morton). The error due to S, given by $\delta I = (\partial I/\partial S)\,\delta S$, is ±9·4 mm, or ±4% of the total loss.

(*c*) *The error in* $(1-p-p_t)$

The difference between the two values derived for p (0·26 and 0·32, Gash and Morton) suggests that the error in this parameter may be relatively large. However, an error of ±30% in p, which covers both values, is equivalent to an error of ±14% in $(1-p-p_t)$ and results in an error in the interception loss, given by $\delta I = \{\partial I/\partial(1-p-p_t)\}\delta(1-p-p_t)$, of 8·9 mm, or ±3% of the total loss.

This analysis ignores error due to the assumption of equivalence between rainday and rainfall event. The assumption was made partly for simplicity but also to avoid the need for any subjective assessment of what constitutes a rainfall event. Disregarding this, the error resulting from all the other variables considered is 19·7 mm, or 8% of the total.

6. Discussion and conclusions

The model described in this paper in effect separates the interception loss resulting from evaporation during rainfall, from that which results from the structure of the canopy. Following some simplifying assumptions, this is accomplished analytically rather than numerically as in the Rutter model, although conceptually the models are similar. The agreement between observation and estimation appears to be sufficiently good, in the case studied, to justify the additional assumptions and simplifications which have been made. Further tests will, however, be required before these assumptions can be considered generally justifiable.

As it has been applied here, the model suffers from the same disadvantage as the Rutter model, in that it requires an input of hourly meteorological data. Although rainfall and rainfall duration are measured at climatological and agrometeorological stations in many areas, the hourly values of the meteorological parameters required to calculate the evaporation are rarely measured above forests, except in research projects. There are some sixty meteorological stations in the United Kingdom where hourly observations are made, but these are generally associated with aviation or marine activities and therefore tend to be at considerable distances, and different altitudes, from the major areas of forest. These are mainly inland, upland regions. In any practical application of the model it will therefore be necessary to use empirical values of at least $\bar{E}$ and possibly $\bar{R}$.

Little is known about the seasonal, areal or altitudinal variation of $\bar{E}$, although some preliminary information can be obtained from Fig. 1, where a seasonal variation is indicated. However, it should be noted that in an analysis of a similar parameter for a site in the south of England during 1966 and 1967, Rutter and Morton (1977) found no such marked seasonal trend, which may therefore be associated with the anomalous meteorological conditions during the summers of 1975 and 1976. A preliminary study of evaporation from an upland forest in central Wales has given a value for $\bar{E}$ of 0·12 mm h^{-1}, which compares with the value of 0·19 mm h^{-1} obtained for Thetford Forest. These two sites represent extremes of climate likely to be found in forested areas of Great Britain; so if this value for central Wales is correct, it is likely that the variation in $\bar{E}$ may be sufficiently small for adequately accurate estimates of evaporation to be made, with little regard for variation in $\bar{E}$. The sensitivity of the interception loss to $\bar{E}$ will vary with rainfall climate, being more sensitive in regions of long rainfall duration. However, in the case studied, Thetford Forest, to obtain an accuracy of $\pm 15\%$ in interception loss (equivalent to approximately 5% in net rainfall) would, with the estimated accuracy in the other variables, require $\bar{E}$ to be estimated to an accuracy of $\pm 37\%$. This relative insensitivity to $\bar{E}$ is illustrated in Fig. 3, where estimates made using the values of $\bar{E}$ derived for each individual period can be compared with those derived using fixed means for the whole period. The consequent possibility of using empirical estimates of $\bar{E}$ to estimate interception loss solely from rainfall data, suggests that the problem of estimating the evaporation of rainfall intercepted by forests may have a relatively simple solution.

Thom and Oliver (1977) have suggested that total regional evaporation (interception plus transpiration) may be calculated from daily measurements made at climatological and agrometeorological stations, using a version of the Penman equation, modified by the introduction of two additional factors. The first is related to the surface roughness of the vegetation, while the second represents the combined effect of the average surface resistance of the vegetation and the duration of canopy wetness. In practice this latter parameter requires an independent estimate of the interception loss (Gash 1978), which, it was suggested, might be obtained from the extrapolation of previous catchment analyses or measurements of interception loss. However, the model presented here allows, in the case of forests, the attractive possibility of obtaining this required estimate of rainfall interception over the same period and for the same site as those being used to calculate the transpiration.

Acknowledgments

I am indebted to N. F. Cowell and C. R. Lloyd for their help in the collection and analysis of the Thetford data and to W. J. Shuttleworth for his constructive criticism at all stages in the preparation of this paper. Thanks are also due to Professor P. G. Jarvis, Professor A. J. Rutter, J. B. Stewart and A. S. Thom for their helpful comments during the final stages of preparation and to referees for their helpful criticism of the first draft of this paper.

The work described in this paper was carried out with funds provided by the Department of the Environment under contract number DGR 480/35.

References

Blake, G. J. 1975 The interception process. In *Prediction in catchment hydrology*, Eds. T. G. Chapman and F. X. Dunin, Australian Academy of Science, Canberra, 59–81.

Gash, J. H. C. 1978 Comment on the paper by A. S. Thom and H. R. Oliver, On Penman's equation for estimating regional evaporation, *Quart. J. R. Met. Soc.*, **104**, 532–533.

Gash, J. H. C. and Morton, A. J. 1978 An application of the Rutter model to the estimation of the interception loss from Thetford Forest, *J. Hydrol.*, **38**, 49–58.

Gash, J. H. C. and Stewart, J. B. 1977 The evaporation from Thetford Forest during 1975, *Ibid.*, **35**, 385–396.

Helvey, J. D. and Patric, J. H. 1965 Canopy and litter interception of rainfall by hardwoods of eastern United States, *Water Resour. Res.*, **1**, 193–206.

Horton, R. E. 1919 Rainfall interception. *Mon. Weath. Rev.*, **47**, 603–623.

Jackson, I. J. 1975 Relationships between rainfall parameters and interception by tropical forest, *J. Hydrol.*, **24**, 215–238.

Leyton, L., Reynolds, E. R. C. and Thompson, F. B. 1967 Rainfall interception in forest and moorland, *Int. Symp on Forest Hydrology*, Eds. W. E. Sopper and H. W. Lull, Pergamon Press, Oxford, 163–178.

Monteith, J. L. 1977 Resistance of a partially wet canopy: whose equation fails?, *Boundary Layer Met.*, **12**, 379–383.

Rutter, A. J. 1975 Chapter 4. The hydrological cycle in vegetation. In *Vegetation and the atmosphere* Vol. 1, Ed. J. L. Monteith, Academic Press, London, 111–154.

Rutter, A. J., Kershaw, K. A., Robins, P. C. and Morton, A. J. 1971 A predictive model of rainfall interception in forests. I: Derivation of the model from observations in a plantation of Corsican Pine, *Agr. Met.*, **9**, 367–384.

Rutter, A. J. and Morton, A. J. 1977 A predictive model of rainfall interception in forests. III: Sensitivity of the model to stand parameters and meteorological variables, *J. Appl. Ecol.*, **14**, 567–588.

Rutter, A. J., Morton, A. J. and Robins, P. C. 1975 A predictive model of rainfall interception in forests. II: Generalization of the model and comparison with observations in some coniferous and hardwood stands, *Ibid.*, **12**, 367–380.

Shuttleworth, W. J. 1976 Experimental evidence for the failure of the Penman–Monteith equation in partially wet conditions, *Boundary Layer Met.*, **8**, 81–99.

1978 A simplified one-dimensional theoretical description of the vegetation–atmosphere interaction, *Ibid.*, **14**, 3–27.

Stewart, J. B. 1977 Evaporation from the wet canopy of a pine forest, *Water Resour. Res.*, **13**, 915–921.

Strangeways, I. C. 1972 Automatic weather stations for network operation, *Weather*, **27**, 403–408.

Thom, A. S. 1971 Momentum absorption by vegetation, *Quart. J. R. Met. Soc.*, **97**, 414–428.

1972 Momentum, mass and heat exchange of vegetation, *Ibid.*, **98**, 124–134.

Thom, A. S. and Oliver, H. R. 1977 On Penman's equation for estimating regional evaporation, *Ibid.*, **103**, 345–357.

Thom, A. S., Stewart, J. B., Oliver, H. R. and Gash, J. H. C. 1975 Comparison of aerodynamic and energy budget estimates of fluxes over a pine forest, *Ibid.*, **101**, 93–105.

Zinke, P. J. 1967 Forest interception studies in the United States, *International Symposium on Forest Hydrology*, Eds. W. E. Sopper and H. W. Lull, Pergamon Press, Oxford, 137–161.

Reprinted from JOURNAL OF APPLIED METEOROLOGY, Vol. 18, No. 5, May 1979
American Meteorological Society
Printed in U. S. A.

Has the Priestley-Taylor Equation Any Relevance to Forest Evaporation?

W. JAMES SHUTTLEWORTH AND IAN R. CALDER

Institute of Hydrology, Wallingford, Oxon, UK

(Manuscript received 11 July 1978, in final form 2 February 1979)

ABSTRACT

Long-term evaporation measurements are expressed in the Priestley-Taylor (1972) "potential evaporation" framework for a spruce forest in Plynlimon, Wales, and a Scots Pine forest in Norfolk, England. The results are used to illustrate the possibility of significant variability in evaporation from forest vegetation in response to precipitation input, and so provide a warning against the indiscriminate use of the Priestley-Taylor formula. A tentative suggestion is made regarding a possible role for potential evaporation in the forest environment.

1. Introduction

Penman (1948) demonstrated that in the absence of any control at the earth's surface, evaporation depends only on meteorological conditions. The particular combination of meteorological parameters can be considered the basis of a metric, used to relate evaporation measurements made at different locations and in different conditions by allowing for the variation due to this meteorological control. In an attempt to provide a more convenient metric, Priestley and Taylor (1972) simplified the Penman equation by redefining potential evaporation E_p as

$$E_p=\frac{\alpha}{\lambda}\frac{\Delta}{\Delta+\gamma}(Q^*-G), \qquad (1)$$

where α is a constant, Q^* the net radiation, G the soil heat flux, Δ the gradient of saturated vapor pressure with temperature and γ the psychrometric constant. Using this expression with data gathered from several different oceanic and saturated land surfaces, these authors found that the actual evaporation E_a could be simply equated to potential evaporation,

$$E_a = E_p, \qquad (2)$$

with the constant α in the order of 1.2–1.3. Several other evaporation data sets have since been similarly expressed (e.g., McNaughton and Black, 1973; Davies and Allen, 1973; Thompson, 1975; Stewart and Rouse, 1977; Mukammal and Neumann, 1977), tending to reinforce the use of Eq. (2), with a value of $\alpha=1.27$, as a means of estimating actual evaporation in potential conditions.

The overall effect of this work is to perpetuate and propagate the idea that actual evaporation, when averaged over several days or weeks, is a simple and easily calculable function of temperature and radiant energy, independent of precipitation input and only weakly related to vegetation cover. The validity and usefulness of Eq. (2) as a practical description of surface evaporation flux for use in large-scale meteorological applications is not the concern of this paper. The authors are concerned that the indiscriminate use of the Priestley-Taylor hypothesis generates complacency regarding the importance of "surface" control in the evaporation process, and that this complacency might be transferred from the meteorologist to the hydrologist and water engineer who, tempted by computational convenience, will adopt it as a universal description of evaporation loss.

This paper compares estimates of potential evaporation using the Priestley-Taylor equation with two measured evaporation data sets for tall (forest) vegetation. The comparison is made both when water is freely available on the surface of the vegetation, and in conditions of transpiration when at least one of the forests represents "potential" conditions in the, perhaps erroneous, but often accepted sense that there is no significant soil moisture stress. The object is to provide general evidence on the variability of evaporation, and in particular to provide evidence on the divergence of forest evaporation from the Priestley-Taylor metric. An attempt is made to provide a nonpredictive but concise description of these data by developing an equation which contains some elements of the Priestley-Taylor metric but which acknowledges the primary change in surface status, that between wet and dry conditions.

0021-8952/79/050639-08$06.00

2. Experimental data

The data sets used in this analysis are extensive evaporation measurements for two mature forest stands in the United Kingdom. The experimental sites are located in the extreme west and east of the country and therefore might be considered indicative of the change in evapotranspiration for this vegetation type in response to a precipitation difference of the order of 1800 mm year^{-1} between the sites.

a. The Plynlimon data set

The data set consists of simultaneous measurements of evaporation and meteorological variables extending on a continuous basis from February 1974 to September 1976. The experimental site, at Plynlimon, Powys, central Wales, is located close to the center of the Institute of Hydrology's experimental catchment (Clarke and McCulloch, 1976), which is extensively forested with Sitka and Norway Spruce.

The evaporation measurements were made with a "natural" lysimeter described by Calder (1976), containing a sample of 26 trees. The representativeness of the lysimeter has been established by comparison with the total evaporation loss from the whole catchment (Calder, 1976); while successful modeling of the detailed variation in measured evaporation in terms of a physically based model (Calder, 1977) gives increased confidence in the experimental data. In parallel with this lysimetric measurement of total evaporation loss, a separate measurement was made of the loss due to rainfall interception by the forest (Calder and Rosier, 1976).

For the early part of the experiment the meteorological variables used in calculating the Priestley-Taylor definition of potential evaporation were measured on a routine basis with a pair of automatic weather stations (Strangeways, 1972) located in a small grassland clearing 400 m from the lysimeter. During this period the net radiation into the forest was estimated by multiplying that measured over grass by 1.1; this average correction factor was obtained as part of a separate study (Roberts, 1977, private communication), and validated by the installation of a single, automatic weather station above the forest canopy, which was used as the primary measurement during the latter part of the experiment.

No measurements of soil heat flux were available at Plynlimon; however, soil heat flux is low beneath a forest—usually a negligible part of the total energy budget (Sinclair *et al.*, 1975). For the purpose of this study, in which evaporation losses are calculated over long periods, the integrated heat flux is assumed negligible.

b. The Thetford data set

Precise measurements of evaporative and meteorological variables were gathered, as part of an extensive study of forest micrometeorology, on a discontinuous basis over the period 1972–76. Several examples of the use of these data already exist in the literature, as do descriptions of the experimental apparatus and analysis techniques used (e.g., Oliver and Oliver, 1973; Stewart and Thom, 1973; Gash and Stewart, 1975; McNeil and Shuttleworth, 1975; Thom *et al.*, 1975). The experimental days were selected on the basis of operational convenience irrespective of meteorological conditions from spring to autumn: while they might be considered a "random" sample of measurements of evaporation over the periods of the field study, it should be remembered that such field studies were not carried out in winter, so that the resulting data are only truly representative of spring, summer and autumn conditions. Measurements of the evaporative loss by rainfall interception, however, are available separately all year round (Gash and Morton, 1978).

Evaporation was determined by the energy budget/Bowen ratio method with gradients of temperature and humidity measured above the forest with accurate, quartz-crystal psychrometers (Gash and Stewart, 1975) arranged as profiles (Stewart and Thom, 1973) until 1973, and used in a thermometer interchange system (McNeil and Shuttleworth, 1975) since that time. The meteorological parameters required to compute potential evaporation were available as part of the data necessary for the evaporation measurement. Only days with continuous data extending at least from dawn to dusk were used in this study: data were more usually available for the complete 24 h period. In total 198 days of data were used from this data set.

3. Results

The comparison between actual measured evaporation and that calculated from Eq. (2) with $\alpha=1.27$ is presented in Fig. 1 as cumulative sums for days of experimental data. The beginning and end of successive years are marked to give an indication of time of collection and amount of data available for each year. The separation into wet and dry periods in this diagram is not meant to be definitive, merely instructive; it was made for the Plynlimon data on the basis of the occurrence of rainfall during the preceding 3 h, and for the Thetford data on the basis of in-canopy wetness sensors (Stewart, 1977). It should be remembered that the Thetford data consists of discontinuous daily samples with a seasonal frequency distribution biased toward summer conditions, hence the preponderance of dry canopy data. The results presented are *not the cumulative evaporation loss from Thetford forest for complete years.*

The calculation of potential evaporation was carried out on an hourly basis *for daylight hours* (net radiation positive). This decision reflects the inconsistent availability of nighttime data in the Thetford data set, and a desire to make the most favorable comparison possible

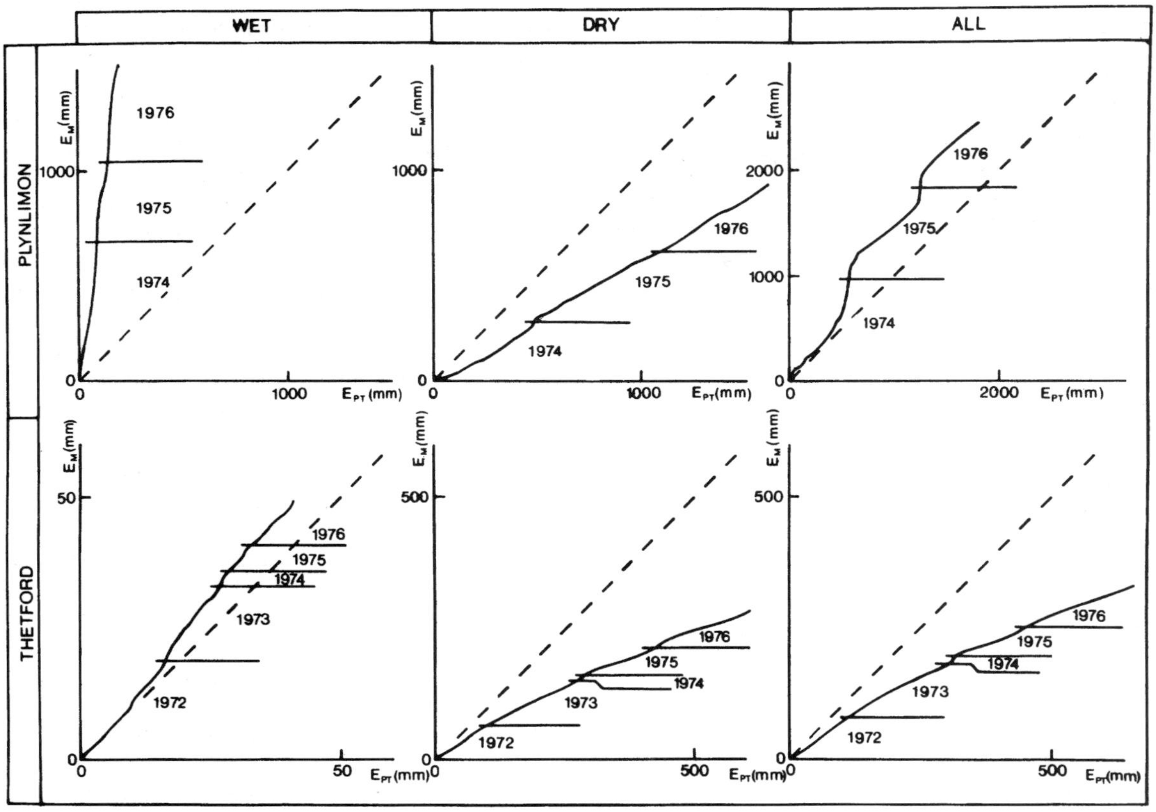

FIG. 1. Comparison between measured evaporation E_M and E_{PT} calculated from Eq. (2) with $\alpha = 1.27$, expressed as cumulative sums over the several years of data collection for the Plynlimon and Thetford evaporation data sets. A separate comparison is made for both wet and dry periods, and for the total loss in all conditions. (Note. The results given for Thetford are *not* the cumulative evaporation loss for complete years.)

with the Priestley-Taylor prediction. The predictive ability of Eq. (2) is best in the absence of advective effects; at night systematic sensible heat advection occurs routinely to maintain the outward net radiation flux. In general, the evaporation process plays a minor role in the total energy transfer at night, and a radiation-based evaporative flux prediction would be less likely to have validity in such conditions. Estimates were made of the effect of restricting the calculation to daylight hours for the two data sets; including the (negative) nighttime contribution reduced the cumulative prediction given by Eq. (2) by about 10–15%, worsening the comparison for the Plynlimon site and improving it for the Thetford site by this amount.

4. Discussion

Although attempts were made to optimize the comparison by removing nighttime periods with sensible heat advection, no attempt was made to restrict the comparison further to daylight periods with zero advection, nor was attention given to soil wetness. In their original work Priestley and Taylor (1972) drew attention to the limited descriptive power of Eq. (2) in advective conditions and over unsaturated soils. Although some attention has since been paid to the importance of soil moisture as a surface control (e.g., Mukammal and Neumann, 1977), less attention has been given to the possibility of long-term, medium-scale advection.

We make no apology for refusing to impose restrictions on these data to further optimize the agreement between measurement and the Priestley-Taylor prediction. It is the essence of our thesis that for tall vegetation, with large atmospheric exchange coefficients, the extent of the control on evaporation exerted by the surface wetness status is considerable, and that meteorologically based evaporation metrics such as those of Priestley-Taylor or Penman, which pay minimal attention to surface conditions, are of limited use. It is apparent from Fig. 1 that in both these data sets, the status of the surface (whether it is wet or dry) is the primary control on evaporation rate. When the vegetation is dry, physiological control systematically reduces the evaporation rate below that predicted by the Priestley-Taylor equation even when no soil moisture stress is present, as demonstrated by the Plynlimon data. In fact, Calder (1978) found no evidence of an additional reduction in evaporation rate in response to the soil moisture deficits of 200 mm (corresponding to soil tensions of 6 bars) observed at this site in the drought of 1976. When the vegetation

TABLE 1. Effective values of α in Eq. (1) relating total measured evaporation for the two forest evaporation data sets.

Location and year		Wet conditions	Dry conditions	All conditions
Plynlimon,	1974	9.33	0.74	2.08
	1975	8.57	0.73	1.44
	1976	11.52	0.76	1.69
	All	9.69	0.74	1.72
Thetford,	1972	1.44	0.81	0.89
	1973	1.79	0.62	0.67
	1974	(1.76)	(0.95)	(1.04)
	1975	1.55	0.50	0.53
	1976	1.16	0.62	0.64
	All	1.50	0.64	0.69

Notes: These values only apply to the sample of daily evaporation measurements used in this analysis: the Thetford data set in particular has a strong sample bias toward the middle of the year.
The 1974 Thetford data are limited and the values of α less reliable than for other years.

is wet, the complete removal of all surface control means that it is possible for the evaporation process to exploit the large atmospheric exchange coefficients and obtain access to the considerable sensible heat already stored, or presently being released by the precipitation process, in the lower levels of the atmosphere.

It is quite possible that meteorologists unfamiliar with forest micrometeorology, in general, and these data, in particular, might be surprised by the routine and long-term observation of sensible heat advection to a wet forest. The conventional attitude, that evaporation is largely a radiation-controlled process, is of course easily justified for short vegetation from more fundamental descriptions such as those provided by Monteith (1965) or Shuttleworth (1976, 1978). At the same time and on the same basis, it is also fairly easy to understand why evaporation from tall vegetation, with large atmospheric exchange coefficients, is more intimately related to the atmospheric vapor pressure deficit (e.g., Stewart and Thom, 1973; Thom and Oliver, 1977). Indeed, for basic theoretical reasons, the simple experimental observation of a finite atmospheric humidity deficit near the surface of forest vegetation in wet canopy conditions very often implies the presence of sensible heat advection, since the incident radiation is often low at such times.

The Plynlimon data presented here are particularly useful in demonstrating an extreme example of large-scale, medium-term advection. In wet canopy conditions the surface obtains typically only 20% of its total energy input in the form of radiant energy; indeed, the effect is so large that even when averaged over several years, the total evapotranspiration loss from the forested area is approximately 12% higher than the total radiant energy input.

An explanation of the process which allows such large-scale energy advection may be provided by consideration of the meteorological conditions which normally prevail during wet conditions. At Plynlimon, as with most upland forested catchments in the United Kingdom, the most common type of rainfall is that brought about by frontal activity. This activity is generally associated with a well-mixed, deep boundary layer which has a large transverse wind speed, typically 20 m s^{-1} (see e.g., Browning *et al.*, 1975). Under such conditions the interactive feedback between the air mass moving over a wet forest of medium size ($\sim$10 km) and the evaporation flux is slight. Air movement over forests with dimensions of the order of 100 km would be necessary before the cooling and humidification of the boundary layer was sufficient to significantly reduce evaporation rates. In fact, such simple horizontal air movements probably rarely occur over these distances because the latent heat released by condensation in clouds is associated with large-scale ascent, thereby producing an effective sink for the cold moist air. Indeed, the mechanism of frontal rainfall can only be maintained by rapid advection in response to this vertical motion.

Clearly, with tall vegetation the behavior of the whole planetary boundary layer is important and it could be dangerous to base the prediction of *total* evaporation loss merely on empirical relationships with radiation, temperature and soil moisture status. Such an approximate empirical relationship may be possible as a rather approximate estimator of (dry canopy) transpiration; indeed, on the basis of Fig. 1 and Table 1, the present data might suggest the use of Eq. (1) in the form

$$E_T \approx (0.72\pm0.07)\frac{\Delta}{\lambda(\Delta+\gamma)}(Q^*-G) \tag{3}$$

as a long-term, first-order estimate of this *component* for both forests, the value of α being the mean and standard deviation of the yearly values weighted by the number of "data days" available. However, the total evaporation loss also depends critically on the frequency and duration of precipitation and in particular on the fraction lost as a result of interception by the forest canopy. Forest evaporation can therefore be almost as variable as precipitation.

In a recent paper Thom and Oliver (1977) have attempted to modify the original Penman approach to take account of differing vegetation types, and the modification of surface control in response to precipitation amounts. However, Gash (1978) has pointed out that their approach, in fact, is equivalent to expressing actual evaporation E_a as

$$E_a = E_T + I(1-C_T), \tag{4}$$

where E_T is a physically realistic transpiration estimate and I the interception loss. The term C_T is a correction to take account of the fact that the transpiration estimate is also made while the evaporation of intercepted water is still in progress. Its value has been calculated as 0.07 for Thetford Forest (Gash and Stewart, 1977).

It is obviously also possible to attempt a similar modification of the Priestley-Taylor metric by using Eq. (3) as an upgraded estimate of forest transpiration. Within this last assumption the value of C_T for Plynlimon is given as the ratio (α_D/α_W), where α_D is the value of α [Eq. (1)] in dry conditions and α_W the value in wet conditions, for the Plynlimon data. It can be seen from Table 1 that C_T is again approximately 0.07, and Eq. (4) can therefore be rewritten in the form

$$E_a \approx (0.72 \pm 0.07)\frac{\Delta}{\lambda(\Delta+\gamma)}(Q^* - G) + 0.93I \qquad (5)$$

as a first-order description of these data.

It is possible that Eq. (5) might be of some predictive use in high rainfall areas, given a physically based, predictive model of the interception component such as those of Rutter (Rutter *et al.*, 1971) or Gash (1979). In such conditions the total evaporation loss is less sensitive to the empirical transpiration estimate. The predictive use of Eq. (5) in dry areas, where transpiration is the major loss, is limited by its empirical content.

It is possible to develop Eq. (5) further, with an additional rather dramatic decrease in its predictive accuracy. Gash (1979) has created an analytic model of forest interception based on physical principles similar to those used in the numerical model of Rutter *et al.* (1971). In so doing he demonstrates that there is at least some physical justification for expressing the interception loss from a forest in terms of a regression on precipitation input, *but* that the regression coefficients are likely to be variable because they depend on the mean rainfall rate and mean wet canopy evaporation rate for the particular location, and the structure of the particular canopy. Direct measurements of interception in these two forests (Calder, 1976, 1977; Gash and Stewart, 1976; Gash and Morton, 1978), summarized in Fig. 2, suggest a rather approximate relationship of the form

$$I \approx (0.29 \pm 0.08)P, \qquad (6)$$

where P is the precipitation input.

In the Priestley-Taylor framework, therefore, the equation which gives a first-order approximation to evapotranspiration in rough (20%) agreement with the long-term, *total* evaporation loss for the two forests considered in this paper has the form

$$E_a = (0.72 \pm 0.07)\frac{\Delta}{\lambda(\Delta+\gamma)}(Q^* - G) + (0.27 \pm 0.08)P. \qquad (7)$$

No claims are made for regarding the predictive ability of this empirical expression outside the experimental data on which it is based. The suggested errors are probably only realistic when describing medium-term average evaporation; the expression is rather a poor description of short-term evaporation loss and is there-

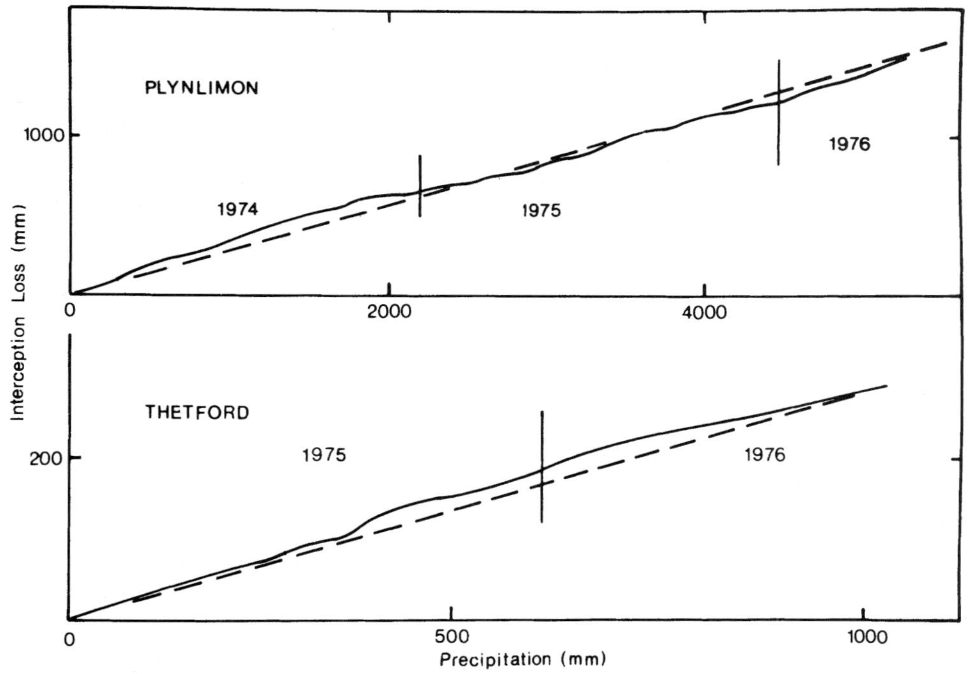

FIG. 2. Cumulative loss of water, as a result of rainfall interception by the forest canopy, expressed in terms of cumulative precipitation input for coniferous forest in Plynlimon and Thetford. The dashed line represents the relationship $I = 0.29\,P$ in both cases.

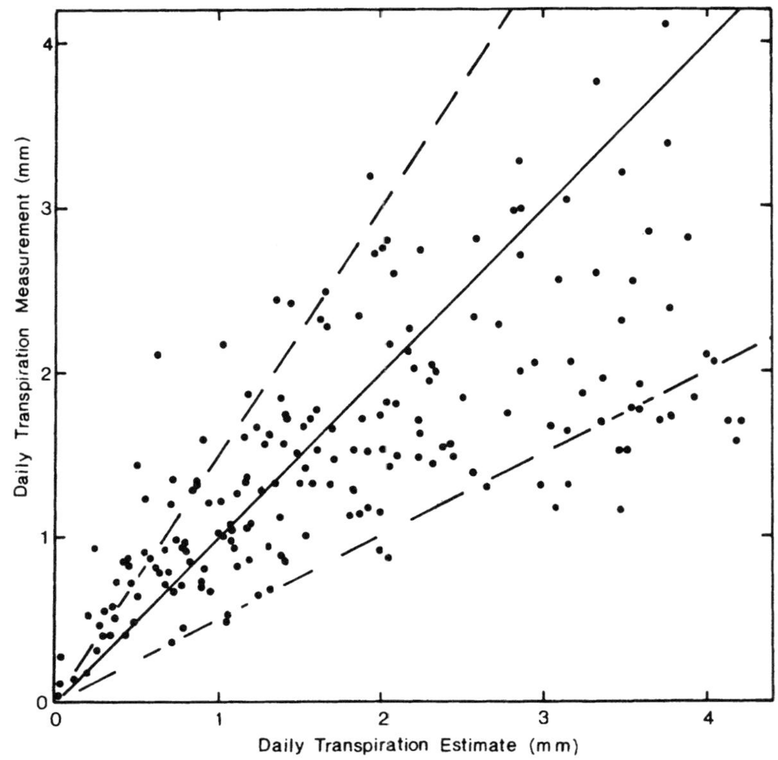

FIG. 3. Comparison between the daily measurement of evaporation made over Thetford Forest compared with a daily estimate made using Eq. (3) in the text. The full line represents equality, the dashed lines a variation of 50% around equality.

fore unlikely to be of much use in short-term management. This is apparent in Fig. 2, as variations in the relationship between interception and precipitation. But it is more obvious in Fig. 3, which is a comparison between a daily transpiration estimate, given by Eq. (3), and measured daily transpiration for Thetford Forest. (The lysimetric measurement used in the Plynlimon data does not permit a similar comparison for that data set.) In the light of this short-term variability, the value of $\alpha=1.05$ found by McNaughton and Black (1973) from 16 dry days of evaporation data, collected over Douglas fir in British Columbia, clearly cannot be regarded as inconsistent with these data; particularly since their value is based on a 24 h comparison. For the reasons outlined earlier the value of $\alpha=0.72$ reported here is the long-term average value based on a daytime comparison; the value deduced from a 24 h comparison would be in the order 10–20% higher, depending on overnight cloud conditions and the amount of sensible heat advected to maintain the night time radiation loss.

Eq. (7) does provide at least a *working description* of the long-term total evaporation loss for both sites as illustrated in Fig. 4, which makes a comparison between measured evaporation and that calculated using this equation. The dashed line indicates equality, while the dotted lines indicate the estimated errors expected for a description using this expression. The agreement for the Plynlimon data set is obviously much closer than that for the Thetford data set. This reflects, in part, the fact that the empiricism implicit in Eq. (7) is biased toward Plynlimon because that is where the weight of data lies; and in part, the fact that a greater fraction of the evaporation from Thetford occurs as the more variable transpiration component. It is probably worth pointing out that most of the shortfall in measured evaporation from Thetford occurs in the summers of 1975 and 1976, within an 18-month period which was the driest on record in the United Kingdom.

5. Concluding remarks

This paper makes no comment on the relative merits of the Priestley-Taylor and the Penman formulas as a basis for evaporation prediction: the two seem to have some conceptual similarity and often produce numerically similar results (at the 5–10% level), but the Priestley-Taylor version is easier to apply. Neither do we suggest the alternative use of Eqs. (5) or (7) as a generally valid description of forest evapotranspiration. Indeed, we reiterate our warning against the indiscriminate use of any simplistic equation with a large empirical content in this application. Eq. (7) does provide at least a reasonable long-term description of the two rather different forest evaporation data sets

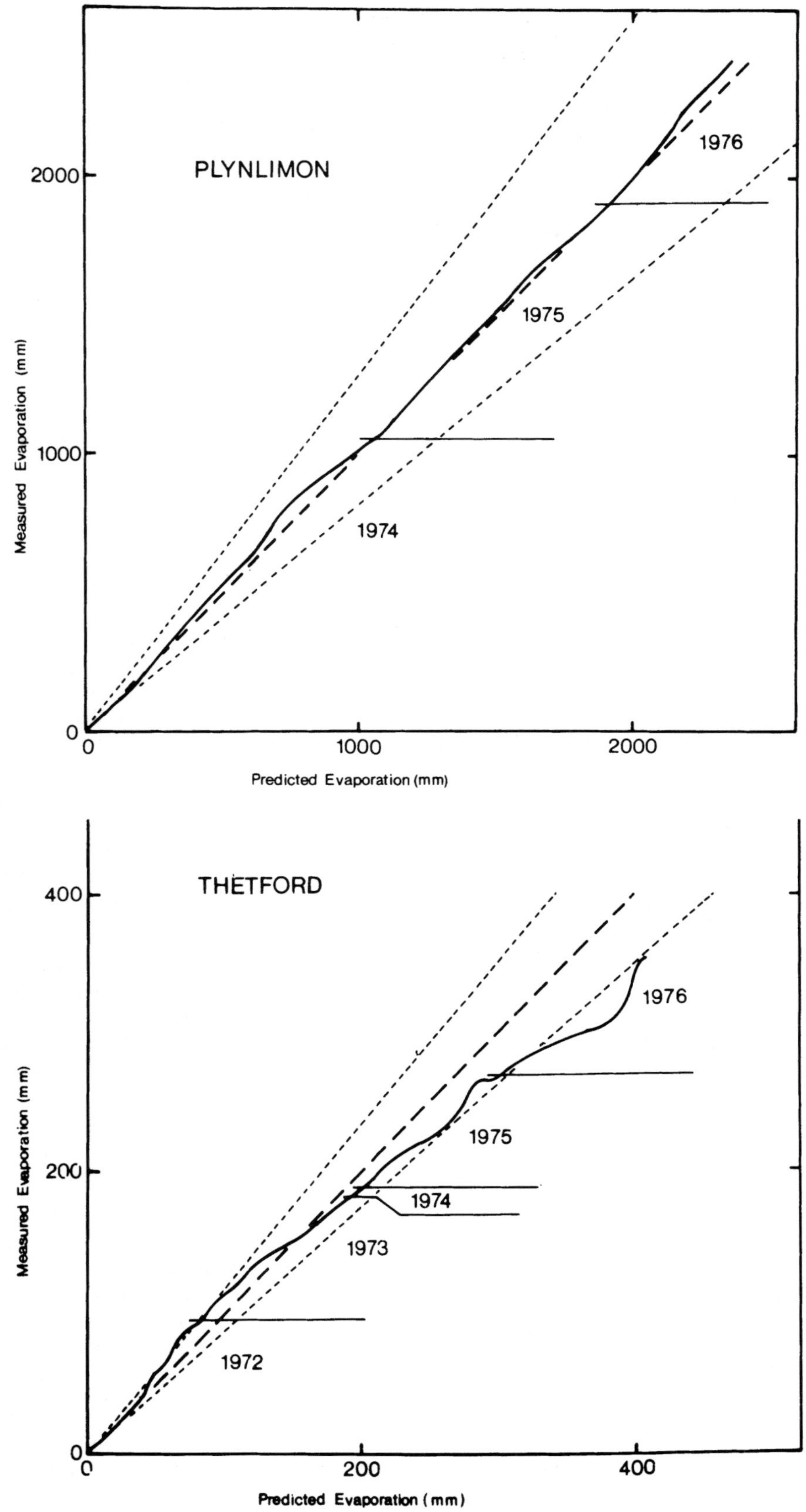

FIG. 4. Comparison between measured evaporation and a calculation based on Eq. (7) for the Plynlimon and Thetford data. The long-dashed line represents equality, the short-dashed lines indicate the expected errors.

presented here—something which is not possible using the simpler Priestley-Taylor metric.

It remains our fundamental belief that evapotranspiration in general, and forest evapotranspiration, in particular, are processes subject to physical and physiological control; and that the laws of physics are not influenced by human desire for computational simplicity. Our purpose is to make explicit the possible variability of forest evaporation as a result of its close dependence on surface controls and the real possibility of medium-scale, long-term advection in high rainfall areas.

Acknowledgments. This paper is a distillation of a great deal of effort by a considerable number of people at the Institute of Hydrology. The authors would particularly like to thank John Stewart, Howard Oliver, John Gash, Chris Moore, David McNeil, Colin Lloyd, Sylvia Oliver, Nick Cowell, Paul Rosier, Philip Hill and Richard Harding who all contributed in some part to this paper.

We would also like to thank the many other people at the Institute who have provided help, and in particular the Director, Dr. J. S. G. McCulloch, whose foresight and consistent support were in no small measure responsible for the existence of these data.

REFERENCES

Browning, K. A., C. W. Pardoe and F. F. Hill, 1975: The nature of orographic rain at wintertime cold fronts. *Quart. J. Roy. Meteor. Soc.*, **101**, 333–352.

Calder, I. R., 1976: The measurement of water losses from a forested area using a "natural" lysimeter. *J. Hydrol.*, **30**, 311–325.

——, 1977: A model of transpiration and interception loss from a spruce forest in Plynlimon, central Wales. *J. Hydrol.*, **33**, 247–265.

——, 1978: Transpiration observations from a spruce forest and comparisons with predictions from an evaporation model. *J. Hydrol.*, **38**, 33–47.

——, and P. T. W. Rosier, 1976: The design of large plastic sheet net rainfall gauges. *J. Hydrol.*, **30**, 403–405.

Clarke, R. T., and J. S. G. McCulloch, 1976: Recent work in comparison of the effect of alternative used (coniferous forest, upland pasture) on water catchment behavior. Presented at the Conference of Water, Space and Amenity Commission, London, May 1975.

Davies, J. A., and C. D. Allen, 1973: Equilibrium, potential evaporation from cropped surfaces in southern Ontario. *J. Appl. Meteor.*, **12**, 649–657.

Gash, J. H. C., 1978: Comment on the paper by A. S. Thom and H. R. Oliver "On Penman's equation for estimating regional evaporation." *Quart. J. Roy. Meteor. Soc.*, **104**, 532–533.

——, 1979: An analytic model of rainfall interception by forests. *Quart. J. Roy. Meteor. Soc.*, **105**, 43–56.

——, and J. B. Stewart, 1975: The average surface resistance of a pine forest derived from Bowen ratio measurements. *Bound.-Layer Meteor.*, **8**, 453–464.

——, and ——, 1977: The evaporation from Thetford Forest during 1975. *J. Hydrol.*, **35**, 385–396.

——, and A. J. Morton, 1978: An application of the Rutter model to the estimation of the interception loss from Thetford Forest. *J. Hydrol.*, **38**, 49–58.

McNaughton, K. G., and T. A. Black, 1973: A study of evapotranspiration from a Douglas fir forest using the energy balance approach. *Water Resour. Res.*, **9**, 1579–1590.

McNeil, D. D., and W. J. Shuttleworth, 1975: Comparative measurements of energy fluxes over a pine forest. *Bound.-Layer Meteor.*, **9**, 297–313.

Monteith, J. L., 1965: Evaporation and Environment. *Symp. Soc. Exp. Biol.*, **19**, 205–234.

Mukammal, E. I., and H. H. Neumann, 1977: Application of the Priestley-Taylor evaporation model to assess the influence of soil moisture on the evaporation from a large weighing lysimeter and class A pan. *Bound.-Layer Meteor.*, **12**, 243–256.

Oliver, S. A., and H. R. Oliver, 1973: A computer method of automatic acquisition of wind speed data. *J. Phys. E., Sci. Instrum.*, **6**, 401–403.

Penman, H. L., 1948: Natural evaporation from open water, bare soil and grass. *Proc. Roy. Soc. London*, **A193**, 120–145.

Priestley, C. H. B., and R. J. Taylor, 1972: On the assessment of surface heat flux and evaporation using large scale parameters. *Mon. Wea. Rev.*, **100**, 81–92.

Rutter, A. J., K. A. Kershaw, P. C. Robins and A. J. Morton, 1971: A predictive model of rainfall interception in forests. I. Derivation of the model from observations in a plantation of Corsican pine. *Agric. Meteor.*, **9**, 367–84.

Shuttleworth, W. J., 1976: A one-dimensional theoretical description of the vegetation-atmosphere interaction. *Bound.-Layer Meteor.*, **10**, 273–302.

——, 1978: A simplified one-dimensional theoretical description of the vegetation-atmosphere interaction. *Bound.-Layer Meteor.*, **14**, 3–27.

Sinclair, T. R., L. H. Allen and E. R. Lemon, 1975: An analysis of errors in the calculation of energy flux densities above vegetation by a Bowen-ratio profile method. *Bound.-Layer Meteor.*, **8**, 129–139.

Stewart, J. B., 1977: Evaporation from the wet canopy of a pine forest. *Water Resour. Res.*, **13**, 915–921.

——, and A. S. Thom, 1973: Energy budgets in a pine forest. *Quart. J. Roy. Meteor. Soc.*, **99**, 154–170.

Stewart, R. B., and W. R. Rouse, 1977: Substantiation of the Priestley-Taylor Parameter $\alpha = 1.26$ for potential evaporation in high latitudes. *J. Appl. Meteor.*, **16**, 649–650.

Strangeways, I. C., 1972: Automatic weather stations for network operation. *Weather*, **27**, 403–408.

Thom, A. S., and H. R. Oliver, 1977: On Penman's equation for estimating regional evaporation. *Quart. J. Roy. Meteor. Soc.*, **103**, 345–357.

Thom, A. S., J. B. Stewart, H. R. Oliver and J. H. C. Gash, 1975: Comparison of aerodynamic and energy budget estimates of fluxes over pine forest. *Quart. J. Roy. Meteor. Soc.*, **101**, 93–105.

Thompson, J. R., 1975: Energy budgets for three small plots—substantiation of Priestley and Taylor's large-scale evaporation parameter, 1975. *J. Appl. Meteor.*, **14**, 1399–1401.

EVAPORATION AND ENVIRONMENT

By J. L. MONTEITH
Rothamsted Experimental Station, Harpenden, Herts.

A turgid leaf exposed to bright sunshine can transpire an amount of water several times its own weight during a summer day. Rapid evaporation is sustained by a supply of heat from the atmosphere and by a movement of water within the plant preventing the desiccation of leaf tissue. In analysis, the need for energy and the need for water have often been dissociated. Meteorologists investigating the energetics of transpiration have assumed that leaves behave like pieces of wet, green blotting paper, and plant physiologists have demonstrated mechanisms for the conduction of water at arbitrary rates unrelated to the physics of the environment.

This paper describes progress towards a reconciliation of parallel concepts in meteorology and physiology. The path for the diffusion of water vapour from leaf cells to the free atmosphere is divided into two parts, one determined primarily by the size and distribution of stomata, and the other by wind speed and the aerodynamic properties of the plant surface. Diffusive resistances for single leaves and for plant communities are established from measurements in the laboratory and in the field and are then used:

(i) to predict relative rates of evaporation from leaves with wet and dry surfaces;

(ii) to investigate the dependence of transpiration rate on wind speed and surface roughness;

(iii) to demonstrate that the relation between transpiration rate and leaf area is governed by stomatal closure in leaves well shaded from sunlight;

(iv) to calculate maximum rates of transpiration for different crops and climates.

A final section on the advection of dry air stresses the importance of physiological restraint on the rate of transpiration from an irrigated field surrounded by dry land.

FIRST PRINCIPLES

The evaporation of water is like a commercial transaction in which a wet surface sells water vapour to its environment in exchange for heat. For each gram of water evaporated at 20° C., the surface demands 585 calories of heat and several forms of payment are acceptable. When

Paper D8 is reprinted from pages 205–234 of *The State and Movement of Water in Living Organisms* (Proc. 19th Symp. Soc. Exp. Biol., Swansea 1964), by permission of The Society for Experimental Biology.

water evaporates from plants, the environment can supply heat by solar radiation, by turbulent transfer from the atmosphere, or by conduction from the soil. A few elementary principles of thermodynamics are needed to relate the state of the environment to the rate of evaporation from a free water surface, an isolated leaf, or from a crop canopy.

In problems of natural evaporation, the state of a given mass of air can be described by its temperature T and its vapour pressure e. The total heat content of the air is the sum of a sensible heat content, depending on temperature, and a latent heat content, depending on vapour pressure. When temperature changes by δT and vapour pressure by δe, approximations discussed by Brunt (1939) lead to the relation

$$\frac{\delta e}{\text{change in latent heat}} = \frac{\gamma \delta T}{\text{change in sensible heat}}, \tag{1}$$

where γ, a function of the physical properties of dry air, is 0·66 mb. ° C.$^{-1}$ at 20° C. and 1 atm. pressure.

Consider a mass of air isolated from all external sources of heat. Any spontaneous change in latent heat content must be balanced by an equal and opposite change in sensible heat content. In particular, when the air is unsaturated and contains a small amount of liquid water, the water will evaporate (increasing the latent heat content of the air) and the air will cool (decreasing the sensible heat content of the air). The process stops when the air becomes saturated at a temperature T', known as the wet-bulb temperature, and from equation 1, the equation for the redistribution of energy is

$$e_s(T') - e = \gamma(T - T'), \tag{2}$$

where $e_s(T')$ is the saturation vapour pressure at T'.

Further increase of vapour pressure can be produced only by adding an amount of heat $\underset{\sim}{Q}$ that is shared between an increase of sensible heat (increasing T) and an increase of latent heat (increasing e). While liquid water remains in the system, small changes are related by

$$\delta e_s = \Delta \delta T, \tag{3}$$

where Δ, the change of saturation vapour pressure with temperature, increases with temperature. (At 0° C., Δ = 0·5 mb. ° C.$^{-1}$; at 25° C. Δ = 1·9 mb. ° C.$^{-1}$.) Combining equations 1 and 3,

$$\text{increase of latent heat} = \frac{\Delta \underset{\sim}{Q}}{\Delta + \gamma}, \tag{4a}$$

$$\text{increase of sensible heat} = \frac{\gamma \underset{\sim}{Q}}{\Delta + \gamma}. \tag{4b}$$

The rate of evaporation from a wet surface with temperature T' can now be calculated from the rate of increase in the latent heat content of surrounding air at temperature T. If the specific heat of air is ρc (cal. cm.$^{-3}$), then 1 cm.^{3} of unsaturated air cooling from T to T' ° C. will supply $\rho c(T-T')$ calories of heat for evaporation. If r_a is the time in which 1 cm.^{3} of air exchanges heat with 1 cm.^{2} of surface, $1/r_a$ is a rate of exchange in cubic centimetres of air per square centimetre of surface per second. The rate at which air supplies heat to the surface is

$$\lambda E_1 = \frac{\rho c(T-T')}{r_a} = \frac{\rho c D}{r_a}, \tag{5}$$

where D is the wet-bulb depression, λ is the latent heat of water (cal. g.$^{-1}$) and E_1 is the evaporation rate (g. cm.$^{-2}$ sec.$^{-1}$).

When water evaporates into *saturated* air, latent heat must be taken up at a rate

$$\lambda E_2 = \frac{\Delta H}{\Delta + \gamma} \tag{6}$$

from equation (4*a*), where H (cal. cm.$^{-2}$ sec.$^{-1}$) is the *rate* of heat supply.

Fig. 1 shows how the evaporation from a wet surface may be found by combining equations (5) and (6). In this example, air at 13° C. and 9 mb. vapour pressure (point W) passes over a wet surface at 16° C. and 18·2 mb. saturated vapour pressure (point Y). In terms of thermodynamics, the

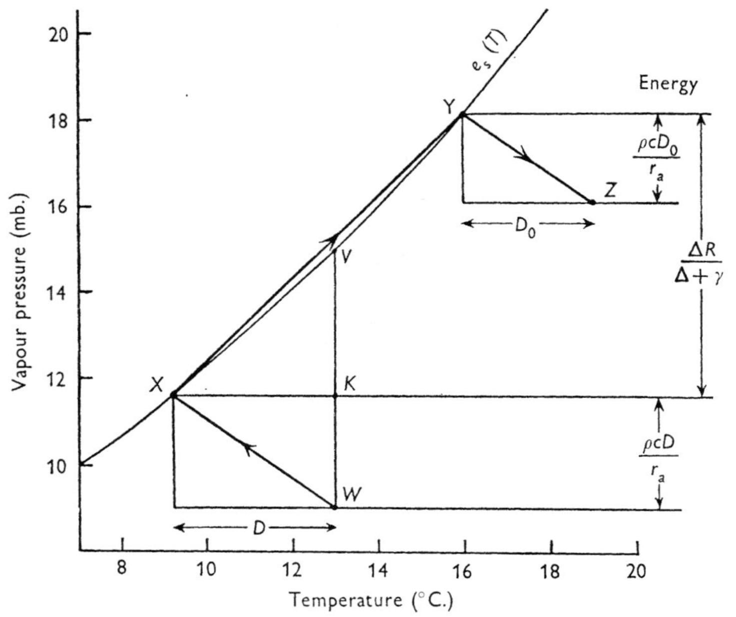

Fig. 1. Graphical representation of evaporation formulae (see text).

amount of energy needed to change air from state W to state Y is completely independent of the path chosen from W to Y and of the infinite possibilities the route WXY is the most revealing. Starting from W, the air is saturated by cooling to its wet-bulb temperature (9·2° C.) along a line WX with slope $-\gamma$ and the heat released for evaporation is λE_1 (equation 5). Maintaining saturation, the air is now heated from state X to state Y where X and Y are both on the saturation vapour pressure curve $e_s(T)$. Approximating to this curve by the chord XY with slope Δ, the corresponding latent heat of evaporation is λE_2 (equation 6). The temperature-dependent function Δ is evaluated at the mean of the wet-bulb temperature of the air and the surface temperature.

The total latent heat of evaporation is

$$\lambda E = \lambda(E_1 + E_2) = \frac{\Delta H}{\Delta+\gamma} + \frac{\rho c D}{r_a}. \qquad (7)$$

For air at W, the relation between the wet-bulb depression D and the saturation vapour pressure deficit is given by

$$\frac{e_s(T)-e}{D} = \frac{JW}{KX} = \frac{JK}{KX} + \frac{KW}{KX} = \Delta' + \gamma,$$

where Δ' is evaluated at the mean of wet- and dry-bulb air temperatures. Assuming $\Delta \doteq \Delta'$, equation (7) may be written

$$\lambda E = \frac{\Delta H + \rho c \{e_s(T)-e\}/r_a}{\Delta+\gamma}. \qquad (8)$$

Equation (8) was first derived by Penman (1948) and equation (7) is a variant preferred by Slatyer & McIlroy (1961). Neither equation is valid for a surface where the vapour pressure is less than the saturation vapour pressure at surface temperature. For example, the water in a leaf evaporates at the surfaces of cell walls surrounding sub-stomatal cavities, and reaches the outer surface of the leaf by molecular diffusion through stomata and through the cuticle. When a leaf has a uniform temperature T_o, the vapour pressure of air in contact with wet cell walls is usually very close to the saturation vapour pressure $e_s(T_o)$. At a dry leaf surface, the vapour pressure e_o is always less than $e_s(T_o)$ during transpiration, and surface air is never saturated. In Fig. 1, the unsaturated air in contact with a leaf is given a wet-bulb depression D_o and its state is represented by point Z. To reach Z from Y, the air is de-saturated by condensation of water vapour and latent heat is released at a rate

$$-\lambda E_3 = \frac{\rho c D_o}{r_a}.$$

The total rate of evaporation for the path WZ is

$$\lambda E = \lambda(E_1+E_2+E_3) = \frac{\Delta H}{\Delta+\gamma}+\frac{\rho c(D-D_o)}{r_a}. \tag{9}$$

Some theoretical implications of this expression were examined by Slatyer & McIlroy but its practical exploitation is severely limited by the difficulty of measuring or calculating the wet-bulb depression of air at a leaf surface. This difficulty may be resolved by eliminating surface parameters as follows.

Anticipating later discussion, the same value of r_a is valid for the diffusion of heat and water vapour. When e_o is the vapour pressure at a leaf surface, the transpiration rate is proportional to $(e_o-e)/r_a$ and, by analogy with Ohm's law, (e_o-e) is a potential difference maintaining a current or flux of water vapour across an external diffusion resistance r_a in the air surrounding a leaf. Similarly, the rate of diffusion *within* a leaf is proportional to $\{e_s(T_o)-e_o\}/r_l$, where r_l is the internal resistance of the leaf. This resistance is provided by stomatal pores, by the cuticle, and by the cell walls during severe water stress (see the contribution by Meidner, p. 185). In a turgid leaf with ample water for transpiration, the resistance of cell walls is negligible and the cuticle presents a resistance much larger than the parallel resistance of stomata. The total leaf resistance is almost equal to the resistance of the stomata alone and is governed by stomatal sizes and populations (Milthorpe, 1962).

When the rates of diffusion within the leaf and in the external air are equal

$$E\left(\frac{\gamma\lambda}{\rho c}\right) = \frac{e_s(T_o)-e_o}{r_l} = \frac{e_o-e}{r_a}, \tag{10}$$

where the constant $\gamma\lambda/(\rho c)$ preserves the consistency of units, and by rearranging terms

$$e_s(T_o)-e = \left(1+\frac{r_l}{r_a}\right)(e_o-e). \tag{11}$$

Equation (11) shows that a leaf with internal resistance r_l can be treated formally as a free water surface if the vapour pressure difference (e_o-e) is replaced by $\{e_s(T_o)-e\}/(1+r_l/r_a)$. Alternatively, the psychrometer constant γ can be replaced by a modified value

$$\gamma^* = \gamma\left(1+\frac{r_l}{r_a}\right). \tag{12}$$

Returning to equation (8), the latent heat of transpiration from a leaf becomes

$$\lambda E = \frac{\Delta H + \rho c\{e_s(T) - e\}/r_a}{\Delta + \gamma^*}, \tag{13}$$

as derived by Penman (1953) with somewhat different symbolism.

The alternative equations (9) and (13) show how the rate of evaporation from a wet surface or from a leaf depends on four weather parameters: net heat from external sources, usually radiation; temperature, humidity, and wind speed. When the leaf resistance is constant, evaporation increases (i) linearly with radiation, (ii) linearly with saturation deficit or wet-bulb depression, (iii) with air temperature when radiation and saturation deficit are constant. The variation with wind speed is more complex. An increase of wind speed increases the rate of exchange of air between the environment and the surface and so increases $1/r_a$. From equation (9), evaporation rate increases with wind speed when $D > D_o$. This condition is *always* satisfied by a free water surface evaporating into unsaturated air

$$(D > 0, \quad D_o = 0)$$

and will *sometimes* be satisfied by a leaf. Evaporation rate is independent of wind speed when $D = D_o$, i.e. when water evaporates into saturated air ($D = D_o = 0$) or when wet-bulb depression is the same at the leaf surface and in the surrounding air. Then the evaporation rate is $\lambda E = \Delta H/(\Delta + \gamma)$. Finally, the rate of transpiration from a leaf will decrease with increasing wind speed when $D < D_o$.

Physically, the dependence of transpiration rate on wind speed depends on relative changes of surface temperature and the rate of exchange ($1/r_a$). Provided H is positive, an increase of wind speed always decreases leaf temperature. By itself, this cooling tends to decrease losses of sensible and latent heat, but this effect is compensated by an increase in ($1/r_a$). When this increased rate of exchange more than compensates for the decrease of temperature, the loss of sensible heat is forced at the expense of latent heat and transpiration rate decreases with wind speed. Conversely, when an increase in the rate of exchange fails to compensate for cooling, the loss of latent heat is forced at the expense of sensible heat and transpiration increases with wind speed.

EXTERNAL RESISTANCE

(i) *Single leaves*

When a potential difference of one volt is maintained across an electrical resistance, the value of the resistance in ohms is the time in seconds for the passage of unit charge. As a diffusive resistance, r_a is the time for unit volume of air to exchange heat (or water vapour) with unit area of surface. This time depends partly on the speed of the air-stream and partly on the geometry of the surface. To compare measurements of r_a for surfaces of different size, it is convenient to work with two non-dimensional parameters (d/Kr_a) and (ud/ν) where d = characteristic length of surface, e.g. diameter of a circular disk or width of a narrow rectangle (cm.); K = molecular diffusion coefficient for heat or water vapour (cm.^{2} sec.$^{-1}$); ν = kinematic viscosity (cm.^{2} sec.$^{-1}$); u = wind speed (cm. sec.$^{-1}$). When K is the molecular diffusion coefficient for heat transfer, (d/Kr_a) is a Nusselt

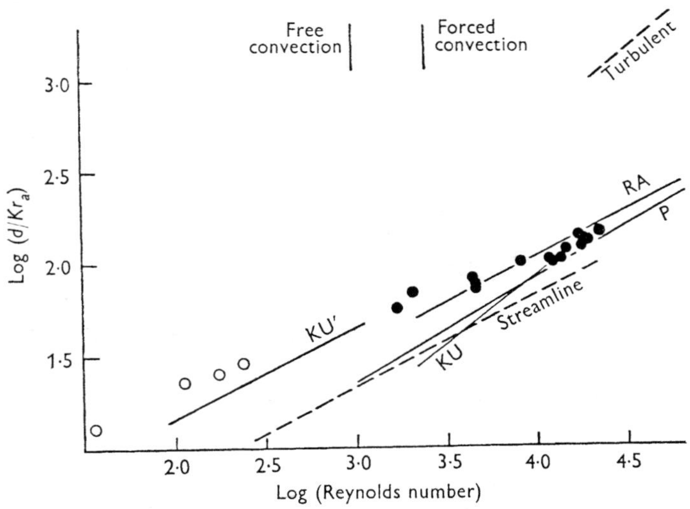

Fig. 2. Non-dimensional exchange parameter (d/Kr_a) as function of Reynolds number. Both scales are logarithmic. ●, Martin (1943); ○, Slatyer & Bierhuizen (1964); KU and KU′, Kuiper (1961); RA, Raschke (1956); P, Powell (1942).

number and is the ratio of the characteristic length d to the effective thickness (Kr_a) of a laminar boundary layer for the molecular diffusion of heat. When K is the diffusion coefficient for water vapour, (d/Kr_a) is an analogous number for vapour transfer. The Reynolds number (ud/ν) is proportional to the ratio of inertial to viscous forces in the air near the surface. When the surface is a flat plate and the Reynolds number is less than 2×10^4, viscous forces predominate and maintain laminar streamline flow, but when the number exceeds 2×10^4 the flow tends to become turbulent (Fishenden & Saunders, 1957). The dashed lines in Fig. 2 show an abrupt increase in Nusselt number, and hence a decrease in r_a, with the onset of turbulence. The full points are calculated from measurements of the evaporation from squares of filter paper (Martin, 1943), and the lines P

14-2

and KU show the mean of wind-tunnel measurements by Powell (1942) and Kuiper (1961). The line RA was derived from measurements by Raschke (1956) on the heat exchange of an *Alocasia* leaf. All three sets of measurements (and a fourth set by de Vries & Venema (1954) omitted for clarity) are close to the streamline relationship for flat plates and (d/Kr_a) is an order of magnitude less than for turbulent flow at the same Reynolds number.

On the left-hand side of the diagram, open circles are recent measurements by Slatyer & Bierhuizen (1964) at wind speeds less than 3 cm. sec.$^{-1}$, and KU' is the mean of measurements by Kuiper at wind speeds less than 20 cm. sec.$^{-1}$. Kuiper attributed the discontinuity between KU and KU' to the transition from streamline to turbulent flow with increasing wind speed. It is unlikely that this transition occurred at such a small Reynolds number (2×10^3) or that the Nusselt number decreased by an order of magnitude after the onset of turbulence. More plausibly, the two sets of measurements at slow wind speeds (KU' and open circles) are characteristic of a régime of free convection where transfer processes are governed by a downward flow of relatively cool dense air from the surface of a transpiring leaf, or by the upward movement of relatively warm air when the leaf is heated by radiation.

When Raschke measured the exchange of heat and water vapour at the surface of leaves in the open air, the diffusion resistances for heat and vapour were identical within the limits of experimental error. Powell found that the resistance for the diffusion of water vapour from wet surfaces in a wind tunnel was 0·9 of the diffusion resistance for heat. This small difference is negligible in most problems of heat balance and evaporation. Slatyer & Bierhuizen report mean values of r_a for heat that are only 0·4 of the resistance for vapour transfer at the same wind speed, and this much larger difference can be attributed to the exchange of heat by buoyancy in the régime of free convection. During the day, at levels in vegetation where the wind speed exceeds 50 cm. sec.$^{-1}$, the exchange of heat, water vapour, and carbon dioxide at leaf surfaces is governed by the laws of forced convection and, in principle, all three rates of transfer can be calculated from the same external resistance.

Reviewing all the measurements summarized in Fig. 2, the line RA is probably the most representative of leaves in the open air. This line gives the relation

$$r_a = 3{\cdot}4\, d(ud/\nu)^{-0{\cdot}5} \qquad (14)$$
$$= 1{\cdot}3\, \sqrt{(d/u)},$$

whereas the results of de Vries & Venema give

$$r_a = 20\, d(ud/\nu)^{-0{\cdot}7}. \qquad (15)$$

For Reynolds numbers between 10^3 and 3×10^4, the two equations give similar values of r_a lying between 0·2 and 0·02 sec. cm.$^{-1}$. Treating a leaf as two similar surfaces in parallel, external resistance is half the resistance for a single surface, and Fig. 3*a* shows values calculated from equation (14) for a range of wind speeds and leaf widths.

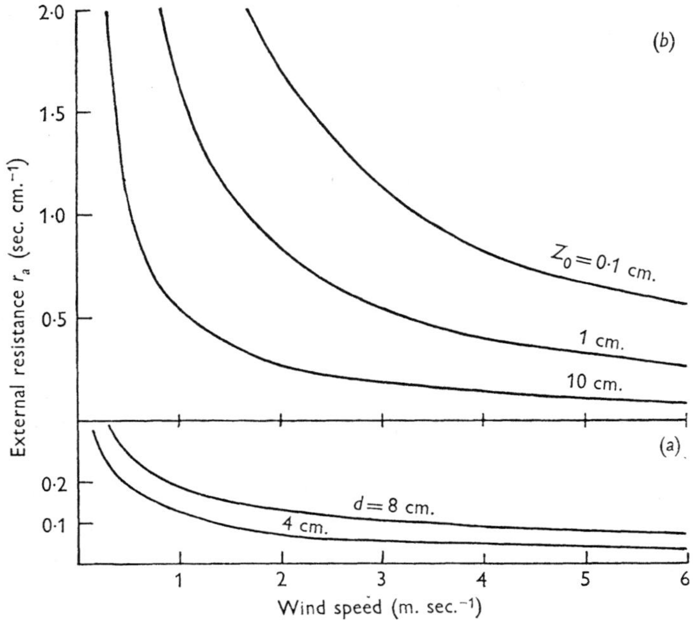

Fig. 3. (*a*) External resistance of single leaves with characteristic length *d* (equation 14); (*b*) external resistance of crop canopy with roughness constant z and wind speed measured at 2 m. above zero plane d' (equation (17*b*)).

(ii) *Uniform crop surfaces*

Diffusion by atmospheric turbulence governs processes of exchange between the atmosphere and the surface of a crop. Over extensive, uniform surfaces, measured diffusion coefficients of water vapour and momentum are equal over a wide range of atmospheric stability (Priestley, 1959) and profiles of vapour pressure and wind speed are the same shape (Robinson & Rider, 1951; Rider, 1954). When vapour pressure *e* and wind speed *u* are measured at several heights *above* a crop canopy, *e* can be plotted against *u* to give a straight line intercepting the axis $u = 0$ at $e = e_o$. Penman & Long (1960) showed that the latent heat of evaporation is

$$\lambda E = \frac{k^2 \rho c (e_o - e) u}{\{\ln (z - d')/z_o\}^2}, \tag{16}$$

where *u* and *e* are measured at the same convenient height z above the crop, and k (=0·41) is a constant of proportionality. The zero plane height

d' is chosen to make wind speed proportional to $\ln(z-d')$, and the roughness parameter z_o is found by plotting u against $\ln(z-d')$ to give an intercept $\ln z_o$ on the axis $u = 0$. By definition, the external or aerodynamic resistance between the crop surface and the air at height z is then

$$r_a = \frac{\rho c}{\gamma\lambda}\frac{e_o - e}{E} \quad (17a)$$

$$= \frac{\{\ln (z-d')/z_o\}^2}{k^2 u}, \quad (17b)$$

and the variation of r_a with roughness and wind speed is shown in Fig. 3*b*. Businger (1956) incorporated equation (17*b*) in Penman's equation for evaporation and the success of this modification was later confirmed by Tanner & Pelton (1960*a*).

When the roughness length of a surface is constant, its aerodynamic resistance is inversely proportional to wind speed. In practice, many crops get smoother, and roughness decreases as wind speed increases (Monteith, 1963*a*; Stoller & Lemon, 1963). For these surfaces, there is a range of wind speeds within which the resistance is almost constant.

The derivation of an aerodynamic resistance from a logarithmic wind profile is strictly valid only when the change of temperature with height is close to the dry adiabatic lapse rate. In practice, this usually means that the surface is only a fraction of a degree Centigrade hotter or cooler than the air at screen height. When the temperature difference is several degrees and the wind speed is a few metres per second or less, the rate of mixing of the air by buoyancy is comparable with mixing by mechanical turbulence. When temperature decreases with height so that the air is unstable, the logarithmic wind profile exaggerates the increase of wind with height, and r_a is overestimated when a value of surface roughness derived in neutral conditions is used in equation (17*b*). Conversely, when temperature increases with height and the air is stable, r_a is underestimated. Resistances can be corrected for buoyancy from functions of the wind and temperature gradients (Monteith, 1963*a*) or from fluxes of sensible heat and momentum (Monteith, Szeicz & Yabuki, 1964).

Table 2 contains values of r_a for several types of vegetation, calculated where possible from the ratio of a vapour pressure gradient to an evaporation rate. When the air is unstable this ratio gives a 'true' resistance, smaller than the value estimated from wind speed and surface roughness. For example, the smoothest surface in Table 2 is short grass with

$$z_o = 0{\cdot}1 \text{ cm}.$$

In neutral conditions, the resistance calculated from equation (17*b*) is about 1 sec. cm.$^{-1}$, but the 'true' value of r_a derived from equation (17*a*) is only 0·5 sec. cm.$^{-1}$.

For several farm crops, the aerodynamic resistance lies between 0·2 and 0·5 sec. cm.$^{-1}$. It is notoriously difficult to measure representative wind profiles over forests, and curious values emerge from attempts to determine roughness length. A roughness of $z_o = 290$ cm. seems large for pine trees only 500 cm. tall (Baumgartner, 1956), but the corresponding resistance of 0·06 sec. cm.$^{-1}$ is consistent with measured vapour pressure gradients and estimated transpiration rates.

INTERNAL RESISTANCE

(i) *Single leaves*

The diffusion resistance of leaf stomata can be calculated from stomatal sizes and populations with the help of simplifying assumptions about the shape of the stomatal wall. Penman & Schofield (1951) derived a resistance of 0·4 sec. cm.$^{-1}$ for sunflower leaves and Penman & Long (1960) quote a figure of 1·4 sec. cm.$^{-1}$ for wheat. Bange (1953) calculated a minimum resistance of 0·8 sec. cm.$^{-1}$ for *Zebrina* leaves. Experimental values of *total* leaf resistance tend to be larger than these theoretical values for the stomatal path alone.

There are three ways of estimating leaf resistance r_l from transpiration rates. The most fundamental is also the most difficult because it needs supplementary measurements of surface temperature T_o and vapour pressure e_o. Given these measurements, r_l is calculated from equation 10 rewritten as

$$r_l = \left(\frac{\rho c}{\gamma\lambda}\right)\frac{e_s(T_o)-e_o}{E}. \qquad (10a)$$

In a second method avoiding the measurement of vapour pressure at the leaf surface, the rate of transpiration E from a leaf is compared with the rate of evaporation E_p from a piece of wet filter paper of the same size exposed in the same air stream. If the paper behaves like a free water surface with surface temperature T_p,

$$E = \frac{\rho c}{\gamma\lambda}\frac{e_s(T_o)-e}{r_l+r_a}, \qquad (18a)$$

$$E_p = \frac{\rho c}{\gamma\lambda}\frac{e_s(T_p)-e}{r_a}. \qquad (18b)$$

Elimination of r_a then gives

$$r_l = \frac{\rho c}{\gamma\lambda}\left[\frac{e_s(T_o)-e}{E}-\frac{e_s(T_p)-e}{E_p}\right]. \qquad (19)$$

In a simpler third method, r_a is calculated from leaf size and wind speed using equation (14) or Fig. 3*a*. When the reflectivity of the leaf and a similar piece of wet filter paper are equal, both surfaces absorb the same amount of radiation and equations (12) and (13) give

$$E/E_p = \frac{\Delta+\gamma}{\Delta+\gamma^*}, \tag{20}$$

$$r_l = r_a\left(1+\frac{\Delta}{\gamma}\right)\left(\frac{E_p}{E}-1\right). \tag{21}$$

When stomata are fully open in bright light, the leaf resistances of several species lie between 1 and 3 sec. cm.$^{-1}$ (Table 1), an order of magnitude greater than the external resistance at wind speeds met in the field.

Table 1. *Leaf resistances estimated from transpiration rates*

Species	r_l (sec. cm.$^{-1}$)	Reference
Turnip, sugar beet	1·5–1·7	Gaastra (1959)
Bean, tomato	2·3–3·3	Kuiper (1961)
Xanthium	0·8–1·7	Mellor *et al.* (1964)
Cotton	0·9–1·3	Slatyer & Bierhuizen (1964)
Barley	1·0–2·0	Armstrong (unpublished)

(ii) *Uniform crop surfaces*

The external resistance of a crop can be found from wind, temperature and humidity profiles above the crop. When these are the same shape, plotting temperature (and vapour pressure) against wind speed at a series of heights gives a straight line from which $T = T_o$ at $u = 0$ can be read (and similarly $e = e_o$). Then by analogy with the equation defining the internal resistance of a single leaf, the surface resistance of a crop canopy may be written

$$r_s = \frac{\rho c}{\gamma\lambda}\left\{\frac{e_s(T_o)-e_o}{E}\right\}. \tag{10b}$$

Although it will seldom be possible to identify T_o and e_o with conditions at a specific level within the canopy, the parameters r_a and r_s allow the equations of heat and vapour flux derived for a single leaf to be applied to a plant community. In particular, the rate of transpiration from a crop in given weather can be calculated from equation (13).

When evaporation from the soil surface is negligible, r_s is expected to be a plant parameter revealing diurnal and seasonal changes of stomatal resistance. When evaporation from the soil is comparable with transpiration, r_s combines the resistance of crop leaves with the capillary resistance to the diffusion of water vapour in soil pores. The surface resistance is inversely proportional to a wetness parameter $\lambda E/\{e_s(T_o)-e_o\}$ proposed by Covey (1959).

Resistances were estimated from evaporation measurements at Rothamsted and were incorporated in a theoretical model of crop transpiration and photosynthesis (Monteith, 1963*a*) that was adversely criticized by Tanner (1963). He objected that the derivation of r_s was invalid when the sources and sinks of heat, water vapour, and momentum were set at different levels within a crop canopy, but this criticism is irrelevant when r_s (and r_a) are determined from measurements *above* the canopy where established profiles are the same shape. When profiles are distorted by small divergences of flux or by effects of buoyancy, resistances can sometimes be estimated from measurements very near the surface where these anomalies are smallest. It has also been argued that r_s is not a purely physiological parameter because it includes an external resistance between transpiring leaves and air moving through the canopy. If so, this external component should change with wind speed but recent measurements at Rothamsted show no such effect.

Table 2 summarizes values of surface resistance calculated from evaporation rates and profiles using equation (10*b*) (Rothamsted, Kew, Davis, O'Neill), or from components of the energy balance and weather above the canopy using equation (13) (Hancock, Munich). Most of these values are the mean resistance for each hour during daylight, weighted by net radiation. At Kew and O'Neill, only 3 or 4 hr. were available each day. The resistance for Cambridge is an unweighted mean from a set of 23 profiles randomly timed over 5 days. From these profiles, Pasquill (1950) established that the ratio of actual transpiration rate to potential transpiration from a wet surface at the same temperature was 0·41, and a corresponding surface resistance is derived from equation (11). Linacre, Palmer & Trickett (1964) calculated the resistance of crops under glass from measurements of surface temperature.

Minimum values of surface resistance are between 0·3 and 0·5 sec. cm.$^{-1}$, roughly a third to a quarter of the minimum resistance for individual leaves (Table 1). Oversimplifying, the effective surface resistance of a transpiring crop is equivalent to the stomatal resistance of three or four leaf layers in parallel, each layer containing unit leaf area.

The surface resistance of barley increased from a minimum of 0·3 sec. cm.$^{-1}$ in mid-June to 0·7 sec. cm.$^{-1}$ at the end of July. During this period, the total leaf area index (leaves and sheaths) decreased from 10 to 6, partly from natural senescence and partly because many of the lower leaves were attacked by mildew. The increase of resistance with time may be caused by

Table 2. *Weather and resistances*

Surface	Height (cm.)	Site	Dates	Reference	Evaporation E (mm. hr.$^{-1}$ × 100)	Net heat H/λ (mm. hr.$^{-1}$ × 100)	Mean air temperature (° C.)	Mean vapour pressure (mb.)	Mean wind speed (m. sec.$^{-1}$)	z_0 (cm.)	Resistances (sec. cm.$^{-1}$) r_a	r_i	r_s	$(\lambda E/H) \times 100$
Grass														
Timothy and meadow fescue	60–70	Rothamsted	18. v. 1961	Monteith (1963*a*)	25	33	12·2	9·6	2	4	0·2	0·37	0·5	76
Rough pasture with some clover	5–10	Cambridge	8–12. v. 1947	Pasquill (1950)	—	—	—	—	3	0·7	0·4	—	0·5	—
Lawn	1	Kew	21. vi. 1949	Robinson & Rider (1951)	20	41	21·5	13·0	3	0·1	0·5	0·83	4·1	49
Rye-grass	—	Davis, California	28. v. 1960	Monteith (1963*b*)	48	60	21·6	13·2	2	5 (?)	0·3	0·57	0·5	80
			1. vi. 1960		83	60	32·1	10·3	7	5 (?)	0·1	1·69	1·1	138
Natural prairie	6	O'Neill, Nebraska	13. viii. 1953	Lettau & Davidson (1957)	37	55	30·6	16·9	6	0·9	0·3	1·32	2·1	67
			25. viii. 1953		11	58	27·8	19·2	8	0·9	0·2	0·84	7·0	19
			31. viii. 1953		9	51	29·8	20·0	6	0·9	0·3	1·17	15	18
Alfalfa–brome mixture	15	Hancock, Wisconsin	24. ix. 1957	Tanner & Pelton (1960*b*)	38	42	16·8	11·2	5	3	0·2	0·50	0·4	90
Barley														
Proctor	60	Rothamsted	26. vi. 1963	Szeicz (unpublished)	28	28	15·1	11·2	4	5	0·3	0·64	0·3	100
	105		23. vii. 1963		25	32	21·0	17·3	2	13	0·2	0·72	0·7	78
Beans	95	Rothamsted	15. vi. 1961	Monteith (1963*b*)	27	29	17·0	12·8	1	9	0·3	0·62	0·5	93
	95		21. vi. 1961		25	40	17·9	11·6	1	7	0·2	0·60	1·1	63
	135		25. vii. 1961		15	32	20·9	14·9	1	3	0·2	0·82	2·3	47
Rice														
Paddy field	—	—	—	Covey (1959) from Kaneseki (1956)	—	—	—	—	—	—	—	—	0·5	—
Pine forest	550	Munich	28. vi.–7. vii. 1952	Baumgartner (1956)	60	90	27·7	12·3	2	290 (?)	0·06	0·62	0·9	67
Glasshouse crops														
Beans	18	Griffith, Australia	—	Linacre *et al.* (1964)	—	—	—	—	—	—	—	0·2	0·3	—
Cotton	18											0·2	0·3	—
Saltbush	18											0·3	0·8	—

a decrease in total leaf area, by an increase in the resistance of individual leaves, or by a combination of both effects. As a further complication, the soil moisture deficit increased from 0·5 to 5 cm., but the maximum deficit was probably too small to restrict the uptake of water by the deepest roots.

Eliminating differences in soil moisture by irrigation, Fritschen & van Bavel (1964) found that daily transpiration from Sudan grass (*Sorghum vulgare* var. *sudanensis*) 75 cm. tall was 70% more than the transpiration from a neighbouring stand 140 cm. tall that had begun to flower. This difference reveals an increase of surface resistance that may be associated with the development of inflorescences or with an increase of stomatal resistance in leaves that had reached maturity.

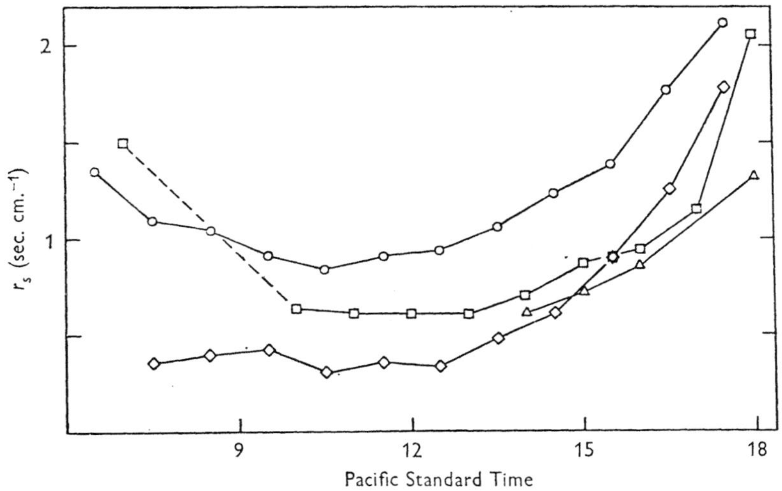

Fig. 4. Diurnal variation of crop surface resistance at Davis, California (Pacific Standard Time). ◇, 28 May 1960; △, 30 July 1962; ○, 1 June 1960 □, 31 July 1962.

The increase of surface resistance at O'Neill during August 1953 is correlated with a decrease in soil moisture in the absence of rain. On 13 August when the soil moisture at 10 cm. depth was 6% of wet weight, the surface resistance was 2 sec. cm.$^{-1}$, but, after 18 days without rain, soil moisture decreased to 3% and the resistance increased to about 15 sec. cm.$^{-1}$. From the same measurements, Halstead (1954) showed that the fraction of net radiation used for evaporation decreased linearly with soil water content.

Diurnal changes of surface resistance are superimposed on these seasonal changes. Because the uptake of water by plant roots lags behind the loss of water from leaves, water stress increases during the day and the increase of surface resistance shown in Fig. 4 suggests that stomata were gradually closing in response to this stress. This interpretation is consistent with an increase of surface resistance when transpiration rate is increased. When the mean rate was 0·48 mm. hr.$^{-1}$, the resistance was 0·5 sec. cm.$^{-1}$. Three days later, hot dry winds forced the rate of evaporation to 0·84 mm. hr.$^{-1}$, and the corresponding increase of resistance to 1·1 sec. cm.$^{-1}$ may be a response of self-preservation stimulated by an increase in the water deficit of leaf tissue. The decrease of resistance after sunrise may demonstrate a slow opening of stomata in response to increasing light intensity; or an increase of permeability in plant roots as the soil gets warmer near the surface.

EVAPORATION FROM WET AND DRY SURFACES

The rate of evaporation E_o from a surface thoroughly wetted by rain or dew can be calculated from equation (8). If E is the rate of transpiration in the same weather when the surface is dry, the relative transpiration rate is

$$\frac{E}{E_o} = \frac{\Delta+\gamma}{\Delta+\gamma(1+r_s/r_a)}, \qquad (20a)$$

and this fraction depends only on air temperature and the ratio r_s/r_a. (Equation (20a) is valid when wet and dry leaves reflect the same fraction of solar radiation, a condition usually satisfied unless the sun is very near the horizon.) Fig. 5 shows how the relative transpiration rate changes with resistance. Taking a constant minimum leaf resistance of 2 sec. cm.$^{-1}$, the ratio r_s/r_a ranges from 5 to 20 when r_a ranges from 0·4 to 0·1 sec. cm.$^{-1}$.

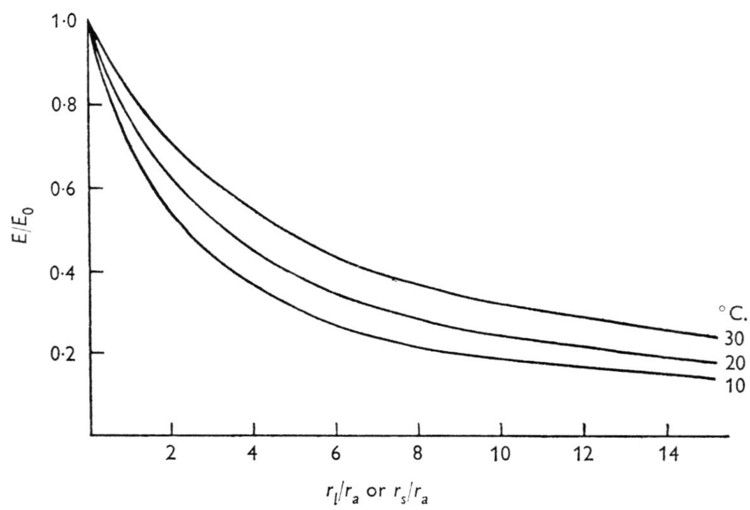

Fig. 5. Change of relative evaporation rate E/E_o with ratio r_l/r_a or r_s/r_a and with temperature.

At 20° C. the relative transpiration rate lies between 14 and 40%. The absolute rate of transpiration for *single leaves* is sensitive to changes in stomatal resistance and is relatively independent of wind speed. In very still air when r_a is very large, E approaches E_o and then the physiological control of transpiration is negligible. This was illustrated by Bange (1953), who found that the rate of transpiration from *Zebrina* leaves increased almost linearly with stomatal width in 'still' air, whereas, in 'wind', transpiration rates were independent of stomatal width between 10 and 20 μ.

Ehlig & Gardner (1964) found that when soil moisture potential reached 16 atm., transpiration from pepper, sunflower and cotton grown under glass decreased to about 20% of the maximum rate of transpiration in wet soil. Because this decrease depends on the *ratio* r_s/r_a and not on r_s alone, the results of these and similar experiments are not necessarily relevant to field conditions.

When *field crops* are supplied with adequate water, the minimum ratio r_s/r_a is between 1 and 2, and at 20° C. the corresponding ratio of E/E_o is between 0·76 and 0·62. This range is consistent with measurements by Frankenberger (1960), who calculated the ratio of actual transpiration to the evaporation from the same crop in similar weather immediately after the leaves were wetted by rain. On many rainless days, the rate of transpiration from grass, potatoes, sugar beet and rye was between 0·6 and 0·8 E_o.

McIlroy & Angus (1964) were unable to detect an increase in the rate of evaporation from grass immediately after the leaves were wetted by irrigation, a result predictable as follows. The grass was mown to a height of 6–10 cm. and the roughness length would be near $z_o = 1$ cm. From equation (17*b*), at a mean wind speed of 2·5 m. sec.$^{-1}$, the value of r_a is 0·7 sec. cm.$^{-1}$, for *neutral* conditions, but internal evidence shows that during most daylight hours air temperature increased with height. In this stable air, the resistance must have exceeded 0·7 sec. cm.$^{-1}$ (p. 214) and may often have reached 1–2 sec. cm.$^{-1}$. Because irrigation was very frequent, the surface resistance was probably about 0·3 sec. cm.$^{-1}$. Assuming a maximum value of $r_s = 0·3$, Fig. 5 shows that the difference between E and E_o would be a few per cent at most and too small to detect. Similar measurements reported by McMillan & Burgy (1960) can be interpreted in the same way.

By contrast, the ratio r_s/r_a for a pine forest is 15 and the transpiration rate is about one-fifth of the evaporation rate from leaves wetted by rain and exposed to the same weather. The rapid loss of intercepted rain after a shower may be an important factor in the water economy of trees growing in dry regions, and in the yield of water from forested catchments.

LATENT HEAT AND AVAILABLE ENERGY

With values of r_s and r_a established for a range of crops, maximum rates of transpiration will now be related to prevailing weather. It is convenient to write equation (13) non-dimensionally as

$$\frac{\lambda E}{H} = \frac{(\Delta/\gamma)+r_i/r_a}{(\Delta/\gamma)+(r_s+r_a)/r_a}, \tag{22}$$

where

$$r_i = \rho c \frac{e_s(T)-e}{\gamma H} \tag{23}$$

is a property of the atmosphere with the dimensions of a diffusive resistance. When $r_i = r_s+r_a$, $\lambda E = H$, showing that r_i is a critical value of (r_s+r_a) at which the net supply of heat from external sources is used exclusively to evaporate water. Then the flux of sensible heat is zero and the system is isothermal because it contains no gradients of temperature. This theorem identifies r_i as an 'isothermal resistance' proportional to the ratio of saturation deficit to net available heat. Table 2 shows that this ratio increases from a minimum of about 0·4 sec. cm.$^{-1}$ in the relatively humid climate of south-east England to a maximum of about 1·3 sec. cm.$^{-1}$ in the much drier climate of the western U.S.A.

(i) *Dependence of $\lambda E/H$ on r_a*

To find how $\lambda E/H$ changes with external resistance, equation (22) is differentiated with respect to r_a assuming constant radiation, temperature and humidity. The evaporation rate is independent of r_a when

$$r_s = \left(1+\frac{\gamma}{\Delta}\right) r_i \tag{24}$$

and substitution of this ratio in equation (22) gives

$$\frac{\lambda E}{H} = \frac{\Delta}{\Delta+\gamma}.$$

Fig. 6 shows this result graphically for weather with $r_i = 0{\cdot}6$ sec. cm.$^{-1}$, $\Delta = 1$ mb. °C.$^{-1}$. At $r_s = (1+\gamma/\Delta)\, r_i = 1$ sec. cm.$^{-1}$, the evaporation rate is independent of r_a and hence of wind speed and surface roughness. From equation (24), the Bowen ratio (sensible heat flux: latent heat flux) is then (γ/Δ). When r_s exceeds 1 sec. cm.$^{-1}$, the evaporation rate decreases with increasing wind speed and roughness, and the Bowen ratio is more than (γ/Δ). Conversely, when r_s is less than 1 sec. cm.$^{-1}$, evaporation increases with wind speed and roughness and the Bowen ratio is less than (γ/Δ).

In laboratory studies of transpiration, r_i is usually large because H is small without intense radiation. Then $r_s < (1+\gamma/\Delta)\, r_i$ and transpiration rates increase with wind speed. To demonstrate the opposite effect, Mellor, Salisbury & Raschke (1964) exposed leaves of *Xanthium* to an intense

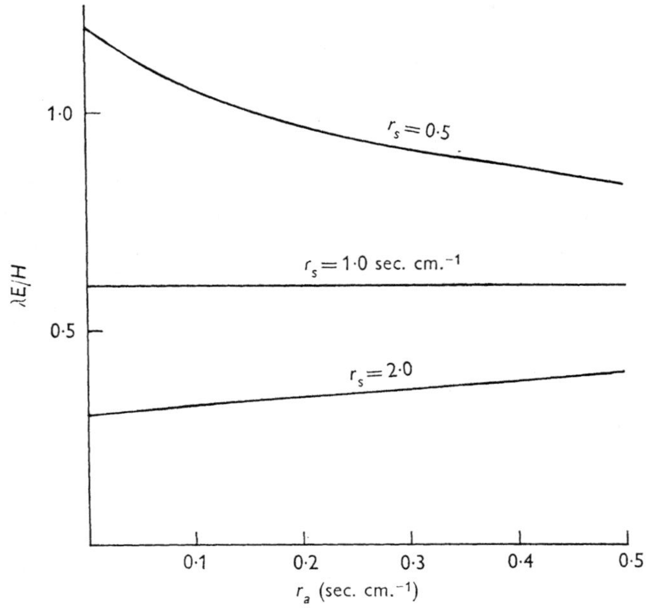

Fig. 6. Fraction of net heat used for evaporation as function of r_a and r_s (or r_l), assuming Δ = 1 mb.° C.$^{-1}$ and r_i = 0·6 sec. cm.$^{-1}$.

source of light (about 1 cal. cm.$^{-2}$ min.$^{-1}$) in air with a saturation deficit of only 0·8 mb. The calculated value of r_i is only 0·2 sec. cm.$^{-1}$, and presumably r_s exceeded $(1+\gamma/\Delta)\, r_i$. The transpiration rate decreased by a factor of two when wind speed increased from 0·2 to 4 m. sec.$^{-1}$.

The dependence of transpiration rate on wind speed is difficult to establish in the field because r_s, r_i and temperature all vary diurnally. Ladefoged (1963) found that the rate of transpiration from beech trees was uncorrelated with wind speed between 1 and 7 m. sec.$^{-1}$. Very roughly, the isothermal resistance corresponding to mean weather was 0·3 sec. cm.$^{-1}$ and, from equation (24), r_s = 0·5 sec. cm.$^{-1}$, consistent with values in Table 2.

(ii) *Dependence of $\lambda E/H$ on r_s*

Because Δ and r_a appear both in the numerator and in the denominator of equation (22), $(\lambda E/H)$ is insensitive to changes of wind speed and temperature. This leaves only two variables, r_s and r_i, to determine the fraction of net available heat used for evaporation. The values of $(\lambda E/H)$ in Table 2 show a clear inverse correlation with r_s alone and Fig. 7 shows that they fit a straight line

$$\frac{\lambda E}{H} = \left(\log \frac{25}{r_s}\right)^{\frac{1}{2}}. \tag{25}$$

There are two reasons why $(\lambda E/H)$ should appear independent of r_i. First, saturation deficit decreases, and so r_i decreases, when air blows from a hot, dry surface (large r_s) to a relatively cool, moist surface (small r_s). This process introduces a correlation between the equilibrium values of r_i and r_s some distance from the boundary. Near the boundary, there is a

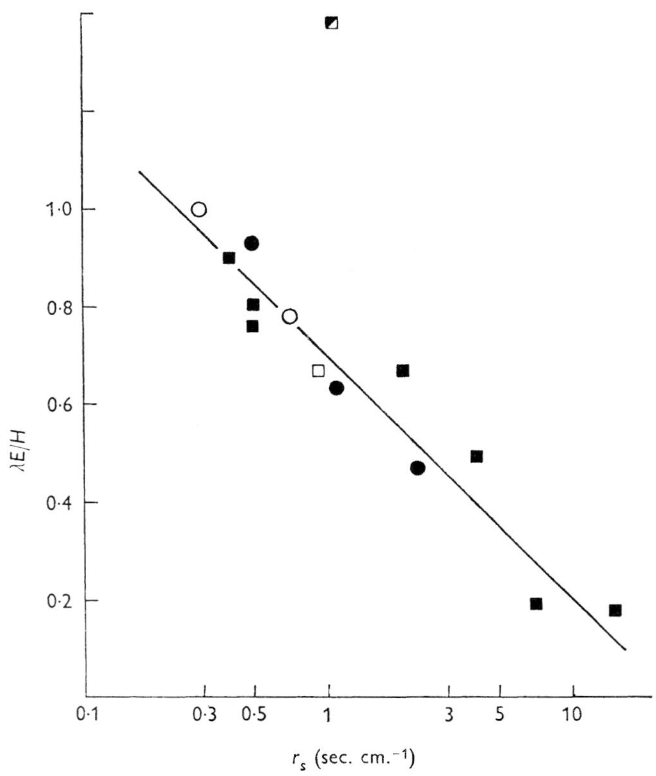

Fig. 7. Fraction of net heat used for evaporation as function of surface resistance r_s Measurements from Table 2. ■, Grass; ◩, grass, Davis, 1 June 1960; □, pine forest; ●, beans; ○, barley.

lack of equilibrium illustrated by the anomalous point for Davis on 1 June 1960, and examined later in more detail. A second and more obvious reason for the good correlation of $(\lambda E/H)$ and r_s is that the range of r_i is much smaller than the range of r_s.

The numerical constant in equation (25) may depend on the choice of measurements that establish it and further work may show that different surfaces have their own characteristic constants. Meanwhile, the equation can be used cautiously to estimate the internal resistance of closed crop canopies from measurements of $(\lambda E/H)$. For example, pineapple is a crop that uses water conservatively for transpiration but its internal resistance has never been measured. Ekern (1964) found that the latent heat of transpiration was about 12% of net radiation and, if equation (25) is valid, $r_s = 15$ sec. cm.$^{-1}$.

TRANSPIRATION AND LEAF AREA

In previous sections, E was the total flux of water vapour from a field crop, the sum of transpiration from all the leaves intercepting radiation and evaporation from the surface of the underlying soil. Within the canopy, the vertical flux increases with height at a rate that was measured in wheat by Saito (1962) and was calculated by Penman & Long (1960) from a simple model. This model will now be extended to include a profile of net radiation determining the energy available for transpiration and a profile of visible radiation (in the range 0·4–0·7 μ) determining stomatal aperture.

To relate the attenuation of radiation to the distribution of leaves, the foliage of a crop is divided into a set of horizontal leaf layers, each containing unit leaf area index. Within each layer, the distribution of leaves can be described by the distribution of angles with respect to a horizontal plane or, more simply, by a bulk parameter s which is the fraction of radiation passing straight through the layer without striking a leaf. This fraction ranges from 0·4 for crops with horizontal leaves (kale, clover) to 0·8 for crops with more nearly vertical leaves (cereals, grasses) (Monteith, 1965). In each layer, a fraction $(1-s)$ of incident radiation is intercepted by leaves whose transmission coefficient is τ. It can be shown that, in the nth layer below the top of the canopy, the intensity on a *horizontal* surface is

$$I_n = \{s+(1-s)\,\tau\}^{n-1} I_0,$$

but the mean intensity of radiation on *leaf* surfaces is only $(1-s)\,I_n$, the amount of radiation intercepted by unit leaf area.

When a crop with $s = 0{\cdot}8$, $\tau = 0{\cdot}1$ is exposed to bright sunshine with a visible intensity of 0·4 cal. cm.$^{-2}$ min.$^{-1}$, the mean intensity of light on leaf surfaces is 0·08 cal. cm.$^{-2}$ min.$^{-1}$ in the first layer (with unit leaf area index) and 0·07 cal. cm.$^{-2}$ min.$^{-1}$ in the second layer. Gaastra (1959) and Kuiper (1961) found that the leaf resistances of several field crops increased rapidly with decreasing light intensity below 0·08 cal. cm.$^{-2}$ min.$^{-1}$ because stomata began to close. Comparison of these figures suggests that the amount of visible light reaching crop leaves in the field may often fail to reach the intensity needed for maximum stomatal opening.

This hypothesis will now be tested from profiles of radiation, temperature, humidity and wind speed available from a field of barley on 23 July 1963 (Long, Monteith, Penman & Szeicz, 1964). The transmission coefficient for visible radiation was 0·07 and *s* was 0·8. The resistance of

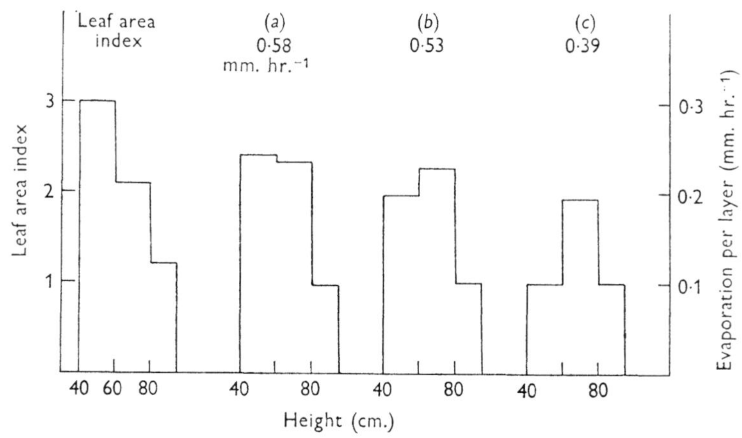

Fig. 8. Calculated transpiration from a stand of barley, 23 July 1963, in three layers, 40 to 60, 60 to 80, 80 to 95 cm.: (*a*) with T, e, u, r_s constant; (*b*) with T, e, u varying with height, r_s constant; (*c*) with T, e, u, r_s varying with height. Figures above each section are total transpiration.

barley leaves was assumed equal to the resistance of Kuiper's bean leaves at the same light intensity. Between 09.00 and 16.00 hours G.M.T., the measured mean rate of transpiration was 0·37 mm. hr.$^{-1}$, and Fig. 8 shows how estimates of this amount are improved as more of the relevant physical parameters are taken into account. In all the estimates the measured change of net radiation with height is included, but, in the first, values of temperature, humidity, and wind speed at a height of 2 m. were assumed to exist at all levels and all stomata were assumed to be fully open. For these conditions, the transpiration rate was 0·58 mm. hr.$^{-1}$ and the largest contribution came from the lowest layer of leaves, between 40 and 60 cm. Putting in the measured values of temperature, humidity and wind, but continuing to assume fully open stomata, the total transpiration calculated is 0·53 mm. hr.$^{-1}$, still too big, with the lowest layer contributing 20% less than before. Finally, putting in values of stomatal resistance estimated from measurements of light intensity, the total flux was 0·39 mm. hr.$^{-1}$, in good agreement with the measurement for the whole canopy.

Fig. 9 illustrates a more general relationship for crops with s = 0·4 and 0·8. The transmission of leaves was assumed to be 0·05 for visible radiation and 0·25 for both solar and net radiation because their profiles are usually similar in shape. Above the canopy, intensities were 0·6 cal. cm.$^{-2}$ min.$^{-1}$ net radiation and 0·45 cal. cm.$^{-2}$ min.$^{-1}$ visible radiation, and the following values were assumed constant throughout the canopy: Δ = 1 mb. °C.$^{-1}$, r_i = 0·6 sec. cm.$^{-1}$, r_a = 0·2 sec. cm.$^{-1}$. The left-hand side of Fig. 9 shows that transpiration rates reach a maximum at a leaf area index of 3 when

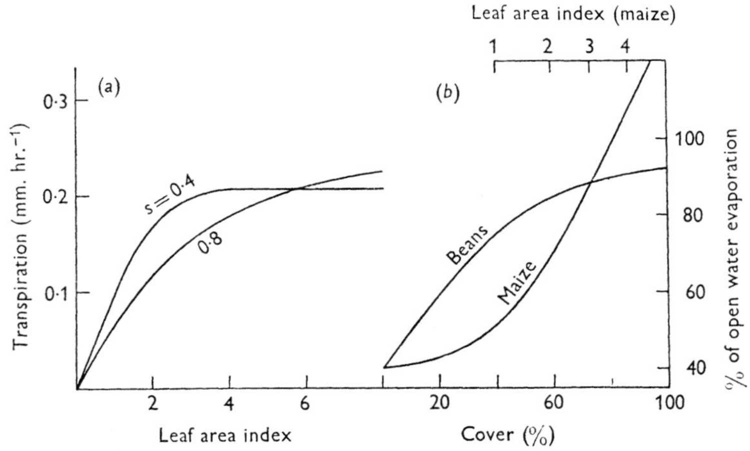

Fig. 9. (*a*) Calculated transpiration for crops with s = 0·4 and 0·8, as function of leaf area index; (*b*) measured evaporation (soil and plants) as a fraction of open water evaporation plotted as function of visual estimate of cover or of leaf area index. (After Cackett & Metelerkamp, 1963, 1964.)

s = 0·4 but increase beyond an index of 6 when s = 0·8. The right-hand side of the figure shows comparable measurements by Cackett & Metelerkamp (1963, 1964) on beans and maize. The two curves may reflect differences in leaf arrangement between beans (small s) and maize (large s). In particular, an increase in the rate of transpiration from maize beyond a leaf area index of 4 is consistent with the theoretical prediction for a crop with s = 0·8. The calculations for Fig. 9*a* assume that evaporation from the soil is negligible. If the soil were wet, the total rate of evaporation would be larger and would increase less rapidly with leaf area index.

POTENTIAL TRANSPIRATION

Much of the evidence reviewed by Penman (1963) supports his hypothesis that vegetation with its roots in moist soil and with leaves forming a complete canopy transpires at a maximum or potential rate determined primarily by weather. From discussion in the last section, a canopy is 'complete' when the transpiration of its lowest leaves is restricted by stomatal closure. When leaves are predominantly horizontal, the minimum leaf area needed to form a complete canopy may be somewhat less than the minimum area at which the soil appears to be fully shaded (Fig. 9). Cereals and other crops with leaves more nearly vertical may not transpire at their maximum rate until the leaf area exceeds the minimum for full shading. Results in Table 2 imply that farm crops satisfying the conditions for potential transpiration have a minimum surface resistance between 0·3 and 0·5 sec. cm.$^{-1}$. On the slender evidence of one set of observations, the minimum resistance of a pine forest is about 0·9 sec. cm.$^{-1}$.

Sets of resistances were chosen to estimate the seasonal variation of potential transpiration from three contrasting surfaces in two climates: very short grass, a tall farm crop, and a pine forest, in south-east England and in California. Table 3 shows the choice of roughness length, surface resistance and reflectivity. Rationalizing differences between surfaces, the roughness length increases by a factor of 50 from short grass to tall crop, and by a further factor of 50 from crop to forest. The surface resistance for the forest is twice the resistance for grass and crop. Heat stored in the soil and in the products of photosynthesis was assumed to be 10% of net radiation so that H is 90% of net radiation. Differences in net radiation absorbed by pine forests and by short, green vegetation were calculated from reflexion coefficients for solar radiation given by Ångström (1925).

Assuming that weather at London Airport and Kew Observatory is representative of the Thames Valley, records were taken from the Monthly Weather Report for 1963, a year in which the distribution of solar radiation during summer was close to the long term average for the decade 1954–63. Net radiation absorbed during daylight was estimated empirically from solar radiation and cloudiness (Monteith, 1965). For the much drier climate of the Sacramento Valley in northern California, corresponding records from a U.S. Weather Bureau Summary were available for 1961 and radiation figures for Davis were kindly supplied by W. O. Pruitt and H. B. Schultz. Fig. 10 shows the seasonal variation of monthly transpiration in the Thames Valley and annual totals are in Table 3.

Relative transpiration rates depend on the relative values of r_s, r_a and r_i. Moving from the short grass to the tall crop, r_a decreases (because z_o

increases) whereas r_s and r_i stay constant in a given climate. Because the surface resistance is less than the critical value $(1+\gamma/\Delta)\,r_i$ (p. 222), transpiration rate increases with increasing roughness. Moving from the tall crop to the forest, there is a further decrease of r_a, a decrease of r_i (because reflectivity decreases) and an increase of r_s. The stomatal resistance is now only slightly smaller than $(1+\gamma/\Delta)\,r_i$ so the increase of roughness from 5 to

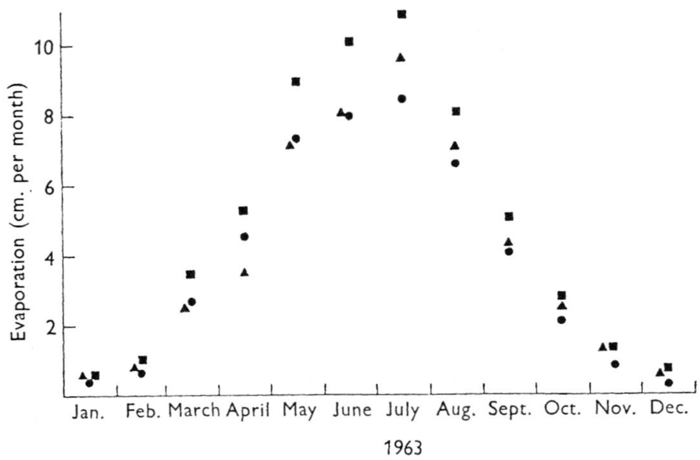

Fig. 10. Estimated annual variation of potential transpiration in Thames Valley, 1963. Some symbols are displaced horizontally to avoid overlapping. ●, Short grass; ■, tall crop; ▲, pine forest.

Table 3. *Assumed crop parameters and annual transpiration*

	reflectivity %	z_0 (cm.)	r_i (sec. cm.$^{-1}$)	r_a (sec. cm.$^{-1}$)	r_s (sec. cm.$^{-1}$)	Δ/γ	E (cm. yr.$^{-1}$)	$\lambda E/H$ (%)
			Thames Valley					
Short grass	25	0·1	0·8	1·1	0·5	1·3	47	76
Tall crop	25	5·0	0·8	0·36	0·5	1·3	58	95
Pine forest	15	250	0·7	0·025	1·0	1·3	48	69
			Sacramento Valley					
Short grass	25	0·1	1·1	1·1	0·5	2·4	148	96
Tall crop	25	5·0	1·1	0·36	0·5	2·4	187	117
Pine forest	15	250	0·9	0·025	1·0	2·4	182	100

250 cm. increases the transpiration rate very little. The decrease of reflectivity *per se* increases transpiration by about 12%. Finally, the increase of surface resistance more than outweighs the changes in reflectivity and roughness and makes transpiration from the forest less than from a tall crop and only slightly greater than from short grass.

The estimate predicts that, in the Thames Valley, vegetation with the roughness of a tall crop and with access to soil moisture throughout the year will use about 95% of available heat to transpire water. This prediction is supported by measurements during summer at Rothamsted, where net radiation was close to the latent heat of transpiration from cereals, sugar beet, and beans. Elsewhere, net radiation was within 20% of latent heat losses from grass (House, Rider & Tugwell, 1960), irrigated alfalfa-brome (Tanner & Lemon, 1962), Bermuda grass (*Cynodon* sp.: van Bavel & Harris, 1962), maize (Graham & King, 1961), and rice (Uchijima, 1961). By inference from Fig. 7, all these crops have a similar minimum surface resistance in the range 0·2–0·5 sec. cm.$^{-1}$ and, as a first approximation to guide the application of water by irrigation, the energy for potential transpiration is equal to the net income of radiation. Environments in which this rule breaks down are examined in a final section.

ADVECTION

When air with a water vapour content χ g. cm.$^{-3}$ moves horizontally with wind speed u cm. sec.$^{-1}$, water vapour is transported at a rate $u\chi$ g. cm.$^{-2}$ sec.$^{-1}$. This transport is an 'advective' flow, present throughout the

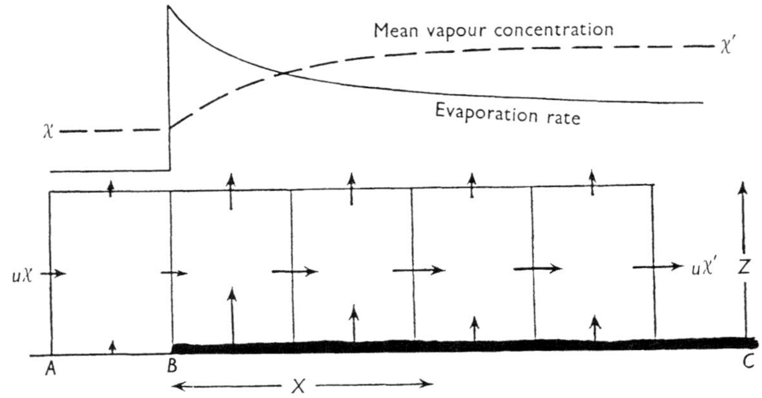

Fig. 11. Diagram representing the advection of water vapour and change in evaporation rate when air blows from a relatively dry surface *AB* to a moist surface *BC*. Fluxes of water vapour are proportional to lengths of arrows.

atmosphere because u and χ are always finite. In most meteorological estimates of evaporation, the vertical flux of water vapour a few metres *above* the surface is assumed equal to the flux *at* the surface. This assumption is not valid when $u\chi$ changes horizontally. For example, when air moves from an underlying dry surface to a moister surface, the concentration of water vapour increases and the increase in the horizontal flux of water vapour is often loosely referred to as 'local advection'. In Fig. 11, *AB* is a section of dry surface (e.g. bare soil) and *BC* is a wet surface (e.g. an irrigated crop). The atmosphere is represented in two dimensions by a layer of height Z divided into arbitrary vertical sections, and fluxes of water vapour are represented by arrows whose length is proportional to flux. To satisfy the conservation of water vapour, the total length of arrows entering any vertical section or box must equal the total length of departing arrows.

The box on the extreme left shows a state of equilibrium over the dry surface where χ is the mean concentration of water vapour between the surface and height Z. Here, the advective flux $u\chi$ entering from the left is equal to the flux leaving on the right and the vertical flux at height Z must, therefore, be equal to the rate of evaporation from the surface. When the air begins to move over the wetter surface, the evaporation rate increases, so χ increases and the horizontal arrows entering and leaving the second box show the increase of advective flux. To satisfy the conservation of vapour, the vertical flux at height Z must now be *less* than the evaporation rate at the surface. Moving towards the right of the diagram, the evaporation rate tends to decrease as the air becomes moister and equilibrium is re-established when the mean concentration of vapour reaches a new value χ'. In the fifth box, the horizontal flux is constant at $u\chi'$ and the vertical flux at Z is once more equal to the rate of evaporation at the surface.

The maximum height Z' at which the vertical flux is equal to the evaporation rate is the depth of a boundary layer characteristic of the surface and increasing with downwind distance X. The relationship between Z' and X has not yet been established by experiment but measurements by Dyer & Pruitt (1962) suggest that Z'/X is much less than 1/50. Theoretical analysis for a neutral atmosphere (Dyer, 1962) predicts that the vertical flux will be 95% of the evaporation rate at a height $Z = X/500$.

Local rates of evaporation at the surface can be estimated from net radiation and from temperature, humidity, and wind speed measured *within* the boundary layer. When weather is measured *above* this layer, Penman's equation (13) underestimates evaporation rates because the vapour pressure profile is not logarithmic. Abdel-Aziz, Taylor & Ashcroft (1964) found that the mean rate of transpiration calculated from equation (13) was only 73% of the actual rate of transpiration from a field of irrigated lucerne in Utah.

The leaves of an irrigated crop surrounded by dry land are often cooler than the air because the latent heat of transpiration exceeds the energy available from net radiation and the deficit of energy is made good by a downward transfer of sensible heat from warmer air to cooler surface. This is one of the *effects* of the advection of dry air over a wet surface, not a *cause* of anomalous evaporation rates as some writers imply. The relation between heat balance and advection was well illustrated by measurements of transpiration from a field of irrigated Bermuda grass 15 m. square. When the surrounding land was wetted by rain, latent heat of evaporation λE was about 0·8 of net radiation but, during 12 rainless days, the transpiration rate increased as the surroundings became drier until λE was almost twice the net radiation. On a much larger scale, daily values of λE for cotton were 1·75 times net radiation although the upwind fetch over a uniformly cropped and irrigated area exceeded 10 miles. Both these experiments were quoted by Tanner & Lemon (1962).

Downwind from the boundary of a wet area, λE decreases as the air becomes cooler and moister. The rate of decrease was predicted by Philip (1959), who solved equations for the diffusion of heat and water vapour from a wet surface ($r_s = 0$). In a numerical example (Rider & Philip, 1960), λE was 1·24, 0·98, 0·78 and 0·42 times net radiation at downwind distances of 1, 10, 10^2 and 10^5 m. Rider, Philip & Bradley (1963) measured changes of χ with height (to 1·5 m.) and with distance (to 18 m.) over irrigated grassland when the upwind fetch over dry tarmac was 350 m. Measuring the divergence of horizontal flux, they found that the ratio of latent heat to the net supply of radiation decreased from 3·7 averaged over the first metre of grass to 1·7 averaged over the first 16 m., but these figures were not compared directly with estimates from theory.

A surprisingly small change of transpiration rate was reported for irrigated clover by Millar (1964). Downwind from the edge of the field, the transpiration rate, measured with a set of lysimeters, decreased by 10% from 14 to 470 cm., and was constant from 5 to 14 m. At mid-day, plants near the boundary began to wilt and the turgidity of leaves increased downwind to 14 m. at least. Other workers have shown that transpiration is often restricted by the rate at which plant roots extract moisture from the soil. Even when soil moisture tension is only 1 atm., the surface resistance of grass and maize begins to increase when the rate of transpiration exceeds 4 mm. day^{-1} (Makkink & van Heemst, 1956; Denmead & Shaw, 1962). Associated physiological symptoms are loss of leaf turgidity, stomatal closure, and possibly some drying of cell walls. When crops are grown in an arid environment, transpiration rates often exceed 4 mm. day^{-1} and surface resistance increases with transpiration rate as shown in Fig. 4. The correlation of transpiration rate and leaf turgidity in Miller's experiment suggests that the downwind change of evaporation rate expected for a wet surface was largely offset by the downwind change of surface resistance. Progress here needs a technique for measuring leaf resistances directly and a modification of theory to include a surface resistance that changes with transpiration rate.

REFERENCES

ABDEL-AZIS, M. H., TAYLOR, S. A. & ASHCROFT, G. L. (1964). *Agron. J.* **56**, 139.

ÅNGSTRÖM, A. (1925). *Geogr. Ann., Stockh.*, **7**, 323.

BANGE, G. G. J. (1953). *Acta bot. neerl.* **2**, 255.

BAUMGARTNER, A. (1956). *Ber. dtsch. Wetterdienstes*, Nr. 28, Bd. **5**.

BAVEL, C. H. M. VAN & HARRIS, D. G. (1962). *Agron. J.* **54**, 319.

BRUNT, D. (1939). *Physical and Dynamical Meteorology*. Cambridge University Press.

BUSINGER, J. A. (1956). *Neth. J. agric. Sci.* **4**, 77.

CACKETT, K. E. & METELERKAMP, H. R. R. (1963). *Rhod. J. agric. Res.* **1**, 18.

CACKETT, K. E. & METELERKAMP, H. R. R. (1964). *Rhod. J. agric. Res.* **2**, 35.

COVEY, W. G. (1959). Ph.D. Thesis, Agricultural and Mechanical College of Texas.

DENMEAD, O. T. & SHAW, R. H. (1962). *Agron. J.* **45**, 385.

DYER, A. J. (1962). Second Annual Report under USAEPG Contract DA-36-039-SC-80334, University of California, Davis.

DYER, A. J. & PRUITT, W. O. (1962). Second Annual Report under USAEPG Contract DA-36-039-SC-80334, University of California, Davis.

EHLIG, C. F. & GARDNER, W. R. (1964). *Agron. J.* **56**, 127.

EKERN, P. C. (1964). Cyclostyled Research Report of the Pineapple Research Institute, Hawaii.

FISHENDEN, M. & SAUNDERS, O. A. (1957). *Heat Transfer*. London: Oxford University Press.

FRANKENBERGER, E. (1960). *Ber. dtsch. Wetterdienstes*, Nr. 73, Bd. **10**.

FRITSCHEN, L. J. & VAN BAVEL, C. H. M. (1964). *Agron. J.* **56**, 201.

GAASTRA, P. (1959). *Meded. LandbHoogesch., Wageningen*, **59**, 1.

GRAHAM, W. G. & KING, K. M. (1961). *Proc. Soil Sci. Soc. Amer.* **25**, 158.

HALSTEAD, M. H. (1954). *Johns Hopkins University, Publications in Climatology*, **7**, 326.

HOUSE, G. J., RIDER, N. E. & TUGWELL, C. P. (1960). *Quart. J. R. met. Soc.* **86**, 215.

KANESEKI, S. (1956). *J. agric. Met., Tokyo*, **12**, 30.

KUIPER, P. J. C. (1961). *Meded. LandbHoogesch., Wageningen*, **61**, 1.

LADEFOGED, K. (1963). *Physiol. Plant.* **16**, 378.

LETTAU, H. H. & DAVIDSON, B. (1957). *Exploring the Atmosphere's First Mile*, vol. 2. London: Pergamon Press.

LINACRE, E. T., PALMER, J. H. & TRICKETT, E. S. (1964). *Agric. Met.* **1**, 165.

LONG, I. F., MONTEITH, J. L., PENMAN, H. L. & SZEICZ, G. (1964). *Met. Rdsch.* **17**, 97.

MCILROY, I. C. & ANGUS, D. E. (1964). *Agric. Met.* **1**, 201.

MCMILLAN, W. D. & BURGY, R. H. (1960). *J. geophys. Res.* **65**, 2389.

MAKKINK, G. G. & VAN HEEMST, H. D. J. (1956). *Neth. J. agric. Sci.* **4**, 1.

MARTIN, E. (1943). *Publ. Carneg. Instn*, **550**, 1.

MELLOR, R. S., SALISBURY, F. B. & RASCHKE, K. (1964). *Planta*, **61**, 56.

MILLAR, B. D. (1964). *Aust. J. agric. Res.* **15**, 85.

MILTHORPE, F. L. (1962). In *Arid Zone Research*, vol. 16. Paris: UNESCO.

MONTEITH, J. L. (1963*a*). In *Environmental Control of Plant Growth*, p. 95. New York: Academic Press.

MONTEITH, J. L. (1963*b*). Second Annual Report under USAEPG Contract DA-36-039-SC-80334, University of California, Davis.

MONTEITH, J. L. (1965). *Ann. Bot., Lond.* (in the Press).

MONTEITH, J. L. (1965). *Weather* (in preparation).

MONTEITH, J. L., SZEICZ, G. & YABUKI, K. (1964). *J. appl. Ecol.* **1**, 321.
PASQUILL, F. (1950). Air Ministry, Met. Res. Cttee, M.R.P. 579.
PENMAN, H. L. (1948). *Proc. roy. Soc* A, **202**, 143.
PENMAN, H. L. (1953). *Rep. 13th int. hort. Congr.* **2**, 913.
PENMAN, H. L. (1963). *Vegetation and Hydrology.* Farnham Royal, England: Technical Communication No. 53, Commonwealth Bureau of Soils.
PENMAN, H. L. & LONG, I. F. (1960). *Quart. J. R. met. Soc.* **86**, 16.
PENMAN, H. L. & SCHOFIELD, R. K. (1951). *Symp. Soc. exp. Biol.* **5**, 115.
PHILIP, J. R. (1959). *J. Met.* **16**, 535.
POWELL, R. W. (1942). *Trans. Instn chem. Engrs, Lond.*, **18**, 136.
PRIESTLEY, C. H. B. (1959). *Turbulent Transfer in the Lower Atmosphere.* Chicago: University of Chicago Press.
RASCHKE, K. (1956). *Arch. Met., Wien*, **7**B, 240.
RIDER, N. E. (1954). *Phil. Trans.* A, **246**, 481.
RIDER, N. E. & PHILIP, J. R. (1960). *Ass. Int. de Hydrologique Scient.* **53**, 421.
RIDER, N. E., PHILIP, J. R. & BRADLEY, E. F. (1963). *Quart. J. R. met. Soc.* **89**, 507.
ROBINSON, G. D. & RIDER, N. E. (1951). *Quart. J. R. met. Soc.* **77**, 375.
SAITO, T. (1962). *J. agric. Met., Tokyo*, **17**, 101.
SLATYER, R. O. & BIERHUIZEN, J. F. (1964). *Aust. J. biol. Sci.* **17**, 115.
SLATYER, R. O. & MCILROY, I. C. (1961). *Practical Microclimatology.* Paris: UNESCO.
STOLLER, J. & LEMON, E. R. (1963). U.S.D.A. Agricultural Res. Service, Production Research Report, no. 72.
TANNER, C. B. (1963). In *Environmental Control of Plant Growth*, p. 141. New York: Academic Press.
TANNER, C. B. & LEMON, E. R. (1962). *Agron. J.* **54**, 207.
TANNER, C. B. & PELTON, W. L. (1960*a*). *J. geophys. Res.* **65**, 3391.
TANNER, C. B. & PELTON, W. L. (1960*b*). *Soils Bulletin*, 2. Department of Soils, University of Wisconsin, Madison, Wis.
UCHIJIMA, Z. (1961). *Bull. nat. Inst. agric. Sci., Tokyo*, A, no. 8, 243.
VRIES, D. A. DE & VENEMA, H. J. (1954). *Veg. acta Geobot.* **5**, 225.

Phil. Trans. R. Soc. Lond. B. **273**, 593–610 (1976) [593]
Printed in Great Britain

The interpretation of the variations in leaf water potential and stomatal conductance found in canopies in the field

BY P. G. JARVIS†
Department of Botany, University of Aberdeen, St Machar Drive, Aberdeen, AB9 2UD

Attempts to correlate values of stomatal conductance and leaf water potential with particular environmental variables in the field are generally of only limited success because they are simultaneously affected by a number of environmental variables. For example, correlations between leaf water potential and either flux of radiant energy or vapour pressure deficit show a diurnal hysteresis which leads to a scatter diagram if many values are plotted. However, a simple model may be adequate to relate leaf water potential to the flow of water through the plant.

The stomatal conductance of illuminated leaves is a function of current levels of temperature, vapour pressure deficit, leaf water potential (really turgor pressure) and ambient CO_2 concentration. Consequently, when plotted against any one of these variables a scatter diagram results. Physiological knowledge of stomatal functioning is not adequate to provide a mechanistic model linking stomatal conductance to all these variables. None the less, the parameters describing the relationships with the variables can be conveniently estimated from field data by a technique of non-linear least squares, for predictive purposes and to describe variations in response from season to season and plant to plant.

1. INTRODUCTION

The development of the diffusion porometer and the pressure chamber has enabled large numbers of measurements of stomatal conductance (k_s) and xylem pressure potential (Ψ_x) to be made in the field, in recent years. Because the measurements are easily made, very many values have been obtained simply to describe stomatal behaviour and water status in plants. Convenient methods are required to reduce these measurements to manageable proportions and to obtain useful information about plant properties from them. One approach which has been tried is to relate the measurements empirically to the environmental variables using simple (e.g. Millar, Jensen, Bauer & Norum 1971) or multiple (e.g. Smart & Barrs 1973) regression techniques. However, as a general rule it would seem more promising to try to use mechanistic models, the parameters of which would contribute useful information about properties of the plants.

In addition, values of Ψ_x and k_s are sometimes used as aids in the interpretation of concurrent variations in the rates of photosynthesis and evaporation by leaves or canopies, or as inputs to models to predict photosynthesis and evaporation by leaves or canopies, in particular seasons and weather conditions (see, for example, Monteith 1965, Acock, Thornley & Warren Wilson 1971, Reed, Hammerly, Dinger & Jarvis 1976). For these purposes, values of k_s and Ψ_x may be obtained from measurement on the foliage at the time for which the values are needed. However, measurement of both of these plant variables in the field is laborious, if made by intermittent sampling with porometer and pressure chamber, or requires careful maintenance of

† Present address: Department of Forestry and Natural Resources, University of Edinburgh, Kings Buildings, Mayfield Road, Edinburgh, EH9 3JU

complex apparatus, if made continuously with assimilation chamber and psychrometer. Furthermore, measurement of k_s by diffusion porometer can be made only when the foliage is dry, and long-term enclosure may lead to changes in the physiological properties of the leaves. Consequently, measurements of Ψ_x and k_s in the field are usually made intensively over selected periods of a few hours or days as shown in figures 1 and 2.

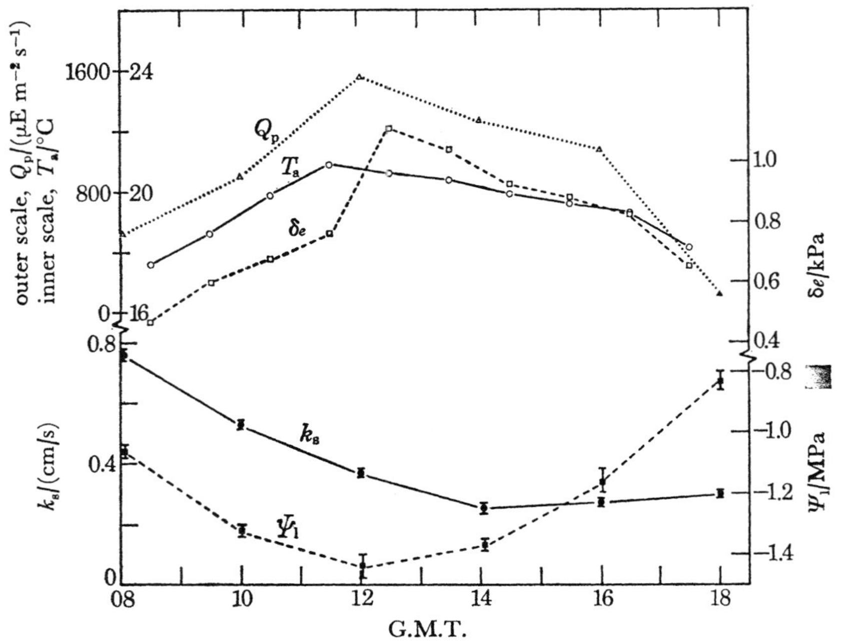

FIGURE 1. Diurnal variation in stomatal conductance (k_s), measured with a diffusion porometer, and leaf water potential (Ψ_l), measured with a pressure chamber, of current year's shoots of Sitka spruce (*Picea sitchensis*) in the upper part of a forest canopy of 25 year old trees on 15 August 1973. Each point is the mean of five measurements; two standard errors of the mean are shown. Concurrent variations in photon flux density (Q_p), air temperature (T_a) and vapour pressure difference (δe) are shown. Data from Watts, Neilson & Jarvis (1976).

Measurements of this frequency provide a considerable amount of detailed information over short periods. However, a similar amount of information is needed throughout the year for the interpretation of measurements of photosynthesis and evaporation, or for the calculation of rates of photosynthesis and evaporation from weather and plant variables. An alternative to very frequent measurement of k_s and Ψ_x in the field is to predict them from measurements of the weather and certain simple plant variables using models containing the parameters derived from occasional periods of measurement in the field. This paper describes two simple models for these purposes.

2. Leaf water potential

The leaf water potential (Ψ_l) is the driving force for the movement of liquid water through the plant. Consequently, the values of Ψ_l attained in the field should be interpreted in relation to the volume flux of water through the plant and the characteristics of the pathway of water transfer from soil to leaf.

The average water potential of the leaf mesophyll (Ψ_l) is nearly identical with the xylem pressure potential in many species, because the solute concentration of the xylem water is low in the extremities of the plant, and equilibration of water potential between leaf mesophyll cells and the xylem is rapid. In this paper, the two will be considered as synonymous.

The pathway of water movement through the plant is a hydraulic system containing frictional resistances to the flow of water. These frictional resistances occur at the internal surfaces of the vessels and tracheids, in simple and bordered pits, in cell walls, in membranes and in the symplasm. The movement of liquid water along a series of frictional resistances requires a difference in potential and this difference is largest when the flow is largest and where the resistance is highest. It is often said that the water deficits in leaves, or the reduction in Ψ_l, occur because the loss of water in transpiration exceeds absorption by the roots. It is, of course, true that the water balance of a cell can change only if the influxes and effluxes of water are unequal. However, to suppose that low water potentials occur in leaves because transpiration from the leaves exceeds absorption by the roots is to obscure the essential nature of water movement through the plant. Under conditions of *steady-state* transpiration from a plant, when absorption of water by the roots equals losses from the leaves, a potential difference must exist between leaves and soil to move water through the frictional resistances. This difference in potential will be larger when transpiration is higher, if the resistances to flow are constant. Water deficits arise because the water potential of cells adjacent to the flow pathway tends to equilibrate with that in the pathway by the movement of water out of the cells. The size of the water deficit depends upon the moisture characteristic of the cells, the resistance to movement of water out of the cells and the water potential in the flow pathway.

For an inelastic, hydraulic system with frictional resistance to flow but no capacitance, a simple one dimensional model of leaf water potential is

$$\Psi_l = \Psi_s - \Sigma q_l R_l - h\rho g, \qquad (1)$$

where the subscripts l and s indicate leaf and soil, respectively, and q_l and R_l are the partial, series-linked flows and resistances in the pathway traversed by a molecule of water from soil to leaf (Richter 1973); h is the height above the reference level, ρ the density of water and g the acceleration due to gravity. This model suggests that there should be a unique relation between Ψ_l and the rate of flow of water through the plant for a plant with particular flow path characteristics, provided that the resistance to flow remains constant.

The partial flows in a branched system are not easily measured, whereas the rate of evaporation of water (E) from whole plants in the field may be derived from measurements of the components of the leaf or canopy energy balance. Consequently, (ER_p) is sometimes substituted for ($\Sigma q_l R_l$) in equation (1) as an approximation, namely:

$$\Psi_l = \Psi_s - ER_p - h\rho g. \qquad (2)$$

R_p is an effective resistance for the entire pathway of parallel and series-linked conduits between the source of water in the soil and the leaves; it depends strongly upon the development and branching of the vascular tissue into a system of parallel-linked conduits (Jarvis 1975). Investigation of the relationship between Ψ_l and the rate of evaporation of water from the leaves should be informative about the properties of this pathway. In addition, the dependence of Ψ_l on evaporation rate should provide a means for rationalizing the apparent effects of many environmental variables on Ψ_l, and provide a means for the prediction of Ψ_l from a knowledge of the environmental conditions.

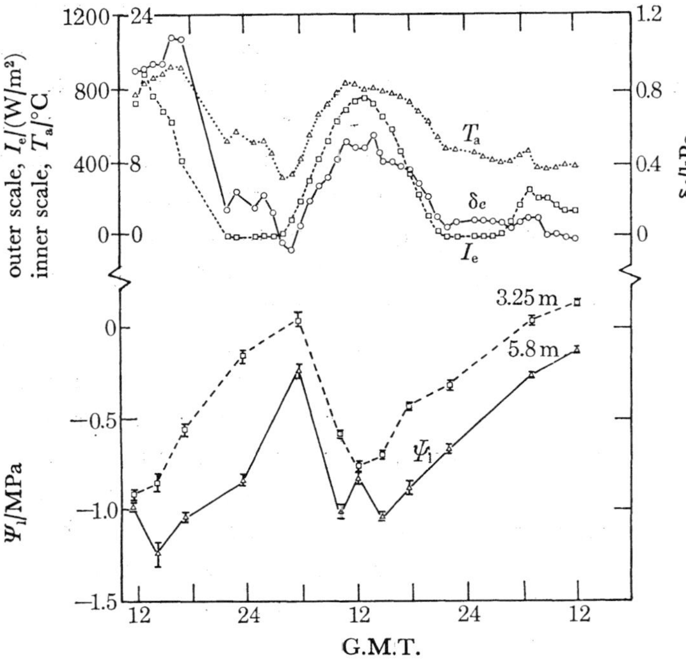

FIGURE 2. Diurnal variation in leaf water potential (Ψ_l) at two heights above the ground in a forest canopy, as in figure 1 (2–4 June 1970). Concurrent variations in irradiance (I_e), T_a and δe are shown. (Redrawn from Hellkvist, Richards & Jarvis 1974, Figure 5.)

Many observations have shown that Ψ_l falls from a high value in the morning to a low value in the middle of the day and then rises again in the afternoon as shown in figure 2 (Klepper 1968, Carr 1971, Goode & Higgs 1973, Jackson 1974). The value of Ψ_l appears to follow air temperature (T_a), vapour pressure deficit (δe) and irradiance (I_e). For example, simple linear regressions of Ψ_l on δe or I_e have been found to account for 70 to 95 % of the variation in Ψ_l over short periods of time (Millar, Jensen, Bauer & Norum 1971; Smart & Barrs 1973).

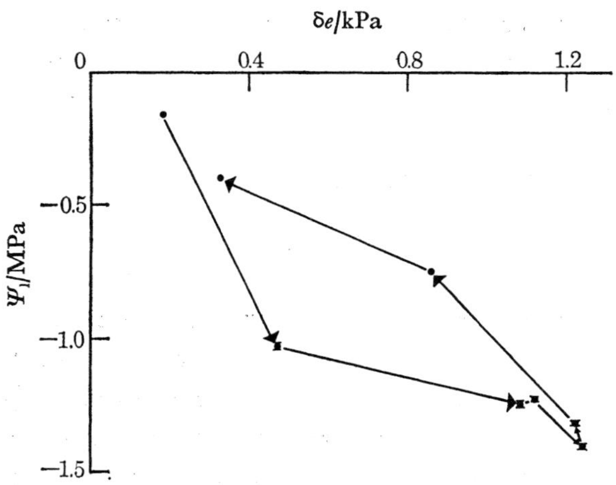

FIGURE 3. The relation between the water potential (Ψ_l) of leaves on hedgerow apple trees (Cox's orange pippin; 26 July 1973) and the vapour pressure difference (δe) during the course of a day. The arrows mark the course of time from dawn to dusk. Each point is the mean of five measurements. Two standard errors of the mean are indicated when they are larger than the symbol. (Redrawn from Landsberg *et al.* 1975, Figure 5.)

However, plots of the diurnal values of Ψ_l against δe (figure 3) or available energy (A) (figure 4) usually show marked hysteresis. This hysteresis results from the simultaneous dependence of Ψ_l on both δe and A. When both increase and decrease in phase together, little or no hysteresis occurs but when the changes in δe lag behind the changes in A, as usually occurs, marked hysteresis results. This hysteresis can introduce appreciable variation into simple regressions of Ψ_l on δe. Multiple regression analysis allows for this with the result that the variation in T, δe and I_e may account for up to 96% of the variation in Ψ_l (Smart & Barrs 1973). However, it would seem more promising to attempt to interpret Ψ_l as a function of these environmental driving variables in terms of their effects on the flow of water through the plant as specified by equation (2).

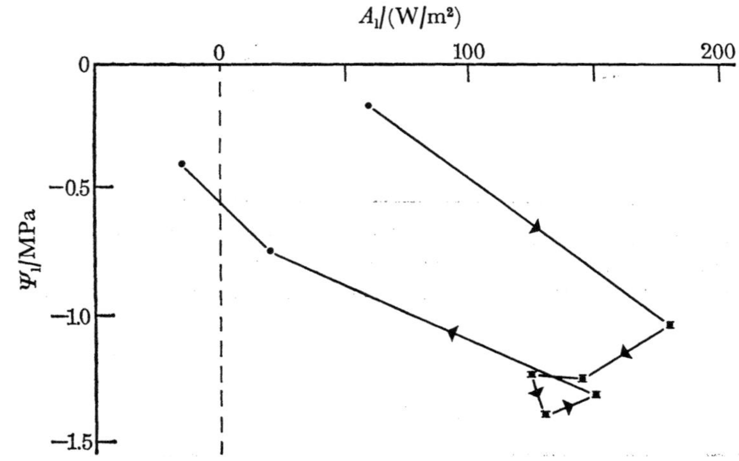

FIGURE 4. The relation between the water potential (Ψ_l) of leaves on hedgerow apple trees and the available energy (A_l) during the course of a day, as in figure 3. A_l is the net radiation absorbed per unit leaf area. (Redrawn from Landsberg *et al.* 1975, Figure 6.)

The Monteith (1965) modification of the Penman equation shows that E is not linearly dependent on any one environmental variable but that:

$$E = \frac{sA + c_p \rho \delta e k_a}{\lambda(s + \gamma(1 + k_a/k_s))}. \tag{3}$$

The rate of evaporation per unit area is a strong function of both A and δe and a weaker function of windspeed, through its effect on the boundary layer conductance (k_a), and of air temperature, through its effect on s (the slope of the relation between saturation vapour pressure and temperature). λ, c_p, ρ and γ are the latent heat of vaporization of water, the specific heat of air at constant pressure, the density of air and the psychrometric constant, respectively; they are all weak functions of temperature. E also strongly depends on the stomatal conductance (k_s) and consequently is affected by all the environmental and physiological variables which affect k_s (see §3).

Consequently it is not to be expected that Ψ_l will be a simple linear function of A, δe or T. However, the same values of Ψ_l plotted in figures 3 and 4 do show a close linear relation with evaporation rate (figure 5). In a similar treatment to that used here, Elfving, Kaufmann & Hall (1972) obtained curvilinear relations. However, they neglected the energy term in the evaporation equation, assuming that E was given by $(\delta e\, k_a)/(1 + k_a/k_s)$. From equation (2), the slope of the line in figure 5 gives the effective resistance of the pathway to water transfer,

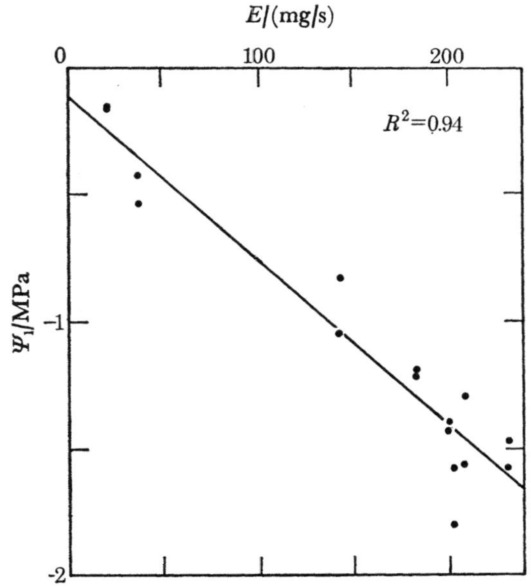

FIGURE 5. The relation between leaf water potential (Ψ_l) of leaves on hedgerow apple trees and the rate of evaporation per tree (E). E was calculated from the energy balance of the leaves by using equation (3) and measured values of A, k_s, k_a, δe and air temperature, and the leaf area per tree. (Redrawn from Landsberg *et al.* 1975, Figure 7.)

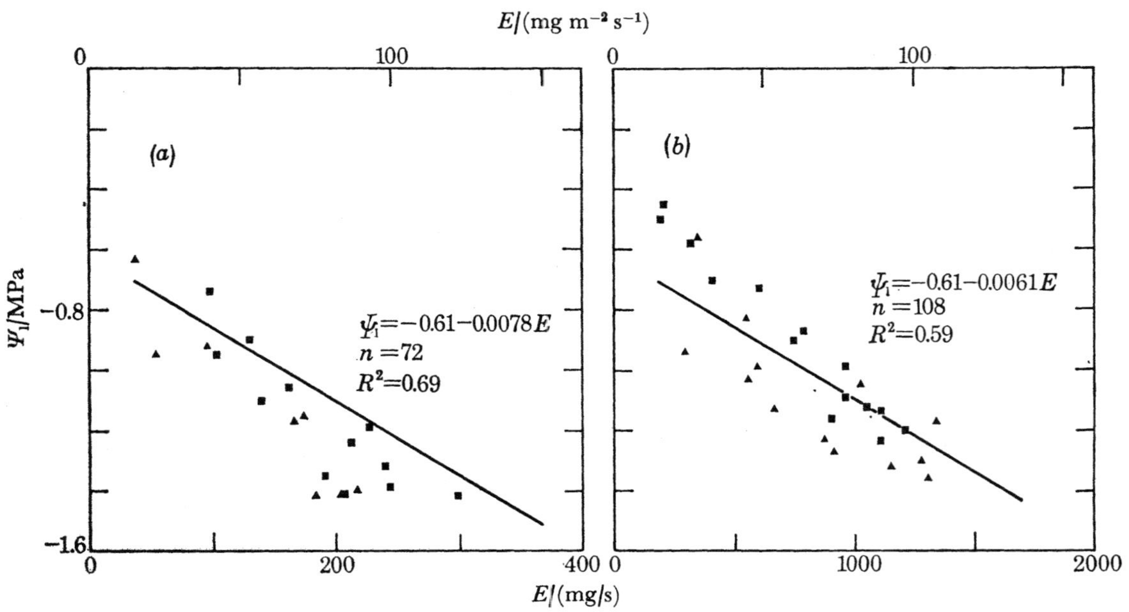

FIGURES 6*a* AND *b*. The relation between leaf water potential (Ψ_l) and evaporation rate (E) from (*a*) a 25 year old stand of Sitka spruce at Fetteresso forest (June–September 1973, 4100 trees/ha, leaf area index *ca.* 10) and (*b*) a 43 year old stand of Scots pine at Thetford forest (June 1973, 800 trees/ha, leaf area index *ca.* 4.3). The points are from measurements made on two separate days at each site. The regression lines have been calculated from 8 and 9 days measurements, respectively. Each point is based on five measurements of Ψ_l, the standard error of which was less than 0.05 MPa. E was measured by the Bowen ratio method by (*a*) P. G. Jarvis, G. B. James & B. Davidson and (*b*) Gash & Stewart (1975) respectively. The measurements of Ψ_l were made by (*a*) W. R. Watts & R. E. Neilson and (*b*) J. M. Roberts, respectively.

and the intercept on the ordinate gives an effective source water potential. Figures 6*a* and *b* show similar relations between Ψ_l and E for Sitka spruce (*Picea sitchensis*) and for Scots pine (*Pinus sylvestris*) on two different days, and regression lines based on measurements over eight or nine days. The ranges of the evaporation rates (expressed per unit ground area) and of the water potentials, and the parameters of the regression lines, are similar at the two sites. It should be emphasized that the spruce and pine data were collected completely independently by different groups of people at sites *ca.* 800 km apart. I am greatly indebted to Dr J. M. Roberts and Mr J. B. Stewart of the Institute of Hydrology for the provision of the Scots pine data.

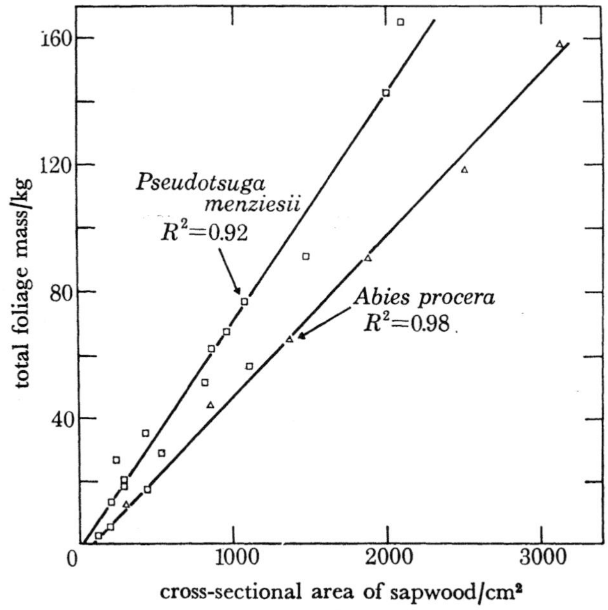

FIGURE 7. The relation between the mass of foliage on a tree and the cross-sectional area of the sapwood at breast height for Douglas fir and noble fir trees of a range of sizes. (Redrawn from Grier & Waring 1975, Figure 1.)

At the Fetteresso Forest site, there were approximately 4100 stems/ha (hectare), whereas at Thetford there were approximately 800 stems/ha. Thus at equal evaporation rates per unit ground area, there was five times as large a volume flow through the Scots pine trees as through the Sitka spruce trees, for approximately the same drop in water potential. Thus the resistance to flow in the pine trees was only about one fifth of the resistance to flow in the spruce trees. This difference occurred despite the fact that the pine trees (top height of *ca.* 15.5 m) were approximately 4 m taller than the spruce trees (top height of *ca.* 11.5 m). As a possible explanation, I suggest that the lower flow resistance of the pine trees may result from a larger cross-sectional area of conducting sapwood than in the spruce trees. This might possibly be a species difference: *Pinus ponderosa*, for example, has a much larger sapwood cross-sectional area than several species of spruce or fir (Lassen & Okkonen 1969). However, it may equally well be the result of homeostasis in the development of sapwood by the trees. It is most unlikely that the spruce trees, after thinning to the same area density as the pines, will have much lower Ψ_l as a result of the higher rates of evaporation per tree. Instead, R_p will almost certainly fall, as a result of the growth of additional sapwood cross-sectional area in the stem and branches, concurrent with an increase in the amount of leaves in the canopy, as found by Grier & Waring

1975 (figure 7). This should result in the maintenance of Ψ_l within approximately the same limits (Jarvis 1975, figure 8).

Hysteresis is also evident sometimes in the relation between Ψ_l and E (figure 8) because Ψ_l shows a tendency to peak later in the day than E. This might result from differential errors in the determination of E in the morning and afternoon, because of differences in the size of the gradients of temperature and humidity. Alternatively, there may be a real change in the flow resistance of the tree during the course of the day, higher resistances to flow occurring in the afternoon than in the morning. This could result from the fall in water potential during the day causing contraction in the diameter of the tracheids and displacement of the torus in the bordered pits (see Gregory & Petty 1973). Another plausible explanation is that the hysteresis is caused by a shift downwards in the sources of available stored water in the plant itself during the day (Jarvis 1975). When transpiration begins, water will tend to be withdrawn simultaneously from all sources in the plant and soil because of the hydraulic nature of the pathway.

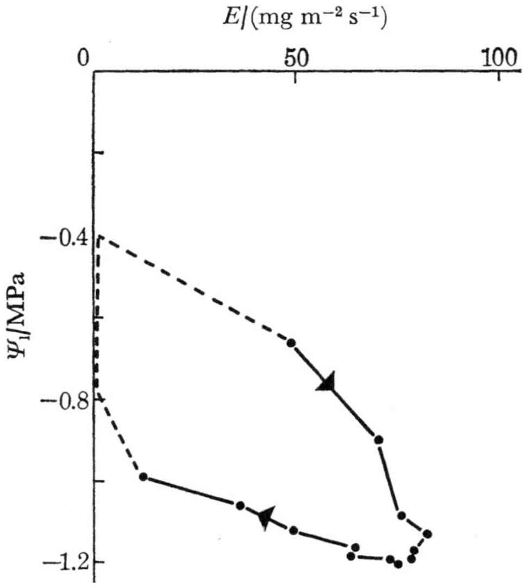

FIGURE 8. An extreme example of hysteresis in the relation between leaf water potential (Ψ_l) and evaporation rate (E) for Scots pine at Thetford Forest. Details as in figure 6*b*. (Data provided by J. B. Stewart & J. M Roberts.)

Initially it is to be expected that most water will move to the liquid-air interface in the leaf from the nearest sources of most readily available water, such as the cells in the leaves, branches and sapwood in the tree crown region. Since these cells are sources of water of finite volume, their capacity to supply water decreases as it is withdrawn and as their water potential tends towards that of the xylem. Consequently a front of water potential and water deficit spreads down through the plant appearing at lower levels somewhat attenuated and with a phase shift. In small herbaceous plants with little storage capacity, both the attenuation and the phase shift may be negligible, but in the stems of trees they are considerable (figure 9). In the afternoon all water transpired must either be withdrawn from storage in the lower levels of the plant or from the soil. Consequently, lower leaf water potentials are likely to be required to maintain a particular transpiration rate than in the morning.

While the explanations suggested above are conjectural, it is apparent that analysis of the values of Ψ_l in those terms provides stimulation for investigation of additional properties of the trees of a more causal nature. For example, it becomes important to know the storage capacity of different tissues in different parts of the plant and the amounts of water coming out of the various stores during the course of the day. More needs to be known about the role of the sapwood as a conducting and storage tissue and possible homeostasis in the cross-sectional area of the sapwood consequent upon changes in the area density of trees. In addition, the model adequately summarizes the interacting effects of radiation, temperature and vapour pressure deficit on Ψ_l; it enables the reduction and interpretation of field measurements of Ψ_l; and it may also be useful in the prediction of stomatal conductance (see § 3) or evaporation rate.

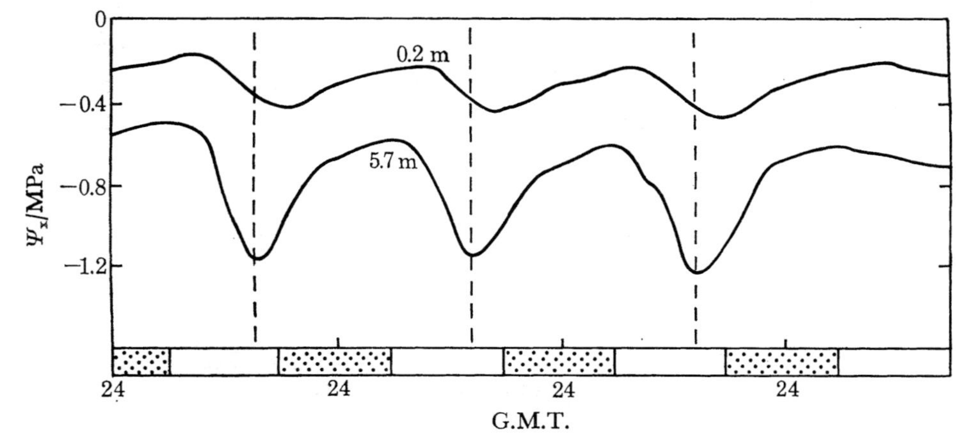

FIGURE 9. Diurnal variation in the xylem pressure potential (Ψ_x) at just above the ground and at the base of the crown in a Sitka spruce tree at Fetteresso Forest during three sunny days (18–21 August 1971). (Redrawn from Richards 1973, Figure 3.5.)

3. STOMATAL CONDUCTANCE

The conductance of illuminated stomata depends upon the current level and history of several environmental variables (Meidner & Mansfield 1968; Neilson & Jarvis 1975). A satisfactory conceptual model to include the known effects of the environmental variables upon k_s depends upon an adequate knowledge of the mechanism of stomatal action at the cellular level. Unfortunately, at the present time the mechanism is not understood in sufficient detail to enable us to specify each partial process and how it is affected by the environmental variables. Consequently, it is not yet possible to interpret the effects of environmental variables on k_s with a mechanistic model of stomatal action.

Nonetheless, a phenomenological model to predict k_s can be made from experimentally determined relations between k_s and the environmental variables. The general form of these relations has been established. The parameters in the relations depend upon the physiological condition of the plants, the previous weather and the season. Consequently, they must be determined at intervals to establish their seasonal course so as to enable calculation of the rate of evaporation and photosynthesis throughout the year. The seasonal variation in the parameters describes the changing adaptation of k_s to the environment. The model described here was developed in close collaboration with Dr K. L. Reed, Department of Natural Resources, Olympia, Washington, U.S.A. A more complete presentation of the model will be made elsewhere.

The variables

The steady state stomatal aperture or conductance depends on photochemical processes and is a function of the quantum flux density (Meidner & Mansfield 1968). The k_s of illuminated stomata is also a function of ambient CO_2 concentration (Meidner & Mansfield 1968), leaf-air vapour pressure difference (Lange, Lösch, Schulze & Kappen 1971), leaf temperature (Neilson & Jarvis 1975) and leaf water status (Stålfelt 1955). We shall restrict attention to these five variables although leaf temperature and leaf water status also depend on several additional environmental variables which influence the leaf energy balance and supply of water to the leaves.

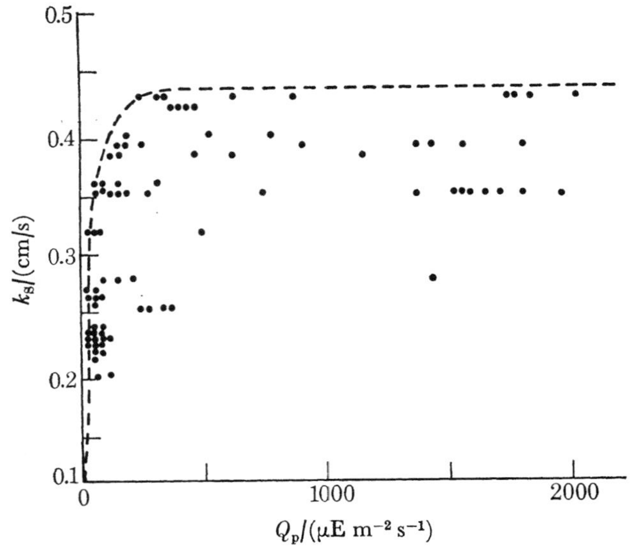

FIGURE 10. A scatter diagram of measurements of stomatal conductance (k_s) of current years shoots of Sitka spruce, in the upper part of the canopy at Fetteresso Forest, plotted against the photon flux density (Q_p) at the time of measurement (Spring, 1973). The dashed line indicates the probable upper limit of the observations. (Data from Watts, Neilson & Jarvis 1976.)

Leaf water potential has been taken as an indication of water status because values are readily available, although it is by no means the most relevant water status variable (see, for example, Turner 1974). Since k_s depends on these five major variables, field measurements of k_s do not usually show a clear relationship with any one of the variables. On many occasions k_s is reduced below the value expected, at a particular value of the independent variable, as the result of the influence of one or more of the other variables. This results in a scatter diagram when k_s is plotted as a function of any one independent variable (figures 10 and 11). Provided that enough measurements have been made adequately to cover the variable space, the upper limit of the scatter diagram indicates the response to the particular independent variable, when the other variables are not limiting. However, this objective is rarely achieved in the field and the upper limit is usually discontinuous and irregular. Consequently, the form of the functional relations between k_s and each driving variable is best determined in controlled environments. Once the form of the relations is known, the parameters in the relations can be determined from a suitable body of field data by using non-linear least squares (e.g. Reed *et al.* 1976). With these parameters, k_s can then be estimated from the weather on other occasions.

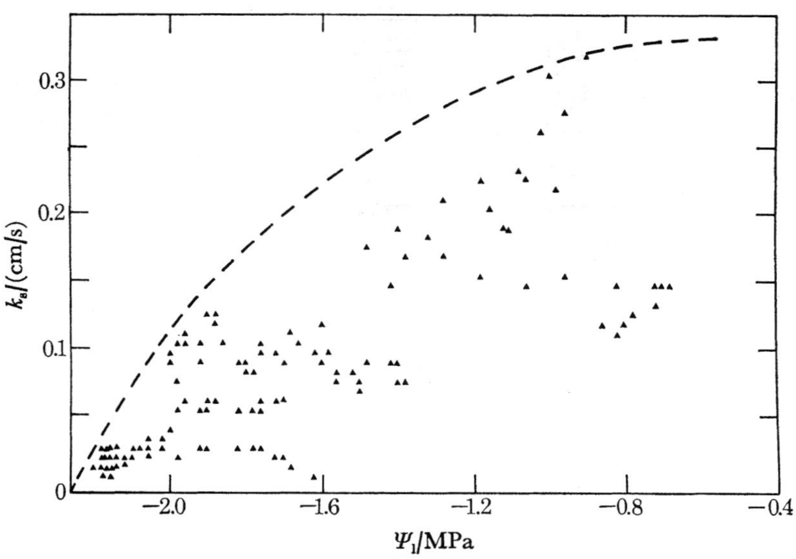

FIGURE 11. A scatter diagram of measurements of stomatal conductance (k_s) of current years shoots of Douglas fir, in the middle of the canopy at the Cedar River Forest site, plotted against the leaf water potential (Ψ_l) at the time of measurement (Autumn, 1973). The dashed line indicates the probable upper limit of the observations. (Data from Leverenz 1974.)

The functions

The observed relation between k_s and photon flux density, Q_p, is similar in form to the relation between CO_2 influx and Q_p and can be approximated by a similar hyperbola (e.g. Turner 1973), with the difference that k_s may have a finite positive value in the dark (figure 12):

$$k_s(Q_p) = b_1 b_2 (Q_p - q)/\{b_1 + b_2(Q_p - q)\}, \tag{4}$$

where $q = b_{10}/b_1$. b_1 is the asymtotic value of k_s at infinite Q_p and b_2 is dk_s/dQ_p at $Q_p = 0$. b_{10} is the value of k_s in the dark and is given by the intercept on the ordinate. b_{10} is introduced to allow for the stomata being open at night, and is not intended to be a cuticular conductance. We have assumed in the following equations that when the stomata are closed the leaf conductance is zero because field data are generally inadequate to define a cuticular conductance.

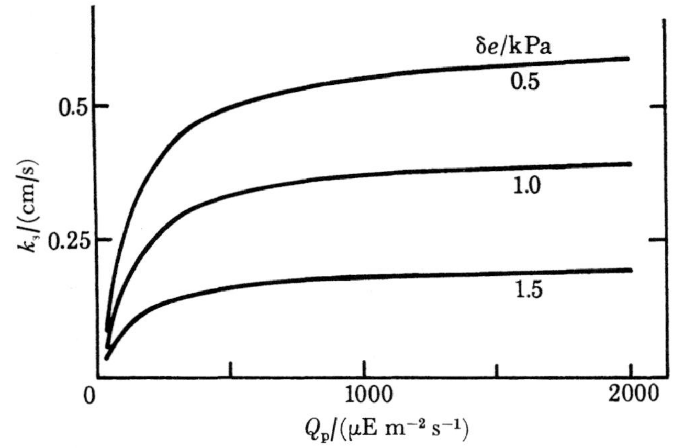

FIGURE 12. The relation between stomatal conductance (k_s) and photon flux density (Q_p), at three levels of vapour pressure difference (δe) as given by equations (4) and (9). Other conditions assumed: T_a = 10 °C; Ψ_l = −0.5 MPa; C_a = 300 cm^3/m^3.

The dependence of k_s on leaf temperature (Sharpe 1973, Neilson & Jarvis 1975) can be represented by the following function (figure 13):

$$k_s(T) = b_3(T-T_l)(T_h-T)^{b_4} \quad (0 \leqslant k_s(T) \leqslant 1), \tag{5}$$

where $$b_3 = 1/(T_0-T_l)(T_h-T_0)^{b_4}$$

and $$b_4 = (T_h-T_0)/(T_h-T_l).$$

$k_s(T)$ is a dimensionless conductance varying between a value of 1 at T_0 and zero at T_l and T_h, the low and high leaf temperatures respectively, at which the curve meets the abscissa.

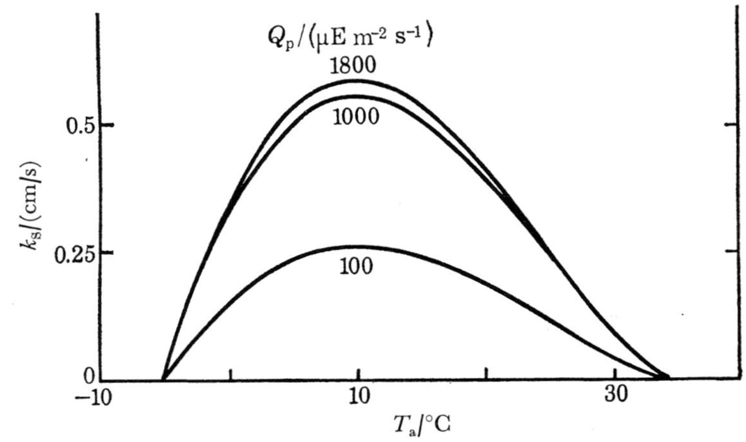

FIGURE 13. The relation between stomatal conductance (k_s) and air temperature (T_a) at three levels of photon flux density (Q_p) as given by equations (5) and (9). Other conditions assumed: $\delta e = 0.5$ kPa; $\Psi_l = 0.5$ MPa; $C_a = 300\ cm^3/m^3$.

A linear reduction in k_s with increasing vapour pressure difference is assumed (Lange, Lösch, Schulze & Kappen 1971, Schulze *et al.* 1973, Camacho, Hall & Kaufmann 1974, Watts, Neilson & Jarvis 1976, Neilson & Jarvis 1975):

$$k_s(\delta e) = 1 - b_5\delta e \quad (0 \leqslant k_s(\delta e) \leqslant 1), \tag{6}$$

where b_5 is the slope of the relation.

The observed responses of k_s to decreasing Ψ_l include linear proportionality, two sloping straight lines, a threshold region followed by a sloping linear decline, and a continuous curvilinear decline (see, for example, Kanemasu & Tanner 1969, Hansen 1971, Biscoe 1972, Beadle *et al.* 1973, Turner 1974). All of these responses can be more or less approximated by a negative exponential relation (figure 14):

$$k_s(\Psi_l) = 1 - e^{-b_6\delta\Psi} \quad (0 \leqslant k_s(\Psi_l) \leqslant 1), \tag{7}$$

where $\delta\Psi = \Psi_l - \Psi_m$ and Ψ_m is the value of Ψ_l at which the relation extrapolates to $k_s = 0$.

The response of k_s to ambient CO_2 concentration (C_a) (Meidner & Mansfield 1968) is assumed to consist of a threshold region between 0 to 100 cm^3/m^3 followed by a linear decline to a minimal value at 1000 cm^3/m^3 (figure 15):

$$k_s(C_a) = \begin{Bmatrix} 1 & C_a < 100 \\ 1 - b_7C_a & 100 < C_a < 1000 \\ b_8 & C_a > 1000 \end{Bmatrix} \quad (0 \leqslant k_s(C_a) \leqslant 1), \tag{8}$$

where $b_7 = (1-b_8)/900$ and b_8 is the minimal value of k_s at values of C_a larger than $1000\ cm^3/m^3$.

Interactions

We are uncertain about the extent of the interactions between the variables. We know for example, that k_s of illuminated leaves is reduced by high values of temperature, leaf–air vapour pressure difference, CO_2 concentration and water stress, acting independently. However, we do not know the extent to which k_s is reduced by these variables acting in concert.

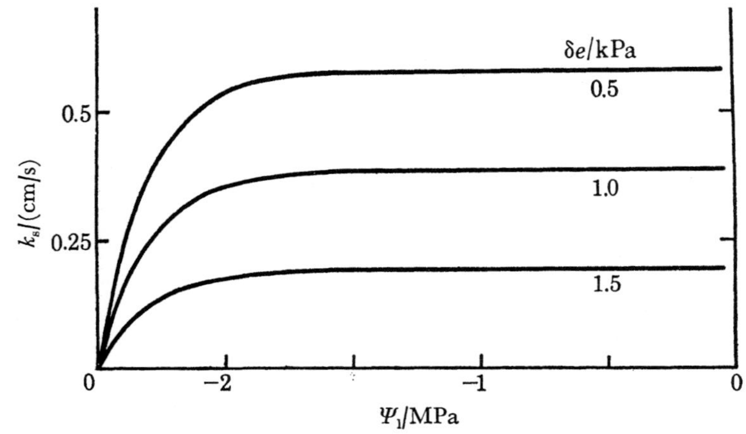

FIGURE 14. The relation between stomatal conductance (k_s) and leaf water potential (Ψ_l) at three levels of vapour pressure difference (δe) as given by equations (7) and (9). Other conditions assumed: $T_a = 10\ °C$; $Q_p = 1800\ \mu E\ m^{-2}\ s^{-1}$; $C_a = 300\ cm^3/m^3$.

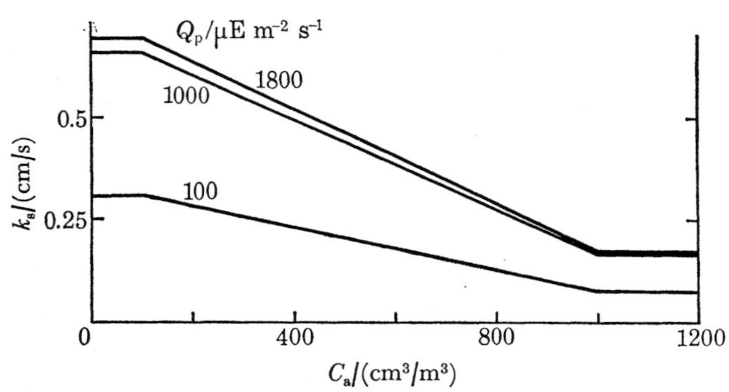

FIGURE 15. The relation between stomatal conductance (k_s) and ambient CO_2 concentration (C_a) at three levels of photon flux density (Q_p) as given by equations (8) and (9). Other conditions assumed: $\delta e = 0.5$ kPa; $\Psi_l = -0.5$ MPa; $T_a = 10\ °C$.

The simplest hypothesis, which we have adopted, is that the k_s obtained is the result of complete expression of the influence of all the variables without any synergistic interactions, namely:

$$k_s\,(Q_p, T, \delta e, \Psi_l, C_a) = k_s\,(Q_p)\,k_s\,(T)\,k_s\,(\delta e)\,k_s\,(\Psi_l)\,k_s\,(C_a). \tag{9}$$

That is to say, if k_s is reduced to 80% of maximum by the prevailing photon flux and to 80% of maximum by the prevailing temperature, the resulting k_s will be 64% of maximum k_s. Further experiments are needed to show whether this hypothesis is adequate.

Determination of parameters

The parameters in the model are b_1, b_2, T_l, T_0, T_h, b_5, b_6, Ψ_m and b_7. They are estimated by the simultaneous solution of equations (4) and (8) using nonlinear least squares regression analysis (Draper & Smith 1966). The technique used is identical with that used by Reed *et al.* (1976) to determine the parameters in a CO_2 uptake model and has the same limitations discussed by them. In particular, some bias in the estimated parameters results if the variables do not adequately fill the variable space as shown in figure 16. Such coverage is difficult to obtain in the field where photon flux, temperature and vapour pressure difference, for example, tend to be correlated. Consequently, sampling in the field and the selection of data for the analysis must be arranged deliberately to provide data which are uniformly distributed as far as is possible. If this is not possible, the shoot environment in the field must be artificially controlled to break the correlation between the weather variables and to provide combinations of infrequently occurring conditions in the 'corners' of the variable space.

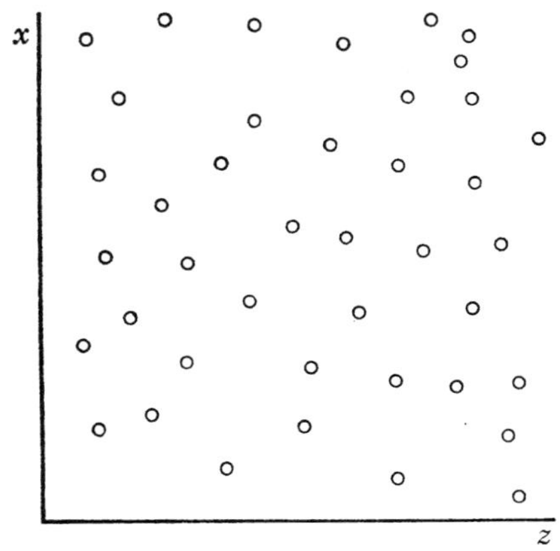

FIGURE 16. A diagram to show the ideal distribution of measurements of stomatal conductance in relation to two driving variables *x* and *z*. The measurements are more of less uniformly distributed throughout the variable space.

Application

The model has been used to interpret two sets of measurements of k_s which were collected independently on different species by different techniques.

Data set I: 80 measurements of k_s on current year shoots of Sitka spruce (*Picea sitchensis*) in the upper part of the canopy made with a null-balance, diffusion porometer (Beardsell, Davidson & Jarvis 1972) by W. R. Watts at Fetteresso Forest, U.K., between 25 April–17 May 1973. Concurrent measurements included the driving variables Q, T, δe and Ψ_l but not C_a which was taken as constant.

Data set II: 113 measurements of k_s on current year shoots of Douglas fir (*Pseudotsuga menziesii*) in the middle part of the canopy made with an assimilation chamber (Leverenz 1974) by J. Leverenz at the Cedar River Forest, Washington, U.S.A., between 8 August to 11 September, 1973. Concurrent measurements included all the driving variables.

The parameters determined are listed in table 1. There was insufficient range of variation in some of the variables to allow determination of the bracketed parameters which were fixed at the indicated values. The model accounted for 51 % of the variation in data set I and 73 % of the variation in data set II.

There are substantial differences in the derived parameters between data sets I and II. The Sitka spruce (data set I) had a higher maximum k_s (b_1), a lower temperature optimum for k_s (T_0), a steeper rise in k_s with increasing photon flux density (b_2), a steeper decline in k_s with increasing vapour pressure difference (b_5) and a more abrupt decline with decreasing leaf water potential (b_6). These differences in the parameters may result from differences between the species but more probably describe differences in the physiological condition of the trees in the spring and autumn at the two sites.

TABLE 1. VALUES OF THE PARAMETERS DERIVED FROM APPLICATION OF THE MODEL

quantity	symbol	data set I	data set II	unit
maximum k_s	b_1	0.43	0.30	$cm\ s^{-1}$
dk_s/dQ_p; $Q_p = 0$	b_2	0.0296	0.0135	$(cm\ s^{-1})/(\mu E\ m^{-2}\ s^{-1})$
k_s; $Q_p = 0$	b_{10}	0	0.01	$cm\ s^{-1}$
low temperature for $k_s = 0$	T_l	(−5.0)	(−5.0)	°C
temperature for max. k_s	T_0	9.0	20.0	°C
high temperature for $k_s = 0$	T_h	(35.0)	(45.0)	°C
$dk_s/d(\delta e)$	b_5	−0.26	−0.12	kPa^{-1}
$dk_s/d\Psi_l$; $k_s = 0$	b_6	−40	−0.93	MPa^{-1}
Ψ_l; $k_s = 0$	Ψ_m	(−2.4)	(−2.4)	MPa
dk_s/dC_a	b_7	(0)	(0)	$(cm^3/m^3)^{-1}$
	R^2	0.51	0.73	

TABLE 2. RANGE OF ENVIRONMENTAL VARIABLES IN DATA SETS I AND II

	data set I	data set II
photon flux density/$(\mu E\ m^{-2}\ s^{-1})$	15–1958	34–1831
temperature/°C	7.0–10.0	12.6–42.8
vapour pressure deficit/kPa	0.28–0.52	0.09–5.26
leaf water potential/MPa	0.71–1.62	0.72–2.20
ambient CO_2 concentration/(cm^3/m^3)	300	310–343

The rather low proportion of the variation accounted for in data set I might be attributed to inadequacies in the model. However, plots of the residuals (predicted k_s − measured k_s) with respect to the driving variables show that the positive and negative erros are, in general, equally distributed each side of zero (see figures 16 and 17, for example). Consequently, the low values of R^2 are more likely to reflect inadequacies in the distribution of the data than inadequacies in the model. Both sets of field results have pronounced limitations. The range of the variables T, δe and Ψ_l in data set I is very small (table 2). Whilst a broad range of a variable does not necessarily imply an ideal distribution for parameter estimation, a narrow range is unlikely to be satisfactory. The ranges of the variables are much wider in data set II but certain of the variables are very strongly correlated one with another: extreme values of Q, T, δe and Ψ_l all occur together. Consequently, k_s is small at even moderate levels of any one of the driving variables. For example, k_s does not exceed 0.1 cm s^{-1}, at photon flux densities above 400 $\mu E\ m^{-2}\ s^{-1}$, or at temperatures higher than 19 °C. Application of the CO_2 uptake model of Reed *et al.* (1975) to similar poorly distributed sets of data also results in low values of R^2, whereas values of R^2 in excess of 0.95 have been obtained with data more uniformly distributed through the variable space. A preliminary test of this model using steady state measurements of stomatal conductance made in the laboratory on forest shoots or potted plants (Ludlow & Jarvis 1971, Neilson *et al.* 1972, Cornic & Jarvis 1972, Neilson & Jarvis 1975), has shown that the model can account for 95% of the variation in the data. Consequently we are now developing the

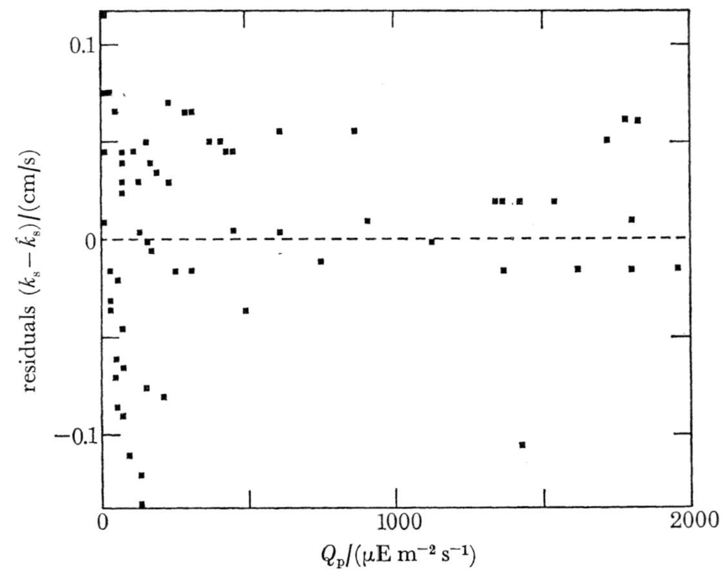

FIGURE 17. The distribution of the differences between stomatal conductance predicted from equations (4) to (9) ($\hat{k}_s$) and measured values of stomatal conductance (k_s) in relation to photon flux density (Q_p). Measurements made at Fetteresso Forest, Spring 1973, on current year's shoots of Sitka spruce in the upper part of the canopy. (Data from Watts, Neilson & Jarvis 1976.)

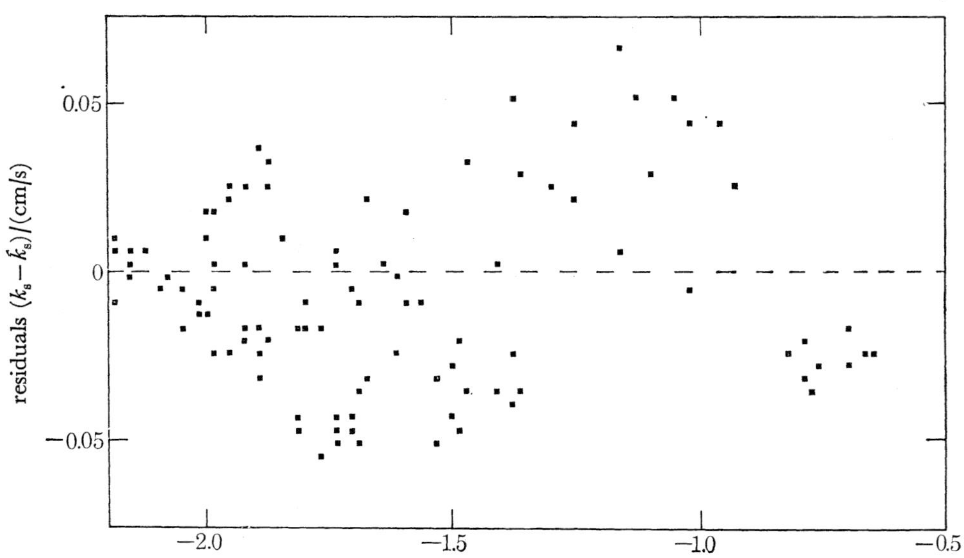

FIGURE 18. The distribution of the differences between stomatal conductance predicted from equations (4) to (9) ($\hat{k}_s$) and measured values of stomatal conductance (k_s) in relation to leaf water potential (Ψ_l). Measurements made at Cedar River Forest, Autumn 1973, on current year's shoots of Douglas fir in the middle of the canopy. (Data from Leverenz 1974.)

use of this kind of analysis to derive seasonal variation in stomatal properties from measurements of k_s made in the canopy at Fetteresso over the last three years.

Interpretation of the response of stomata to environmental variables in this way is practically useful, in that the parameters can be used to make predictions, but it is not wholly satisfactory. The parameters have limited physiological meaning because the model is descriptive rather than mechanistic. Nonetheless, the parameters provide a better description of the properties of the stomata than can be obtained from multiple regression coefficients. Furthermore, as the model evolves to become more mechanistic, it should become possible to interpret the parameters at the process level. We consider that approach outlined here to be a useful interim way of using measurements made in the field to describe the changing properties of the stomata.

I am very grateful for the generous provision of unpublished results from Dr W. R. Watts and Dr R. E. Neilson of this Department and Mr J. B. Stewart and Dr J. M. Roberts of the N.E.R.C. Institute of Hydrology. Their permission to use their results is greatly appreciated. Specific acknowledgement is made at the appropriate places in the text or figure legends. I am also very grateful to Dr K. L. Reed of the State of Washington's, Department of Natural Resources, Olympia, for his part in the development of the model of the influence of environmental variables on stomatal conductance and to Mr J. W. Leverenz for access to his primary data.

References (Jarvis)

Acock, B., Thornley, J. H. M. & Warren Wilson, J. 1971 Photosynthesis and energy conversion. In *Potential crop production* (ed. P. F. Waring & J. P. Cooper), pp. 43–75. London: Heinemann.

Beadle, C. L., Stevenson, K. R., Neumann, H. H., Thurtell, G. W. & King, H. M. 1973 Diffusive resistance, transpiration and photosynthesis in single leaves of corn and soybean in relation to leaf water potential. *Can. J. Pl. Sci.* **53**, 537–544.

Beardsell, M. F., Jarvis, P. G. & Davidson, B. 1972 A null balance porometer suitable for use with leaves of many shapes. *J. appl. Ecol.* **9**, 677–690.

Biscoe, P. V. 1972 The diffusion resistance and water status of leaves of *Beta vulgaris*. *J. exp. Bot.* **23**, 930–940.

Camacho, S. E., Hall, A. E. & Kaufmann, M. R. 1974 Efficiency and regulation of water transport in some woody and herbaceous species. *Pl. Physiol.* **54**, 169–172.

Carr, K. V. 1971/72 The internal water status of the tea plant (*Camellia sinensis*): some results illustrating the use of the pressure chamber technique. *Agric. Meteorol.* **9**, 447–460.

Cornic, G. & Jarvis, P. G. 1972 Effects of oxygen on CO_2 exchange and stomatal resistance in Sitka spruce and maize at low irradiances. *Photosynthetica* **6**, 225–239.

Draper, N. R. & Smith, H. 1966 *Applied regression analysis*. New York, London & Sydney: Wiley.

Elfving, D. C., Kaufmann, M. R. & Hall, A. E. 1972 Interpreting leaf water potential measurements with a model of the soil-plant-atmosphere continuum. *Physiologia Pl.* **27**, 161–168.

Gash, J. H. C. & Stewart, J. B. 1975 The average surface resistance of a pine forest derived from Bowen ratio measurements. *Boundary-layer Meteorol.* **8**, 453–464.

Goode, J. E. & Higgs, K. H. 1973 Water, osmotic and pressure potential relationships in apple leaves. *J. hort. Sci.* **48**, 203–215.

Gregory, S. L. & Petty, J. A. 1973 Valve action in bordered pits. *J. exp. Bot.* **24**, 763–767.

Grier, C. C. & Waring, R. H. 1975 Estimating Douglas fir and noble fir foliage mass from sapwood area. *For. Sci.* (In the Press.)

Hansen, G. K. 1971 Photosynthesis, transpiration and diffusion resistance in relation to water potential in leaves during water stress. *Acta Agric. Scand.* **21**, 163–170.

Hellkvist, J., Richards, G. P. & Jarvis, P. G. 1974 Vertical gradients of water potential and tissue water relations in Sitka spruce trees measured with the pressure chamber. *J. appl. Ecol.* **11**, 637–668.

Jackson, D. K. 1974 The course and magnitude of water stress in *Lolium perenne* and *Dactylis glomerata*. *J. agric. Sci. Camb.* **82**, 19–27.

Jarvis, P. G. 1975 Water transfer in plants. In *Heat and mass transfer in the environment of vegetation* (ed. D. A. de Vries), pp. 369–394. Washington D.C.: Scripta.

Kanemasu, E. T. & Tanner, C. B. 1969 Stomatal diffusion resistance of snap beans. I. Influence of leaf-water potential. *Pl. Physiol.* **44**, 1547–1552.

Klepper, B. 1968 Diurnal pattern of water potential in woody plants. *Pl. Physiol.* **43**, 1931–1934.

Landsberg, J. J., Beadle, C. L., Biscoe, P. V., Butler, D. R., Davidson, B., Incoll, L. D., James, G. B., Jarvis, P. G., Martin, P. J., Neilson, R. E., Powell, D. B. B., Slack, E. M., Thorpe, M. R., Turner, N. C., Warrit, B. & Watts, W. R. 1975 Diurnal energy, water and CO_2 exchanges in an apple (*Malus pumila*) orchard. *J. appl. Ecol.* **12**, 659–684.

Lange, O. L., Lösch, R., Schulze, E-D. & Kappen, L. K. 1971 Responses of stomata to changes in humidity. *Planta, Berl.* **100**, 76–86.

Lassen, L. E. & Okkonen, E. A. 1969 Sapwood thickness of Douglas fir and five other western softwoods. U.S.D.A. Forest Service Research Paper FPL 124, 1–16.

Leverenz, J. W. 1974 Net photosynthesis as related to shoot hierarchy in a large dominant Douglas fir tree. M.Sc. thesis, University Washington, Seattle.

Ludlow, M. M. & Jarvis, P. G. 1971 Photosynthesis in Sitka spruce (*Picea sitchensis* (Bong.) Carr.). I. General characteristics. *J. appl. Ecol.* **8**, 925–953.

Meidner, H. & Mansfield, T. A. 1968 *Physiology of stomata*. Maidenhead: McGraw Hill.

Millar, A. R., Jensen, R. E., Bauer, A. & Norum, E. B. 1971 Influence of atmospheric and soil environmental parameters on the diurnal fluctuations of leaf water status of barley. *Agric. Meteorol.* **8**, 92–105.

Monteith, J. L. 1965 Evaporation and environment. *Symp. Soc. exp. Biol.* **19**, 205–234.

Neilson, R. E. & Jarvis, P. G. 1975 Photosynthesis in Sitka spruce (*Picea sitchensis* (Bong.) Carr.). VI. Response of stomata to temperature. *J. appl. Ecol.* **12**, 879–891.

Neilson, R. E., Ludlow, M. M. & Jarvis, P. G. 1972 Photosynthesis in Sitka spruce (*Picea sitchensis* (Bong.) Carr.). II. Response to temperature. *J. appl. Ecol.* **9**, 721–745.

Reed, K. L., Hamerley, E. R., Dinger, B. E. & Jarvis, P. G. 1976 An analytical model for field measurements of photosynthesis. *J. appl. Ecol.* (In the press.)

Richards, G. P. 1973 Some aspects of the water relations of Sitka spruce. Ph.D. thesis, University of Aberdeen.

Richter, H. 1973 Frictional potential losses and total water potential in plants: a re-evaluation. *J. exp. Bot.* **24**, 983–994.

Schulze, E-D., Lange, O. L., Kappen, L., Buschbom, U. & Evenari, M. 1973 Stomatal responses to changes in temperature at increasing water stress. *Planta, Berl.* **110**, 24–42.

Sharpe, P. J. H. 1973 Adaxial and abaxial stomatal resistance of cotton in the field. *Agron. J.* **65**, 570–574.

Smart, R. E. & Barrs, H. D. 1973 The effect of environmental and irrigation interval on leaf water potential of four horticultural species. *Agric. Meteorol.* **12**, 333–346.

Stålfelt, M. G. 1955 The stomata as a hydrophotic regulator of the water deficit of the plant. *Physiologia Pl.* **8**, 572–593.

Turner, N. C. 1973 Illumination and stomatal resistance to transpiration in three field crops. *Ecology and conservation* **5**, 63–68.

Turner, N. C. 1974 Stomatal behavior and water status of maize, sorghum and tobacco under field conditions. II. At low soil water potential. *Pl. Physiol.* **53**, 360–365.

Watts, W. R., Neilson, R. E. & Jarvis, P. G. 1976 Photosynthesis in Sitka spruce (*Picea sitchensis* (Bong.) Carr.). VIII. Forest canopy measurements of stomatal conductance and $^{14}CO_2$ uptake. *J. appl. Ecol.* (In the press.)

A Model Predicting Stomatal Conductance and its Contribution to the Control of Photosynthesis Under Different Environmental Conditions.

J. Timothy Ball, Ian E. Woodrow, and Joseph A. Berry
Department of Plant Biology, Carnegie Institution of Washington,
290 Panama St., Stanford, CA 94305. (CIW-DPB) Pub. No. 953)

In the past, stomatal responses have generally been considered in relation to single environmental variables in part because the interactions between factors have appeared difficult to quantify in a simple way. A linear correlation between stomatal conductance (*g*) and CO_2 assimilation rate (*A*) has been reported when photon fluence was varied and when the photosynthetic capacity of leaves was altered by growth conditions, provided CO_2, air humidity and leaf temperature were constant (1). Temperature and humidity are, however, not consistent in nature. Lack of a concise description of stomatal responses to combinations of environmental factors has limited attempts to integrate these responses into quantitative models of leaf energy balance, photosynthesis, and transpiration. Moreover, this lack has hindered progress toward understanding the stomatal mechanism. We have taken a multi-variant approach to the study of stomatal conductance and we show that under many conditions the responses of stomata can be described by a set of linear relationships. This model can be linked to models of leaf carbon metabolism and the environment to predict fluxes of CO_2, H_2O and energy. In this paper, we show how the model of conductance can be linked to a description of CO_2 assimilation as a function of intercellular CO_2 (whether empirical or the output of a model) to predict the distribution of flux control between the stomata and leaf "biochemistry" under conditions in a gas-exchange cuvette.

Methods

CO_2 assimilation (*A*) and conductance (*g*) were measured as we varied conditions over the following wide range: photosynthetically active photon fluence (*I*) from 0.1 to 2.0 mmol m^{-2} s^{-1} the concentration (mole fraction) of CO_2 at the leaf surface (c_s) from 100 to 1000 μmol mol^{-1}, temperature (*T*) from 20 to 35°C, and the diffusion gradient for water vapor between the intercellular air spaces and the leaf surface (w_s) from 2 to 30 mmol mol^{-1}. We express gas concentrations external to the leaf in terms of mole fraction at the leaf surface (interior to the boundary layer) so that the results apply to any boundary layer conditions.

Results

A linear relationship between *A* and g when I was varied is apparent in our data but when other factors are varied the relationship of g to *A* is less straight forward (Fig. 1a). Several patterns are, however, present when CO_2 or humidity are varied while all other factors are held constant.

Stomatal conductance tends to decline with increasing CO_2 concentration while assimilation increases. It has been shown that the intercellular mole fraction CO_2 (c_i) changes linearly with ambient mole fraction CO_2 (1). When air humidity is decreased stomata tend to close while relatively little change in assimilation rate occurs. Humidity responses at different values of *I* through which *A* and *g* vary widely converge toward a single relationship where the ratio *A/g* is a linear function of humidity (2).

The data of Fig. 1a may be replotted showing stomatal conductance to water vapor (g_{sw}) varies directly with assimilation rate (*A*) scaled by relative humidity at the leaf surface (h_s), and divided by the mole fraction CO_2 at the leaf surface (c_s). This is a simple empirical model of stomatal conductance which can be expressed as

$$g_{sw} = kA\frac{h_s}{c_s} \qquad (1)$$

This model accounts for variations in stomatal conductance when, as in nature, several environmental factors are changing at once. The slope constant of the model (*k*) represents the

Biggens, J. (ed.), Progress in Photosynthesis Research, Vol. IV. ISBN 90 247 3453 3

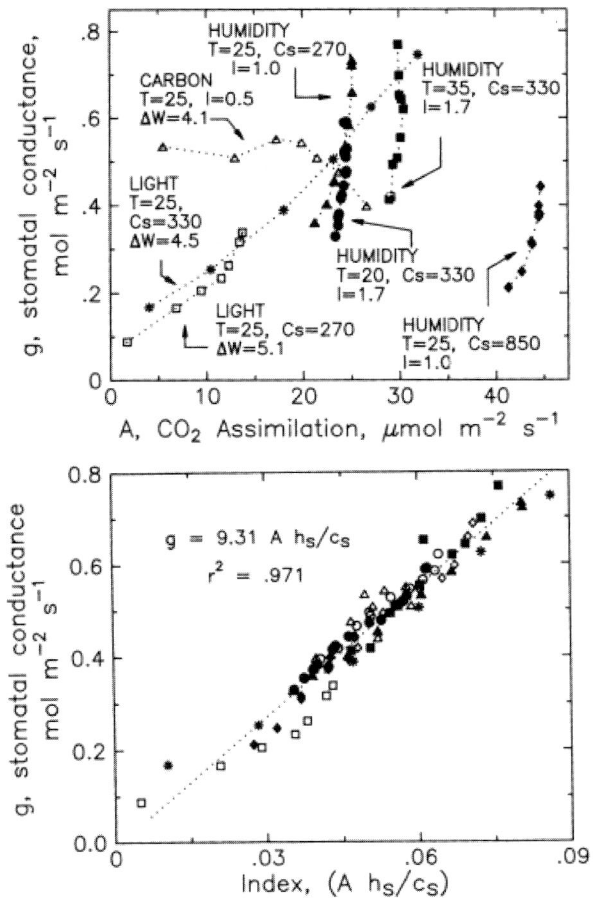

Figure 1. (A) Patterns in the relationship of g to A as humidity, light, and CO_2 concentration were changed under different environmental conditions as labeled in the figure.
(B) A simple empirical model of stomatal conductance in *Glycine max*. The linear relationship of conductance to a combined environmental-physiological index: the product, net CO_2 assimilation times relative humidity at the leaf surface divided by the CO_2 concentration at the leaf surface. The model (dotted line) is an unconstrained regression (intercept not significantly different from the original) of the data shown here. A = net CO_2 assimilation rate (μmol m^{-2} s^{-1}), h_s = decimal relative humidity (w_s/w_i, unitless) and c_s is mole fraction times 10^6 (μmol mol^{-1}). The data are those in Figure 1a (represented by the same symbols).

composite sensitivity of stomatal conductance to assimilation, CO_2 concentration, humidity and temperature. The intercept of the regression through the data of Fig. 1 is not significantly different from the origin. Regression lines through data from other species have different slopes and some have slightly positive y-intercepts (Ball and Berry unpublished).

Of particular interest is the interaction of temperature, humidity and the assimilation rate in controlling conductance. The stomatal response to humidity has been presumed to be a function of the driving force of transpiration (ΔW), the gradient of water vapor concentration between the leaf and the air (4). Varying humidity while the leaf is held at different temperatures shows g_{sw} is less sensitive to ΔW at higher temperatures (Fig. 2a). The change in sensitivity may relate to the large change in saturation mole fraction water vapor (w_{sat}) with temperature. Relative humidity at the leaf surface (h_s) scales w_s for changes in w_{sat}: (h_s=1 - w_s/w_{sat}). Scaled both by w_{sat} and by A, the response of stomatal conductance to humidity converges to a single relationship (Fig. 2b). From this we conclude that stomata respond to relative humidity. The response of g_{sw} to temperature include responses to relative humidity and to assimilation rate.

There are a number of ways that this model of stomatal behavior can be used including simulation of leaf responses in nature. We describe below a less obvious use of the model, namely how it can be used to estimate stomatal "limitation" of photosynthesis.

<u>Discussion</u>: Application of the empirical model to the calculation of the control of CO_2 flux.

A number of expressions for the stomatal limitation of photosynthesis have been suggested (4)

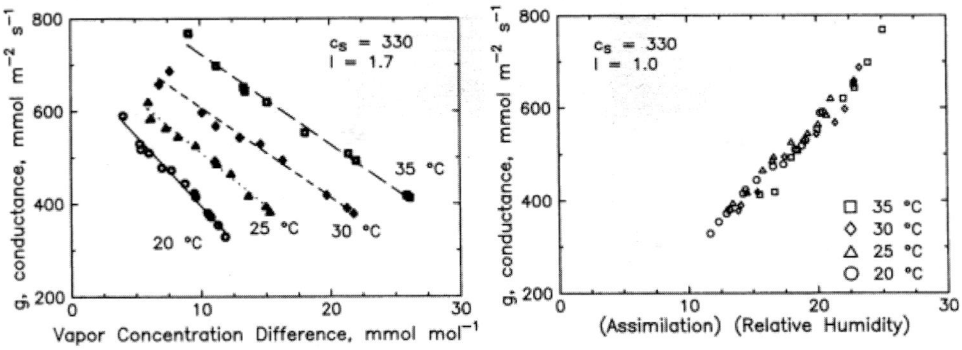

Figure 2. The interacting effects of temperature and humidity on stomatal conductance. *I* was 1.7 mmol m^{-2} s^{-1} and c_s was 330 μmol m^{-2} s^{-1}.
(A) Stomatal conductance as a function of vapor concentration difference at temperatures ranging from 20 to 35°C.
(B) Stomatal conductance as a function of net CO_2 assimilation rate times relative humidity of the sub-boundary-layer air (w_s/w_i) at 20, 25, 30, and 35 degrees (°)

but these generally fail to adequately consider non-linearities of the photosynthetic system and/or would attribute possible feedback from non-stomatal components of the system (e.g. leaf biochemistry) to stomata. The analysis of metabolic control outlined by Kacser and Burns (5) is ideally suited to quantitative descriptions of the control of fluxes in complex systems.

Control analysis defines the control coefficient of system component *P* with respect to flux *V* as

$$C_P^V = \frac{\partial V \cdot P}{\partial P \cdot V} = \frac{\partial \ln V}{\partial \ln P} \qquad (2)$$

In the present case we consider the flux of CO_2, *A*, through a two system: the stomatal (*s*) and the "biochemistry" (*c*). Since the sum of all control coefficients in a system is unity the control coefficient for one component can be thought of as roughly the proportion of total "limitation" at that step (5). The "biochemistry" component here is comprised of all aspects of carbon metabolism downstream of c_i.

The expression for the control of stomata with respect to CO_2 assimilation, derived in (6) becomes

$$C_S^A = [1 - \frac{(c_s - c_i)g' - g_{sc} - \frac{E}{2}}{A'}] \qquad (3)$$

(where g_{sc} is stomatal conductance to CO_2, *E* is transpiration rate, *g* and *A'* are the slopes of g_{sc} and *A* as function of c_i, respectively) when leaf temperature is held constant, boundary layer conductance is large and known, and w_s and c_s calculable, --typical conditions for many gas exchange experiments.

If one ignores the influence of the efflux of water from a leaf on the diffusion of CO_2 into the leaf using the approximation $A = g_{sc}\,[c_s\text{-}c_i]$ and substitutes for *A* in Eq. 1 the empirical model of conductance can be rearranged to

$$\frac{c_i}{c_s} = 1 - \frac{1.6}{h_s k} \qquad (4)$$

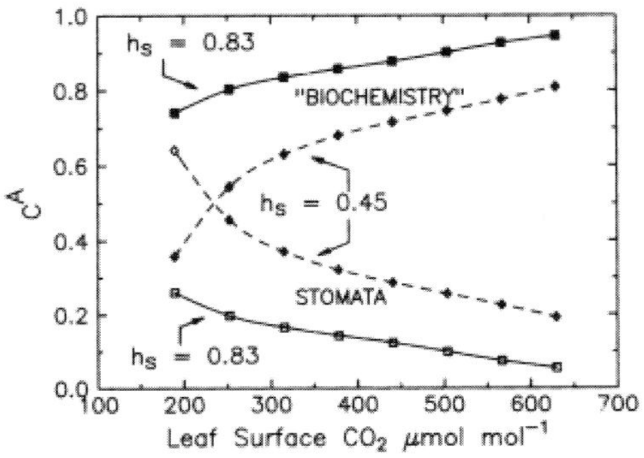

Figure 3. The distribution of control of CO_2 assimilation between the "biochemistry" and the stomata as a function of the mole fraction CO_2 at the leaf surface. The relative humidities of the air at the leaf-surface was 0.83 (solid lines) and 0.45 (dashed lines) at leaf temperature of 25°C.

Obtaining an expression for *g'* from the empirical model and substituting into Equation 3

$$C_S^A = A'\left[\frac{kAh_s}{c_s(h_s k - 1.6)} + \frac{kAh_s}{1.6c_s} + \frac{E}{2}\right]^{-1} \qquad (5)$$

This equation is parameterized with the slope of an empirical relationship as in Fig. 1b, measured or modeled values of *A* and *A'* and external conditions h_s and c_s. A more complete analysis of stomatal control, applicable in nature, that considers control of evaporation, convective heat transfer, radiative exchange, as well as CO_2 fixation and that includes control due to the boundary layer as well as stomata and "biochemistry" is available (6).

The distribution of flux control with changes in mole fraction CO_2 at the leaf surface at two humidities (Fig. 3) illustrates the interaction of two environmental variables in the "limitation" of photosynthesis. At low external CO_2 concentrations diffusion of CO_2 through stomata constitutes a large portion of the control but as CO_2 concentration increases the "biochemistry" tends toward saturation clearly dominating the rate of CO_2 fixation. At low humidity stomata can constitute the major control of assimilation and can retain significant control to higher CO_2 concentrations. This rise in stomatal control of *A* is accompanied by a reduction in *A* (Fig. 1) and reflects a "cost" in terms of carbon gain attached to the "benefit" of reducing water loss.

Conclusion

The empirical model of stomatal conductance presented here provides new fundamental insight into the nature and interaction of stomatal responses particularly with regard to humidity. We have suggested several ways that the model can be used including the calculation of the control coefficients of stomata and biochemistry with regard to assimilation. This approach to quantifying "limitation" has several advantages including: expandability to include additional system components (e.g. boundary-layer, or more specific aspects of the "biochemistry") and ability to illustrate interactions between and the distribution of control among all components of an integrated system (6).

Acknowledgments

IEW is supported by a Harkness Fellowship from the Commonwealth Fund (New York) and CSIRO (Australia) Post-doctoral Award. The Authors would like to thank the Commonwealth Fund and the McKnight Foundation for financial support and Dr. J.A.G. Peet for assistance.

References

1. Wong, S.C., Cowan, I.R. and Farquhar, G.D. (1978) Plant Physiol. 62, 670-74.
2. Ball, J.T., and Berry, J.A. (1982) Carnegie Inst. Washington Yrbk. 81, 88-92.
3. Lösch, R. (1979) in, Structure, Function, and Ecology of Stomata (Sen. D.N., Chawan, D., Bansal, R. eds.) pp. 189-216.
4. Jones, H.G. (1985) Plant, Cell, and Environ. 8, 95-104.
5. Kacser, H., and Burns, J.A. (1973) Symp. Soc. Exp. Biol. 27, 65-104.
6. Woodrow, I.E., Ball, J.T., and Berry, J.A. (1986) Proceed. VII Int'l. Photosynthesis Congress, in press.

Quart. J. R. Met. Soc. (1985), **111**, pp. 839–855 551.573:551.584.4:556.132.6

Evaporation from sparse crops–an energy combination theory

By W. JAMES SHUTTLEWORTH and J. S. WALLACE
Institute of Hydrology, Wallingford, Oxon.

(Received 25 July 1984; revised 24 January 1985)

SUMMARY

A one-dimensional model is adopted to describe the energy partition of sparse crops. Theoretical development of this model yields a combination equation which describes evaporation in terms of controlling resistances associated with the plants, and with the soil or water in which they are growing. The equation provides a simple but physically plausible description of the transition between bare substrate and a closed canopy. Although the aerodynamic transfer resistances for incomplete canopies have, as yet, no experimental justification, typical values, appropriate to a specimen agricultural crop and soil, are shown to have limited sensitivity in the model. Processes which require further study if the equation is to be used to calculate evaporation throughout a crop season are also discussed.

1. INTRODUCTION

Previous steps in the development of a physically based model of the vegetation–atmosphere interaction (e.g. Shuttleworth 1976, 1978) explicitly treat the vegetation as a closed, stable canopy of uniform structure. They emphasize the interaction of the vegetation, with fluxes arising at the soil surface introduced as an unspecified, and implicitly small, input to the model (Shuttleworth 1979). In this paper this theoretical work is reinterpreted and developed into the situation of sparse crops, where the use of a one-dimensional model has less obvious justification. In describing such crops the soil and plant components must carry equal status, since they can be of similar size and their relative importance can change significantly with crop cover.

The philosophy of this paper is to make minimum concession to the more obvious three-dimensional structure of sparse and row crops. Accordingly a one-dimensional model of the interaction is adopted to derive a combination equation, which can provide a physically plausible transition between the bare substrate and closed canopy limits. The equation is expressed in terms of conceptual resistances now familiar to the micro-meteorologist and plant physiologist: canopy resistance and boundary layer resistance etc; it also requires the less familiar concept of a surface resistance for bare soil (Monteith 1981). In the later sections of the paper typical values of these resistances are used to illustrate how energy partition varies between crops of the same height, but with different leaf areas.

2. THE ONE-DIMENSIONAL MODEL

The problem of sparse crops is approached using the knowledge that the two asymptotic limits (bare substrate and a closed canopy) can both be represented by a one-dimensional model. Moreover it is acknowledged that, in describing the closed canopy situation, models which represent the canopy interaction as occurring as a single source (Monteith 1965) have increasing acceptance. Such models represent a practical compromise between physical rigour and field application. The basic assumption, that there is numerical similarity between bulk stomatal resistance and an integration of component stomatal resistances in dry conditions (Monteith 1965), has been tested experimentally (Black *et al.* 1970; Szeicz *et al.* 1973; Tan and Black 1976), numerically (Sinclair *et al.* 1971) and explored theoretically (Shuttleworth 1976).

Implicit in this successful assumption is the idea that the real three-dimensional nature of a crop can be ignored in terms of its practical consequences. It is also assumed that aerodynamic mixing within the crop is sufficiently good to allow the hypothetical existence of a 'mean canopy airstream' (Thom 1972) which can be described by meteorological parameters such as temperature, humidity and wind speed. These assumptions are necessary if progress is to be made towards providing a description which is general (in the sense that it is not merely a description of a particular crop at a particular time). Previous models based on these assumptions have been proposed and tested (Black *et al.* 1970; Szeicz *et al.* 1973) even in closed canopies, where the aerodynamic interaction within the canopy is minimized.

The degree of aerodynamic mixing in sparse and row-structured crops is likely to be greater than that in closed canopies. Adopting the Monteith assumption for such crops is therefore arguably more plausible in terms of mixing in the vertical. It does, however, require a reconsideration of the scale relevant to the horizontal averaging process, and will involve additional uncertainty regarding the consequences of persistent features within the aerodynamic mixing pattern in the horizontal plane (see, for example, Arkin and Perrier 1974).

Any one-dimensional description assumes horizontal uniformity, but in practice recognizable three-dimensional features of the crop (individual plants and rows, and persistent aerodynamic mixing features) are always present. If a description of the Monteith type is to be used it is necessary that the elements of which the model is composed (e.g. energy fluxes, stomatal resistance, etc.) are defined as horizontal averages over area scales in which persistent features occur in sufficient numbers to allow such averaging. A one-dimensional description is clearly not relevant to horizontal scales less than this.

In this study a simple two-component structure is maintained and energy partition treated as occurring at 'the crop' and 'the soil'. This simplification appears particularly arbitrary in the case of row crops since it involves, for instance, no distinction between soil beneath the vegetation and that between the rows. The presence of a defined three-dimensional structure clearly cannot be totally ignored in certain aspects of the interaction. For example, in the case of interception of solar radiation, row orientation may affect the effective absorption coefficient relevant at the scale of the horizontal

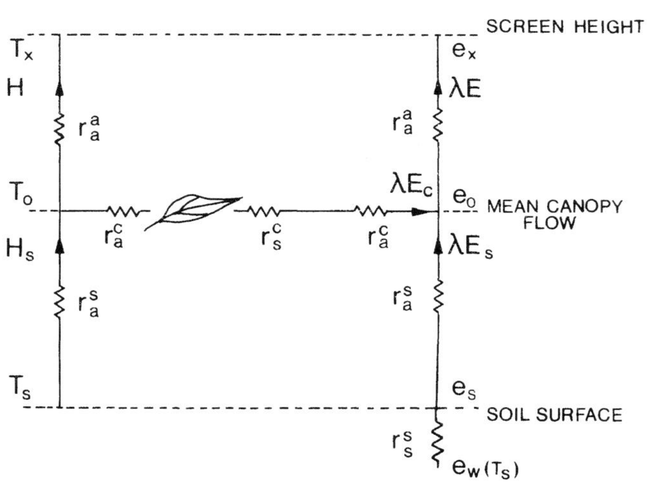

Figure 1. Schematic diagram of a one-dimensional description of energy partition for sparse crops. The nomenclature used is given in section 3(*a*).

averaging. The diurnal cycle in solar altitude reduces the distinction between the daily total absorption by such organized structures and that of randomly distributed sparse vegetation. Nonetheless, the realistic philosophy in dealing with row crops is to treat separately such aspects of the problem, but then to return to an averaging scale over which a one-dimensional model is assumed to apply.

The model adopted and developed in this paper is illustrated in Fig. 1. It incorporates not only the now familiar concept of a bulk stomatal resistance for the vegetation, r_s^c, but also the less familiar concept of a surface resistance at the substrate surface, r_s^s. The physical origin of this surface resistance is less obvious for soil than it is for vegetation but its *mathematical* definition is precise. Its presence reflects the fact that the layer of air adjacent to the soil surface is not necessarily saturated unless the soil surface is wet. If soil evaporation is λE_s, and the temperature and vapour pressure at the soil surface T_s and e_s respectively, then the surface resistance of the soil is defined by the equation

$$r_s^s = (\rho c_p/\gamma)\{e_w(T_s) - e_s\}/\lambda E_s \tag{1}$$

where $e_w(T_s)$ is the saturated vapour pressure at temperature T_s and the other quantities are defined in section 3(a). Monteith (1981) interprets this resistance in conceptual terms by describing evaporation from a drying soil as occurring from wet soil below a dry soil layer of increasing thickness, treated as isothermal. This provides a description qualitatively consistent with observation. Although this obvious over-simplification leads to problems in physical interpretation (Fuchs and Tanner 1967), Eq. (1) defines an entity which can form the subject of empirical models. Such modelling is not discussed further in the present paper.

The model described in Fig. 1 adopts the concept of a bulk boundary layer resistance, r_a^c, which controls transfer between the surface of the vegetation and the canopy air stream (Thom 1972). Vertical transport is controlled by two further aerodynamic resistances. The first, r_a^a, is the transfer resistance between the hypothetical mean canopy flow and the reference height, x, above the crop. The second, r_a^s, is the aerodynamic resistance encountered by the energy fluxes leaving the substrate before they are incorporated into the mean canopy flow. For simplicity in this analysis it is assumed that the various aerodynamic resistances are identical for sensible and latent heat. Making this assumption simplifies the formalism but does not alter the derivation in any fundamental way.

It is worth remembering at this point that the so called 'aerodynamic resistance' generally employed when describing energy partition with a combination equation (Monteith 1965) is obtained by the addition of the component resistances described in the previous paragraph. In this way the aerodynamic resistance used in a combination equation describing a closed canopy (with no soil evaporation) is $r_a^a + r_a^c$, while that which would be used to describe evaporation from the substrate is $r_a^a + r_a^s$. The relevant 'surface' resistances in these two situations are r_s^c and r_s^s. The objective in the following sections is to derive a combination equation, descriptive of both plant and substrate evaporation, but which asymptotes towards simpler combination equations involving the aerodynamic and surface resistances relevant in limiting situations.

3. THEORETICAL DEVELOPMENT

(a) *Nomenclature*

A, A_s	Total energy flux leaving the complete crop, the substrate, as sensible and latent heat per unit ground area ($\mathrm{W\,m^{-2}}$)
C	Extinction coefficient of the crop for net radiation (dimensionless)

c_p	Specific heat at constant pressure ($J\,kg^{-1}K^{-1}$)
d	Zero plane displacement of crop with complete canopy cover ($L = 4$) (m)
D	Vapour pressure deficit at reference height, $e_w(T_x) - e_x$ (mb)
D_0	Vapour pressure deficit at canopy source height, $e_w(T_0) - e_0$ (mb)
e_0, e_s	Vapour pressure at canopy source height, soil surface, (mb)
e_x	Vapour pressure at reference height (mb)
$e_w(T)$	Saturated vapour pressure at temperature T ($T = T_x$, T_0, T_s) (mb)
G	Soil heat flux ($W\,m^{-2}$)
h	Crop height (m)
H, H_s	Sensible heat flux from the complete crop, substrate, ($W\,m^{-2}$)
k	von Kármán's constant (dimensionless)
K	Eddy diffusion coefficient (m^2s^{-1})
K_h	Eddy diffusion coefficient at top of canopy (m^2s^{-1})
L	Projected area of leaf per unit ground area (leaf area index) (dimensionless)
n	Eddy diffusivity decay constant in a crop with complete canopy cover ($L = 4$) (dimensionless)
P	Biochemical storage of energy in the crop below reference height ($W\,m^{-2}$)
r_a^a	Aerodynamic resistance between canopy source height and reference level ($s\,m^{-1}$)
r_a^c	Bulk boundary layer resistance of the vegetative elements in the canopy ($s\,m^{-1}$)
r_a^s	Aerodynamic resistance between the substrate and canopy source height ($s\,m^{-1}$)
r_b	Mean boundary layer resistance per unit area of vegetation ($s\,m^{-1}$)
r_s^c	Bulk stomatal resistance of the canopy ($s\,m^{-1}$)
r_s^s	Surface resistance of the substrate ($s\,m^{-1}$)
r_{ST}	Mean stomatal resistance ($s\,m^{-1}$)
$r_a^a(0)$	Value of r_a^a for bare substrate ($s\,m^{-1}$)
$r_a^a(\alpha)$	Value of r_a^a for crop with complete canopy cover ($L = 4$) ($s\,m^{-1}$)
$r_a^s(0)$	Value of r_a^s for bare substrate ($s\,m^{-1}$)
$r_a^s(\alpha)$	Value of r_a^s for crop with complete canopy cover ($L = 4$) ($s\,m^{-1}$)
R_n	Net radiation flux into the complete crop ($W\,m^{-2}$)
R_n^s	Net radiation flux into the substrate ($W\,m^{-2}$)
S	Physical storage of energy in the atmosphere and crop below reference height ($W\,m^{-2}$)
T_0	Air temperature at canopy source height (°C)
T_s	Temperature of the substrate surface (°C)
T_x	Air temperature at reference height (°C)
u	Wind speed at the reference height ($m\,s^{-1}$)
u_*	Friction velocity ($m\,s^{-1}$)
x	Reference height above the crop where meteorological measurements are available (2 m in this analysis)
z	Height (variable) (m)
z_0	Roughness length of crop with complete canopy cover ($L = 4$) (m)
z_0'	Roughness length of the bare substrate (m)
Δ	Mean rate of change of saturated vapour pressure with temperature, $\{e_w(T_x) - e_w(T_0)\}/(T_x - T_0)$ ($mb\,K^{-1}$)
γ	Psychrometric 'constant' ($mb\,K^{-1}$)
λE	Latent heat flux from the complete crop ($W\,m^{-2}$)
λE_c	Latent heat flux from the plant canopy ($W\,m^{-2}$)

λE_s Latent heat flux from the substrate ($\mathrm{W\,m^{-2}}$)
ρ Density of air ($\mathrm{kg\,m^{-3}}$)

(b) *The energy budget*

All combination equations rely on drawing up an energy budget between the outward fluxes of sensible and latent heat and the energy available in other forms. In this problem two budgets are drawn up, one at the substrate surface and one for the complete crop. The sum of the above-canopy fluxes of sensible heat, H, and latent heat, λE, is the available energy, A, and is given by

$$A = \lambda E + H \tag{2}$$

$$= R_n - S - P - G \tag{3}$$

where R_n is the incoming net radiation, S and P are the physical and biochemical energy storage terms, and G is the heat conduction into the substrate. In a similar way, the energy available at the substrate, A_s, is given by

$$A_s = \lambda E_s + H_s \tag{4}$$

$$= R_n^s - G \tag{5}$$

where R_n^s is the net radiation at the substrate surface.

In drawing up such energy budgets it is clearly necessary to consider average values of the several components defined over horizontal scales which involve significant numbers of the identifiable crop and soil features. In general R_n^s is less than R_n and A_s less than A. In the limit of bare substrate A and A_s are equal.

(c) *In-canopy deficit*

By analogy with Ohm's law for the electrical analogue shown in Fig. 1, the difference in vapour pressure and temperature between the level of mean canopy flow and reference height can be written in terms of resistance and flux as

$$e_x - e_0 = -\lambda E\, r_a^a \gamma / \rho c_p \tag{6}$$

and

$$T_x - T_0 = -H r_a^a / \rho c_p . \tag{7}$$

Introducing the definition of Δ into the expression for the vapour pressure deficit at the canopy source height, D_0, gives

$$D_0 = e_w(T_x) - \{e_w(T_x) - e_w(T_0)\} - e_0$$

and substituting Eqs. (7), (6) and (2), yields a relationship between D_0 and D such that

$$D_0 = D + \{\Delta A - (\Delta + \gamma)\lambda E\} r_a^a / \rho c_p . \tag{8}$$

(d) *The sparse crop combination equation*

In the model illustrated (Fig. 1), the evaporation from the substrate, λE_s, and from plants in the canopy, λE_c, can be separately calculated from equations of the Penman–Monteith type, thus

$$\lambda E_s = (\Delta A_s + \rho c_p D_0 / r_a^s)\{\Delta + \gamma(1 + r_s^s / r_a^s)\}^{-1} \tag{9}$$

$$\lambda E_c = \{\Delta(A - A_s) + \rho c_p D_0 / r_a^c\}\{\Delta + \gamma(1 + r_s^c / r_a^c)\}^{-1} . \tag{10}$$

The total evaporation from the crop, λE, is the sum of these two, and it can be shown (see appendix) that D_0 can be eliminated and the resultant equation arranged in the form

$$\lambda E = C_c PM_c + C_s PM_s \tag{11}$$

where PM_c and PM_s are terms each similar to the Penman–Monteith combination equations which would apply to evaporation from a closed canopy and from bare substrate respectively. They have the form

$$PM_c = \frac{\Delta A + \{\rho c_p D - \Delta r_a^c A_s\}/(r_a^a + r_a^c)}{\Delta + \gamma\{1 + r_s^c/(r_a^a + r_a^c)\}} \tag{12}$$

$$PM_s = \frac{\Delta A + \{\rho c_p D - \Delta r_a^s (A - A_s)\}/(r_a^a + r_a^s)}{\Delta + \gamma\{1 + r_s^s/(r_a^a + r_a^s)\}}. \tag{13}$$

The coefficients C_c and C_s are given by the expressions

$$C_c = \{1 + R_c R_a / R_s (R_c + R_a)\}^{-1} \tag{14}$$

and

$$C_s = \{1 + R_s R_a / R_c (R_s + R_a)\}^{-1} \tag{15}$$

where

$$R_a = (\Delta + \gamma) r_a^a \tag{16}$$

$$R_s = (\Delta + \gamma) r_a^s + \gamma r_s^s \tag{17}$$

$$R_c = (\Delta + \gamma) r_a^c + \gamma r_s^c. \tag{18}$$

It can be easily seen that Eq. (11) has correctly defined asymptotic limits. If there is no substrate evaporation, r_s^s and hence R_s are infinite, PM_s is zero in Eq. (11), and C_c is unity. If, in addition, there is no sensible heat flux from the substrate, H_s and hence A_s are zero, and Eq. (11) reduces to the conventional Penman–Monteith equation describing closed canopy evaporation with no substrate interaction. In a similar way, if there is no canopy present r_s^c and hence R_c are infinite, and $A = A_s$. Equation (11) reduces to a conventional form, describing substrate evaporation with the Penman–Monteith equation involving a surface resistance applicable to the substrate.

In the more general intermediate situation, when both substrate and canopy evaporation occur, Eq. (11) provides a physically plausible description of the total evaporation λE. Once calculated, this can be substituted into Eq. (8) to compute D_0, and in this way the component fluxes λE_c and λE_s calculated from Eqs. (9) and (10) if required.

(e) *The leaf area dependence of surface resistance*

The assumptions and observations made in Shuttleworth (1979) and associated papers (Shuttleworth 1976; 1978) are implicit in Eq. (11). In the context of the present problem a point of particular relevance is the fact that the mean boundary layer resistance of the canopy, r_a^c, and the bulk stomatal resistance of the canopy, r_s^c, are both 'surface' resistances, influenced by the surface area of the vegetation present. They vary inversely with the total leaf area of the vegetative elements present (Shuttleworth 1976). This is important when considering the effect of changes in leaf area index on energy partition and it is convenient in this case to rewrite these two surface resistances in the form

$$r_s^c = r_{ST}/2L \tag{19}$$

$$r_a^c = r_b/2L \tag{20}$$

where r_{ST} is the mean stomatal resistance (of amphistomatous leaves) and r_b the mean boundary layer resistance, both expressed per unit surface area of vegetation. L is the leaf area index of the canopy, i.e. the projected area of the vegetation per unit ground area. In this way these two resistances are high in very sparse canopies and, all other things being equal, decrease inversely as the area of vegetation per unit ground area increases. In practice this is just a first approximation since r_b may itself be influenced by changes in mean canopy wind speed, and r_{ST} may change in response to differences in shading.

The resistance r_s^s is also a surface resistance and should therefore be divided by the area of exposed substrate per unit ground area. In practice the area occupied by plant stems is likely to be a small fraction of ground area. Any litter present on the ground can be regarded as part of the substrate.

4. Vegetation density variations

In this section the response of Eq. (11) is explored when applied to crops of the same height but with different leaf areas, which are subject to a specified atmospheric demand. For the purposes of comparison, calculations are carried out assuming *measured* meteorological variables are available above the crops at a height, x, of 2 m. In this way it is assumed that any density-related interaction between the crop and the meteorological variables, e.g. albedo changes, can be treated separately (we return to this point later). Calculations are made for soil and water substrates, using a specification of the component resistances drawn from the literature, and taken as applying to a 'typical' agricultural crop.

(*a*) *Model specification*

(i) *Available energy*. The physical and biochemical storage terms, S and P, are ignored in Eq. (3). Since net radiation during daylight hours is primarily determined by direct radiation, it has been found experimentally (e.g. Ross 1981) that the radiation reaching the soil surface, R_n^s, can be calculated using a Beer's law relationship of the form

$$R_n^s = R_n \exp(-CL) \tag{21}$$

where C is the extinction coefficient of the crop for net radiation, chosen arbitrarily as 0·7 (see for example Monteith 1973). It is convenient here to ignore variations in C which may occur in response to structural differences in crops of different density, although this could (if known) be included in specific cases. In the present calculation the heat conduction into the substrate, G, is arbitrarily set to 20% of the radiation received at the substrate surface, R_n^s, and therefore G also changes with crop density.

(ii) *Mean stomatal resistance*. The mean stomatal resistance, r_{ST}, is taken as 400 s m^{-1}. It follows from Eq. (19) that, for a leaf area index, L, of 4, the bulk stomatal resistance, r_s^c, is 50 s m^{-1}. This value is typical of a fully grown agricultural crop (see Wallace *et al.* 1981).

(iii) *Mean boundary layer resistance*. Measurements of mean boundary layer resistance, r_b, generally have significant scatter and exhibit some dependence on in-canopy wind speed. The value of r_b for stands of vegetation of different density is therefore uncertain. Typical values measured in the field are in the order 25 s m^{-1} (see Denmead 1976; Uchijima 1976) and this value is assumed here. The corresponding bulk boundary layer resistance is 3 s m^{-1} for a leaf area index of $L = 4$. In practice this resistance is only significant in Eq. (11) when acting in combination with the much larger bulk stomatal resistance; errors in its assumed value are shown later to be of limited numerical importance.

(iv) *Surface resistance of the substrate*. The description of substrate evaporation in terms of a surface resistance is somewhat novel in this paper, and, in consequence, typical values are difficult to specify. Calculations are performed for three values of r_s^s: 0, 500 and 2000 s m^{-1}. The first value, 0, corresponds to a substrate of wet soil or free water, whilst the last, 2000, is arguably typical of fairly dry soil—theoretically, it corresponds to molecular diffusion through a 1·5 cm thick layer of dry sandy soil (Fuchs and Tanner 1967). The third value, 500, is chosen merely as an intermediate value; on a unit area basis it is in the same order as that of dry vegetation.

(v) *Eddy diffusion resistance.* Clearly crop density affects the size of the aerodynamic resistances r_a^a and r_a^s, but the quantitative response of within-canopy aerodynamic transfer to differing leaf area index is perhaps the least understood aspect of micrometeorology, and likely to remain so in the foreseeable future. In these circumstances we have assumed the simplest possible model in which r_a^a and r_a^s vary linearly with leaf area index between the values associated with their two limits, namely bare substrate and a complete canopy cover (which we arbitrarily assume corresponds to $L = 4$). In fact, as we demonstrate later, the quantitative effect of this simple treatment has a limited effect on component and total evaporation fluxes.

In crops with complete cover the effective source, at which mean canopy air stream conditions are assumed to apply, is defined to occur at a height $(d + z_0)$ in the crop, where d is zero plane displacement and z_0 is crop roughness length. Monteith (1973) relates d and z_0 to crop height for the fully developed crop through the expressions

$$d = 0{\cdot}63h \qquad z_0 = 0{\cdot}13h. \tag{22}$$

In sparser crops we assume that the effective source height remains fixed at this fraction of crop height. It follows that in this exercise, where crop height is held constant, the effective height at which the plant components of sensible and latent heat arise is assumed to be independent of crop density. For simplicity, stability effects are ignored here and the eddy diffusion coefficients describing the vertical movement of heat and water vapour are equated to those for momentum.

Above the fully developed crop ($L \geqslant 4$), the eddy diffusion coefficient, K, is given by

$$K = ku_*(z - d) \qquad (z > h) \tag{23}$$

where k is von Kármán's constant, z is height and u_* is the friction velocity, which, in conditions of neutral atmospheric stability, is given by the expression

$$u_* = ku/\ln\{(x - d)/z_0\} \tag{24}$$

where u is the wind speed at the reference height x. It is also assumed that, in the closed canopy, the eddy diffusion coefficient decreases exponentially with height, thus

$$K = K_h \exp\{-n(1 - z/h)\} \tag{25}$$

where K_h is the value of K at the top of the crop, $ku_*(h - d)$; we use a value of $n = 2{\cdot}5$ which is typical of the agricultural crop being specified (Monteith 1973). Using these several assumptions and performing an integration over the height ranges 0 to $d + z_0$ and $d + z_0$ to x respectively, it is possible to write

$$r_a^s(\alpha) = \frac{\ln\{(x - d)/z_0\}}{k^2 u} \frac{h}{n(h - d)} [\exp n - \exp[n\{1 - (d + z_0)/h\}]] \tag{26}$$

$$r_{\mathrm{a}}^{\mathrm{a}}(\alpha) = \frac{\ln\{(x-d)/z_0\}}{k^2 u}\left[\ln\{(x-d)/(h-d)\} + \frac{h}{n(h-d)} \times \times \left[\exp[n\{1-(d+z_0)/h\}] - 1\right]\right]. \tag{27}$$

For a crop height (h) of 0·3 m and wind speed (u) of $2\,\mathrm{m\,s^{-1}}$, this gives $r_{\mathrm{a}}^{\mathrm{s}}(\alpha) = 128\,\mathrm{s\,m^{-1}}$ and $r_{\mathrm{a}}^{\mathrm{a}}(\alpha) = 42\,\mathrm{s\,m^{-1}}$.

With a bare substrate, computation is simpler. Here the aerodynamic resistances are given by

$$r_{\mathrm{a}}^{\mathrm{s}}(0) = \ln(x/z_0')\ln\{(d+z_0)/z_0'\}/k^2 u \tag{28}$$

$$r_{\mathrm{a}}^{\mathrm{a}}(0) = \ln^2(x/z_0')/k^2 u - r_{\mathrm{a}}^{\mathrm{s}}(0) \tag{29}$$

where z_0' is the effective roughness length of the substrate. For bare soil z_0' is commonly taken as 0·01 m (see Van Bavel and Hillel 1976) and for simplicity here differences in the surface roughness between wet soil and free water substrates are neglected. The values of these resistances in the present situation with $u = 2\,\mathrm{m\,s^{-1}}$ are $r_{\mathrm{a}}^{\mathrm{s}}(0) = 49\,\mathrm{s\,m^{-1}}$ and $r_{\mathrm{a}}^{\mathrm{a}}(0) = 34\,\mathrm{s\,m^{-1}}$ respectively.

Since we do not yet know exactly how $r_{\mathrm{a}}^{\mathrm{a}}$ and $r_{\mathrm{a}}^{\mathrm{s}}$ will vary, we assume they have a linear relationship between their asymptotic limits. Thus

$$\left.\begin{aligned} r_{\mathrm{a}}^{\mathrm{a}} &= \tfrac{1}{4}Lr_{\mathrm{a}}^{\mathrm{a}}(\alpha) + \tfrac{1}{4}(4-L)r_{\mathrm{a}}^{\mathrm{a}}(0) \\ r_{\mathrm{a}}^{\mathrm{s}} &= \tfrac{1}{4}Lr_{\mathrm{a}}^{\mathrm{s}}(\alpha) + \tfrac{1}{4}(4-L)r_{\mathrm{a}}^{\mathrm{s}}(0) \end{aligned}\right\} \quad 0 \leqslant L \leqslant 4 \tag{30}$$

$$\left.\begin{aligned} r_{\mathrm{a}}^{\mathrm{a}} &= r_{\mathrm{a}}^{\mathrm{a}}(\alpha) \\ r_{\mathrm{a}}^{\mathrm{s}} &= r_{\mathrm{a}}^{\mathrm{s}}(\alpha). \end{aligned}\right\} \quad L > 4 \tag{31}$$

It is shown later that the exact forms of these relationships are usually of limited numerical importance in calculations of evaporation. Implicit in the above equations is the assumption that the roughness length and zero plane displacement of crops with intermediate cover ($0 < L < 4$) vary between the values appropriate to complete cover (z_0 and d) and bare soil (z_0' and zero) and are therefore *not* a fixed fraction of crop height. However, the effective source height of the energy fluxes from the vegetation (mean canopy flow) *is* assumed to be a fixed fraction of crop height.

(*b*) *Model predictions*

Calculations are carried out for the following meteorological conditions: $R_{\mathrm{n}} = 400\,\mathrm{W\,m^{-2}}$; $D = 0, 10, 20$ mb; $T_x = 25\,°\mathrm{C}$; and $u = 2\,\mathrm{m\,s^{-1}}$. Such meteorological conditions might be considered typical for midday in the middle of a growing season at a subtropical site. However, the objective is not to make detailed predictions for particular meteorological conditions, it is rather to illustrate the general features of the theoretical treatment described.

(i) *Free water substrate.* The situation in which $r_{\mathrm{s}}^{\mathrm{s}} = 0$ is in some regards a particular case. It might be considered to represent the behaviour of paddy rice or crops which are (over) watered by trickle irrigation. Results for this situation are illustrated in Fig. 2.

Figure 2(a) illustrates total crop evaporation rates for the meteorological conditions specified above, for crops of different density, defined by their leaf area index. These rates are compared with the energy available to the whole crop, and that available to the substrate. The energy available to the system increases slightly with crop cover, while that of the substrate falls monotonically. This behaviour merely reflects the assumptions regarding G and R_n^s made in section 4(*a*)(i). The total evaporation rate varies considerably with vapour pressure deficit, but for a given atmospheric demand is fairly independent of crop cover (±9%). The rate dips around $L = 1$ when radiation capture by the plants is significant, but their bulk stomatal resistance is still quite large.

Figure 2(b) illustrates the fraction of the available energy partitioned by the vegetation, $(A - A_s)/A$, and the fractional contribution made to total evaporation by the plants, $\lambda E_c/\lambda E$, for different atmospheric deficits. The fraction of the total evaporation generated by the crop is fairly insensitive to deficit and noticeably less than the fraction of radiation intercepted.

(ii) *Substrate resistance dependence*. Figure 3 illustrates the effect of changes in the surface resistance of a soil substrate for crops of different density. Calculations are presented for an atmospheric vapour pressure deficit of 20 mb; the effect at different deficits is qualitatively similar, although the actual value of the evaporation rate changes.

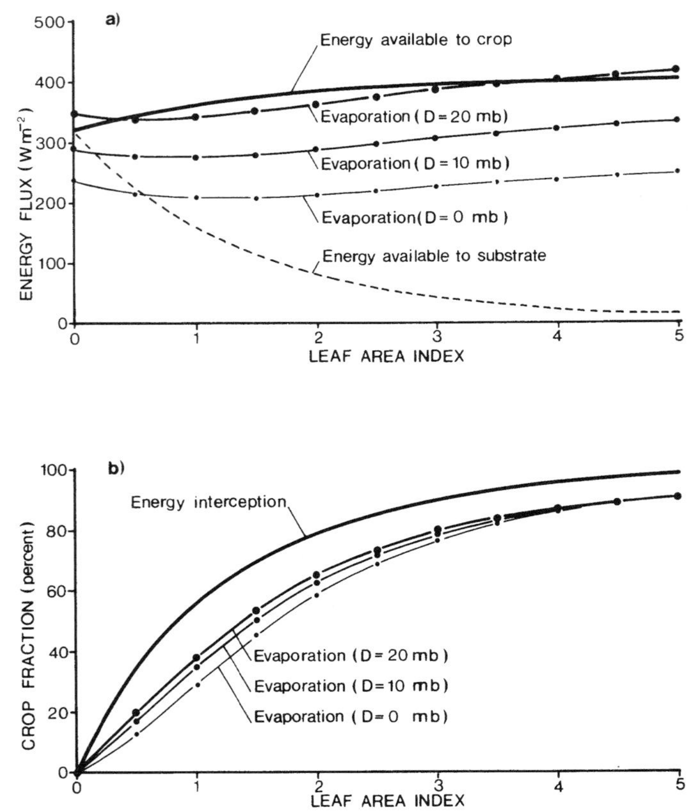

Figure 2. (a) Energy available to the crops and their free water substrates expressed as a function of L, compared with computed total crop evaporation rates for the model and conditions described in the text, with vapour pressure deficits of 0, 10 and 20 mb

(b) Fraction of total evaporation originating from the plants expressed as a function of L, computed for the model and conditions described in the text, with vapour pressure deficits of 0, 10 and 20 mb and a free water substrate. The fraction of energy intercepted by the vegetation is also shown for comparison. (R_n, T_x, u, C, n, r_s^s, r_{ST}, r_b, x, h and z_0' held constant)

The total evaporation rate of sparse crops is significantly altered by the condition (i.e. surface resistance) of the soil substrate (Fig. 3(a)). The contribution to total evaporation made by plants is also sensitive to r_s^s, and can easily exceed the fraction of energy intercepted by the canopy when leaf area index is low ($L < 2$) and soil surface resistance high. In this situation some of the energy incident on the soil is transferred as sensible heat to the canopy and utilized there for transpiration.

(c) *Model sensitivity*

In this section the sensitivity of the calculations made using the sparse crop combination equation (Eq. (11)) to the assumptions made in section 4(*a*) is explored when the model is used with surface resistances appropriate to sparse crops growing in soil.

(i) *The parametrization of aerodynamic resistance.* Calculations of evaporation rate and the fraction of evaporation arising in the crop are presented in Tables 1 and 2 respectively. The values presented are for a vapour pressure deficit of 20 mb, a stomatal resistance of 400 s m^{-1} and a soil surface resistance of 500 s m^{-1} with the assumptions made in 4(*a*), except that extreme changes are made in the parametrization of aerodynamic resistance.

Tables 1(a) and 2(a) illustrate the effect of halving and doubling the assumed value of mean boundary layer resistance. Clearly the model, and the physical process it

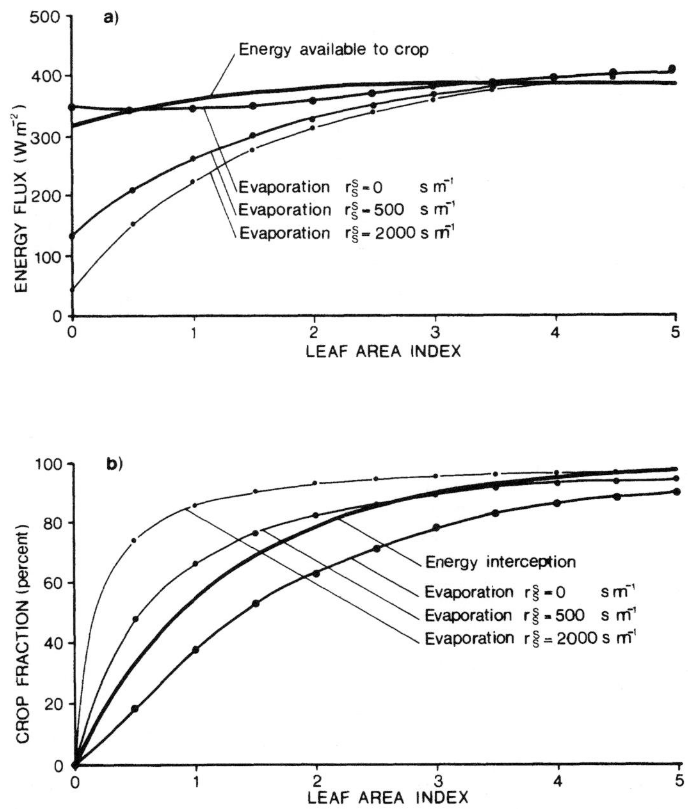

Figure 3. (a) Computed total crop evaporation rates expressed as a function of L for the model and conditions described in the text, with substrate surface resistances of 0, 500 and 2000 s m^{-1}, compared with the energy available to the crop.
(b) Fraction of total evaporation originating from the plants expressed as a function of L computed for the model and conditions described in the text with substrate surface resistances of 0, 500 and 2000 s m^{-1}. (R_n, T_x, u, D, C, n, r_{ST}, r_b, x, h and z_0' held constant)

TABLE 1. CALCULATED TOTAL CROP EVAPORATION RATES ($W m^{-2}$) FOR $D = 20$ mb $r_s^s = 500 s m^{-1}$ AND $r_{ST} = 400 s m^{-1}$ WITH CHANGES IN THE PARAMETRIZATION OF AERODYNAMIC RESISTANCE (SEE TEXT).

Model change	Leaf area index						
	0	0·5	1·0	1·5	2·0	3·0	4·0
(a) $r_b = 12·5 s m^{-1}$	135	210	263	302	331	370	394
No change	135	209	261	300	329	368	392
$r_b = 50 s m^{-1}$	135	207	259	297	325	364	387
	(0)	(1·4)	(1·5)	(1·7)	(1·8)	(1·6)	(1·8)
(b) $n = 1·25$	135	207	259	299	329	370	396
No change	135	209	261	300	329	368	392
$n = 5·0$	135	223	270	303	328	362	382
	(0)	(7·7)	(4·2)	(1·3)	(0·3)	(2·2)	(3·6)
(c) Cover	164	221	265	299	326	365	392
No change	135	209	261	300	329	368	392
Bare	135	206	259	300	332	378	409
	(21·5)	(7·2)	(2·3)	(0·3)	(1·8)	(3·5)	(4·3)

Numbers in brackets are the full range difference in the two perturbed rates expressed as a percentage of the rate given by the unmodified model.

TABLE 2. FRACTION OF TOTAL EVAPORATION ORIGINATING FROM THE PLANTS (PER CENT) CALCULATED FOR $D = 20$ mb, $r_s^s = 500 s m^{-1}$ AND $r_{ST} = 400 s m^{-1}$ WITH CHANGES IN THE PARAMETRIZATION OF AERODYNAMIC RESISTANCE (SEE TEXT).

Model change	Leaf area index						
	0	0·5	1·0	1·5	2·0	3·0	4·0
(a) $r_b = 12·5 s m^{-1}$	0	48·2	66·7	76·6	82·8	89·7	93·3
No change	0	47·9	66·5	76·4	82·5	89·5	93·2
$r_b = 50 s m^{-1}$	0	47·4	66·0	75·9	82·1	89·2	92·9
	(0)	(1·7)	(1·0)	(0·9)	(0·8)	(0·6)	(0·4)
(b) $n = 1·25$	0	48·7	67·3	76·9	82·6	88·8	92·0
No change	0	47·9	66·5	76·4	82·5	89·5	93·2
$n = 5·0$	0	43·3	62·5	74·4	82·3	91·2	95·4
	(0)	(11·3)	(7·2)	(3·3)	(0·4)	(2·7)	(3·6)
(c) Cover	0	44·4	64·6	75·7	82·4	89·7	93·2
No change	0	47·9	66·5	76·4	82·5	89·5	93·2
Bare	0	48·7	67·3	76·9	82·6	88·8	91·9
	(0)	(9·0)	(4·1)	(1·6)	(0·2)	(1·0)	(1·4)

Numbers in brackets are the full range difference in the two perturbed values expressed as a percentage of the fraction given by the unmodified model.

describes, are rather insensitive to the value of r_b: changing the value by a factor two changes λE and $\lambda E_c/\lambda E$ by 2% or less.

The effect of halving and doubling the constant n, which describes the exponential decay in eddy diffusivity through a fully developed crop, Eq. (25), is illustrated in Tables 1(b) and 2(b). The magnitude of the response is in the order of 5%, the proportionately largest effect in sparse canopies.

Tables 1(c) and 2(c) test the sensitivity of the model to extreme changes in the parametrization of r_a^a and r_a^s as a function of the leaf area index, L, given in Eqs. (30) and (31). Calculations are made with these aerodynamic resistances held at their complete cover and bare substrate limits. Again the magnitude of the response is typically in the order of 5%. The effect on total evaporation rate of using the complete cover resistance in bare substrate conditions is greater than this, 21·5%, but corresponds to miscalculating the aerodynamic resistance by a factor of two in conditions of high vapour pressure deficit.

(ii) *Net radiation absorption.* Table 3 presents calculations of (a) evaporation rate and (b) the fraction of evaporation originating from the crop for $D = 20$ mb, $r_{ST} = 400\ \mathrm{s\ m^{-1}}$ and $r_s^s = 500\ \mathrm{s\ m^{-1}}$, with other parameters as in section 4(*a*) except that C is altered. Calculations are made for $C = 0{\cdot}7$, as elsewhere in the analysis, and for $C = 0{\cdot}5$ and $0{\cdot}9$. Such a range in radiation absorption coefficient is not atypical of that found for real crops. With these assumptions the response of total evaporation rate to changes in C is small, less than 1%; that in the plant fraction of this evaporation is larger, 5–10%.

Some care is necessary in interpreting these particular results. The calculations presented in Table 3 represent the behaviour of sparse crops growing in soil with a surface resistance $r_s^s = 500\ \mathrm{s\ m^{-1}}$ and a stomatal resistance $r_{ST} = 400\ \mathrm{s\ m^{-1}}$ (consistent with the rest of this section). However, in this situation we get a minimal response of evaporation to changes in C. This parameter directly controls the fractional absorption of radiation by the plants and is therefore a driving mechanism in the initial routing of energy for partition by the surface resistances of the plants and soil. In this calculation r_s^s and r_{ST} are in the same order and this tends to suppress the response of the total evaporation rate to changes in C. Changes in the fraction of evaporation originating from the plants are more affected by changes in C.

(iii) *Mean stomatal resistance.* Figure 4 illustrates how the value of r_{ST}, and hence r_s^c, controls transpiration in crops with different density. The total evaporation rate changes significantly, though not of course proportionally, when r_{ST} is halved and doubled, and the proportion of evaporation originating at the plants also changes, especially in the sparser crops.

5. Concluding remarks

The assumption (made in section 4(*a*)(v)) that the effective source height of the crop component of energy flux remains fixed at the value of $z_0 + d$ relevant to complete cover,

TABLE 3. (a) Total crop evaporation rate ($\mathrm{W\ m^{-2}}$) and (b) fraction of total evaporation originating from the plants (percent), calculated for $D = 20$ mb, $r_s^s = 500\ \mathrm{s\ m^{-1}}$ and $r_{ST} = 400\ \mathrm{s\ m^{-1}}$, for changes in the net radiation extinction coefficient $C = 0{\cdot}5, 0{\cdot}7, 0{\cdot}9$.

Calculated parameter	Leaf area index						
	0	0·5	1·0	1·5	2·0	3·0	4·0
(a) Evaporation							
$C = 0{\cdot}5$	135	208	260	298	327	366	390
$C = 0{\cdot}7$	135	209	261	300	329	368	392
$C = 0{\cdot}9$	135	209	262	301	330	369	392
	(0)	(0·5)	(0·8)	(1·0)	(0·9)	(0·8)	(0·5)
(b) Plant fraction							
$C = 0{\cdot}5$	0·0	45·5	63·2	72·9	79·1	86·7	91·0
$C = 0{\cdot}7$	0·0	47·9	66·5	76·4	82·5	89·5	93·2
$C = 0{\cdot}9$	0·0	50·1	69·1	78·9	84·8	91·1	94·1
	(0)	(9·6)	(8·9)	(7·9)	(6·9)	(4·9)	(3·3)

Numbers in brackets are the full range difference in the two perturbed values expressed as a percentage of those given with $C = 0{\cdot}7$.

irrespective of crop density, should not be misinterpreted as implying that z_0 and d are themselves assumed constant with changing crop density. These roughness parameters do of course change in the present model and tend to the values for bare soil (z'_0 and zero) in the limit. The assumption made here is, rather, that the plant components of the energy fluxes can be considered as arising at a particular fraction of the crop height

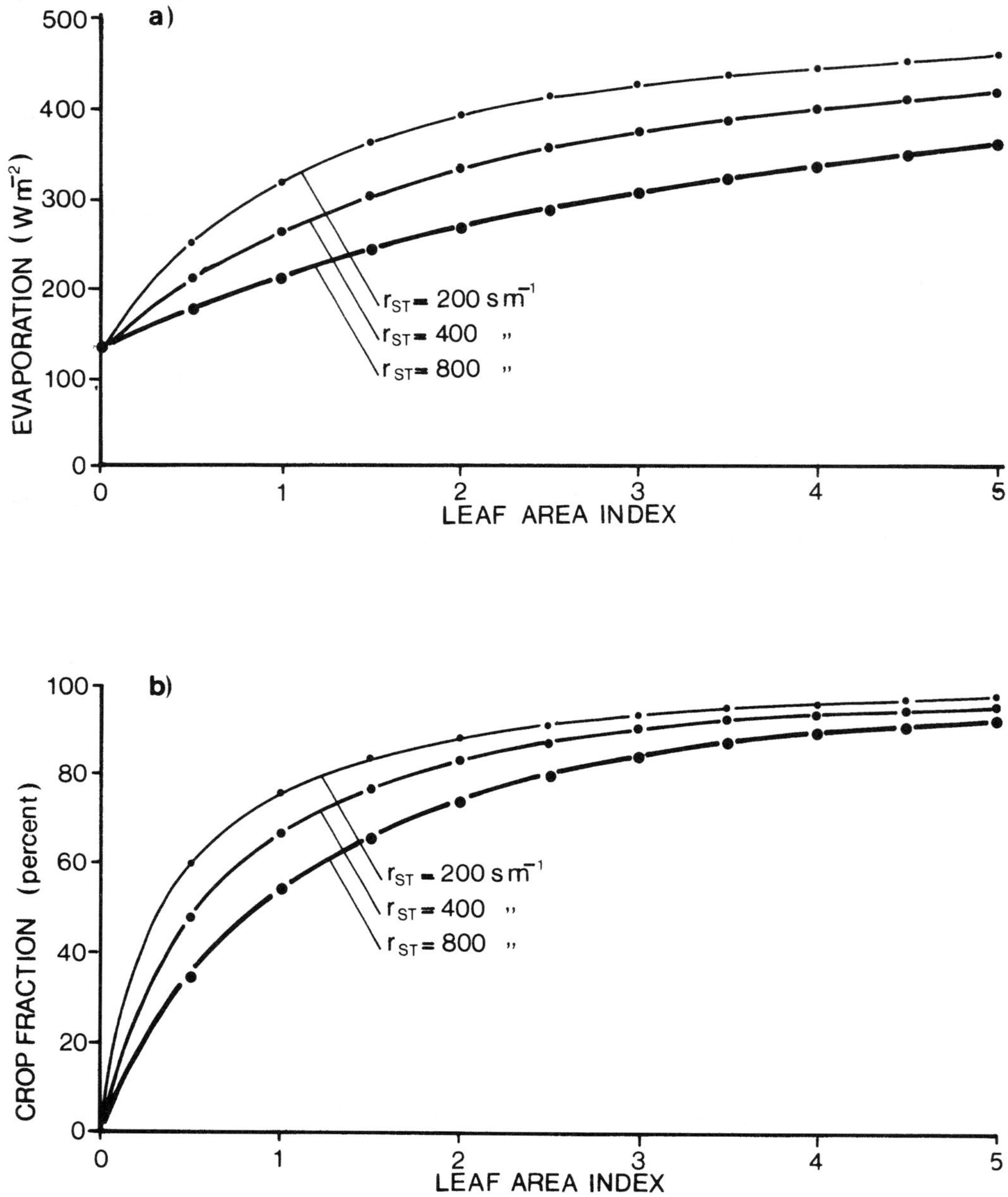

Figure 4. (a) Computed total crop evaporation rates expressed as a function of L for the model and conditions described in the text with mean stomatal resistances of 200, 400 and 800 s m^{-1}.
(b) Fraction of total evaporation originating from the plants expressed as a function of L computed for the model and conditions described in the text with mean stomatal resistances of 200, 400 and 800 s m^{-1}. (R_n, T_x, u, D, C, n, r_s^s, r_b, x, h and z'_0 held constant)

(0·76h; Eq. (22)), which is specified by the value of $z_0 + d$ relevant to a closed canopy, and remains fixed at this fraction of h as vegetation density changes.

The assumption that the crop flux source height is a constant fraction of crop height involves approximation, as indeed does the additional implicit assumption that a single source level is appropriate for both heat and vapour for all crop densities, and that this is coincident with the effective sink of momentum in a closed canopy. Moreover, we have chosen to simplify the present calculations and presentation by ignoring the effect of stability in calculating components of aerodynamic resistance (although it would be fairly simple to include an iterative correction to allow for these using empirical stability functions above the crop or soil). These several approximations in the present model serve to exacerbate and illustrate a more basic lack of understanding of how aerodynamic transfer resistances evolve as crops grow. In the light of this, the limited sensitivity of the present theoretical description to extreme changes in the parametrization of aerodynamic resistance (see section 4(c)(i)) is an important feature of the present paper.

The current example calculations of the variation in energy partition with crop density at fixed crop height (presented in sections 4(b) and (c)) should not be regarded as describing the variation in energy partition as a crop grows, even though this may eventually be the most likely use of the model presented here. The aim of the present paper is to suggest a mathematical scheme which can be used to calculate evaporation when measurements or submodels of the crop height, leaf area, stomatal and substrate resistance, net radiation interception and soil heat flux are available. The difficulties involved in providing such crop-specific submodels are not underestimated. Some are clearly interrelated, for example, leaf area influences soil heat flux and net radiation (through albedo); and, apart from the direct effect on bulk stomatal resistance, leaf area could also have indirect effects on biological control by changing the radiation loading on individual leaves. Nonetheless the present paper represents an attempt to provide a framework through which such submodels may be combined to calculate energy partition. This may ultimately yield a more accurate method for calculating evaporation from sparse crops and, hence, a better understanding of how stomatal control takes over from soil conditions as crop cover increases.

Acknowledgments

We are pleased to acknowledge the financial support of the U.K. Overseas Development Administration during the writing of this paper.

Appendix

Derivation of the sparse crop combination equation

Introducing Eq. (8) into Eqs. (9) and (10) gives

$$\lambda E_{\mathrm{s}} = \frac{\Delta A_{\mathrm{s}} + (\rho c_p/r_{\mathrm{a}}^{\mathrm{s}})[D + \{\Delta A - (\Delta + \gamma)\lambda E\}r_{\mathrm{a}}^{\mathrm{a}}/\rho c_p]}{\Delta + \gamma(1 + r_{\mathrm{s}}^{\mathrm{s}}/r_{\mathrm{a}}^{\mathrm{s}})} \tag{A1}$$

$$\lambda E_{\mathrm{c}} = \frac{\Delta(A - A_{\mathrm{s}}) + (\rho c_p/r_{\mathrm{a}}^{\mathrm{c}})[D + \{\Delta A - (\Delta + \gamma)\lambda E\}r_{\mathrm{a}}^{\mathrm{a}}/\rho c_p]}{\Delta + \gamma(1 + r_{\mathrm{s}}^{\mathrm{c}}/r_{\mathrm{a}}^{\mathrm{c}})}. \tag{A2}$$

The total evaporation flux, λE, is given by $\lambda E = \lambda E_s + \lambda E_c$. Adding Eqs. (A1) and (A2) gives

$$\lambda E = \frac{\Delta A_s r_a^s + \rho c_p[D + \{\Delta A - (\Delta + \gamma)\lambda E\}r_a^a/\rho c_p]}{(\Delta + \gamma)r_a^s + \gamma r_s^s} + \frac{\Delta(A - A_s)r_a^c + \rho c_p[D + \{\Delta A - (\Delta + \gamma)\lambda E\}r_a^a/\rho c_p]}{(\Delta + \gamma)r_a^c + \gamma r_s^c}. \quad \text{(A3)}$$

Multiplying (A3) by the product of the two denominators and collecting terms in λE gives

$$\begin{aligned}\lambda E[\{(\Delta + \gamma)r_a^s + \gamma r_s^s\}\{(\Delta + \gamma)r_a^c + \gamma r_s^c\} + (\Delta + \gamma)r_a^a \times \\ \times \{(\Delta + \gamma)r_a^c + \gamma r_s^c\} + (\Delta + \gamma)r_a^a\{(\Delta + \gamma)r_a^s + \gamma r_s^s\}] \\ = (\Delta A_s r_a^s + \rho c_p D + \Delta A r_a^a)\{(\Delta + \gamma)r_a^c + \gamma r_s^c\} + \\ + \{\Delta(A - A_s)r_a^c + \rho c_p D + \Delta A r_a^a\}\{(\Delta + \gamma)r_a^s + \gamma r_s^s\}.\end{aligned} \quad \text{(A4)}$$

If we define

$$R_s = (\Delta + \gamma)r_a^s + \gamma r_s^s \quad \text{(A5)}$$

$$R_c = (\Delta + \gamma)r_a^c + \gamma r_s^c \quad \text{(A6)}$$

$$R_a = (\Delta + \gamma)r_a^a \quad \text{(A7)}$$

and substitute these into (A4) we get

$$\begin{aligned}\lambda E(R_s R_c + R_c R_a + R_s R_a) = \{\Delta A(r_a^a + r_a^s) + \rho c_p D - r_a^s\Delta(A - A_s)\}R_c + \\ + \{\Delta A(r_a^a + r_a^c) + \rho c_p D - r_a^c \Delta A_s\}R_s.\end{aligned} \quad \text{(A8)}$$

Now

$$R_s + R_a = (\Delta + \gamma)(r_a^a + r_a^s) + \gamma r_s^s \quad \text{(A9)}$$

and

$$R_c + R_a = (\Delta + \gamma)(r_a^a + r_a^c) + \gamma r_s^c \quad \text{(A10)}$$

so we can write Eq. (A8) as

$$\lambda E(R_s R_c + R_c R_a + R_s R_a) = PM_s R_c(R_s + R_a) + PM_c R_s(R_c + R_a) \quad \text{(A11)}$$

where

$$PM_s = [\Delta A + \{\rho c_p D - \Delta r_a^s(A - A_s)\}/(r_a^a + r_a^s)][\Delta + \gamma\{1 + r_s^s/(r_a^a + r_a^s)\}]^{-1} \quad \text{(A12)}$$

and

$$PM_c = [\Delta A + (\rho c_p D - \Delta r_a^c A_s)/(r_a^a + r_a^c)][\Delta + \gamma\{1 + r_s^c/(r_a^a + r_a^c)\}]^{-1}. \quad \text{(A13)}$$

So Eq. (A11) becomes

$$\lambda E = C_c PM_c + C_s PM_s \quad \text{(A14)}$$

providing

$$C_s = R_c(R_s + R_a)/(R_s R_c + R_c R_a + R_s R_a) = \{1 + R_s R_a/R_c(R_s + R_a)\}^{-1} \quad \text{(A15)}$$

$$C_c = R_s(R_c + R_a)/(R_s R_c + R_c R_a + R_s R_a) = \{1 + R_c R_a/R_s(R_c + R_a)\}^{-1}. \quad \text{(A16)}$$

Equation (A14) is the desired sparse crop combination equation. The contributions λE_c and λE_s can now be computed from Eqs. (9) and (10) with D_0 given by Eq. (8).

References

Arkin, G. F. and Perrier, E. R. 1974 Vorticular air flow within an open row crop canopy. *Agric. Meteorol.*, **13**, 359–374

Black, T. A., Tanner, C. B. and Gardner, W. R. 1970 Evaporation from a snap bean crop. *Agron. J.*, **62**, 66–69

Denmead, O. T. 1976 'Temperate Cereals', in *Vegetation and the Atmosphere*, Vol. 2. (Ed. J. L. Monteith) Academic Press, London pp. 1–31

Fuchs, M. and Tanner, C. B. 1967 Evaporation from drying soil. *J. Appl. Meteorol.*, **6**, 852–857

Monteith, J. L. 1965 Evaporation and the environment. *Symp. Soc. Expl. Biol.*, **19**, 205–234

1973 *Principles of Environmental Physics*. Edward Arnold, London.

1981 Evaporation and surface temperature. *Quart. J. R. Met. Soc.*, **107**, 1–27

Ross, J. 1981 *The radiation regime and architecture of plant stands.* (Tasks for vegetation sciences, 3; ed. Helmut Leith) Dr W. Junk Publishers. The Hague

Shuttleworth, W. J. 1976 A one-dimensional theoretical description of the vegetation–atmosphere interaction. *Boundary-Layer Meteorol.*, **10**, 273–302

1978 A simplified one-dimensional theoretical description of the vegetation–atmosphere interaction. *ibid.*, **14**, 3–27

1979 Below-canopy fluxes in a simplified one-dimensional theoretical description of the vegetation–atmosphere interaction. *ibid.*, **17**, 315–331

Sinclair, T. R., Allen, L. H. and Stewart, D. W. 1971 'A simulation model for crop environmental interactions and its use in improving crop productivity'. Proc. 1971 Summer Comput. Simul. Counc., La Jolla, California

Szeicz, G., Van Bavel, C. H. M. and Takami, S. 1973 Stomatal factor in the water use and dry matter production by sorghum. *Agric. Meteorol.*, **12**, 361–389

Tan, C. S. and Black, T. A. 1976 Factors affecting the canopy resistance of a Douglas fir forest. *Boundary-Layer Meteorol.*, **10**, 475–488

Thom, A. S. 1972 Momentum, mass and heat exchange of vegetation. *Quart. J. R. Met. Soc.*, **98**, 124–134

Uchijima, Z. 1976 'Maize and Rice', in *Vegetation and the Atmosphere*, Vol. 2 (Ed. J. L. Monteith) Academic Press, London, pp. 33–64

Van Bavel, C. H. M. and Hillel, D. I. 1976 Calculating potential and actual evaporation from a bare soil surface by simulation of concurrent flow of water and heat. *Agric. Meteorol.*, **17**, 453–476

Wallace, J. S., Batchelor, C. H. and Hodnett, M. G. 1981 Crop evaporation and surface conductance calculated using soil moisture data from central India. *ibid.*, **25**, 83–96

E Evaporation as a Component of the Global Climate System – Commentary

During the 1970s, a handful of visionary meteorologists realised that the numerical weather prediction models that had previously been developed for forecasting the weather a day or so ahead, could also be run for many years ahead to simulate the *statistics* of weather, i.e. climate. This led to a new branch of meteorological science in which these so-called General Circulation Models (GCMs) were used to predict the climatic impacts of major land-use change or change in the chemical composition of the atmosphere. The average residence time of water vapour in the atmosphere is only about a week. Consequently, forecasts of the weather a few days ahead have limited sensitivity to the accuracy of the estimates of the evaporation from the land surfaces represented in weather prediction models. However, when meteorological models are run to predict changes in climate for at least months, but more usually years and perhaps centuries ahead, the correct representation of the water vapour input to the atmosphere by evaporation becomes critical. Recognition of this model sensitivity to land surface inputs created the need for new descriptions of the land–atmosphere exchanges of water vapour, sensible heat and momentum (and later, CO_2) between continental surfaces and the modelled atmosphere in the form of sub-models within GCMs. The creation of these sub-models, often called Land Surface Parameterizations (LSPs), but sometimes called Soil–Vegetation–Atmosphere Transfer Schemes (SVATS), also generated the need for new data to ensure that they correctly represented the whole of the Earth's land surface. The papers reproduced in this section have been selected to reflect the growth of this important new interface between hydrological science and atmospheric science and, in the final paper, also the growth in understanding needed to parameterize land surface processes at the large scale of individual GCM grid squares.

E1 Evaporation in a global climate model

Dickinson, R. E., Henderson-Sellers, A., Kennedy, P. J. & Wilson, M. F. (1986) Biosphere Atmosphere Transfer Scheme (BATS) for the NCAR Community Climate Model.

The Biosphere–Atmosphere Transfer Scheme (or BATS) was the first GCM land surface parameterization specifically designed to model the surface energy and water fluxes from the whole of the world's land surface that attempted to reproduce the behaviours of a geographically prescribed range of soil and vegetation types, or *biomes*. Such properties as aerodynamic surface roughness, albedo, and the plant physiological and soil controls on evaporation were all included in a way which is realistic at the local scale, although in practice LSPs are often used to describe the interactions of whole grid squares in the GCM, with areas of 10^4–10^6 km^2. This is because such models, when embedded in GCMs, must balance realism and complexity on the one hand, against the need for efficient computing on the other. This point should be borne in mind when reading the paper. It describes in detail how BATS achieved this balance between realism and computational efficiency, and set a precedent in doing so. It is important to see the model as a whole, which is why the Series Editor has indulged us with the space to reproduce the complete report. An updated version of BATS is still widely used today, and this report is still widely cited as the precursor for this whole field.

As the number of GCMs in use increased, more LSPs were developed; the SiB model developed by Sellers *et al.* (1986) was another notable pioneer. In fact, there was an amazing proliferation of models in the late 1980s and early 1990s. However, despite many of the models being based on similar basic physics, an intercomparison (Project for Intercomparison of Land-surface Schemes PILPS; Henderson-Sellers *et al.*, 1993) initially revealed a worryingly large range of predictions from the same driving data. This drew attention to the point that LSPs needed to be validated and calibrated against observations if they are to be credible.

Once realistic LSPs had been introduced, it became possible to begin using GCMs to test the sensitivity of climate to large-scale changes in land use. Several experiments, for example, studied the hypothetical removal of vegetation from the Sahel and the deforestation of Amazonia. GCMs are still used for this purpose today, but currently there is greater emphasis on using them to predict the climatic impacts of increased greenhouse gas concentrations. In this application, the correct representation of the surface fluxes is vitally important if the vegetation–atmosphere impacts and feedbacks are to be predicted with accuracy as the chemical composition of the atmosphere changes.

E2 Evaporation data to support climate modelling

Shuttleworth, W. J., Gash, J. H. C., Lloyd, C. R., Moore, C. J., Roberts, J., Marques, A. de O., Fisch, G., Silva, V. de P., Ribeiro, N. M. G., Molion, L. C. B., de Abreu Sa, L. D., Nobre, J. C., Cabral, O. M. R., Patel, S. R. & de Moraes, J. C. (1984)
Eddy correlation measurements of energy partition for Amazonian forest.

As mentioned above, in the 1980s global climate models were being used to assess the potential impact of large-scale tropical deforestation on regional and perhaps global climate. However, previous observational evaporation studies had almost entirely been in temperate latitudes – to serve the needs of water resources and agriculture in the then developed world. Consequently, there were almost no data for tropical forests to calibrate the representation of this biome and the parameters describing tropical forest in LSPs were little more than educated guesses. Perhaps inevitably, therefore, the first field study specifically mounted to support climate modelling, which is described in this paper, took place in the Brazilian Amazon. The measurements were made in the Reserva Ducke, north of Manaus, as a collaboration between Brazilian and British scientists. This paper was the first of many to be produced describing the results of this project, but it captures the essence of the first important new result, namely that tropical rain forest has a finite surface resistance and exercises stomatal control to conserve water despite the abundant supply of water. Further results from the study are summarised by (Shuttleworth, 1988a, 1989). The data from this project were subsequently widely used to recalibrate and/or validate the representation of tropical rainforest by LSPs (e.g. Sellers, *et al.*, 1989; Shuttleworth & Dickinson, 1989).

The study described in this paper, as well as being the first motivated by the need to improve climate models, was also the first to demonstrate and be facilitated by the then new capability to make routine measurements of surface energy and momentum fluxes using the eddy correlation technique (see B4, B5). This same improvement in measurement capability subsequently allowed the hydrometeorological community to move away from short, single-site measurement campaigns, to multi-participant and usually multinational mesoscale field experiments in support of improving GCMs, with multiple simultaneous measurements over several different vegetation types (e.g. André *et al.*, 1988; Goutorbe *et al.*, 1997), and then eventually to studies which involved long-term continuous measurements of surface exchanges.

E3 Land-surface/atmosphere interaction at the GCM grid scale

de Bruin, H. A. R. (1989) Physical aspects of the planetary boundary layer.

The simultaneous measurements of surface energy and water exchanges at multiple sites, atmospheric boundary layer sounding, and aircraft measurements during mesoscale field experiments (see above), and the associated mesoscale meteorological modelling activity, generated new understanding of land surface–atmosphere coupling. Most significantly, it was realised that the nature of this coupling depends on the length scale of the heterogeneity present in the underlying land surface cover. Although this paper by Henk de Bruin primarily aims to provide a comprehensive review of physical aspects of the atmospheric boundary layer, we select it as the benchmark paper in this field because, in the concluding sections, it was the first to recognise that there are two different types of surface heterogeneity for horizontal terrain. Both need to be recognised when representing land surface–atmosphere interactions. Note that, as a result of publication delays in de Bruin's paper, this realization is sometimes wrongly ascribed to one of the present authors (Shuttleworth, 1988b).

The key point that de Bruin introduces here is that, if the land surface is patchy and heterogeneous at length scales of less than about 10 km, then each patch of vegetation can reasonably be assumed to interface separately with the atmosphere at some *blending height*, with surface fluxes then moving upwards into the well-mixed portion of the overlying atmospheric boundary layer (ABL), which is typically 0.5–3.0 km deep in daytime conditions. However, if the structure of the underlying surface is such that there is heterogeneity in the surface cover with a length scale of tens of kilometres, the ABL recognises and responds to this heterogeneity in surface exchanges. Additional mesoscale circulations are then generated which can substantially enhance mixing and convection in the atmosphere, perhaps generating additional clouds and/or precipitation. The result, that land surfaces with both small-scale heterogeneity and, paradoxically, with uniform vegetation cover can be assumed to interface with the well-mixed portion of a spatially uniform ABL at an assumed blending height, greatly simplifies representation of the land surface–atmosphere interaction at the grid scale used in GCMs.

References

André, J.-C., Goutorbe, J.-P., Perrier, A., Becker, F., Bessemoulin, P., Bougeault, P., Brunet, Y., Brutsaert, W., Carlson, T., Cuenca, R., Gash, J., Gelpe, J., Hildebrand, P., Lagouarde, P., Lloyd, C., Mahrt, L., Mascart, P., Mazaudier, C., Noilhan, J., Ottle, C., Payen, M., Phulpin, T., Stull, R., Shuttleworth, J., Schmugge, T., Taconet, O., Tarrieu, C., Thepenier, R.-M., Valencogne, C., Vidal-Madjar, D. & Weill, A. (1988) HAPEX-MOBILHY: first results from the special observing period. *Ann. Geophys.* **6**, 477–492.

Dickinson, R. E. & Henderson-Sellers, A. (1988) Modelling tropical deforestation: a study of GCM land-surface parameterizations. *Quart. J. Royal Met. Soc.* **114**, 439–462.

Goutorbe, J.-P., Lebel, T., Dolman, A. J., Gash, J. H. C., Kabat, P., Kerr, Y. H., Monteny, B., Prince, S. D., Stricker, J. N. M., Tinga, A. & Wallace, J. S. (1997) An overview of HAPEX-Sahel: a study in climate and desertification. *J. Hydrol.* **188–189**, 4–17.

Henderson-Sellers, A., Yang, Z.-L. & Dickinson, R. E. (1993) The Project for Intercomparison of Land-surface Parameterization Schemes. *Bull. Am. Met. Soc.* **74**, 1335–1349.

Sellers, P. J., Mintz, Y., Sud, Y. C. & Dalcher, A. (1986) A simple biosphere model (SiB) for use within general circulation models, *J. Atmos. Sci.* **43**, 305–331.

Sellers, P. J., Shuttleworth, W. J., Dorman, J. L., Dalcher, A. & Roberts, J. (1989) Calibrating the simple biosphere model for Amazonian tropical forest using field and remote sensing data. Part 1: Average calibration with field data. *J. Appl. Met.* **28**, 727–759.

Shuttleworth, W. J. (1988a) Evaporation from Amazonian rain forest. *Proc. Royal Soc. London B* **233**, 321–346.

Shuttleworth, W. J. (1988b) Macrohydrology – the new challenge for process hydrology. *J. Hydrol.* **100**, 31–56.

Shuttleworth, W. J. (1989) Micrometeorology of temperate and tropical forest. *Phil. Trans. Royal Soc. London B* **324**, 299–334.

Shuttleworth, W. J. & Dickinson, R. E. (1989) Tropical deforestation: comment on the discrepancy between recent GCM and observational results. *Quart. J. Royal Met. Soc.* **115**, 1177–1179.

NCAR/TN-275+STR
NCAR TECHNICAL NOTE

December 1986

Biosphere-Atmosphere Transfer Scheme (BATS) for the NCAR Community Climate Model

R. E. Dickinson
A. Henderson-Sellers
P. J. Kennedy
M. F. Wilson

ATMOSPHERIC ANALYSIS AND PREDICTION DIVISION

NATIONAL CENTER FOR ATMOSPHERIC RESEARCH
BOULDER, COLORADO

CONTENTS

CONTENTS—Continued

LIST OF TABLES

LIST OF FIGURES

LIST OF APPENDICES

PREFACE

Since early 1977, various members of the Atmospheric Analysis and Prediction (AAP) Division, at the National Center for Atmospheric Research (NCAR), and other collaborators have been developing a comprehensive model of land–surface processes suitable for use, initially with the previous grid–point General Circulation Model, and now with the spectral Community Climate Models (CCM0 and CCM1). Special emphasis has been given to describing properly the role of vegetation in modifying the surface moisture and energy budgets. The result of these efforts has been incorporated into a boundary package, referred to as the Biosphere–Atmosphere Transfer Scheme (BATS)—a piece of software of about three thousand lines of code written in SI units that runs as an offline version or coupled to the CCM. It includes (i) assignment of land type and soil information to each model grid square, (ii) calculation of soil, snow or sea–ice surface temperature in response to net surface heating and depending on soil or snow heat capacity and thermal conductivity, (iii) calculation of soil moisture, evaporation, and surface and groundwater runoff, (iv) specification of vegetation cover in terms of fractional ground shading and relative areas of transpiring and nontranspiring plant surfaces inferred from the land–use type, (v) surface albedo in terms of soil moisture, vegetation cover, and snow cover, including the shading of snow by vegetation, (vi) plant water budget including dew, intercepted precipitation, and transpiration as limited by stomatal resistance and soil dryness, (vii) surface drag coefficients as a function of bulk Richardson number and vegetation cover, and (viii) determination of foliage temperature in response to energy–balance requirements and consequent fluxes of heat and moisture from the foliage to canopy air. Scientific studies are being pursued to better establish the importance of various aspects of land–surface processes for climate and climate models. The purpose of this report is thus to describe the physical processes, current numerical parameterizations, and some of the code structure of this boundary package.

Robert E. Dickinson
September 1986

ACKNOWLEDGMENTS

Authors A. Henderson-Sellers and M. F. Wilson were visitors to the AAP Division and the Advanced Study Program (ASP), respectively, at NCAR during the period of this work. A. Henderson-Sellers's permanent affiliation is with the Department of Geography, the University of Liverpool, United Kingdom, and M. F. Wilson is with the British Meteorological Office, Bracknell, United Kingdom. R. E. Dickinson and P. J. Kennedy are members of the AAP Division at NCAR. The author thanks D. Williamson, J. Hack, F. Giorgi, and M. Verstraete for reviewing the manuscript and A. Modahl for its preparation.

1. INTRODUCTION

The purposes of the biosphere–atmosphere transfer scheme (BATS), as coupled with the NCAR Community Climate Model (CCM), are to (a) calculate the transfers of momentum, heat, and moisture between the earth's surface and atmospheric layers, (b) determine values for wind, moisture, and temperature in the atmosphere, within vegetation canopies, and at the level of surface observations, and (c) to determine (over land and sea ice) values of temperature and moisture quantities at the earth's surface. Included in the latter are the determination of the moisture content of soil, the excess rainfall that goes into runoff, and the physical state of the moisture at the surface, *i.e.*, whether it is snow or water. To carry out these calculations, it is necessary to prescribe a predominant land–surface category for each surface grid point, and for those categories that have significant vegetative cover, to calculate the exchanges of moisture and energy among the vegetation and the soil and the air.

The earth's surface receives and emits various kinds of energy. The most important of these, from the viewpoint of physical processes, are (a) solar radiation, absorbed after undergoing atmospheric alterations and reflections determined by the radiative transfer subroutine and the surface albedo; (b) infrared radiation, emitted from the surface according to $\epsilon\sigma_s T^4$, where ϵ is the thermal emissivity, σ_s is the Stefan–Boltzmann constant, and T is the ground or vegetation temperature (downward atmospheric flux is obtained from the detailed radiative transfer calculation of the CCM or is parameterized); (c) sensible heat flux, calculated in the absence of vegetation by an aerodynamic transfer formula as proportional to the difference between surface and overlying boundary–layer air temperatures; and (d) latent heat flux calculated similarly to (c) but proportional to specific humidity differences. The net energy received at the earth's surface melts snow, or warms the surface, or is conducted downward to be stored in lower layers.

Over a long-enough time period (a few days or more), the effect of heat storage within unfrozen soil becomes negligible. Over shorter times, it may be as important as the various energy fluxes. Besides this "thermal inertia," which depends on soil moisture and composition as well as on time scale, other major complications in model parameterization involve the detail to which surface processes are parameterized and, in particular, the manner in which the ratio of sensible to latent heat fluxes is established.

One important feature of realistic climatic processes is the large diurnal variations of fluxes of sensible and latent heat over land surfaces. Indeed, even the sign of these quantities usually varies from day to night. These fluxes are strongly nonlinear functions of the static stability of the lowest atmospheric layers, and surface moisture is, in turn, a strongly nonlinear function of surface temperature, so that significant errors can be made in calculating these fluxes using diurnally averaged temperatures. Furthermore, the actual diurnal variations of temperatures at and near the ground are very significant properties of climate, as well as depending on other climatic properties such as soil moisture, intensity of incident solar flux, and convection in the boundary layer. Thus, a valid simulation of these diurnal temperature variations is one of the major design objectives of many current GCMs.

Examination of the processes important for determining diurnal variation of temperatures at the surface suggests that the presence and nature of vegetation are important factors in this problem.

The physical characteristics of vegetation most important here are (a) absorption of solar radiation and consequent shading of the ground, (b) exchanges of sensible and latent heat with the atmosphere over a total canopy surface area that may be an order of magnitude larger than that of the ground, and (c) the presence of canopy surface moisture. This moisture results from the film of water formed by dew or by intercepted rainfall; moisture beyond that which the foliage can hold drips to the ground. Over dry areas of the foliage, vapor pressure is controlled by stomatal openings (as parameterized by a stomatal resistance factor).

For example, over forested areas, interception of rainfall by leaves and re-evaporation can remove most of the precipitation in a light drizzle. Furthermore, much of the transfer of moisture from the ground to the air is through leaf transpiration. Snowfall and frost are also significantly affected by vegetative cover.

The present document revises the previous description (Dickinson *et al.*, 1981) and is intended to complement and explain the physics of the BATS code rather than serve as a substitute for it. It is divided as follows: Section 2 describes how land-cover and soils data were determined for the model and lists parameters dependent upon these classifications; it also describes specification of land-surface albedo. Section 3 summarizes what is involved in executing the code of the BATS surface physics package linked to the CCM. Section 4 describes how soil temperature is calculated. Section 5 gives the determination of soil moisture and snow cover for a nonvegetated grid square. Section 6 summarizes the calculation of momentum drag coefficients including the modifications needed for vegetation or for sea ice. Section 7 describes how the model calculates sensible and latent fluxes over a nonvegetated surface and how the model interpolates between ground and lowest model level to obtain air temperature and mixing ratios at the level of surface observations (Stephenson-screen level). Section 8 then describes what parameters are used to represent vegetation and how the model vegetation stores water. Section 9 summarizes the calculation of soil moisture and snow cover in the presence of vegetation.

We also include an appendix that summarizes some of the code notation and briefly describes the flow of operations in the code. Figure 1 shows conceptually many of the processes included in the boundary parameterizations.

Although our treatments of surface processes are much more elaborate than those of conventional GCMs, they are still conceptually much simpler than the most elaborate models available for individual processes, in particular for soil hydrology, plant water budgets, and snow physics. It is hoped that the present descriptions are sufficiently detailed to permit firm evaluations of the sensitivity of the climate system to various surface processes and hence to establish where further work and more careful parameterizations may be needed.

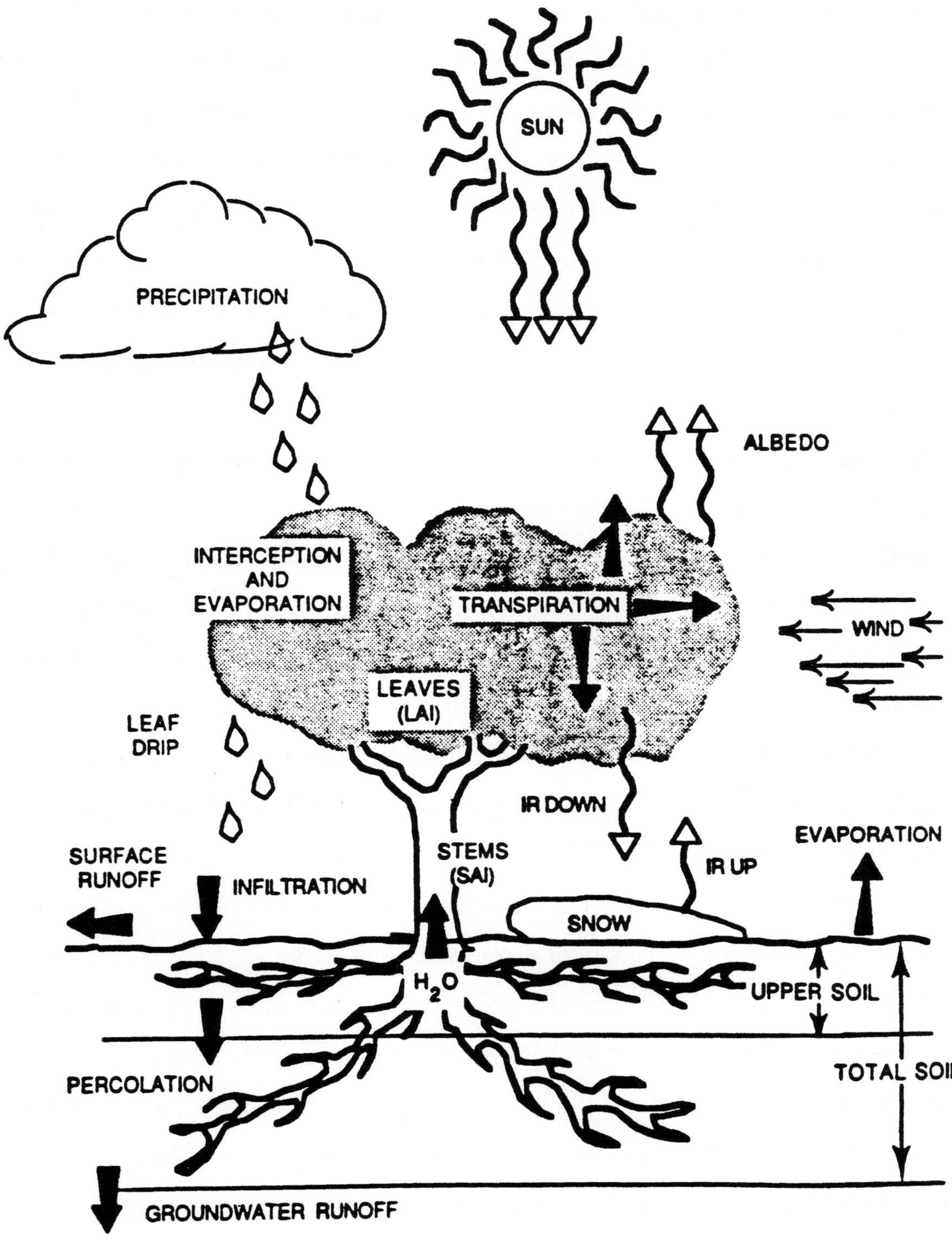

Figure 1. Schematic diagram illustrating the features included in the land-surface parameterization scheme used here.

4

a. Mathematical Symbols

For the convenience of the reader, we summarize here all the mathematical symbols defined and used in the text.

α_f	Albedo of the vegetative canopy
α_{IR}	Snow albedo for near-infrared radiation
α_{IRD}	Snow albedo for diffuse near-infrared radiation
α_{IRO}	= 0.65, albedo of new snow for near-infrared solar radiation with solar zenith angle less than 60°
α_V	Snow albedo for visible solar radiation
α_{VD}	Snow albedo for diffuse visible solar radiation
α_{VO}	= 0.95, albedo of new snow for visible solar radiation with solar zenith angle less than 60°
$\Delta\alpha_g$	Increase of albedo due to dryness of soil surface
Δt	Model time step
β	Fraction of transpiration from top soil layer
ϵ	Thermal emissivity of ground
κ	R/C_p
λ	Wavelength of radiation
λ_s	Thermal diffusivity coefficient
ρ_a	Density of surface air
ρ_i	Density of ice relative to water
ρ_s	Density of subsurface soil layer
ρ_{sw}	Density of snow relative to water
ρ_w	Volume of liquid water per unit volume of soil, weighted toward top layer as defined by Eq. (18a)
ρ_{wsat}	Porosity or saturated soil water density
σ	Ratio of pressure p to surface air pressure p_s
σ_f	Fractional foliage cover for each grid point
σ_s	Stefan–Boltzmann constant
τ_{SNOW}	Nondimensional age of snow

5

τ_0	Time constant used in calculation of age of snow
τ_1	Period of heating (in soil temperature calculation) $= 8.64 \times 10^4$ seconds $= 1$ day
ϕ	Soil water suction (negative potential)
ϕ_{MAX}	Soil water suction for permanent wilting of plants
ϕ_0	Soil water suction for saturated soil
Υ_{ro}	Maximum transpiration that can be sustained
Υ_w	Rate of transfer of water by diffusion to the upper soil layer from the lower; subscript zero denotes rate in the absence of gravity
a	Fraction of sea ice covered by leads
c_i	= Specific heat of ice $\simeq 0.45\ c_w$
c_s	Specific heat of subsurface layer
c_{sw}	$= 0.49\ c_w$, specific heat of snow for unit snow density
c_w	$= 4.186 \times 10^6$ J m^{-3} K^{-1}, specific heat of water
$c_{A,F,V}$	Various bulk conductances as defined in the text
c_1	$= 2\pi^{1/2} = 3.5449$ (in ground temperature calculation)
c_2	$= 2\pi$ (in ground temperature calculation)
c_3	= rate of subsoil relaxation = 0.2 (in soil temperature calculation)
d_{ie}	Equivalent water depth of snow
d_{ice}	Depth of ice
d_s	$= 10\ S_{cv}/\rho_{sw} = 10 \times$ snow depth
d_1	$= (k_s\tau_1)^{1/2}$ is soil depth influenced by periodic heating
f	Ratio of soil evaporation to potential evaporation
f_d	Fraction of vegetation that is green and unwetted
f_w	Fraction of wet foliage
f_{SNOW}	Fraction of grid square covered by snow
$f(ZEN)$	Increase of snow albedo due to solar zenith exceeding 60°

6

g	Acceleration due to gravity
h_s	$= (S_g + (F_{IR}^{\downarrow} - F_{IR}^{\uparrow}) - F_s - L_{v,s}F_q - L_f S_m)$, heat balance at the surface
k	von Karman constant
k_s	Soil or snow thermal conductivity
p_s	Surface air pressure
p_1	Lowest model level air pressure
q	Specific humidity
q_1	Lowest model level water–vapor mixing ratio
q_{af}	Water–vapor mixing ratio of the air within the foliage
q_g	Soil surface water–vapor mixing ratio
$q_{g,s}$	Saturated mixing ratio at soil surface temperature
q_s	Water–vapor mixing ratio inside the leaves
r_a	Aerodynamic resistance factor $(1/(C_f U_{af}))$
r_s	Stomatal resistance factor
$r_{\ell a}$	Resistance to heat or moisture transfer through the laminar boundary layer at the foliage surface
r_F	$= r_{\ell a}/(\sigma_f L_{SAI})$, bulk foliage resistance
r_v	Average resistance for transfer of water vapor from foliage due to stomatal and aerodynamic resistance of the foliage surface
r_1	Rate of "snow aging" due to grain growth from vapor diffusion
r_2	Rate of "snow aging" due to melt water
r_3	Rate of "snow aging" from dirt and soot
r''	Fraction of potential evaporation from a leaf
s	Volume of water in soil divided by volume of water at saturation
s_i	Soil water as defined above in layer i
s_w	Soil water as defined above at permanent wilting point

t	Time
u_1	Lowest model level west–to–east wind component
u^*	Friction velocity (ρu^{*2} = surface momentum flux)
v_1	Lowest model level south–to–north wind component
z_0	Roughness length
z_1	Height of lowest model level
A_{LBG}	Albedo for bare soil
A_{LBL}	Albedo of plants for infrared radiation
A_{LBS}	Albedo of plants for short–wave radiation
A_{LBGO}	Albedo for wet bare soil
B	Soil parameter defining change in soil water potential and hydraulic conductivity with soil water, a function of soil texture
C_f	Coefficient of transfer between foliage and air in the foliage
C_p	Specific heat of air
C_s	Constant in snow albedo calculation (= 0.2)
C_D	Surface drag coefficient
$C_{D,F}$	Average momentum transfer coefficient over the grid square in the presence of vegetation
C_{DH}	Aerodynamic transfer coefficient for heat
C_{DN}	Drag coefficient for neutral stability
C_{DVEG}	Aerodynamic drag coefficient for the canopy
C_{DW}	Aerodynamic transfer coefficient for water vapor
C_N, C_S	Constants in snow albedo calculation (= 0.5, 0.2)
$C(ZEN)$	Cosine of the solar zenith angle
D	Average diffusivity for water flow through soil
D_f	Characteristic dimension of foliage elements in direction of wind flow
D_s	Rate of excess snow dropping from leaves per unit land area

D_w	Rate of excess water dripping from leaves per unit land area
D_{min}	Minimum value of soil diffusivity
D_{max}	$= B\phi K/\rho_{wm}$, maximum value of soil diffusivity
$E_{a,f,g}$	Water flux per unit land area; f refers to origin at foliage, g at ground, and a to the total flux; a superscript N, in Eq. (70)–(71), refers to the flux at the Nth iteration for leaf temperature
E_{tr}	Transpiration
E_{trmx}	Maximum transpiration the vegetation can sustain
E_D	Evaporation demand as determined by the air–ground humidity difference
F_c	Heat conduction through snow–covered sea ice or bare sea ice
F_s	Atmospheric sensible heat flux from ground to atmosphere
F_q	Moisture flux from ground to atmosphere
F_{qm}	Moisture flux through the wet surface that the soil can sustain, *i.e.*, maximum as limited by diffusion
F_{vi}	Visible radiation incident at the top of the canopy
F_{vl}	Visible radiation absorbed by the lower canopy
F_{vu}	Visible radiation absorbed by the upper canopy
F_{AGE}	Snow age factor
$F_{IR}^{\downarrow}$ - $F_{IR}^{\uparrow}$	Net long–wave radiation flux from atmosphere to bare ground
$F_{SEAS}(T)$	Seasonal variation of vegetation cover with temperature (Eq. 1b)
G	Net water applied to the surface $(P_r + S_m - F_q)$ in absence of vegetation
G_{HEAT}	Heat conduction rate from surface to underlying reservoir
$H_{a,f,g}$	Heat flux per unit land area to the atmosphere; f refers to origin at foliage, g at ground, a to total flux

K	Hydraulic conductivity of soil
K_0	Saturated soil hydraulic conductivity, equivalent to downward flow rate for saturated soil due to gravity
K_{VEG}	Land use/vegetation type in model
L_d	Unwetted fraction of leaf–stem area free to transpire
L_f	Latent heat of fusion
L_s	Latent heat of sublimation
L_v	Latent heat of evaporation
$L_{v,s}$	Either L_v or L_s, depending on whether T_g is greater than freezing or not
L_w	Ratio of wetted leaf stem area to total leaf stem area exchanging water with the atmosphere
L_{AI}	Leaf area index
L_{SAI}	Leaf–stem area index, sum of L_{AI} and S_{AI}
L_{SAIW}	Total vegetation surface that exchanges water with the atmosphere
P	Precipitation rate
P_r	Rate of precipitation falling as rain
P_s	Rate of precipitation falling as snow
Q_c	Latent heat release
R	Gas constant
R_f	Dependence of the stomatal resistance factor on solar radiation
R_g	Leakage down to subsoil and groundwater reservoirs
Ri_B	Surface bulk Richardson number
R_m	Net radiation that would be absorbed by the vegetation if the foliage temperature $= T_{g1}$
R_n	Net radiation absorbed by the vegetation
R_s	Surface runoff

R_{ti}	Fraction of roots in the soil layer
R_w	Total runoff $(R_s + R_g)$
R_{OUGH}	Assigned roughness of vegetation types in model
$S^{\downarrow}$	Incident solar flux
S_f	Dependence of the stomatal resistance factor on temperature
S_g	Solar radiation absorbed over bare ground at earth's surface
S_m	Rate of snow melt
S_{cv}	Snow cover (liquid water equivalent)
S_{cv}/ρ_{sw}	Average snow depth
S_{gf}	Solar flux absorbed by vegetation
S_{sw}	Surface soil water in upper layer of soil in meters (maximum value = S_{swmax})
S_{tw}	Total water in rooting zone of soil (maximum value = S_{twmax})
S_{AI}	Stem area index
S_0	S_{tw}/S_{twmax}
S_1	S_{sw}/S_{swmax}
T_a	Air temperature at 10 m over ocean, 1.3 m over grass for land
T_{af}	Temperature within the foliage layer
T_c	Reference temperature for deciding whether precipitation is rain or snow
T_f	Temperature of foliage
T_{g1}	Surface soil temperature
T_{g2}	Subsurface temperature (depth of about 0.2 m)
T_{g3}	Deep soil temperature (depth of about 2 m = annual average)
T_m	= 273.16 K, the melting or freezing temperature of pure water
T_B	Approximate freezing point of sea water
T_1	Air temperature of lowest model layer $\times(p_1/p_s)^{-\kappa}$
U_{af}	Magnitude of wind within the foliage layer
U_c	Subgrid-scale horizontal convection velocity

V_a	$= (u_1^2 + v_1^2 + U_c^2)^{1/2}$, strength of wind at the anemometer level
W_{dew}	Total water stored by canopy per unit land area
W_{DMAX}	Maximum water the canopy can hold
W_{LT}	Soil dryness (or plant wilting) factor
W_{m1}	Rate of melting (freezing if negative) of surface soil water
W_{m2}	Rate of melting (freezing if negative) of subsurface soil water
Z_0	Depth of total active soil layer, varies typically from 0.5 to 2 m in thickness as a function of vegetation cover and/or land use
Z_1	Depth of upper soil layer, fixed at 0.10 up to now

2. LAND TYPES, SOIL TYPES, FOLIAGE COVER, AND SURFACE ALBEDO

This section discusses model data fields that are read into the CCM upon initialization and the calculation of surface albedo that is done in a separate subroutine linked closely to the radiation subroutine and the initialized fields.

The surface processes occurring in the model depend, in principle, on the nature of the land surface, including its vegetative cover and soil properties. Because of the coarse model resolution, a high level of detail is not possible. However, the following parameterizations appear consistent with the model resolution and objectives: (a) Land surfaces are broken into vegetative types and separately into soil types. (b) These vegetative types, along with soil moisture and season, determine the (i) fractional vegetative cover over a grid square, (ii) plant and soil albedos, (iii) surface roughness, and (iv) soil thermal and hydrologic properties.

The land-use and soils data used with BATS have been derived from two global archives. The methods of land-type and soil-information assignment are described in the following two subsections.

Reflection of solar radiation is a strong function of wavelength for many surfaces. Furthermore, the spectrum of solar radiation reaching the surface depends strongly on the optical properties of the atmosphere. Therefore, use of spectrally integrated surface albedos may be too inaccurate. The model radiative calculations use a crude spectral decomposition of solar radiation by distinguishing between wavelengths shorter than or longer than 0.7 μm. Plants, in particular, have much lower albedos for visible than for near-infrared solar radiation, because they need the visible radiation for photosynthesis but, otherwise, must avoid overheating. By contrast, snow has much higher visible than near-infrared albedos, and these albedos depend significantly on solar zenith angles. The current parameterizations for vegetative and soil surface albedos are also described in this section. The computation of surface albedos, and hence absorbed solar radiation, is handled externally to the rest of the BATS package in subroutine ALBEDO which is linked to the radiation scheme in the CCM. Snow, if present, must also be considered in some detail because of its highly variable albedo and because vegetation can "mask" the snow surface. The albedo of the snow is parameterized by a measure of the snow age. This parameterization is also described at the end of this section.

a. Land-Type Assignment

Currently, two global land-surface archives are designed for incorporation into climate models—the vegetation and cultivation data of Matthews (1983, 1984) and the land use and soils data of Wilson (1984), described in Wilson and Henderson-Sellers (1985). Both land-use data archives were used in the construction of the land-type data set for the CCM. The implementation of the soils data of Wilson (1984) is described in the following subsection.

Matthews (1983, 1984) defines 31 classes of natural land type and prescribes the intensity of cultivation for each 1°-$\times$-1° grid element. Wilson (1984) also uses base resolution of $1^\circ \times 1^\circ$ but defines two land-cover types for each land element—primary $\geq 50\%$ of the grid square, and secondary $< 50\%$ but $\geq 25\%$. She uses 53 land-cover classes including a number of cultivation categories. To simplify somewhat the CCM land-cover specification, both data archives were reduced to percentages of the 18 more basic land-cover classes shown in Table 1. Examples of the role of these different use classes are given in Fig. 2.

Table 1. Vegetation/Land-Cover Types

1. Crop/mixed farming
2. Short grass
3. Evergreen needleleaf tree
4. Deciduous needleleaf tree
5. Deciduous broadleaf tree
6. Evergreen broadleaf tree
7. Tall grass
8. Desert
9. Tundra
10. Irrigated crop
11. Semi-desert
12. Ice cap/glacier
13. Bog or marsh
14. Inland water
15. Ocean
16. Evergreen shrub
17. Deciduous shrub
18. Mixed Woodland*

*The mixed woodland class was added later, as described in the following text.

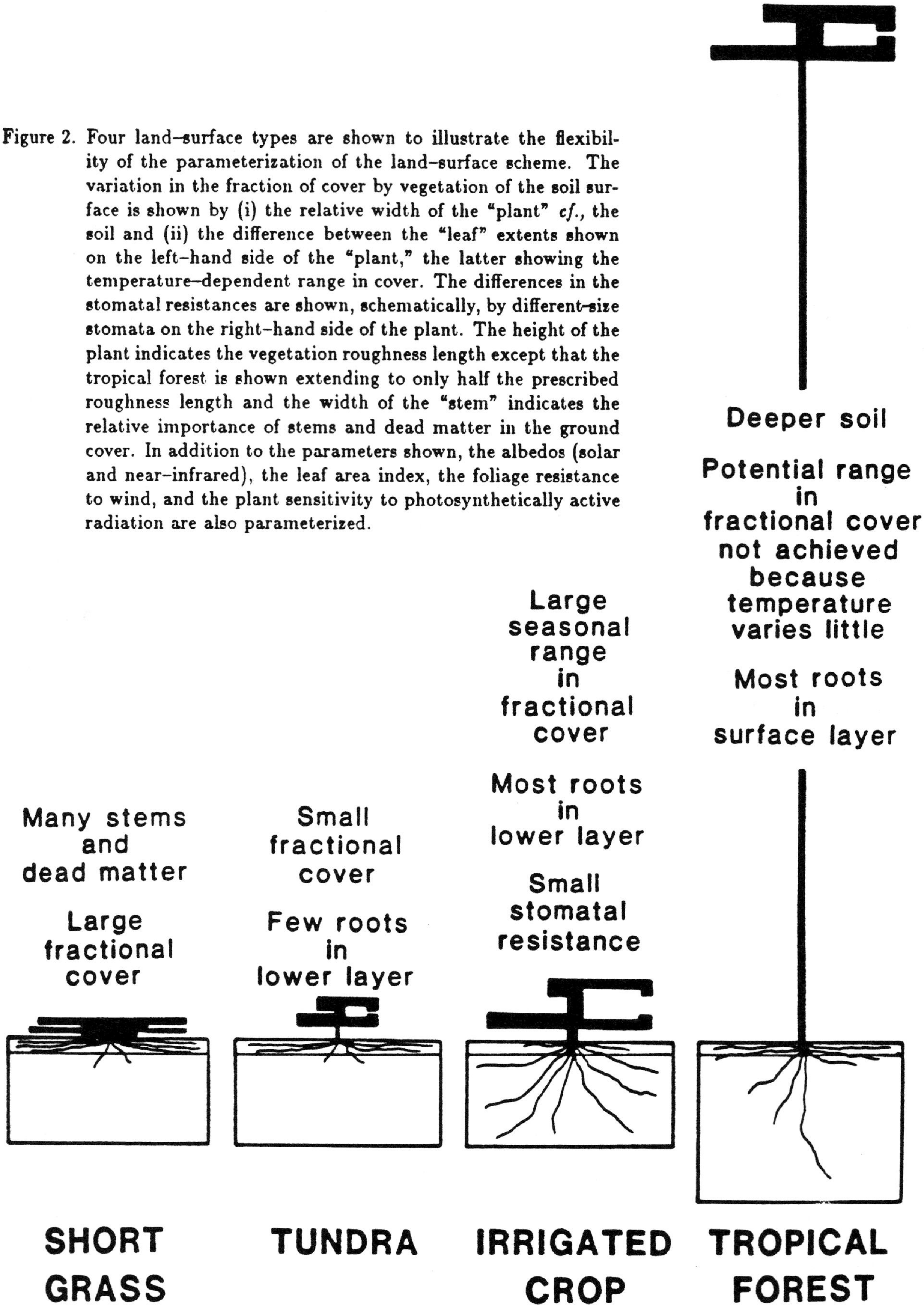

Figure 2. Four land–surface types are shown to illustrate the flexibility of the parameterization of the land–surface scheme. The variation in the fraction of cover by vegetation of the soil surface is shown by (i) the relative width of the "plant" *cf.*, the soil and (ii) the difference between the "leaf" extents shown on the left–hand side of the "plant," the latter showing the temperature–dependent range in cover. The differences in the stomatal resistances are shown, schematically, by different–size stomata on the right–hand side of the plant. The height of the plant indicates the vegetation roughness length except that the tropical forest is shown extending to only half the prescribed roughness length and the width of the "stem" indicates the relative importance of stems and dead matter in the ground cover. In addition to the parameters shown, the albedos (solar and near–infrared), the leaf area index, the foliage resistance to wind, and the plant sensitivity to photosynthetically active radiation are also parameterized.

Data sets at the model base resolution of approximately 4.5° × 7.5° were derived from each of the 1°–sets and initially in terms of percentages of the 17 basic types in each grid element. Each grid element was assigned a dominant type except when use of type 18 (mixed woodland) seemed more appropriate. The dominant land types inferred from each data set were compared and if the classifications agreed, the result was incorporated into the final CCM data set. At some model grid squares, differences occurred between the Matthews and the Wilson data sets, and two alternative or combined steps were taken. First, an average (of the two percentage components) was established and second, a comparison was made between this average and the ecotype classification of Olson *et al.* (1983), and subjective assessment was made as to what classification to use. The resultant archive is of 18 land–use classes defined on the model mesh. These 18 classes of land cover are used to define a wide variety of land surface, hydrological, and vegetation properties as described below.

For each grid point, a value K_{VEG} is read into the model: $K_{VEG} = 0$ for all ocean grid points; $K_{VEG} = 1$ for all points with arable/mixed farming, etc. The assigned roughness length R_{OUGH} (given in Table 2) is used as the aerodynamic roughness (see section 8), and the fraction of a grid square covered by snow f_{SNOW} is inferred according to the formula,

$$f_{SNOW} = f_{SN}/(1 + f_{SN}),$$

with

$$f_{SN} = 0.1\ S_{cv}/(\rho_{sw} \times R_{OUGH}).$$

The quantity $S_{cv}/(\rho_{sw})$ is the average snow depth and $R_{OUGH}/0.1$ is about half the average vertical extent of the surface objects. The quantity f_{SNOW}, calculated separately for the vegetated and nonvegetated parts of a grid square, determines what fraction of the area has the albedo of snow versus that of no snow. Over vegetated areas, all vegetation covered by the snow pack is neglected by reducing the fraction of vegetation cover defined in subsection (c) by the factor $(1 - f_{SNOW})$.

TABLE 2. VEGETATION/LAND COVER PARAMETERS

Parameter	Land Cover/Vegetation Type*																	
	1	2	3	4	5	6	7	8	9	10	11	12	13	14	15	16	17	18
a) Maximum fractional vegetation cover	0.85	0.80	0.80	0.80	0.80	0.90	0.80	0.0	0.60	0.80	0.10	0.0	0.80	0.0	0.0	0.80	0.80	0.80
b) Difference between maximum fractional vegetation cover and cover at temperature of 269 K	0.6	0.1	0.1	0.3	0.3	0.5	0.3	0.0	0.2	0.6	0.1	0.0	0.4	0.0	0.0	0.2	0.3	0.2
c) Roughness length (m)	0.06	0.02	1.0	1.0	0.8	2.0	0.1	0.05	0.04	0.06	0.1	0.01	0.03	0.0024	0.0024	0.1	0.1	0.8
d) Depth of the total soil layer (m)**	1.0	1.0	1.5	1.5	2.0	1.5	1.0	1.0	0.5	1.0	1.0	1.0	1.0	1.0	1.0	1.0	1.0	2.0
e) Depth of the upper soil layer (m)**	0.1	0.1	0.1	0.1	0.1	0.1	0.1	0.1	0.1	0.1	0.1	0.1	0.1	0.1	0.1	0.1	0.1	0.1
f) Rooting ratio (upper to total soil layers)	3	8	10	10	10	12	8	9	4	3	8	5	5	5	5	5	5	10
g) Vegetation albedo for wavelengths <0.7μm	0.10	0.10	0.05	0.05	0.08	0.04	0.08	0.20	0.09	0.08	0.17	0.80	0.06	0.07	0.07	0.05	0.08	0.06
h) Vegetation albedo for wavelengths ≥0.7μm	0.30	0.30	0.23	0.23	0.28	0.20	0.30	0.40	0.26	0.28	0.34	0.60	0.18	0.20	0.20	0.23	0.28	0.24
i) Minimum stomatal resistance (s m^{-1})	150	250	250	250	250	250	250	250	250	250	250	250	250	250	250	250	250	250
j) Maximum LAI	6	2	6	6	6	6	6	0	6	6	6	0	6	0	0	6	6	6
k) Minimum LAI	0.5	0.5	5.0	1.0	1.0	5.0	0.5	0.0	0.5	0.5	0.5	0.0	0.5	0.0	0.0	5.0	1.0	3.0
l) Stem (& dead matter) area index	0.5	4.0	2.0	2.0	2.0	2.0	2.0	0.5	0.5	2.0	2.0	2.0	2.0	2.0	2.0	2.0	2.0	2.0
m) Inverse square root of leaf dimension ($m^{-1/2}$)	10	5	5	5	5	5	5	5	5	5	5	5	5	5	5	5	5	5
n) Light sensitivity factor (m^2 W^{-1})	0.01	0.01	0.03	0.03	0.03	0.03	0.01	0.01	0.01	0.01	0.01	0.01	0.01	0.01	0.01	0.01	0.01	0.03

*See definitions in Table 1.

**Soil depths in code are in mm as are all water storages to make the conversion factor 1.0 between water amounts and SI energy fluxes.

A number of the land–surface parameters defined as a function of land–cover type should, strictly, be made functions of soil type. However, for the present implementation, rooting ratios and upper and total soil depths (illustrated in Fig. 1) remain a function of land–cover type only (Table 2). A number of parameters do vary with soil type and are described in the next subsection.

b. Soil–Information Assignment

Wilson (1984) classifies soil color, texture, and drainage using the FAO Soil Map of the World (FAO/UNESCO, 1974) as the data source. Soil information is archived at 1°–×–1° resolution in three color classes, three texture classes, and three drainage classes. A soil–type archive has been constructed from these data on the R–15 (a 40–×–48 mesh related to 15 latitudinal and 15 longitudinal wave numbers in a rectangular array) model grid. Eight color classes were derived with the scale stretched at the light end. Twelve texture classes, ranging from 1 = very coarse (equivalent to sand) to 12 = very fine (equivalent to heavy clay), were assigned by averaging the textures from the 1°–×–1° data set. Texture class 6 is comparable to a loam soil. Three drainage classes were assigned by selecting the dominant drainage class from the 1°–×–1° data set except when the proportion of poor plus impeded drainage exceeded the proportion of free drainage. In that case, the model grid element was assigned a classification of poor drainage.

Soil properties have been associated with the twelve texture classes; dry and saturated albedos in two wavelength regions have been associated with the eight color classes. The drainage classes have not yet been implemented.

c. Vegetation Cover

Vegetation modifies surface processes by (a) shading the ground from sky radiation, (b) increasing the surface roughness and consequent exchanges of heat, moisture, and momentum, and (c) controlling water loss from dry leaves through transpiration processes. The first two effects are purely mechanical and do not require live leaf surfaces. Therefore, it is useful to define vegetation cover as that fraction of the ground shaded by both transpiring and nontranspiring plant surfaces. The latter includes not only stems, trunks, branches, etc., but also dead leaves, either on live plants, on dead plants, or as litter. Conversely, the nonvegetated area is that which appears as bare ground as seen, *e.g.*, by looking down from a satellite averaged over various viewing angles. The prescription of vegetation is then completed by defining leaf area and stem area indices, L_{AI} and S_{AI}, as the projected area of transpiring and nontranspiring surface per unit area of vegetated land. These are listed in the parameter table (Table 2) and are used in section 8. Since there are no satisfactory data sources for the 18 land-cover classes as defined, we derived our values by intelligent guessing, guided by various micrometeorological texts, *e.g.*, Geiger (1965), Monteith (1975), and other literature, in particular, Federer (1971) for leafless hardwood forests.

The fractional vegetation cover σ_f is determined for each grid point and at each time step as a function of the subsoil temperature T_{g2}, the land-cover type K_{VEG}, and the seasonal range of vegetation cover $\Delta\sigma_f$ listed above.
We take

$$\sigma_f = \sigma_f \text{ summer, if } T_{g2} > 298.0 \text{ K},$$

$$\sigma_f = \sigma_f \text{ summer} - \Delta\sigma_f \text{ seasonal, if } T_{g2} < 273.0 \text{ K},$$

$$\sigma_f = \sigma_f \text{ summer} - [1 - F_{SEAS}(T_{g2})]\Delta\sigma_f \text{ seasonal,} \tag{1a}$$

where

$$F_{SEAS}(T) = 1 - 0.0016 \times (298.0 - T)^2 \tag{1b}$$

is a factor that ranges from 1.0 at 298.0 K to 0.0 at 273.0 K and 323 K.

d. Vegetation and Soil Albedos

For each of the land grid points, three other variables are defined in subroutine ALBEDO—visible solar albedo of vegetation ($\lambda < 0.7\mu$m), near-infrared albedo of vegetation ($\lambda > 0.7\mu$m), and soil albedo. The values for the albedo of vegetation were determined from a variety of sources, in particular Monteith (1975, 1976), but also with reference to Monteith (1959), Kung *et al.* (1964), Barry and Chambers (1966), Federer (1968), Oguntoyinbo (1970), Stewart (1971), Tucker and Miller (1977), Rockwood and Cox (1978), Kriebel (1979), Fuller and Rouse (1979), and Kukla and Robinson (1980).

Gates *et al.* (1965) give spectral albedos of individual leaves. The near-infrared albedo for vegetation cannot be inferred from albedos of individual leaves since the high reflectivity (~50%) and transmissivities of these surfaces imply that radiative transfer considerations are dominant (Dickinson, 1983). Our division into visible and near-infrared fluxes is largely based on the data of Tucker and Miller (1977), Kriebel (1979), and Fuller and Rouse (1979).

The albedo of the nonvegetated soil surface was determined from a variety of sources, *e.g.*, Kondratyev (1969), Condit (1970), Idso *et al.* (1975), and Tucker and Miller (1977). The detailed distributions of the soil albedo depend upon soil type and soil wetness. However, for vegetation cover σ_f of 0.80 or more, relatively little short-wave radiation reaches the ground so that these parameterizations become secondary.

The values for vegetation albedo presently used in subroutine ALBEDO are listed in Table 2. We have not been able to find much data on the albedos of branches and brown vegetation as contrasted to green (besides Federer, 1968, 1971), but we assume that it is the same as green. It appears that trunks and branches may have lower albedos by as much as 5%, that red or brown leaves may have higher albedos by as much as 5%, and that the spectral variation of albedo will differ from that of green leaves.

The albedo for bare soil A_{LBG} is taken to be

$$A_{LBG} = A_{LBGO} + \Delta\alpha_g(S_{sw}),$$

where A_{LBGO} is the albedo for a saturated soil and where the increase of albedo due to dryness of surface soil is given for $\lambda < 0.7\mu$m as a function of the ratio of surface soil water content S_{sw} to the top layer depth Z_1,

$$\Delta\alpha_g(S_{sw}) = 0.01(11 - 40S_{sw}/Z_1) > 0. \quad (2)$$

This formulation is chosen so that soil albedos range in a nonlinear fashion between the saturated and dry values listed above. The term S_{sw} becomes small ($\leq$0.03 m) before the soil albedo shows a significant increase. Moisture is retained around the soil grains until ~80% dryness occurs. The soil albedos for $\lambda > 0.7\mu$m are twice those for $\lambda < 0.7\mu$m. Dry and saturated soil albedos for the eight color classes used are shown in Table 3.

e. Snow Albedos

Snow albedos are determined from the following equations and inferred from the calculations of Wiscombe and Warren (1980) and the snow model and data of Anderson (1976):

$$\alpha_V = \alpha_{VD} + 0.4f(ZEN)[1 - \alpha_{VD}], \tag{3}$$

$$\alpha_{IR} = \alpha_{IRD} + 0.4f(ZEN)[1 - \alpha_{IRD}], \tag{4}$$

where α_V = albedo for $\lambda < 0.7\mu$ m, α_{IR} = albedo for $\lambda > 0.7\mu$ m, and the subscript D denotes diffuse albedos as given by

$$\alpha_{VD} = [1 - C_S F_{AGE}]\, \alpha_{VO}, \tag{5a}$$

$$\alpha_{IRD} = [1 - C_N F_{AGE}]\, \alpha_{IRO}, \tag{5b}$$

$$C_S = 0.2,\ C_N = 0.5,$$

and

α_{VO}	= 0.95, the albedo for visible radiation incident on new snow with solar zenith angle less than 60°
α_{IRO}	= 0.65, the albedo of new snow for near-infrared solar radiation with solar zenith angle less than 60°
$f(ZEN)$	= factor between 0.0 and 1.0 giving increase of snow visible albedo due to solar zenith angle exceeding 60°
$C(ZEN)$	= cosine of the solar zenith angle
F_{AGE}	= a transformed snow age defined below and used in this section to give the fractional reduction of snow albedo due to snow aging for solar zenith angle less than 60°

The following parameterizations are used:

$$f(ZEN) = \frac{1}{b}\left[\frac{b+1}{1+2b\,C(ZEN)} - 1\right],\ f(ZEN) = 0 \text{ if } C(ZEN) > 0.5. \tag{6}$$

TABLE 3. SOIL PARAMETERS

I/FUNCTIONS OF TEXTURE

Parameter	Texture Class (from sand (1) to clay (12))											
	1	2	3	4	5	6	7	8	9	10	11	12
a) Porosity (volume of voids to volume of soil)	0.33	0.36	0.39	0.42	0.45	0.48	0.51	0.54	0.57	0.60	0.63	0.66
b) Maximum soil suction (m)	0.03	0.03	0.03	0.2	0.2	0.2	0.2	0.2	0.2	0.2	0.2	0.2
c) Saturated hydraulic conductivity ($mm\ s^{-1}$)	0.2	0.08	0.032	0.013	8.9×10^{-3}	6.3×10^{-3}	4.5×10^{-3}	3.2×10^{-3}	2.2×10^{-3}	1.6×10^{-3}	1.1×10^{-3}	0.8×10^{-3}
d) Ratio of saturated thermal conductivity to that of loam	1.7	1.5	1.3	1.2	1.1	1.0	0.95	0.90	0.85	0.80	0.75	0.70
e) Exponent "B" defined in Clapp & Hornberger (1978)	3.5	4.0	4.5	5.0	5.5	6.0	6.8	7.6	8.4	9.2	10.0	10.8
f) Moisture content relative to saturation at which transpiration ceases	0.095	0.128	0.161	0.266	0.300	0.332	0.378	0.419	0.455	0.487	0.516	0.542

II/FUNCTIONS OF COLOR

Parameter	Color (from light (1) to dark (8))							
	1	2	3	4	5	6	7	8
a) Dry soil albedo								
$< 0.7\ \mu m$	0.23	0.22	0.20	0.18	0.16	0.14	0.12	0.10
$\geqslant 0.7\ \mu m$	0.46	0.44	0.40	0.36	0.32	0.28	0.24	0.20
b) Saturated soil albedo								
$< 0.7\ \mu m$	0.12	0.11	0.10	0.09	0.08	0.07	0.06	0.05
$\geqslant 0.7\ \mu m$	0.24	0.22	0.20	0.18	0.16	0.14	0.12	0.10

Equation (6) has the property for all b that it vanishes at $C(ZEN) = 0.5$ and is unity at $C(ZEN) = 0$ (sun on the horizon); b is adjustable to best available data—for now $b = 2.0$.

Snow albedo decreases with time due to growth of snow grain size and accumulation of dirt and soot. We parameterize the decrease term F_{AGE} by

$$F_{AGE} = \tau_{SNOW}/[1 + \tau_{SNOW}]. \tag{7}$$

The nondimensional age of snow τ_{SNOW} is incremented as follows:

$$\Delta\tau_{SNOW} = \tau_0^{-1}(r_1 + r_2 + r_3)\Delta t, \tag{8}$$

where $\tau_0^{-1} = 1 \times 10^{-6}\ \mathrm{s}^{-1}$,

$$r_1 = \exp\left[5000\left(\frac{1}{273.16} - \frac{1}{T_{g1}}\right)\right],$$

$$r_2 = r_1^{10} \leq 1,$$

and

$$r_3 = 0.01 \quad \text{over Antarctica}, \quad = 0.3 \text{ elsewhere}.$$

The term r_1 represents the effect of grain growth due to vapor diffusion, the temperature dependence being essentially proportional to the vapor pressure of water.

The term r_2 represents the additional effect near and at freezing of melt water and r_3 the effect of dirt and soot.

A snowfall of 0.01 m liquid water is assumed to restore the surface age, hence albedo, to that of new snow. Since the precipitation in one model time step will generally be less than that required to so restore the surface when it snows for a given time step, we reduce the snow age by a factor depending on the amount of the fresh snow in m, ΔP_s, as follows:

$$\tau_{SNOW}^{N+1} = (\tau_{SNOW}^{N} + \Delta\tau_{SNOW})(1 - 100\Delta P_s)\tau_{SNOW} > 0, \tag{9}$$

where $\Delta\tau_{SNOW}$ is defined as in Eq. (8).

23

3. COUPLING OF THE BATS PACKAGE TO THE REST OF THE MODEL AND SUMMARY OF CALCULATIONS (AS OF SEPTEMBER 1986)

The CCM subroutine for surface calculations is modified by a Cray UPDATE modpack that links the BATS boundary package to the CCM. Because the land versus ocean distributions of the prescribed data sets may not match those of the CCM, checks are made to provide soil and vegetation data to any CCM points that are classified land in the CCM but ocean in the land-use data set. All the needed CCM fields are passed into a data structure adopted from the previous NCAR third-generation GCM. The lowest model temperature is multiplied by σ^{κ} for use as a "potential temperature." Instantaneous precipitation is calculated by differencing the accumulated precipitation at two time levels. A routine named SOLBDC is called to obtain soil hydrological properties as a function of the soil-data fields. The routine BNDRY is then called to enter the main boundary package. Finally, the fields are transferred back into a form suitable for the CCM. Note, in particular, that we pass heat, moisture, and momentum fluxes evaluated for surface temperatures present at the finish of the boundary package for use in the vertical diffusion routine.

Subroutine BNDRY calls individual physical process subroutines and evaluates parameters common to several routines, as indicated in Fig. 3 and Appendix C. In particular, it provides the relative soil moisture from the model moisture and maximum soil moisture storage. It calls subroutine DRAG to obtain transfer coefficients between the lowest model layer and the surface (canopy plus ground). These coefficients depend on the lowest model altitude, surface roughness z_0—determined from surface cover—and a bulk Richardson number calculated from the strength of the lowest model-layer winds and the temperature difference between surface and lowest model layer.

The vegetation part of the code is only executed for grid squares with vegetation cover greater than 0.01. A mean wind within the canopy is obtained from the mean wind outside the canopy times the square root of the drag coefficient

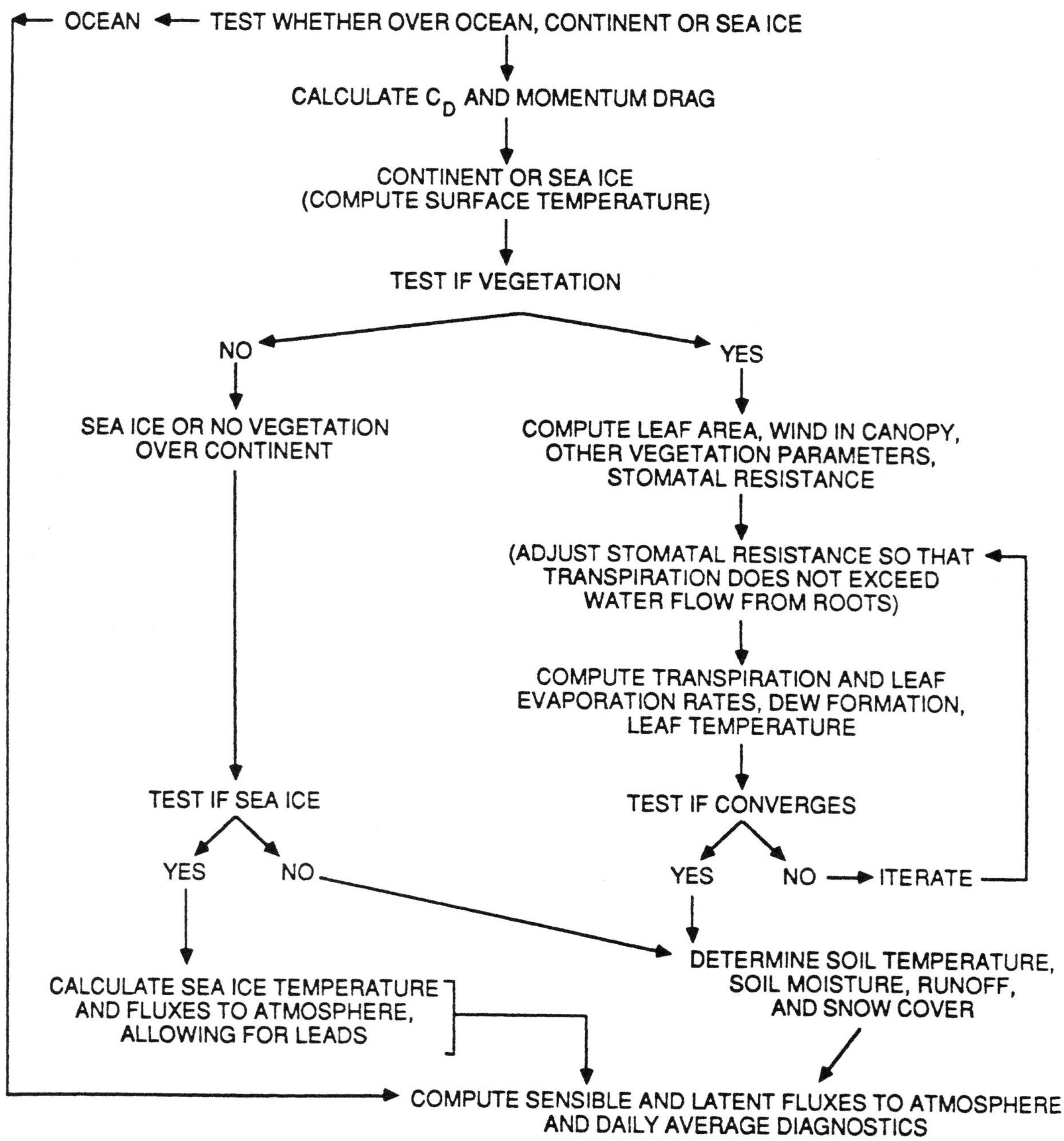

Figure 3. Flow diagram showing major features of the surface parameterization scheme employed in the NCAR CCM.

25

which equals the friction velocity u^*. The coefficient of transfer of heat and momentum from leaves is calculated. Foliage water is modified by intercepted rainfall. The temperature of the foliage (leaves) is calculated. Any rain or snow intercepted by leaves in excess of their maximum capacity is determined as falling to the ground and saved for soil-water or snow-budget calculations.

Returning to a calculation for all surfaces, we calculate rain or snow incident on the ground (minus any that was intercepted by the foliage) and partition soil evaporation into that from soil water and that from overlying snow. Routines are called to calculate the sea ice or the ground temperature and the budgets of snow cover and soil water. The updated temperatures, soil moisture, and foliage transpiration are used to determine net fluxes of heat and momentum from the surface to the lowest atmospheric model layer.

A surface-air temperature T_a is estimated for comparison with observations by using boundary-layer theory to interpolate between model layer one temperature and surface (ground, canopy-air or ocean) temperatures. The surface-air temperature is evaluated at 10 m over ocean and at 1.3 m over land, where it is assumed measured within a plot of short grass, embedded in the given cover type. The approximate formula used is

$$T_a = \widetilde{T}_{af} + \Delta T,$$

where

$$\widetilde{T}_{af} = T_{af}, \quad \text{if vegetated,}$$

$$\widetilde{T}_{af} = T_g, \quad \text{if nonvegetated,}$$

$$\Delta T = (T_1 - \widetilde{T}_{af}) z_{DELT},$$

$$z_{DELT} = \text{minimum of } (1., b/[\log(z_1/z_0)]^2),$$

where $z_0 = 0.01$ m, $b = 4.9C_D/C_{DN}$, where C_D is the drag coefficient at z_1 and C_{DN} is its value for neutral conditions.

The BATS code was coupled to a version of CCM0B that included several of the improvements in vertical diffusion adopted for CCM1 and some other changes

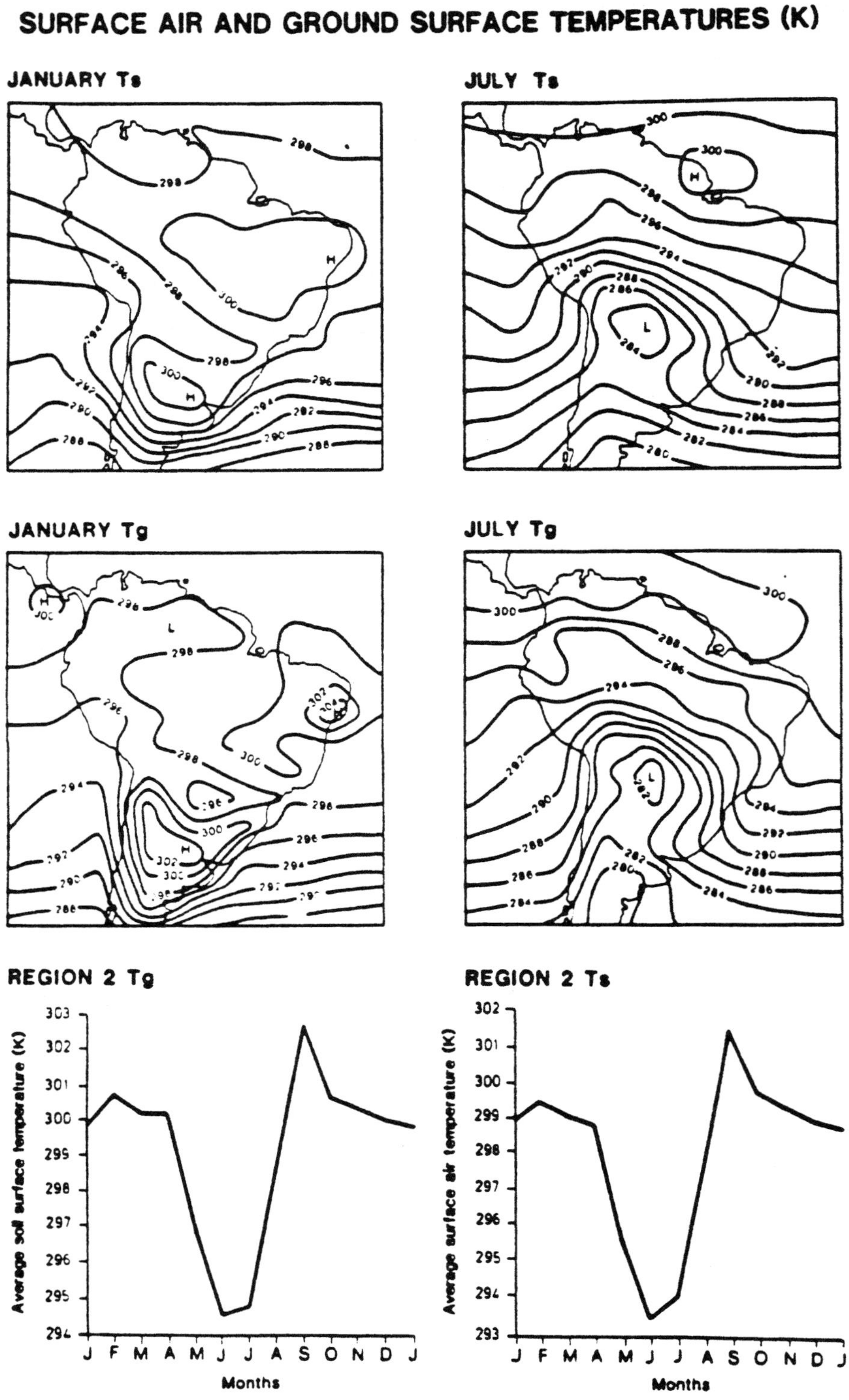

Figure 4. Mean January and July surface air and soil temperatures over South America for the BATS parameterization scheme run in CCM0B. Region 2 refers to the average over the Amazon region south of the equator (Dickinson and Henderson-Sellers, 1986).

27

that are not to be used further. Several seasonal cycles were carried out for a control as well as a 13-month Amazon deforestation simulation. An example of results from the control simulation is shown in Fig. 4. Over the next year, the previous control simulations will be repeated with CCM1, only modified by BATS so that cleaner comparisons of its effects on the atmospheric simulations can be made.

Further examples of results using the BATS parameterization may be found in the recent papers of Henderson-Sellers *et al.* (1986), Wilson *et al.* (1986a,b), and Dickinson and Henderson-Sellers (1986).

4. SOIL TEMPERATURE

The model follows Deardorff (1978) in using a "force-restore" method for the subsoil temperature reservoir. The soil and subsurface temperatures are obtained from

$$\frac{\partial T_{g1}}{\partial t} = \frac{c_1 h_s}{\rho_s c_s d_1} - \frac{c_2(T_{g1} - T_{g2})}{\tau_1}, \tag{10a}$$

$$\frac{\partial T_{g2}}{\partial t} = -c_3\left[\frac{(T_{g2} - T_{g1})}{\tau_1} + c_4(T_{g2} - T_{g3}) + Q_{sf}\right], \tag{10b}$$

where

T_{g1} = Surface soil and/or snow temperature (also referred to as T_g)

T_{g2} = Subsurface soil and/or snow temperature

T_{g3} = Fixed annual mean deep soil temperature

t = Time

c_1 $= 2\pi^{1/2} = 3.5449$

h_s $= (S_g + (F_{IR}^{\downarrow} - F_{IR}^{\uparrow}) - F_s - L_{v,s}F_q - L_f S_m - L_f W_{m1})$ (10c

c_2 $= 2\pi = 6.28$

c_3 = Rate of subsoil relaxation; we assume = 0.2. Note: This factor depends on how deep a soil thermal reservoir we consider (which is somewhat arbitrary). Our choice was dictated by the requirement that T_{g2} have a seasonal cycle but not a significant diurnal one and correspond to a thermal reservoir of 1 to 2 m soil

c_4 = Damping of soil surface temperature to annual mean value. The parameter c_4 is set to zero everywhere except over permafrost where T_{g3} is set to freezing T_m.

τ_1 = Period of heating = 8.64×10^4 seconds (= 1 day)

where

S_g = Solar flux absorbed over bare ground at earth's surface

$F_{IR}^{\downarrow} - F_{IR}^{\uparrow}$ = Net IR (long wave) flux from atmosphere to bare ground

F_s = Atmospheric sensible heat flux from ground to atmosphere

F_q = Atmospheric moisture flux from ground to atmosphere

$L_{v,s}$ = Latent heat of evaporation or sublimation

L_f = Latent heat of fusion

Q_{sf} = $L_f W_{m2} c_1 / (\rho_s c_s d_1 c_2)$ = Rate of subsoil temperature change because of melting or freezing

S_m = Rate of snow melt

W_{m1} = Rate of melting (negative for freezing) of surface soil water

W_{m2} = Rate of melting (negative for freezing) of subsurface soil water

$\rho_s c_s$ = Specific heat of subsurface layer per unit mass

d_1 = $(k_s \tau_1)^{1/2}$ = Soil depth influenced by a periodic heating (about 0.2 m for a typical soil)

where

k_s = soil or snow thermal conductivity ($m^2\ s^{-1}$)

While Eq. (10) is intended primarily for a diurnal cycle of soil heating, the coefficient c_4 is chosen so as to give the proper temperature–amplitude seasonal forcing.

According to Nakano and Brown (1972) and further references they give, the fraction of soil water that is a frozen state at temperatures below 0°C depends on soil type, temperature, and soil moisture. As a crude but simple parameterization, we assume that the fraction of frozen soil is linear for temperatures from freezing

29

$T = T_m$ to $T = T_m - 4°$C and that 0.15 of the soil water density remains unfrozen at temperatures colder than that. Then the heat required for melting (freezing) ground to be used in (10a–10c) is 0.0 if $T \geq T_m$, or $T \leq T_m - 4°$. Otherwise,

$$W_{m1} = 0.25(S_{sw} - F_{ru})\frac{\partial T_{g1}}{\partial t}, \qquad S_{sw} > F_{ru}, \tag{11a}$$

$$W_{m2} = 0.25(S_{tw} - S_{sw} + F_{ru} - F_{ra})\frac{\partial T_{g2}}{\partial t}, \quad S_{tw} - S_{sw} > (F_{ra} - F_{ru}). \tag{11b}$$

F_{ru} and F_{ra} are the water in the upper and lower soil layers, respectively, that is too bound to the soil particles to freeze. These quantities are taken to be 0.15 of the depth of the upper and lower soil layers, respectively. The factor of 0.25 is the inverse of soil freezing range of 4°, and both this factor and the range should be adjusted together.

Define

$$\widetilde{h}_s = (h_s + L_f W_{m1})^N + 2C_g T_{g1}^N, \tag{11c}$$

where C_g is the coefficient of latent heat transfer from ground to air. Adding the last term in (11c) to h_s and subtracting the same term at $N+1$ below in Eq. (12) have the effect of including the sensible heat flux at the $N+1$ time level which is a more stable finite–differencing than evaluating F_s at the N'th time level.

The force–restore finite differencing for Eqs. (10)–(11) becomes

$$\widetilde{C}T_{g1}^{N+1} = \widetilde{C}T_{g1}^{N} - \frac{\Delta t}{2\tau_1}c_2(T_{g1}^{N+1} + T_{g1}^{N}) + \widetilde{B}\Delta t[\widetilde{h}_s - C_g(T_{g1}^{N} + T_{g1}^{N+1})] + c_2\frac{\Delta t}{\tau_1}T_{g2}^{N}, \tag{12}$$

where

$$\widetilde{B} = c_1/(\rho_s c_s d_1),$$

$$\widetilde{C} = 1.0 + 0.25(S_{sw} - F_{ru})L_f\widetilde{B}\delta(T_m, T_m - 4.0°),$$

$$\delta(T_a, T_b) = \begin{cases} 1., & \text{if} \quad T < T_a, \text{ and } T > T_b\ , \\ 0, & \text{otherwise}\ . \end{cases}$$

If we set

$$A = 0.5c_2\Delta t/\tau_1 + \widetilde{B}C_g\Delta t; \;\; B = (\widetilde{B}\widetilde{h}_s + \frac{c_2}{\tau_1}T_{g2}^N)\Delta t,$$

30

then

$$T_{g1}^{N+1} = \frac{(\widetilde{C} - A)T_{g1}^{N} + B}{\widetilde{C} + A}, \tag{12b}$$

where ΔT is the same time step as in the GCM. Equation (10b) is solved in the same manner except there is no term corresponding to h_s as in the above. Note that Eqs. (10a)–(10b) do not exactly conserve energy due to time variations in d_1 and c_s. The consequent error is believed negligible.

Details in the implementation of Eqs. (10)–(12) depend on the nature of the surface as now discussed.

a. Over Soil

Thermal conductivity and heat capacity of soil depend primarily on its moisture and organic content. From the theory of de Vries (1963) and Fig. 3.1 of de Vries (1975), we infer the following approximate formula for a loam soil:

$$\rho_s c_s = (0.23 + \rho_w) c_w, \tag{13}$$

$$k_s = \lambda_s / (0.23 + \rho_w), \tag{14a}$$

$$\lambda_{s\ell} = \frac{(2.9\rho_w + 0.04)\lambda_c}{(1 - 0.6 \times \rho_w)\rho_w + 0.09}, \tag{14b}$$

where $\lambda_c = 10^{-7}$ m^2 s^{-1}, ρ_w is the volume of liquid water per unit volume of soil as defined by Eq. (27b), and $c_w = 4.186 \times 10^6$ J m^{-3} K^{-1} is the specific heat of water. From this formula, the thermal conductivity of all soil types is given by

$$\lambda_s = S_{KRAT}^{i} \times \lambda_{s\ell}, \tag{14c}$$

where S_{KRAT}^{i} is the ratio of the thermal conductivity of texture type i to that of loam. The values of S_{KRAT}^{i} are listed in Table 1.

b. Over Snow–Covered Land

The computation is similar to bare land except that the constants are for snow (*e.g.*, Corps of Engineers, 1956; Yen, 1969; Anderson, 1976), *i.e.*,

$$k_s = k_{so} \rho_{sw}, \tag{15a}$$

where $k_{so} = (7.0 \times 10^{-7}/0.49)$ m^2 s^{-1} ,

$$\rho_s c_s = \rho_{sw} c_{sw}, \tag{15b}$$

where

$$c_{sw} = 0.49 c_w$$

is the specific heat of snow for unit snow density, numerically equal to the specific heat of ice per unit mass, and where ρ_{sw} is the density of the snow relative to water. Typically, for fresh snow, $\rho_{sw} = 0.1$ and for old surface snow, $\rho_{sw} = 0.3$.

For determining snow density, we use

$$\rho_{sw} = 0.1 + 0.3 F_{AGE}, \tag{15c}$$

where F_{AGE} is the snow age factor given by Eq. (7).

To allow for fractional bare ground, weighted means of snow and ground conductivities and of heat capacities are used, with equal weight for 0.1 m of snow, *i.e.*,

$$k_s = \frac{d_s k_{sSNOW} + k_{sGROUND}}{1 + d_s},$$

where $d_s = 10 S_{cv}/\rho_{sw}$ and S_{cv} is the snow cover (liquid water equivalent in m). With 0.1 m of snow, *i.e.*, $d_s = 1$, the penetration of the daily temperature wave in the snow is approximately the depth of the snow. The term $\rho_s c_s$ is averaged likewise. A more precise averaging does not seem warranted because of the likely unevenness of the snow distribution.

c. Over Bare Sea Ice or Snow-Covered Sea Ice

Over sea ice, the diurnal component of heat storage within the ice is neglected and replaced by steady conduction of heat from the underlying ocean. Hence, the sum of the components of energy fluxes becomes

$$\frac{\partial T_g}{\partial t} = \frac{h_s}{[0.5(c_{sw} S_{cv} + w_t c_i d_{ice})]}, \tag{16}$$

where $c_i = 0.45 c_w$ and d_{ice} is the depth of ice, specified as a function of season, or computed, and $w_t = 1.4\rho_{sw} d_{ice} \div (S_{cv} + 1.4 \rho_{sw} d_{ice})$,

$$h_s = S_g + F_{IR}^{\downarrow} - F_{IR}^{\uparrow} - F_s - L_v F_q - L_f S_m + F_c. \tag{17}$$

The square–bracketed term in the denominator of (16) is an effective heat capacity of the snow and ice, assuming a temperature profile through the snow and ice controlled by steady–state heat conduction. The heat conduction through snow–covered sea ice or bare sea ice, following Maykut and Untersteiner (1971) and Semtner (1976), is given by

$$F_c = \frac{K_{snow}(T_B - T_g)}{1 + (K_{snow}/K_{ice})},$$

where

$$K_{snow} = k_s \rho_{sw}^2 c_{sw}/S_{cv},$$

$$K_{snow}/K_{ice} = 1.4\rho_{sw}^2 d_{ice}/S_{cv},$$

$$T_B = -1.8°C, \qquad (18)$$

approximate freezing point of sea ice. Inclusion of leads in the ice is discussed in section 6c.

5. SOIL MOISTURE AND SNOW COVER IN THE ABSENCE OF VEGETATION

For the soil moisture–snow cover specification, the earth's surface is divided into two regions—(1) oceanic regions (nonsea–ice–covered and sea–ice–covered regions), and (2) continental regions with and without snow cover. For the nonsea–ice–covered oceanic regions, the surface temperature T_g is prescribed from observational data in the standard model. For other regions, the computation of T_g depends on the current conditions of snow cover, soil moisture, type of surface, and temperature of the first layer of the atmosphere.

a. Precipitation (Rain and Snow)

The rainfall and latent heat release (Q_c) in each atmospheric layer depend in a complicated fashion on layer humidity and precipitation in overlying layers (the latter determining the phase of the precipitation).

The precipitation rate at the ground (P) is obtained as the sum of net precipitation from each layer. This precipitation is assumed to fall as snow P_s, if for the lowest model layer ($\sigma = 0.991$) $T_1 \leq T_c$, or as rain P_r if $T_1 > T_c$ (Auer, 1974), where $T_c = T_m + 2.2°$.

$$P_s = P,\ P_r = 0,\ \text{if } T_1 \leq T_c, \qquad (19)$$

$$P_s = 0,\ P_r = P,\ \text{if } T_1 > T_c.$$

b. Soil Moisture Budget

Moisture incident on the ground either infiltrates the soil or is lost to surface runoff. We consider two parameters to represent soil moisture:

S_{tw} = Total water in the rooting zone depth D_0 of soil

S_{twmax} = Maximum total soil water

S_{sw} = Surface soil water representing water in the upper layer (depth D_1) of soil

S_{swmax} = Maximum upper soil water

The rooting zone depth D_0 and the upper soil depth D_1 are functions of land–cover type and are listed in Table 2.

Both S_{tw} and S_{sw} gain the same amount of water from rainfall P_r and lose the same amount from evaporation F_q and surface runoff R_s, since these occur at the soil surface. Unique to S_{sw} is the term denoted Υ_w representing the transfer of water to the upper layer from the rest of the soil column. Unique to S_{tw} is the leakage down to subsoil and ground water reservoirs (representing the bulk of the runoff) and it is denoted R_g. The equations for S_{tw} and S_{sw} in the absence of vegetation are written

$$\frac{\partial S_{sw}}{\partial t} = G - R_s + \Upsilon_w, \qquad (20)$$

$$\frac{\partial S_{tw}}{\partial t} = G - R_s - R_g, \qquad (21a)$$

where

$$G = P_r + S_m - F_q. \qquad (21b)$$

G equals the net water applied to the surface, P_r = rainfall, S_m = snowmelt, and F_q = evaporation. Negative F_q represents dew formation. The terms F_q, R_s, R_g, and Υ_w are parameterized on the basis of a multilayer soil model (Dickinson, 1984). This parameterization is described in the remainder of this section.

c. Infiltration and Percolation to Ground Water

In principle, each grid square should have a distribution of soil types as established by its climate, vegetation, geology, etc. However, for now we assume a single soil type for each grid square and with the following properties, mostly dependent upon the soil texture and listed in Table 3 (*e.g.*, Campbell, 1974; Clapp and Hornberger, 1978).

(a) Porosity = P_{ORSL} (see Table 3), *i.e.*, at saturation 1 m^3 of soil holds P_{ORSL} (m^3) of water.

(b) Soil water suction (negative potential): $\phi = \phi_o s^{-B}$ where ϕ_o values are listed in Table 3; B ranges from 3.5 to 10.8 (Table 3), and s = volume of water divided by volume of water at saturation. For example, if $s = 0.3$, $\beta = 5$, and $\phi_o = 0.2$ m, $\phi = 200$ m of tension, the value at which transpiration is assumed to cease, as indicated in Table 3, line f. Note that $s = \rho_{sw}/P_{ORSL}$.

(c) Hydraulic conductivity: $K_w = K_{wo} s^{2B+3}$, with values for K_{wo} ($\mathrm{m\,s^{-1}}$) listed in Table 3 which represents the flow rate for saturated soil due to gravity.

Water represented by s diffuses through the soil with a diffusivity

$$D = -K_w \partial\phi/\partial s = K_{wo} B s^{B+2}. \tag{22}$$

Besides the diffusive movement, there is gravitational drainage which dominates the flow for large–enough length scales. This provides the subsoil drainage expression,

$$R_g = K_{wo} s^{2B+3}, \tag{23}$$

e.g., at $s = 0.6$ and $K_{wo} = 4 \times 10^{-4}\ \mathrm{m\,s^{-1}}$, $R_g = 0.5 \times 10^{-8}\mathrm{m\,s^{-1}} (= 0.5\ \ \mathrm{mm\,d^{-1}})$.

d. Evaporation

The evaporative terms F_q and the transfer between the upper soil layer and below are difficult to parameterize with sufficient generality. The current expressions are based on the behavior of a soil column that is initially at field capacity and dried by a diurnally varying potential evaporation applied at the surface. Based on multilayer soil model integrations and theoretical arguments, we adopt the following parameterizations:

$$F_q = \text{minimum of } (F_{qp}, F_{qm}), \tag{24}$$

where F_{qp} = potential evaporation, and F_{qm} = maximum moisture flux through the wet surface that the soil can sustain. F_{qp} is calculated using Eq. (39) with $f_g = 1$ as the evaporation from a wet surface with the same aerodynamic characteristics as the soil surface, and

$$F_{qm} = C_k D s_0/(Z_0 Z_1)^{1/2}, \tag{25a}$$

where Z_0 is the depth of the soil active layer (between 0.5 and 2 m), Z_1 is the depth of the surface soil layer (restricted to be between 1- and 20-cm thickness), and where D is an average soil diffusivity defined by

$$D = 1.02 D_{max} s_1^B s_0^2 (s_0/s_1)^{B_f}, \tag{25b}$$

using

$$D_{max} = B\phi_0 K_0/\rho_{wm}, \tag{25c}$$

where ρ_{wm} is the fraction of saturated soil filled by water, and K_0 and ϕ_0 the maximum conductivity and soil suction of Table 3, and defining

$$B_f = 5.5 - 0.8B[1 + 0.1(B-4)\log_{10}(K_0/K_R)], \tag{25d}$$

with $K_R = 10^{-5}$ m s^{-1}. The term C_k in Eq. (25a) is

$$C_k = (1 + 1550 D_{min}/D_{max}) \left[\frac{B - 3.7 + 5/B}{B+5}\right]. \tag{25e}$$

Equation (25) was tested for values of ϕ_0 from 0.1 to 1.0 m s^{-1}. The nominal value of D_{min} is 10^{-9} m^2 s^{-1}, and the expression was tested for values of D_{min} from 10^{-8} m^2 s^{-1} to 10^{-10} m^2 s^{-1}. The general structure of the terms in (25a–e) is indicated from dimensional analysis and physical reasoning, but their detailed structure was inferred from trial-and-error numerical integrations. Since potential evaporation rarely exceeds 4×10^{-7} m s^{-1}, soil much wetter than field capacity will generally evaporate at the potential rate.

During the process of evaporation in response to daytime heating, the soil moisture distribution approaches a self-similar profile. For such a profile, the water loss from the top layer is proportional to the water loss from the total active layer, with a proportionality factor that is just the ratio of the water contents within these two layers, *i.e.*, $(s_1 Z_1)/(s_0 Z_0)$. In that case, the down-gradient rate of transfer of water from the lower soil layer to the surface layer, denoted Υ_w, is given by

$$\Upsilon_w = \Upsilon_{w0} = C_k D(s_0 - \varepsilon s_1)/(Z_0 Z_1)^{1/2}, \tag{26a}$$

where $\varepsilon = Z_1/Z_0$ is the ratio of the depth of the upper layer to the depth of the total active soil layer. Although Eq. (26a) with $\varepsilon = Z_1/Z_0$ works well for the assumed regime of daily evaporation, it gives unrealistic results during infiltration when s_1 exceeds s_0. With infiltration, it is more realistic to assume that the downward transfer by diffusion is proportional to $(s_0 - s_1)$. To interpolate between these two limits, we assume

$$\varepsilon = \frac{Z_1/Z_0}{\text{maximum of } (Z_1/Z_0 \text{ or } (1 - s_1/s_0) \div (1 - \hat{s}_1/\hat{s}_0))}\ , \tag{26b}$$

where $\hat{s}_1/\hat{s}_0$ is the ratio of s_1 to s_0 during evaporation with a self-similar profile. For simplicity, we use typical values appropriate to $B = 5$, $K_0 = 10^{-5}$ m s^{-1}, $Z_1 = 0.1$ m and $Z_0 = 1$ m, giving $\hat{s}_1/\hat{s}_0 = 0.8$. An additional gravitational drainage term is also included so that the model water transfer between layers is evaluated from

$$\Upsilon_w = \Upsilon_{w0} - (s_1/s_0)^{B+1.5} R_g, \tag{26c}$$

where R_g is given by Eq. (23). The geometric mean of s_1 and s_0 has been assumed for the value of s at the interface between the upper and lower layer. The possibility of a gradient-driven upward water flow from an underlying saturated zone is neglected. With use of Eqs. (26a–c), water in the surface layer is generally transferred downward as rapidly as generated by rainfall of GCM grid square intensity, for soils other than heavy clay. Hence, water does not pond very frequently on the surface from saturation of the surface layer. It is thus necessary to assume additional surface runoff near saturation, tuned to get the observed surface runoff (globally about half the total runoff), or to assume a subgrid-scale variation of rainfall intensity.

e. Surface Runoff

During periods of heavy rainfall or snowmelt and high soil moisture, much of the water incident on natural surfaces does not penetrate to ground-water reservoirs but, rather, flows immediately into streams and rivers.

The classical hydrology description is that such fast runoff flows on the surface in sheets or rivulets (overland flow). However, surface runoff is now described as occurring primarily over the fraction of a grid square where the soil has become saturated due to a high water table or to impermeable surface soil (the so-called variable source area). In some localities, near-surface flows, guided by underlying impermeable layers or subground air channels, are also important. Due to the complex nature of surface runoff processes, it is not possible to model them in detail. Thus, the surface runoff is usually inferred as a function of the rainfall history over a basin by analyzing the flood peaks in stream and river hydrographs. Although, in principle, there is no unique decomposition of a hydrograph into surface and ground-water runoff, in practice it is not difficult to devise such approximate divisions. L'vovich (1979), in particular, has provided annual average global maps of surface and ground-water runoff. He finds, on the average, that about half the total runoff is surface runoff. The physical basis for modeling hydrology on the small scale is reviewed in Kirby (1979). Available models for runoff simulation are summarized in Fleming (1975).

Guided by the criteria that there should be small surface runoff at the soil moisture of field capacity and complete surface runoff at saturated soil, we have parameterized surface runoff R_s by

$$\begin{aligned} R_s &= (\rho_w/\rho_{wsat})^4 G, \quad T_{g2} \geq 0°C, \\ &= (\rho_w/\rho_{wsat}) G, \quad T_{g2} < 0°C, \end{aligned} \tag{27a}$$

38

where ρ_{wsat} is the saturated soil water density and ρ_w is the soil water density weighted toward the top layer, as defined by

$$\rho_w = \rho_{wsat} \frac{(s_o + s_1)}{2}, \tag{27b}$$

and G is defined by (21b). For negative $G, R_s = 0$. If the subsurface temperature is below freezing, the fraction of surface runoff is increased. This allows crudely for the effect of frozen ground.

f. Snow Cover

The most detailed model of snow energy balance and melt processes to date has been given by Anderson (1976). He models, in particular, water and energy transfer and density changes throughout the snow column. By contrast, we model explicitly only the snow–surface processes. There is no explicit distinction between subsurface snow versus soil temperature, *i.e.*, T_{g2} refers, in principle, to a subsurface snow temperature after more than a few centimeters of liquid equivalent snow have accumulated. The most serious conceptual errors occur during time of snow melt or rainfall on a snowpack. We put water on the snow surface directly into the soil, whereas real melt or rain water has to percolate through the snow pack and may refreeze. We also implicitly neglect melting at the bottom of the snow pack due to heat conducted from the ground (ground melt) unless this heat reaches the top snow surface.

If it is snowing or if there is snow cover, we first check to see if T_g is 0.0°C, and if so, compute the snow melt rate before computing the surface temperature. The snow melt rate S_m^N, when positive, is given by

$$S_m^N = L_f^{-1} \times [h_s^N - G_{HEAT}], \tag{28}$$

where h_s^N is the expression given by (10c) without the snowmelt (*i.e.*, $h_s = S_g + F_{IR}^{\downarrow} - F_{IR}^{\uparrow} - F_s - L_s F_q$) and where G_{HEAT} is the heat conduction to the underlying reservoir, which is either snow or ground, depending on the snow depth, that is consistent with Eq. (10a), *i.e.*, $G_{HEAT} = c_2 \rho_s c_s d_1 [(T_{g1} - T_{g2})/c_1 \tau_1]$.

39

In calculating F_s and F_q over melting snow, we take T_g to be 0.0°C. Also, $L = L_s$ is the latent heat of sublimation of water ($L_s = 2.834 \times 10^6$ J kg^{-1}) and L_f is the latent heat of fusion of water ($L_f = 3.34 \times 10^5$ J kg^{-1}).

If S_m calculated from Eq. (28) exceeds the snow cover, S_m is set to the snow cover and the ground temperature is calculated from Eqs. (10a and c), including the heat loss required to melt the snow. If snow remains after melting the snow amount given by Eq. (28), we set

$$T_g = 0^\circ \text{ C}.$$

The snow cover is updated from

$$\frac{\partial S_{cv}}{\partial t} = P_s - F_q - S_m, \tag{29}$$

where S_{cv} is the snow–cover amount measured in terms of liquid water content, P_s is the snow precipitation rate, and F_q equals the rate of sublimation.

6. DRAG COEFFICIENTS

In the CCM0, the drag coefficient C_D is a prescribed constant. However, in reality, it is quite variable (*e.g.*, *cf.* Garratt, 1977). Thus, in the BATS scheme, C_D is calculated as a function of C_{DN}, the drag coefficient for neutral stability, and Ri_B is the surface bulk Richardson number, *i.e.*,

$$C_D = f(C_{DN}, Ri_B), \tag{30}$$

where C_{DN} is the drag coefficient for neutral stability, Ri_B is the surface bulk Richardson number,

$$Ri_B = \frac{g z_1 (1 - T_{g1}/T_a)}{V_a^2}, \tag{31}$$

where $V_a^2 = u_1^2 + v_1^2 + U_c^2$ with T_{g1} the surface soil (or snow or ice) temperature and T_1, u_1, v_1 the air temperature $\times$ $(p_s/p_1)^\kappa$ and wind components at z_1, where z_1 is the height of the lowest model level, g = acceleration due to gravity, and

$$\begin{aligned} U_c &= 0.1 \text{ m s}^{-1}, \quad \text{if } T_{g1}/T_1 < 1, \\ &= 1.0 \text{ m s}^{-1}, \quad \text{if } T_{g1}/T_1 > 1. \end{aligned} \tag{32}$$

40

Also, modifying slightly the formulae given in some notes of Deardorff (personal communication), we take for $Ri_B < 0$

$$C_D = C_{DN}(1 + 24.5(-C_{DN}Ri_B)^{1/2}), \tag{33a}$$

and for $Ri_B > 0$

$$C_D = C_{DN}/(1 + 11.5Ri_B). \tag{33b}$$

In reality, the C_D value for heat is somewhat larger under unstable conditions and somewhat smaller under stable conditions. This refinement did not seem to be numerically large enough to warrant its inclusion. A similar formulation is now used in the new CCM1.

The neutral drag coefficient is obtained from mixed-layer theory as

$$C_{DN} = \left[\frac{k}{\ln(z_1/z_0)}\right]^2, \tag{34}$$

where $k = 0.40$, the von Karman constant, and z_0 is the roughness length. For water surfaces, we take $z_0 = 2.3 \times 10^{-4}$ m so that $C_{DNW} = 0.0014$. For bare land, we take $z_0 = 10^{-2}$ m so that $C_{DNL} = 2.4\ C_{DNW}$.

a. Vegetation

Over vegetated grid squares, the neutral drag coefficient is estimated by a linear combination for drag coefficients or for vegetation over bare soil or over snow. We assume that the snow coefficient has the same value as that of the ocean, *i.e.*, C_{DNW}. If $C_{D,FN}$ denotes the neutral drag coefficient over the grid square, then

$$C_{D,FN} = \sigma_f C_{FN} + (1 - \sigma_f)(S_{cv}C_{DNW} + (1 - S_{cv})C_{DNL}), \tag{35}$$

where C_{FN} is the local drag coefficient over vegetation, determined using Eq. (34) and $z_0 = z_{0V}$ tabulated for the given vegetation type, and S_{cv} is the fraction of ground covered by snow. The vegetation fraction has been corrected for snow covering vegetation as follows:

$$\sigma_f = (1 - W_T)\sigma_{fo}, \tag{36}$$

where σ_{fo} is the fraction of vegetation in the absence of snow and

$$W_T = W_{T0}/(1 + W_{T0}), \tag{37a}$$

$$W_{T_0} = \text{snowdepth}/(10.z_{0V}). \tag{37b}$$

The term S_{cv} is calculated from $S_{cv} = S_{cv0}/(1 + S_{cv0})$, S_{cv0} = snow depth/(0.1 m). Snow depth is calculated from the liquid–water content of snow divided by the model–calculated snow density. These formulae are based on the assumption that half the vegetation is covered by snow for snow depth of 10 z_{0V}, as in (37b), and half of bare ground for a snow depth of 0.01 m, assuming $z_{0V} \gg 0.01$ m, as it is for the input data.

The form of area averaging that has been assumed would be more appropriate for spatially separated vegetated and nonvegetated regions within the grid square. By contrast, a sparsely but uniformly vegetated area can exert considerably more drag than a more heavily vegetated area.

b. Leads over Sea Ice

Over sea ice, it is important to allow for leads. This is done by prescribing a fraction of surface a covered by leads, over which the assumed water temperature = 1.8°C, and at sea level the saturated mixing ratio of leads = 3.3×10^{-3} kg kg^{-1}. Drag coefficients are calculated separately for the sea ice and for lead surfaces and then are linearly combined according to the relative fraction of the lead.

For lead factors for sea ice, we assume

a = 0.02 for Arctic, latitude < 80°N,

a = 0.01 for 80°–85°N and a = 0.005 within 5° of the North Pole,

and

a = 0.04 for Antarctica.

Observations have indicated that Arctic sea ice is more compact than Antarctic sea ice.

7. SENSIBLE AND LATENT HEAT FLUXES OVER ICE OR A BARE SURFACE AND CORRESPONDING VALUES FOR ATMOSPHERIC TEMPERATURE AND HUMIDITY AT THE LEVEL OF A STEPHENSON SCREEN

The sensible and latent heat fluxes over water, sea ice, or bare surfaces are obtained using the momentum drag coefficients obtained in the previous section as follows:

$$F_s = \rho_a C_p C_D V_a (T_{g1} - T_a), \tag{38}$$

where ρ_a is the surface air density, C_D the aerodynamic drag coefficient for heat, C_p the specific heat for air, and V_a the wind speed. Similarly, the moisture flux (from the surface) to the atmosphere F_q is given by

$$F_q = \rho_a C_D V_a f_g (q_g - q_a), \tag{39}$$

where q_g is the saturated mixing ratio at the temperature of the surface (ground, snow, ice, or water), q_a is the mixing ratio of the model lowest level, and f_g is a wetness factor, which has the value of 1.0 except for diffusion-limited soil surfaces, where it is defined by the ratio of actual to potential ground evaporation, *i.e.*,

$$f_g = F_q / F_{qp}.$$

8. ENERGY FLUXES WITH VEGETATION

Our treatment of vegetation within the CCM is not dissimilar to the usual one-layer formulation in the micrometeorological literature ascribed to Monteith (*e.g.*, Thom and Oliver, 1977), except that we consider separate ground-energy equations and separate resistances for transfer from above the canopy to air within the canopy and from air within the canopy to the foliage surfaces and allow for partial wetting of the canopy. In applying a self-consistent model rather than observations as is usually done, it is necessary to derive variables within the canopy in terms of lowest atmospheric model-layer variables. At each land grid point, we prescribe a fractional vegetation cover σ_f as described in section 2 (Deardorff, 1978; Shuttleworth, 1978).

a. Parameterization of Foliage Variables

The one-sided surface area of vegetation per unit area of ground consists of transpiring surfaces specified by a leaf area index, *i.e.*, (L_{AI}) and nontranspiring surfaces (including dead vegetation) specified by a stem area index (S_{AI}). The S_{AI} is a constant for each land type, whereas the L_{AI} has a seasonal variation, using the same dependence on subsoil temperature as used for σ_f,

$$L_{AI} = L_{AI}^{MIN} + F_{SEAS}(T_{g2}) \times (L_{AI}^{MAX} - L_{AI}^{MIN}), \tag{40}$$

where the seasonal factor $F_{SEAS}(T)$ is defined by Eq. (1b). We denote the sum of L_{AI} and S_{AI} the L_{SAI} (leaf stem area index), *i.e.*, $L_{SAI} = L_{AI} + S_{AI}$. To include evaporation from wetted stems and leaves, we follow Deardorff in defining the fractional area of the leaves covered by water as

$$\widetilde{L}_W = \left(\frac{W_{dew}}{W_{DMAX}}\right)^{2/3}, \tag{41}$$

where W_{dew} is the total water intercepted by the canopy and W_{DMAX} is the maximum water the canopy can hold, as defined in the next subsection. The same expression is used for the stems. The fraction L_d of foliage surface free to transpire is then defined by

$$L_d = (1.0 - \widetilde{L}_w)L_{AI}/L_{SAI}. \tag{42}$$

The values presently used for defining L_{AI} and S_{AI} are listed in the land–cover type parameter list for section 2. The grassland values were obtained from Ripley and Redman (1976). Other values are based on a wide variety of inputs. The $S_{AI} = 0.5$ of class 1 (cropland) corresponds to the 0.1 used by Deardorff (1978) for S_{AI}/L_{AI}.

Also needed is the magnitude of wind within the foliage layer taken to be

$$U_{af} = V_a C_D^{1/2}. \tag{43}$$

b. Vegetation Storage of Intercepted Precipitation and Dew

When it rains, the surfaces of vegetation become covered with a film of water before drip through and stem flow carry water to the ground. This water can then

44

reevaporate to the air, but at the same time transpiration is suppressed over wet green leaves. Similarly, the formation of nighttime dew can keep foliage cool in the morning and suppress transpiration. Typical values for reevaporation of intercepted rainfall are in the range of 10 to 50% of rainfall, depending primarily on rainfall intensity. The suppression of transpiration by wet leaves has been less studied but is probably also significant. Snowfall is also intercepted by foliage, and frost formation on foliage commonly occurs. These are of somewhat less significance for the water budget because of lower evapotranspiration rates at low temperatures. Hence, it is reasonable to assume that vegetation storage of solid water is the same as liquid water. In doing so, we ignore the larger initial water storage of snow interception and its frequently more rapid removal by blowoff. We assume a maximum water storage of 0.0002 m $\times$ L_{SAI}. The water stored per unit land-surface area is calculated from the incident precipitation and difference between transpiration and water flux to the plant surface, *i.e.*,

$$\frac{\partial W_{dew}}{\partial t} = \sigma_f P - E_f + E_{tr}. \tag{44}$$

If $W_{dew} > W_{DMAX} = 0.0002$ m $\times \sigma_f L_{SAI}$, then W_{dew} is set equal to W_{DMAX} and the excess leaf moisture is added to the precipitation on the soil, either as water or snow, depending on whether or not the snowfall criteria are satisfied. More general drip formulae are discussed by Massman (1980).

c. Foliage Fluxes

We first discuss the evaporation from wet foliage. The water flux from dry foliage follows similar considerations but, in addition, the resistance to water flux by stomata needs consideration. The water on wet foliage (leaves and stems) evaporates according to

$$E_f^{WET} = A_f \rho_a r_{\ell a}^{-1} (q_f^{SAT} - q_{af}), \tag{45}$$

where A_f is the projected area of the wet surface, which for the whole wet canopy is $\sigma_f \times L_{SAI}$, q_f^{SAT} is the saturation water-vapor mixing ratio at temperature of the foliage T_f, q_{af} is the water-vapor mixing ratio of air within the canopy, and

$r_{\ell a}$ is the aerodynamic resistance to moisture and heat flux of the foliage molecular boundary layer per unit foliage projected area. Equation (45), if negative, gives the rate of accumulation of dew on foliage whether wet or not.

The conductance for heat and vapor flux from leaves is given by

$$r_{\ell a}^{-1} = C_f \times (U_{af}/D_f)^{1/2}, \tag{46}$$

where $C_f = 0.01$ m s$^{-1/2}$, D_f is the characteristic dimension of the leaves in the direction of wind flow, and U_{af} is the magnitude of the wind velocity incident on the leaves. Equation (46) corresponds to models of laminar boundary layers past flat plates (as discussed in Gates, 1980). Equation (45) gives the evaporation at each surface within the canopy; however, a different definition of Eq. (46) would be needed to apply it to stems and needles. It is also assumed that Eq. (45) can be applied to the canopy as a whole.

Similar to (45), the heat flux from the foliage is given by

$$H_f = \sigma_f L_{SAI} r_{\ell a}^{-1} \rho_a C_p (T_f - T_{af}). \tag{47}$$

Two parameters are needed to evaluate Eqs. (45) and (47), *i.e.*, the foliage temperature T_f and the water mixing ratio q_{af} for air within the canopy.

The flux E_f from foliage that is only partly wet, with wetted fraction denoted $\widetilde{L}_w$, now follows from Eq. (45) as

$$E_f = r'' E_f^{WET}, \tag{48}$$

where

$$r'' = 1 - \delta(E_f^{WET}) \left[1.0 - \widetilde{L}_w - L_d \left(\frac{r_{\ell a}}{r_{\ell a} + r_s} \right) \right], \tag{49}$$

where r_s = 'stomatal' resistance to be discussed further and where δ is a step function and is one for positive argument and zero for negative argument. The fraction of wetted area over nontranspiring foliage is assumed to be the same as that for transpiring foliage.

Transpiration E_{tr} occurs only from dry leaf surfaces and is only outward,

$$E_{tr} = \delta(E_f^{WET}) L_d \left(\frac{r_{\ell a}}{r_{\ell a} + r_s} \right) E_f^{WET}. \qquad (50)$$

The above formulation neglects the likely small temperature differences between wet and dry surfaces, as well as the effects of several other possible inhomogeneities. For example, many plants, such as hardwood trees, have leaf stomata that occur only on the underside of leaves, whereas leaf water may be stored largely on the upper surface of the leaves.

d. Stomatal Resistance

The term stomatal resistance, as used here, refers to the total mechanical resistance encountered by diffusion from inside a leaf to outside. This term is sometimes referred to as leaf resistance to distinguish it from the resistance due to the stomata alone. Water vapor inside leaves is maintained at or very near its saturated value, for otherwise the mesophyllic cells of the leaf would desiccate and the leaf wilt. The stomata are pores which, when open, are the main conduits for transpired water. Hence, the net resistance r_s to water passing from the inside to the outside of the leaf depends largely on the size, distribution, and degree of opening of these stomata. However, some water diffusion also occurs through leaf cuticles, which can be the primary route for transpiration when the stomata are closed. In general, the opening of the stomata, and hence r_s, changes with various environmental parameters, including inability of roots to supply adequately the transpiration demand.

The stomatal resistance factor is taken to be

$$r_s = r_{smin} \times R_\ell \times S_\ell \times M_\ell. \qquad (51)$$

The factors on the right-hand side are discussed, *e.g.*, by Hinckley *et al.* (1978); R_ℓ gives the dependence of r_s on solar radiation. Studies have been made of the dependence of r_s on visible fluxes (*cf. e.g.*, Hinckley *et al.*, 1978, Fig. 3; Watts, 1977, Figs. 10–11; Denmead and Millar, 1976, Fig. 1). These studies show that a minimum stomatal resistance is reached from between 10% of full sun (approximately true for the trees studied) and full sun as indicated by Denmead for wheat. Full sun corresponds to about 500 W m^{-2} of incident visible solar radiation ($\lambda < 0.7\mu m$). The factor R_ℓ varies between about 1 for overhead sun and r_{smax}/r_{smin} for night-time, where r_{smax} is the cuticular resistance of the leaves. It is assumed to be of the form

$$R_\ell = \frac{1+f}{f + r_{smin}/r_{smax}}, \tag{52}$$

where $f = F_v/F_{vc}$ with $r_{smax} = 5000$ s m^{-1}, F_v = flux of visible solar radiation, and F_{vc} the visible solar flux for which R_ℓ is about double its minimum value. Present model values are $F_{vc} = 30$ W m^{-2} for trees, and $F_{vc} = 100$ W m^{-2} for grasslands and crops. The dependence of stomatal resistance on vapor pressure deficit is not included because it appears to be highly variable from species to species (mechanistic models for the dependence of stomatal resistance on vapor pressure deficit have been reviewed by Farquhar, 1978). Hence, r_{smin}, which takes different values for different kinds of vegetative cover, is intended to apply for average vapor pressure deficit conditions.

Equation (52) is averaged over the different leaves in the canopy which receive different amounts of radiation. The canopy is divided into an upper and lower fraction and it is assumed that 0.75 of the radiation is absorbed by the upper canopy and 0.25 is absorbed by the lower canopy, and that the leaf stem area indexes of upper (L_{SAIU}) and lower (L_{SAIL}) canopy are each one-half of the total leaf stem area index, *i.e.*, $L_{SAIU} = L_{SAIL} = 0.5L_{SAI}$, and hence

$$F_{vu} = 0.75F_{vi}/L_{SAIU}, \tag{53a}$$

$$F_{vl} = 0.25F_{vi}/L_{SAIL}, \tag{53b}$$

where F_{vi} is the visible radiation incident at the top of the canopy. Equation (52) is applied using F_{vu} and then F_{vi} for F_v and the resulting conductivities averaged to get R_ℓ in the model. Singh and Szeicz (1980) have described a similar division of canopy into sunlight and shaded parts in determining stomatal resistance.

The seasonal temperature factor $S_\ell = F_{SEAS}(T_f)$, as given by Eq. (1b). The moisture factor M_ℓ depends on the soil moisture and the ability of plant roots to take water readily from the soil for a given level of root moisture. Initially, M_ℓ = 1, and if the plants' transpiration exceeds a maximum value, depending on soil moisture as described below, M_ℓ is increased so that the transpiration is maintained at the maximum value. Finally, if r_s as given by Eq. (51) exceeds r_{smax}, it is set to r_{smax}.

e. Root Resistance

The transpiration rate calculated from Eq. (50) must be consistent with the maximum transpiration that the vegetation can sustain as defined below. If E_{tr} is found to exceed E_{trmx}, then r_s is redetermined so that $E_{tr} = E_{trmx}$, the maximum transpiration available from the plant. The determination of this maximum is based on a simplification of the SPAC (Soil Plant Atmosphere Continuum) model of Federer (1979). For recent reviews of root–water models, see Hillel (1980) and Molz (1981). The plant water uptake in each soil layer is limited by the difference between soil and the leaf potential divided by an effective resistance. This effective resistance depends on the total length of root per unit area and the internal plant resistance per unit root length. When the soil is dry enough, the diffusion of water from the soil to the roots also contributes to this resistance. Lumping various terms together yields the following formula,

$$E_{trmx} = \Upsilon_{r0} \Sigma_i R_{ti} (1 - W^i_{LT}), \tag{54}$$

where Υ_{r0} is the maximum transpiration that can be sustained. The sum is over soil layers, each designated by subscript i, R_{ti} is the fraction of roots in a given soil layer, and W^i_{LT} is a soil dryness (or plant wilting) factor. We have taken $\Upsilon_{r0} = 2 \times 10^{-7}$ m s^{-1} $\times \sigma_f \times S_{EASB}$, where S_{EASB} is a factor that is 1 during the growing season and drops to 0 when the soil is frozen. For typical vegetation, $\sigma_f = 0.8$, and close to field capacity, $W^i_{LT} = 0.05$ so that typically for well–watered growing season vegetation, $\Upsilon_{r0} = 1.5 \times 10^{-7}$ m s^{-1}, corresponding to a maximum latent energy flux of about 380 W m^{-2}. Observed latent fluxes from unwetted foliage under intense solar radiation rarely exceed 400 W m^{-2}. Federer (1979, *cf.* his Fig. 1), modeling data for birch and maple forests, finds maximum transpiration of about 1.5×10^{-7}m s^{-1}, although with sufficient root capacity the transpiration could exceed 2.0×10^{-7} m s^{-1}.

The negative of the leaf potential can be approximated by the maximum value it takes before leaf desiccation, since the leaf potentials of many plants approach this value under any significant water stress. The term W^i_{LT} then is $(\phi - \phi_0)$ divided by $(\phi_{max} - \phi_0)$, where ϕ is the soil water suction and ϕ_{max} is the maximum value of negative of potential of leaves before desiccation. Hence, for soil suction of the form given in section 5c,

$$W^i_{LT} = \frac{s_i^{-B} - 1}{s_w^{-B} - 1}, \tag{55}$$

where s_i is the soil water of the i'th layer and s_w the soil water for which transpiration goes to zero. The current model value is $s_w = 0.125$. Normally, soils of clay-like texture would have larger values and sandy-textured soils lower values. The factor W^i_{LT} ranges from 0 at saturation to 1 at the "permanent wilting point."

The ratio of maximum water extraction, given by Eq. (54) for a given soil layer, to the maximum water extraction for the total active soil layer multiplied by the fraction of roots in the given layer gives the fraction of total transpiration that is taken from the given layer. This permits a variable distribution of roots between the upper soil layer and the total active soil column. The parameter $ROOTF$ as given in Table 2 is the ratio of the density of the roots in the upper layer to the density of roots in the total column. Thus, *e.g.*, for a 1-m total layer and 20-cm upper layer, if $ROOTF = 5$, then 50% of the roots is in the upper soil layer.

f. Energy Balance of Plant Canopy and Soil

The air within the canopy has negligible heat capacity and so heat flux from the foliage H_f and from the ground H_g must be balanced by heat flux to the atmosphere H_a, *i.e.*,

$$H_a = H_f + H_g, \tag{56}$$

where flux to the atmosphere is given by

$$H_a = \rho_a C_p C_D V_a (T_{af} - T_a), \tag{57}$$

where C_p is the specific heat of air, V_a is the magnitude of atmospheric wind above the canopy, and C_D is the aerodynamic bulk transfer coefficient between canopy air and atmosphere above, assumed to be the same for heat and moisture as for momentum. The flux from the ground is assumed to be

$$H_g = \rho_a C_p C_D [(1 - \sigma_f) V_a + \sigma_f U_{af}](T_{g1} - T_{af}). \qquad (58)$$

Equations (56)–(58) are solved for T_{af}, *i.e.*,

$$T_{af} = (c_A T_a + c_F T_f + c_G T_{g1})/(c_A + c_F + c_G), \qquad (59)$$

where

$$c_A = C_D V_a, \qquad (60)$$

$$c_F = \sigma_f L_{SAI} r_{\ell a}^{-1}, \qquad (61)$$

$$c_G = C_D [(1 - \sigma_f) V_a + \sigma_f U_{af}] \qquad (62)$$

are conductances for heat flux, to the atmosphere above the canopy, and from foliage, and the ground, respectively. Similarly, the canopy air is assumed to have zero capacity for water vapor storage so that the flux of water from the canopy air E_a balances the flux from the foliage E_f and the flux from the ground, E_g, *i.e.*,

$$E_a = E_f + E_g, \qquad (63)$$

where E_f was defined earlier in Eq. (48), and

$$E_a = \rho_a c_A (q_{af} - q_a), \qquad (64)$$

$$E_g = \rho_a c_G f_g (q_{g,s} - q_{af}), \qquad (65)$$

where $q_{g,s}$ is the saturated ground water vapor concentration and f_g is a wetness factor, defined as the ratio of actual to potential ground evaporation as obtained from the soil moisture parameterization. Equation (63) is solved for q_{af} using the definitions of E_a, E_f, E_g, *i.e.*,

$$q_{af} = (c_A q_a + c_V q_f^{SAT} + c_G f_g q_{g,s}) \div (c_A + c_V + f_g c_G), \qquad (66)$$

51

where $c_V = r'' c_F$ is the average conductance of foliage to water vapor flux.

g. Leaf Temperature

The final budget equation required to obtain the transpiration from vegetation is an equation of foliage energy conservation, *i.e.*,

$$R_n(T_f) = LE_f(T_f) + H_f(T_f), \tag{67}$$

where R_n is the net radiation absorbed by the foliage. Equation (67) is solved for foliage temperature T_f by Newton–Raphson iteration. In the model formulation, the emissivity of the ground and foliage is assumed to be unity; the foliage is assumed to have zero heat capacity, and photosynthetic and respiratory energy transformations are neglected. The exchange of thermal infrared (longwave) radiation from canopy to ground is approximated by

$$4\sigma T_{g1}^3 (T_f - T_{g1}),$$

and net exchange from canopy to sky by

$$4\sigma T_{g1}^3 (T_f - T_{g1}) - F_{RLNK},$$

where F_{RLNK} as calculated in the radiation routine is $\sigma T_{g1}^4 - F_{LW}$, the net loss of long-wave radiation from a bare surface. Absorbed solar radiation S_{qf} is obtained in the radiation routine from incident solar radiation and canopy albedo as obtained from subroutine ALBEDO.

The net radiation absorbed by vegetation per unit area F_{RAD} becomes

$$F_{RAD} = S_{F1} - 8\sigma\sigma_f T_{g1}^3 T_f, \tag{68}$$

where

$$S_{F1} = \sigma_f (S_{qf} - F_{RLNK} + 8\sigma T_{g1}^4). \tag{69}$$

The previously defined resistances are calculated iteratively as T_f changes. In particular, it is necessary to reevaluate the sensible heat flux to the atmosphere and the

evapotranspiration E_f (*cf.* Eqs. 45–48) through its dependence on foliage temper**ature** saturation mixing ratio q_f^{SAT} and through q_{af} dependence on C_D (Eqs. 61, **66**).

Thus, we expand

$$H_f^{N+1} = H_f^N + \delta H(T_f^{N+1} - T_f^N), \tag{70}$$

$$E_f^{N+1} = E_f^N + \delta E^N(T_f^{N+1} - T_f^N), \tag{71}$$

where

$$\delta H = \frac{\partial H_f^N}{\partial C_D}\frac{\partial C_D}{\partial T_f}, \tag{72}$$

$$\delta E = \frac{\delta E_f^N}{\partial q_f^{SAT}}\frac{\partial q_f^{SAT}}{\partial T_f} + \frac{\partial E_f^N}{\partial C_D}\frac{\partial C_D}{\partial T_f}. \tag{73}$$

The expression q_f^{SAT}, as all saturation mixing ratios, is calculated using

$$q^{SAT} = \frac{0.622 \times p^{SAT}}{p - 0.378 p^{SAT}}, \tag{74}$$

with p^{SAT} calculated from Tetens' formula

$$p^{SAT} = 611 \exp\left\{\frac{A(T - T_m)}{T - B}\right\}, \tag{75}$$

where

$$A = 21.874(T \leq T_m), 17.269(T > T_m),$$

and

$$B = 7.66(T \leq T_m), 35.86(T > T_m),$$

where $T_m = 273.16$.

Hence,

$$\frac{\partial q_f^{SAT}}{\partial T_f} = \frac{A(T_m - B)}{(T_f - B)^2} q^{SAT}. \tag{76}$$

We calculate T_f iteratively, as

$$T_f^{N+1} = \frac{S_f + c_H(c_A T_a + c_G T_g) + L_v(T_f^N \delta E - E^N) + T_f^N \delta H}{8\sigma_f T_g^3 + c_H(c_A + c_G) + L_v \delta E + \delta H}, \tag{77}$$

53

taking

$$c_H = c_F/(c_A + c_F + c_G). \tag{78}$$

Because of the discontinuous first derivative of C_D at neutral stability, non-converging oscillations about neutral stability are possible. Thus, we assume that if the sign of the stability switches between two iterations $C_D = C_{DN}$, the neutral stability drag coefficient can then be used. The detailed expressions for δH, δE in the code are too complicated to be repeated here and have not been thoroughly checked. However, they ensure convergence, and since they drop out of the convergent solution, their exact form is not important.

The T_f's are constrained to change by no more than 5° in one iteration. Let

$$\Delta^{N+1} = T^{N+1} - T^N,$$

and

$$\widetilde{\Delta} = \text{larger of}(\Delta^{N+1}, \Delta^N).$$

The iteration is stopped for $\widetilde{\Delta} < 0.001$ or after six iterations have been carried out.

After completion of the T_f calculation, we obtain

$$\Delta T_{a,f} = T_a - T_{af},$$

$$\Delta q_{a,f} = q_a - q_{af},$$

for use in obtaining sensible and latent heat fluxes to the atmosphere, and update the leaf water budget by adding the difference between the foliage evapotranspiration and its water vapor flux to the atmosphere. Leaf water can either increase because of dew formation or decrease because of evaporation of leaf water.

9. SOIL MOISTURE AND SNOW COVER WITH VEGETATION

Finally, we note that, in the presence of vegetation the soil moisture and snow cover, Eqs. (20), (21a), and (29) become

$$\frac{\partial S_{sw}}{\partial t} = P_r(1 - \sigma_f) - R_s + \Upsilon_w - \beta E_{tr} - F_q + S_m + D_w, \tag{79a}$$

$$\frac{\partial S_{tw}}{\partial t} = P_r(1 - \sigma_f) - R_w - E_{tr} - F_q + S_m + D_w, \qquad (79b)$$

$$\frac{\partial S_{cv}}{\partial t} = P_s(1 - \sigma_f) - F_q - S_m + D_s, \qquad (79c)$$

where β = fraction of transpiration from the top soil layer, D_w is the rate of excess water dripping from leaves per unit land area, D_s is the corresponding rate at which excess snow falls from the leaves, as mentioned following Eq. (44), and $R_w = R_s + R_g$ is the total runoff.

APPENDIX A. Fortran Symbols for Some Parameters in Model Code

(Units are in parentheses; no units indicate a dimensionless parameter)

AGE	Factor to reduce visible snow albedo due to age of snow
AGE1	Snow aging factor due to crystal growth
AGE2	Snow aging factor due to surface melting
AGE3	Snow aging factor due to accumulation of other particles
ALBGL	Albedo, soil, near-infrared, direct
ALBGLD	Albedo, soil, near-infrared, diffuse
ALBGS	Albedo, soil, visible, direct
ALBGSD	Albedo, soil, visible, diffuse
ALBL	Albedo, vegetation, near-infrared, direct
ALBLD	Albedo, vegetation, near-infrared, diffuse
ALBS	Albedo, vegetation, visible, direct
ALBSD	Albedo, vegetation, visible, diffuse
ALBZN	Zenith angle correction for albedo
BSW	Soil diffusivity parameter (= B in Clapp & Hornberger, 1978)
CDR	Surface drag coefficient
CDRN	Surface drag coefficient for neutral stability
CDRX	Surface drag coefficient for ice with leads
CF	Heat transfer coefficient from leaves
CH20	Specific heat of water (J m^{-3} K^{-1})
CICE	Specific heat of ice (J m^{-3} K^{-1})
CLND	Proportionality factor (to the ocean) for drag coefficient over land
COSZRS	Cosine of solar zenith angle (radians)
CROUGH	Proportionality factor (to the ocean) for drag coefficient over vegetation
CZF	Correction term for new snow albedo for large solar zenith angle
DENSI	Ratio of snow density to fresh snow density
DEPA	Depth of active soil column (mm)
DEPU	Depth of top soil layer (mm)
DEWMX	Maximum allowed depth of water on leaves (mm)
DMAX	Maximum soil diffusivity (m^2 s^{-1})
DTLAF	Difference between leaf temperature and temperature of air in foliage
EF	Moisture flux from foliage (mm s^{-1})
EFPOT	Potential evaporation rate from foliage (mm s^{-1})
EMS	Atmospheric emissivity
ETR	Transpiration rate (mm s^{-1})
ETRC	Maximum possible transpiration, given the soil moisture (mm s^{-1})

56

EVMX0	Maximum evaporation at saturation (mm s^{-1})
FACT	Turbulent removal flux coefficient
FB	Fractional intercepted visible radiation absorbed in lower canopy
FC	Light sensitivity factor for crops and grasses (m^2 W^{-1})
FDRY	Fraction of foliage that is green and dry
FEVPG	Evaporative heat flux from ground (mm s^{-1})
FLMX	Maximum rate of infiltration allowed by soil (mm s^{-1})
FSEAS	Function for temperature dependence of, vegetation cover
FSENG	Sensible heat flux from ground (W m^{-2})
FT	Fractional intercepted visible radiation absorbed in upper canopy
FWET	Fraction of foliage covered by water
GWATR	Net input of water to the soil surface (mm)
GWMX	Maximum soil moisture content (mm)
HS	Net energy flux into the surface (W m^{-2})
HSF	Sensible heat from leaves (W m^{-2})
HTVP	Latent heat of vaporization of water (or sublimation) (J kg^{-1})
LVEG	Vegetation type
PHI0	Minimum soil suction (mm)
PORSL	Soil porosity ratio
QSAT	Saturated specific humidity (kg kg^{-1})
RA	Aerodynamic resistance factor for leaves
RAP	Potential rate of evaporation (mm s^{-1})
RHOSW	Density of snow relative to water
RIB	Surface bulk Richardson number
RLAI	Inverse of sum of leaf area index and stem area index
RMAX0	Maximum stomatal resistance (s m^{-1})
RNOF	Summed runoff (= RSUR + RSUBS) (mm s^{-1})*
ROTF	Ratio of soil moisture extraction from top to total, when fully saturated
ROUGH	Aerodynamic roughness length (m)
ROUGHG	Proportionality factor (to the ocean) for drag coefficient over total surface
RS	Stomatal resistance (s m^{-1})
RSCS	Heat capacity of soil (J kg^{-1})

*Surface and total runoff terms are accumulated at the end of subroutine WATER so that output values are in mm day^{-1}.

RSMIN	Minimum stomatal resistance (s m^{-1})
RSUBS	Subsurface drainage (mm s^{-1})*
RSUR	Surface runoff (mm s^{-1})*
RTRT	Ratio RATW1/RATW0 = WATU/WATT
SABVEG	Radiation absorbed by vegetation (W m^{-2})
SAI	Stem area index
SCOND	Term proportional to leaf transpiration water loss
SCOR	Snow surface drag coefficient correction factor
SDROP	Snow falling off vegetation
SEASB	Seasonal factor depending on lower soil temperature and controlling LAI and root water uptake
SEASF	Difference between VEGC and fractional cover at 269 K
SG	Solar flux absorbed by bare ground (W m^{-2})
SIGF	Fractional vegetation cover
SK	Soil heat conductivity (J m^{-1})
SM	Rate of snowmelt
SNAL0	New snow albedo, visible
SNAL1	New snow albedo, near-infrared
SOLIS	Incident visible radiation, used for stomatal resistance (W m^{-2})
SQF	Solar flux absorbed by leaf (W m^{-2})
SWILT	Soil moisture content at which permanent wilting occurs (mm)
SWTRT	Average soil moisture per unit volume of soil, weighted toward top
TA	Temperature of air above canopy (K)
TAF	Temperature of air in canopy (K)
TAU1	Length of day (s)
TG	Temperature of soil at surface (K)
TGB	Temperature of subsoil
TM	A reference temperature = TS − 2.2 (K)
TRANS0	Movement of water from below to upper soil layer (mm s^{-1})
TRANS1	Net drying out of upper layer (mm s^{-1})
TRSMX	Maximum transpiration rate at saturation allowing for fractional vegetation cover and seasonal temperature dependence (mm s^{-1})
TRSMX0	Maximum rate of transpiration with saturated soil, 100% vegetation, optimum temperature (mm s^{-1})
TS	Temperature of air above canopy (K)
TVEG	Vegetation type
UAF	Velocity of air within foliage (m s^{-1})
VA	Wind speed above canopy (m s^{-1})
VEGC	Maximum fractional cover of vegetation

58

WATA	Average of WATT and WATU
WATT	Ratio of total soil moisture to maximum allowed
WATU	Ratio of upper soil moisture to maximum allowed
WILTR	Fractional soil moisture at which permanent wilting occurs
WLTTB	Wilting factor for total soil
WLTUB	Wilting factor for upper soil
WT	Fraction of grid square covered by snow
XKMX	Maximum soil hydraulic conductivity (mm s^{-1})
XLA	Maximum leaf area index
XLAI	Leaf area index
XLAI0	Minimum leaf area index
XRUN	Leaf drip (mm s^{-1})

59

APPENDIX B. Pointers Defining Location of Boundary Subroutine Variables*

NPK	Surface pressure
NDRAGK	Drag or surface stress
NDELTK	Temperature difference between above-canopy air and foliage air
NDELQK	Specific humidity difference between above-canopy air and foliage air
NZBK	Surface elevation
NTGK	Temperature of the soil surface
NTSK	Temperature of the air above canopy
NQGK	Specific humidity at soil surface
NQSK	Specific humidity above canopy
NUSK	Westerly wind above canopy
NVSK	Southerly wind above canopy
NLDOCK	Land/ocean mask (0 = ocean, 1 = land, 2 = sea ice)
NSWTK	Total soil water
NSCVK	Snow cover, water equivalent
NRNOK	Total runoff
NRNOSK	Surface runoff
NTGBK	Temperature of lower soil layer
NSICEK	Thickness of sea ice
NSAGK	Snow age, nondimensional
NLDEWK	Depth of water on foliage
NTLEFK	Temperature of foliage
NEVPAK	Moisture flux, averaged over history-write interval
NSENAK	Sensible heat flux, averaged over history-write interval
NTGAK	Soil surface temperature, averaged over history-write interval
NTSAK	Above-canopy air temperature, averaged over history-write interval
NPRCAK	Precipitation, averaged over history-write interval
NFRLAK	Net infrared energy flux, averaged over history-write interval
NFRSAK	Net absorbed solar energy flux, averaged over history-write interval
NGWETK	Soil wetness factor, *i.e.*, ratio of evaporation to that from a wet soil
NTMINK	Minimum above-canopy temperature, over history-write interval
NTMAXK	Maximum above-canopy temperature, over history-write interval
NVEGK	Fractional vegetation cover
NEVPRK	Moisture flux to atmosphere
NSENTK	Sensible heat flux to atmosphere
NZ1K	Altitude of lowest atmospheric level
NFRLK	Net infrared energy flux at surface
NFRSK	Net absorbed solar energy flux
NTVEGK	Vegetation type (see Table 1)

*Listed in order of occurrence in the common block BPNT.

NRHSK	Density of surface air
NSSWK	Water in upper soil
NPBPK	Accumulated carbon uptake
NRFEK	Number of rainfall events
NLREK	Accumulated duration of rainfall events
NQSAK	Specific humidity above canopy, averaged over history–write interval
NRESPK	Accumulated carbon uptake of soil and vegetation minus carbon loss by respiration
NPRCPK	Instantaneous precipitation rate
NIRCPK	Intercepted precipitation rate
NPRDK	Currently not used
NTAFK	Temperature of the air within foliage
NSFK	Solar energy flux

APPENDIX C: Calling Diagram for Boundary Subroutine

Note: Diagram below is for a stand-alone boundary package. In link-up with CCM, appropriate driver and output functions are assumed by the CCM.
Order of calling is top to bottom, and left to right. Subroutine names are capitalized.

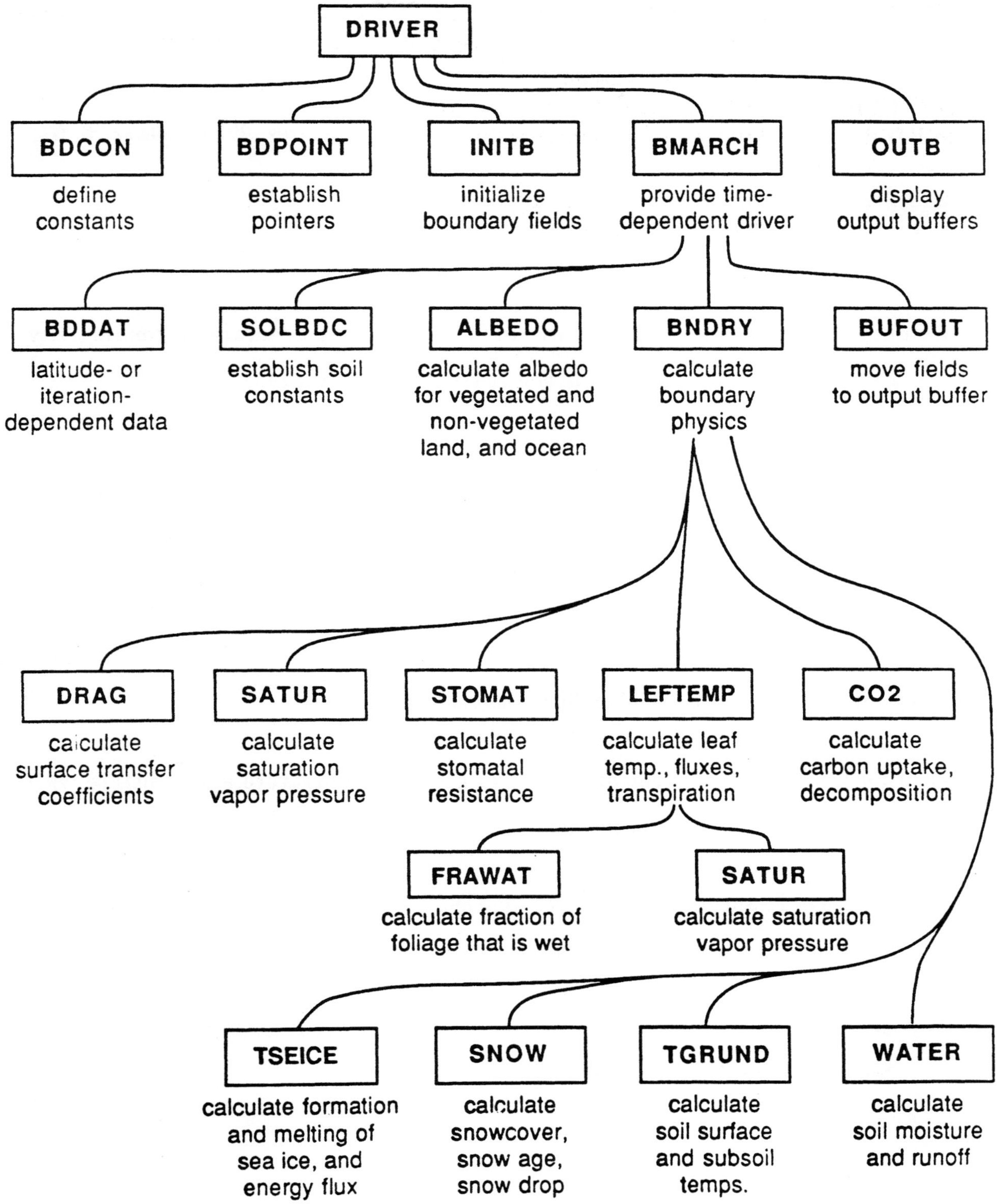

REFERENCES

Anderson, E. A., 1976: *A Point Energy and Mass Balance Model of a Snow Cover.* Office of Hydrology, National Weather Service, 150 pp.

Auer, A., 1974: The rain versus snow threshold temperatures. *Weatherwise,* p. 67.

Barry, R. G. and R. E. Chambers, 1966: A preliminary map of summer albedo over England and Wales. *Quart. J. Roy. Meteor. Soc.,* **92**, 543–548.

Campbell, G. S., 1974: A simple method for determining unsaturated conductivity from moisture retention data. *Soil Science,* **117**, 311–314.

Clapp, R. B. and G. M. Hornberger, 1978: Empirical equations for some soil hydraulic properties. *Water Resources Research,* **14**, 601–604.

Condit, H. R., 1970: The spectral reflectance of American soils. *Photogrammetric Engineering,* **36**, 955–966.

Corps of Engineers, 1956: *Snow Hydrology,* 437 pp.

Deardorff, J., 1978: Efficient prediction of ground temperature and moisture with inclusion of a layer of vegetation. *J. Geophys. Res.,* **83**, 1889–1903.

Denmead, O. T. and B. D. Millar, 1976: Field studies of the conductance of wheat leaves and transpiration. *Agron. J.,* **68**, 307–311.

de Vries, D. A., 1963: Thermal properties of soils. In *Physics of Plant Environment.* North Holland Publishing Co., 594 pp.

____, 1975: Heat transfer in soils. In *Heat and Mass Transfer in the Biosphere, Part I*. D. A. de Vries and N. H. Afgan, Eds., Scripta Book Co., Amsterdam, 594 pp.

Dickinson, R. E., 1983: Land-surface processes and climate—Surface albedos and energy balance. *Advances in Geophysics,* **25,** 305–353.

____, 1984: Modeling evapotranspiration for three-dimensional global climate models. In *Climate Processes and Climate Sensitivity.* J. E. Hansen and T. Takahashi, eds., American Geophysical Union, Washington, DC, 58–72.

____, R. E., J. Jaeger, W. M. Washington, and R. Wolski, 1981: *Boundary Subroutine for the NCAR Global Climate Model.* NCAR Technical Note/TN-173+1A, National Center for Atmospheric Research, Boulder, CO, 75 pp.

____ and A. Henderson-Sellers, 1986: Modelling tropical deforestation: A study of GCM land-surface parameterizations. *Quart. J. Roy. Meteor. Soc.,* to be submitted.

FAO/UNESCO, 1974: *Soil Map of the World.* FAO, Paris, France.

Farquhar, G. D., 1978: Feedforward responses of stomata to humidity. *Ann. J. Plant Physiol.,* **5**, 787–800.

Federer, C. A., 1968: Spatial variation of net radiation, albedo, and surface temperature of forests. *J. Appl. Meteor.,* **7,** 789–795.

____, 1971: Solar radiation absorbed by leafless hardwood forests. *Agricultural Meteorol.,* **9,** 1971/1972, 3–20.

____, 1979: A soil-plant atmosphere model for transpiration and availability of soil water. *Water Resources Research,* **15,** 555–562.

Fleming, G., 1975: *Computer Simulation Techniques in Hydrology.* Elsevier, New York, NY, 333 pp.

Fuller, S. P. and W. R. Rouse, 1979: Spectral reflectance changes accompanying a post–fire recovery sequence in a subarctic spruce lichen woodland. *Remote Sensing of Environment,* **8**, 11–23.

Garratt, J. R., 1977: Review of drag coefficients over oceans and continents. *Mon. Wea. Rev.,* **105**, 915–929.

Gates, D. M., 1980: *Biophysical Ecology.* Springer–Verlag, New York, NY, 611 pp.

____, M. J. Keegan, J. C. Schleter, and V. R. Weidner, 1965: Spectral properties of plants. *Appl. Optics,* **4**, 11–20.

Geiger, R., 1965: *The Climate Near the Ground.* Harvard University Press, Cambridge, MA, 611 pp.

Henderson–Sellers, A., M. F. Wilson, and R. E. Dickinson, 1986: Zero–dimensional sensitivity studies with the NCAR CCM land surface parameterization scheme. *Proc. International Satellite Land-Surface Climatology Project (ISLSCP) Conference,* Rome, Italy, 2–6 December 1985, European Space Agency SP–248.

Hillel, D., 1980: *Applications of Soil Physics.* Academic Press, New York, NY, 385 pp.

Hinckley, T. M., J. P. Lassoie, and S. W. Running, 1978: Temporal and spatial variations in the water status for forest trees. *Forest Science Monograph,* **20**, Society of American Foresters, 72 pp.

Idso, S. B., R. D. Jackson, R. J. Reginato, B. A. Kimball, and F. S. Nakayama, 1975: The dependence of bare soil albedo on soil water content. *J. Appl. Meteor.,* **14**, 109–113.

Kirby, M. J., Ed., 1979: *Hillslope Hydrology.* Wiley, New York, NY, 338 pp.

Kondratyev, K. Ya., 1969: *Radiation in the Atmosphere.* Academic Press, New York, NY, 912 pp.

Kriebel, K. T., 1979: Albedo of vegetated surfaces: Its variability with differing irradiances. *Remote Sensing of the Environment,* **8**, 283–290.

Kukla, G. J. and D. Robinson, 1980: Annual cycle of surface albedo. *Mon. Wea. Rev.,* **108**, 56–67.

Kung, E. C., R. A. Bryson, and D. H. Lenschow, 1964: Study of a continental surface albedo on the basis of flight measurements and structure of the earth's surface cover over North America. *Mon. Wea. Rev.,* **92**, 543–564.

L'vovich, M. I., 1979: *World Water Resources and Their Future.* American Geophysical Union, Washington, DC, 415 pp.

Massman, W. J., 1980: Water storage on forest foliage: A general model. *Water Resources Research,* **16**, 210–216.

Matthews, E., 1983: Global vegetation and land use: New high resolution data bases for climate studies. *J. Clim. Appl. Meteor.,* **22**, 474–487.

____, 1984: Prescription of land–surface boundary conditions in GISS GCMII and Vegetation, land–use and seasonal albedo data sets: Documentation of archived data tape. *NASA Technical Memos 86096 and 86107,* NASA, Goddard Institute for Space Studies, New York, NY, 20 pp. and 9 pp.

Maykut, G. and N. Untersteiner, 1971: Some results from a time–dependent thermodynamic model of sea ice. *J. Geophys. Res.,* **76**, 1550–1575.

Molz, F. J., 1981: Models of water transport in the soil–plant system: A review. *Water Resources Research,* **17**, 1245–1260.

Monteith, J. L., 1959: The reflection of short–wave radiation by vegetation. *Quart. J. Roy. Meteor. Soc.,* **85**, 386–392.

Monteith, J. L., Ed., 1975: *Vegetation and the Atmosphere, Vol. I, Principles.* Academic Press, New York, NY, 278 pp.

Monteith, J. L., 1976: *Vegetation and the Atmosphere, Vol. II, Case Studies.* Academic Press, New York, NY, 439 pp.

Nakano, Y. and S. Brown, 1972: Mathematical modeling and validation of the thermal regimes in tundra soils, Barrow, Alaska. *Arctic and Alpine Research,* **4**, 19–38.

Oguntoyinbo, J. S., 1970: Reflection coefficient of natural vegetation, crops and urban surfaces in Nigeria. *Quart. J. Roy. Meteor. Soc.* **96**, 430–441.

Olson, J. S., J. A. Watts, and L. J. Allison, 1983: Carbon in live vegetation of major world ecosystems. U.S. Department of Energy, DOE/NBB–0037, No. TR004, U.S. Department of Energy, Washington, DC, 152 pp.

Ripley, E. A. and R. E. Redman, 1976: Grassland. In Monteith, *loc. cit.,* p. 351.

Rockwood, A. A. and S. K. Cox, 1978: Satellite inferred surface albedo over northwestern Africa. *J. Atmos. Sci.,* **35**, 513–522.

Semtner, A. J., 1976: A model for the thermodynamic growth of sea ice in numerical investigations of climate. *J. Phys. Oceanogr.,* **6**, 379–389.

Shuttleworth, W. J., 1978: A simplified one–dimensional theoretical description of the vegetation–atmosphere interaction. *Boundary-Layer Meteor.,* **14**, 3–27.

Singh, B. and G. Szeicz, 1980: Predicting the canopy resistance of a mixed hardwood forest. *Agricultural Meteor.*, **21**, 49–58.

Stewart, J. B., 1971: The albedo of a pine forest. *Quart. J. Roy. Meteor. Soc.*, **97**, 561–564.

Thom, A. S. and H. R. Oliver, 1977: On Penman's equation for estimating regional evaporation. *Quart. J. Roy. Meteor. Soc.*, **103**, 345–357.

Tucker, C. J. and L. D. Miller, 1977: Soil spectra contributions to grass canopy spectral reflectance. *Photogrammatic Engineering and Remote Sensing*, **43**, 721–726.

Watts, W. R., 1977: Field studies of stomatal conductance. In *Environmental Effects in Crop Physiology*. J. J. Landsberg and C. V. Cutting, eds., Academic Press, New York, NY, 388 pp.

Wilson, M. F., 1984: The construction and use of land surface information in a general circulation climate model. Unpublished Ph.D. thesis, University of Liverpool, United Kingdom, 346 pp.

____, A. Henderson-Sellers, R. E. Dickinson, and P. J. Kennedy, 1986a: Sensitivity of the biosphere–atmosphere transfer scheme (BATS) to the inclusion of variable soil characteristics. *J. Climate Appl. Meteor.*, in press.

____, ____, ____, and ____, 1986b: Investigation of the sensitivity of the land-surface parameterization of the NCAR community climate model in regions of tundra vegetation. *J. Climatology*, submitted.

____ and ____, 1985: A global archive of land cover and soils data for use in general circulation climate models. *J. Climatology*, **5**, 119–143.

Wiscombe, W. J. and S. G. Warren, 1980: A model for the spectral albedo of snow. I. Pure snow. *J. Atmos. Sci.*, **37**, 2712–2733.

Yen, Y.–C., 1969: Recent studies in snow properties. In *Advances in Hydrosciences*, **5**, Academic Press, New York, NY, 305 pp.

Quart. J. R. Met. Soc. (1984), **110**, pp. 1143–1162 551.551.8:551.573:551.584.41

Eddy correlation measurements of energy partition for Amazonian forest

By W. JAMES SHUTTLEWORTH, JOHN H. C. GASH, COLIN R. LLOYD, CHRISTOPHER J. MOORE and JOHN ROBERTS
Institute of Hydrology, Wallingford, Oxon, U.K.

ARI De O. MARQUES FILHO, GILBERTO FISCH, VICENTE De PAULA SILVA FILHO and MARIA De NAZARÉ GÓES RIBEIRO
Instituto Nacional de Pesquisas da Amazônia, Manaus, Amazonas, Brasil

LUIZ C. B. MOLION, LEONARDO D. De ABREU SÁ and J. CARLOS A. NOBRE
Instituto de Pesquisas Espaciais, São José dos Campos, São Paulo, Brasil

OSVALDO M. R. CABRAL
Centro Nacional de Pesquisa da Serigueira e Dende – EMBRAPA, Manaus, Amazonas, Brasil

SUKARAN R. PATEL
Universidade Federal da Paraíba, Campina Grande, Paraíba, Brasil

and

J. CARVALHO De MORAES
Universidade Federal do Pará, Belém, Pará, Brasil

(Received 2 February 1984; revised 12 June 1984)

SUMMARY

Measurements of energy partition for Amazonian forest made with novel eddy correlation equipment are presented for eight dry days in September 1983. These are interpreted to provide estimates of the aerodynamic and surface resistance for this vegetation type. Daily total evaporation for a transpiring canopy accounts for 70% of the available radiant energy, and is two thirds of conventional estimates of potential evaporation. The results are used to provide an initial calibration of a simple, physically based model of daily evaporation for Amazonian rain forest.

1. INTRODUCTION

Progress in forest micrometeorology over the past fifteen years has been such that the acquisition, and even to some extent the interpretation, of data has become routine. Finance and practicability have dictated that the overwhelming majority of this work has occurred in temperate latitudes (e.g. Black and McNaughton 1971; Stewart and Thom 1973; McNeil and Shuttleworth 1975; Gay and Fritschen 1979). At the same time, there has been increasing awareness that tropical rain forest, and the continental rain forest of the Amazon basin in particular, may have an important role in global climatology. Energy partition measurements for this important vegetation type represent a significant gap in the literature.

This paper describes results from the first field season of a major Anglo–Brazilian collaborative study of the micrometeorology and plant physiology of Amazonian rain forest. Several intensive periods of data collection from micrometeorological instrumentation are planned over the next few years. Data collection during the first field season was limited by the need to first erect a meteorological tower and set up a field site in this difficult environment.

Arguably the most important single result from past studies of forest evaporation is the realization that interception loss, the evaporation of free water from the leaf surface during and after rain, and transpiration require separate description for this vegetation type (e.g. Shuttleworth and Calder 1979). This is a direct consequence of the different magnitudes of the transfer resistances involved. During transpiration the latent heat flux is subject to a surface resistance, intimately related to stomatal control, which is generally at least an order of magnitude greater than the aerodynamic transfer resistance. When intercepted water is evaporating from a wet canopy this resistance is effectively short circuited and evaporation can alter dramatically.

It remains to be seen how important the interception component is in the context of evaporation from tropical rain forests, but its direct measurement and modelling have proved tractable elsewhere (e.g. Gash *et al.* 1980; Calder 1977), and its detailed study forms part of the experimental programme of the collaboration described above. On the other hand, the measurement and modelling of transpiration can be realistically studied only using micrometeorological techniques. In this paper we concentrate on this last aspect and present data corresponding to dry canopy conditions. These data are for eight dry days: 6, 7, 9, 10, 17, 18, 25 and 27 September 1983.

2. Experimental site and background meteorology

The measurements were made using a 45 m scaffolding tower at a site, 2°57′S 59°57′W, selected as representative of the natural vegetation and regional topography in the Ducke Reserve, 25 km from Manaus, Amazonas. The plant density is high, up to 3000 stems ha^{-1}, but less than 10% have girths of 200 mm or more. The forest canopy is of very varied species and typical of undisturbed natural forest. It extends with no obvious sub-storeys to a height of 35 m, with occasional emergent trees reaching 40 m, one upwind and quite close to the experimental tower (within 20 m). Rooting in the yellow laterite soil is shallow, with a dense root mat down to 150 mm.

The topography is gently undulating with valleys several tens of metres deep occurring at about 300 m intervals. The topography of the canopy top is modulated by the differential growth of the vegetation. The tower is sited near the top of a broader than average ridge. The eddy correlation and profile measurements described in this paper assume that the fluxes are one dimensional and an adequate fetch is required for this assumption. The tower has fetches of undisturbed forest extending many kilometres in most directions. The minimum fetch occurs over a narrow range of angles downwind of the site; here the species content of the forest has been modified by human intervention about 1 km from the tower.

The climatological average rainfall pattern for this region exhibits a marked seasonal dependence, with a monthly maximum of 270 mm in March and a minimum of 40 mm in August. The data presented here were collected during September 1983, towards the end of a 'dry season' in which rainfall was anomalously high (150 mm per month). Over the period of data collection the specific humidity remained fairly constant in the range 16 to 19 $g\,kg^{-1}$, while specific humidity deficit above the canopy ranged from (2 ± 2) $g\,kg^{-1}$ at night to (10 ± 2) $g\,kg^{-1}$ in the early afternoon. The average temperature was in the order 27 °C with a daily range of 7 to 10°C. Wind speeds were low, around $(2\pm1)\,m\,s^{-1}$ with a strong easterly bias. Hourly average solar radiation exceeded 900 $W\,m^{-2}$ on rare occasions but peak daily values of 500 to 700 $W\,m^{-2}$ were more common. Net radiation varied between $-40\,W\,m^{-2}$ at night and an average daytime maximum around 500 $W\,m^{-2}$.

3. Instrumentation

The evaporation, sensible heat and momentum fluxes presented in this paper were measured with a 'Hydra', a battery-powered eddy correlation, flux measuring device developed by the Institute of Hydrology. The instrument was mounted on a pole above the tower at a height of 48·4 m. Measurements of the other meteorological variables used in this analysis were obtained using initially one, and later two, automatic weather stations mounted at the top of the tower at a height of 45 metres. These instruments are visible in Fig. 1, an aerial photograph of the tower. Towards the end of the experiment data from other instrumentation became intermittently available and are used in this paper. In particular for two of the days studied some temperature and humidity gradients were obtained with a Thermometer Interchange System (TIS). This last instrumentation has been described in detail by McNeil and Shuttleworth (1975), and this description is not repeated here.

Figure 1. Aerial photograph of the scaffolding tower. The Hydra is mounted vertically above it at a height of 48·4 m, the two automatic weather stations are at the top of the tower at a height of 45 m.

(*a*) *The Hydra*

(i) *Hardware components*. The Hydra system comprises a vertical sonic anemometer (Shuttleworth *et al.* 1982), a single beam, 2·7 μm, infrared absorption hygrometer (Moore 1983), a 50 μm thermocouple thermometer, and two Gill propeller anemometers mounted orthogonally in the horizontal plane. These sensors are interfaced to a battery-powered RCA 1802 CMOS microprocessor system which is integral to the device, and provides on-line calculation of fluxes, variances and mean values. The data are logged on a solid state store (G.K. Instruments) which, in this experiment, was removed and interrogated with a PET microcomputer every few days.

Manufacturers calibrations were assumed for the Gill propellers, but the other sensors were calibrated. The calibration of the infrared hygrometer exhibits a marked temperature dependence (see Moore 1983), and was calibrated against an EG & G Type 992 dew point hygrometer over a wide range of temperatures in a controlled temperature environment chamber. This chamber was also used to calibrate the thermocouple thermometer. The sonic anemometer was calibrated by extensive comparison against a Kaijo-Denki DAT 311 sonic anemometer, with both anemometers operating in field conditions. This empirical calibration was deemed necessary in view of defects in the Hydra anemometer performance, related in part to limited cosine response (Shuttleworth *et al.* 1982), and, possibly, in part to the interaction between the Hydra framework and the wind field. Because of these effects, the calibration of the anemometer, and in consequence the flux measurements produced by the Hydra, could well be prone to systematic error in the order of 5% due to sonic anemometer calibration alone. Calibration defects in other sensors are believed to be small, in the order of one or two per cent, but could contribute to further potential error. Probably more important is the possibility of flux underestimation due to sensor exposure, sensor separation, and the frequency limitations imposed by sensor time constants and the averaging procedures used to evaluate the eddy flux. In the context of the present paper, there is clearly a need to check the performance of this novel device in this application. Such investigation receives detailed attention in later sections of this paper.

The Hydra system cannot provide reliable measurements during rain, when the hygrometer, sonic anemometer and thermocouple are wet, and the sensors generally take about an hour to dry out after a rain storm. Moreover the on-line software, outlined in the next section, uses running auto-regressive means with long time constants to identify the fluctuating components of the measured meteorological variables. The version of the software used in this experiment did not contain subroutines to provide rapid initialization of these auto-regressive means after a discontinuity in sensor availability, i.e. after initial 'turn on' or a heavy rainstorm. For this reason data produced by the software within three hours of such an event are not used in the present analysis. This fact was in part responsible for the decision to restrict the present analysis to completely dry days.

(ii) *On-line calculations.* The real-time software used in the Hydra microprocessor system is described in detail elsewhere (Lloyd *et al.* 1984).

In outline, the program takes samples of the fluctuating meteorological variables at a frequency of 10 Hz and computes the evaporation (E), sensible heat flux (H) and friction velocity (u_*) from the expressions

$$E = \langle w\rho_v\rangle - \langle w\rangle\langle\rho_v\rangle \tag{1}$$

$$H = \rho c_p(\langle wT\rangle - \langle w\rangle\langle T\rangle) \tag{2}$$

and

$$u_*^2 = \langle wu\rangle - \langle w\rangle\langle u\rangle \tag{3}$$

where w is the vertical wind velocity, ρ_v absolute humidity, ρ air density, c_p the specific heat of air at constant pressure, T temperature and u the wind velocity in the direction of the mean wind. In these expressions $\langle\ \ \rangle$ symbolizes a moving average calculated for temperature, for example, by the expression

$$\langle T_i\rangle = a\langle T_{i-1}\rangle + (1-a)T_i \tag{4}$$

where T_i is the present temperature sample, $\langle T_{i-1}\rangle$ the previous moving average and $\langle T_i\rangle$ the present resultant value of the moving average. The procedure is a digital analogue of a low pass electrical filter of time constant τ, and the weighting function a is given by the expression

$$a = \exp(-\Delta t/\tau) \tag{5}$$

where Δt is the time interval between inputs. The calculation of the variance of fluctuating variables is analogous to Eqs. (1), (2) and (3), and the variance of temperature, for example, is given by the expression

$$\sigma_T^2 = \langle T^2\rangle - \langle T\rangle^2. \tag{6}$$

The infrared absorption hygrometer is of the type first described by Hyson and Hicks (1975), in which humidity fluctuations are given by the ratio of the fluctuating signal to its mean component. In the Hydra this division is performed digitally. The calculation of u_* requires knowledge of u, the wind speed in the direction of the mean wind. The program identifies this direction from suitably defined moving averages of the two orthogonal measured components, and provides the required wind vector by resolving in the direction of the mean wind. In addition to performing these calculations, the software monitors outputs generated by the sonic anemometer and hygrometer which indicate their functional status, and logs this as part of the data output.

An important consequence of calculating the flux, variance and mean according to the above prescription is that the resulting values are delayed with respect to real time. This effect is apparent in the data and a correction applied to take account of it is discussed in the next section.

There is some doubt over the most appropriate value of the parameter τ (and hence a) which should be used in all these calculations. It is necessary that it should be long enough to include most of the low frequency contributions in the co-spectra between vertical wind speed and the other fluctuating meteorological variables. At the same time very long time constants may make the calculations susceptible to contamination by instrument-generated low frequency correlations between the measured variables.

Previous experience suggests that time constants of at least 20 minutes are required, but, in view of the uncertainty, tests were carried out to investigate the suitability of different time constants in this environment. On some days two identical microprocessor systems were used in parallel to analyse the same sensor inputs using different values of the parameter a, corresponding to time constants of 18·75 and 31·25 minutes. Figure 2 shows a typical result, and illustrates the evaporation and latent heat flux obtained on 27 September 1983. The different time constants give two effects: the daily cycle in the measured flux with the longer time constant is delayed (by about 12 minutes); and is also slightly larger. Both of these effects are apparent in Fig. 2 but apply more generally.

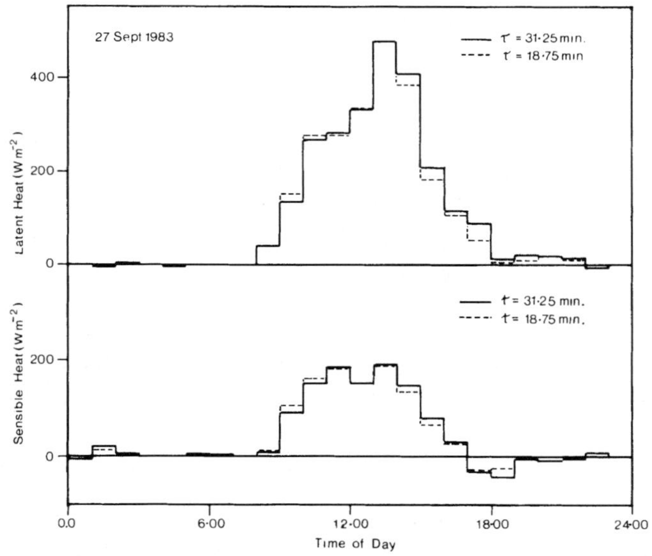

Figure 2. Measured values of latent and sensible heat obtained from the Hydra with two computer systems operating in tandem using different time constants, τ, in the autoregressive mean calculations. The full line corresponds to $\tau = 31{\cdot}25$ minutes, the dashed line $\tau = 18{\cdot}75$ minutes.

The integrated increase in measured flux given by the longer time constant over the four complete days for which comparative data were available was 2%. The longer time constant, 31·25 minutes, was used for all the results given in this paper.

(iii) *Off-line corrections.* The sonic anemometer calibration has a slight temperature dependence corresponding to the change in the speed of sound with temperature (Shuttleworth *et al.* 1982), while the calibration of the infrared hygrometer varies in a complex but defined way with both temperature and absolute humidity (Moore 1983). The data produced by the Hydra were passed through a calibration analysis to take account of these effects.

An adjustment has to be made to the measured latent heat to compensate for the temperature contamination of the humidity fluctuations which results from variations in atmospheric and sensor temperature (see Moore 1983). Corrections were also made for the effect described by Webb *et al.* (1980), who showed that density changes in the atmosphere, due to changes in temperature and humidity, induce small changes in the effect of a given vertical wind velocity. The joint effect of these corrections in the present data is some minor redistribution of measured flux during the day at the $10\,\mathrm{W\,m^{-2}}$ level. The effect on integrated daily latent heat flux is negligible, typically less than 0·5%.

The measurement of horizontal wind speed made with two orthogonal Gill propeller anemometers is subject to the errors associated with this sensor, and in particular to stalling at low wind speed and defects in cosine response. The on-line software used in the Hydra assumes a constant calibration for these sensors. Retrospective, first-order corrections were made to those turbulence parameters (e.g. u_*, $\overline{u}$) which involve a measurement of horizontal wind speed. The correction procedure follows Drinkrow (1972) in expressing the response curve $R(\theta)$ of the Gill anemometer in the form

$$R(\theta) = \cos\theta - 0{\cdot}085\sin(2|\theta|) + 0{\cdot}036\sin(4|\theta|) \qquad (7)$$

where θ is the angle between the wind vector and the anemometer axis. For each hour the standard deviation of vertical wind speed and the mean wind vector were used to estimate the r.m.s. angle between the wind and the horizontal plane, and the two measured mean horizontal components used to provide an initial estimate of the orientation in the horizontal plane. A simple iterative procedure based on Eq. (7) was then used to derive correction factors for the two horizontal axes and related turbulence parameters.

It is recognized that the Hydra instrumentation provides a less than perfect measurement of u_* as a result of defects in sensor performance and shortcomings in the corrections made for these defects. It is not possible to make simple corrections for anemometer stalling and, in the light and variable wind conditions prevalent during this experiment, this could well result in significant error, probably a reduction in all wind-speed-related measurements. Previous experience with this device (Wallace *et al.* 1984) suggests systematic underestimation in u_*, possibly as high as 30%, though more probably less than this. Moreover, the measurement of u_* using eddy correlation techniques is in any case sensitive to errors associated with the distortion of mean wind flow around supporting structures and the instruments themselves. Estimates of the error in u_* for the Kansas data (Businger *et al.* 1971) range between 7 and 33%, varying with the model used and the assumed effective size of tower obstructions (Wieringa 1980; Wyngaard *et al.* 1981; and Wieringa 1981). The results based on measurements of u_* presented below exhibit significant experimental 'noise', and in addition could well be subject to systematic errors, possibly as large as 30%. It is perhaps fortunate that in the context of the present paper, and indeed in the broader context of energy partition from transpiring forests, large errors in the description of aerodynamic transfer have little effect.

Earlier mention was made of the fact that the application of autoregressive means in the real-time analysis carried out in the Hydra software has the effect of introducing a phase delay into the measurements it produces. The effect is in many respects analogous to the application of an electrical voltage to a simple RC filter, although the analogy is not perfect for flux and variance calculations because there can be some time dependence in the second term of Eqs. (1), (2), (3), and in particular of Eq. (6). We develop this analogy and make a first-order correction for the most important frequency component, that corresponding to the daily cycle. It is easily shown that, at low frequencies, the output of a simple low pass filter is subject to a phase delay with respect to the input which is equivalent to a time slip equal to the time constant, τ. A first-order correction was made to take account of this in the Hydra output. Accordingly, the value of (say) evaporation, E_i', relevant to the ith hour in a particular day was calculated from the expression

$$E_i' = E_i(60 - \tau)/60 + E_{i+1}\tau/60 \tag{8}$$

where E_i and E_{i+1} are the latent heat fluxes calculated on-line for hours i and $i + 1$ respectively, and τ is the time constant in minutes. Application of this procedure introduces an element of smoothing into the data.

(*b*) *Automatic weather stations* (*AWS*)

The experiment draws data from initially one and, after 8 September, two automatic weather stations (Didcot Instruments). These stations were used to provide the hourly average measurements of temperature, wet bulb depression, net radiation and wind speed used in the analysis. Independent measurements of these meteorological variables were intermittently available from other micrometeorological instrumentation and were used for intercomparison and verification when this occurred. In particular, the automatic weather stations were non-standard in that they included aspirated psychrometers. A comparison between the psychrometers and the accurate quartz crystal psychrometers used in the TIS, confirmed that the AWS psychrometers were adequately aspirated. The measurements of temperature and wet bulb depression provided by the AWS are, however, prone to experimental noise in the order 0·2°C.

The anemometers used in the AWS have metal cups of sturdy design and are prone to stalling errors at low wind speed (less than $0{\cdot}5\,\mathrm{m\,s^{-1}}$), and overspeeding at all speeds. In part such overspeeding is due to the fact that wind speed is *a priori* greater than the wind vector. Bernstein (1967) indicates that the size of this last effect is estimated by the expression

$$s = \bar{u}(1 + \sigma_v^2/2\bar{u}^2) \tag{9}$$

where s is the wind speed, $\bar{u}$ the mean wind speed vector and σ_v^2 the variance of the transverse wind speed. Preliminary estimates of this contribution to overspeeding, based on measurements of σ_v^2 obtained with the Hydra in daytime conditions, suggest an increase in the order of 10 to 15% at wind speeds around $2\,\mathrm{m\,s^{-1}}$. Mechanical overrun will further contribute to the overspeeding error. Despite these several shortcomings, the AWS wind speed is used throughout this analysis. It is the only measurement consistently available throughout the experiment, and is in any case typical of the type of wind speed measurement commonly available in climatological data, with which future energy partition models will necessarily be used. Its use does, however, further contribute to experimental noise and possible systematic uncertainties in the wind-speed-based results presented here.

A comparison was made between the two net radiation measurements provided by the AWS systems and independent measurements made with two Funk net radiometers which were available for part of the experimental period. Integrated daily values of net radiation agree to within systematic errors in the order 5%, but hourly values exhibit scatter greater than this, typically 10 to 15%. The hourly average value provided by the two stations is based on twelve individual measurements made at five-minute intervals and this sampling procedure is in significant part responsible for the experimental noise observed in the data. To minimize the effect, the mean values of net radiation, wind speed, temperature and wet bulb depression from the two weather stations are used in this analysis whenever both were available. Only one station was operational on 6 and 7 September and the results for these two days are perhaps less reliable because of this.

4. Results

(*a*) *Integrated daily flux*

Assuming horizontal homogeneity, the energy budget of the forest vegetation can be written as:

$$A = \lambda E + H \tag{10}$$

$$= R_n - G - S - P \tag{11}$$

where A is the available energy per unit area; λE is the latent heat flux out of the forest; H is the sensible heat flux out of the forest; R_n is the net all-wavelength radiation; G is the energy flux into the ground; S is the equivalent energy flux into storage; P is the net rate of energy absorption per unit area by the combined processes of photosynthesis and respiration, estimated to be of the order of a few per cent of R_n and neglected here.

The presence of a storage term in this energy budget adds considerable complexity to the measurement and modelling of the daily cycle in forest energy partition. Early attempts to model this component simply (e.g. Stewart and Thom 1973) proved incapable of describing sustained nighttime radiation (Thompson 1979). More recent, sophisticated computer models (e.g. Goudriaan 1979) suggest that storage can be a very significant contribution to the energy budget, both in terms of magnitude and duration. The size, complexity and density of tropical forest is such that energy storage is likely to be a particularly important component, but difficult to calculate.

By adopting an eddy correlation measurement technique we successfully avoid the need to make the calculation, but the presence of the term complicates validation of the measurement to some extent. Accordingly we choose to assume that, to first order, the storage term, S, and indeed the soil heat flux, G, in Eq. (11) are approximately zero over an integrated daily cycle. Within this assumption, the cumulative net radiation over a complete day should equal the cumulative sum of the eddy correlation measurements of latent and sensible heat, i.e. $\Sigma R_n = \Sigma A = \Sigma(\lambda E + H)$. It should be remembered that this check is not completely definitive in that it cannot rule out the possibility of a fortuitous numerical cancellation between daytime and nighttime errors.

In Fig. 3 we present the comparison of cumulative net radiation and the cumulative sum of outgoing energy flux for the eight complete days considered in this analysis. The agreement is satisfactory at about a 5% level, but integrated over all eight days indicates a flux loss in the eddy correlation measurement of $6{\cdot}3\,\mathrm{MJ\,m^{-2}}$ in a total radiant input of $96{\cdot}5\,\mathrm{MJ\,m^{-2}}$. Two days, 6 and 10 September, are significantly worse than the other six and no definite reason is advanced for this, although wind directions on 10 September and the early morning of 6 September, when most of the shortfall occurs, were such as to leave the Hydra sensors poorly exposed.

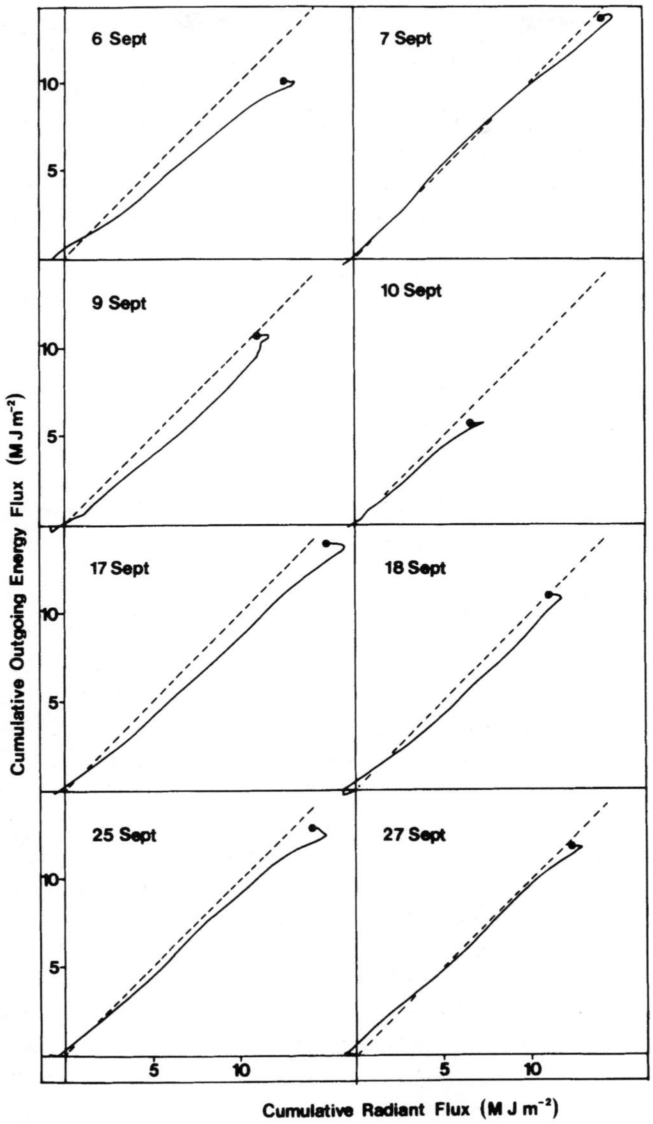

Figure 3. Comparison of the cumulative net radiation and the cumulative sum of latent and sensible heat measured by the Hydra for the eight fine days studied in this analysis. The 1:1 line is shown dashed.

(*b*) *Energy fluxes*

Figure 4 illustrates the daily energy exchange cycle for the eight fine days considered in this analysis. In most respects the behaviour is similar from day to day. There is a fairly constant partition of the available energy during daylight hours as latent and sensible heat, often some sustained evaporation into the early evening, and outgoing radiation in the order 30 to 40 W m^{-2} as the forest cools during the night. The suppression

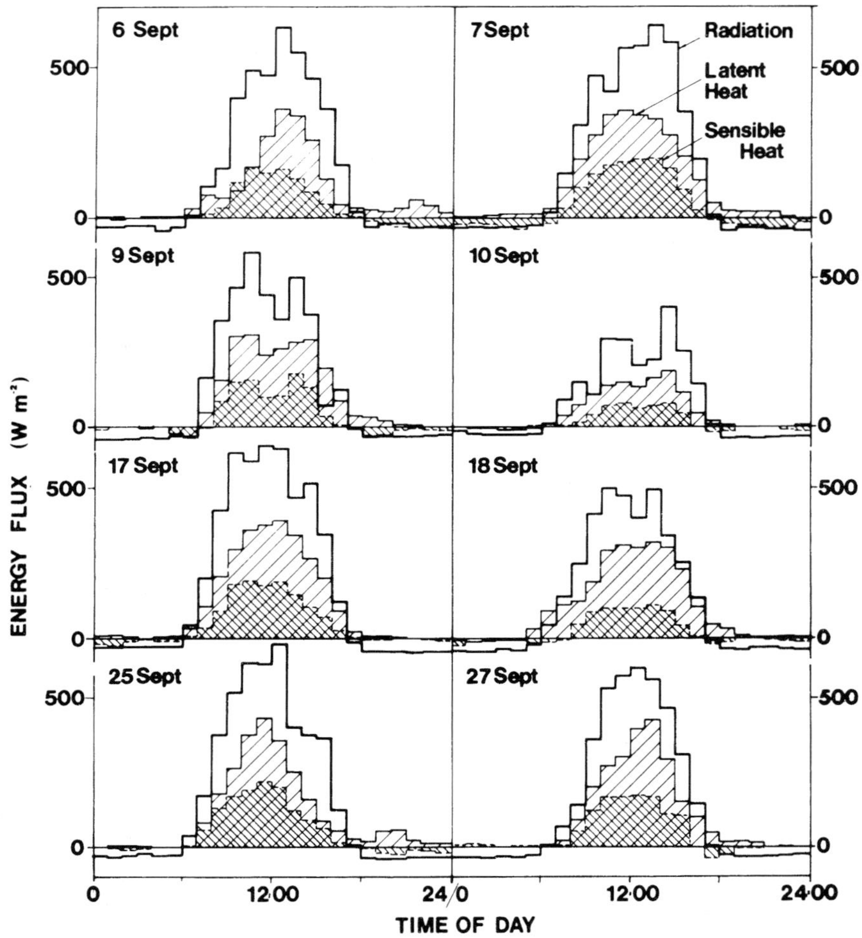

Figure 4. Daily variation in measured net radiation, latent heat and sensible heat for the eight fine days studied in this analysis. In all cases the thick line indicates net radiation, and latent heat and sensible heat flux correspond to the lightly shaded and darkly shaded portions respectively.

of turbulent interaction between the forest and the atmosphere at night, typified by the low energy fluxes, is a feature of the data and applies to all the turbulence variables. In fact the detailed behaviour of measured turbulent fluxes at night is difficult to interpret, and it is suspected that hourly measurements produced by the Hydra are subject to apparently haphazard fluctuations in the order 10 W m^{-2} which are of instrumental origin, perhaps related to correlations in the long term drift of the component sensors. No doubt such noise also occurs during the day but is less apparent: presumably the effect on integrated flux measurements is small.

Table 1 presents the integrated daily flux of radiant energy, latent and sensible heat for the eight fine days together with the 'evaporative fraction' defined in two ways:

$$\alpha = \Sigma\lambda E/\Sigma(\lambda E + H) \quad (12)$$

and

$$\alpha' = \Sigma\lambda E/\Sigma R_n. \quad (13)$$

The value of this parameter is reasonably similar for the eight days studied, the value

TABLE 1. INTEGRATED DAILY ENERGY FLUXES FOR THE EIGHT FINE DAYS CONSIDERED IN THE ANALYSIS, TOGETHER WITH THE EVAPORATIVE FRACTION DEFINED BY Eqs. (12) AND (13).

Date (Sept. 1983)	Integrated daily energy flux ($MJ\,m^{-2}$)			Evaporative fraction	
	Radiation	Latent heat	Sensible heat	α	α'
6	12·58	7·59	2·50	0·752	0·603
7	14·12	10·32	3·41	0·751	0·731
9	10·94	8·13	2·62	0·756	0·743
10	6·67	4·28	1·43	0·739	0·642
17	14·89	10·20	3·74	0·731	0·685
18	10·99	9·18	1·92	0·827	0·835
25	14·09	9·05	3·82	0·701	0·642
27	12·27	8·64	3·33	0·722	0·704
Mean	12·06	8·42	2·85	0·747	0·698

of α' is generally lower than α because of the integrated eddy flux loss mentioned in section 4(a). It is also more variable from day to day since it includes the additional variability in this loss factor. However, the implication of Table 1 is clear: the daily evaporative fraction for transpiring tropical forest for these data is around 0·7, corresponding to a Bowen ratio, $\beta\,(=H/\lambda E)$, of 0·43. We return to this important result later.

Earlier mention was made of the fact that on two days, 25 and 27 September, a Thermometer Interchange System (TIS) of the type described by McNeil and Shuttleworth (1975) was operational for significant portions of the day. It is possible to use this to provide some confirmation of the above result. The system provided two differential measurements of temperature and humidity between the height ranges 35·6 to 39·3 m and 40·9 to 44·6 m. Under the assumption that the fluxes of sensible and latent heat obey a one-dimensional diffusion equation, and that the eddy diffusivity for heat and water vapour are equal, the ratio of the differential temperature and humidity measurements provides a measure of the Bowen ratio in steady state conditions (see McNeil and Shuttleworth for details). The evaporative fraction is directly calculable from this ratio through the expression $\alpha = (1+\beta)^{-1}$. Hourly measurements made in this way tend to be noisy because the differential gradients are small.

Figure 5 illustrates the daily trend in α calculated for the two portions of the TIS and compares this with that calculated from the eddy correlation measurements for daylight hours when relevant measurements were available. Although these results are subject to considerable experimental noise, they are in broad agreement, with the TIS results perhaps tending to favour an even lower evaporative fraction.

(c) *Aerodynamic resistance estimates*

The aerodynamic resistance for momentum transfer, r_A, is given by the expression

$$r_A = u/(u_*)^2 \tag{14}$$

where u is the measured wind speed and u_* is the friction velocity. In neutral conditions (Thom 1975) it is usual to write

$$u_*/u = k \ln^{-1}\{(z-d)/z_0\} \tag{15}$$

where k is von Kármán's constant (assumed equal to 0·41), z is the measurement height of the wind speed u, and z_0 and d are the roughness length and zero plane displacement of the vegetation.

Our purpose in this section is to provide a working description of r_A for use in

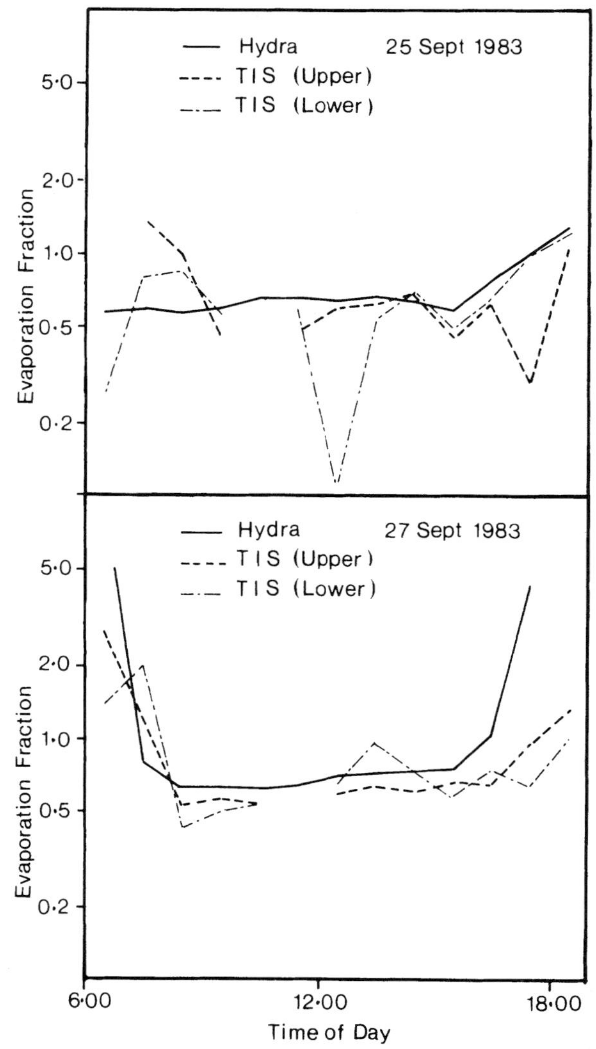

Figure 5. The variation in evaporative fraction, $\alpha(=\lambda E/(H+\lambda E))$, for daylight hours on 25 and 27 September as measured by the Hydra (———) and the upper (– – –) and lower (– · – · –) portions of the Thermometer Interchange System.

deriving surface resistance and in later discussion. For this purpose we reference the aerodynamic transfer resistance to the wind speed at 45 m measured by the automatic weather stations for the reasons mentioned in section 3(*b*). Measurements of u_* at 48·5 m are drawn from the Hydra and we assume no divergence in momentum flux between the measurement heights of these two instruments.

There is reason to believe that aerodynamic resistance to the transfer of energy flux differs from that for momentum in non-neutral conditions (e.g. Stewart and Thom 1973), and all aerodynamic resistances are subject to stability corrections. In practice this complexity is of very little numerical importance in modelling evaporation from forests for reasons which become clear in the following sections. Consequently such models generally assume simple formulations corresponding to neutral stability (e.g. Gash *et al.* 1980), and some assume a constant value for this resistance (e.g. Gash and Stewart 1977). In accordance with this philosophy we present data for daytime conditions, when the vast majority of flux transfer occurs, regardless of atmospheric stability. The data are, however, heavily selected to correspond to wind directions in which the two Gill propeller anemometers are reasonably well exposed and where their mutual interference is minimized. Only mean wind directions which are within 45° of the bisecting angle between the two propellers are considered. In fact the large variability in wind direction in this environment is such that, even with this selection, some interference cannot be ruled out in the hourly average values considered. The data are further selected to exclude hours when they are unreliable by virtue of their timing relative to initial turn-on of the Hydra system, or sonic anemometer loss during rain.

Measured values of u_* are plotted against AWS wind speed in Fig. 6(a), together with a line of linear regression constrained to pass through the origin. The gradient of this line is 0·175 which, from Eq. (15), corresponds to a value of $(z - d)/z_0 = 10{\cdot}4$. In

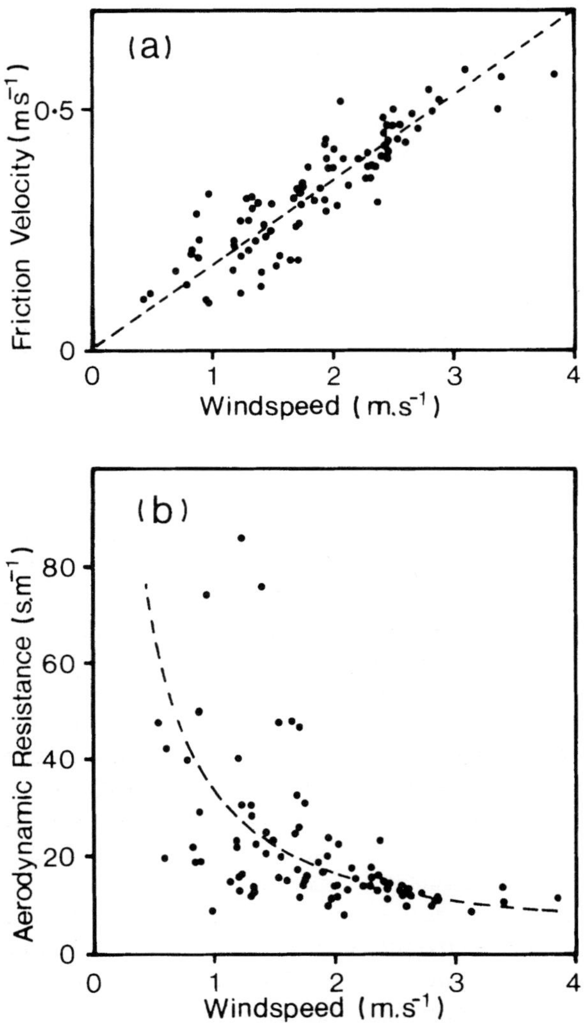

Figure 6. (a) Values of u_* measured with the Hydra as a function of u measured with the automatic weather station. The dashed line is a line of linear regression constrained to pass through the origin.

(b) Values of aerodynamic resistance, r_A, deduced from measurements of u_*, made with the Hydra, and u, made with the automatic weather station, expressed as a function of u. The dashed line is the curve $r_A = 33/u$ used as a working description in this analysis.

previous studies of momentum transfer to forest canopies, attempts have been made to refer observed values of d and z_0 to canopy height h, but the results are extremely variable. The values given in Table V of Jarvis *et al.* (1976), when interpreted for this experiment by setting $h = 35\,\text{m}$ and $z = 1{\cdot}29h$, give estimates of $(z - d)/z_0$ in the range 6 to 20, with a mean value of 10·1.

Assuming the value 10·4 for $(z - d)/z_0$ and combining Eqs. (14) and (15) gives the function

$$r_A(u) = 33/u \tag{16}$$

which we adopt as a working description of these data for the purpose of this analysis. Figure 6(b) compares this function with values of r_A calculated from the individual measurements using Eq. (14). The results are subject to considerable scatter particularly at low wind speeds when problems associated with propeller and cup anemometer stalling, directional variability and fractional overspeeding errors are most important. Errors here are in the order of 100%; the dependence at higher wind speed is slightly better defined but scatter is still in the order of 50%. We claim no better accuracy than this for Eq. (16). Although most of the scatter in Figs. 6(a) and (b) is probably of instrumental origin, large experimental noise in the measurement of the aerodynamic properties of forests is in fact an extremely common, though rarely discussed, phenomenon. It is probable that the absence of steady state conditions and the (relative) proximity of measurement systems to the surface make some contribution.

(*d*) *Surface resistance*

On rearranging the Penman–Monteith equation (Monteith 1965), the surface resistance, r_s, is given by

$$r_s = \{(\Delta'\lambda/c_p)\beta - 1\}r_A + \rho\lambda\{q_w(T) - q\}/\lambda E \tag{17}$$

where Δ' is the slope of the saturated specific humidity curve at mean air temperature; ρ is the density of air; λ is the latent heat of vaporization of water; c_p is the specific heat of air at constant pressure; $q_w(T)$ is the saturated specific humidity at T, the air temperature at height z; q is the specific humidity at height z; and other quantities are defined earlier in the text.

Values of surface resistance were calculated from Eq. (17) for each hour using Eq. (16) to estimate r_A. In this way a value of r_A is consistently available from the AWS wind speed measurement. The alternative, direct measurements of r_A using Eq. (14) has restricted availability as a result of wind direction constraints on the u_* measurement. For the eight days considered the surface resistance follows a similar daily trend and the eight-day median value is presented in Fig. 7. The resistance falls rapidly with daybreak and then stays fairly constant around $130\,\text{s}\,\text{m}^{-1}$ until early afternoon, after which time it rises progressively, increasing by almost an order of magnitude by early evening.

It is worth noting at this stage that the coefficient of r_A in Eq. (17) is small. For Bowen ratios around 0·43 and mean temperatures of 28 °C, a 50% change in a typical value of r_A ($18\,\text{s}\,\text{m}^{-1}$) only causes a 3% change in a typical value of r_s ($130\,\text{s}\,\text{m}^{-1}$).

The daily trend in surface resistance shown in Fig. 7 is in qualitative agreement with some separate, direct measurements of stomatal resistance made in parallel with these data. The physiological cause of this behaviour is not clear at this stage. There could well be some response to vapour pressure deficits, since these are higher in the afternoon, but the behaviour is more suggestive of a response to a daily increase in leaf water potential. The rapid decrease at dawn and increase at dusk could well reflect a dependence on radiation level.

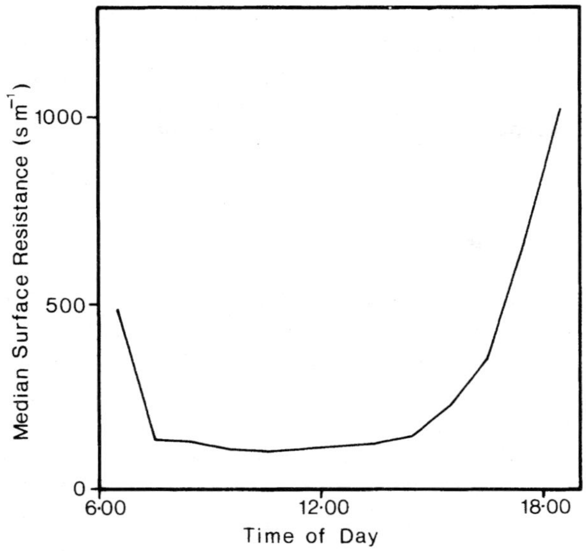

Figure 7. The median value of surface resistance, r_s, for the eight fine days considered in this analysis for daylight hours.

5. Discussion

The data presented in section 4 provide an initial specification of the aerodynamic and surface controls on energy partition for Amazonian forest. Assuming this site is representative, they are arguably typical of natural forest in mid-continent Amazonia in conditions when soil moisture is freely available. The results may be of interim use in hydrological and meteorological energy partition models of dry canopy evaporation in conjunction with short-term meteorological variables, measured (or model specified) near the forest canopy. Their usefulness in this role is, however, presently curtailed by the limited time period and soil moisture conditions they describe, and, equally importantly, by the present lack of an experimental calibration of model parameters in an equivalent short-term model of wet canopy (interception) loss. Extending the descriptive range of the dry canopy specification presented here, and providing a calibration of detailed wet canopy models remain important objectives of the collaborative study described in section 1. We therefore postpone further discussion of this potentially important application of these data.

All models of energy partition are ultimately empirical, their complexity, physical realism and descriptive accuracy are conditioned by the data requirement involved in their use (Shuttleworth 1979, 1983; Stewart 1983). The past absence of any measurements against which more complex but physically realistic models could be calibrated in the Amazon basin, has necessarily required the speculative use of energy partition models with a high empirical content. Moreover there is, in this region, a shortage of short-term, above-canopy meteorological data with which complex models could in any case be operated—a shortage which is unlikely to be rectified in the near future. In this situation, there is a clear need to interpret the present data in the framework of existing, empirical energy partition models, and if necessary to modify these. We concentrate on this aspect in the remainder of this section.

(*a*) *Comparison with estimation formulae*

In the phraseology of classical hydrology, the experimental situation in which the data presented in this paper were gathered represents 'evaporation from a well-watered crop'. The meteorologist and hydrologist schooled in the concept of potential evaporation might therefore be surprised by the result (given in section 4(*b*)), that the fraction of incoming radiant energy used in evaporation for a transpiring forest canopy is around 0·7. To the forest hydrometeorologist this result is entirely plausible, and not unexpected (cf. Shuttleworth and Calder 1979). It reflects the fact that transpiring forest canopies exert significant surface control even when water is freely available in the soil.

To emphasize this point we present, in Table 2, values for some of the more commonly used estimation equations, together with the water equivalent of total incoming

TABLE 2. WATER EQUIVALENT OF RADIATION, ACTUAL EVAPORATION AND THE CALCULATED RATES DEFINED IN THE TEXT: PENMAN EVAPORATION (λE_P), THOM–OLIVER EVAPORATION (λE_{TO}), PRIESTLEY–TAYLOR EVAPORATION (λE_{PT}) AND EQUILIBRIUM (λE_E) FOR THE FINE DAYS CONSIDERED IN THE ANALYSIS. ALL NUMBERS ARE mm OF WATER PER DAY.

Date (Sept. 1983)	Radiation equivalent	Actual evaporation	Calculated evaporation rates			
			λE_P	λE_{TO}	λE_{PT}	λE_E
6	5·16	3·11	5·21	5·35	5·02	3·99
7	5·80	4·22	6·12	6·51	5·57	4·46
9	4·49	3·32	4·74	5·04	4·32	3·27
10	2·74	1·75	2·71	2·74	2·54	2·02
17	6·12	4·18	6·38	6·72	6·00	4·76
18	4·52	3·76	4·62	4·79	4·38	3·48
25	5·79	3·70	5·78	5·87	5·63	4·46
27	5·04	3·54	4·85	4·84	4·90	3·89
Mean	4·96	3·45	5·05	5·24	4·80	3·80

radiation and the measured values of actual evaporation. If Q_N is the integrated radiation input, D the daily average specific humidity deficit, and u the daily average wind speed ($\mathrm{m\,s^{-1}}$); then, expressed in SI units and referenced to specific humidity (rather than vapour pressure), these estimation equations take the following form:

Penman evaporation (Penman 1948)

$$\lambda E_P = \{\Delta' Q_N + \rho c_p D f(u)\}/(\Delta' + c_p/\lambda) \tag{18}$$

where

$$f(u) = 0{\cdot}004(1 + 0{\cdot}54u); \tag{19}$$

Thom–Oliver evaporation (Thom and Oliver 1977), with values of the constants suggested for general use,

$$\lambda E_{TO} = \{\Delta' Q_N + \rho c_p 2{\cdot}5 D f(u)\}/(\Delta' + 2{\cdot}4 c_p/\lambda); \tag{20}$$

Priestley–Taylor evaporation (Priestley and Taylor 1972),

$$\lambda E_{PT} = 1{\cdot}26\Delta' Q_N/(\Delta' + c_p/\lambda); \tag{21}$$

equilibrium evaporation, the steady state, asymptotic limit approached by a closed atmosphere over a large expanse of vegetation with fixed surface resistance (McNaughton and Jarvis 1983),

$$\lambda E_E = \Delta' Q_N/(\Delta' + c_p/\lambda). \tag{22}$$

Inspection of Table 2 reveals that all the estimates of potential evaporation λE_{P} λE_{TO}, λE_{PT} and even λE_{E} exceed actual evaporation in dry canopy conditions, and the first two exceed the available radiation input. The potential evaporation estimates, which agree to within 5%, exceed actual evaporation by a factor 1·5. Equilibrium evaporation is a much closer approximation to measured transpiration.

(*b*) *Towards a practical estimate of daily evaporation*

Given the results presented in section 4, i.e. high measured values of surface resistance and comparatively low values of aerodynamic resistance, the conclusion drawn from work in temperate forest (cf. Shuttleworth and Calder 1979), that transpiration and the evaporation of intercepted rainfall must be estimated separately, applies equally to Amazonian rain forest. Gash (1978) demonstrated that the approach advocated by Thom and Oliver (1977) is in essence the Penman–Monteith equation (Monteith 1965) written in a form suitable for daily application. We follow this philosophy and write, as a working model of daily evaporation for forest surfaces, the equation

$$\lambda E_{\mathrm{F}} = \frac{\Delta' Q_{\mathrm{N}} + \rho c_p D/r_{\mathrm{A}}}{\Delta' + (c_p/\lambda)(1 + r_{\mathrm{sd}}/r_{\mathrm{A}})} + I(1 - c) \tag{23}$$

where I is a model calculation (or measurement) of interception loss; r_{sd} is the effective daily value of dry canopy surface resistance in well watered conditions; and c compensates for including transpiration in wet conditions, and is given by

$$c = \frac{\Delta' + c_p/\lambda}{\Delta' + (c_p/\lambda)(1 + r_{\mathrm{sd}}/r_{\mathrm{A}})}. \tag{24}$$

In their original formulation Thom and Oliver recommend the equation

$$r_{\mathrm{A}} = 53 \ln^2\{(z - d)/z_0\}\}f(u) \tag{25}$$

as an estimate of aerodynamic resistance. In the present context we prefer to combine Eqs. (14) and (15) and adopt

$$r_{\mathrm{A}} = \ln^2\{(z - d)/z_0\}/k^2 u \tag{26}$$

as a working description of the aerodynamic resistance, with $(z-d)/z_0 = 10{\cdot}4$. The values of r_{A} given by these two expressions differ significantly at low wind speeds but the consequences of this difference in estimates of transpiration in Eq. (23) is small, typically in the order 2%.

The first term in Eq. (23) describes transpiration and the present data provide an initial calibration of the empirical constant r_{sd}. Taking mean values of the measured variables, measured over the eight days considered in this analysis, and using Eq. (23), to calculate the value of r_{sd} which gives the evaporative fraction 0·7 suggested by Table 1, we deduce that the effective daily value of dry canopy resistance for this forest is $215\,\mathrm{s\,m^{-1}}$. The fact that the effective daily value of dry canopy resistance is significantly greater than the measured midday values presented in section 4(*d*) serves as a reminder that r_{sd} is an empirical parameter linked to daily average values. Evaporation, which largely occurs during the day, is implicitly assumed to occur more slowly over twenty-four hours, and the effective resistance is increased.

The correction factor $(1 - c)$ in Eq. (23) compensates for the fact that some transpiration is calculated even under wet conditions (cf. Gash 1978). Using the effective dry canopy resistance $r_{\mathrm{sd}} = 215\,\mathrm{s\,m^{-1}}$ in Eq. (24) gives a value of $(1 - c) = 0{\cdot}71$. This is less than that typical of temperate forests, e.g. 0·93 (Shuttleworth and Calder 1979), and this in part reflects the fact that the rates of evaporation in dry and wet canopy conditions are more comparable for tropical forests since mean temperatures are higher.

The second term in Eq. (23) describes the evaporation from the wet forest canopy and implies the separate existence of a direct measurement or sub-model of the interception loss, I. Gash (1979) created an analytic model of interception based on physical principles similar to those used in the numerical model of Rutter *et al.* (1971). In so doing he demonstrated that there is at least some physical justification for expressing the interception loss from a forest in terms of a regression on precipitation input. The regression coefficients depend on mean rainfall rate and mean wet canopy evaporation rate, and on the canopy storage capacity. For temperate forests the interception component can make an important and possibly dominant contribution to total evaporation. Preliminary estimates (see later) suggest that interception is of less relative importance in this environment and in this period contributes only about 30% of total evaporation. In this way the errors involved in rather a coarse model of I, in Eq. (23), have a proportionately reduced percentage error in estimates of λE_F.

In the present study parallel measurements of interception loss were made during September and early October 1983 at a site adjacent to the meteorological tower. Measurements of the rainfall amount reaching the ground were made with 16 raingauges which were relocated randomly along a linear transect after each rainstorm. The total interception loss was 32·7 mm in a total precipitation, P, of 193·25 mm (16·9%). Using $I = 0{\cdot}169\,P$ as a working model of interception loss during September 1983, and taking meteorological data from the automatic weather stations, Eq. (23) predicts a mean daily evaporation of 3·73 mm for this period. This can be compared with the mean daily estimates $\lambda E_P = 4{\cdot}27$ mm, $\lambda E_{PT} = 4{\cdot}15$ mm, $\lambda E_{TO} = 4{\cdot}30$ mm and $\lambda E_E = 3{\cdot}29$ mm, given by Eqs. (18), (20), (21) and (22) applied over the equivalent period. Over these, atypically wet, days in September 1983, an estimated 48% of the precipitation falling on the site returned to the atmosphere by evaporation.

The measurements of transpiration given in section 5(*a*), and the estimates of total evaporation given in this section, together typify the general behaviour of forest evaporation at a catchment scale (cf. Shuttleworth and Calder 1979). When the canopy is dry, evaporation proceeds at rates which are significantly less than those suggested by estimation formulae developed and calibrated over short vegetation (λE_P, λE_{PT} and λE_{TO}), rates which are more comparable to equilibrium evaporation (λE_E). However, during periods when rainfall is common, as in the present data, the interception loss tends to compensate (or even over-compensate) for this. The best understanding of forest micrometeorology currently available, calibrated with the dry canopy data presented here, and merged with preliminary estimates of interception loss, suggests that evaporation from Amazonian rain forest will:

(i) approach, and possibly exceed, accepted estimates of potential evaporation in wet months (rainfall 270 mm per month);
(ii) fall to 70% of potential rates, or less, in dry months (rainfall 30 mm per month).

Acknowledgments

The results presented here were obtained in the first year of an extensive programme of research into Amazonian micrometeorology supported by the Natural Environment Research Council, UK, the Conselho Nacional de Desenvolvimento Cientifico and Tecnologico (CNP_q), Brasil, and the British Council. The authors are pleased to acknowledge this support. The project received a great deal of practical and administrative help from members of staff at the Institute of Hydrology, the Instituto Nacional de Pesquisas da Amazonia and Centro Nacional de Pesquia da Seringueira e Dende (EMBRAPA), all of which are gratefully recognised. We would particularly like to thank Robin Clarke for his help in setting up the project, Howard Oliver and David McNeil for their work in preparing the equipment, and Francisco de Assis Saturnino for his help in tower and site construction.

REFERENCES

Bernstein, A. B. 1967 A note on the use of cup anemometer in wind profile experiments. *J. Appl. Met.*, **6**, 280–286

Black, T. A. and McNaughton, K. G. 1971 Psychrometric apparatus for Bowen ratio determination over forest. *Bound.-Layer Meteorol.*, **2**, 246–254

Businger, J. A., Wyngaard, J. C., Izumi, Y. and Bradley, E. F. 1971 Flux-profile relationships in the atmospheric surface layer. *J. Atmos. Sc.*, **28**, 190–201

Calder, I. R. 1977 A model of transpiration and interception losses from a spruce forest in Plynlimon, central Wales. *J. Hydrol.*, **33**, 247–265

Drinkrow, R. 1972 A solution to the paired Gill-anemometer response function. *J. Appl. Meteorol.*, **11**, 76–80

Gash, J. H. C. 1978 Comment on the paper by A. S. Thom and H. R. Oliver: 'On Penman's equation for estimating regional evaporation'. *Quart. J. R. Met. Soc.*, **104**, 532–533

1979 An analytical model of rainfall interception by forests. *ibid.* **105**, 43–56

Gash, J. H. C. and Stewart, J. B. 1977 The evaporation from Thetford Forest during 1975. *J. Hydrol.*, **35**, 385–396

Gash, J. H. C., Wright, I. R. and Lloyd, C. R. 1980 Comparative estimates of interception loss from three coniferous forests in Great Britain. *ibid.* **48**, 89–105

Gay, L. W. and Fritschen, L. J. 1979 An energy budget analysis of water use by Saltcedar. *Water Resour. Res.*, **15**, 1589–1592

Goudriaan, J. 1979 'MICROWEATHER simulation model, applied to a forest', pp. 47–57 in *Comparison of forest water and energy exchange models*, (Swen Halldin, Ed), International Society of Ecological Modelling, Copenhagen

Hyson, P. and Hicks, B. B. 1975 A single-beam infrared hygrometer for evaporation measurement. *J. Appl. Meteorol.*, **14**, 301–307

Jarvis, P. G., James, G. B. and Landsberg, J. J. 1976 'Coniferous Forest', in *Vegetation and the atmosphere*, Vol. II (J. L. Monteith, Ed.), Academic Press (London)

Lloyd, C. R., Shuttleworth, W. J., Turner, M. and Gash, J. H. C. 1984 A microprocessor system for eddy-correlation. *Agric. and For. Meterol.* (in press)

McNaughton, K. G. and Jarvis, P. G. 1983 'Predicting effects of vegetation changes on transpiration and evaporation', Chapter 1, in *Water deficits and plant growth*, Vol VII, (T. T. Kozolowski, Ed.) Academic Press (New York)

McNeil, D. D. and Shuttleworth, W. J. 1975 Comparative measurements of energy fluxes over a pine forest. *Bound.-Layer Meteorol.*, **9**, 297–313

Monteith, J. L. 1965 Evaporation and environment. *Symp. Soc. Exp. Biol.*, **19**, 205–234

Moore, C. J. 1983 On the calibration and temperature behaviour of single-beam infrared hygrometers. *Bound.-Layer Meteorol.*, **25**, 245–269

Penman, H. L. 1948 Natural evaporation from open water, bare soil and grass. *Proc. Roy. Soc.*, A, **193**, 120–145

Priestley, C. H. B. and Taylor, R. J. 1972 On the assessment of surface heat flux and evaporation using large scale parameters. *Mon. Weath., Rev.*, **100**, 81–92

Rutter, A. J., Kershaw, K. A., Robins, P. C. and Morton, A. J. 1971 A predictive model of rainfall interception in forests. I: Derivation of the model from observations in a plantation of Corsican pine. *Agric. Meteorol.*, **9**, 367–384

Shuttleworth, W. J. 1979 'Evaporation'. Inst. Hydrol. Rept. No. 56. Wallingford

1983 'Evaporation models in the global water budget', pp. 147–171 in *Variations in the Global Water Budget* (A. Street-Perrott, M. Beran and R. Radcliffe, Ed.). D. Reidel Publishing Company, Dordrecht

Shuttleworth, W. J. and Calder, I. R. 1979 Has the Priestley–Taylor equation any relevance to forest evaporation? *J. Appl. Met.*, **18**, 639–646

Shuttleworth, W. J., McNeil D. D. and Moore, C. J. 1982 A switched continuous-wave sonic anemometer for measuring surface heat fluxes. *Bound.-Layer Meteorol.*, **23**, 425–448

Stewart, J. B. 1983 A discussion of the relationships between the principle forms of the combination equation for estimating crop evaporation. *Agric. Meteorol.*, **3**, 111–127

Stewart, J. B. and Thom, A. S. 1973 Energy budgets in pine forests. *Quart. J. R. Met. Soc.*, **99**, 154–170

Thom, A. S. 1975 'Momentum, mass and heat exchange of plant communities', chapter 3, in *Vegetation and the atmosphere*, Vol. I (J. L. Monteith, Ed.). Academic Press (London)

Thom, A. S. and Oliver, H. R. 1977 On Penman's equation for estimating regional evaporation. *Quart. J. R. Met. Soc.*, **103**, 345–357

Thompson, N. 1979 Turbulence measurements above a pine forest. *Bound.-Layer Meteorol.*, **16**, 293–310

Wallace, J. S., Lloyd, C. R., Roberts, John and Shuttleworth, W. J. 1984 A comparison of methods for estimating aerodynamic resistance of heather in the field. *Agric. and For. Meteorol.*, (in press)

Webb, E. K., Pearman, G. T. and Leuning, R. 1980 Correction of flux measurements for density effects due to heat and water vapour transfer. *Quart. J.R. Met. Soc.*, **106**, 85–100

Wieringa, J. 1980 A revaluation of the Kansas Mast influence on measurements of stress and cup anemometer overspeeding. *Bound.-Layer Meteorol.*, **18**, 411–430

1981 A reply to Wyngaard *et al.* 1981. *ibid.*, **22**, 251–255

Wyngaard, J. C., Businger, J. A., Kaimal, J. C. and Larson, S. E. 1981 Comments on 'A revaluation of the Kansas Mast influence on measurements of stress and cup anemometer overspeeding'. *ibid.*, **22**, 245–250

Estimation of Areal Evapotranspiration (Proceedings of a workshop held at Vancouver, B.C., Canada, August 1987). IAHS Publ. no. 177, 1989.

Physical aspects of the planetary boundary layer with special reference to regional evapotranspiration

H.A.R. de Bruin
Wageningen Agricultural University
Department of Physics and Meteorology
Duivendaal 2, 6701 AP Wageningen
The Netherlands

Abstract A brief review of the physical processes related to the vertical transport of water vapour into and within the planetary boundary layer (PBL) is presented. The main features of the stable and near neutral PBL are discussed with special reference to regional evapotranspiration. It is shown that through interactions between the surface fluxes and the PBL, evapotranspiration and the saturation deficit are dependent variables. Consequently, the Penman-Monteith equation is a descriptive rather than predictive formula. Future models of the regional evapotranspiration will have to take account of the effects of surface inhomogeneities on the dynamics of the PBL, and the formation of fog and clouds.

Aspects physiques de la couche limite planétaire avec référence spéciale à l'évapotranspiration

Résumé On a rapidement passé en revue les processus physiques qui se rapportent au mouvement vertical de la vapeur d'eau vers, et à l'intérieur même, de la couche limite planétaire (Clp). On a présenté les principales caractéristiques de la Clp qui est stable et quasiment neûtre en se référant plus spécialement à l'évapotranspiration régionale. On a ainsi montré que l'évapotranspiration régionale et le déficit de saturation sont des variables interdépendantes à cause des interactions qui existent entre les flux de la surface et de la Clp. Tout cela fait de l'équation de Penman-Monteith une formule plutôt descriptive que prédictive. Les prochains modèles d'évapotranspiration devront tenir compte des effets de l'inhomogénéité de la surface sur la dynamique de la Clp et sur la formation du brouillard et des nuages.

NOTATION

c	entrainment constant (Eq.13)	–
c_p	specific heat of air at constant pressure	J kg^{-1} K^{-1}
g	acceleration of gravity	m s^{-2}

h	height of the PBL	m
k	von Karman constant	–
q	specific humidity	–
q_m	specific humidity of the well-mixed PBL	–
$q_s(T)$	saturation specific humidity at temperature T (and pressure p)	–
q_t	specific humidity of stable air at z = h	–
r_a	aerodynamic resistance for transport of heat and wàter vapour of the air layer between the surface and reference height z	$s\ m^{-1}$
r_s	canopy or surface resistance	$s\ m^{-1}$
s	slope of the saturation specific humidity versus temperature curve	K^{-1}
t^*	time constant (Eq.10)	s
u_*	friction velocity	$m\ s^{-1}$
w^*	convective velocity scale (Eq.3)	$m\ s^{-1}$
A	$= Q^*-G$, available energy	$W\ m^{-2}$
D	saturation (specific humidity) deficit ($D = q_s(T)-q$)	–
D_{eq}	equilibrium saturation deficit (Eq.9)	–
D_m	$= q_s(\theta_m)-q_m$ (Eq.5)	–
D_t	$= q_s(\theta_t)-q_t$ (see text)	–
E	evapo(transpi)ration	$kg\ m^{-2}\ s^{-1}$
E_{eq}	equilibrium evapo(transpi)ration (Eq.11)	$kg\ m^{-2}\ s^{-1}$
G	soil heat flux density	$W\ m^{-2}$
H	sensible heat flux density	$W\ m^{-2}$
H_v	virtual heat flux density	$W\ m^{-2}$
L	Obukhov length (Eq.2)	m
Q^*	net radiation	$W\ m^{-2}$
γ	psychrometric constant($= c_p/\lambda$)	K^{-1}
γ^*	$=\gamma(1+r_s/r_a)$	K^{-1}
γ_q	slope of the q profile of stable air above the well-mixed PBL	m^{-1}
γ_θ	slope of the θ profile of stable air above the well-mixed PBL	$K\ m^{-1}$
τ	surface flux of momentum	$N\ m^{-2}$
Δq	'jump' in q profile at z = h	–

$\Delta\theta$	'jump' in θ profile at z = h	K
λ	latent heat of vaporization of water	J kg^{-1}
ρ	density of air	kg m^{-3}
θ	potential temperature	K
θ_m	potential temperature of the well-mixed PBL	K
θ_t	potential temperature of the stable air at z=h	K

INTRODUCTION

In most models for evapo(transpi)ration, this quantity (E) is described as a function of routine meteorological variables. The best-known and, probably, most successful example is the Penman-Monteith equation (Monteith, 1981):

$$\lambda E = \frac{s\ A + \rho\ c_p\ D/r_a}{s + \gamma\ (1 + r_s/r_a)} \qquad (1)$$

As early as 1963 Bouchet realised that the Penman-equation (Penman, 1948), on which (1) is based, is a *descriptive* rather than a *predictive* formula, since E is expressed in *dependent* variables. In particular, the saturation deficit (D) is correlated with E due to the fact that surface evaporation (which is generally accompanied by surface heating) affects directly the temperature and humidity of the lowest part of the atmosphere, and thus influences D. In this way Bouchet (1963) argued that the Penman-equation is not able to predict the 'potential' evapotranspiration rate, especially if the surface is very dry.

In fact, the interaction between surface evaporation and the atmosphere leads to an additional relationship between E and D (and other parameters occurring in the right-hand side of (1) such as net radiation). Recently, it has been recognised by evaporation researchers that such a relationship can be obtained only if models for the planetary boundary layer (PBL) are included in evaporation theory. In the last decade much progress has been made in PBL-modelling. An important 'driving-force' behind this research is the air pollution problem with which many countries are faced. In PBL research an opposite development took place: it was found that the evolution of the PBL strongly depends on the surface heating or cooling, which - through the surface energy balance - is directly related to evaporation. As a result, in most PBL models the relevant PBL parameters are described in terms of the surface fluxes of heat and water vapour, where the latter are considered independent boundary conditions. A good example is the model for the well-mixed PBL used by Driedonks (1981, 1982), who showed that the height of the PBL evolves approximately proportionally to the square-root of the time integral of the surface heat flux.

The developments in evaporation and PBL research described above took place almost independently. Results obtained in both disciplines were combined only recently, e.g. Pielke (1984) and Holtslag (1987). It should be noted that in the last five years or so, climate modellers have recognised the significance of a proper description of evaporation, e.g. Sellers *et al.* (1986).

It is the objective of this paper to describe briefly the most important aspects of the PBL as far as these are relevant for regional evapotranspiration. The paper is meant as an introduction for those evaporation researchers who are not familiar with PBL-theory.

THE PBL

General

The PBL is the layer of the atmosphere that is directly affected by the Earth's surface, e.g. through its roughness or by surface heating. In most cases the flow within the PBL is turbulent. Generally, the PBL is capped by an inversion. It is common to distinguish between the unstable and the stable PBL depending on the sign of the virtual heat flux density at the surface, $H_v = H + 0.07\ \lambda E$; if $H_v > 0$ (H_v upwards) the PBL is unstable. In the lowest part of the PBL (the surface layer), Monin-Obukhov similarity theory applies (Panofsky & Dutton, 1984). Here the Obukhov-length (L) plays a dominant role and is given by

$$L = \frac{-T\ u_*^3}{kg\ H_v}\ \rho c_p \qquad (2)$$

The depth of the surface layer is typically one tenth of the PBL height (h). Here we will confine ourselves to horizontally homogeneous conditions. Also, condensation of water vapour within the PBL is excluded. Following Holtstag & Nieuwstadt (1986) the (turbulent) state of the (dry) PBL can be categorized schematically in terms of z/h and h/L, where z is the height above the ground.

The unstable PBL

As noted before, the unstable PBL is formed if at the surface the virtual heat flux (H_v) is upwards. In this case, buoyancy enhances the mechanically generated turbulence caused by wind shear. The unstable PBL can be in two different states (Fig. 1):

a) a near neutral upper layer, when $0 < -h/L < 10$, or

b) a mixed layer where $-h/L > 10$.

Both layers are capped by the entrainment or interfacial layer (described later). Also, it is seen from Fig. 1 that the surface layer transforms into the free convection layer in the region right of the line $-z/L = 1$.

It is common in research on atmospheric turbulence to use a limited number of scaling parameters, which are grouped as dimensionless numbers. For the different regions the most important scaling parameters are shown in Fig. 1.

The near neutral upper layer forms during strong winds and with overcast conditions. Over land it occurs much less frequently than over sea. Scaling parameters are z and h and the surface fluxes of momentum and (virtual) heat (τ and H_v, respectively). Its height may also depend upon the Coriolis parameter f (Tennekes, 1982). Surprisingly, less is known about the near neutral upper layer than about the mixed layer. This is partly because this layer seldom forms over land.

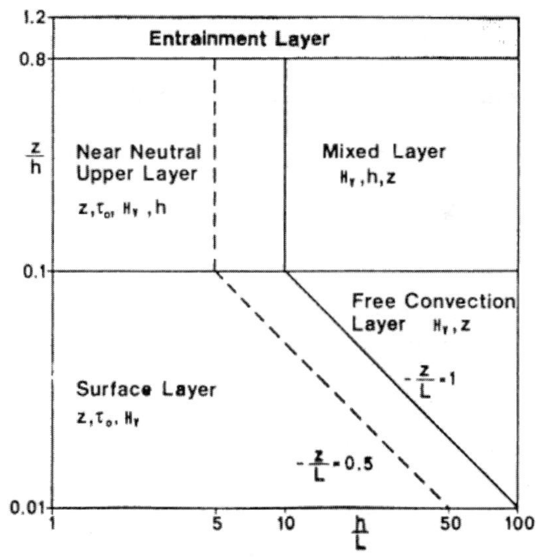

FIG. 1. Definition of scaling regions in the unstable PBL (after Holtslag & Nieuwstadt, 1986).

The mixed layer is much better understood. The most important scaling parameters are H_V, h and z. The convective velocity scale w* is often used to characterize the state of the mixed layer, with

$$w^* = \left(\frac{g\ H_V\ h}{T\ \rho c_p}\right)^{1/3} \tag{3}$$

A schematic representation is given in Fig. 2 of the mean temperature profile under convective conditions according to the three-layer model of Deardorff (1979). The vertical profile of the heat flux is also shown. Large convective eddies cause almost perfect mixing, so that θ is constant with height in the well-mixed layer between z ≈ 0.1 h and the interfacial or entrainment layer around z = h. These eddies are depicted in Fig. 3, which is an artist's representation of the unaveraged, local structure of the convective PBL (after Wyngaard, 1985).

Not much is known of the interfacial layer. Fortunately, it appears that its existence has little effect for many practical calculations and can be ignored. For this case the profiles of potential temperature and sensible heat have the form as shown in Fig. 4. The well-mixed layer is capped by an inversion, with a sudden jump in θ occuring at z = h. Above the well-mixed layer the air is stable. This layer is described by dθ/dz = γ_θ = constant. Essentially, the convective PBL evolves in time since the turbulent eddies described above intrude into the stable air aloft. In this way, warmer air is entrained into the mixed layer, while h increases. In summertime, the mixed layer grows typically from about 200 m in the early morning to about 1-2 km in the late afternoon.

The mixed layer is very important for evaporation studies since most evaporation occurs during mixed layer formation. Therefore, most evaporation models that include a description of the PBL refer to the well-mixed PBL. So far, we have not discussed the profile of (specific) humidity (q) within the convective PBL. Often, it is taken to be similar to the θ-profile, as presented in Fig. 4. However, observations (Andre *et al.*, 1979; Brutsaert, 1987) show a different behaviour (Fig. 5). The reason for this is that at the top, dry air, and at the ground, moist air, are brought into the mixed layer. Apparently, convective mixing is not vigorous enough to destroy the vertical gradient in the q-profile. Note that in the case of temperature, there is a positive heat flux into the mixed layer both at the top and at the surface. Thus it is much easier for convective mixing to maintain a vertical θ-profile.

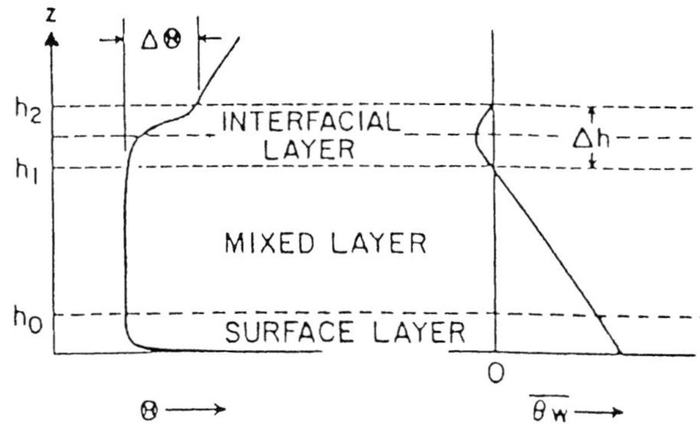

FIG. 2. A three-layer model of the well-mixed PBL (after Deardorff, 1979).

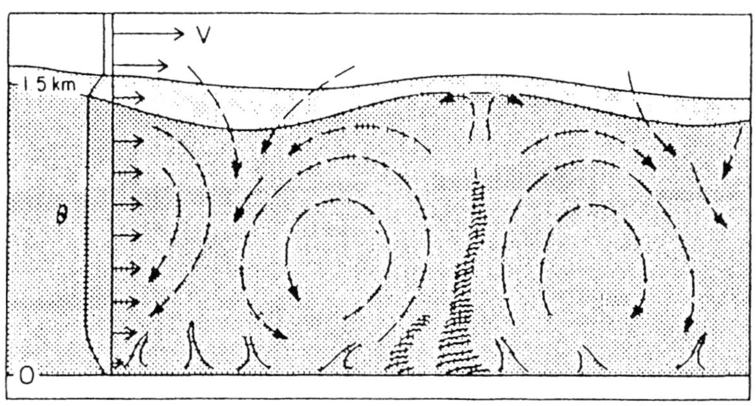

FIG. 3. An artist's representation of the unaveraged, local structure of the convective PBL (after Wyngaard, 1986).

Monin-Obukhov similarity theory applies in the surface layer (Businger, 1973; Panofsky & Dutton, 1984). This scheme appears to be very successful in the unstable surface layer, although recent research has revealed that besides z and the surface fluxes of virtual heat and momentum, appearing in the Monin-Obukhov theory, the surface roughness also is a scaling parameter (Wyngaard, 1985). It is noted that the flux-profile relationships proposed by Dyer & Hicks (1970) or Businger *et al.* (1971), which usually apply to the unstable surface layer, do not approach the theoritical so-called free convection limit. According to free convection theory (Monin & Yaglom, 1971) the surface sensible heat flux becomes proportional to $(\delta\theta/\delta z)^{3/2}$ at conditions of low wind speed. The Dyer-Hicks function does not have this behaviour. Businger (1973) gives an explanation for this feature. In the case of free convection, the surface layer is called the free convection layer, and H_v and z are the most important scaling parameters. Often the turbulent velocity scale, $w_f = (g\ H_v\ z/T\rho\ c_p)^{1/3}$, is used in the free convection layer (Tennekes, 1970). Note that τ does not play a role in this layer.

In summary it can be stated that the unstable PBL is primarily driven by surface heating, which in turn is determined by (solar) radiation and by evaporation.

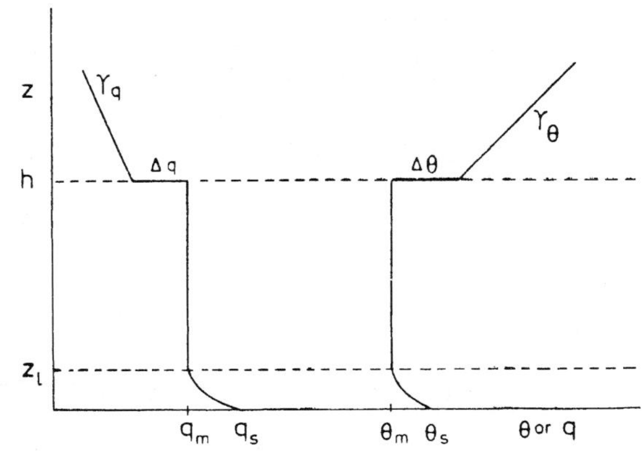

FIG. 4. Profiles of potential temperature (θ) and specific humidity (q) in a well-mixed PBL.

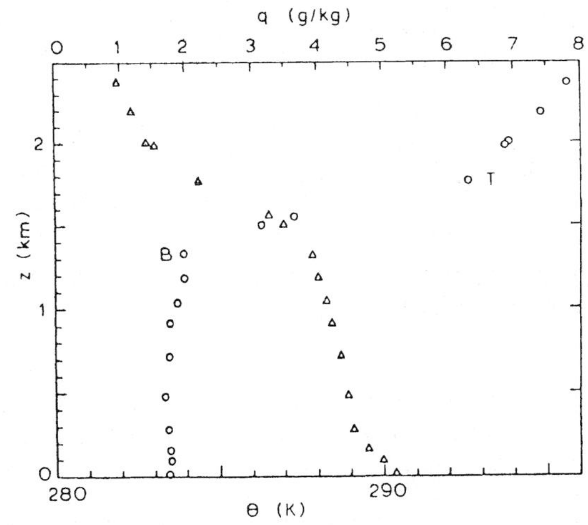

FIG. 5. An example of mean potential temperature (circles) and specific humidity (triangles) (after Brutsaert, 1987).

The stable PBL

Usually a stable PBL forms at night. The state of turbulence in the stable PBL is primarily determined by H_v and τ. H_v is now downwards, the surface is cooler than the overlying air, the stratification is stable, and mechanically generated turbulence is now suppressed by buoyancy. It appears that wind speed and cloud cover are the dominant meteorological forcing variables. For clear skies the outgoing longwave radiation greatly exceeds the incoming radiation, the net radiation at the surface is negative and the surface cools. If wind speed increases, mechanically-induced turbulent mixing is enhanced and warmer air from higher levels is brought down to the surface. This process tends to decrease surface cooling. Cloudiness increases the longwave incoming radiation, decreasing net radiation loss and surface cooling.

Evaporation at nighttime shows a peculiar behaviour. Due to radiative cooling, the surface may cool to below dew point. The ensuing condensation on the surface gives rise to an inverted vapour lapse so that a downward water vapour flux ($E < 0$) can occur, known as dew fall (Oke, 1978). However, this phenomenon is confined to a limited range of wind speeds. For very low wind speeds, turbulence is not powerful enough to maintain a vertical flux and dew fall vanishes. On the other hand, if the wind speed is too large, the surface is heated by a downward heat flux, which tends to destroy the water vapour inversion. For that reason E is usually upwards at higher wind speeds.

Net radiation is also affected by wind speed. In still air, the downward turbulent heat flux is small and the surface can cool to a low temperature. At higher wind speeds warmer air from higher levels is brought down and the surface temperature is warmer. Thus the outgoing longwave radiation tends to increase with increasing wind speed. As a result the net radiation decreases (becomes more negative) with increasing wind speed.

The features of the surface energy balance during nighttime are schematically depicted in Fig. 6. This shows an example of the output of a model for the nocturnal surface layer recently presented by Holtslag & de Bruin(1988).

Fig. 7 shows the different regions that can be distinguished in the stable PBL according to Holtslag & Nieuwstadt (1986). In addition, an artist's representation of the local structure of the nocturnal PBL is presented in Fig. 8 (Wyngaard, 1985). Besides the stable surface layer discussed above and the near neutral upper layer, there are three other regions:

a) the local scaling region

b) the intermittancy area and

c) the region in-between called the z-less scaling region.

Surprisingly, it is found that the profiles of wind and temperature within almost the entire stable PBL can be described using Monin-Obukhov similarity theory (Holtslag, 1984; Holtslag & de Bruin, 1988). This is in spite of the fact that the theory, strictly speaking, is confined to the surface layer, which is very thin during nighttime. Also it is noted that the stability function proposed by Webb (1970), which often is used for stable conditions (Panofsky & Dutton, 1984), is only applicable for $(z/L) \leq 0.5$. If $(z/L) > 0.5$, the function proposed by Holtslag & de Bruin(1988), based on the work of Hicks (1976) and Carson & Richards (1978), is recommended.

Usually evaporation in stable conditions is small and, moreover, knowledge of the stable PBL is far from complete. Therefore, we will not consider the stable PBL in more details. The reader is referred to monographs such as Nieuwstadt & van Dop (1982).

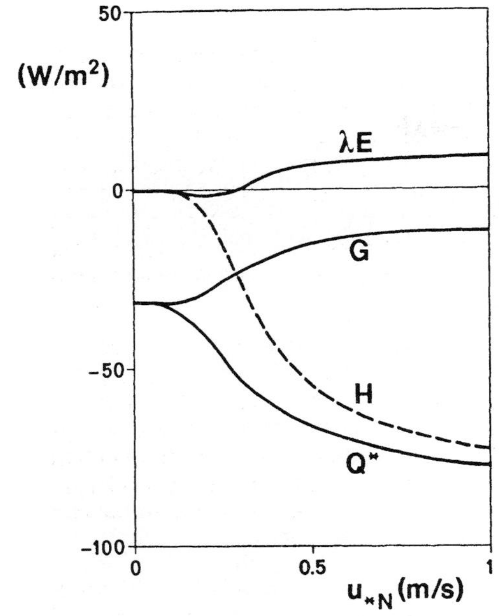

FIG. 6. The variation of net radiation (Q^*), sensible (H), latent (λE) and soil (G) heat flux density with wind speed. Note that u_{*N} is the friction velocity derived from the observed wind speed at 10 m and the logarithmic wind profile.

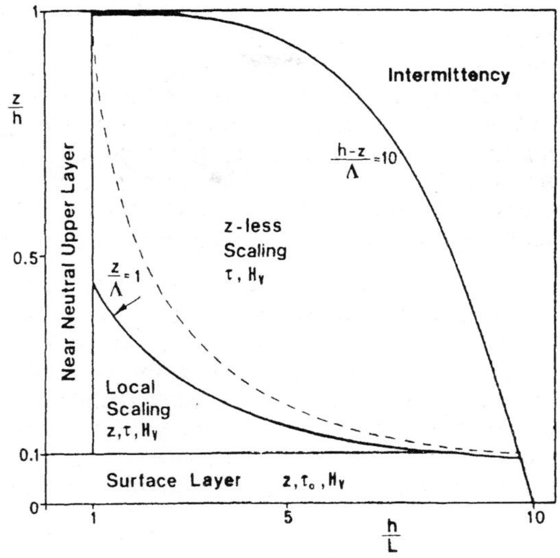

FIG. 7. Definition of scaling regions in the stable PBL (after Holstag & Nieuwstadt, 1986).

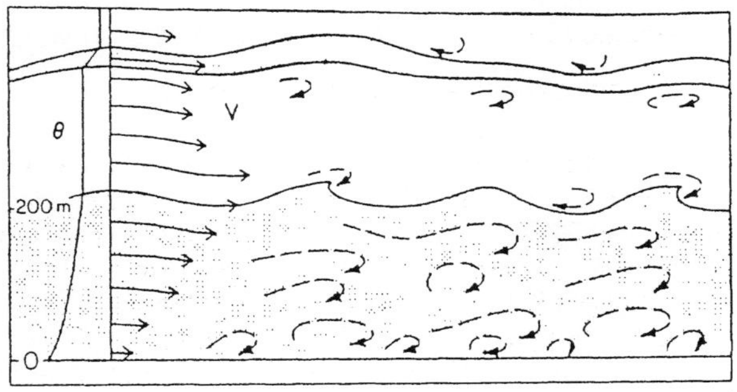

FIG. 8. An artist's representation of the local structure of the nocturnal PBL (after Wyngaard, 1986).

INTERACTION BETWEEN E AND THE PBL

General

After the brief, general discussion of the different turbulent states of the PBL we now return to the problem denoted by Bouchet (1963), i.e. how the surface evaporation is related to the saturation deficit of the overlying PBL. This relationship will be illustrated using two relatively simple PBL-models. Both cases refer to the well-mixed (convective) PBL.

The closed box model

To illustrate the relationship that exists between the surface fluxes of sensible heat and water vapour on the one hand and the temperature and humidity of the air near the ground on the other, we first consider the simple case where the PBL is assumed to be a closed box. This closed box model has been used by Perrier (1980) and in a somewhat different form by McNaughton (1976). It is described by McNaughton & Jarvis (1983) and Brutsaert (1986). It is assumed that the PBL is well-mixed above the surface layer, implying that θ and q are constant with height. Gradients of θ and q are allowed within the surface layer. The Penman-Monteith equation applies between the ground and the top of the surface layer. The depth of the surface layer is typically one tenth of the PBL-height. The PBL is capped by an inversion, which is assumed to act as an impermeable lid for heat and water vapour.

If sensible heat and water vapour are supplied at the surface, the potential temperature (θ_m) and specific humidity (q_m) of the well-mixed PBL will increase according to

$$\frac{d\theta_m}{dt} = \frac{H}{\rho c_p\ h} \text{ and } \frac{dq_m}{dt} = \frac{E}{\rho h} \tag{4}$$

We define $D_m = q_s(\theta_m) - q_m$, (5)

where $q_s(\theta_m)$ is the saturated specific humidity at θ_m. D_m is a measure of the saturation deficit of the PBL. Differentiating (5) and combining the result with (4) yields

$$\rho c_p h \frac{dD_m}{dt} = sH - \gamma\lambda E = s\,A - (s + \gamma)\,\lambda E \tag{6}$$

where s is calculated at θ_m, $\gamma = c_p/\lambda$ and $A = Q^* - G = H + \lambda E$. From (6) it is seen how D_m and E are interrelated in the case of the closed-box model. This result can be combined with the Penman-Monteith equation, applied to the entire surface layer,

$$\lambda E = \frac{sA}{s + \gamma^*} + \frac{\rho c_p D_m / r_a}{s + \gamma^*} \tag{7}$$

where $\gamma^* = \gamma(1 + r_s/r_a)$. λE can be eliminated from (6) and (7) leading to a simple first order differential equation for D_m that can be written as

$$\frac{dD_m}{dt} = \frac{D_{eq} - D_m}{t^*} \tag{8}$$

where the *equilibrium saturation deficit* (D_{eq}) is defined by

$$D_{eq} = \frac{Ar_a}{\rho c_p} \frac{s\gamma^* - s\gamma}{s + \gamma} \tag{9}$$

and the *time constant* is given by

$$t^* = \frac{hr_a\,(s + \gamma^*)}{s + \gamma} \tag{10}$$

D_{eq} can be regarded as a forcing function. If D_{eq} and t^* are constant in time, D_m will approach D_{eq} and dD_m/dt tends to zero. Then it follows from (6) that λE reaches its *equilibrium* rate λE_{eq} defined by

$$\lambda E_{eq} = \frac{s}{s + \gamma} A \tag{11}$$

Hence, for a closed PBL the Priestley & Taylor (1972) equation is obtained with $\alpha = 1$.

Although this example is not realistic, since h is not constant, it illustrates clearly that E and D_m are interrelated. The final result is that E tends to E_{eq}, so that E is determined primarily by net radiation (usually $Q^* >> G$).

A mixed-layer model including entrainment

In reality, the PBL height is not constant as assumed in the previous section. Turbulent eddies, created within the well-mixed layer (primarily by surface heating), intrude into the stable air aloft and the PBL will grow. As a consequence, this air, which is relatively warm and dry, is entrained into the well-mixed layer, affecting θ_m and q_m.

According to Tennekes (1973) and others, (4) has to be replaced by

$$\frac{d\theta_m}{dt} = \frac{H}{\rho c_p h} + \frac{\Delta\theta}{h}\frac{dh}{dt} \tag{12a}$$

$$\frac{dq_m}{dt} = \frac{E}{\rho h} + \frac{\Delta q}{h}\frac{dh}{dt} \tag{12b}$$

where $\Delta\theta$ and Δq (Fig. 4) are the 'jumps' at $z = h$ in the θ and q profiles, respectively. The last terms of (12) describe the effect of the growth of the PBL height and the resulting entrainment of warmer and drier air.

Because the entrainment process is primarily determined by surface heating, the last term of (12a) appears to be proportional to the second, so that

$$\frac{d\theta_m}{dt} = \frac{(1 + c)H}{\rho c_p h} \tag{13}$$

where $c \approx 0.2$.

Let θ_t and q_t be the value of θ and q of the stable air at $z = h$. Then $\theta_t = \theta_m + \Delta\theta$ and $q_t = q_m + \Delta q$ (usually $\Delta q < 0$), and defining $D_t = q_s(\theta_t) - q_t$, it can be shown from (12) that

$$\frac{dD_m}{dt} \approx \frac{sH - \gamma\lambda E}{\rho c_p h} + \frac{D_t - D_m}{h}\frac{dh}{dt} \tag{14}$$

This equation is equivalent to (6) of the closed-box model with h constant. The last term of (14) describes the effect of entrainment. Note that neither D_t nor h are constant. According to Driedonks (1981)

$$\frac{dh}{dt} \approx \frac{1 + 2c}{(1 + c)\gamma\theta}\frac{d\theta_m}{dt} \tag{15}$$

under convective conditions, provided $h \geq 2\,h_0$ (h_0 is the initial value of h just after sunrise). From (13) to (15) it can be shown that

$$\rho c_p h \frac{dD_m}{dt} = (s + s^*)\,H - \gamma\lambda E \tag{16}$$

where $s^* = (D_t - D_m)(1 + 2c)/(\gamma_\theta\ h)$.

To illustrate the difference with the closed-box model we consider the case where $dD_m/dt \rightarrow 0$. Then λE approaches

$$\lambda E = \frac{A}{1 + \gamma/(s + s^*)} \tag{17}$$

Defining a parameter α (Priestley & Taylor, 1972) such that

$$\lambda E = \alpha\frac{s}{s + \gamma}A \tag{18}$$

it is seen that now α approaches

$$\alpha = \frac{s + s^*}{\gamma + s + s^*} \cdot \frac{s + \gamma}{s} \tag{19}$$

According to (19) α can be smaller or greater than one. If $s^* > 0$, $\alpha > 1$. In that case $D_t > D_m$, i.e. the stable air has a larger specific humidity deficit than the well-mixed layer. On the other hand, $\alpha < 1$ if $D_t > D_m$.

It should be stressed that (19) is presented here only to illustrate the differences with the closed-box model. It is not generally applicable. To solve the set of equations listed above an additional equation for the surface fluxes is needed, e.g. the Penman-Monteith equation. In this way, one arrives at a set of coupled differential equations, which can be solved only if the initial and boundary conditions are known. It is outside the scope of this paper to discuss this matter in detail.

de Bruin(1983) and McNaughton & Spriggs (1986) present the mixed layer dynamics for special cases. Some results of the paper by de Bruin(1983) are presented in Fig. 9 where the calculated daytime variation of α is shown. It can be seen that around noon $\alpha \approx 1.3$ if $r_s \approx 0$; $\alpha \approx 1$ if $r_s \approx 60$-90 s m^{-1} and $\alpha < 1$ if $r_s > 100$ s m^{-1}. This is in good agreement with observations. Note that the more complete approach by McNaughton & Spriggs (1986) yields similar results (Jarvis & McNaughton, 1986). For practical calculations, the above result is important. Namely, it appears that a Priestley-Taylor type of estimate can be used for the surface fluxes E and H. This result can be used for rough estimates of evaporation. Note that for most crops r_s shows a significant diurnal variation. Usually, it is approximately constant in the morning, while it increases sharply in the afternoon. Recently, Brunet & Raupach (personal communication) found that the model for the well-mixed PBL described above is insensitive to this effect. For practical calculations, a mean value for r_s can be used. Finally, it is noted that recently authors such as Ten Berge (1986) and Pan & Mahrt (1987) coupled models for the PBL and the surface flux. These authors considered the case of bare soil. In this case the Penman-Monteith equation cannot be used.

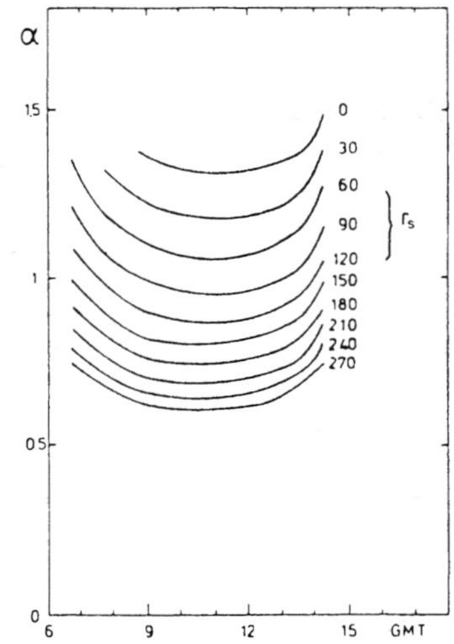

FIG. 9. Diurnal variation of the Priestley-Taylor parameter α for different surface resistances, r_s (after de Bruin, 1983).

Surface inhomogeneities

Until now we have confined ourselves to horizontally uniform terrain. By coupling PBL models with models for surface evaporation a problem arises. Most natural surfaces are uniform only on a scale of, say, 100 m (for simplicity we ignore hilly or mountainous regions). During daytime, the PBL has a height of, typically, 500-1000 m, corresponding to a horizontal scale of at least 10 km. The Earth's surface is seldom homogeneous on this scale. It must be concluded that in real life the coupling of the well-mixed PBL model with the Penman-Monteith equation, strictly speaking, is not permitted. It is interesting to note that all climate and weather modellers are faced with the problem that the land surface is rarely uniform on a grid-scale.

In my opinion future work on PBL and regional evaporation must be focussed on this inhomogeneity problem. Two types of surface inhomogeneities can be distinguished for horizontal terrain:

small scale inhomogeneities

These affect the surface layer, but do not affect the PBL. An example is an agricultural area in the Netherlands consisting of relatively small fields. It is to be expected that the large PBL eddies cause horizontal mixing such that the PBL "feels" one mean surface flux of heat, momentum, etc. The problem is how to determine these mean fluxes.

larger scale inhomogeneities

These inhomogeneties effect directly the dynamics of the PBL. An example is a lake with a diameter of, say, 20 km surrounded by uniform land. In this case the PBL "feels" two surface fluxes, one over land and one over water. Recently, Nieuwstadt (personal communication) considered the dynamics of the depth of such a non-homogeneous well-mixed PBL. He found that the behaviour of such a system can differ significantly from the uniform case.

SUMMARY AND CONCLUDING REMARKS

In this paper, a brief survey is given of the behaviour of the planetary boundary layer with special reference to evapotranspiration on a regional scale. It is meant as an introduction for those evaporation researchers who are not familiar with PBL-theory. It is shown that, through interactions between the surface fluxes and the PBL, evapotranspiration and saturation deficit are dependent variables. As a result, the Penman-Monteith equation is a descriptive formula, which is not able to predict, for example, the effects of changes in land management on a regional scale. Can this question be tackled by including PBL physics in our evaporation models? In my opinion, prediction of effects of changes in land management can be made only if we do so. However, the PBL models used so far by evaporation researchers are much too simple for this purpose. Future research will have to take into account such features as the effects of surface inhomogeneities on the dynamics of the PBL, and the formation of fog and clouds within the PBL.

REFERENCES

Andre, J.C., Lacarrere, P. & Mahrt, L.J. (1979) Sur la distribution verticale de l'humidité dans une couche limite convective. *J. Rech. Atmos.* 13, 135-146.

Bouchet, R.J. (1983) Evapotranspiration réele et potentielle, signification climatique. *Proc. IASH* 62, 134-142.

Brutsaert, W. (1986) Catchment-scale evaporation and the atmospheric boundary layer. *Wat. Resour. Res.* 22, 39S-45S

Brutsaert, W. (1987) Nearly steady convection and the boundary-layer budget of water vapour and sensible heat. *Boundary-Layer Meteorol.* 39, 283-300.

Businger, J.A. (1973) Turbulent transfer in the atmospheric surface layer. In: *Workshop on Micrometeorology,* (D.A. Haugen, ed.), 67-100. Am. Met. Soc., Boston.

Businger, J.A., Wyngaard, J.C., Izumi, Y. & Bradley, E.F. (1971) Flux profile relationships in the atmospheric surface layer. *J. Atmos Sci.* 28, 181-189.

Carson, D.J. & Richards, P.J.R. (1978) Modelling surface turbulent fluxes in stable conditions. *Boundary-Layer Meteorol.* 14, 67-81.

Deardorff, J.W. (1979) Prediction of convective mixed-layer entrainment for realistic capping inversion structure. *J. Atmos. Sci.* 36, 424-436.

de Bruin, H.A.R. (1983) A model for the Priestley-Taylor parameter α. *J. Clim. Appl. Meteorol.* 22, 572-578.

Driedonks, A.G.M. (1981) Dynamics of the well-mixed atmospheric boundary layer, *Scientific Report* W.R. 81-2, K.N.M.I., de Bilt, the Netherlands.

Driedonks, A.G.M. (1982) Models and observations of the growth of the atmospheric boundary layer. *Boundary-Layer Meteorol.* 23, 283-306.

Dyer, A.J. & Hicks, B.B. (1970) Flux-gradient relationships in the constant flux layer. *Quart. J. Roy. Met. Soc.* 96, 715-721.

Hicks, B.B. (1976) Wind profile relationships from the "Wangara" experiment. *Quart. J. Roy. Met. Soc.* 102, 531-551.

Holtslag, A.A.M. (1984) Estimates of diabatic wind speed profiles from near surface weather observations. *Boundary-Layer Meteorol.* 29, 225-250.

Holtslag, A.A.M. & Nieuwstadt, F.T.M. (1986) Scaling the atmospheric boundary layer. *Boundary-Layer Meteorol.* 36, 201-209.

Holtslag, A.A.M. (1987) Surface fluxes and boundary layer scaling: models and applications. Ph.D. Thesis, Agricultural University of Wageningen, The Netherlands.

Holtslag, A.A.M. & de Bruin, H.A.R. (1988) Applied modelling of the nighttime surface energy balance over land (Submitted to *J. Clim. Appl. Meteorol.*)

Jarvis, P.G. & McNaughton, K.G. (1986) Stomatal control of transpiration: Scaling up from leaf to region. *Adv. Ecological Res.* 15, 1-49. Academic Press, London.

McNaughton, K.G. (1976) Evaporation and advection. I. Evaporation from extensive homogeneous surfaces. *Quart. J. Roy. Met. Soc.* 102, 181-191.

McNaughton, K.G. & Jarvis, P.G. (1983) Predicting effects of vegetation changes on transpiration and evaporation. In: *Water Deficit and Plant Growth, Vol. VII.* (T.T. Koslowski, ed.), 1-47. Academic Press, New York.

IAHS, the International Association of Hydrological Sciences

IAHS is a nonprofit-making nongovernmental scientific organization committed to serving the science of hydrology and the worldwide community of hydrologists. IAHS, established in 1922, now has more than 4500 members from 124 countries.

The IAHS objective is to promote hydrology as part of the geosciences and as the basis for water management applications. The primary aims are to:

- initiate, facilitate, and coordinate research into hydrological problems requiring international cooperation
- provide for international discussion, comparison, and publication of hydrological research
- support hydrological sciences in developing countries
- bridge the gap between science and practice, including providing training opportunities and materials

International scientific commissions deal with surface water, groundwater, snow and ice hydrology, erosion and sedimentation, water quality, water resources, land–atmosphere systems and specific techniques, while working groups address particular issues. The Predictions in Ungauged Basins (PUB) Decade, 2003–2012, is a major initiative.

IAHS fulfils its mission of worldwide dissemination of research results and practice, and raises funds, by publishing and selling *Hydrological Sciences Journal* and books. Details of all publications are available at the web site, including abstracts of published papers. IAHS supports hydrological science in financially disadvantaged countries by: providing publications free of charge to libraries in those countries; and, offering discounts on publications purchased by members and contributing towards their costs when attending IAHS meetings.

Membership of IAHS is *free of charge* and open to anyone wishing to participate in IAHS activities, such as IAHS symposia and workshops.

IAHS Ltd is a UK registered charity (no. 1078635); all revenues are used to support the aims of the Association. Also, IAHS is one of the seven associations that comprise the International Union of Geodesy and Geophysics (IUGG, *www.iugg.org*).

IAHS AISH

www.iahs.info

Hydrological Sciences Journal

IF 1.606

Hydrological Sciences Journal (*HSJ*) is the official journal of IAHS—published by hydrologists for hydrologists. It is one of the oldest (published since 1956) and the most international of all the water-related journals. *HSJ* aims to provide a forum for original papers and significant developments in hydrology. The **impact factor** of *HSJ* is **1.606** (ISI Journal Citations Report for 2005). The scope of the Journal includes:

- the hydrological cycle on the Earth
- surface water, groundwater, snow and ice, in all their physical, chemical and biological processes, their interrelationships, and their relationship to geographical factors, atmospheric processes and climate, and Earth surface processes including erosion and sedimentation
- hydrological extremes and their impacts
- measurement, mathematical representation and computational aspects of hydrological processes
- hydrological aspects of the use and management of water resources and their change under the influence of Man's activities
- water resources systems, including the planning, engineering, management and economic aspects of applied hydrology

Information about all IAHS publications, including the contents and abstracts of published papers, is available at ***www.iahs.info***.

Journal subscriptions can be ordered from IAHS Press (an *order form* is available on request, or from the web site). IAHS members benefit from substantial discounts; IAHS membership is free of charge.

***HSJ* Backfile, vols 1–47** *(1956–2002)*
The complete *HSJ* backfile is now available online, free of charge at ***www.iahs.info***

Orders *Please send Journal orders, enquiries and submissions to:*

Mrs Frances Watkins
IAHS Press, Centre for Ecology and Hydrology
Wallingford, Oxfordshire OX10 8BB, UK
frances@iahs.demon.co.uk
tel.: + 44 1491 692405
fax: + 44 1491 692448/692424

Scales in Hydrology and Water Management

Echelles en hydrologie et gestion de l'eau

Edited by Ioulia Tchiguirinskaia, Mike Bonell & Pierre Hubert

IAHS Publication 287 (*September 2004*)
ISBN 978-1-901502-62-6; 170 + x pp.; **Price £36.95**

www.iahs.info

A joint IAHS / UNESCO-IHP publication

Schlumberger

Financial sponsorship for this publication was provided by Schlumberger Water Services

Space and time scales are fundamental for both scientific and operational hydrology issues. The question of how different scales interact and how we can transport results from one scale to another is basic to hydrological understanding and strategically important for decision-making.

Two questions dominate water resources management: how much remains for the future? and what is the response to environmental stresses? To answer them, a proper grasp of the interactions at different scales between the atmosphere, lithosphere, hydrosphere and biosphere, and their relations to agricultural, urban and industrial activities is needed.

This publication presents a collection of essays by engineers and scientists currently tackling the issues of scale dependence in hydrological phenomena and water resources management.

Please send book orders and enquiries to:

Mrs Jill Gash
IAHS Press, Centre for Ecology and Hydrology
Wallingford, Oxfordshire OX10 8BB, UK

jilly@iahs.demon.co.uk
tel.: + 44 1491 692442
fax: + 44 1491 692448/692424

See the IAHS website for information about the International Association of Hydrological Sciences (IAHS), membership, meetings, and publications, including abstracts of papers published since 2000.

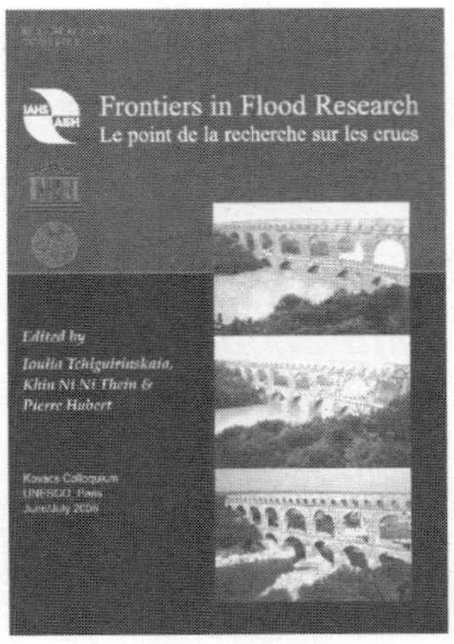

Frontiers in Flood Research

Le point de la recherche sur les crues

edited by Ioulia Tchiguirinskaia
CEREVE, Ecole Nationale des Ponts et Chaussée, France

Khin Ni Ni Thein and
UNESCO Division of Water Sciences, Paris, France

Pierre Hubert
UMR Sisyphe, CIG, Ecole des Mines de Paris, France

A joint IAHS/UNESCO-IHP publication

The abstracts of the papers in this volume can be seen at: www.iahs.info with information about other IAHS publications and IAHS activities.

IAHS Publication 305
(*Jun e 2006*)
ISBN 978-1-901502-63-3
212 + xii pp price £47.00

Major floods are the most chronic and costly natural hazards in many countries. Despite new developments in flood related sciences and rigorous implementation of hazard-reduction policies, damages from flood disasters still continue to escalate. Losses resulting from floods extend far beyond economic hardship. Consequently, growing international attention focuses on innovative flood research and on integrated flood risk management.

The contributors to this volume, international experts in current flood research, were invite speak at the 2006 Kovacs Colloquium, a biannual event organized by the International Hydrological Programme (IHP) of UNESCO and IAHS. This book presents their papers a new information and insights into key issues of flood research. Scientific understanding of flood phenomenon and of its potential impacts from the perspective of the engineering, operational, economic, social and ecological aspects of flood research, and with a focus c integrated flood risk management, is provided. The contributions are listed below, and the abstracts can be seen at the IAHS website

---

Large Sample Basin Experiments for Hydrological Model Parameterization:

Results of the Model Parameter Experiment – MOPEX

Edited by Vazken Andréassian, Alan Hall, Nanée Chahinian & John Schaake

IAHS Publication 307 (*2006*) 978-1-901502-73-2 348+ iv pp, plus a DVD with data, price £70.00

This volume contains 25 contributions, many of which follow from presentations made at the last two MOPEX workshops: July 2004 at ENGREF, Paris, France, and April 2005 at the IAHS Assembly in Foz do Iguaçu, Brazil. An objective is to show how valuable it is to work on large data sets in hydrological modelling.

- The first section provides an introduction to the goals of the MOPEX project, and presents the databases that were used by the workshop participants. Much of the data is published on the DVD accompanying this volume.
- The second section groups four review papers that were not presented during the MOPEX workshops, but solicited especially for this volume, in order to provide alternative views on the use of large sample basin experiments in hydrology.
- The third and the fourth sections present model parameterization experiments based on samples of a large number of basins: in the third, the focus is on the databases that were gathered specifically for the MOPEX program, while the fourth presents regionalization and parameterization studies based on other large hydrometeorological databases.
- The final section compiles the most recent results of the project, and discusses MOPEX in a wider context.

The abstracts of the papers in this volume can be seen at: **www.iahs.info** *with information about other IAHS publications and IAHS activities.*

Please send book orders and enquiries to:

Mrs Jill Gash
IAHS Press, Centre for Ecology and Hydrology
Wallingford, Oxfordshire OX10 8BB, UK

jilly@iahs.demon.co.uk
tel.: + 44 1491 692442
fax: + 44 1491 692448

See the IAHS website for information about the International Association of Hydrological Sciences (IAHS), membership, meetings, and publications.